Traditional and Folk Herbal Medicine

Recent Researches

— *Volume 2* —

The Editor

Dr. Vijay Kumar Gupta, Ph.D., FLS, London (born 1953-) Chief Scientist, CSIR- Indian Institute of Integrative Medicine, Jammu, India. He did his M.Sc. (1975) and Ph.D. (1979) in Zoology both from University of Jammu, Jammu-India. His research capabilities are substantiated by his excellent work on histopathology, ecology and reproductive biology of fishes, turtles, birds and mammals, which has already got recognition in India and abroad.

Dr. Gupta has to his credit more than 100 scientific publications and review articles which have appeared in internationally recognized Indian and foreign journals. Founder fellow, life member and office bearer of many national societies, academies and associations. He has successfully completed a number of research/consultancy projects funded by government, private and multinational agencies. His current areas of interest are histopathology, toxicology, pre-clinical safety pharmacology, reproductive efficacy studies of laboratory animals and biodiversity.

He is the Series Editor of the recently published multi-volume set of books, "**Comprehensive Bioactive Natural Products (Vols. 1-8)**", published by M/S Studium Press, LLC, USA. He is also Editor-in-Chief of the books, "**Utilisation and Management of Medicinal Plants**", "**Medicinal Plants: Phytochemistry, Pharmacology and Therapeutics (Vols. 1 and 2)**", "**Traditional and Folk Herbal Medicine**", "**Natural Products: Research Reviews**", "**Bioactive Phytochemicals: Perspectives for Modern Medicine**", "**Perspectives in Animal Ecology and Reproduction (Vols. 1-9)**" and "**Animal Diversity, Natural History and Conservation (Vols. 1-3)**". The Editor-in-Chief of the American Biographical Institute, USA, has appointed him as *Consulting Editor* of *The Contemporary Who's Who*. Dr. Gupta also appointed as Nominee for the *Committee for the Purpose of Control and Supervision of Experiments on Animals* (CPCSEA, Govt. of India). The *Linnaean Society of London, U.K.* has awarded fellowship to him in November 2009 in recognition of his contribution towards the cultivation of knowledge in Science of Natural History. Recently, Modern Scientific Press, USA has nominated Dr. Gupta as the Editor of the *International Journal of Traditional and Natural Medicine*.

Traditional and Folk Herbal Medicine

Recent Researches

— *Volume 2* —

Editor
V.K. Gupta
Chief Scientist
CSIR-Indian Institute of Integrative Medicine
Canal Road, Jammu – 180 001,
India

2014
Daya Publishing House®
A Division of
Astral International Pvt. Ltd.
New Delhi – 110 002

ISBN 9789351301080

Published by : **Daya Publishing House®**
A Division of
Astral International Pvt. Ltd.
– ISO 9001:2008 Certified Company –
4760-61/23, Ansari Road, Darya Ganj
New Delhi-110 002
Ph. 011-43549197, 23278134
E-mail: info@astralint.com
Website: www.astralint.com

Laser Typesetting : **Classic Computer Services**, Delhi - 110 035

Printed at : **Replika Press Pvt. Ltd.**

PRINTED IN INDIA

Editorial Board

Prof. Ian Fraser Pryme

Department of Biomedicine,
University of Bergen,
Jonas Lies vei 91, N-5009 Bergen, NORWAY
Tel: +47-55-586438
Fax: +47-55-586360
E-mail: ian.pryme@biomed.uib.no

Dr. George Qian Li

Herbal Medicines Research and Education Centre,
Faculty of Pharmacy, The University of Sydney,
NSW 2006, AUSTRALIA
Tel: 612 9351
Fax: 9351 8638
E-mail: george.li@sydney.edu.au

Dr. Julia Serkedjieva

Institute of Microbiology,
Bulgarian Academy of Sciences,
26, Academician Georgy Bonchev St.,
1113 Sofia, BULGARIA
Tel: +359 2 979 31 85
Fax: +359 2 870 01 09
E-mail: jserkedjieva@microbio.bas.bg

UNIVERSITY OF OXFORD

Department of Primary Care Health Sciences

*Postal Address: Radcliffe Observatory Quarter,
Woodstock Road, Oxford, OX2 6GG*

*Visitors' Address: 2nd floor, New Radcliffe House,
Walton Street, Jericho, Oxford, OX2 6NW*

Tel: +44 (0)1865 289300, Fax: +44 (0)1865 289287

email: merlin.willcox@phc.ox.ac.uk

Foreword

Traditional herbal medicine is important for several reasons. First and foremost, it has been the mainstay of health care for the human species since time immemorial. This has only changed in industrialized countries, and only in the last century or so. Even now, many modern medicines are directly extracted from, or are derived from natural plant products. It would be impossible to list them all here, but this book contains several examples. Artemisinin, the most potent antimalarial drug ever discovered, is extracted from *Artemisia annua*, which has been used as a medicine in China for at least 2000 years. Metformin, the most commonly prescribed antidiabetic drug, is derived from *Galega officinalis*, a traditional herbal medicine for diabetes. Another traditional medicine for diabetes, the Madagascan periwinkle (*Catharanthus roseus*), was found to contain vincristine and vinblastine, which have revolutionized the treatment of childhood leukaemia.

However the herbs which have been "transformed" into modern drugs remain a small minority of the global traditional herbal pharmacopoeia, which contains many thousands of plant species. Many of these have barely been researched, but continue to be used in their traditional forms. There remains a treasure-trove of new medicines waiting to be discovered and developed, either as modern medicines or as standardized herbal medicines – or both. Many patients may yet benefit from the fruits of such research

Yet at the same time, the increasing globalization of modern medicine is resulting in the loss of precious traditional knowledge, as many traditional healers' children are no longer interested in learning from them. It is said that when a healer dies without passing on his or her knowledge, it is as if a library has burnt down, losing its valuable contents forever. Furthermore every year more medicinal plants are becoming endangered or even extinct through over-harvesting, unsustainable use, or destruction of their natural habitat for short-term profits.

Reversing this trend will not be easy. It requires us all to make a concerted effort to show the value of the natural world to the long-term survival of humanity, and to develop new and better treatments for the many illnesses which still afflict mankind. This book contributes to that effort, by discussing the medicinal importance of many traditional remedies. When humans realize that the traditional knowledge of our grandparents, and the wild plants in our environments, may hold the key to our own survival, we may start to respect and preserve the world we have inherited from our grandparents, and which we must pass on to our grandchildren.

Dr. Merlin Willcox MRCGP

Clinical Researcher

Preface

Plants had been used for medicinal purposes long before recorded history. Ancient Chinese and Egyptian papyrus writings describe medicinal uses for plants as early as 3,000 BC. Indigenous cultures used herbs in their healing rituals, while others developed traditional medical systems in which herbal therapies were used. Researchers found that people in different parts of the world tended to use the same or similar plants for the same purposes.

Folk medicine is significant source of Ayurvedic, Unani, Traditional Chinese Medicine (TCM) and Medical herbalism. Folk medicine incorporates crude medicinal herbs, decoctions and infusions and syrups. Folk medicine is still practiced by some vendors, hakims and vaids in remote areas and some folk preparations are of surprisingly high curative value. A large proportion of the population in a number of developing countries still relies on traditional practitioners, including traditional birth attendants, herbalists and bonesetters and on local medicinal plants to satisfy their primary health care needs. Vincristine and vinblastine, the potent anti cancer drugs were derived from folk plant, *Vinca rosea* (periwinkle) used traditionally for the treatment of diabetes mellitus. Today herbal medicine is making dramatic comeback and scientists are turning to natural products for answer to ailments like cancer, Aids, hepatitis and rheumatoid arthritis.

The World Health Organization (WHO) estimates that 4 billion people, 80% of the world population, presently use herbal medicine for some aspect of primary health care. WHO notes that of 119 plant-derived pharmaceutical medicines, about 74% are used in modern medicine in some ways that correlate directly with their traditional use as medicines by native cultures. India having approximately eight percent of world's biodiversity, has the potential of becoming a major global player in screening and developing herb-based formulations, medicines and products. About 17000 plant species are said to exist in India of which 7500 plant species find mention in the recorded folklore of India. Of these only 1700 have been documented in 'Ayurveda' and classical healing treatise.

In the early 19th century, when chemical analysis first became available, scientists began to extract and modify the active ingredients from plants. Later, chemists began making their own version of plant compounds and, over time, the use of herbal medicines declined in favor of drugs. Almost one fourth of pharmaceutical drugs are derived from botanicals. In spite of spectacular advances in modern medicine including genetic engineering, biotechnology, organ transplant, designer drugs etc., a large population in the developing countries still rely on herbal and other complementary medicines for primary health care.

In the present venture an attempt has been made to put all the information on traditional and folk herbal medicine from different sources across the globe in the form of a book. This volume includes nineteen articles and cover a wide array of current research dealing with both fundamental and applied research in the field. The articles are highly authoritative, thought provoking and contributed by eminent scientists in their respective fields.

This volume like the previous one shall definitely prove very useful to the researchers who are interested in seeking knowledge/information pertaining to the traditional and folk herbal medicine. We are thankful to the publisher M/S Astral International Pvt. Ltd. New Delhi for publishing the book well in time without compromising the quality of the contents and the printing. Contributions from our esteemed contributors are also greatly acknowledged and errors if any are owned by the editor.

Dr. V.K. Gupta, *FLS*

Contents

Traditional and Folk Herbal Medicine:
Recent Researches Vol. 2 (2014)
Editor: **V.K. Gupta**
Published by: **DAYA PUBLISHING HOUSE, NEW DELHI**

Pages **1–44**

1

Prospects, Paradigms and Drawbacks of Current Research on Herbal Medicines

Md. Atiar Rahman[1,2] and Md. Shahidul Islam[1*]

ABSTRACT

Phytotherapeutic agents which are standardized preparations consisting of complex mixtures of active ingredients from one or more plants or plant materials in crude or processed state are known as herbal medicine. An uprising interest has been paid on the research of traditional herbal medicines since last couple of decades due to their incredible pharmacological activities, economic viability and less side effects. Apart from this, multidrug resistance in human pathogenic microorganisms and synthetic drug-dependence provided impetus to explore traditional medicines although the burden of the affordability of synthetic drugs was emphasized. Currently herbal medicines are widely spoken as a green medicine for their safe and dependable health care paradigms and anti-degradative environmental roles. They are now expanded over the world beyond India and China, overcoming the cutbacks in healthcare research, quality development and molecular uses, where they have been used primitively. About 65-80 per cent of the world's population depends essentially on folk herbals medicines for their primary health care. A significant growth of phytotherapeutic market ($20 billion per annum in European and USA only) over the last 15 years reflects the expansion of use of traditional medicines. But in general, a lack of quality information and scientific evidences on these medicines are still a question. The trend in domestication, production, biotechnological and genetic research to ensure efficacy, safety and quality of herbal medicines are yet to be

1 Department of Biochemistry, School of Life Sciences, University of KwaZulu-Natal (Westville Campus), Durban 4000, South Africa

2 Department of Biochemistry and Molecular Biology, University of Chittagong, Chittagong-4331, Bangladesh

* *Corresponding author*: E-mail: islamd@ukzn.ac.za

rationalized. This chapter will discuss comprehensively the concurrent research on harmonization and improvement in common and molecular uses and studies, safety evaluation, prevalence and factors associated with the use, self-management and combination of scientific evidences of traditional herbal medicines in the era of new generation synthetic drugs.

Keywords: Folk medicine, Herbal medicine, Traditional medicine, Efficacy, Health care.

Introduction

Herbal medicine is an essential part of traditional medicine in almost every culture. Traditional herbal medicine has been practiced in China since thousands of years. The effectiveness and use of herbal medicines have been increased sharply in last few decades in the United States and other western countries for the treatment of many chronic diseases due to their incredible pharmacological activities, economic viability and harmless effects (Prashant *et al.*, 2008). They have been recognized especially in those disease conditions where few effective treatments have emerged from modern pharmaceutical research (Ehling, 2001). In recent years, scientists in academic and government research institutions and in the biotechnology and pharmaceutical industries have increasingly explored herbal medicines as a rich source for modern, molecular target-specific drug discovery (Marko, 2001; Goldman, 2001).

Modern synthetic medications for life threatening diseases like AIDS, cancer and chronic diseases like diabetes have been attributed to the unbearable economic burden substantiated to death in many cases. Currently 40 million people are infected with Human Immunodeficiency Virus (HIV) (UNAIDS, 2007) and more than 20 million deaths have been attributed to Acquired Immunodeficiency Syndrome (AIDS) since the onset of the HIV pandemic. The introduction of highly active antiretroviral therapy (HAART) has led to reduce the AIDS-related morbidity and mortality (MoH, 2004). However the management of HAART side effects has remained a challenge in many resource limited settings (UNAIDS, 2008). Similarly, huge numbers of cancer patients being incapable to afford the cost of modern synthetic anticancer drugs depend upon the herbal medicines especially in China.

Apart from this, worldwide multidrug resistance in human pathogenic microorganisms, intolerance and dependence of synthetic drugs provided impetus for exploring the folkloric traditional drugs in modern world although the early researchers emphasized the economic downturn and burden of affordability of synthetic drugs. Herbal or traditional or complementary and alternative medicines are generally considered as non-toxic compared to synthetic medicines and drugs. Currently they are wide-spoken as green medicine, the safe and dependable health care paradigms and anti-degradative environmental tools. According to WHO, about 65-80 per cent of the world's population depends essentially on folkloric herbal drugs for primary healthcare. A marked worldwide growth has seen in the phytotherapeutic market over the last 15 years which reflects the expansion of use of folkloric drugs. For the European and USA markets alone, this was reached about $7 billion and $5 billion per annum, respectively. Even Malaysian annual sales for traditional medicines, from 2000 to 2005, increased from US$ 385 million (RM 1 billion) to US$ 1.29 billion (RM 4.5 billion). Thus most of the large pharmaceutical companies paid more attention to herbal medicine research (Chew *et al.*, 2012).

Despite the long history of herbal medicine products and their promising effects in treating diseases including cancer, several major problems such as toxicity and the efficacy of most of the herbal recipes, ambiguous active ingredients of herbals, unknown ingredients and molecular mechanisms to some extent may have prevented their wide acceptance in the biomedical and pharmaceutical community in the United States and other western countries (Fang *et al.*, 2004).

However, recent reports showed that herbal drugs and herbal researches have minimized some of the drawbacks with research in healthcare, quality development, market growth and molecular uses. This chapter will discuss the general trends of herbal drugs, new paradigm context of herbal drugs, economic and healthcare contribution in life-threatening diseases and the use of herbals for wide range of diseases. Fundamental efficacy and safety measures of herbal drugs, modernization of herbal drugs will also be discussed at the ending section of this chapter.

Herbals at a Glance

Definition of Herbal

Phytotherapeutic agents which are standardized preparations consisting of complex mixtures of active ingredients from one or more plants or plant materials in crude or processed state are known as herbal medicine. Herbal medicines are the synthesis of therapeutic experiences of generations of practicing physicians with indigenous systems of medicine for over hundreds of years. In spite of the great advances observed in modern medicine, in recent decades, herbal drugs are still making a large contribution to the healthcare system. Over the past decades, interest in drugs derived from higher plants, especially the phytotherapeutic ones, has increased expressively. It is estimated that about 25 per cent of all modern medicines are directly or indirectly derived from higher plants (Farnsworth and Morris, 1976; De Smet, 1997; Cragg *et al.*, 1997; Shu, 1998). About 60 per cent of currently available antitumoral and antimicrobial drugs on the market and most of those in the late stages of clinical trials are derived from natural products, mainly from higher plants. Most of the herbal sources are abundant in tropical countries although medicinal plants are distributed worldwide.

Herbal Drugs in the View of WHO

The World Health Organization (WHO) has recently defined traditional medicine (including herbal drugs) as comprising therapeutic practices that have been in existence, often for hundreds of years, before the development and spread of modern medicine and are still in use today (Gansser,1964). Herbal drugs constitute only those traditional medicines which primarily use medicinal plant preparations for therapy. The earliest recorded evidence of herbal use in Indian, Chinese, Egyptian, Greek, Roman and Syrian texts back to about 5000 years, however the classical Indian texts include *Rigveda, Atherveda, Charak Samhita* and *Sushruta Samhita*. The herbal or traditional medicines medicaments have, therefore, been derived from rich traditions of ancient civilizations and scientific heritage.

Till 1991, the World Health Organization (WHO) has not systematically evaluated traditional medicines, when they developed guidelines for the assessment of herbal medicine, where suggestions for herbal medicine standardization are outlined and the scenario and perceptions of herbal medicine were discussed. According to the WHO, because of poverty and lack of access to modern medicine, about 65-80 per cent of the world's population which lives in developing countries depends essentially on plants-based traditional medicines and traditional healers for primary health care. Currently, the major pharmaceutical companies have demonstrated much interest on investigating higher plants as a source of new modern medicine and also for the development of standardized phytotherapeutic agents with proven safety measures (De Smet, 1997; Brevoort, 1995; Blumenthal, 1999). Herbal medicinal preparations are normally very popular in developing countries with a history of tradition of use of medicinal plants and also in some developed countries such as Germany, France, Italy and the United States where appropriate guidelines for registration of such medicines are available (Eisenberg *et al.*, 1993; Fisher and Ward, 1994; Roberts and Tyler, 1998).

Herbal Medicines Characterized as Phytotherapeutic Agents

Phytotherapeutic agents or phytomedicines are standardized herbal preparations consisting of complex mixtures of one or more plants which are used in most countries for the management of various diseases. Herbal drugs might contain plant parts or plant materials as active ingredients in the crude or processed state plus certain excipients, *i.e.*, solvents, diluents or preservatives (Akerele, 1993; Bulletin of the World Health Organization, 1998). But combinations with chemically defined active substances or isolated constituents are not considered to be herbal medicines. That is why, homeopathic preparations are not considered as herbals although they contain plant materials. Usually, the active principles responsible for their pharmacological actions are unknown and phytotherapeutic agents is the fact that they normally do not possess an immediate or strong pharmacological action which makes herbals not to be used as a medicine for an emergency treatment. However, in contrast to modern medicines, herbal medicines are frequently used to treat chronic diseases. Some major differences and drawbacks of herbal medicines compared to the synthetic medicines are:

1. The active principles are frequently unknown;
2. Standardization, stability and quality control are feasible but not easy;
3. The availability and quality of raw materials are frequently problematic;
4. Well-controlled double-blind clinical and toxicological studies to prove their efficacy and safety are rare;
5. They have a wide range of therapeutic use and are suitable for chronic treatments;

6. The occurrence of undesirable side effects seems to be less frequent with herbal medicines, but well-controlled randomized clinical trials have revealed that the herbal medicines may also have side effects;
7. They are usually cheaper than synthetic drugs.

Herbal Drugs Characterized as Zootherapeutic Agents

Animals (and their derived products) are essential ingredients in the preparation of many traditional remedies (Table 1.1). Animals and products derived from their organs have constituted part of the inventory of medicinal substances used in various cultures since ancient times (Alves *et al.*, 2007; Alves *et al.*, 2008; Lev, 2003). Despite its prevalence in traditional medical practices worldwide, research on medicinal animals has often been neglected compared to medicinal plants. The more emphasis was given to medicinal plants due to the diversity of species as well as acceptability by people compared to animal originated medicines (Alves *et al.*, 2011). Additionally, in many aspects, plants are easier to collect, store and trade. However, some recent publications have shown the importance of zootherapy in various socio-cultural environments around the world, and examples of the use of animal-derived remedies which can be found in many urban, semi-urban and remote localities many parts of the world, particularly in developing countries (Mahawar and Jaroli, 2008; V´azquez *et al.*, 2006; Walston, 2005). In Brazil, many species of animals have been used for medicinal purposes since colonial times, with widely disseminated therapeutic alternatives (Alves, 2008; Moura and Marques, 2008; Ferreira *et al.*, 2009). Many people still use animal-derived medicines as an alternative or supplement to western health care (Alves and Rosa, 2006; Alves and Rosa, 2007; Alves *et al.*, 2009). In a recent review, it has been reported that the 290 different animal species are being used in traditional folk medicine in Brazil (Sodeinde and Soewu, 1999). This number is certainly underestimated since the amount of studies on the theme is very limited and most of the medicinal animals comes from studies conducted in coastal areas and Amazon region (Alves *et al.*, 2009; Costa-Neto and Marques, 2000; Figueiredo, 1994). However, many other parts of the world have also been practiced in animal sources as remedial agents (Silva, 2008).

Classes of Herbal Drugs

According to European Pharmacopoeia (Supplement, 2000), herbal drug preparations are obtained by several step processes such as extraction, distillation, expression, fractionation, purification, concentration and fermentation. Herbal drugs are normally marketed as standardized preparations in the form of extracts, liquid, solid (powdered extract), or viscous preparations (Keller, 1996; Blumenthal *et al.*, 1998). Some other forms like essence (herbal cosmetics), infusion (herbal infusion tea), tincture (liquid herbal), poultice (raw plant material), decoctions, fatty oils, essential oils and expressed juice are also available in the market. They are prepared by maceration, percolation or distillation (volatile oils). Ethanol, water, or mixtures of ethanol and water are used for the production of fluid extracts. Solid or powered extracts are prepared by evaporation of the solvents used in the process of extraction of the raw materials.

Table 1.1: Categories of diseases treated with zootherapeutic remedies in surveyed area Queimadas, Paraiba State), according to the Brazilian Centre for the Classification of Diseases (Alves *et al.*, 2009).

Categories	*Diseases and Illnesses Mentioned by Respondents*	*Total*
1	Teething, inflammation, hoarseness, cracks in the sole of the feet, general pain, to assist children who take longer than usual to start walking, children that speak with lateness, itching, problems of navel, "hard nerve" and healing of umbilical cord of newborn baby	11
2	Erysipelas, oral mycosis, athlete's foot, measles, rubella, chickenpox, tuberculosis, warts, whooping cough and mumps, moniliasis	11
3	Asthma, bronchitis, effusion, catarrh, cough, flu, sore throat, sinusitis and tonsillitis	9
4	Poor digestion	1
5	Burns, wounds, muscle strain, luxation, distension and sprains	6
6	Arthritis, arthrosis, backache, osteoporosis and rheumatism	5
7	Hemorrhage, cardiac problems, thrombosis and cerebral hemorrhage	7
8	Menstrual problems and urinary infection	2
9	Alcoholism, snakes bite, insect bite, to suck a splinter out of skin or flesh, swelling and bleeding	6
10	Boils, skin spots and acne,	3
11	Problems in nerves	1
12	Tumors	1
13	Deafness, earache	2
14	Diabetes, weakness and malnutrition	3
15	Sexual impotence	1
16	Conjunctivitis and pterygium	2
	Total	68

(1) Undefined diseases; (2) some infections and parasitic diseases; (3) respiratory system; (4) digestive system; (5) injuries, poisoning and other consequences of external causes; (6) osteomuscular system and conjunctive tissue; (7) circulatory system; (8) urinogenital system; (9) external causes of morbidity and mortality; (10) skin and subcutaneous tissue; (11) nervous system; (12) Symptoms, signs and abnormal findings from medical and laboratorial examination, not categorized in other part or section; (13) ear (middle and inner ear) and mastoid apophysis; (14) diseases of the endocrine glands, metabolism and nutrition; (15) mental and behavioural perturbations; (16) ophthalmological diseases.

Soft extracts are semi-solid preparations obtained by evaporation or partial evaporation of the solvent used for the preparation. **Dry extracts** are solid preparations obtained by evaporation of the solvent used for their production. Dry extracts usually have a loss on drying or water content is not greater than 5 per cent (*m/m*). **Liquid extracts** are liquid preparations of which, in general, 1 part by mass or volume is equivalent to 1 part by mass of the herbal drug or animal matter. These preparations are adjusted, if necessary, so that they satisfy the requirements of the content of the solvent and, where applicable, for the constituents. **Tinctures** are liquid preparations which are usually obtained using either 1 part of herbal drug or animal matter and 10 parts of extraction solvent or 1 part of herbal drug or animal matter and 5 parts of extraction solvent (Gaedcke and Steinhoff, 2003). Although the fluid extract and tinctures are similar but the difference between them is that tinctures are not concentrated but a diluted preparation.

Extraction of Herbal Drug

The comminuted, mostly dried parts of the plant and the extraction solvent are combined in the extractor. This mixture is allowed to stand for several hours depending on the extract ion process. Generally, extraction with additional extraction solvent is repeated until no more constituents are transferred to the extraction solution. In the next step, the extraction solution (miscella, eluate) is separated from the herbal drug residue. Since it contains the extractable herbal matter, the eluate represents an important intermediate product during the total herbal processing.

The uses of solvents for herbal drug extractions are based on the type of herbals are to be prepared. If the extraction solvent is ethanol or water or a mixture of both, the preparations resulting after filtration are liquid extracts (fluid or tincture, depending on the level of concentration). If the extraction is carried out with vegetable oils (*e.g.*, rape oil or olive oil) oily macerates are obtained which are usually filled in soft gelatin capsules. If the resulting preparation is a dry extract, processing of the eluate continues with cautious evaporation of the majority of the extraction solvent in the vacuum-evaporator. The resulting aqueous soft extract is evaporated to dryness using suitable drying aggregates, *e.g.* spray dryer, belt dryer or spray belt dryer. In most cases, suitable technical excipients such as maltodextrin, lactose or silicium dioxide are needed to be added. After grinding and sieving, flowable powder is obtained which can be further processed to tablets or coated tablets or filled in soft or hard gelatin capsules (Gaedcke and Steinhoff, 2003).

Herbal Medicines: A New Paradigm

It is not enough to stand on the ancient belief of Chinese and Indian herbal medicine, which is being used for thousands of years, in order to support its continued use in the new paradigm of integrative medicine. Herbalists are entering an era in medicine where our beliefs must be supported by experimental and clinical researches. These research data are used to translate our knowledge of holistic diagnostics and medical intervention into a biomedical language that both patients and allied health practitioners can get benefit of it. Clinical research is highly dependent on having access to quality herbal products that provide positive results for our patients and success in our clinical trials to reach a new paradigm of the use of herbal medicines. The major challenge for the herbalists is to provide incentives to the patients to utilize herbal medicines in the countries dominated by evidence-based pharmaceuticals.

Evidence based medicine is a term developed by Sackett (1996). He suggested that in order to guide therapeutic decisions that produce positive clinical outcomes, we must integrate individual clinical expertise with the best available external clinical evidence from systematic research that is patient centered. The philosophy of traditional medical care is already patient centered. Herbalists and traditional healers possess a rich history of clinical expertise based on volumes of clinical case studies. If we have to integrate into a biomedical model, an understanding of how to prove our herbal medicines are efficacious using research is mandatory. The primary investigators in herbal research should be acupuncturists (MacPherson *et al.*, 2008). They are in the best position to understand the unique qualities, applications and therapeutics of the herbs. Armed with research data and herbal clinical expertise, acupuncturists will be in a unique position to educate their patients and other biomedical practitioners in optimizing research design and prescribing effective treatment protocols. Acupuncturists will be seen taking their position as leaders in herbal medicine and collaborative partners in an integrative medical model. Success in clinical trials can only confirm the quality herbal products are being utilized.

Quality control is also necessary to improve the quality of the randomized controlled trials for herbal medicines (Leung *et al.*, 2006). This will help to rely upon herbal manufacturers to provide access to quality herbal products and should demand that these medicines are not only safe for the patients, but also stand up to the highest levels of good manufacturing practices (GMP) (Leung *et al.*, 2006). Advances in biomedical science techniques are also allowed to create novel therapeutic agents from the herbal world (Pan *et al.*, 2010). Acupuncturists can be instrumental in bridging the gap in integrated research for new drug therapies. Conversely, the temptation to sacrifice quality for profits lingers in front of hungry entrepreneurs eager to cash in on the baby boomers Wellness Revolution (Pilzer, 2002). Acupuncturists will be the gatekeepers of patient safety, utilizing only those products proven to meet the standards of an integrated medical culture. Access to quality herbal products for the patient is also challenged by the current governmental regulatory and insurance based policies. Unlike pharmaceutical drugs that qualify for third-party insurance reimbursement, equal access of herbs to insurance coverage has not become a reality. This will make the inclusion of herbal therapy into mainstream medicine difficulty. Acupuncturists must be vigilant in leading the agenda to enlist legislative support for herbal coverage in governmental health plans. This can be accomplished through education; dispelling myths of ineffective, unsafe and inferior quality herbs. Instead, acupuncturists as professionals should promote the responsible use of GMP herbal products and document the reduction in expenditures that their interventions make versus comparative biomedical drug therapy.

It is widely known that herbs are safe and effective, without convenient dosage forms, appealing aesthetics, competitive pricing to drug therapy and easy availability, however mainstream patients are not likely to "go herbal." If the herbal product is difficult to prepare, swallow or administer, patient compliance with therapy declines despite of results. Patients are more likely to take medicines that look appealing, easy to consume and pleasant in taste. The ability to manufacture aesthetically appealing herbal products that are widely available at a competitive cost will promote patient usage.

Acupuncturists can no longer wait for integrative medicine to come to the doors.

Why Do We Need a New Paradigm: From United States Health Care Crisis

Although each country's health care system is unique, being influenced by cultural, political and economic factors, some of the trend seen in the United States can illustrate the general characteristics of the current medical model around the world. For more than a decade, the United States system has been under scrutiny. Patients, policy-makers, the medical community, health insurance companies and social commentators have been vocal about the shortcomings of the current system and need for health care reform. The reasons for their discontent are numerous and multifaceted. The United States system has been shown to be the most expensive and inadequate system in the developed world (Angell, 1999; Starfield, 2000). According to a recent WHO analysis, even though United States spends the highest amount of money per person on health care (about US$ 4000 per person ill in 1997), in overall quality, its care ranks 37th in the world (Labders, 2000). Despite US$1.3 trillion worth of health care expenditures (about 14 per cent of ODIP), the number of uninsured grew to 43 million (16 per cent of the population) in 1998 (Iglehart, 1999)· Questions are also being raised about the invasiveness, adequacy, and safety of the current system. Suffering caused by chronic and age-related diseases is on the rise and some common diseases continue to take their toll. Still more of a person's health care money is spent in attempts to prolong life at all costs, rather than on other health care needs such as prevention, treatment of chronic, stress-related illness and improvement of quality of life. The emphasis on crisis care has resulted in development of highly technical invasive treatments and potent drugs that raise the risk of iatrogenesis. Alarmingly, as many as 20-30 per cent of patients receive contraindicated and/or unnecessary care (Starfield, 2000). Recent reports estimate the number of deaths due to medical errors to be between 44,000 and 98,000 Americans. An additional 106,000 deaths a year are attributed to non-error, adverse effects of medications, raising questions about the overall safety of purified, potent drugs (Kohn *et al.*, 2000). Although the data are available, this circumstances exist more or less in most of the countries, the deteriorating situation in health care management can only be eradicated through the application of safe, non- or less toxic, economically viable drugs.

Prospects of Traditional Medicine Researches

Market Size of Herbal Drugs

It has been reported that pharmaceutical sales worldwide are forecast to grow from 600 billion U.S. dollars in 2010 to over 950 billion U.S. dollars by 2013 (Pan *et al.*, 2010). Advances in biomedical science techniques will allow the creation of novel therapeutic agents from the herbal world (Pan *et al.*, 2010).During the past decades, public interest in natural therapies, namely herbal medicine, has increased dramatically not only in developing countries but mainly in industrialized countries (De Smet, 1997; Blumenthal, 1999; Grünwald, 1995). This has increased the international trade in herbal medicine enormously and has attracted most of the pharmaceutical companies, including the multinationals. Until a few years ago, only small companies had interest in the marketing of herbal medicines. Currently, most large multinational companies are interested in commercializing herbal drugs. It is estimated that the European market alone reached about $7 billion in 1997. The German market corresponds to about 50 per cent of the European market, about $3.5 billion which represents about $42.90 per capita. This market is followed by France, $1.8 billion; Italy, $700 million; the United Kingdom, $400 million; Spain, $300 million; the Netherlands, about $100 million. European herbal medicines are distributed under 6 basic therapeutic categories: cardiovascular, 27.0 per cent; respiratory, 15.3 per cent; digestive, 14.4 per cent; tonic, 14.4 per cent; hypnotic/sedative, 9.3 per cent; topical, 7.4 per cent; others, 12.0 per cent. The herbal medical database indicates that the herbal medicine markets in Asia and Japan reach $2.3 and 2.1 billion, respectively (Grünwald, 1995). However, in no other country has the herbal medicine marketplace grown more than it has in the in USA. A few years ago, this was a non-existent category of medicine. The US herbal medicine market reached about $3.2 billion in 1996, and is expected to reach $5 billion in 1999 (Blumenthal, 1999; Roberts and Tyler, 1998). According to a national survey, about 60 million Americans over 18 years of age use herbal drugs to treat colds, burns, headaches, allergies, rashes, depression, diarrhea and menopause, among others. Each American spends about $54 a year buying these medicines. Currently, most medicinal herbs, such as *Aloe vera*, *Panax quinquefolius* (American ginseng), Echinacea, *Alium sativum*, *Ginkgo biloba*, *Serenoa repens* (saw palmetto), *Valeriana officinalis*, etc., are cultivated in the USA and exported to Europe and the orient. As a natural consequence, many large companies have introduced a line of herbal products into their sales (Blumenthal, 1999; Roberts and Tyler, 1998). According to Grünwald (1995), the phytomedicine market has grown at an expressive rate worldwide since 1985 (from 5 to 18 per cent a year).

In Australia, the report states that according to industry stakeholders, herbs that will generate the highest possible return for growers are *E. purpurea*, skullcap and goldenseal. Based on current international pricing for dry plant material *E. purpurea* A$40-45/kg, skullcap A$42/kg and goldenseal A$140-200/kg (as of April 2005) it appears attractive for Australian growers to supply these herbs. Despite current high prices, the unpredictability of international prices will make it difficult to secure grower interest. For example, when Australian growers were supplying skullcap 3-4 years ago at a cost base of A$22/kg for dry plant material, growers from North America were supplying at A$16/kg. When domestic production of skullcap ceased, prices have been gradually rising to its current level of A$42/kg. Prospects paper Australia. (Gee Yap, 2006)

In Malaysia, from 2000 to 2005, annual sales for traditional medicines increased from US$ 385 million (RM 1 billion) to US$ 1.29 billion (RM 4.5 billion)(Aziz and Tey, 2009). The prevalence and factors associated with its use is largely unknown, although the use is believed to be widespread as a complementary therapy (Aziz and Tey, 2009).

Herbal Spectrum in Health Care

Traditional medicine and medicinal plants, in general, continue to be a powerful source of new drugs, now contributing about 90 per cent of the newly discovered pharmaceuticals. Traditional medicine continues to provide health coverage for over 80 per cent of the world population, especially in the developing world. The past and the present are all full of living examples of discoveries of drugs, ranging from anticancer, antiasthma, antidiabetic, antihypertensives and many others which owe their origin to traditional medicine. The world is currently facing a number of health challenges, partly due to expand boundaries of diseases as a result of global warming, improved means of transportation and increased migration of people. It is also acknowledged that the change of lifestyle in the developing world, where people are now leading sedentary lives, and have greatly changed the types of food essentially consuming more refined foods which led to increase in the prevalence of glucose intolerance (King and Rewers, 1993). Likewise, emerged number of new diseases such as HIV/AIDS, Ebola, and human spongiform encephalopathy (Budka *et al.*, 1995; VanEverbroeck *et al.*, 2000; Brown *et al.*, 2004)that don't yet have a cure. Re-emergence of diseases like tuberculosis and the emergence of pathogens that are resistant to the currently available chemotherapeutic agents, is yet another challenge for pharmaceutical industries. The living example is the emergence of multi-drug resistant (MDR) *Mycobacterium tuberculosis*. Malignant tumors and some important parasites, like *Plasmodium falciparum* and *Candida albicans* are also developing resistance, thus making management of diseases more and more difficult. Amidst this crisis, the synthetic chemist is long recognized to have failed to address these health challenges. The only rational alternative is therefore, traditional medicine, which supports more than 80 per cent of the population in developing countries, and it is now widely accepted that traditional medicines are more affordable, less toxic, and have a wide acceptance around the world (Cowan, 1999; WHO, 2002).

Nature has been a source of medicinal treatments for thousands of years, and plant-based systems continue to play an essential role in the primary health care. It is estimated that 25 to 50 per cent of all current pharmaceuticals are derived from plants (Cowan, 1999). In fact, it is now believed that plant based systems contribute 90 per cent of the newly discovered pharmaceuticals. Nature has provided many of the effective anticancer agents in current use, such as the microbially derived drugs, dactinomycin, bleomycin, and doxorubicin, and the plant-derived drugs vinblastine, irinotecan, topotecan, etoposide, and paclitaxel (Cragg *et al.*, 1997). Traditional medicine, has particularly, been a good source of these plant derived drugs. A number of healthcare benefits of herbal drugs are discussed below:

Effectiveness of Herbal Drugs in the Context of Traditional Chinese Medicine (TCM)

The remarkable longevity and current popularity of herbal drugs implies its clinical efficacy. However, to incorporate it into mainstream health care, many are looking for a systematic proof of efficacy from vigorous research studies. In the last few decades, China has been a pioneer in investigating TCM and has provided a wealth of research data demonstrating the efficacy of its concepts and techniques. Unfortunately, these studies often are not available in English and/or do not have sufficient research strength. At the same time, studies conducted in the Western world frequently concentrate on investigating a single therapeutic technique outside the TCM framework (Hui, 1999). Despite these limitations, reviews of TCM research yield suggestive evidence of efficacy of all the major therapeutic modalities: acupuncture, acupressure, herbal medicine, *Tai Chi* and *Qi Gong*. In 1997, National

Institutes of Health (NIH) consensus determined that, based on available studies, acupuncture should be considered effective for a number of conditions such as post-operative and chemotherapy nausea or vomiting, and post-operative dental pain. Acupuncture has also recommended as an adjunct treatment or an acceptable alternative for addiction, stroke rehabilitation, asthma, menstrual cramps, tennis elbow, carpal tunnel syndrome, lower back pain, headache and pain syndromes such as fibromyalgia and myofascial pain (NIH Consensus Conference, 1998). Concurrently, basic research studies have elucidated some neurophysiological mechanisms implicated in acupuncture, such as release of endogenous opioids and stimulation of the limbic system (Avants, 2000; Hui, 2000). Similarly, there is growing knowledge of herbs and their biological effects, extensive research of Chinese, Korean and Kampo formulas has shown that many herbs have unique immunomodulating, cardiovascular, antiviral, anticancer and other beneficial effects. Clinical research of herbal medicine has demonstrated herbal efficacy in the areas of gastroenterology, hepatology, reproductive endocrinology, dermatology, neurology and psychiatry. In randomized controlled trials, traditional herbal preparations have been shown to be effective for hepatocellular carcinoma, eczema, vascular dementia and irritable bowel syndrome (Bendoussan *et al.*, 1998; Eskinazi *et al.*, 1999).

On the other hand, less research has been done on the other therapeutic techniques, acupressure, *Tai chi* and *Qi Gong*, but some data are available. Recent studies on massage therapy have shown benefits for a spectrum of conditions, such as chronic pain syndromes, chronic fatigue syndrome, fibromyalgia and others (Braverman and Schulman, 1999; Ernst, 1999). *Tai Chi* exercise has been shown to improve balance in elderly subjects and it may help decrease blood pressure (Young *et al.*, 1999). *Qi Gong* practice may have beneficial effects in patients with asthma, diabetes and cancer (Sancier, 1996).

Herbals as an Antidiabetic Drug

In ancient time, *Galega officinalis* (L.) was used ethnomedically for the treatment of diabetes. Phytochemical work led to the isolation of the antihyperglycaemic compound galegine from the plant. Further structure activity studies and modification of galegine led to the synthesis of metformin and eventually the other biguanide oral hypoglycaemic agents (Sneader, 1985). Harunganin was recently isolated from a Tanzanian plant, *Harungana madagascariensis* and is now patented in the USA for treatment of diabetes (Inman and Luo, 1998). Other compounds isolated from traditionally used plants and patented for treatment of diabetes include cryptolepine (Bierer *et al.*, 1998a,b; Luo *et al.*, 1998), maprouneacin (Carney *et al.*, 1999), vismin (Inman and Luo, 1998), and quinones SP18904 and SP18905 (Luo *et al.*, 1999). The most interesting discovery was nordihydroguaiaretic acid (ndga) (Luo *et al.*, 1998) which, besides being active orally in diabetic mice, also lowered cholesterol levels. However, the recent discoveries on the antidiabetic compounds are vanillic acid derivative and its sulfate adduct isolated from green algae, *Cladophora socialis* (Chlorophyceae), showed potent inhibition of protein tyrosine phosphatase 1B (PTP1B), an important enzyme in regulating the insulin receptor (Jia *et al.*, 2003; Hays *et al.*, 2008). Cinnamaldehyde was identified as the compound responsible for anti-diabetic activity in *Cinnamonum zeylanicum* Blume (Lauraceae) (Jung *et al.*, 2006).

Three flavonoids, apigenin-5-O-[a-L-rhamnopyranosyl-(1/4)-6-O-b-D-acetylglucopyranoside] (Semple *et al.*, 2008), apigenin (Washburn, 2009), and apigenin-5-O-[a-L-rhamnopyranosyl-(1/4)-6-O-b-D-glucopyranoside] (Hardman *et al.*, 2010) were found to be responsible for the anti-hyperglycemic effect of the ethanol extract of the leaves of *Cephalotaxus sinensis*. A significantly increased level of glucose transporter GLUT-4 was also seen from mice adipocytes treated with this extract. Two bis (catechol glycoside) esters were isolated from the leaves of *Dodecadenia grandiflora* (Lauraceae) showed significant antihyperglycemic activity in STZ-induced diabetic rats (Gershell, 2005; Carpino *et al.*, 2010). Two main bioactive compounds, kraussianone-1 and kraussianone-2, from the roots of *Eriosema kraussianum* (Fabaceae) were studied for their hypoglycaemic properties (Sears, 2010; Anuradha, 2010).Two acylated kaempferol-3-O-a-L-rhamnopyranosides, from *Machilus philippinense* Merr. (Lauraceae) were isolated by bioassay-guided fractionation and found to inhibit a-glucosidase type IV(Cox *et al.*, 2010). Hung *et al.* (2012) has illustrated number of natural antidiabetic compounds other than mentioned in this section in an outstanding review.

Herbals as an Anticancer Drug

The plant *Catharanthus rosea* which was used in Madagascar for the treatment of diabetes was later observed that the extracts of this plant suppressed the bone marrow. Follow up of this observation led to the discovery and the isolation of the anticancer compounds vincristine, vinblastine, and vinorelbine. The compound taxol which was first isolated from the Pacific yew, *Taxus brevifolia*, was found to have anticancer activity. This has now resulted in the synthesis of the analogues paclitaxcel and docetaxel, which now constitute a new class of broad spectrum anticancer compounds. Plants of the family of Combretaceae are widely used in traditional medicine in African countries (Fyhrquist *et al.*, 2004; Moshi and Mbwambo, 2005). However, there are mainly four groups of plant-derived anti-cancer drugs in clinical use. They are vinca alkaloids (*e.g.*, vinblastine and vincristine), epipodophyllotoxin derivatives (*e.g.*, etoposide and teniposide), camptothecin derivatives (*e.g.*, topotecan and irinotecan) and taxanes (*e.g.*, paclitaxel) (Da Rocha *et al.*, 2001; Efferth *et al.*, 2007). All of these natural products are also traditionally used and recorded as poisonous herbs (Wang *et al.*, 2012). They have been produced through sophisticated technologies with a "Fighting fire with fire" framework which represents a distinctive thinking style of Chinese practitioners and this idea has set its proper position in cancer therapy (Figure 1.1).These drugs show strong therapeutic action in different types of cancers although their toxicities or side effects simultaneously exist and cannot be ignored Gambogic Acid, secreted by various *Garcinia* species including *Garcinia hanburyi* Hook. F. in Southeast Asia, has been demonstrated to be a potent anti-cancer agent with diverse molecular targets through different mechanisms. It is also an important anti-cancer drug candidate for its excellent cytotoxic potential against various malignant tumors (Zhao *et al.*, 2010b). GA presents potent effects on growth inhibition and G0/G1 cell cyclearrest in chronic myeloid leukemia cells via interruption of steroid receptor coactivator-3 (SRC-3) (Li *et al.*, 2009). The plants of this family are now a source of combretastatins, a new class of antimitotic drugs that also inhibit angiogenesis and are currently being developed for the treatment of

Figure 1.1: The research framework of PCHM is built up from many aspects, more sophisticated technologies and more suitable research methodologies are needed (Wang *et al.*, 2012).

cancer. Work done in our laboratory has shown that compounds, from this family hold a promise for new anticancer drugs. Triptolide from *Tripterygium wilfordii* Hook F. was first reported in 1972 as antileukemic (Kupchan *et al.*, 1972), the anti-tumor activity of triptolide has been demonstrated in various tumor models, including promyelocytic leukemia, human hepatocellular carcinoma, cervical adenocarcinoma, pancreatic carcinoma, multiple myeloma (MM), cholangiocarcinoma, and oral cancer cells (Liu, 2011). Tripterine, also known as celastrol, has different chemical structures from triptolide but shares several common properties (Salminen *et al.*, 2010). Celastrol possesses anti-tumor activity against a broad spectrum of tumors, both *in vitro* and *in vivo* (Kannaiyan *et al.*, 2011), which attributes to its multiple effects on apoptosis, cell cycle arrest, angiogenesis, and metastasis. Many signaling molecules are involved in the biological activity of celastrol, including the NF-αB (Sethi *et al.*, 2007), activation of caspase family proteins (Yang *et al.*, 2011a), VEGF receptors (Pang *et al.*, 2010), heat shock proteins, potassium and calcium channels (Sun *et al.*, 2006) and the immunoglobulin Fc epsilon receptor I (Fc_RI) (Kim *et al.*, 2009). In addition, celastrol also shows a synergistic effect when combined with chemotherapeutic agents, including radiation induced suppression of the proliferation of melanoma, ovarian cancer and lung cancer (Lee *et al.*, 2011).

Herbals as an Anti-HIV Drug

Despite the advances in anti-retroviral therapy which has transformed HIV/AIDS from a fatal to a manageable chronic disease, increasing viral drug resistance, side effects and uneven access to anti-retroviral drugs remain a considerable therapeutic challenge. As a consequence of these shortcomings and partly based on the fact that HIV/AIDS gives rise to opportunistic infections whose symptoms have been managed in Africa by traditional healers for centuries by using herbal medicines (Leteane *et al.*, 2012).The most detrimental and life threatening diseases in the 20th century, no doubt, is HIV-borne diseases AIDS. The use of traditional medicines for the management of HIV/AIDS patient is directed to three main areas; direct effect on the virus, immunostimulant activity, and treatment of opportunistic infections. A number of plants have now been identified from random screening or from traditional systems of treatment that interfere with nearly all stages of the viral life cycle. Compounds have been reported that inhibit virus-cell fusion, reverse transcription, virus adsorption, and proteolytic cleavage (Uchiumi *et al.*, 2003). A good number of plants contain these compounds, such as *Phyllanthus amarus* (Notka *et al.*, 2004), *Phyllanthus niruri* (Naik and Juvekar, 2003), *Salvia miltiorrhiza* (Abd-Elazem *et al.*, 2002), *Shepherdia argentea, Viola yedoensis, Arctium lappa, Epimedium grandiflorum, Glycyrrhiza uralensis* and *Castanospermum australe* (WHO, 1989), and *Croton tiglium* originate from traditional medicine. Other traditionally used medicines that have shown anti-HIV activity are *Aspilia pluriseta* and *Rumex bequaertii* (Cos *et al.*, 2002). In a recent study done on 21 plant species belonging to 14 families that are used in Ethiopian traditional medicine, extracts of four plants, *Bersama abyssinica* Fresen, *Combretum paniculatum* Vent., *Dodonaea angustifolia* L.f., and *Ximenia americana* L. displayed anti-HIV activity against both HIV-1 and HIV-2 strains (Asres *et al.*, 2001). Similarly screening of plants used in traditional medicines of other countries has led to identification of interesting anti-HIV activity (Piras *et al.*, 1997).

The most recent success story is the discovery of (+)-Calanolide A, a novel non-nucleoside HIV-1- reverse transcriptase inhibitor (NNRTI), which was first isolated from a Malaysian plant, *Calophyllum lanigerum* (Creagh *et al.*, 2001).The anti-HIV properties of Calanolide A show that it is effective towards multiple NNRTI-resistant mutations that include the highly resistant Y181C mutation. Clinical viral isolates that are resistant to nucleoside analogues have been shown to be hypersensitive to the antiviral effects of Calanolide A. In laboratory experiments Calanolide A also works synergistically towards HIV when used with other anti-HIV drugs. The drug is being developed by Sarawak MediChem and is now undergoing phase III clinical trials. In a recent report, it has been shown that Calanolide A is also active against both drug-sensitive and drug-resistant *Mycobacterium tuberculosis* (Xu *et al.*, 2004). *Sutherlandia frutescens* and *Lobostemon trigonu* are two plants that are marketed in South Africa for the treatment of HIV/AIDS patients. A recent study proved that these two plants have inhibitory effects against HIV-1 reverse transcriptase activity (Harnett *et al.*, 2005).

Clinicians treating HIV+ patients with Sutherlandia tablets produced by the South African company Phyto Nova have reported increase in CD4 and reduction of viral load. Sutherlandia is also reported to have apoptotic activity against human cancer cell lines (Chinkwo, 2005), thus suggesting as an additional potential for dealing with opportunistic malignancies in HIV patients.

Few other plant sources like *Taraxacum officinale*, (dandelion) extract showed strong activity against HIV-1 RT and inhibit both the HIV-1 vector and the hybrid-MoMuLV/MoMuSV retrovirus replication with less side effects (Han *et al.*, 2011). The root and stem bark of two other popular medicinal plants *Cassia sieberiana* and *Cassia abbreviata* showed direct inhibitory effects on the HIV-1c replication at its binding and entry stage (Leteane *et al.*, 2012). Two extracts, one from *Euphorbiaceae, Trigonostema xyphophylloides* (TXE) and one from *Dipterocarpaceae, Vatica astrotricha* (VAD) inhibited HIV-1 replication and syncytia formation in CD4+ Jurkat cells, and had little adverse effects on host cell proliferation and survival. These two extracts during the infection significantly blocked infection of the reporter virus (Park *et al.*, 2009). A lectin, BanLec, isolated from banana inhibits HIV-1 infection by binding to the glycosylated viral envelope and blocking cellular entry (Michael *et al.*, 2010).In Africa, herbal medicines are often used as primary treatment for human immunodeficiency virus (HIV) related problems. Concurrent use of traditional herbal medicines (THM) with antiretroviral drugs (ARVs) is widespread among HIV infected patients. However, the extent of THM use is not known in most settings in Sub-Saharan Africa (Namuddu *et al.*, 2011).

Herbals as an Anti-bacterial Drug

The concerns are being raised by several international organizations dealing with public health, for the emergency of bacterial resistance as a major health and social problem (French, 2010; Spellberg *et al.*, 2008).In fact multi-resistant bacteria endanger with increasing frequency of the successful outcome of antibiotic therapy of synthetic drugs and the fact that the spread of drug resistant organisms has largely outpaced antibiotic research and development. In fact, from the period 1983-1987 to

2003–2008, the FDA approval of new antibiotics has been decreased up to 70 per cent, with 16 drugs approved in the former and only 5 in the latter period. This trend of reduction has been constantly running for the last 30 years (Spellberg *et al.*, 2008). The strategies to search for the new plant-derived antibiotics have been the subject of the recent reviews (Moellering, 2011). In the quest of new antibacterial molecules, screening for the new leading compounds is a common step. The compounds from natural origin still provide a high number of interesting structures, even in this era of combinatorial chemistry. Plants have already yielded even the compounds with inhibiting activities against Gram positive bacteria acting as efflux pumps inhibitors (EPI) (Tegos *et al.*, 2002). That is why the use of medicinal and herbal remedies to treat infectious diseases is common in many countries (Ankli *et al.*, 2002). Hundreds of plant extracts have been using since a long time as safe and effective antimicrobial agents, which treat, therapeutically and prophylactically, however a wide variety of bacterial infections still a matter of concern for biomedical research (Mohebi *et al.*, 2011). Medicinal plants play an important role for the management of different microbial infections (Shinwari *et al.*, 2009) although the plants are not tested for either for microbiological or foreign materials contamination to assure the quality (Kruti *et al.*, 2011). However, these medicines can destroy microbes that cause certain contagious diseases and should be used in conjunction with modern medicines and antibiotics in order to obtain a better effect.

People use herbs to treat different diseases because they are cheap and effective, although doctors are often reluctant to prescribe them because of knowledge deficiency, real concerns about product safety, patients liability, and the presence of pathogens and compounds in the products that may be dangerous to health (Ernst, 2002; Hussain *et al.*, 2009). However, antimicrobial properties of medicinal plants are being increasingly reported from different parts of the world (Saxena, 1997). Many researchers have examined the uses of medicinal plants, but only a few studies have tested these Ethno-botanical findings in a laboratory setting to confirm the real antimicrobial properties of these plants (Bhattarai *et al.*, 2008; Shakya *et al.*, 2008). Medicinal plants can provide a wealth of antimicrobial agents, and hundreds have already been investigated for biological activities. In this mixed context, our experience shows that there are a number of plants with a strong potential for offering solutions to resolve of bacterial, viral, protozoa, and other opportunistic infections.

Herbals as an Anti-fungal Drug

Plants are rich source of bioactive secondary metabolites such as tannins, terpenoids, saponins, alkaloids, flavonoids, and other compounds, which have been reported to have antifungal properties *in vitro*. Since the plant kingdom provides a number of leading compounds of novel structures so a wide-scale investigation of species has been considered in the recent years. Therefore, the research on natural products and compounds derived from natural products has accelerated recently. A series of molecules with antifungal activity against different strains of fungus have been found in plants from several recent studies. These molecules may be used directly or considered as a precursor for developing better molecules (Arif *et al.*, 2009). Recently, a number of studies have reported the antifungal activity of phenolic compounds from natural sources. The site(s) and number of hydroxyl groups on the phenol group are thought to be related to their relative toxicity to microorganisms, with evidence that increased hydroxylation results in increased toxicity. The mechanisms thought to be responsible for phenolic toxicity to microorganisms include enzyme inhibition by the oxidized compounds, possibly through reaction with sulfhydryl groups or through more nonspecific interactions with the proteins (Arif *et al.*, 2009).

Another study on the mechanism of action of diterpenic and phenolic compounds isolated from *Euphorbia* species showed that these compounds modulate a different extent of azole antifungal resistance mediated by Pdr5p, Snq2p, and Cdr1p (Ferreira *et al.*, 2006). A number of phenolics, flavonoids, coumarins, quinones, saponins, xanthones, alkaloids, lectins and polypeptides, terpenoids and essential oils from several plant sources have been shown to have antifungal activities. Amentoflavone from *Selaginella tamariscina* and four other compounds eupomatenoid-3, eupomatenoid-5, conocarpan, and orientin, from *Piper solmsianum* exhibited strong antifungal action against dermatophytes with a potency as high as the standard antifungal drug ketoconazole (Campos *et al.*, 2005). Along with other 5,8-dihydroxyumbelliprenin from *Fsafetida foetida* showed strong antifungal activity against the dermatophytes. The naphthoquinones kigelinone, isopinnatal, dihydro-a-lapachone, and lapachol from *Kigelia pinnata*were reported for antifungal activity (Binutu *et al.*, 1996). Spirostanol steroidal saponins isolated from the roots of *Smilax medica*, together with the smilagenin 3-O-b-Dglucopyranoside and disporoside A exhibited antifungal activity against the human pathogenic yeasts *C. albicans*, *C. glabrata*, and *C. tropicalis* (Sauton *et al.*, 2005). Two out of eight saponins from *Tribulus terrestris* were reported to have promising antifungal activity against fluconazole-resistant *Candida* strains (MIC_{80} = 4.4 µg/ml), *C. neoformans* (MIC_{80} = 10.7 µg/ml), and *C. krusei* (MIC_{80} = 8.8 µg/ml) (Zhang *et al.*, 2005). Xanthone Caledonixanthone E isolated from the stem bark of *Calophyllum caledonicum* has been reported as a strong antifungal agent. Alkaloids from *Datura metel* and *Aniba panurensis* showed strong activities against *Aspergillus* and *Candida* in both *in vitro* and *in vivo* conditions. A lectin and a polypeptide isolated from the roots of *Astragalus mongholicus* exerted antifungal activity towards various fungi (Yan *et al.*, 2005). A novel antifungal peptide, cucurmoschin, isolated from the seeds of the black pumpkin inhibited mycelial growth in the fungi (Wang and Ng, 2003). An antifungal protein, AFP-J purified from potato tubers *Solanum tuberosum*, strongly inhibited yeast fungal strains, including *C. albicans*, *Trichosporon beigelii*, and *S. cerevisiae* (Park *et al.*, 20005). From the above-mentioned studies, it is clear that a number strong anti-fungal agent are available in the natural products of kingdom which can be isolated and purified in order to manufacture potent anti-fungal medicines.

Herbals as an Anti-malarial Drug

Malaria is a common problem in many parts of the world including Africa, South America and some parts of Asia. A number of plant based traditional medicines are currently being used for the treatment of malaria associated symptoms. In ancient time, the Cinchona bark was used for the treatment of malaria by South American Indians. This led to the isolation of quinine in the 17th century. Based on the structure of quinine other analogues have been synthesized. Mefloquine is a quinoline methanol with a structure similar to that of quinine. Many other antimalarial drugs were synthesized based on the structure of quinine include the 4-aminoquinolines, 8-aminoquinolines (primaquine and tefenaquine), and the aryl alcohols (halofantrine, lumefantrine, and pyronaridine). Recently, a number of antimalarial agents such asquinghasou, the antimalarials artemisinin, dihydroartemisinin, artemether, and arteether have been discovered from the Chinese traditional medicine (Meshnick *et al.*, 1996). A number of other plants have shown antimalarial activity and have potential for discovery of future antimalarial drugs. Recently *Agecerious corniculatum* (Sundaram *et al.*, 2012), Quassinoid compound Simalikalactone D from *Quassia amara* L. (Bertani *et al.*, 2012), *Aspilia africana* (Christian *et al.*, 2012) *Ficus platyphylla* (Shittu *et al.*, 2011) have shown antimalarial effects in experimental control mice.

Herbals as an Anti-asthmatic Drug

Asthma is a chronic disease characterized by acute exacerbation of coughing, dyspnoea, and wheezing and chest tightness. Numbers of medicinal plants have been used for the treatment of asthmatic condition since a long time. Khellin from *Amni visnaga* was used in USA as a bronchodilator however nausea and vomiting was the major long-term side effects of this plant products. As an attempt to synthesize khellin analogues with fewer side effects the antiasthmatic drug chromolyn (Sodium chrogycate) was discovered (Sneader, 1985). Sodium chromogylycate is now used to stabilize the membranes of macrophages and prevent degranulation, and hence release of mediators of broncho-constriction in asthmatic patients. Apart from the active isolations of plants, extracts of *Adhatoda vasica, Albizzia lebbeck, Artemicia caerulenscens, Boswellia serrata, Calotropis gigantea, Calotropis procera, Cedrus deodara, Clerodendron serratum, Curcuma longa, Eugenia caryophylis, Eleocarpus spharicus, Inula racemosa, Ocimum sanctum, Picrorrhiza kurroa, Piper longum, Sarcostemma brevistigma, Solanum xanthocarpum, Tephrosia purpurea, Tinospora cordifolia, Tylophora asthmatica, Vitex negundohave* been used as anti-asthmatic remedy in the different parts of the World (Ismail, 2010).

Herbals as a Molecular Tools for Treatment

As early as 1882, investigation of Chinese Medicine (CM) led to the discovery of ephedrine from *Ephedra sinica* (Ma-Huang). This compound and its numerous synthetic analogues are used today as effective anti-asthmatic agents. A lot of scientists are trying to enter into the new era of herbal regimes. Lee (2000) reported that he has been designing following new bioactive natural products in his Natural Products Library and their analogues as chemotherapeutic agents:

1. Plant anti-tumour agents–novel plant-derived cytotoxic anti-tumour principals and analogues.
2. Novel anti-tumour etoposide analogues–design and synthesis of novel analogues related to the anti-cancer drug etoposide with better pharmacological profiles.
3. Development of anti-AIDS agents- plant-derived anti-HIV compounds. Two classes of modified natural products from betulinic acid (*Szyigium claviflorum*) and the coumarin suksdorfin (*Lomatium suksdorfii*).

4. Novel anti-malarial agents–anti-malarial analogues related to qinghaosu.
5. Anti-viral hepatitic drugs–design and synthesis of analogues related to DDB, a clinically useful anti-viral hepatitic drug.
6. New non-steroidal anti-inflammatory and anti-arthritis agents.
7. New anti-fungal agents.
8. Novel anti-viral agents–inhibitors of HSV I and II and HCMV.
9. Combinatorial chemistry–focused combinatorial libraries based on bioactive natural products models.
10. Chinese medicines–discovery and development of active principals, fractions, and formulations from traditional Chinese herbal medicines.

In the area of anti-cancer drug design, the Natural Product Laboratory(NPL) has discovered more than 1000 new cytotoxic anti-tumour principals and their derivatives or synthetic analogues from medicinal herbs in the past nearly three decades. Cytotoxic extracts, fractions and compounds are screened first in an initial in-house human tumour cell line (HTCL) panel and then followed through bioactivity-directed fractionation and isolation (BDFI), focusing on one or two of the most active bioassays systems. Promising samples are confirmed effective in forms the evaluation of National Cancer Institute(NCI) *in vitro* HTCL and *in vivo* xenograft assays (Boyd, 1993). In this manner, BDFI of Chinese medicinal herbs, coupled with rational structural modification and synthesis, have resulted in more than 100 new cytotoxic anti-tumour compounds and their analogues that are of current interest to NCI and have shown confirmed activity in NCI's *in vitro* primary human tumour cell lines bioassay (Figure 1.2).

Etoposide is a semisynthetic derivative of the natural product podophyllotoxin, which is isolated from *Podophyllum peltatum*. However, their differences include both structural changes at the 4- and 40-position and their mechanism of action: podophyllotoxin targets tubulin, while etoposide targets the enzyme DNAtopo II (Cragg and Suffness, 1988). Etoposide and the related thiophene analogueteniposide are widely used to treat various cancers, including leukaemias and lymphomas; however, myelosuppression, drug resistance, and poor bioavailability remain as problems associated with their use (O'Dwyer and Alonso, 1984; Issell *et al.*, 1984).

Evaluation of the Safety and Efficacy of Herbal Medicines

Depending on the particular country and existing legislation, herbal products used for diagnosis, cure, mitigation, treatment, or prevention of diseases is normally regarded as drugs. However, in some countries, including United States, botanical products are marketed as dietary supplement. Other countries treat the herbal preparations as drugs, and to be registered these products for testing their safety and clinical efficacy. However, so far, few programs have been established to study the safety and efficacy of herbal medicines as originally proposed by the WHO Guide lines for the assessment of herbal medicines (Zeisel, 1999; Petrovick *et al.*, 1999; Keller, 1996).

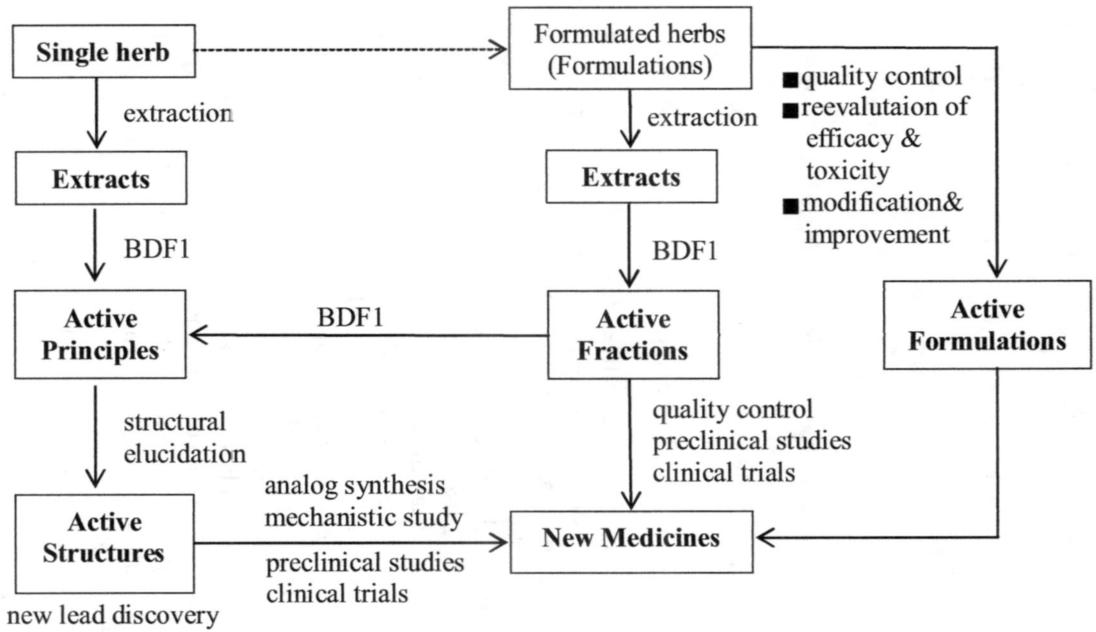

Figure 1.2: Principal concepts of research on new medicinals (Lee KH, 2000). BDF1 denotes bioactivity-directed fractionation and isolation.

Although clinical trials with herbal drugs are feasible, a survey of the specialized literature reveals that few well-controlled double-blind (placebo-controlled) trials have been carried out with herbal medicines. Several recent reviews published in the Annals of Internal Medicine, the Journal of the American Medical Association (JAMA), the British Medical Journal, the Lancet, and the British Journal of Clinical Pharmacology, among others, confirm the feasibility of herbal medicines. However, some drawbacks have been focused strong in these reviews on the way of the establishment of herbal drugs, for example:

1. Lack of standardization and quality control of the herbal drugs used in clinical trials;
2. Use of different dosages of herbal medicines;
3. Inadequate randomization in most studies, and patients were not properly selected;
4. Numbers of patients in most trials are insufficient for the attainment of statistical significance;
5. Difficulties in establishing appropriate placebos because of the tastes, aromas, etc;
6. Wide variations in the duration of treatments using herbal medicines.

As a consequence of such difficulties, few herbal drugs have been studied adequately and well-controlled double-blind clinical trials to prove their true safety and efficacy. However, a large number of clinical trials have been performed only with some herbal drugs, including the extract of *Ginkgo biloba* (used for the treatment of CNS and cardiovascular disorders) (Brautigam *et al.*, 1998) and *Hypericum perforatum* (St. John's wort, used as an antidepressant) (Josey and Tackett, 1999). It is important to emphasize that most such clinical studies have received the same criticisms as those mentioned above. Although the clinical trials have shown that these herbal drugs are quite safe and devoid of serious side effects, their clinical efficacy still requires well-controlled randomized double-blind studies. Other botanical drugs such as *Panax ginseng* (ginseng) herbs used as a tonic, *Tanacetun parthenium* (feverfew) used to treat migraine headache, *Allium sativum* (garlic) used to lower low-density protein cholesterol and some cardiovascular disturbances, *Matricaria chamomilla* (chamomile) recommended as a carminative, anti-inflammatory and antispasmodic, *Silybium marianun* (milk thistle) used for repairing liver function including cirrhosis, *Valeriana officinalis* (valerian) used as a sedative and sleeping aid, *Piper methysticum* (Kava kava) used as an anxiolytic, *Aesculus hippocastanum* (horse chestnut) used for the treatment of chronic venous insufficiency, *Cassia acutfolia* (Senna) and *Rhamnus purshiana* (cascara sagrada) which are used as laxatives, *Echinacea purpura* (Echinacea) used as an anti-inflammatory and immunostimulant, *Arnica montana* (arnica) used to treat post-traumatic and post-operative conditions, and *Serenoa repens* (saw palmetto) used for the treatment of benign prostatic hyperplasia are all herbal medicines with the largest worldwide market, and although they have been evaluated in different clinical trials, additional well-controlled and appropriate randomized clinical trials are still needed in order to prove their authentic efficacy. However, the situation will change very quickly because the increase in the world market for medicinal herbs has attracted most of the largest pharmaceutical companies, including some multinationals, and some of them have recently acquired small companies specialized in phytotherapeutic agents (Blumenthal,1999). Since people have accumulated experience and possess expertise in drug development, and mainly because they have the money to carry out the clinical trials, great progress in herbal medicine should occur in the near future. The general idea that herbal drugs are very safe and free from side effects is false. Plants have hundreds of constituents and some are very toxic such as the most cytotoxic anti-cancer plant-derived drugs, digitalis, the pyrrolizidine alkaloids, ephedrine, phorbol esters, etc. However, the adverse effects of most herbal drugs are relatively less frequent when the drugs are used properly compared with synthetic drugs, but well controlled clinical trials confirm that they really existed (Bagheri *et al.*, 1998; De Smet, 1995).

Drawbacks to Use Herbal Drugs and Herbal Research

Problems Related to Herbal Drugs Use

The use of herbal products is complicated by numerous problems as summarized below:

Lack of Scientific Evidence of Safety and Efficacy

Most herbal products have not been scientifically evaluated; thus, information is limited about their pharmacokinetics, pharmacodynamics, efficacy, and safety. Randomized controlled trials are the best way to determine efficacy and safety; however, controlled trials for most herbal products are not available, which are also not demanded by consumers or health care practitioners, and not required by regulatory agencies. Belief in the safety of complementary and alternative medicine (CAM) products remains unsubstantiated, and even when data are available, findings are often questionable because of a lack of consistency in research methods, small sample size, absence of placebo group, inclusion of healthy volunteers or low-risk populations without co-morbid conditions, lack of standardization of supplements being investigated, and absence of data about drug–herb interactions. Thus, findings for specific herbal products are often limited for making decisions about the efficacy or safety.

Lack of Regulatory Oversight

Despite the lack of scientific evidence about the efficacy and safety of herbal products, more than 30,000 dietary supplements are sold in the U.S., with an additional 1,000 new supplements introduced annually. Most patients believe that the government oversees the safety of CAM; but the fact is that the only requirement is for the manufacturer to send a copy of the product label to the U.S. Food and Drug Administration (FDA). A new dietary supplement or new formulation can be introduced and marketed overnight, without significant restriction, despite containing new, experimental, and unregulated herbal ingredients. The Dietary Supplement, Health and Education Act of 1994 provides some oversight but nonetheless allows untested herbal products on the market (Sheehan and Atherton, 1992). Unfortunately, many supplements contain ingredients or contaminants with adverse effects or interactions with other drugs. If these supplements were subjected to pre-market safety testing, many would no doubt to be banned.

The FDA has identified several supplements, including ephedra, chaparral, comfrey, lobelia, and yohimbe, that have caused serious harm, yet it took 9 years to get ephedra alone removed from the market (Bensoussan *et al.*, 1998). The other products are still available for sale in stores and over the internet. Restricting a specific substance or group of substances is not particularly helpful because manufacturers can simply introduce new formulations. From 1995 to 1997, hundreds of cases of ephedra toxicity were reported to the FDA as causing stroke, myocardial infarction, seizure, and death from cardiac arrhythmia (McNamara, 1996). Side effects occurred in otherwise healthy young adults those who were used ephedra for energy boost and weight loss (Laakmann *et al.*, 1998). As a result, ephedra was eventually banned by the FDA, but other structurally related sympathomimetic amines emerged to take its place in weight-loss and energy-enhancement products, such as synephrine, which is now the most popular ephedra replacement. It has a similar pharmacology and structure. Internet resources describe synephrine as an herbal supplement that stimulates fat metabolism, suppresses appetite, and boosts caloric expenditure without side effects. Synephrine inhibits intestinal cytochrome P450 and increases blood levels of many drugs. Like ephedra and caffeine, the combination of synephrine and caffeine produces arrhythmogenesis, stroke, CHF, and seizures (Haller and Benowitz, 2000). The general public regards herbal products as natural. Thus, there is a widespread yet false perception that herbal products are safe.

Lack of Quality Control

The wide variations in herbal supplement manufacturing techniques, storage methods, and lot-to-lot variability require urgent attention from the FDA and other regulatory agencies. The U.S. Pharmacopeia has been developing quality control standards, but supplement manufacturers are not required to follow these guidelines. Such latitude allows cheaply manufactured and even expired herbal products to be sold without restriction. The variation in herbal sources and the inconsistent labeling of ingredients continue to be the major concerns. More than 40 per cent of all herbal products fail to contain as much of their active ingredients as claimed on their labels (Harris poll, 2008) or substitute cheaper ingredients for more expensive ones (Brody, 2009). Contamination with heavy metals, adulteration with pharmaceuticals, and prohibited animal and plant ingredients are repeatedly found in herbal products (Saper, 2008). Intentional and unintentional adulteration with antibiotics, nonsteroidal anti-inflammatory drugs, hormones, and heavy metals is common. Untested and unregulated steroids are sold to the public, including minors (Corns, 2003). In 2007, the FDA issued 9 safety alerts warning consumers to stop using 13 brands marketed as dietary supplements because testing identified undocumented prescription medications in them (Corns, 2003). Nine contained erectile dysfunction drugs such as sildenafil or tadalafil, 3 contained lovastatin, and 1 contained the weight-loss drug sibutramine. The ramifications of using herbal supplements are sometimes discerned through subsequent medical tests or treatment.

Aristolochic acidnephropathy has been reported in patients who took herbal supplements without knowing that the manufacturer had substituted *Aristolochia* for another herb (Debelle*et al.*, 2008). In addition to being a renal toxin, aristolochic acid is on the World Health Organization's list of human carcinogens. It is banned from use in several European countries, but is still readily available in the U.S. (Corns, 2003; Debelle *et al.*, 2008). The FDA guidelines for manufacturers mandate that they avoid contaminating products with other herbs, pesticides, heavy metals, or prescription drugs (Nortier *et al.*, 2000). Nonetheless, products have been repeatedly reported to contain adulterant compounds and to put unknowing consumers at risk of adverse side effects and drug interactions.

Public Misinformation

Unethical marketing techniques have led to false advertisements about the safety and efficacy of the herbs reaching the public today, more so than at any other time. Herbal products are promoted in magazines, newspapers, and books, on radio and television, and through the internet. Herbs are advocated for treatment on the basis of unproven, word-of-mouth traditions and beliefs. Some promotional materials even claim that the featured product is doctor recommended, the world's most powerful, patented, or now presented without a prescription (even if it had never been prescribed). In a study of internet marketing of herbal products, Morris and Avorn (2003) found that at least 81 per cent of internet sites made 1 or more health claims; with more than 50 per cent claiming to treat, prevent, diagnose, or cure specific diseases despite regulations barring such statements (Guidance for industry, 2009). The sale of herbal supplements can be halted only if the product has been proven dangerous and documentary proof has been submitted to the Secretary of the Department of Health and Human Services. However, the FDA does not routinely test food supplements, and has to demonstrate that a dietary supplement is "unsafe," before it can take action to restrict the product's use or removal from the marketplace (Guidance for industry, 2009). The burden of proof for demonstrating that a supplement is safe must be placed on manufacturers rather than the government or consumers, and consumer protection laws must be instituted.

Lack of Knowledge about Herb-Drug Interactions by Patients and Health Care Providers

Devastating effects due to a lack of understanding about adverse herb–drug interactions can occur in vulnerable populations, including elderly persons, children, pregnant women, immunocompromised people (*e.g.*, those with human immunodeficiency virus or transplant recipients), and those with hepatic or renal dysfunction. Most (64 per cent) of the patients with a diagnosis of atrial fibrillation, CHF, or ischemic heart disease who were attending a cardiovascular clinic reported concomitant use of alternative therapies and prescription drugs (Morris and Avorn, 2003). More than one-half (58 per cent) took supplements that had potential interactions with warfarin, amiodarone, sotalol, or digoxin. In a geriatric clinic, 46 per cent of the patients were taking supplements with anticoagulation properties. Most (73 per cent) were also taking a prescription anticoagulant and were unaware of potential interactions. Findings from another study indicated that 1 in 6 adults taking conventional prescription medications reported concomitant use of at least 1 CAM product during the preceding week. More than 40 per cent of the respondents said that taking herbal medicine along with prescription medication was more useful than taking either alone (FDA, 2010).The common herb-drug interactions are summarized in Table 1.2.

Table 1.2: Herbs-Drug Interactions: common herbs (Tachjian *et al.*, 2010).

Herb	*Drug or Drug Class*	*Interaction or Other Comments*
Comfrey	Phenobarbital	Increases metabolism of comfrey, producing a lethal metabolite from pyrrolizidine that results in severe hepatotoxicity
Danshen	Anticoagulant or antiplatelet agents	Increases bleeding due to additive effects
	Digoxin	Increases side effects of digoxin
Echinacea	Amiodarone or ibutilide	Increases QT interval
	Statins, fibrates, niacin	Increases risk of hepatotoxic effects
	Antidiabetes drugs	Increases blood glucose
		Decreases effectiveness of oral hypoglycaemic agents
	Class IA and class III antiarrhythmics	Increases QT interval
	Beta-blockers	Decreases effects of beta-blockers, leading to hypertension and tachycardia
	Monoamine oxidase inhibitors	Hypertension
Evening primrose oil	Phenobarbital	Decreases seizure threshold
Garlic	Aspirin, clopidogrel, warfarin, or heparinoid drugs	Increases bleeding risk
Ginkgo biloba	Antidiabetes drugs	Increases hypoglycaemia
	Aspirin	Increases bleeding
	Warfarin	Inhibits PAF hemorrhage
Ginseng	Antidiabetes drugs	Increases hypoglycaemia
	Digoxin	Interferes with digoxin assay, leading to falsely increased levels
	Warfarin	Decreases effectiveness of warfarin
	Phenelzine sulfate	Headache
		Irritability
		Insomnia
Hawthorn	Digoxin	Increases effects of digoxin
	Calcium-channel blockers or nitrates	Increases vasodilatory effects
Kava	Alprazolam	Increases CNS depression
		Increases effects of alcohol
Licorice	Spironolactone	Increases effects of spironolactone

Contd...

Table 1.2–*Contd...*

Herb	*Drug or Drug Class*	*Interaction or Other Comments*
Saw palmetto	Anticoagulant or antiplatelet agents	Increases bleeding
Soy milk	Warfarin	Decreases effectiveness of warfarin
St. John's wort	Digoxin	Decreases serum digoxin concentration
	Clopidogrel	Increases activity of clopidogrel
		Increases bleeding
	Warfarin	Decreases warfarin bioavailability and effectiveness
	Simvastatin	Decreases effectiveness of simvastatin
	Paroxetine	Nausea
		Lethargy
		Incoherence
	Class IA and III antiarrhythmic agents	Decreases effectiveness (precipitating arrhythmias)
	Cyclosporine	Decreases cyclosporine concentration due to increased clearance (transplant rejection)
	Theophylline	Decreases serum concentration
	Indinavir	Decreases serum concentration (treatment failure in HIV patients)
Yohimbine	Clonidine, Guanabenz	Decreases blood pressure reduction effect of centrally active agents
	CNS stimulants (*e.g.*, amphetamines)	Risk of hypertensive crisis due to its monoamine oxidase inhibitor activity
	ACE inhibitors	Decreases effectiveness of ACE inhibitors
	Hypertension	Beta-blockers
		Decreases effectiveness of beta-blockers
		Hypertension
		Increases heart rate

Under-Reporting of Adverse Drug Reactions

Most consumers are unlikely to attribute health problems to a supplement that they assume it safe and natural, and are reluctant to report an adverse effect from a self-medicated substance. Supplement manufacturers also rarely report serious adverse effects to the FDA despite federal requirements. Between 1990 and 1994, fewer than 10 of more than 2,500 reports of adverse effects made to the FDA came from an herbal product manufacturer. A report from the inspector general of the Department of Health and Human Services revealed that >1 per cent of adverse reactions caused by dietary supplements are reported to the FDA (Cohen *et al.*, 2002). Recent FDA post-marketing guidance for the supplement industry mandates the reporting of all serious adverse events to the FDA within 15 days (Kuo *et al.*, 2004).

Measures to be taken to Improve the Quality of Herbal Drugs

Raw Material

The microbiological quality assurance of herbal medicines must necessarily involve a proper stance regarding the Good Agricultural Practice. Researchers have even recommended the cultivation indoors, with full control over the climate and the presence of pathogens, and the possibility of genetic enhancement combined with biotechnological studies, indicating a tendency to domestication instead of harvesting from the wild(Blumenthal *et al.*, 1998; Calixto, 2000). The cultivation of medicinal plants may bring to the achievement of desirable objectives related to the eight Millennium Development Goals proposed by the United Nations (UN) and addressed by the World Health Organization (WHO, 2009) again indicate the strategic importance of these products to the various traditional communities. These communities include family farmers, indigenous communities, Maroons, and other groupings in a vulnerable socioeconomic position, which have a historical knowledge of the properties and use of various medicinal plants (Yoney *et al.*, 2010; Leonti, 2011). For the success of this strategy it is necessary to ensure these communities with the access to this information and know-how, since they supply raw materials. This connection apparently seems to be unattainable due to social, anthropological and cultural incompatibilities. But what we have observed in practice is the gradual development of market relations between the major urban centers and the more reserved communities, which generally seek to avoid contact with the outside elements. This fact demonstrates the tendency of many of these communities to open channels for knowledge and values exchange in order to facilitate the development and quality of life without their identity loss.

Sampling and Sample Preparation

Initial sample or raw material should be analyzed and assayed for microbial contamination of herbal preparation. Initially 10 ml or 10 g of sample are needed for the performance of the assays. In case of raw materials, an interesting strategy to homogenize the sample and facilitate the interaction with the culture medium used, is the previous hand-grinding with a properly sterilized mortar and pestle in a sterile environment such as a laminar flow biological safety cabinet. So, when analysis objects are leaves, stems or other plant constituents in their raw state, the sample preparation becomes important. On the other hand, thinner powders interact well with the solutions used in the homogenization and in the adjustment to near-neutral pH (pH = 7.0), favoring the growth of microorganisms present in the sample. Therefore, the pharmaceutical form may provide different strategies for sample preparation. When there is doubt about the presence of substances with inhibitory microbial growth a preparatory test for assessing the inhibitory capacity can be used. Carry out a transfer of each of the viable strains of *Staphylococcus aureus, Pseudomonas aeruginosa, Salmonella, Escherichia coli* and *Candida albicans* to 5 ml of soyabean casein broth and incubate at 30°C-35°C for 24 hours for bacteria and at 20°C- 25°C for 24 hours in the case of *Candida albicans*, making decimal dilutions in buffered peptone solution to obtain a suspension containing 50 to 100 viable cells. Use 1 ml of the microorganisms' inoculum separately in the presence or absence of the sample and incubate again as described for bacteria and fungi. If there is a massive growth of microorganisms, microbiological assays may be undertaken; in case of little or no growth of viable strains it is necessary to use inactivating agents.

Viable Microorganisms Count

After preliminary tests, pour 10 ml or 10 g sample into a container containing 90 ml of soybean casein broth, nutrient broth, buffered peptone broth or other appropriate diluents.

It is important to attempt to the fact that the pH should be near-neutral and the characteristics of the sample, such as raw materials from forest origin, may contain higher amounts of fungi and therefore prone to high levels of aflatoxins. So, an appropriate medium would be one that favours the growth of fungi capable of producing these substances. From this first dilution, transfer 1 ml to tubes containing 9 ml of the used diluent, for example, soybean casein broth, so as to obtain serial dilutions of 10-1, 10-2 and10-3. In some cases the number of dilutions can be increased to improve the counting. From each dilution, transfer volumes of 0.1 to 0.5 ml to Petri dishes in duplicate containing Soybean casein agar or Sabouraud-Dextrose Agar for counting bacteria and fungi, respectively. The spreading is accomplished with the aid of Drigalski spatula or asterile glass rod. In case of bacteria, the incubation should be at 30°C-35 °C for 48 to 72 hours; for fungi, temperatures from 20°C to 25 °C for 5-7 days at least. Antibiotics are commonly used in the culture medium in order to inhibit the growth of undesirable microorganisms. Therefore, amphotericin B is added to the culture medium for bacterial growth (Soybean Casein Agar or Nutrient Agar) and chloramphenicol to the culture medium for fungi growth (Sabouraud-Dextrose Agar or Potato Agar).

WHO's Recommendations in Quality Control

To ensure the quality of vegetable raw materials, in 1992, the WHO recommends that pharmacopoeia specifications for the plant materials which must include:

1. Botanical name, with reference to the authors;
2. Specification of the part used;
3. Morphological description, macro and microscopic examination;
4. Determination of particle size distribution;
5. Determination of total ash or sulfated ash (residue on ignition) and acid insoluble ash;
6. Determination of extractable matter;
7. Determination of water and loss on drying;
8. Determination of essential oils;
9. Identification by thin layer chromatography;
10. Quantitative determination of active constituents;
11. Limit tests for heavy metals;
12. Determination of pesticide residues.

Harmonizing Traditional and Modern Medicine

The current interest in traditional and complementary medicine in the United States is attracting attention in many parts of the community–the health care industry, governmental agencies, media and the public. An increasing number of insurers and managed care organizations are providing benefits for traditional medicine, a majority of U.S. medical schools now offer courses covering traditional medicine and, as Eisenberg's national studies have revealed, more people are using complementary therapies (Wetzel *et al.*, 1998). To facilitate research on the effectiveness of alternative therapies, the National Center for Complementary and Alternative Medicine (NCCAM) received a budget of US$50 million in 1999. Recognizing the need to encourage quality and quantity of scientific information on botanicals, as well as develop a systematic evaluation of safety and efficacy of dietary supplements, two research centers were also established this year to investigate the biological effects of botanicals (Eisenberg *et al.*, 1993). Many patients are using traditional and modern medical paradigms concurrently, creating a need for the appropriate and smooth merger of the two medicines. The theories and techniques of traditional medicine encompass most practices classified as complementary medicine in the United States, and have become increasingly important in the health care system (Eisenberg *et al.*, 1998).Traditional medicines are affordable, low tech, safe and effective when used appropriately. Ongoing research around the world on acupuncture, herbs, massages (NIH, 1999) and Tai-Chi (Field *et al.*, 1997; Sunshine *et al.*, 1996) has shed light on some of the theories and practices of traditional medicine. Evidence derived from vigorous research designs well as patient demand are fueling the merger of traditional medicine with modern medicine at the clinical level, while more academic researchers and institutions are becoming interested in the potential of integrating these two healing traditions.

Modernization of Herbal Drugs

Modern technologies being used or under development for the quality control (QC) of herbal medicines include: microscopic analyses (morphological investigation and powder methods), chromatographic methods for fingerprint profiling [thin-layer chromatography, high-performance liquid chromatography, liquid chromatography and gas chromatography (TLC, HPLC,LC, GC), GC- and LC-mass spectrometry (GC-MS and LCMS), LC/MS/MS and capillary electrophoresis (CE)], DNA polymorphism methods to analyze DNA sequences and establish relationships between genomic DNA fragment patterns and ratios of active components, and biological methods or bioassays. Modern instrumental quantification, identification, and comparison have been documented increasingly, for example, for licorice (by HPLC) (Hosoya, 1985), ginseng (by positive ion MS) (Shibata, 1985) and Uncariae powder (Gou-Teng-San) (by 3D-HPLC) (Wang *et al.*, 1999). The latter prescription is used in exceptionally nervous patients, who exhibit, among various symptoms, insomnia (Shu-Wei Hsu's 'Pen ShihFang', Sung Dynasty) and vascular dementia (Wang *et al.*, 1999). Such technical modernization of herbal medicine is crucial to promote standardization and establish QC through qualitative and quantitative analysis of active constituents and bioactivity. In addition, efforts should be made to guarantee safety by evaluation of toxicity and side effects and to ensure efficacy based on clinical scientific observation, including double blind clinical trials, rather than anecdotal evidence only. Similarly, dietary supplements must meet all high quality assurance standards. Modern analytical and pharmacological methods are being used to research and document the properties (not only the quantitative composition, but also the pharmacological action) of herbal formulations in comparison with their individual components. The following three approaches for the development of new, effective and safe world-class drug scan be envisioned. Active principals (bioactive compounds) approach, Active fractions approach, Active formulations approach. Figure 1.1 illustrates graphically that, now and in the future, not only bioactive principals (lead compounds or leads), but also active fractions and active formulations, must be investigated to continue the legacy of herbal medicine, as well as the uses of other medicinal herbs, for drug development in the 21st century.

Conclusions

Despite of a number of limitations herbal drugs have been exploring in the world healthcare market as a competitive and cheaper alternative to expensive synthetic drugs. According to the recent data and publications, the use of plant originated natural medicines or herbals are increasing at a significant rate in all over the world and many multinational pharmaceutical industries are funding for broad band on the safety and efficacy of herbal medicine research. Although it is anticipated that the popularity of herbal medicines will be increased day by day but care should be taken on herbal material preparations, processing, storage, herbal manufacturing, quality control, evaluation of toxicity, safety and efficacy, drug-interactions as well as the harmonization of herbal and synthetic drugs. The patients and consumers also should be more careful before taking the herbal medicines or food supplements. It is recommended to make themselves familiar with the safety and efficacy of herbal medicines before they start to consume or use anyone of them. At a time, traditional healers can be trained by natural product scientists those who are working with the pharmacological and toxicological effects of medicinal plants and/or natural medicines and subsequent patients can consult with a traditional healer before taking a herbal medicine.

Acknowledgements

This work was supported by a competitive research grant from the Research Office, University of KwaZulu-Natal (UKZN), Durban; an incentive grant for rated researchers and a grant support for women and young researchers from the National Research Foundation (NRF), Pretoria, South Africa. The first author of this work was received a Postdoctoral Fellowship from the Research Office of UKZN, Durban, South Africa

References

Abd-Elazem, I.S., Chen, H.S., Bates, R.B., and Huang, R.C. (2002). Isolation of two highly potent and non-toxic inhibitors of human immunodeficiency virus type 1 (HIV-1) integrase from *Salvia miltiorrhiza*. *Antiviral Research*, **55**: 91-106.

Akerele, O. (1993). Summary of WHO guidelines for the assessment of herbal medicines. *Herbal Gram*, **28**: 13-19.

Alves, R.R.N., and Rosa, I. L.(2006). From cnidarians to mammals: the use of animals as remedies in fishing communities in NE Brazil. *Journal of Ethnopharmacology*, **107**: 259–276.

Alves, R.R.N., Rosa, I.L., and Santana, G. G. (2008).The role of animal-derived remedies as complementary medicine in Brazil. *BioScience*, **57**: 949–955.

Alves, R.R.N. (2008). Animal-based remedies as complementary medicine in Brazil. *Forschende Komplementarmedizin*, **15**: 226–227.

Alves, R.R.N., Mendonc, L. E. T., Confessor, M. V. A., Vieira, W. L. S., and Lopez, L. C. S. (2009). Hunting strategies used in the semi-arid region of northeastern Brazil. *Journal of Ethnobiology and Ethnomedicine*, **5**: 12.

Alves, R.R.N., Barbosa, J.A.A., Santos, S.L.D.X., Souto, W.M.S., and Barboza, R.R. D. (2011).Animal-based remedies as complementary medicines in the semi-arid region of Northeastern Brazil. *Evidence-Based Complementary and Alternative Medicine*, **2011**: 1-15.

Angell, M. (1999). The American health care system revisited–A new series. *New England Journal of Medicine*, **340**: 48.

Ankli, A., Heinrich, M., Bork, P., Wolfram, L., Bauerfeind, P., Brun, R., Schmid, C., Weiss, C., Bruggisser, R., Gertsch, J., Wasesch, A.M., and Sticher, O. (2002). Yucatec Mayan medicinal plants: evaluation based on indigenous uses. *Journal of Ethnopharmacology*, **79**: 43–52.

Anuradha, C.V. (2009). Amino acid support in the prevention of diabetes and diabetic complications. *Current Protein and Peptide Science*, **10**: 8–17.

Arif, T., Bhosale, J.D., Kumar, N., Mandal, T.K., Bendre, R.S., Lavekar, G.S., and Dabur, R. (2009). Natural products – antifungal agents derived from plants. *Journal of Asian Natural Products Research*, **11**: 621–638.

Asres, K., Bucar, F., Kartinig, T., Witvrouw, M., Pannecoupue, C., and De Clercq, E. (2001). Antiviral activity against human immunodeficiency virus type 1 (HIV-1) and type 2 (HIV-2) of ethnobotanically selected Ethiopian medicinal plants. *Phytotherapy Research*, **15**: 62-69.

Avants, S.K., Margolin, A., Holford, T.R., and Kosten, T.R.(2000). A randomized controlled trial of auricular acupuncture for cocaine dependence. *Archieves of Internal Medicine*, **160**: 2305-2312.

Aziz, Z., and Tey, N.P.(2009). Herbal medicines: Prevalence and predictors of use among Malaysian adults. *Complementary Therapies in Medicine*, **17**: 44–50.

Bagheri, H., Broué, P., Lacroix, I., Larrey, D., Olives, J.P., Vaysse, Ph., Ghisolfi, J., and Montastru, J.L. (1998). Fulminant hepatic failure after herbal medicine ingestion in children. *Térapie*, **53**: 77-83.

Bensoussan, A., Talley, N.J., Hing, M., Menzies, R., Guo, A., and Ngu, M. (1998). Treatment of irritable bowel syndrome with Chinese herbal medicine: a randomized controlled trial. *Journal American Medical Association*, **280**: 1585-9

Bertani, S., Houël, E., Jullian, V., Bourdy, G., Valentin, A., Stien, D., and Deharo, E.(2012). New findings on Simalikalactone D, an antimalarial compound from *Quassia amara* L. (Simaroubaceae). *Experimental Parasitology*, **130**: 341-7.

Bhattarai, S., Chaudhary, R.P., and Taylor,R.S.L. (2008).Screening of selected ethnomedicinal plants of Manang district, Central Nepal for antibacterial activity.*Ethnobotany*, **20**: 9-15.

Bierer, D.E., Dubenko, L.G. Zhang, P., Lu, Q., Imbach, P.A., Garofalo, A.W., Phuan, P., Fort, D.M., Litvak, J., Gerber, R.E., Sloan, B., Luo, J., Cooper, R., and Reaven, G.M. (1998b). Antihyperglycemic activities of cryptolepine analogues: an ethnobotanical lead structure isolated from *Cryptolepis sanguinolenta*. *Journal of Medicinal Chemistry*,**41:** 2754-2764.

Binutu, O.A., Adesogan,K.E., and Okogun, J.I. (1996).Antibacterial and antifungal compounds from *Kigelia pinnata*.*Planta Med.*, **62**: 352-3.

Blumenthal, M., Brusse, W.R., Goldberg, A., Gruenwald, J., Hall, T., Riggins, C.W., and Rister, R.S. (1998). The Complete German Commission E Monographs.Therapeutic Guide to Herbal Medicines. *The American Botanical Council*, Austin, TX, USA.

Blumenthal, M. (1999). Herb industry sees mergers, acquisitions, and entry by pharmaceutical giants in 1998. *HerbalGram*, **45**: 67-68.

Braverman, D.L., and Schulman, R.A. (1999). Massage techniques in rehabilitation medicine. *Physical Medicine and Rehabilitation Clinics of North America*, **3**: 631-49.

Brevoort, P. (1995). The U.S. botanical market: An overview. *HerbalGram*, **36**: 49-59.

Brown, P., Gibbs, Jr., Rodgers-Johnson, P., Asher, D.M., Sulima, M.P., Bacote, A., and Gajdusek, D.C. (2004). Human spongiform encephalopathy: The National Institute of Health series of 300 cases of experimentally transmitted disease. *Annals of Neurology*, **35**: 513-529.

Boyd, M.R. (1993). In: Neiderhuber JE, ed. Current Therapy in Oncology. Philadelphia, PA: BC Decker, pp.11-22.

Brautigam, M.R.H., Blommaert, F.A., Verleye, G., Castermans, J., Steur, E.N.H.J., and Kleijnen, J. (1998). Treatment of age-related memory complaints with *Ginkgo biloba* extract: a randomized double blind placebo-controlled study. *Phytomedicine*, **5**: 425-434.

Brody, J.E. (2009). Americans gamble on herbs as medicine. The New York Times. Available at: http: //query.nytimes.com/gst/fullpage.html?res_9B0D EFDA123B F93AA35751C0A96F958260. Accessed January 6.

Bulletin of the World Health Organization (1998). Regulatory situation of herbal medicines.A worldwide review. Geneva, 1-43.

Budka, H., Aguzzi, A., Brown, P., Brucher, J.M, Bugiani, O., Collinge, J., Diringer, H., Gullotta, F., Haltia, M., and Hauw, J.J. (1995). Neuropathological diagnostic criteria for Creutzfeldt-Jakob disease (CJD) and other human spongiform encephalopathies (prion diseases). *Brain Pathology*, **5**: 319-322.

De Campos, M.P., Cechinel Filho, V., Da Silva, R.Z., Yunes, R.A., Zacchino, S., Juarez, S., Bella Cruz, R.C., and Bella Cruz, A.(2005). Evaluation of antifungal activity of *Piper solmsianum* C. DC.var. *solmsianum* (Piperaceae). *Biol Pharm Bull.*, **28**: 1527-30.

Calixto, J. B. (2000). Efficacy, safety, quality control, marketing and regulatory guidelines for herbal medicines (phytotherapeutic agents). *Brazillan Journal of Medicinal and Biological Research*, **33**: 179-189. 2000.

Carpino, P. A., and Goodwin B. (2010). Diabetes area participation analysis: a review of companies and targets described in the 2008–2010 patent literature, *Expert Opinion on Therapeutic Patients*, **20**: 1627–51.

Chinkwo, K.A. (2005). *Sutherlandia frutescens* extracts can induce apoptosis in cultured carcinioma cells. *Journal of Ethnopharmacology*,**98**: 163-170.

Chew, A.L., Jessica, J.J.A., and Sasidharan, S. (2012). Antioxidant and antibacterial activity ofdifferent parts of *Leucas aspera*. *Asian Pacific Journal of Tropical Biomedicine*, **2**: 176-180.

Christian, A.G., Mfon, A.G., Dick, E.A., David-Oku, E., Linus, A.J., and Chukwuma, E.B. (2012). Antimalarial potency of the leaf extract of *Aspilia africana* (Pers.) C.D. Adams. *Asian Pacific Journal of Tropical Medicine*, **5**: 126-9.

Cohen, R.J., Ek, K., and Pan, C.X. (2002). Complementary and alternative medicine (CAM) use by older adults: a comparison of self-report and physician chart documentation. The *Journal of Gerontology Series A Biological Sciences and Medical Sciences*, **57**: M223–7.

Corns, C.M. (2003). Herbal remedies and clinical biochemistry. *Annals of Clinical Biochemistry*, **40**: 489 –507.

Cos, P., Hermans, N., De, B.T., Apers, S., Sindambiwe, J.B., Witvrouw, M., De, C.E, Vanden, B.D., Pieters, L., and Vlietinck, A.J. (2002). Antiviral activity of Rwandan medicinal plants against human immunodeficiency virus type-1 (HIV-1). *Phytomedicine*, **9:** 62-68.

Costa-Neto, E. M., and Marques, J. G. W. (2000). Faunistic resources used as medicines by artisanal fishermen from Siribinha Beach, State of Bahia, Brazil. *Journal of Ethnobiology*, **20**: 93–109.

Cowan, M.M. (1999). Plant products as antimicrobial agents. *Clinical MicrobiologyReviews*, 12: 564-582.

Cox, M.E., Rowell, J., Corsino, L., and Green, J. B. (2010). Dipeptidyl peptidase-4 inhibitors in the management of type 2 diabetes: safety, tolerability, and efficacy. *Drug, Healthcare and Patient Safety*, **2**: 7–19.

Cragg, G.M., Newman, D.J., and Snader, K.M. (1997).Natural products in drug discovery and development.*Journal of Natural Products*, **60**: 52-60.

Cragg, G., and Suffness, M. (1988). Metabolism of plant-derived anticancer agents. *Pharmacol. Ther.*, **37**: 425-61.

Creagh, T., Ruckle, J.L., Tolbert, D.T., Giltner, J., Eiznhamer, D.A., Dutta, B., Flavin, M.T., and Xu, Z. (2001). Safety and pharmacokinetics of single doses of (+)-calanolide A, a novel, naturally occurring nonnucleoside reverse transcriptase inhibitor, in healthy, human immunodeficiency virus negative-negative human subjects. *Antimicrobial Agents and Chemotherapy*, **45:** 1379-1386.

Da Rocha, A.B., Lopes, R.M., and Schwartsmann, G.(2001). Natural products in anticancer therapy.*Current Opinion in Pharmacology*,**1**: 364–369.

De Smet, P.A.G.M. (1997). The role of plant derived drugs and herbal medicines in healthcare. *Drugs*, **54**: 801-840.

Debelle, F.D., Vanherweghem, J.L., and Nortier, J.L.(2009). Aristolochic acid nephropathy: a worldwide problem. *Kidney International*,**74**: 158–69.

Eisenberg, D.M., Kessler, R.C., Foster, C., Norlock, F.E., Calkins, D.R., and Thomas, L. Delbanco, T.L. (1993). Unconventional medicine in the United States. *New England Journal of Medicine*, **328**: 246-252.

Efferth, T., Li, P.C., Konkimalla, V.S., and Kaina, B. (2007).From traditional Chinese medicine to rational cancer therapy. *Trends in Molecular Medicine*,**13**: 353–361.

Ehling, D. (2001). Oriental medicine: An introduction. *Alternative Therapies in Health and Medicine*,**7**: 71-82.

Ernst, E. (1999). Massage therapy for low back pain: a systematic review. *Journal of Pain and Symptom Management*, **17**: 65-9.

Ernst, E. (2002). Toxic heavy metals and undeclared drugs in Asian herbal medicines. *Pharmacological Sciences*, **23**: 136-139.

Eskinazi, D., Blumenthal, M., Farnsworth, N., and Riggins, C.W. (1999). Botanical Medicine: Efficacy, Quality Assurance, and Regulation. Mary Ann Liebert, Inc. Publisher, New York.

Ferreira, F.S., Brito, A.V., Ribeiro, S. C., Saraiva, A. A. F., Almeida W. O., and Alves R. R.N. (2009). Animal-based folk remedies sold in public markets in Crato and Juazeiro do Norte, Cear´a, Brazil. *BMC Complementary and Alternative Medicine*, **9**: 17.

Ferreira, M.J.U., Duarte, N., Ko³aczkowski, M., and Michalak, K. (2006). Antifungal and multidrug resistance modulatory effects of diterpenic and phenolic compounds. *Planta Medica*, **23**: 72-77.

Fang, X., Shao, S., Zhang, H., and Wang, S. (2005). CHMIS-C: A comprehensive herbal medicine information system for cancer. *Journal of Medicinal Chemistry*, **48**: 1481-1488.

Farnsworth, N.R., and Morris, R.W. (1976). Higher plants-the sleeping giant of drug development. *American Journal of Pharmaceutical Education*, **148**: 46-52.

French, G. L. (2010).The continuing crisis in antibiotic resistance. *International Journal of Antimicrobial Agents*, **36**: S3.

Field, T., Henteleff, T., Hernandez-Reif, M., Marting, E., Mavunda, K., Kuhn, C., and Schanberg, S. (1997b). Children with asthma have improved pulmonary functions after massage therapy. *Journal of Pediatrics*, **132:** 854-858.

Figueiredo N. (1994). Os 'bichos' que curam: os animais e a medicina 'folk' emBel´em do Par´a," *Boletim doMuseu Paraense Em´ýlio G¨oeldi*, **10**: 75–91.

Fisher, P., and Ward, A. (1994). Complementary medicine in Europe. *British Medical Journal*, **309**: 107-111.

Fyhrquist, P., Mwasumbi, L., Haeggstrom, C.A., Vuorela, H., Hiltunen, R., and Vuorela, P. (2004). Antifungal activity of selected species of *Terminalia*, *Pteleopsis* and *Combretum* (Combretaceae) in Tanzania. *Pharmaceutical Biology*, **42**: 308-317.

Gaedcke, F. W., and Steinholf, K. B. (2003).Herbal Medicinal Products. Stuttgart, Germany: CRC Press, ISBN: 0-8493-1023-7.

Gansser, A. (1994). Geology of the Himalayas, Interscience, New York.

Gee, Yap. (2006). Assessment of market potential for selected medicinal herb products, Rural Industries Research and Development Corporation. RIRDC Publication No 06/084. 2006.

Gershell, L. (2005). Type 2 diabetes market. *Natural Review of Drug Discovery*, **4**: 367–8.

Goldman, P. (2001). Herbal medicines today and the roots of modern pharmacology. *Annals of Internal Medicine*,**135**: 594-600.

Grünwald, J. (1995). The European phytomedicines market: figures, trends, analysis. *HerbalGram*, **34**: 60-65.

Guidance for industry: botanical drug products. U.S. Department of Health and Human Services. Food and Drug Administration. Center for Drug Evaluation and Research. Available at: http: //www.fda.gov/cder/guidance/4592fnl.htm.

Haller, C.A., and Benowitz, N.L. (2000). Adverse cardiovascular and central nervous system events associated with dietary supplements containing ephedra alkaloids. *New England Journal of Medicine*, **343**: 1833– 8.

Han, H., He, W., Wang, W., and Gao, B.(2011). Inhibitory effect of aqueous dandelion extract onHIV-1 replication and reverse transcriptase activity.*BMC complementary and alternative Medicine*, **11**: 112

Hardman, T. C., Rutherford, P., Dubrey, S. W., and WierzbickiA. S. (2010).Sodium-glucose co-transporter 2 inhibitors: from apple tree to'Sweet Pee'. *Current Pharmaceutical Design*, **16**: 3830–8.

Harnett, S.M., Ooosthuizen, V., and Van de Venter, M. (2005).Anti-HIV activities of organic and aqueous extracts of *Sutherlandia frutescens* and *Lobostemon trigonus*. *Journal of Ethnopharmacology*, **96**: 113-119.

Harris poll shows widespread public ignorance of supplement regulation. Available at: http: //www.supplementquality.com/news/Harris_survey.html. Accessed November 10, 2008.

Hays, N. P., Galassetti, P. R., and Coker, R. H. (2008). Prevention and treatment of type 2 diabetes: current role of lifestyle, natural product, and pharmacological interventions. *Pharmacol. Ther.*, **118**: 181–91.

Hui, K.K. (1999). Harmonizing traditional Chinese and modern Western medicine: a perspective from the United States. In: Consultation Meeting on Traditional and Modern Medicine: Harmonizing the two Approaches. Beijing Nov 1999. Report: 185-189. Manila, World Health Organization Regional Office for the Western Pacific Report, 1999.

Hui, K.K., Liu, J., Makris, N., Gollub, R.L., Chen, A.J., Moore, C.I., Kennedy, D.N., Rosen, B.R., and Kwong, K.K. (2000). Acupuncture modulates the limbic system and subcortical gray structures of the human brain: Evidence from fMRI studies in normal subjects. *Human Brain Mapping*, **9**: 13-25.

Hung,H.Yi., Qian, K., Morris-Natschke, S. L., Hsu, C.S., and Lee K. H.(2012). Recent discovery of plant-derived anti-diabetic natural products. *Natural Product Report*, **29**: 580.

Hussain, J., Khan,A.L., Rehman,N., Hamayun,M., Shinwari,Z.K., Malik, W., and Lee, I.J. (2009). Assessment of herbal products and their composite medicinal plants through proximate and micronutrients analysis. *Journal of Medicinal Plants Research*, **3**: 1072-1077.

Iglehart, J.K. (1999).The American health care system expenditures. *New England Journal of Medicine*, **340**: 70-76.

Inman, W.D., and Reed, M.J. (1997).Triterpenoid compound for the treatment of diabetes. In: U.S. Patent. South San Francisco, CA: Shaman Pharmaceuticals.

Ismail, M.Y.M. (2010). Antiasthmatic herbal drugs areview. *International Journal of Pharmacy and Pharmaceutical Sciences*, **3**: 28-29.

Jia, W., Gao, W., and Tang, L. (2003). Antidiabetic herbal drugs officially approved in China. *Phytotherapy Research*, **17**: 1127–34.

Josey, E.S., and Tackett, R.L. (1999). St. John's wort: a new alternative for depression? *International Journal of Clinical Pharmacology and Therapeutics*, **37**: 111-119.

Jung, M., Park,M., Lee, H. C., Kang, Y. H., Kang, E. S., and Kim, S. K. (2006). Antidiabetic agents from medicinal plants. *Curr. Med. Chem.*, **13**: 1203–18.

Kannaiyan, R., Shanmugam, M.K., and Sethi, G.(2011). Molecular targets of celastrol derived from Thunder of God Vine: potential role in the treatment of inflammatory disorders and cancer. *Cancer Letters*, **303**: 9–20.

Keller, K. (1996). Herbal medicinal products in Germany and Europe: experiences with national and European assessment. *Drug Information Journal*, **30**: 933-948.

Kim, Y., Kim, K., Lee, H., Han, S., Lee, Y.S., Choe, J., Kim, Y.M., Hahn, J.H., Ro, J.Y., and Jeoung, D.(2009). Celastrol binds to ERK and inhibits FcepsilonRI signaling to exert an anti-allergic effect. *European Journal of Pharmacology*, **612**: 131–142.

King, H., and Rewers, M. (1993). Global estimates for prevalence of diabetes mellitus and impaired glucose tolerance in adults. WHO Ad Hoc Diabetes Reporting Group. *Diabetes Care*, **16**: 157-177.

Kohn, L.T., Corrigan, J.M., and Donaldson, M.S. (2000). To err is human: building a safer health system. Committee on quality of health care in America. Institute of Medicine, National Academy Press.

Kruti, P., B. Solanki, K. Maniar, N. Gurav and S. Bhatt. (2011). Natural herbal supplements an assessment of their nutritional value and their phytochemical constituents. *International Journal of Pharmacology and Biological Sciences*, **2**: 420-431.

Kuo, G.M., Hawley, S.T., Weiss, L.T., Balkrishnan, R., and Volk, R.J. (2004).Factors associated with herbal use among urban multiethnic primary care patients: a cross-sectional survey. *BMC Complementary and Alternative Medicine*, **4**: 18.

Kupchan, S.M., Court, W.A., Dailey Jr., R.G., Gilmore, C.J., and Bryan, R.F. (1972). Triptolide and tripdiolide, novel antileukemic diterpenoid triepoxides from *Tripterygium wilfordii*. *Journal of the American Chemical Society*, **94:** 7194–7195.

Labders, S.J. (2000). The world's health care, how do we rank? *American Medical News*, **28**: 5-6.

Laakmann, G., Schüle, C., Baghai,T., and Kieser, M. (1998). St. John's wort in mild to moderate depression: the relevance of hyperforin for the clinical efficacy. *Pharmacopsy chiatry*, **31**: 54-59.

Lee, K.H. (2000). Research and future trends in the pharmaceutical development of medicinal herbs from Chinese medicine. *Public Health Nutrition*,**3**: 515-522

Lee, J.H., Choi, K.J., Seo, W.D., Jang, S.Y., Kim, M., Lee, B.W., Kim, J.Y., Kang, S., Park, K.H., Lee, Y.S., and Bae, S. (2011). Enhancement of radiation sensitivity in lung cancer cells by celastrol is mediated by inhibition of Hsp90. *International Journal of Molecular Medicine*, **27**: 441–446.

Leonti, M. (2011). The future is written: Impact of scripts on the cognition, selection, knowledge and transmission of medicinal plant use and its implications for ethnobotany and ethnopharmacology. *Journal of Ethnopharmacology*,**134**: 542-55.

Leteane, M.M., Ngwenya, B.N., Muzila, M., Namushe, A., Mwinga, J., Musonda, R., Moyo, S, Mengestu, Y.B., Abegaz, B.M., and Andrae-Marobela, K. (2012). Old plants newly discovered: *Cassia sieberiana* D.C. and *Cassia abbreviata* Oliv. Oliv root extracts inhibit *in vitro* HIV-1c replication in peripheral blood mononuclear cells (PBMCs) by different modes of action. *Journal of Ethnopharmacology*, **141**: 48-56.

Leung, K., Bian, Z., Moher, D., Dagenais, S., Li, Y., Liu, L., *et al*. (2006). Improving the quality of randomized controlled trials in Chinese herbal medicine, part III: quality control of Chinese herbal medicine used in randomized controlled trials. *Zhong Xi Yi Jie He Xue Bao*, **4:** 225-32.

Lev, E. (2003). Traditional healing with animals (zootherapy): medieval to present-day Levantine practice. *Journal of Ethnopharmacology*, 85: 107–118.

Li, R., Chen, Y., Zeng, L.L., Shu, W.X., Zhao, F., Wen, L., and Liu, Y. (2009). Gambogic acid induces G0/G1 arrest and apoptosis involving inhibition of SRC-3 and inactivationof Akt pathway in K562 leukemia cells. *Toxicology*,**262**: 98–105

Liu, Q.(2011). Triptolide and its expanding multiple pharmacological functions. *International Immunopharmacology*, **11**: 377–383

Luo, J., Chuang, T., Cheung, J., Quan, J., Tsai, J., Sullivan, C., Hector, R.F., Reed, M.J., Meszaros, K., King, S.R., Carlson, T.J., and Reaven, G.M. (1998). Masoprocol (nordihydroguaiaretic acid): a new antihyperglycemic agent isolated from the creosote bush (*Larrea tridentata*). *European Journal of Pharmacology*, **346**: 77-79.

Luo, J., Cheung, J., Yevich, E.M., Clark, J.P., Tsai, J., Lapresca, P., Ubillas, R.P., Fort, D.M., Carlson, T.J., Hector, R.F., King S.R. Mendez, C.D., Jolad, S.D., and Reaven, G.M. (1999). Novel terpenoid-type quinones isolated from *Pycnanthus angolensis* of potential utility in the treatment of type 2 diabetes. *Journal of Pharmacology and Experimental Therapeutics*,**288**: 529-534.

MacPherson, H., Hammerschlag, R., Lewith, G., and Schnyer, R. (2008). Acupuncture Research. New York: Churchill Livingstone Elsevier.

Marko, D., Schatzle, S., and Friedel, A. (2001). Inhibition of cyclin-dependent kinase 1 (CDK 1) by indirubin derivatives in human tumor cells. *British Journal of Cancer*, **84:** 283-289.

McCluggage, D., and Higdon, P.L. (1999). Holistic Care for Birds: A Manual of Wellness and Healing. Wiley Publishers.

Mahawar, M. M., and Jaroli, D. P. (2008). Traditional zootherapeutic studies in India: a review. *Journal of Ethnobiology and Ethnomedicine*, **4**: 17.

Meshnick, S.R., Taylor, T.E., and Kamwonwongpaisan, S. (1996). Artemisinin and the antimalarial endoperoxides: from herbal remedy to targeted chemotherapy. *Microbiology Reviews*,**60**: 301-315.

Swanson, M.D., Winter, H.C., Goldstein, I.J., and Markovitz, D.M. (2010). A lectin isolated from bananas is a potent inhibitor of HIV replication. *Journal of Biological Chemistry*, **285**: 8646–8655.

Moellering, R.C. Jr. (2011). Discovering new antimicrobial agents. *International Journal of Antimicrobial Agents*, **37**: 2-9.

MoH: Uganda Ministry of Health Report. 2004.

Mohebi, R., Ghafourian, S., Sekawi, Z., Khosravi, A., Galehdari, E.A., Hushmandfar, R., Ranjbar, R., Maleki, A., Mohammadzadeh, M., Rahbar, M., and Sadeghifard, N. (2011).*In vitro* and *in vivo* antibacterial activity of acorn herbal extract against some Gram-negative and Gram-positive bacteria. *Roumanian Archives of Microbiology and Immunology*,**70**: 149-52.

Morris, C.A. (2003). Internet marketing of herbal products. *Journal of the American Medical Association*, **290**: 1505–9.

Moshi, M.J., and Mbwambo, Z.H. (2005). Some pharmacological properties of extracts of *Terminalia sericea* roots. *Journal of Ethnopharmacology*, **97**: 43-47.

Moura, F. D. B. P., and Marques J. G. W. (2008). Folk medicine using animals in the Chapada Diamantina: incidental medicine?" *Ciˆencia and Sa´ude Coletiva,* **13**: 2179–2188,

Namuddu, B., Kalyango, J.N., Karamagi, C., Mudiope, P., Sumba, S., Kalende, H., Wobudeya, E., Kigozi, B.K., and Waako, P. (2011). Prevalence and factors associated with traditional herbal medicine use among patients on highly active antiretroviral therapy in Uganda. *BMC Public Health,***11**: 855.

Naik, A.D., and Juvekar, A.R. (2003). Effects of alkaloidal extract of *Phyllanthus niruri* on HIV replication. *Indian Journal of Medical Sciences,* **57**: 387-393.

NIH Consensus Conference.(1998). Acupuncture. *Journal of the American Medical Association,* **280**: 1518-1524.

Notka, F., Meier, G., and Wagner, R. (2004).Concerted inhibitory activities of *Phyllanthus amarus* on HIV replication *in vitro* and *ex vivo*. *Antiviral Research,* **64**: 93-102.

Nortier, J.L., Martinez, M.C., Schmeiser, H.H., Arlt, V.M., Bieler, C.A., Petein, M., Depierreux, M.F., De Pauw, L., Abramowicz, D., Vereerstraeten, P., and Vanherweghem, J.L. (2000). Urothelial carcinoma associated with the use of a Chinese herb (*Aristolochia fangchi*). *New England Journal of Medicine,* **342**: 1686 – 92.

Pan, S.Y., Pan, S., Yu, Z.L., Ma, D.L., Chen, S.B., Fong, W.F., Han, Y.F., and Ko, K.M. (2010). New perspectives on innovative drug discovery: an overview. *Journal of Pharmacy and Pharmaceutical Sciences,* **13** : 450-71.

Pang, X., Yi, Z., Zhang, J., Lu, B., Sung, B., Qu, W., Aggarwal, B.B., and Liu, M. (2010). Celastrol suppresses angiogenesis-mediated tumor growth through inhibition of AKT/mammalian target of rapamycin pathway. *Cancer Research,***70**: 1951–1959.

Park, Y., Choi, B.H., Kwak, J.S., Kang, C.W., Lim, H.T., Cheong, H.S. and Hahm, K.S.(2005). Kunitz-type serine protease inhibitor from potato (*Solanum tuberosum* L. cv. Jopung). *Journal of Agriculture and Food Chemistry,* **53**: 6491-6.

Park, I.W., Han, C., Song, X., Green, L.A., Wang, T., Liu, Y., Cen, C., Song, X., Yang, B., Chen, G., and He, J.J. (2009).Inhibition of HIV-1 entry by extracts derived from traditional Chinese medicinal herbal plants.*BMC complementary and alternative Medicine,* **9**: 29

Petrovick, P.R., Marques, L.C., and Paula, I.C. (1999). New rules for phytopharmaceutical drug registration in Brazil. *Journal of Ethnopharmacology,* **66**: 51-55.

Pilzer, P. (2002). The Wellness Revolution (Second ed.). Hoboken, New Jersey: John Wiley and Sons, Inc.

Piras, G., Makino, M., and Baba, M. (1997). Shosaiko-to, a traditional Kampo medicine, enhances the anti-HIV-1 activity of lamivudine (3TC) *in vitro*. *Microbiology and Immunology,* **41**: 835-839.

Prashant, K.R., Dolly, J., Singh, K.R., Gupta, K.R., and Watal G. (2008).Glycemic properties of *Trichosanthes dioica* leaves. *Pharmaceutical Biology*, **46**: 894-899.

Roberts, J.E., and Tyler, V.E. (1998). Tyler's Herbs of Choice. The Therapeutic Use of Phytomedicinals. The Haworth Press, Inc., New York.

Sackett, D. (1996). The conscientious, explicit and judicious use of current best evidence in making decisions about the care of individual patients. *British Medical Journal*, **312**: 71-2.

Salminen, A., Lehtonen, M., Paimela, T., and Kaarniranta, K. (2010). Celastrol: molecular targets of Thunder God Vine. *Biochemical and Biophysical Research Communications*, **394**: 439–442.

Sancier, K. (1996). Medical applications of Qi-Gong. *Alternative Therapies*, **2**: 40-46.

Saper, R.B., Phillips, R.S., Sehgal, A., Khouri, N., Davis, R.B., Paquin, J., Thuppil, V., and Kales, S.N.(2008). Lead, mercury, and arsenic in US- and Indian-manufactured ayurvedic medicines sold via the Internet. *Journal of American Medical Association*, **300**: 915–23.

Sato, Y. (1990). Protection effects of *Aloe arborescens* on skin injury by X-irradiation. *Yakmgakn Zasshi*, **110**: 876-884.

Sautor, M., Miyamoto, T., and Lacaille, M.A. (2005).Steroidal saponins from *Smilax medica* and their antifungal activity. *Journal of Natural Products*, **68**: 1489-1493.

Saxena, K. (1997). Antimicrobial screening of selected medicinal plants from India. *Journal of Ethnopharmacology*, **58**: 75-83.

Sears, B., and Ricordi, C. (2011). Anti-inflammatory nutrition as a pharmacological approach to treat obesity. *Journal of Obesity*, 2011: pii431985

Sethi, G., Ahn, K.S., Pandey, M.K., and Aggarwal, B.B. (2007). Celastrol, a novel triterpene, potentiates TNF-induced apoptosis and suppresses invasion of tumor cells by inhibiting NF-kappaB-regulated gene products and TAK1-mediated NF-kappaB activation. *Blood*, **109:** 2727–2735.

Semple, G., Fioravanti, B., Pereira, G., Calderon, I., Choi, J. Uy, K., Xiong, Y., Ren, A., Morgan, M., Dave, V., Thomsen, W., Unett, D. J., Xing, C., Bossie, S., Carroll, C., ChuZ. L., *et al*. (2008). Discovery of the first potent and orally efficacious agonist of the orphan G-protein coupled receptor 119.*Journal of Medicinal Chemistry*, **51**: 5172–5.

Sheehan, M.P., and Atherton, D.J. (1992). A controlled trial of traditional Chinese medicinal plants in widespread non-exudative atopiceczema.*Brazillian Journal of Dermatology*, **126**: 179-84.

Shakya, M.N., Pradhan, R., and Ranjitkar, R. (2008). A preliminary screening of some Nepalese medicinal plants for antimicrobial activity. *Bulletin of Department of Plant Resource*, **30**: 87-94.

Shittu, I., Emmanuel, A., and Nok, A.J. (2011). Antimalaria effect of the ethanolic stem bark extracts of *Ficus platyphylla* Del. *Journal of Parasitology Research*, 618-209.

Shinwari, Z.K., Khan, I., Naz, S., and Hussain, A. (2009).Assessment of antibacterial activity of three plants used in Pakistan to cure respiratory diseases. *African Journal of Biotechnology*, **8**: 7082-7086.

Shu, Y.Z. (1998). Recent natural products based drug development: A pharmaceutical industry perspective. *Journal of Natural Products*, **61**: 1053-1071.

Silva, A. L. (2008).Animais medicinais: conhecimento e uso entre as populac¸˜oes ribeirinhas do Rio Negro, Amazonas, Brasil. *Boletim do Museu Paraense Em´ýlio Goeldi," S´erie Ciˆencias Naturais*, **3**: 343–357.

Sneader, W. (1985). Drug Discovery: The Evolution of Modern Medicines. New York: Wiley.

Sodeinde,O. A., and Soewu D. A. (1999).Pilot study of the traditional medicine trade in Nigeria. *Traffic Bulletin*, **18**: 35–40.

Spellberg, B., Guidos, R., Gilbert, D., Bradley, J., Boucher, H.W., Scheld, W.M., Bartlett, J.G., and Edwards, J. Jr. (2008).The epidemic of antibiotic-resistant infections: a call to action for the medical community from the Infectious Diseases Society of America. *Clinical Infectious Diseases*, **46**: 155-64.

Starfield, B. (2000). Is US health really the best in the world? *Journal of the American Medical Association*, **284**: 483-485

Sun, H., Liu, X., Xiong, Q., Shikano, S., and Li, M. (2006). Chronic inhibition of cardiac Kir2.1 and HERG potassium channels by celastrol with dual effects on both ion conductivity and protein trafficking. *The Journal of Biological Chemistry*, **281**: 5877–5884.

Sundaram, R., Ganesan, R., and Murugesan, G.(2012). *In vitro* antiplasmodial activity of spiro benzofuran compound from mangrove plant of Southern India. *Asian Pacific Journal of Tropical Medicine*,**5**: 358-61.

Sunshine, W., Field, T., Schanberg, S., Quintino, O., Kilmer, T., Fierro, K., Burman, I., Hashimoto, M., McBride, C., and Henteleff, T. (1996). Massage therapy and transcutaneous electrical stimulation effects on fibromyalgia. *Journal of Clinical Rheumatology*, **2**: 18-22.

Tachjian, A., Maria, V., and Jahangir, A. (2010).Use of herbal products and potential interactions in patients with cardiovascular diseases. *Journal of the American College of Cardiology*, **55**: 515-525

Tegos, G., Stermitz, F.R., Lemovskaya, O., and Lewis, K. (2002). Multidrug pump inhibitorsuncover remarkable activity of plant antimicrobials. *Antimicrobial Agents and Chemotherapy*, **10**: 3133–41.

Uchiumi, F., Hatano, T., Itoh, H., Yoshida, T., and Tanuma, S. (2003).Transcriptional suppression of the HIV promoter by natural compounds. *Antiviral Research*, **58**: 89-98.

UNAIDS: Message on the occasion of World AIDS Day 1 December. 2007.

UNAIDS: Joint United Nations Programmc on HIV/AIDS (UNAIDS) (2008). Report on the global AIDS epidemic.

V´azquez, P. E., M´endez, R. M., Guiasc ´ on, O. G. R., and Pi˜nera, E. J. N. (2006). Uso medicinal de la fauna silvestre en los Altos de Chiapas, M´exico. *Interciencia*, **31**: 491–499.

Van Everbroeck, B., Pals, P., Dziedzie, T., Dom, R., Godfraind, C., Sciot, R., Brucher, J.M., Martin, J.J., and Cras, P. (2000). Retrospective study of Creutzfeldt-Jakob disease in Belgium: neurophysiological findings. *Acta Neoropathologica* (Berl), **99**: 358-364.

Walston N. (2005). An overview of the use of Cambodia's wild plants and animals in traditional medicine systems, TRAFFIC Southeast Asia, Indochina.

Wang, S., Wu, Xu., Tan, M., Gong, J., Tan, W., Bian, B., Chen, M., and Wang, Y. (2012). Fighting fire with fire: Poisonous Chinese herbal medicine for cancer therapy. *Journal of Ethnopharmacology*, **140**: 33– 45.

Wang, H.X., and Ng, T.B.(2003). Isolation of cucurmoschin, a novel antifungal peptide abundant in arginine, glutamate and glycine residues from black pumpkin seeds. *Peptides*, **24**: 969-72.

Washburn, W. N. (2009). Evolution of sodium glucose co-transporter 2 inhibitors as anti-diabetic agents. *Expert Opinion on Therapeutic Patients*, **19**: 1485–99.

Wetzel, M.S., Eisenberg, D.M., and Kaptchuk, T.J. (1998).Courses involving complementary and alternative medicine at U.S. medical schools. *Journal of American Medical Association*, **280**: 784-787.

WHO. (2002). Traditional Medicine: Growing Needs and Potential. WHO Policy Perspectives on Medicines. World Health Organization, Geneva pp.1-6.

WHO. (1989). *In vitro* screening of traditional medicines for anti-HIV activity: memorandum from a WHO meeting. *Bulletin of the World Health Organization*, **67,** 613-618.

Xu, Z.Q., Barrow, W.W., Sulling, W.J., Westbrook, L., Barrow, E., Lin, Y.M., and Flavin, M.T. (2004). Anti-HIV natural product (+)- calanolide A is active against both drug susceptible and drug-resistant strains of *Mycobacterium tuberculosis*. *Bioorganic Medicinal Chemistry*,**12**: 1199-1207.

Yan, Q., Jiang, Z., Yang, S., Deng, W., and Han, L.(2005). A novel homodimeric lectin from *Astragalus mongholicus* with antifungal activity. *Archives of Biochemistry and Biophysics*, **442**: 72-81.

Yang, H.S., Kim, J.Y., Lee, J.H., Lee, B.W., Park, K.H., Shim, K.H., Lee, M.K., and Seo, K.I. (2011a). Celastrol isolated from *Tripterygium regelii* induces apoptosis through both caspase-dependent and -independent pathways in human breast cancer cells. *Food and Chemical Toxicology*,**49**: 527–532.

Yoney, A., Prieto, J. M., Lardos, A., and Heinrich, M. (2012). Ethnopharmacy of Turkish-speaking Cypriots in Greater London. *Phytotherapy Research*, **24**: 731-740.

Young, D.R., Appel, L.J., Jee, S., and Miller, E.R. 3rd. (1999).The effects of aerobic exercise and Tai Chi on blood pressure in older people: results of a randomized trial. *Journal of the American Geriatrics Society*, **47**: 277-284.

Zeisel, S.H. (1999). Regulation of "nutraceuticals". *Science*, **285**: 1853-1855.

Zhang, J.D., Cao, Y.B., Xu, Z., Sun, H.H., An, M.M., Yan, L., Chen, H.S., Gao, P.H., Wang, Y, Jia, X.M., and Jiang, Y.Y.(2005). *In vitro* and *in vivo* antifungal activities of the eight steroid saponins from *Tribulus terrestris* L. with potent activity against fluconazole-resistant fungal pathogens. *Biol Pharm Bull.*, **28**: 2211-2215.

Zhao, L., Zhen, C., Wu, Z.Q., Hu, R., Zhou, C.L., and Guo, Q.L.(2010b). General pharmacological properties, developmental toxicity, and analgesic activity of gambogic acid, a novel natural anticancer agent. *Drug and Chemical Toxicology*, **33**: 88–96.

Traditional and Folk Herbal Medicine: Recent Researches Vol. 2 (2014) *Pages* **45–72**
Editor: **V.K. Gupta**
Published by: **DAYA PUBLISHING HOUSE, NEW DELHI**

2

Artemisia annua (Annual Wormwood): A Precious Traditional and Folk Antimalarial Medicinal Plant

Kaliyaperumal Karunamoorthi[1, 2*]

ABSTRACT

The traditional and folk herbal medicine is a long-standing prehistoric practice and heavily relies on herbal remedies for prevention and treatment. Plants medicinal values had been noted for centuries together. The ancient human has prudently validated the phytotherapeutic potentialities of numerous herbs by a long-term observation as well as trial and error experiments. Throughout the human civilization, malaria has been considered as a world's most serious public health problem as well as an old scourge and has ruled humankind for nearly 500,000 years. In the human history, the traditional and folk herbal medicinal system has played the crucial role in the fight against different diseases and ailments particularly malaria saving millions of lives and curing billions of debilitating episodes of malaria. Evidently the world's potent antimalarial chemotherapy both Quinine and Artemisinin have been isolated and identified from cinchona tree bark in 1820 and from the leaves of the Artemisia annua in 1972, respectively.

The Chinese government has initiated antimalarial programme (1965-75) to search a potent antimalarial from their traditional folk herbal medicine and ultimately artemisinin derivatives (ART) (Qinghaosu) has been identied positively. Currently, artemisinin based combination therapy (ACT) is the drug of choice against both drug-resistant and cerebral

1 Unit of Medical Entomology and Vector Control, Department of Environmental Health Science and Technology, College of Public Health and Medical Sciences, Jimma University, Jimma, Ethiopia

2 Research and Development Centre, Bharathiar University, Coimbatore, Tamil Nadu, India

* *Corresponding author*: E-mail: k_karunamoorthi@yahoo.com

malaria causing strains of Plasmodium falciparum and the drug has also been shown to be effective against other infectious diseases such as hepatitis B, schistosomiasis and drug resistance cancer too. It has been estimated that roughly 500 million episodes of clinical malaria are there per year and the majority of which needs to be ideally treated with ACT. In this viewpoint, this chapter becomes more significant and pertains. It shall provide a detailed outline about the traditional system of medicine and the history of traditional and folk herbal medicine. It also describes about A. annua including its physical characteristics, habitats, medicinal properties, medicinal values and other traditional uses. Besides, it also deals about the importance of bioactive molecules, which are responsible for the potential antimalarial properties and the future perspective to formulate the affordable and accessible potent novel antimalarial agents from the traditional and folk herbal medicinal plants in the near future.

Keywords: Traditional and folk herbal medicine, *Artemisia annua*, Antimalarial agent.

Introduction

Arthropod-borne diseases are major causes of morbidity and mortality in many tropical and subtropical countries and principally the devastating nature of malaria is indubitably intolerable (Karunamoorthi and Ilango, 2010). Even centuries ago, malaria was a common disease in the days of Rome; in fact, the term malaria has inherited from the Latin language "**mala aria**". In those days erroneously it was commonly believed to be caused by inhaling the poisonous vapors from stagnant water or marshlands/swamps and therefore the medieval Italians coined the disease "mala-aria" which translates literally as "bad air" ("mala" = bad and "aria" = air). The two terms combining together known as "malaria" which is still used today to describe the disease and the term malaria is used by an Italian scientist by the first time in the history named Dr. Francisco.

Today we know that malaria is a vector-borne infectious disease caused by the protozoan *Plasmodia* parasites. There are four types of *Plasmodium* species namely *P. falciparum*, *vivax*, *malariae* and *ovale*, while the vector carrying and transmitting the disease is of the female *Anopheles* mosquito species (Toure *et al.*, 2004). The two principal forms of malaria are: (1) uncomplicated, which is not life threatening and where the purpose is to cure only the infection; and (2) severe, which includes cerebral forms, where the purpose is to prevent death (WHO, 2006).

The recent World Health Organization (WHO) Malaria Report (2011) estimates that 3.3 billion people were at the risk of malaria in 2010, although of all geographical regions, populations living in sub-Saharan Africa (SSA) have the highest risk of acquiring malaria; among 216 million episodes of malaria in 2010, approximately 81 per cent, or 174 million cases, were reported from the African Region. There were an estimated 655,000 of malaria deaths in 2010, of which 91 per cent were from Africa. However, malaria is both a preventable and curable illness.

Malaria in pregnancy imposes a serious threat to the mother, fetus and neonate, as pregnancy reduces a woman's immunity, increasing the risk of malarial infection. Despite the availability of effective interventions, malaria remains as one of the most important cause of maternal and childhood morbidity and mortality in SSA

(Karunamoorthi *et al.*, 2010). It is also an important cause of stillbirths, low birth weight, and early infant mortality (Menendez, 1995). Malaria kills a poor African child for every 45 seconds and continues to be a major public health problem in the resource-limited countries of Africa, Asia, Latin America and beyond (Karunamoorthi, 2012a). Malaria is a disease of poverty inflicting a serious negative impact on health and socioeconomic development in the poorest countries of the world that cannot afford to succeed (Karunamoorthi, 2012b).

The explosive global economic development, people movement, water projects, climate change, increased and unplanned urbanization have substantially altered the malaria transmission dynamics and pattern too and this in turn requires implementing need-based and community-oriented malaria control strategies (Karunamoorthi, 2012b). Malaria control heavily relies on a limited number of arsenals like artemisinin derivatives (ART) and pyrethroids. At the moment, malaria control and elimination is one of the most serious challenging tasks due to the spread of multi-drug resistant strains of *P. falciparum* (Karunamoorthi, 2011).

The devastating spread of parasite resistance has prompted the worldwide search for new classes of effective antimalarial drugs (O'Neill, 2004). Hope for the containment of the disease now rests largely on a remarkable set of artemisinin drugs developed by Chinese scientists in the 1970s and early 1980s, which rapidly kill the malaria parasites (Ridley, 2003). It is extracted from sweet wormwood (*A. annua*), and is used as the basis for synthesizing several modern antimalarial drugs, such as artesunate and artemether. These artemisinin derivatives are increasingly important in the treatment of drug-resistant malaria, as they are the most potent antimalarials available, rapidly killing all blood stages of the malaria parasite *P. falciparum* (O'Neill, 2004). WHO now recommends treating uncomplicated malaria with artemisinin-based combinations therapies (ACTs).

The History and Origin of ACT

The emergence of chloroquine-resistant *P. falciparum* malaria in Southeast Asia in the 1960s caught the attention of the Chinese government (Jiaxiang, 1991). In 1967, China's President Mao Zedong had insisted that traditional Chinese medicine was a "**Great Treasure House**", a phrase immortalised in "**The Little Red Book**". He generously funded activities aimed at harnessing the potentials of Qing Hao. Chinese government's enthusiasm for optimal use of Qing Hao and other traditional herbs was exemplified by Project 523. During the Vietnam War, Ho Chi Minh asked Mao Zedong for help, because more North Vietnamese soldiers were dying from malaria than from armed conflicts. In late 1967, a total of 523 Chinese scientists from more than 60 institutions, including chemists, ethnopharmacologists, entomologists and clinicians, were recruited by the Chinese government to the antimalaria drug project. One of the group's researchers, Tu Youyou, from the Chinese Academy of Traditional Chinese Medicine led a small group that identified artemisinin as the active compound of *A. annua* in 1972 (Figure 2.1) (Tu, 2004).

Artemisinin, which is effective in treating chloroquine-resistant cases, is quite different from the old generation of anti-malarial drugs because of its peroxide group (Klayman *et al.*, 1984; Teixeira da Silva, 2004). Attempts to derivatise artemisinin

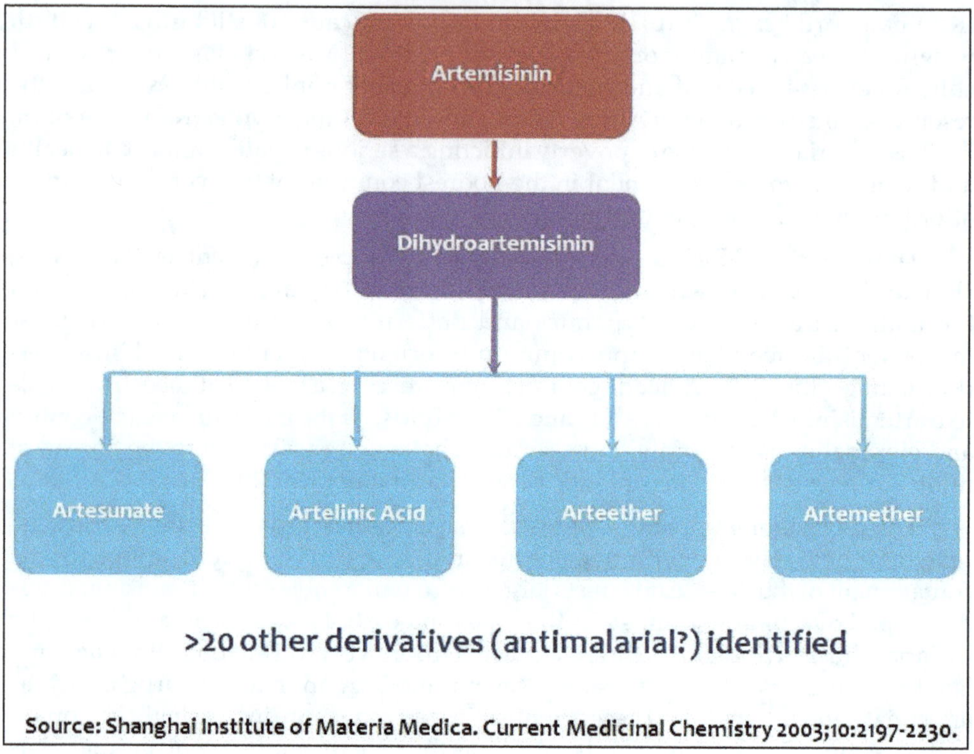

Figure 2.1: The derivatives of artemisinin.

were done even before its chemical structure was fully elucidated. Chemists have successfully semi-synthesized anti-plasmodial analogs, such as dihydroartemisinin, artemether, arteether, artesunate and artelinic acid as given in the figure. The derivatives of artemisinin are widely recommended today (WHO, 2010).

Significance of ACT

The use of drugs in combination is broadly accepted in the treatment of several diseases, such as tuberculosis (TB), Human Immunodeficiency Virus/Acquired Immunodeficiency Syndrome (HIV/AIDS) infection and cancer. The idea behind the therapeutic effect of ACT is that the artemisinin rapidly kills most of the Plasmodium parasites, and those that survive are subsequently killed by a high concentration of the companion drug. One of the main advantages of combination therapy is the likelihood that this strategy will reduce the chances of an infectious agent becoming resistant to the action of the drugs. In the case of malaria, combination therapy has been applied since around 1990. This strategy is being hampered because the *Plasmodium* parasite has developed resistance, as a result of monotherapy, to certain components of currently applied combination drugs. The efficacy and very short half-life (<6 hours) of the artemisinin derivatives make it less likely to develop resistance. Furthermore, the artemisinins have broad stage specificity and can be used to treat both uncomplicated and severe malaria (White, 2004; Yeung *et al.*, 2004).

Besides, artemisinin derivatives are particularly effective in combinations because of their high killing rates (parasite reduction ratios RR circa 10,000 fold per cycle) (White, 1997 and 1999), lack of adverse effects, and absence of significant resistance. In ACTs a 3-day treatment course exposes 2 asexual cycles and so reduces the number of parasites in the body by approximately one hundred million-fold. The gametocytocidal activity of the artemisinin compounds is an important bonus reducing transmissibility and thus further reduces malaria incidence in low-transmission settings (Price *et al.*, 1996; Barnes *et al.*, 2005; Carrara *et al.*, 2006).

Traditional System of Medicine (TSM)

TSM is one of the centuries-old practice and long-serving companion to the human kind to fight against disease and to lead a healthy life. Every indigenous people have been using their unique approach of TMS practice where among, the Chinese, Indian, African TMSs are world-wide renowned (Karunamoorthi *et al.*, 2012). Despite globalization and modernization, still a sizeable fraction of the rural poor purely relies on the traditional medicines as their primary health care modality in the poor-health care settings (Karunamoorthi, 2012c). In rural areas in Africa and elsewhere in developing countries, traditional medicines are often trusted, affordable and accessible, as they are made from locally available plants or other elements (Kitua and Malebo, 2004).

Indeed, the induction of modern health care services has posed immense threat to indigenous health-care practices due to their potential and speedy therapeutic effect. Therefore, in the recent past the TSM are disappearing, displaced, and undervalued by the people. Nevertheless, over the past three decades the worldwide scientific community is extremely inclined towards our long-standing traditional and folk herbal medicine to explore the opportunities to formulate the novel phytotherapeutic agents in the future (Karunamoorthi *et al.*, 2012).

Definitions

Herb

An herb in the botanical sense is a plant lacking a permanent woody stem that produces seeds and owers and that dies down after its growing season. In a traditional medical sense an herb is a small, nonwoody, plant valued for its medicinal, savory, or aromatic qualities. An herbal preparation is, then, a natural remedy derived from herbs.

Traditional Chinese Medicine (TCM)

TCM is an ancient Chinese healthcare system based on the concept of balanced qi (a vital energy believed by practitioners to ow throughout the body that regulates a person's spiritual, emotional, mental, and physical balance). Qi is inuenced by the opposing forces of yin (negative energy) and yang (positive energy). In this system disease is believed to result when the ow of qi is disrupted and yin and yang become unbalanced. Herbal and nutritional therapeutic interventions supplement physical exercises, meditation, acupuncture, and massage in restoring health.

Complement Medicine (CM)

CM is a group of medical and healthcare systems, practices, and products that are not considered at this time to be a part of conventional medicine, primarily because there is insufficiently convincing evidence of their efficacy, but are used together with conventional medicine. Some of these practices employ herbs.

Alternative Medicine (AM)

AM is also a group of medical and healthcare systems with practices and products. These are used in place of conventional medical practices. Dietary supplements and herbals are employed in some of these practices. An example is the use of shark cartilage to treat cancer.

History of Traditional and Folk Herbal Medicine

Plants have been used as a source of medicine throughout history and continue to serve as the basis for many pharmaceuticals used today. The medicinal properties of plants are described already on Assyrian clay tablets dated about 2000 B.C. and documented in the Egyptian culture, the Indian Ayurveda (Patwardhan, 2005), in Traditional Chinese Medicine (TCM)(Kong *et al.*, 2009), and various European documents. Traditional medicines not only provide valuable clues for finding new drugs, but also may help to shift the drug discovery paradigm from 'finding new-entity drugs' to 'combining existing agents', and might even direct the combinations between such agents (Kong *et al.*, 2009; Wagner *et al.*, 2009).

As a result, at the moment, there is a growing acceptance and keen interest towards traditional system of medicines observed worldwide. It has lead many researchers to concentrate on traditional medicine and in particular on traditional antimalarial phytotherapy, since malaria is a major public health issue in the resource-limited settings; where there is a fragile healthcare system, which quite often fails to meet the neediest poor people. The existing potent antimalarial artemisinin combination therapy (ACT) is unaffordable, inaccessible and there are reports suggesting development of resistance too (Karunamoorthi and Tsehaye, 2012).

Traditional and Folk Herbal Medicine and Anti-Malarial Chemotherapy

The use of traditional antimalarial plants is one of the prehistoric age-old customs and it is largely conceived by long-term observation and trial and error mode. Ethnic groups are the repositories of the knowledge on the herbal remedy (Karunamoorthi and Tsehaye, 2012). Over 1,277 plants belonging to 160 families were reported in 2004 to be used traditionally for the treatment of malaria following an extensive survey of the literature (Willcox *et al.*, 2004), and since then the number of species has increased substantially due to the increasing worldwide interest in anti-malarial plants (Rasoanaivo *et al.*, 2011).

The history of anti-malarial chemotherapy is intimately linked with the history of herbal medicinal products. Quinine, the original natural product used in antimalarial chemotherapy was identified from cinchona tree bark, and purified in 1820. The attempt to synthesize quinine led to the development of methylene blue and the dye industry. From this came the classical 4 aminoquinolines and amino-alcohols, such as chloroquine, amodiaquine and mefloquine, which have been the mainstay of malarial treatment over the last century. The natural product naphthoquinone lapichol also was identified as the active ingredient in tree bark used to treat malaria (Wells, 2011). This discovery helped direct the selection of lapinone, which in turn provided the foundation for the discovery of atovaquone, a component of Malarone® and still it is a mainstay of malaria prophylaxis for travellers (Hudson, 1993).

Artemisinin was first isolated from the leaves of the *A. annua* in 1971. The highly unusual endoperoxide group has formed the basis for a range of fully synthetic, longer acting molecules which are now in clinical development (Vennerstrom *et al.*, 2004), and which may be the basis of the next generation of therapies, active against artemisinin-resistant malaria. The pattern is consistent with the initial identification of natural product scaffolds, followed by their modification for clinical use by a combination of medicinal chemistry, formulation development and combination therapy (Wells, 2011).

A. annua: A Gifted Chinese Antimalarial Plant

Qing Hao is a Chinese term used to refer to the antipyretic herb, derived from two plant species commonly found in China's wastelands – *A. annua* and, *Artemisia apiacea*. The plant with the more intense blue-green flowers, called "blue-green *hao*" (*qing hao*) or "fragrant *hao*" (*xiang hao*), is today identified as *A. apiacea*, while the variant with yellow flowers, called *huang hua hao*, is now known to be *A. annua* (Willcox and Bodeker, 2004). *A. annua* is a strongly fragrant, annual herbaceous plant and it is the only natural botanical source for artemisinin (**Qinghaosu**) (Jain *et al.*, 1996; Klayman, 1985). It is native to China, where it is known as **Qinghao** (green herb) and has been used for over 2,000 years to treat symptoms associated with fever and malaria.

The current pharmacopoeia of the People's Republic of China (The People's Republic of China, 1985) describes the dried herb of *A. annua* as a remedy for fevers including malaria and for therapeutic use (Sto¨ ger, 1991). Cinchona has been known in the West for almost 400 years (Lee, 2000a). In contrast, *A. annua*, has been known to the Chinese herbal practitioners for at least 1,000 years but introduced into Western allopathic medicine only in the last 30 years. The major difference between Cinchona and Artemisia is the speed of recovery, with the active principle of the latter, artemisinin, was isolated, synthesised and then modified to produce effective therapeutic entities, largely due to the efficiency of modern chemical techniques (Lee, 2000b). It is vivid that *A. annua*, has been in use world-wide and has its own vernacular names in the regions available. Such names are given below in the Table 2.1.

Taxonomical Position of *A. annua*

Asteraceae is one of the highly medicinal families consisting of a wide range of plants that has colonized in almost all the continents of the world. It possesses 11 subfamilies, the most well known being Chicorioideae and Asteroideae. Artemisia annua belongs to the class of owering plants, Magnoliopsida. The family of Asteraceae is the second largest family of owering plants in the world (Bhakuni *et al.*, 1988 and

1990). Artemisia is a large, diverse genus of plants with between 200 and 400 species, and comprises hardy herbs and shrubs. The taxonomic position of *A. annua* is given below in Table 2.2.

Table 2.1: Vernacular names *of Artemisia annua* in the respective languages.

Sl.No.	*Language*	*Vernacular Name(s)*
1.	Bengali	Nagdoun
2.	Chinese	Caohao, Cao Qinghao, Cao Haozi, Chouhao, Chou Qinghao, Haozi, Jiu Bingcao, Kuhao, San Gengcao, Xianghao, Xiang Qinghao, Xiang Sicao, Xiyehao
3.	English	Annual wormwood, sweet wormwood
4.	French	Armoise annuelle
5.	German	Chinesischer Beifuß
6.	Gujarati	Nagdamani, Sukhdarshan
7.	**Hindi**	Nagdona, Daman
8.	Japanese	Kusoninjin
9.	Korean	Chui-ho, Hwang-hwa-ho, Gae-tong-sook
10.	Malayalam	Makkippuvu, Masipatri
11.	Marathi	Nagdamani, Nagdavan
12.	**Sanskrit**	Nagdamani
13.	Tamil	Makkippu
14.	Telugu	Masipatri
15.	Vietnamese	Thanh cao hoa vàng

Table 2.2: Taxonomic position of *Artemisia annua* L.

Kingdom	*Plantae*–Plants
Subkingdom	*Tracheobionta*–Vascular plants
Super-division	*Spermatophyta*–Seed plants
Division	*Magnoliophyta*–Flowering plants
Class	*Magnoliopsida*–Dicotyledons
Order	Asterales
Tribe	Anthemideae
Family	Asteraceae (Compositae)–Aster family
Genus	Artemisia L.–Sagebrush
Species	*Artemisia annua* L.–Annual wormwood

Origin

Its ancient Chinese name, **Qing Hao**, literally means "**Green herb**" The genus *Artemisia* comprises over 400 species, many of which have an aromatic, bitter taste. *A. annua* is an annual plant, with a crop cycle of six months native to Asia, most probably China. Its occurs naturally as part of a steppe vegetation in the northern parts of Chahar and Suiyuan provinces in China, at 1000 to 1500 m above sea level. In Youyang County, Chongqing, China, the hometown of *A. annua* the plants have high (0.9 per cent) levels of artemisinin. Currently, *A. annua* has been naturalized in many countries (Ferreira *et al.*, 1997).

Historical Importance of *A. annua*

A. annua's pharmacological use began in ancient Chinese medicine (Ferreira *et al.*, 1997; Hsu, 2006). One of the first recorded uses of *A. annua* can be found in the **Chinese Handbook of Prescriptions for Emergency Treatments** of 340 AD where it is recommended for the treatment of fevers. In 1527, a Chinese herbalist, **Li Shi-Zhen,** recommended *A. annua* for the treatment of childhood fevers and a closely related plant *Artemisia apiacea* for the treatment of the then unknown disease that we know today as malaria (Klayman, 1985). Zhang Ji (150–219A.D.) in his classic text **On Cold Damage (Shang Han Lun)** recommends a decoction of *A. annua* to treat fevers with sweating and jaundice (Mitchell *et al.*, 1999). He recommends that the fruits of *Gardenia jasminoides* and roots of rhubarb (*Rheum palmatum*) should be added to the decoction (Willcox *et al.*, 2004a).

Proposed Theories on the Name of Artemisia

There are two theories as to the origin of its name. Ferreira *et al.* (1997) say that it is named after the Greek goddess Artemis, meaning literally "**she who heals sickness**," who was in fact goddess of the hunt, of forests, and was thought to be responsible for sudden death in women (Guirand, 1959). Apparently plants of this genus, probably *Artemisia absinthium*, were used to control pain in childbirth and to induce abortions. Bruce-Chwatt (1982) says that Artemisia was named after **Queen Artemisia of Caria** (Turkey), who lived in the fourth century B.C. She was so aggrieved on the death of her husband and brother, King Mausolus of Halicarnassus, that she mixed his ashes with whatever she drank to make it taste bitter. *A. annua* is so named because it is almost the only member of the genus with an annual cycle (Willcox *et al.*, 2004).

Geographical Distribution

A. annua is widely distributed in the temperate, cool temperate and subtropical zones (mainly in Asia) of the world. It originated from China and grows mainly in the middle, eastern and southern parts of Europe and in the northern, middle and eastern parts of Asia. However, it also grows in the Mediterranean region and countries in North Africa, as well as in south and south-west Asia. In addition, following its settlement in North America from northern Asia, it grows widely in Canada and the United States (Editorial Committee on Flora, 1991).

Physical Characteristics

The wormwood bush can grow to a height of 2 meters, and produces a number of bushy stems that are covered with fine, silky grey-green hairs. Wormwood produces small yellow-green flowers from summer through to early autumn or fall (Gordon, 1980). *A. annua* is a large vigorous weedy annual shrub often reaching more than 2 m

Figure 2.2: *Artemisia* spp. plant.

tall, usually ribbed and single stemmed with alternate branches and stem covered with fine, silky grey-green hairs. Leaves are aromatic, alternate, deeply dissected, glabrous, up to 12 cm long. Leaves contain both 10-celled biseriate trichomes and 5 cell filamentous (T) trichomes. Flowers are tiny nodding (capitula) only 2 or 3 mm across, greenish or yellowish, enclosed by numerous, imbricated bracts displayed in lose panicles. They have bisexual central (disc) florets containing little nectar and pistillate marginal (ray) florets, the latter extruding stigmas prior to the central flowers. Both flowers have synpetalous tubular corolla with the top split into five lobes in the hermaphroditic florets and 2-3 lobes in the pistillate florets with ovaries inferior and unilocular and with the receptacle triangular in shape. Both florets and receptacle bear abundant 10-celled biseriate trichomes; Ttrichomes (filamentous) occur at the pedicel and bracts. Fruit are achene, obovoid, 0.6-1 mm long, faintly nerved and glabrous (Orwa *et al.*, 2009).

Ecology and Biology of *A. annua*

It occurs naturally as part of steppe vegetation in the northern China, at 1000-1500 m altitude. It is usually a naturalized weed of waste places, roadsides, fallow fields and neglected gardens. The plants are longer-lived, more hardy and aromatic when grown in poor dry soils (Orwa *et al.*, 2009). *A. annua* is a determinate short-day plant where non-juvenile plants are very responsive to photoperiodic stimulus and flower about two weeks after induction. The critical photoperiod seems to be about 13.5 hours, but there are likely to be photoperiod x temperature interactions. It is in flower from August to September with mature seeds produced in September and October. In the tropics, flowering is induced when the plants are very small (Orwa *et al.*, 2009). The reproduction of the shrub occurs by insects, self-pollination, and wind distribution (Bown, 1995).

Cultivation of *A. annua*

Artemisinin cannot be synthesized cost effectively, so it is still extracted from *A. annua* aerial parts. Therefore, the science of commercial cultivation of *A. annua*, to maximize artemisinin yields, is already well developed (Laughlin *et al.*, 2002). A few countries are currently cultivating *A. annua* on a large scale, such as China, Kenya, the United Republic of Tanzania and Vietnam. Small-scale cultivation has been undertaken in India and other countries in Africa, south Europe and South America. *A. annua* for industrial use is mainly collected from the wild (WHO, 2006a). It has been introduced for cultivation in high-altitude regions or regions with pronounced cool periods, such as Brazil, Cameroon, Ethiopia, India, Kenya, Mozambique, Myanmar, Tanzania, Thailand, Uganda and Zambia (Simon *et al.*, 1990). The chronology of the global expansion of *A. annua* cultivation is given below in Table 2.3 (Ellman, 2005).

Table 2.3: Chronology of global expansion of *A. annua* cultivation.

Sl.No.	*Location of Cultivation*	*Period*
1.	Origin in China	168 B.C
2.	Moved to Vietnam	1970s
3.	Moved to Europe, Americas and Australia	1980s
4.	Moved to East Africa	1994

Chemical Constituents of *A. annua*

Chinese chemists have devotedly isolated several effective bioactive compounds from this plant since the Vietnam War started in 1971. Artemisinin is extracted from *A. annua* and has four derivatives: dihydroartemisinin, artesunate, artemether and arteether. These four derivatives have approximately five times higher potency than artemisinin, but a shorter shelf-life. Artemisinin and its derivatives work against malaria parasites by being schizontocidal and gametocidal, with rapid action on all blood stages (GMAP, 2008). Therefore, they reduce the disease severity ultimately reducing the mortality and transmitability, which is extremely remarkable.

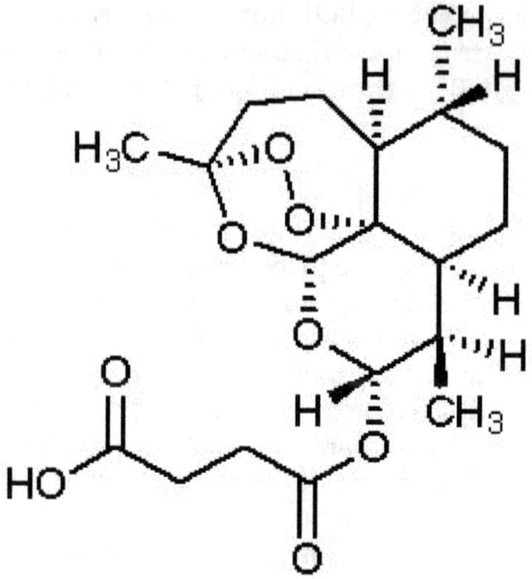

Artesunate

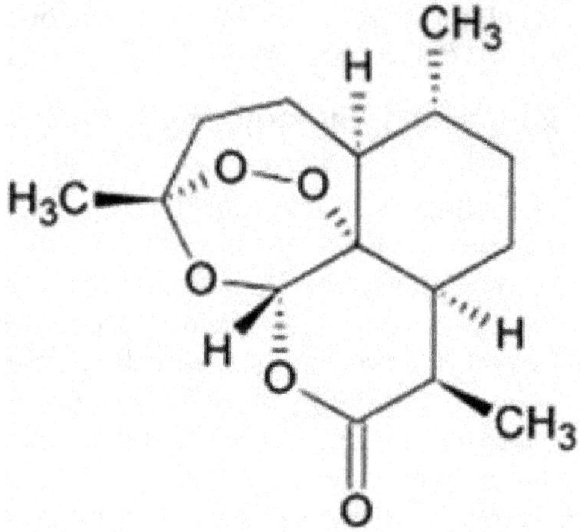

Artemether

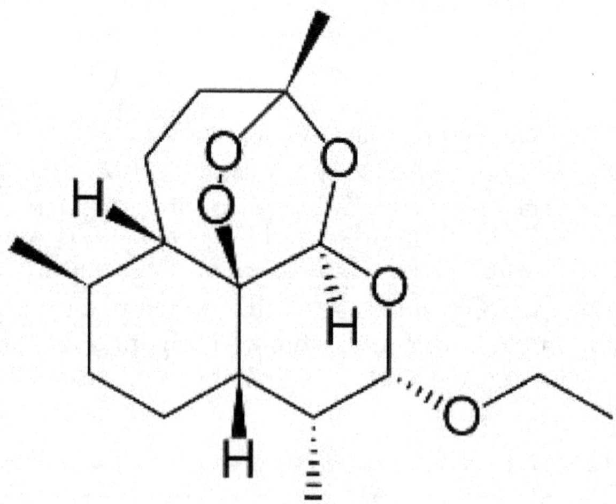

Arteether

Synthetic derivatives that are water soluble (artesunate) and oil soluble (artemether) have been developed to enable intravenous and intramuscular administration, respectively (Van Agtmael *et al.*, 1999). Many secondary metabolites of terpene peroxides are isolated from *A. annua* such as artemisia ketone, artemisinic alcohol, arteannuin B and myrcene hydroperoxide (Bertea *et al.*, 2005; Brown *et al.*, 2003).

Artemisinin Derivatives: A Magic Antimalarial Compounds

Artemisinin is a sesquiterpene lactone, containing an unusual endoperoxide group, which is believed to be responsible for the antimalarial activity of *A. annua.* Unlike quinine-related drugs and antifolate drugs, artemisinin and its derivatives are gametocytocidal and reduce the transmission potential of *falciparum* malaria (Price *et al.*, 1996). Artemisinin is poorly soluble in oil or water, so is usually administered orally, although it can be given rectally (Ashton *et al.*, 1998) and, when suspended in oil, intramuscularly (Titulaer *et al.*, 1990).

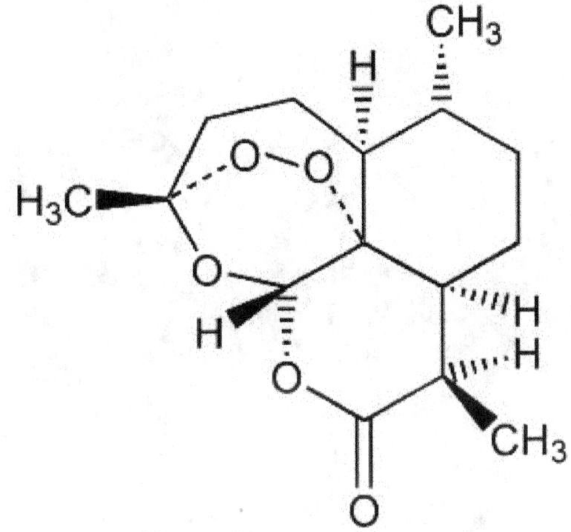

Dihydroartemisinin

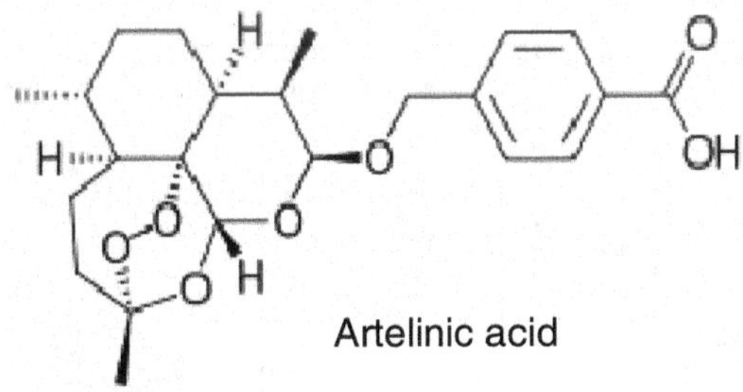

Artelinic acid

Artelinic acid (or its salt, artelinate) is an experimental drug that is being investigated as a treatment for malaria (Bustos *et al.*, 1994). It is a semi-synthetic derivative of the natural compound artemisinin. Artelinic acid has a lower rate of neurotoxicity than the related artemisin derivatives arteether and artemether (Genovese *et al.*, 2000), but is three times more toxic than artesunate (Li *et al.*, 2007).

It is now universally accepted that this family of compounds is among the most powerful antimalarial drugs ever discovered. The pharmacological and clinical evidence is well documented (Wright and Warhurst, 2002; Wilairatana and Looareesuwan, 2002). Reports differ regarding the stage at which the highest content of artemisinin occurs during plant development: just before plant flowering (Liersch *et al.*, 1986; Woerdenbag *et al.*, 1994; Chan *et al.*, 1995), or at the point of full flowering (Singh *et al.*, 1988; Ferreira *et al.*, 1995). Both views suggested a possible linkage between flowering and artemisinin biosynthesis. Artemisinin was found to be extractable from all aerial tissues of the plant, with very high content in the capitula, but not in pollen grains or root tissue. Roots transformed with *Agrobacterium rhizogenic* have also been reported to produce artemisinin (Banerjee *et al.*, 1997; Weathers *et al.*, 2004).

Traditional uses of *A. annua*

Malaria patients treated with a tea from *A. annua*, in a dose corresponding to the recommendations of the Chinese pharmacopoeia, showed a very rapid disappearance

Figure 2.3: The author discusses about the traditional phytotherapuetic potentialities of *Artemisia* spp. with the Ethiopian key informants.

of malaria parasites from the blood, similar to the effect described for pure artemisinin (de Vries and Dien, 1996). Old ineffective drugs were to be abandoned in favor of new formulations based on a compound called artemisinin that could finally reduce the staggering death toll. More than 20 African countries have signed on. But the catch is, there aren't nearly enough of the new drugs to go around (Enserink, 2005). The use of herbal teas prepared from the dried leaves of locally grown *A. annua* is being promoted as an alternative treatment for malaria in areas where people do not have access to, or cannot afford effective anti-malarials such as artemisinin combination therapy (Harborne, 1988).

Orally wormwood is taken for loss of appetite, indigestion and gastrointestinal problems (Blumenthal *et al.*, 1998; BHP, 1996). It is often used in conjunction with other herbs to deal with gallbladder disorders and flatulence. Historically wormwood has been used as a parasitic worm killer, an aphrodisiac, tonic and to induce perspiration (Jellin *et al.*, 2000). Other traditional applications include regulating menstruation and reducing fever (Gruenwald *et al.*, 1998). The daily dose is specified as 4.5 to 9 grams of dried herb to be prepared as a tea infusion with boiling water. 25-30 grams of the dried herb boiled for 30 minutes renders an analgesic and antipyretic decoction to be taken once daily for seven days (WHO, 2006a).

Wormwood is generally regarded as safe when used appropriately and for short durations. Wormwood should not be taken in large amounts or long-term. This herb has been declared unsafe for use during pregnancy due to its uterine and menstrual stimulating effects. Due to the lack of sufficient reliable information, wormwood should not be used while breastfeeding. The leaves of *A. annua* are also burned in China as a fumigant insecticide to kill mosquitoes (Foster and Chongxi, 1992). Some studies have reported insecticidal effects of *A. annua* extract including growth retardation, as well as antifeedant and larvicidal effects (Haghighian *et al.*, 2008; Shekari *et al.*, 2008; Hasheminia *et al.*, 2011).

A. annua in Modern Drug Research and Application

The genus Artemisia belongs to the large family of Asteraceae, encompassing more than 200-400 species. These plants have antimalarial effects and are used as a clinical medicinal plant in many countries (Ridder *et al.*, 2008). Artemisinin and its derivatives are the most potent and rapidly acting antimalarial drugs and the parasite biomass is reduced by 10,000-fold per asexual life cycle, compared to 100- to 1000-fold for other antimalarials. They also decrease gametocyte carriage by 90 per cent, thus reducing transmission of malaria (White *et al.*, 1999). Artemisinin compounds have become the cornerstone of malaria case-management whether used orally for the treatment of uncomplicated malaria episodes or in parenteral or rectal administration for severe malaria (Kindermans *et al.*, 2007).

Artemisinin is currently the best therapeutic against both drug-resistant and cerebral malaria-causing strains of *P. falciparum*. It is now widely used throughout Southeast Asia and Africa, where malaria has become resistant to nearly all anti-malarial drugs (Newton and White, 1999). This new policy was issued to protect the artemisinins and the partner drug from resistance in the wake of reduced efficacy of the well-known once-first-line antimalarials (chloroquine, and sulphadoxine-pyrimethamine) due to resistance (Kindermans *et al.*, 2007).

A. annua and Other Diseases

Artemisinin is an important therapeutic that, along with its derivatives, is a proven treatment for a number of diseases besides malaria (Dhingra *et al.*, 2000), including hepatitis B (Romero *et al.*, 2005), and parasites that cause schistosomiasis (Borrmann *et al.*, 2001; Utzinger *et al.*, 2001). Artemisinin has also been shown to be effective against a variety of cancer cell lines including breast cancer, human leukemia, colon, and small-cell lung carcinomas (Efferth *et al.*, 2001; Singh and Lai, 2001). Furthermore, artemisinin may be especially effective in treating drug resistant cancers (Efferth *et al.*, 2002; Sadava *et al.*, 2002).

In addition, artemisinin is a possible antibacterial agent as well as a natural pesticide (Efferth *et al.*, 2001; Kumar *et al.*, 2000). ART is also used to treat infections of *Pneumocystis carinii*, *Toxoplasma gondii*, human cytomegaloviruses, Herpes simplex viruses, and hepatitis B and C. Artemisinin has also been shown to have in vitro activities against certain tumor cells (Efferth, 2007). Since artemisinin is a relatively safe drug with other anti-microbial activity and no obvious adverse reactions or serious side effects, even for pregnant women (WHO, 2003), it is of particular interest in the possible treatment of a variety of diseases.

Artemisia and AIDS

In African countries a tea infusion prepared from *A. annua* has been used for the treatment of malaria only for the past 10-20 years. Since then, several informal claims in Africa exist that the *A. annua* tea infusions are also able to inhibit HIV. Lubbe *et al.* (2012) has confirmed with *in vitro* studies that *A. annua* is effective in inhibiting HIV. Also of interest is their conclusion that the component artemisinin plays no part in this activity, and that *Artemisia afra*, which in indigenous in Africa from South Africa to Kenya, is also effective in inhibiting HIV.

Livestock (Anti-parasitic and Nutritional Potential)

A. annua and artemisinin uses for the livestock industry are currently in expansion based on current reports of its anti-protozoal, anti-bacterial and antioxidant activities of the plant, its extracts, and its essential oil. Some examples include Babesia (Kumar *et al.*, 2003), Eimeria or coccidiosis (Allen *et al.*, 1997; Arab *et al.*, 2006; Brisibe *et al.*, 2008), and the trematodal blood fluke *Schistosoma* spp. (Xiao and Catto, 1989; Xiao *et al.*, 2000; Lescano *et al.*, 2004), and bacteria (Juteau *et al.*, 2002). Recently, different tissues of the plant have been analyzed for its potential use in animal feed and scored high values for antioxidant capacity (ORAC and total phenolics) and as source of aminoacids, with negligible amounts of anti-nutritive components such as phytates and oxalates (Ferreira and Janick, 2009).

Potential Toxicity

The risk of toxic effects from the use of this plant appears to be limited, since *A. annua* is included in the pharmacopoeia of the People's Republic of China, with recommendations for its dose and therapeutic use, and it may therefore be regarded as one of the established medicines in the present world. Adverse effects from pure artemisinin are minimal (de Vries and Dien, 1996). As with any experimental medication, however, close attention must be given to possible risks and side effects of *A. annua* preparations (Mueller *et al.*, 2000).

Artemisinin has no reported toxicity if taken in recommended doses for short periods in the treatment of malaria (Meshnick, 2002). The pollen is extremely allergenic and a cause of hay fever, as its cousin ragweed (*Ambrosia artemisifolia*). Artemisinin derivatives (dihydroartemisinin, artesunate, artemether, arteether) present better bioavailability and antimalarial activity than artemisinin, but have different safety margins than artemisinin. The bioavailability and half-lives also vary with the delivery system (intramuscular, intra-venous, intra-peritoneal and oral) (Ferreira and Janick, 2009).

Artemisinin has been considered as a relatively safe drug with no obvious adverse reactions or serious side effects, even for pregnant women (Dellicour and Hall, 2007). A recent study (Efferth and Kaina, 2010), however, outlines potential toxic effects of artemisinin when used for long periods and offers suggestions for safe use. A very rare side effect of ART drug is haemoglobinuria (Black water). Cases of severe allergic reaction have also been reported (Leonardi *et al.*, 2001).

Demand of *A. annua* for the Synthesis Artemisinin

Artemisinin is a sesquiterpene lactone isolated from the aerial parts of *A. annua* plants. As indicated earlier, although chemical synthesis of artemisinin is possible, it is not economically feasible (Weathers *et al.*, 2006). A major hurdle in using this compound to treat malaria estimated to cause 300 to 500 million cases and over 1 million deaths each year, worldwide has been producing enough artemisinin to meet world demand (Milhous and Weina, 2010). Efforts to eradicate malaria are an increasing demand for an affordable, high-quality, robust supply of artemisinin (Graham *et al.*, 2010).

Demand for ACTs is predicted to increase dramatically, from around 100 million treatments in 2008 to 250 million by 2015. However, the low yield of Artemisia creates problems for production and there are concerns that artemisinin supplies will not reliably keep pace with the increasing demand for ACTs (CNAP, 2006). Agro-economic analysis indicates that, worldwide, between 17, 000 and 27,000 ha of Artemisia annua are required to satisfy global demand for ACT. An approximate, depending on numerous factors, 4 000 hectares are required for every 100 million ACT treatments. Present acreage stands at an estimated 5,000 hectares. Optimal conditions to grow *A. annua* are as follows in Table 2.4 (Ellman, 2005);

Table 2.4: Ideal conditions for cultivation of *A. annua.*

Cultivation Characteristic	*Parameters*
Longitude	105–115°E
Latitude	25–35°N
Altitude	1000–1500 metres above sea level
Climate	Temperate, subtropical
Rainfall	Reliable for planting/transplanting; 700–1000 mm per annum
Soil	Good drainage, moderate fertility, pH over 5.5
Day length	Short day length triggers flowering
Land availability	Intensive cultivation; land for food and other crops
Labour availability	High demand for transplanting, weeding, harvesting
Profitability	Competitive with returns from other cash crops
Processing facility	Easy access, sufficient capacity
Life-cycle	In China and Vietnam, the life-cycle is completed in ten months under natural conditions, whereas in Europe and America, plants grow more rapidly, completing their life-cycle in just six months.

Graham *et al*. (2010) have paved the way to fast-track breeding varieties of the *A. annua* plant with highly desirable genetic traits. The authors report a genetic map of the plant and identify key loci that could improve agricultural yields, decrease production costs, ensure a steady global supply of the drug, and improve grower confidence in the crop. They have recognized that the yield of artemisinin vary by geographic origin and is inheritable when the super leafy strains possess bountiful glandular trichomes outgrowth structures, where artemisinin is produced and stored by the plant were crossed.

The artemisinin yield from Klayman's 1985 Potomac River variety of *A. annua* was only 0.06 per cent of dry weight of plant material, whereas the Chinese had described varieties in Sichuan province with yields of 0.01 to 0.5 per cent (Klayman, 1985). Even the best of basic crop science, would not keep pace with the growing global health demands and emerging resistance to the antimalarial mainstay, chloroquine (Milhous and Weina, 2010).

Major Challenges and Opportunities

3·3 billion people are at risk of malaria, which is endemic in 106 countries. Between 655 000 and 1·2 million people die every year from *P. falciparum* infection (WHO Malaria Report 2011; Murray *et al.*, 2012). Much of this morbidity and mortality could be avoided if drugs available to patients were efficacious, high qualitied, and used correctly. Children in sub-Saharan Africa and Southeast Asia have the highest risk of contracting and dying from malaria. The global burden of malaria has reduced in the past decade (Johanson *et al.*, 2010), and endemic countries are reliant on the long-term availability of effective antimalarial drugs to maintain this progress (Newman, 2011). Poor-quality antimalarial drugs lead to drug resistance and inadequate treatment, which pose an urgent threat to vulnerable population and jeopardise progress and investments in combating malaria (Nayyar *et al.*, 2012).

Resistance or tolerance to artemisinin derivatives has been described in western Cambodia, and is characterised by slow rates of parasite clearance after treatment (Dondorp *et al.*, 2009). Natural products have been the starting points for new malaria medicines over the last century. By making the most of the potential for clinical data, it is possible that they could continue to influence our thinking for the next century (Wells, 2011). Even though currently ACT serves a pivotal role to combat malaria, they are mostly inaccessible and unaffordable in the resource-limited settings. Besides, there are signs of resistance to artemisinins emerging in Thai-Cambodian border and this has led to explore the low-cost and easily accessible plant species as a potential source of antimalarials (Karunamoorthi and Tsehaye, 2012).

"**In wilderness is the preservation of the world**" Henry David Thoreau (1817-1862). After two thousand years of continuous use, artemisinin has proved to be one of the gems in the "**Great Treasure House**" of Chinese traditional medicine. Its transformation into a mainstream therapeutic agent and drug of choice for malarial treatment worldwide over a period of three decades provides several valuable lessons for complementary medicine practitioners (Awofeso, 2011);

- ☆ There is a need for multidisciplinary collaboration. The full potential of artemisinin was only realized following the work of multidisciplinary Project 523 supported by the Mao-led Chinese government to determine effective anti-malarial chemotherapy. This initiative was pivotal facilitating enhanced understanding of the pharmacology of artemisinin, and in using modern technology to enhance its potency. Collaborations between complementary medicine practitioners and allopathic medicine researchers have so far been weak, and needs to be strengthened, especially in view of the increasing popularity of complementary medicine, and the potential of some traditional remedies to contribute inexpensively towards global health improvement, as is the case with artemisinin.

- ☆ Second, the rise of artemisinin illustrates the need for comprehensive toxicity studies of traditional medicinal plants and on clinical trials of single plants or remedies containing mixtures of plants. Such research should be published in open-access mode on Internet accessible databases, and in English so as to ensure prompt dissemination of findings. If the artemisin study had not been published in an English version of the Chinese Medical Journal, the adoption of the drug as the mainstay of malaria treatment might have been delayed even further.
- ☆ Finally, complementary therapy practitioners should work with allopathic medicine researchers to explore other uses of current medicinal plants through laboratory testing of their pharmacologies, and by experimental studies of efficacy in multiple disease conditions.

Artemisinin has proved to be highly multifunctional in its scope of use. The same may apply to other medicinal plants in current use for single diseases. This point is particularly salient since at least 80 per cent of the worlds' population in developing countries use plant materials as their source of primary health care, and less than 20 per cent of all plant species have been evaluated chemically or biologically (Farnsworth *et al.*, 1985).

Conclusion

Indeed, the traditional and folk herbal medicine is as old as human civilization. The earlier man largely depended on medicinal plants for his basic day to day therapeutic needs. The present scrutiny results shows that the plant kingdom is a repository for the majority of existing modern drugs. Moreover, traditional and folk herbal medicine and the potent antimalarial agents are closely inter-related as the potential antimalarials are derived from the medicinal plants. *A. annua* has been used in the TCM for many centuries to treat malaria related febrile illness and it has been validated by the Chinese scientists only in 1972. Currently, Although, WHO recommends artemisinin to be the drug of choice, with the strain of *P. falciparum*, a sign of drug resistance has been observed.

Reports emerging from resource-poor settings in particular Africa and Southeast Asian countries speaks about the substandard and counterfeit of antimalarial drugs sold from the private drug vendors. It is one of the dangerous practices leading to antimalarial drug resistance, ultimately even the potential ACT will become ineffective. Therefore, in the developing resource-constrained settings, concurrent interventions and a multifaceted approach should be extremely warranted by creating awareness among the poor consumer and health-worker and about the substandard and counterfeit drugs. Actions to be taken towards their cost, is extremely important. Besides, preserving our traditional knowledge on medicinal plants as well as validating the age-old practice by more scientific research and human safety clinical trials must be warranted and measures must be taken to explore on the innovative and novel potent antimalarials in the near future.

Acknowledgements

I would like to thank Mrs. Melita Prakash for her sincere assistance in editing the manuscript. My last but not least heartfelt thanks go to my colleagues from the Department of Environmental Health Science, College of Public Health and Medical Sciences, Jimma University, Ethiopia, for their kind support and cooperation.

References

Allen, P.C., Lydon, J., and Danforth, H.D. (1997). Effects of components of *Artemisia annua* on coccidia infections in chickens. *Poultry Science,* **76**: 1156-1163.

Arab, H.A., Rahbari, S., Rassouli, A., Moslemi, M.H., and Khosravirad, F. (2006). Determination of artemisinin in *Artemisia sieberi* and anticoccidial effects of the plant extract in broiler chickens. *Tropical Animal Health and Production*, **38**: 497-503.

Ashton, M., Duy Sy, N., Van Huong, N., et al. (1998). Artemisinin kinetics and dynamics during oral and rectal treatment of uncomplicated malaria. *Clinical Pharmacology* and *Therapeutics,* **63**: 482-493.

Awofeso, N. (2011) Project 523: transformation of Artemisinin from traditional Chinese medicine to mainstream anti-malaria chemotherapy. *Spatula DD,* **1(2)**: 113-118

Banerjee, S., Zehra, M., Gupta, M.M., and Kumar, S. (1997). Agrobacterium rhizogenesmediated transformation of *Artemisia annua*: production of transgenic plants. *Planta Medica*, **63**: 467-469.

Barnes, K.I., and White, N.J. (2005). Population biology and antimalarial resistance: The transmission of antimalarial drug resistance in *Plasmodium falciparum*. *Acta Tropica*, **94**: 230-240.

Barnes, K.I., Durrheim, D.N., Little, F., Jackson, A., Mehta, U., Allen, E., Dlamini, S.S., Tsoka, J., Bredenkamp, B., Mthembu, D.J., White, N.J., and Sharp, B.L. (2005). Effect of artemether-lumefantrine policy and improved vector control on malaria burden in KwaZulu-Natal, South Africa. *PLoS Medicine*, **2**: e330.

Bertea, C.M., Freije, J.R., van der Woude, H., Verstappen, F.W.A., Perk, L., Marquez, V., De Kraker, J.W., Posthumus, M.A., Jansen, B.J.M., de Groot, A., Franssen, M.C.R., and Bouwmeester, H.J. (2005). Identification of intermediates and enzymes involved in the early steps of artemisinin biosynthesis in *Artemisia annua. Planta Medica*, **71**: 40-7.

Bhakuni, D.S., Goel, A.K., Jain, S., Mehrotra, B.N., and Srimal, R.C. (1990). Screening of Indian plants for biological activity: part XIV. *Indian Journal of Experimental Biology*, **28**: 619-637.

Bhakuni, D.S., Goel, A.K., Jain, S., Mehrotra, B.N., Patnaik, G.K., and Prakash, V. (1988). Screening of Indian plants for biological activity: part XIII. *Indian Journal of Experimental Biology*, **26**: 883-904.

BHP. British Herbal Pharmacopoeia (1996). Fourth Edition. British Herbal Medicine Association Scientific Committee, West Yorks, England.

Blumenthal, M., *et al.* (1998). The Complete German Commission E Monographs: Therapeutic Guide to Herbal Medicines. Austin: American Botanical Council.

Borrmann, S., Szlezak, N., Faucher, J.F., Matsiegui, P.B., Neubauer, R., Biner, R.K., Lell, B., and Kremsner, P.G. (2001). Artesunate and praziquantel for the treatment of *Schistosoma haematobium* infections: a doubleblind, randomized, placebo-controlled study. *Journal of Infectious Diseases*, **184**:1363-1366.

Bown, D. (1995). Encyclopedia of Herbs and Their Uses (RHS). DK Publishing (Dorling Kindersley), London.

Brisibe, E.A., Umoren, U.E., Owai, P.U., and Brisibe, F. (2008). Dietary inclusion of dried *Artemisia annua* leaves for management of cocidiosis and growth enhancement in chickens. *African Journal of Biotechnology*, **7**: 4083-4092.

Brown, G.D., Liang, G.Y., and Sy, L. (2003). Terpenoids from the seeds of *Artemisia annua*. *Phytochemistry*, **64**: 303-323.

Bruce-Chwatt, L.J. (1982). Qinghaosu: a new antimalarial. *British Medical Journal*, **284**: 767-768.

Bustos, M.D., Gay, F., and Diquet, B. (1994). *In-vitro* tests on Philippine isolates of *Plasmodium falciparum* against four standard antimalarials and four Qinghaosu derivatives. *Bulletin of World Health Organization*, **72**: 729-735.

Carrara, V.I., Sirilak, S., Thonglairuam, J., Rojanawatsirivet, C., Proux, S., Gilbos, V., Brockman, A., Ashley, E.A., McGready, R., Krudsood, S., Leemingsawat, S., Looareesuwan, S., Singhasivanon, P., White, N., and Nosten, F. (2006). Deployment of early diagnosis and mefloquine-artesunate treatment of falciparum malaria in Thailand: the Tak Malaria Initiative. *PLoS Medicine*, **3**: e183.

Chan, K.L., Chris, K.H., Jinadasa, S., and Yuen, J.H. (1995). Selection of high artemisinin yielding *Artemisia annua*. *Planta Medica*, **61**: 285-287.

CNAP. (Center Novel Agriculture Product) New hybrid varieties of *Artemisia annua* from the University of York. The CNAP Artemisia Research Project. 2006.

de Vries, P.J., and Dien, T.K. (1996). Clinical pharmacology and therapeutic potential of artemisinin and its derivatives in the treatment of malaria. *Drugs*, **52**: 818-836.

Dellicour, S., Hall, S., Chandramohan, D., and Greenwood, B.D. (2007). The safety of artemisinins during pregnancy: a pressing question. *Malaria Journal*, **6**: 15.

Dhingra, V., Rao, V.M., and Narasu, L. (2000). Current status of artemisinin and its derivatives as anti-malarial drugs. *Life Science*, **66**: 279-300.

Dondorp, A.M., Nosten, F., Yi, P., et al. (2009). Artemisinin resistance in *Plasmodium falciparum* malaria. *New England Journal of Medicine*, **361**: 455-467.

Editorial Committee on Flora of the People's Republic of China of the Chinese Academy of Sciences. (Lin Rong, Lin Yourun, eds.). -NýV

iir×_ [Flora of the People's Republic of China]. Beijing, Science Press, **76(2)**: 1991.

Efferth, T. (2007). Willmar Schwabe Award 2006: antiplasmodial and antitumor activity of artemisinin-from bench to bedside. *Planta Medica*, **73**: 299-309.

Efferth, T., and Kaina, B, (2010). Toxicity of the antimalarial artemisinin and its derivatives. *Critical Reviews* in *Toxicology*, **40**: 405-421.

Efferth, T., Dunstan, H., Sauerbrey, A., Miyachi, H., and Chitambar, C.R. (2001). The anti-malarial artesunate is also active against cancer. *International Journal of Oncology*, **18**: 767-773.

Efferth, T., Olbrich, A., and Bauer, R. (2002). mRNA expression profiles for the response of human tumor cell lines to the anti-malarial drugs artesunate, arteether, and artemether. *Biochemical Pharmacology*, **64**: 617-623.

Ellman, A. (2006). Experiences with cultivation of *Artemisia annua* in Africa. Meeting on the production of artemisinin and artemisinin-based combination therapies. 6-7 June 2005, Arusha, United Republic of Tanzania. WHO/HTM/MAL/ 2006.1113. Available at: whqlibdoc.who.int/hq/2006/WHO_HTM_MAL_ 2006.1113_eng.pdf (accessed on 25th May 2012).

Enserink, M. (2005). Source of new hope against malaria is in short supply. *Science*, **307**: 33.

Farnsworth, N.R., Akerele, O., Bingel, A.S., Soejarto, D.D., and Guo, Z. (1985). Medicinal plants in therapy. *Bulletin of World Health Organization*, **63**: 965-981.

Ferreira, J., and Janick, J. (2009). New Crop FactSHEET Annual Wormwood (*Artemisia annua* L.). Available at: www.hort.purdue.edu/newcrop/cropfactsheets/ artemisia.pdf (accessed on 5th May 2012).

Ferreira, J.F.S., Simon, J.E., and Janick, J. (1995). Relationship of artemisinin content of tissue cultured, greenhouse and field grown plants of *Artemisia anuua*. *Planta Medica*, **61**: 351-355.

Ferreira, J.F.S., Simon, J.E., and Janick, J. (1997). *Artemisia annua*: botany, horticulture, pharmacology (A Review). *Horticultural Reviews*, **19**: 319-371.

Foster, S., and Chongxi, Y. (1992). Herbal Emissaries: Bringing Chinese Herbs to the West. Healing Arts Press, Rochester, VT.

Genovese, R.F., Newman, D.B., and Brewer, T.G. (2000). Behavioural and neural toxicity of the artemisinin antimalarial, arteether, but not artesunate and artelinate, in rats. *Pharmacology Biochemistry and Behavior*, **67(1)**: 37-44.

GMAP. (Global Malaria Action Plan) "Saving Lives, Buying Time: Economics of Malaria Drugs in an Age of Resistance," National Academy of Sciences 2004, Available at: www.nap.edu/catalog/11017.html. Global Malaria Action Plan Report of the 2008 Artemisinin Conference, 8 to 10 October, York, UK. Available at: www.york.ac.uk/org/cnap/artemisiaproject/pdfs/AEconference-report-web.pdf (accessed on 5th May 2012).

Gordon, L.A. (1980). Country Herbal. Devon, England: Webb & Bower (Publishers) Ltd.

Graham, I.A., et al., (2010). The genetic map of *Artemisia annua* L. identifies loci affecting yield of the antimalarial drug artemisinin. *Science*, **327**: 328.

Gruenwald, J, et.al. (1998). PDR for Herbal Medicines. First Edition. Montvale, NJ: Medical Economics Company, Inc.

Guirand, F. (1959). Larousse Encyclopedia of Mythology. Paul Hamlyn, London.

Haghighian, F., Jalali Sendi, J., Aliakbar, A., and Javaherdashti, M. (2008). The growth regulatory, deterrency and ovicidal activity of worm wood (*Artemisia annua* L.) on *Tribolium confusum* Duv. and identification of its chemical constituents by GC-MS. *Pestycydy*, **12(1-2)**: 51-59.

Harborne, J.B. (1988). Introduction to Ecological Biochemistry. London: Academic Press Ltd.

Hasheminia, S.M., Jalali Sendi, J., Talebi Jahromi, K., and Moharramipour, S. (2011). The effect of *Artemisia annua* L. and *Achillea millefolium* L. crude leaf extracts on the toxicity, development, feeding efficiency and biochemical activities of small cabbage *Pieris rapae* L. (Lepidoptera: Pieridae). *Pesticide Biochemistry and Physiology*, **99**: 244-249.

Hsu, E. (2006). The history of Qinghao in the Chinese materia medica. *Transactions of the Royal Society of Tropical Medicine and Hygiene*, **100**: 505-508.

Hudson, A.T. (1993). Atovaquone - a novel broad-spectrum anti-infective drug. *Parasitology Today*, **9**: 66-68.

Jain, D.C., Mathur, A.K., Gupta, M.M., Singh, A.K., Verma, R.K., Gupta, A.P., and Kumar, S. (1996). Isolation of high artemisinin-yielding clones of *Artemisia annua*. *Phytochemistry*, **5**: 993-1001.

Jellin, J.M., Batz, F., and Hitchens, K. (2000). Natural Medicines Comprehensive Database. Third Edition. Stockton, California: Therapeutic Research Faculty.

Jiaxiang, S. (1991). Antimalarial Drug Development in China. National Institutes of Pharmaceutical Research and Development, Beijing, on Behalf of the UNDP/World Bank/WHO Special Programme for Research and Training in Tropical Diseases.

Johanson, E., Cibulskis, R., and Steketee, R. (2010). Malaria funding and resource utilization: the first decade of roll back malaria. March, 2010. Available at: http://www.rbm.who.int/ProgressImpactSeries/docs/RBMMalariaFinancingReport-en.pdf (accessed June 20, 2011).

Juteau, F., Masotti, V., Bessiere, J.M., Dherbomez, M., and Viano, J. (2002). Antibacterial and antioxidant activities of *Artemisia annua* essential oil. *Fitoterapia*, **73**: 532-535.

Karunamoorthi, K. (2011). Vector control: A cornerstone in the malaria elimination campaign. *Clinical Microbiology and Infection*, **17(11)**: 1608-1616. DOI: 10.1111/j.1469-0691.2011.03664.x.

Karunamoorthi, K. (2012a). Global Malaria Eradication: Is It Still Achievable and Practicable? Malaria: Etiology, Pathogenesis and Treatments. Anna Margrét Peterson and Gerald E. Calamandrei (Eds.), ISBN: 978-1-62100-363-2. Nova Science Publishers, New York, USA. (In Press).

Karunamoorthi, K. (2012b). Global malaria burden: Socialomics implications. *Journal of Socialomics* **1**:e108. doi:10.4172/jsc.1000e108.

Karunamoorthi, K. (2012c). Neem oil: biological activities and usage. Recent Progress in Medicinal Plants (RPMP) Vol. 33: Fixed oils and fats of pharmaceutical importance. Govil JN, Bhattacharya S. (Eds). ISBN: 1-933699-23-X. Studium Press LLC., P.O.Box-722200, Houston, Texas-77072, United States of America.

Karunamoorthi, K., and Ilango, K. (2010) Larvicidal activity of *Cymbopogon citratus* (DC) Stapf. and Croton *macrostachyus* Del. against *Anopheles arabiensis* Patton (Diptera: Culicidae), the principal malaria vector. *European Review of Medical and Pharmacological Sciences*, **14(1)**: 57-62.

Karunamoorthi, K., and Tsehaye, E. (2012). Ethnomedicinal knowledge, belief and self-reported practice of local inhabitants on traditional antimalarial plants and phytotherapy. *Journal of Ethnopharmacology*, **141(1)**: 143-150. DOI:10.1016/ j.jep.2012.02.012.

Karunamoorthi, K., Deboch, B., and Tafere, Y. (2010). Knowledge and practice concerning malaria, insecticide-treated net (ITN) utilization and antimalarial treatment among pregnant women attending specialist antenatal clinics. *Journal of Public Health*, **18(6)**: 559-566.

Karunamoorthi, K., Jegajeevanram, K., Jerome, X., Vijayalakshmi, J., and Melita, L. (2012). Tamil Traditional Medicinal System – Siddha: An Indigenous Health Practice in the International Perspectives. *International Journal of Genuine Traditional Medicine*, **2(2):** 1-11. Available at: TANG/www.e-tang.org (accessed on 27th June 2012).

Kindermans, J., Pilloy, J., Olliaro, P., and Gomes, M. (2007). Ensuring sustained ACT production and reliable artemisinin supply. *Malaria Journal*, **6**: 125. doi:10.1186/ 1475-2875-6-125.

Kitua, A.Y., and Malebo, H.M. (2004). Malaria control in Africa and the role of traditional medicine. In Traditional Medicinal Plants and Malaria. Boca Raton: CRC Press;Willcox ML, Bodeker and Rasoanaivo P 3-18.

Klayman, D.L. (1985). Qinghaosu (artemisinin): an antimalarial drug from China. *Science*, **228**:1049-1055.

Klayman, D.L., Lin, A.J., Acton, N., Scovill, J.P., Hoch, J.M., Milhous, W.K., Theoharides, A.D., and Dobek, A.S. (1984). Isolation of artemisinin (Qinghaosu) from *Artemisia annua* growing in the United States. *Journal of Natural Products*, **47**: 715-717.

Kong, D.X., Li, X.J., and Zhang, H.Y. (2009). Where is the hope for drug discovery? Let history tell the future. *Drug Discovery Today*, **14**: 115-119.

Kumar, S., Gupta, A.K., Pal, Y., and Dwivedi, S.K. (2003). *In-vivo* therapeutic efficacy trial with artemisinin derivative, Buparvaquone and Imidocarb Dipropionate against Babesia equi infection in donkeys. *Journal of Veterinary Medical Science*, **65**:1171-1177.

Kumar, S., Khanuja, S.P.S., Kumar, T.R.S., Jain, D.C., Srivastava, S., Bhattacharya, A.K., Saikia, D., Shasany, A.K., Darokar, M.P., and Sharma, R.P. (2000). Method for the use of alpha arteether as an anti-bacterial and anti-fungal agent. US Patent 6127405.

Laughlin, J.C., Heazlewood, G.N., and Beattie, B.M. (2002). Cultivation of *Artemisia annua* L. In Artemisia, Wright, C.W., Ed. Taylor & Francis, London.

Lee, M.R. (2002a). Plants against Malaria Part 1: Cinchona or The Peruvian bark. *The Journal of the Royal College of Physicians* of *Edinburgh*, **32**: 189-196.

Lee, M.R. (2002b). Plants against malaria part 2: *Artemisia annua* (Qinghaosu or the sweet wormwood). *The Journal of the Royal College of Physicians* of *Edinburgh*, **32**: 300-305.

Leonardi, E., Gilvary, G., White, N.J., and Nosten, F. (2001). Severe allergic reactions to oral artesunate: a report of two cases. *Transactions of the Royal Society of Tropical Medicine and Hygiene*, **95**: 182"183.

Lescano, S.Z., Chieffi, P.P., Canhassi, R.R., Boulos, M., and Neto, V.A. (2004). Antischistosomal activity of artemether in experimental *Schistosoma mansoni*. *Revista de Saude Publica*, **38**: 71-75.

Li, Q., Xie, L.H., Johnson, T.O., et al. (2007). Toxicity evaluation of artesunate and artelinate in *Plasmodium berghei* infected and uninfected rats. *Transactions of the Royal Society of Tropical Medicine and Hygiene*, **101(2)**: *104-112*.

Liersch, R., Soicke, H., Stehr, C., and Tullner, H.U. (1986). Formation of artemisinin in *Artemisia annua* during one vegetation period. *Planta Medica*, **52**: 387-390.

Lubbe, A., Seibert, I., Klimkait, T., and van der Kooy, F. (2012). Ethnopharmacology in overdrive: The remarkable anti-HIV activity of *Artemisia annua*. *Journal of Ethnopharmacology*, **141(3)**: 854-859.

Menendez, C. (1995). Malaria during pregnancy: a priority area of malaria research and control. *Parasitology Today*, **11**: 178-183.

Meshnick, S.R. (2002). Artemisinin: mechanisms of action, resistance and toxicity. *International Journal for Parasitology*, **32**:1655–1660.

Milhous, W.K., and Weina, P.J. (2010). The Botanical Solution for Malaria. *Science*, **327**: 207.

Mitchell, C., Ye, F., and Wiseman, N. (1999). Shang Han Lun (On Cold Damage): Translation and Commentaries. Paradigm Publications, Brookline, MA.

Mueller, M.S., Karhagomba, I.B., Hirt, H.M., and Wemakor, E. (2000). The potential of *Artemisia annua* L. as a locally produced remedy for malaria in the tropics: agricultural, chemical and clinical aspects. *Journal of Ethnopharmacology*, **73**: 487-493.

Murray, C.J.L., Rosenfeld, L.C., Lim, S.S., *et al.* (2012). Global malaria mortality between 1980 and 2010: a systematic analysis. *Lancet*, **379**: 413-431.

Nayyar, G.M.L., Breman, J.G., Newton, P.N., and Herrington, J. (2012). Poor-quality antimalarial drugs in Southeast Asia and sub-Saharan Africa. *Lancet Infectious Diseases*, **12**: 488-496.

Newman, R.D. (2011). Learning to outwit malaria. *Bulletin of World Health Organization*, **89**: 10-11.

Newton, P., and White, N. (1999). Malaria: New development in treatment and prevention. *Annual Review of Medicine*, **50**: 179-192.

O'Neill, P.M. (2004). A worthy adversary for malaria. *Nature*, **430**: 838.

Orwa, C., Mutua, A., Kindt, R., Jamnadass, R., and Anthony, S. (2009). *Artemisia annua*. Agroforestree Database: a tree reference and selection guide version 4.0. World Agroforestry Centre, Kenya. Available at: www.worldagroforestry.org/treedb/AFTPDFS/Artemisia_annua.pdf (accessed on 5th May 2012).

Patwardhan, B. (2005). Ethnopharmacology and drug discovery. *Journal of Ethnopharmacology*, **100**: 50-52.

Price, R.N., Nosten, F., Luxemburger, C., ter Kuile, F.O., Paiphun, L., Chongsuphajaisiddhi, T., and White, N.J. (1996). Effects of artemisinin derivatives on malaria transmissibility. *Lancet*, **347**: 1654-1658.

Rasoanaivo, P., Wright, C.W., Willcox, M.L., and Gilbert, B. (2011). Whole plant extracts versus single compounds for the treatment of malaria: synergy and positive interactions. *Malaria Journal*, **10(Suppl 1)**: S4.

Ridder, S.D., Kooy, F.V., and Verpoorte, R. (2008). *Artemisia annua* as a self-reliant treatment for malaria in developing countries. *Journal of Ethnopharmacology*, **120**: 302-314.

Riddle, J.M., and Estes, J.W. (1992). Oral contraceptives in ancient and medieval times. *American Scientist*, **80**: 226-233.

Ridley, R.G. (2003). To kill a parasite. *Nature*, **424**: 888-889.

Romero, M.R., Efferth, T., Serrano, M.A., Castano, B., Macias, R.I., Briz, O., and Marin, J.J. (2005). Effect of artemisinin/artesunate as inhibitors of hepatitis B virus production in an '*in vitro*' system. *Antiviral Research*, **68**: 75-83.

Sadava, D., Phillips, T., Lin, C., and Kane, S.E. (2002). Transferrin overcomes drug resistance to artemisinin in human small-cell lung carcinoma cells. *Cancer Letters*, **179**:151-156.

Shekari, M., Jalali Sendi, J., Etebari, K., Zibaee, A., and Shadparvar, A. (2008). Effects of *Artemisia annua* L. (Asteracea) on nutritional physiology and enzyme activities of elm leaf beetle, *Xanthogaleruca luteola* Mull. (Coleoptera: Chrysomellidae). *Pesticide Biochemistry and Physiology*, **91**: 66-74.

Simon, J.E., Charles, D., Cebert, E., Grant, L., Janick, J., and Whipkey, A. (1990). *Artemisia annua* L.: A promising aromatic and medicinal. In J. Janick and J E Simon (eds.). Advances in new crops. West Lafayette: Timber Press, 522-526.

Singh, A., Vishwakarma, R.A., and Husain, A. (1988). Evaluation of *Artemisia annua* satrins for higher artemisinin production. *Planta Medica*, **54**: 475-476.

Singh, N.P., and Lai, H. (2001). Selective toxicity of dihydroartemisinin and holotransferrin toward human breast cancer cells. *Life Sciences*, **70**: 49-56.

Sto¨ ger, E.A. (Ed.), (1991). Arzneibuch der chinesischen Medizin. Deutscher Apotheker-Verlag, Stuttgart, Germany.

Teixeira da Silva, J.A. (2004). Mining the essential oils of the Anthemideae. *African Journal of Biotechnology*, **3**: 706-720.

The People's Republic of China, (1985). In: Pharmacopoeia, vol. 1. The People's Health Publisher, Beijing.

Titulaer, H.A.C., Zuidema, J., Kager, P.A., Wetsteyn, J.C.F.M., Lugt, C.H.B., and Merkus, F.W.H.M. (1990). The pharmacokinetics of artemisinin after oral, intramuscular and rectal administration to volunteers. *Journal of Pharmacy and Pharmacology*, **42**: 810-813.

Toure, Y.T., Oduola, A.M., and Morel, C.M. (2004). The *Anopheles gambiae* genome: next steps for malaria vector control. *Trends in Parasitology*, **20**: 142-149.

Tu, Y. (2004). The development of new antimalarial drugs: Qinghaosu and dihydro-qinghaosu. *Chinese Medical Journal*, **112**: 976-977.

Utzinger, J., Xiao, S., N'Goran, E. K., Berquist, R., and Tanner, M. (2001). The potential of artemether for the control of schistosomiasis. *International Journal of Parasitology*, **31**:1549-1562.

Van Agtmael, M.A., Cheng-Qi, S., Qing, J.X., Mull, R., and van Boxtel, C.J. (1999). Multiple dose pharmacokinetics of artemether in Chinese patients with uncomplicated falciparum malaria. *International Journal of Antimicrobial Agents*, 12: 151-158.

Vennerstrom, J.L., Arbe-Barnes, S., Brun, R., Charman, S.A., Chiu, F.C., Chollet, J., Dong, Y., Dorn, A., Hunziker, D., Matile, H., McIntosh, K., Padmanilayam, M., Santo Tomas, J., Scheurer, C., Scorneaux, B., Tang, Y., Urwyler, H., Wittlin, S., and Charman, W.N. (2004). Identification of an antimalarial synthetic trioxolane drug development candidate. *Nature*, **430**: 900-904.

Wagner, H., and Ulrich-Merzenich, G. (2009). Synergy research: approaching a new generation of phytopharmaceuticals. *Phytomedicine*, **16**: 97-110.

Weathers, P.J., DeJesus-Gonzalez, L., Kim, Y.J., Souret, F.F., and Towler, M.J. (2004). Alteration of biomass and artemisinin production in *Artemisia annua* hairy roots by media sterilization method and sugars. *Plant Cell Reports*, **23**: 414-418.

Weathers, P.J., Elkholy, S., and Wobbe, K.K. (2008). Artemisinin: the biosynthetic pathway and its regulation in *Artemisia annua*, a terpenoid-rich species. *In Vitro Cellular and Developmental Biology - Plant*, **42(4)**: 309-317.

Wells, T.N.C. (2011). Natural products as starting points for future anti-malarial therapies: going back to our roots? *Malaria Journal*, 10(Suppl 1):S3.

White, N. (1999). Antimalarial drug resistance and combination chemotherapy. *Philosophical Transactions of the Royal Society B: Biological Sciences*, **354**: 739-749.

White, N.J. (1997). Assessment of the pharmacodynamic properties of the antimalarial drugs *in-vivo*. *Antimicrobial Agents* and *Chemotherapy*, **41**: 1413-1422.

White, N.J. (2004). Antimalarial drug resistance. *The Journal of Clinical Investigation*, **113**: 1084-1092

WHO, 2006. Guidelines for the treatment of malaria, WHO/HTM/MAL 2006/108, World, Health Organization, Geneva, Switzerland.

WHO, Assessment of the safety of artemisinin compounds in pregnancy. Available at: http://www.who.int/malaria/cmc_upload/0/000/016/323/artem_pregnancy.html, pp 1–2; 2003. (accessed on 29th May 2012).

WHO. WHO monograph on good agricultural and collection practices (GACP) for *Artemisia annua* L. World Health Organization, Geneva, Switzerland, 2006a.

WHO. World Malaria Report 2011. 2011. World, Health Organization, Geneva, Switzerland. Available at:http://www.who.int/malaria/world_malaria_report_2011/9789241564403_eng.pdf (accessed Jan 20, 2012).

WHO: Guidelines for the treatment of malaria. World, Health Organization, Geneva, Switzerland. 2010.

Wilairatana, P., and Looaresuwan, S. (2002). The clinical use of artemisinin and its derivatives in the treatment of malaria. In Artemisia, Wright, C.W., Ed. Taylor & Francis, London.

Willcox, M., and Bodeker, G. (2004). Traditional herbal medicines for malaria. *British Medical Journal*, **329**: 1156-1159.

Willcox, M., Bodeker, G., Bourdy, G., Dhingra, V., Falquet, J., Ferreira, J.F.S., Graz, B., Hirt, H., Hsu, E., de Magalhães, P.M., Provendier, D., and Wright, C.W. (2004). *Artemisia annua* as a Traditional Herbal Antimalarial. Traditional Medicinal Plants and Malaria.

Willcox, M., Burford, G., and Bodeker, G. (2004a). An overview of ethnobotanical studies on plants used for the treatment of malaria. In Traditional Medicinal Plants and Malaria. Boca Raton: CRC Press; Willcox M, Bodeker G, Rasoanaivo P. 187-197.

Woerdenbag, H.J., Pras, N., Chan, N.G., Bang, B.T., and Bos, R. (1994). Artemisinin related sesquiterpenes, and essential oil in *Artemisia annua* during a vegetative period in Vietnam. *Planta Medica*, **60**: 272–275.

Wright, C.W. and Warhurst, D.C. (2002). The mode of action of artemisinin and its derivatives. In Artemisia, Wright, C.W., Ed. Taylor & Francis, London.

Xiao, S.H., and Catto, B.A. (1989). In vitro and in vivo studies of the effect of artemether on *Schistosoma mansoni*. *Antimicrobial Agents and Chemotherapy*, **33**: 1557-1562.

Xiao, S.H., Booth, M., and Tanner, M. (2000). The prophylactic effects of artemether against *Schistosoma japonicum* infections. *Parasitology Today*, **16**: 122-126.

Yeung, S., Pongtavornpinyo, W., Hastings, I.M., Mills, A.J., and White, N.J. (2004). Antimalarial drug resistance, artemisinin-based combination therapy, and the contribution of modelling to elucidating policy choices. *American Journal of Tropical Medicine and Hygiene* **71**: 179-186.

Traditional and Folk Herbal Medicine: *Pages* **73–114**
Recent Researches Vol. 2 (2014)
Editor: **V.K. Gupta**
Published by: **DAYA PUBLISHING HOUSE, NEW DELHI**

3

Traditional and Folk Herbal Medicine in Czech Republic: A Review

Roman Pavela[1*]

ABSTRACT

In-depth knowledge of the medicinal plants is essential in understanding the manifold utilities of such plants. This paper compares approximately 22 species of fungi belonging to 13 families, 1specie of lichen, and 283 species of plants belonging to 81 families used as popular medicine. Among the plants, the most numerously represented are 10 families comprising 277 species in total and which comprise 54.8 per cent of all species of plants which have been utilized in folk medicine in the Czech Republic. Most of the species belonging to the families Asteraceae, Lamiaceae and Rosaceae with 37, 27, and 22 species, respectively.

Keywords: Czech Republic, Ethnobotany, Folk herbal medicine, Medicinal plants, Plant extracts.

Introduction

Herbal medicine is the oldest form of healthcare known to mankind. Herbs had been used by all cultures throughout history. It was an integral part of the development of modern civilization. Primitive man observed and appreciated the great diversity of plants available to him. The plants provided food, clothing, shelter, and medicine. Much of the medicinal use of plants seems to have been developed through observations of wild animals, and by trial and error. As time went on, each tribe

1 Crop Research Institute, Drnovská 507, 161 06 Praha 6–Ruzyne

* *Corresponding author*: E-mail: pavela@vurv.cz

added the medicinal power of herbs in their area to its knowledgebase. They methodically collected information on herbs and developed well-defined herbal pharmacopoeias. Indeed, well into the 20th century much of the pharmacopoeia of scientific medicine was derived from the herbal lore of native peoples (Schultes, 1962; Planta *et al.*, 2000; Prunerova, 2006).

Within the area of the Czech Republic, the recent tradition of using herbs for medicinal purposes has always followed the development of Western Europe. The first professional Czech writings reflecting herbal medicine utilisation were published in the 1500s (Drabek, 1972; Drabek and Zeidler, 2004; Pavela, 2009). The following centuries witnessed economic and technological developments that positively impacted living standards and health care of the population. However, frequent military conflicts, political instability and regional income disparities during the 19th and 20th centuries have intensified the prevalence of medicinal plants use for treatment and/or prevention of minor ailments, as well as seasonal and chronic illnesses (Novacek, 1998). The use of medicinal plants by the Czech population has recently become a highly discussed issue among pharmacists, particularly with special regard to the harmonization of Czech legislation with European Union standards (Knotek, 2012).

This review sums up information about medicinal plants which have a direct ethno botanical connection to the traditional use of plants in the Czech lands.

Material and Methods

The Czech Republic is a landlocked country, lying in the central part of Europe in the middle of the temperate zone of the northern hemisphere. The country's borders make up neighborhood with Poland, Germany, Austria and Slovakia. The major European watershed passes through the territory of the Czech Republic to separate the basins of the North, Baltic and Black Seas. The Czech Republic can be found on the borderline of two mountain systems. The western and central parts of the country are occupied by the hilly uplands The climate of the Czech Republic is affected by the interaction of oceanic and continental effects. Western winds prevail and intensive cyclone type activities frequently change air masses and bring rather heavy precipitation. The maritime effect is mainly felt in Bohemia, continental climate effects have a bigger impact on Moravia and Silesia.

Literature was included from authors who had either summarized then-renowned literature or oral sayings in a sufficient or hígh-quality manner or from authors who had summarized their own life experience in monographs (Pavela, 2009; Prunerova, 2006; Knotek *et al.*, 2012). Only plants which had their origins or were transplanted in Central Europe (Dostál, 1989) were included. Plants which do not have a historically well-founded tradition in cultivation, such as, *e.g.*, tropical or sub-tropical species of vegetables and spices, were intentionally excluded.

Results and Discussion

Table 3.1 shows 22 species of fungi divided into 13 families, 1 species of lichen, and 283 species of plants (5.7 per cent of all plants in the Czech Republic) divided into 81 families. Among the plants, the most numerously represented are 10 families

Table 3.1: List of the medicinal plants of Czech Republic.

Family	*Botanical Name*	*Part Used*	*Method Use*	*Popular Use*	*References*
Lichen					
Parmeliaceae	*Centraria islandica* (L.) Acharius	Whole lichen	Decoction, extract	Used as an antiphlogosis, also on bronchitis and alimentary tract diseases.	Bodlák *et al.,* 1995; Mackù and Krejca, 1964; Krkavec and Hájková, 1962; Janca and Zentrich, 1995
Fungi					
Hypocreales					
Clavicipitaceae	*Claviceps purpurea* (Fr.) Tul.	Sclerocium	Ectract	Drug acts on sympathicusan and as uterotonic.	Mackù and Krejca, 1964
Hymenochaetales					
Hymenoch-aetaceae	*Inonotus obliquus* (Pers. Et Fr.) Pill.	Sporocarp	Infusion, tea	Drug acts on stomach cancer, also as antiphlogosis and tonic.	Janca and Zentrich, 1994; Kolberár and Endris, 1958
Agaricales					
Agaricaceae	*Agaricus bisporus* (J.E. Lange) Pilát	Sporocarp	Powder	Powder taken for pulmonary diseases, for nettle-rash and eczematous dermatitis, also as hypotensive and for sugar levels degradation.	Bodlák *et al.,* 1995; Janca and Zentrich, 1997; Kolberár and Endris, 1958
Amanitaceae	*Amanita muscaria* (L.: Fr.) Hooker	Sporocarp	Ectract	Used as a cardiotonic and nervous-tonic.	Kolberár and Endris, 1958
Tricholomat-aceae	*Armillaria mellea* (Vahl) P. Kumm.	Sporocarp	Decoction, powder	Powder taken for pulmonary and hepatic diseases, also as a cerebrospinant and digestive.	Kolberár and Endris, 1958
Tricholomata-ceae	*Collybia velutipes* (Curt.: Fr.) Kumm.	Sporocarp	Decoction, powder	Drug acts on immunity system as tonic, also as an antiphlogosis and antidepressant.	Bodlák *et al.,* 1995; Kolberár and Endris, 1958

Contd...

Table 3.1–*Contd...*

Family	*Botanical Name*	*Part Used*	*Method Use*	*Popular Use*	*References*
Coprinaceae	*Coprinus comatus* Fr.	Sporocarp	Ectract	Used as an antidiabetic for sugar levels degradation, also as a hypotensive and for hemorrhoids.	Kolberár and Endris, 1958
Tricholomata-ceae	*Lepista nuda* (Bull.) Cooke	Sporocarp	Infusion, tea	Used as an antidiabetic for sugar levels degradation in blood	Kolberár and Endris, 1958
Pleurotaceae	*Pleurotus dryinus* (Pers.: Fr.) Kummer	Sporocarp	Infusion, powder	Drug acts on lowering of cholesterol-aemia, for consolidation of immunity system and as spasmolysan.	Kolberár and Endris, 1958
Pleurotaceae	*Pleurotus pulmonarius* (Fr.) Quél.	Sporocarp	Infusion, powder	Drug acts on lowering of cholesterol-aemia, for consolidation of immunity system and as spasmolysan.	Kolberár and Endris, 1958
Tricholomata-ceae	*Tricholoma pessundatum* (Fr.) Quél.	Sporocarp	Infusion, tea	Used as an antidiabetic for sugar levels degradation in blood	Bodlák *et al.*, 1995; Kolberár and Endris, 1958
Tricholomata-ceae	*Tricholoma populinum* Lange	Sporocarp	Infusion, tea	Drug acts on immunity system as tonic.	Kolberár and Endris, 1958
Polyporales					
Polyporaceae	*Dendropolyporus umbellatus* (Pers.: Fr.) Jul.	Sporocarp	Tea, powder	Used as a diuretic, anti-laxative, digestive and tonic.	Kolberár and Endris, 1958
Polyporaceae	*Fomes fomentarius* (L.: Fr.) Kick.	Sporocarp	Decoction, powder	Drug acts on rubella and varicella, also as an analgetic and hemostatic.	Bodlák *et al.*, 1995; Kolberár and Endris, 1958
Ganodermata-ceae	*Ganoderma lucidum* (Fr.) Karst.	Sporocarp	Tea, tincture, powder	Drug acts on lowering of cholesterol-aemia, for consolidation of immunity system, also as an antiphlogosis, cardiotonic and for bronchitis.	Kolberár and Endris, 1958
Meripilaceae	*Grifola frondosa* (Dicks.) Gray	Sporocarp	Powder	Used as a cytostatic and tonic for immunity system, also on lowering of cholesterolaemia and antidiabetic.	Kolberár and Endris, 1958

Contd...

Table 3.1–*Contd...*

Family	*Botanical Name*	*Part Used*	*Method Use*	*Popular Use*	*References*
Fomitopsidaceae	*Laricifomes officinalis* (Vill.) Kotl. and Pouzar	Sporocarp	Powder	Used as a laxative, digestive, for bronchitis, also as antiphlogosis.	Bodlák *et al.*, 1995; Kolberár and Endris, 1958
Polyporaceae	*Polyporus squamosus* (Huds.) Fr.	Sporocarp	Tea, powder	Used as a diuretic, anti-laxative, digestive and tonic.	Kolberár and Endris, 1958
Polyporaceae	*Trametes versicolor* (L.: Fr.) Pilat	Sporocarp	Tea, powder	Drug acts on immunity system as tonic, also as an antiphlogosis and treatment for choleplania.	Kolberár and Endris, 1958
Tremellaeles					
Auriculariaceae	*Hirneola auricula-judae* (L.) Berk.	Sporocarp	Macerate, infusion, decoction	Used as a hypotensive, analgetic, hemostatic and tonic.	Bodlák *et al.*, 1995; Kolberár and Endris, 1958
Aphyllophorales					
Hericiaceae	*Hericium erinaceum* (Bull.) Pers	Sporocarp	Extract	Used as a cytostatic and tonic for immunity system.	Kolberár and Endris, 1958
Gasteromycetidae					
Lycoperdaceae	*Langermannia gigantea* (Batsch) Rostk.	Sporocarp, spore	Extract, powder	Used as an adstringent, hemostatic, antiphlogosis, antiaphrodisiac and on dermatitis.	Bodlák *et al.*, 1995; Kolberár and Endris, 1958
Pteridophita					
Equisetaceae	*Equisetum arvense* L.	Whole plants	Infusion, detoction, tincture	Drug used as diuretic for gout, lithemia, antirheumatic, urinary bladder and kidney diseases; externally applied as antiphlogosis, on a dermatitis.	Bodlák *et al.*, 1995; Korbelár, 1981; Mackù and Krejca, 1964; Krkavec and Hájková, 1962; Janca and Zentrich, 1998
Dryopteridaceae	*Dryopteris filix-mas* (L.) Schott.	Roots	Extract, decoction	Extract used as anthelmintic; decoction is externally applied on dermatitis.	Bodlák *et al.*, 1995; Mackù and Krejca, 1964; Krkavec and Hájková, 1962

Contd...

Table 3.1–*Contd...*

Family	*Botanical Name*	*Part Used*	*Method Use*	*Popular Use*	*References*
Lycopodiaceae	*Lycopodium clavatum* L.	Spores	Powder	Internally used as analgetic; externally applied on dermatitis.	Korbelár, 1981; Krkavec and Hájková, 1962; Janca and Zentrich, 1996;
Polypodiaceae	*Polypodium vulgare* L.	Rhizome	Decoction, infusion, tea, powder	Used as expectorant, enteral disinfectant and anthelmintic.	Korbelár, 1981
Spermatophyta					
Gymnospermae					
Pinaceae	*Larix decidua* Mill.	Balsam, bark	Inhalation of decoction, infusion, decoction	Used as an antiphlogosis and adstringent.	Janca and Zentrich, 1995
Pinaceae	*Pinus silvestris* L.	Eyes	Decoction, tincture	Decoction or tincture used for bronchitis and pulmonitis; externally applied on vigilance, neuroses and as antirheumatic.	Bodlák *et al.*, 1995; Korbelár, 1981; Mackù and Krejca, 1964; Krkavec and Hájková, 1962; Janca and Zentrich, 1996
Cupressadae	*Juniperus comunis* L.	Tree; fruits	Tea; ethanol extract; tincture; liqueur	Tea and liqueur used as diureticum and for arthritis, extract and tincture externally applied as desinfection.	Bodlák and Zentrich, 1995; Bodlák *et al.*, 1995; Korbelár, 1981; Mackù and Krejca, 1964
Angiospermae					
Polygonaceae	*Acetosa pratensis* Mill.	Leaves, rhizome	Tea, decoction, infusion	Externally applied in poultices on dermatitis, also as gargles for an aphthae.	Janca and Zentrich, 1996
Ranunculaceae	*Aconitum napellus* L.	Roots	Decoction, extract	Used as cardiotonic	Korbelár, 1981

Contd...

Table 3.1–*Contd...*

Family	*Botanical Name*	*Part Used*	*Method Use*	*Popular Use*	*References*
Araceae	*Acorus calamus* L.	Rhizome	Decoction,	Decoction or extract used as stomachicum, carminative, spasmolysan; antibactericidal, ethanol extract externally used for wounds and as an anti-inflammatory for well perfused legs.	Bodlák *et al.*, 1995; Krkavec and Hájková, 1962; Janca and Zentrich, 1996
Ranunculaceae	*Adonis aestivalis* L.	Herb	Decoction, extract	Used as cardiotonic.	Mackù and Krejca, 1964
Ranunculaceae	*Adonis vernalis* L.	Herb	Decoction, extract	Used as cardiotonic.	Korbelár, 1981
Apiaceae	*Aegopodium podagraria* L.	Rhizome	Decoction, infusion	Used as diuretic, antirheumatic and spasmolysan.	Janca and Zentrich, 1995
Hippocastana-ceae	*Aesculus hippocasta-neum* L.	Flowers, seeds, bark	Decoction, extract, tincture	Drug used as anti-laxative, on well perfuse of coronary vessels, enteritis, headache and fever.	Bodlák *et al.*, 1995; Korbelár, 1981; Krkavec and Hájková, 1962
Rosaceae	*Agrimonia eupatoria* L.	Whole plants	Decoction, extract, tincture, liniment	Decoction, extract and tincture used on inflammatic, curing a liver-complaint, billiary colic and disfuncion of metabolism; decoction and extract externally applied gargled for sore throat and toothache, liniment externally applied in poultices on bruise.	Bodlák *et al.*, 1995; Korbelár, 1981; Krkavec and Hájková, 1962
Asteraceae	*Achillea millefolium* L.	Flower heads	Tea, decoction	Tea and decoction used on rheumatoid arthritis, as a tonic on blood circulation and rhexis, stomachic, hemostatic, hypotention, cardiotonic; infusion and ethanol extract externally used for sores, tumors, wounds and as an anti-inflammatory for swollen.	Bodlák *et al.*, 1995; Korbelár, 1981; Krkavec and Hájková, 1962

Contd...

Table 3.1–*Contd...*

Family	*Botanical Name*	*Part Used*	*Method Use*	*Popular Use*	*References*
Lamiaceae	*Ajuga reptans* L.	Herb	Tincture	Used as a tonic, hypotensive and . antirheumatic	Janca and Zentrich, 1998
Malvaceae	*Alcea rosea* L.	Flower	Decoction	Decoction used as expectonant and . anti-inflammatic	Bodlák *et al.*, 1995; Korbelár, 1981
Rosaceae	*Alchemilla vugaris* L.	Herb	Decoction	Decoction used on gynopathy, gastric ulcers, diabetes; as expectonant and anti-inflammatic; externally applied in poultices on bruises, wounds and venous ulcerations.	Bodlák *et al.*, 1995; Korbelár, 1981; Mackù and Krejca, 1964; Krkavec and Hájková, 1962
Brassicaceae	*Alliaria officinalis* Andrs.	Whole plants	Tea, infusion, decoction	Used as antiphlogosis, tonic on digestive, also as gargle on gingivostomatitis.	Janca and Zentrich, 1995
Liliaceae	*Allium cepa* L.	Bulbs	Decoction, syrup	Decoction and syrup used as antibiotic, expextorant; for bronchytis, to trea influenza and the common cold.	Bodlák *et al.*, 1995; Korbelár, 1981
Liliaceae	*Allium sativum* L.	Bulbs	Decoction, extract, tincture	Drug used as cardiotonic, stoma-chicum, spasmolysan, antiseptic and tonic; on flatulence well perfuse of coronary vessels, enteritis, headache and fever, on sclerosis; externally applied on dermatosis.	Bodlák *et al.*, 1995; Korbelár, 1981
Liliaceae	*Allium ursinum* L.	Leaves	Tincture, wine	Drug used as cardiotonic, stoma-chicum, spasmolysan, antiseptic and tonic; on flatulence well perfuse of coronary vessels, enteritis, headache and fever, on sclerosis; externally applied on dermatosis.	Bodlák *et al.*, 1995; Korbelár, 1981; Mackù and Krejca, 1964; Janca and Zentrich, 1994; Janca and Zentrich, 1998

Contd...

Table 3.1—*Contd...*

Family	*Botanical Name*	*Part Used*	*Method Use*	*Popular Use*	*References*
Betulaceae	*Alnus glutinosa* (L.) Gaertner	Leaves, bark	Infusion, decoction, tincture	Used as an anti-laxative, anti-phlogosis and adstringent.	Janca and Zentrich, 1995
Stellariaceae	*Alsinula media* (L.) Dostál	Herb	Juice, macerate	Juice taken for bronchitis and phthisis, also as diuretic and adstringent; macerate applied externally as an antirheumatic.	Janca and Zentrich, 1996;
Malvaceae	*Althaea officinalis* L.	Whole plants	Decoction, infusion, tea	Used as an expectorant and an antiphlogosis.	Korbelár, 1981; Janca and Zentrich, 1996;
Primulaceae	*Anagallis arvensis* L.	Whole plants	Decoction, infusion, tea	Drug acts as stimulant nervous.	Mackù and Krejca, 1964
Daucaceae	*Anethum graveolens* L.	Fruits	Decoction, crushed fruits	Decoction or crushed fruits used as cardiotonic (prevention on cardiac infarction), spasmolysan, stomachicum, hypotenzive and expectorants.	Bodlák *et al.*, 1995; Korbelár, 1981; Mackù and Krejca, 1964
Boraginaceae	*Anchusa officinalis* L.	Whole plants	Decoction, infusion, tincture, tea	Used as expectorant and an emollient.	Mackù and Krejca, 1964
Fabaceae	*Anthyllis vulneraria* L.	Flowerer	Decoction	Decoction used as tonic, adstringent and stomachicum; externally applied as a dentifrice, on dermatitis and on a woulds as desinfection.	Bodlák *et al.*, 1995; Korbelár, 1981; Mackù and Krejca, 1964
Rosaceae	*Aphanes arvensis* L.	Herb	Decoction	Used as a diuretic, also taken on renal calculi; externally applied for dermatic itch.	Janca and Zentrich, 1998
Asteraceae	*Arctium lappa* L.	Roots	Decoction, extract	Decoction or extract used as anti-diabetic, diuretic and tonic; externally applied as hair-tonic on hair loss and production of dandruff or dermatitis.	Korbelár, 1981; Krkavec and Hájková, 1962; Janca and Zentrich, 1995

Contd...

Table 3.1–*Contd...*

Family	*Botanical Name*	*Part Used*	*Method Use*	*Popular Use*	*References*
Ericaceae	*Arctostaphyllos uva-ursi* (L.) Spreng.	Herb	Decoction, extract	Decoction or extract used as diuretic (for treatment of renal calculi), anti-laxatixe and as antirheumatic and vulnerary.	Bodlák *et al.*, 1995; Korbelár, 1981; Mackù and Krejca, 1964; Janca and Zentrich, 1995
Daucaceae	*Archangelica officinalis* (Moench) Hoffm.	Fruits, roots	Decoction, extract, crushed drug	Decoction, crushed fruits or roots and water extracts used as stomachic, diuretic, carminative and palliation of menstrual ache; externally applied as bathing for effects anti-rheumatiz and sedative.	Bodlák *et al.*, 1995; Korbelár, 1981; Mackù and Krejca, 1964
Aristolochiaceae	*Aristolochia clematitis* L.	Herba	Decoction	Externally applied on dermatitis.	Korbelár, 1981; Janca and Zentrich, 1996
Brassicaceae	*Armoracia rusticana* Gaertner.	Roots	Fresh juice	Fresh juice used as tonic of glands of internal secretion, diuretic, stomachicum, expectorant and analgetic; externally applied as antirheumatic and vulnerary.	Bodlák *et al.*, 1995; Korbelár, 1981; Mackù and Krejca, 1964
Asteraceae	*Arnica montana* L.	Flowers, roots	Decoction, extract, tincture	Decoction, extract or tincture used as antiseptic; externally applied as antirheumatic and vulnerary.	Bodlák *et al.*, 1995; Korbelár, 1981
Asteraceae	*Artemisia abrotanum* L	Herb	Tea, decoction, tincture	Used as a anthelmintic, digestive, stomachic and as drug acts on cholechromeresis.	Mackù and Krejca, 1964; Krkavec and Hájková, 1962
Asteraceae	*Artemisia absinthium* L.	Whole plants	Tea, decoction, tincture	Tea, decoction and tincture used as a vermifuge, stomachic and for cholechromeresis.	Bodlák *et al.*, 1995; Korbelár, 1981; Mackù and Krejca, 1964; Krkavec and Hájková, 1962; Janca and Zentrich, 1995

Contd...

Table 3.1–*Contd...*

Family	*Botanical Name*	*Part Used*	*Method Use*	*Popular Use*	*References*
Asteraceae	*Artemisia dracunculus* L.	Herb	Tea, decoction	Used as a anthelmintic, digestive, stomachic and as drug acts on cholechromeresis.	Mackù and Krejca, 1964
Asteraceae	*Artemisia vulgaris* L.	Whole plants	Tea, decoction	Tea or decoction used as a vermifuge, cholechromeresis and stomachic; externally applied as antirheumatic.	Bodlák *et al.*, 1995; Korbelár, 1981; Mackù and Krejca, 1964
Aristolochiaceae	*Asarum europeaum* L.	Roots	Tincture, decoction, syrup	Used as diuretic, analgerisc and expectorant.	Mackù and Krejca, 1964; Krkavec and Hájková, 1962
Fabaceae	*Astragalus glycyphylos* L.	Herb, roots	Macerate, infusion	Used as a diuretic, sudatorium, hypotensive and sedative; externally applied as a hair-restorer.	Janca and Zentrich, 1998
Apiaceae	*Astrantia major* L.	Whole plants	Infusion, decoction, tincture	Used as digestive and diuretic; externally applied as antiallergic and on dermatitis.	Janca, 1995
Solanaceae	*Atropa belladonna* L.	Roots, leaves	Decoction, infusion, tea, extract	Used as spasmolysan, antiasthmatic and as drug act on nervous systém.	Korbelár, 1981
Lamiaceae	*Ballota nigra* L.	Herb	Decoction, infusion, tea	Drug acts as stimulant nervous.	Mackù and Krejca, 1964
Asteraceae	*Bellis perennis* L.	Flower	Tea, decoction, syrup, tincture	Used as diuretic, antirheumatic and antipyretic.	Bodlák *et al.*, 1995; Korbelár, 1981; Mackù and Krejca, 1964; Krkavec and Hájková, 1962
Berberidaceae	*Berberis vulgaris* L.	Bark from	Tea, infusion roots; fruits	Tea used as diuretic, urinary bladder diseases; infusion for kidney and gall diseases.	Bodlák *et al.*, 1995; Korbelár, 1981
Lamiaceae	*Betonica officinalis* L.	Whole plants	Decoction	Decoction used as tonic and sedation.	Bodlák *et al.*, 1995; Korbelár, 1981; Mackù and Krejca, 1964; Krkavec and Hájková, 1962

Contd...

Table 3.1–*Contd...*

Family	*Botanical Name*	*Part Used*	*Method Use*	*Popular Use*	*References*
Betulaceae	*Betula pendula* Roth.	Leaves	Tea, decoction, syrup	Used as diuretic, antirheumatic and diaphoretic.	Bodlák *et al.*, 1995; Korbelár, 1981; Mackù and Krejca, 1964; Krkavec and Hájková, 1962
Asteraceae	*Bidens tripartita* L.	Herb	Tea, decoction	Used as diuretic, diaphoretic, antirheumatic and externally for treatment on tetter.	Bodlák and Zentrich, 1995
Boraginaceae	*Borago officinalis* L	Herb	Tea, decoction	Tea used as diuretic, diaphoretic, antirheumatic, antipyretic, for treatment on gastrohelcoma and bronchitis; externally applied in poulticed on venous ulcerations, bruises, wounds and sores.	Bodlák *et al.*, 1995; Korbelár, 1981; Mackù and Krejca, 1964
Cucurbitaceae	*Bryonia alba* L.	Roots	Decoction	Seldom used as laxative and diuretic.	Korbelár, 1981
Cucurbitaceae	*Bryonia dioica* Jacq.	Roots	Decoction	Seldom used as laxative and diuretic.	Korbelár, 1981
Asteraceae	*Calendula officinalis* L.	Flower	Tea, decoction, liniment	Tea and decoction used as spasmolysan, antiphlogistic and for cholechromeresis; externally aplliead as linimed on milk-leg, venous ulceration and fungal infections.	Bodlák *et al.*, 1995; Korbelár, 1981; Janca and Zentrich, 1995
Ericaceae	*Calluna vulgaris* L.	Whole plants	Tea, decoction	Used as diuretic, antiphlogestic and tonic.	Bodlák *et al.*, 1995; Korbelár, 1981; Mackù and Krejca, 1964; Krkavec and Hájková, 1962
Brassicaceae	*Capsella bursa-pastoris* L.	Herb	Tea, decoction, tincture	Tea, decoction and tincture used as a hemostatic and doctoring haematopiesis.	Bodlák *et al.*, 1995; Korbelár, 1981; Mackù and Krejca, 1964; Krkavec and Hájková, 1962
Asteraceae	*Carduus acanthoides* L.	Flower heads	Decoction, extract	Decoction or tea used for bronchitis and the common cold, also as diureticum and antilaxation	Bodlák and Zentrich, 1995

Contd...

Table 3.1–*Contd...*

Family	*Botanical Name*	*Part Used*	*Method Use*	*Popular Use*	*References*
Asteraceae	*Carlina acaulis* L.	Roots	Tea	Tea used as a diuretic, spasmolysan; externally applied as an anti-eczematous	Bodlák and Zentrich, 1995; Mackù and Krejca, 1964; Krkavec and Hájková, 1962
Asteraceae	*Carthamus tinctorius* L.	Flowers, seeds	Tea, oil from seeds	Tea or oil used as a laxative, for to treat choleplania and as an antipyretic.	Bodlák *et al.*, 1995; Korbelár, 1981
Daucaceae	*Carum carvi* L.	Fruits	Decoction, extract	Used as a carminative, spasmolysan, stomachic and anthelmintic.	Bodlák *et al.*, 1995; Korbelár, 1981
Fabaceae	*Castanea sativa* Mill.	Leaves, seeds	Extract, tincture	Used as an expectorant and an antiphlogosis.	Janca and Zentrich, 1998
Asteraceae	*Centaurea cyanus* L.	Flower	Infusion	Used as an antiphlogosis, digestive and diuretic; externally applied on conjunctivitis (as eye-lavage).	Bodlák *et al.*, 1995; Korbelár, 1981; Mackù and Krejca, 1964; Krkavec and Hájková, 1962
Gentianaceae	*Centaurium erythraea* Rafn.	Leaves	Decoction, infusion	Used as a digestive and stomachic.	Bodlák *et al.*, 1995; Korbelár, 1981; Mackù and Krejca, 1964; Krkavec and Hájková, 1962; Janca and Zentrich, 1997
Asteraceae	*Chamaemelum nobile* (L.) All.	Flower	Tea, decoction	Internally applied as a carminative, spasmolysan, sudorific and antiphlogosis; externally applied as an antiphlogosis, adstringent and antiseptic.	Bodlák *et al.*, 1995; Korbelár, 1981; Mackù and Krejca, 1964
Oenotheraceae	*Chamaenerion angustifolium* (L.) Scop.	Herb	Decoction	Used as tranquillizer, antidepressant and tonic.	Bodlák *et al.*, 1995; Korbelár, 1981; Mackù and Krejca, 1964
Brassicaceae	*Chamaeplium officinale* (L.) Wallr.	Herb	Decoction, infusion	Used as a diuretic and expectorant.	Janca and Zentrich, 1998
Asteraceae	*Chamomilla recutita* (L.) Rauschert.	Flower	Tea, decoction	Internally applied as a carminative, spasmolysan, sudorific and antiphlogosis; externally applied as an antiphlogosis, adstringent and antiseptic.	Bodlák *et al.*, 1995; Korbelár, 1981; Mackù and Krejca, 1964; Krkavec and Hájková, 1962

Contd...

Table 3.1–*Contd...*

Family	*Botanical Name*	*Part Used*	*Method Use*	*Popular Use*	*References*
Papaveraceae	*Chelidonium majus* L.	Herb	Decoction, extract	Internally applied on a catarrh, bronchitide, hepatic and gall diseases, also as spasmolysan and antiasthmatic; externally applied on tubercle and dermatitis. The latex in the stems is applied to warts in order to destroy them.	Bodlák *et al.*, 1995; Korbelár, 1981; Mackù and Krejca, 1964; Krkavec and Hájková, 1962
Chenopodiaceae	*Chenopodium bonus-henricus* L.	Whole plants	Tea, infusion, decoction	Used as a green food for tonic effect, also as temperate laxative.	Janca and Zentrich, 1995
Chenopodiaceae	*Chenopodium ambro sioides* L.	Whole plants	Infusion	Used as an anthelmintic.	Janca and Zentrich, 1995
Asteraceae	*Cichorium intybus* L.	Whole plants	Decoction, tea	Used as stomachic, laxative, also on cholechromeresis.	Bodlák *et al.*, 1995; Korbelár, 1981; Mackù and Krejca, 1964; Krkavec and Hájková, 1962
Asteraceae	*Cirsium arvense* (L.) Scop.	Flower, leaves	Tincture, infusion, decoction	Used as a cytostatic agent and spasmolysan; externally applied as antirheumatic and on dermatitis.	Janca and Zentrich, 1995
Asteraceae	*Cirsium eriophorum* Scop.	Whole plants	Juice of plants, decoction	Juice or decoction of plants externally applied on dermatosis, ulcer disease; as an anti-inflammatory and as a carcinostatic.	Bodlák *et al.*, 1995; Korbelár, 1981
Ranunculaceae	*Clematis recta* L.	Herb	Powder	Externally applied on dermatitis.	Mackù and Krejca, 1964
Asteraceae	*Cnicus benedictus* L.	Flower heads	Tea	Tea or infusion used as stomachic, citostatic, for relieving gall-secretion and antibiotic	Bodlák and Zentrich, 1995; Bodlák *et al.*, 1995; Korbelár, 1981; Mackù and Krejca, 1964
Liliaceae	*Colchicum autumnale* L.	Seeds, tubers	Extract	Drug used on rheumatic and podagra diseases.	Korbelár, 1981; Mackù and Krejca, 1964
Rosaceae	*Comarum palustre* L.	Whole plants	Decoction,	Used as an antilaxative.	Mackù and Krejca, 1964

Contd...

Table 3.1–*Contd...*

Family	*Botanical Name*	*Part Used*	*Method Use*	*Popular Use*	*References*
Apiaceae	*Conium maculatum* L.	Herb, fruits	Decoction	Externally applied as analgetic.	Korbelár, 1981
Ranunculaceae	*Consolida regalis* S.F. Gray	Whole plants	Decoction, infusion, tea	Used as anthelmintic, diuretic and laxative.	Korbelár, 1981; Mackù and Krejca, 1964
Liliaceae	*Convallaria majalis* L.	Whole plants	Extract	Used as cardiotonic.	Korbelár, 1981; Mackù and Krejca, 1964
Convulvulaceae	*Convolvulus arvensis* L.	Whole plants	Decoction, infusion, tea, extract	Used as laxative and as drug acts on cholechromeresis.	Korbelár, 1981; Mackù and Krejca, 1964
Asteraceae	*Conyza canadensis* (L.) Cronq.	Herb	Decoction	Decoction used for bronchitis, as a diuretic, antirheumatic, anti-laxative and adstringent.	Janca and Zentrich, 1997
Daucaceae	*Coriandrum sativum* L.	Fruits	Decoction	Used as spasmolysan, stomachic, expectorant and carminative.	Bodlák *et al.*, 1995; Korbelár, 1981; Mackù and Krejca, 1964
Corylaceae	*Corylus avellana* L.	Leaves	Decoction, infusion	Used as spasmolysan and diuretic.	Bodlák *et al.*, 1995; Korbelár, 1981; Mackù and Krejca, 1964; Janca and Zentrich, 1995
Asteraceae	*Cota tinctoria* (L.) J.Gay	Herb	Tincture, infusion	Used as an anthelmintic; externally applied as an antirheumatic.	Janca and Zentrich, 1998
Rosaceae	*Crataegus leavigata* (Poir.) D.C.	Flowers, fruits	Tea, tincture	Tea or tincture used for control pressure; tea and fruits was recommended for preventing heart diseases, infusion taken over a long term is used as a blood depurative and tincture used after infarction.	Bodlák and Zentrich, 1995; Bodlák *et al.*, 1995; Korbelár, 1981; Krkavec and Hájková, 1962
Rosaceae	*Crataegus monogyna* Jacq.	Flowers, fruits	Tea, tincture	Used as well as *C. leavigata*	Bodlák and Zentrich, 1995

Contd...

Table 3.1–*Contd...*

Family	*Botanical Name*	*Part Used*	*Method Use*	*Popular Use*	*References*
Rosaceae	*Crataegus oxyocantha* L.	Flowers, fruits	Tea, tincture	Used as well as *C. leavigata*	Bodlák and Zentrich, 1995; Bodlák *et al.*, 1995; Korbelár, 1981; Mackù and Krejca, 1964; Krkavec and Hájková, 1962
Iridaceae	*Crocus sativus* L.	Stamens	Tea, tincture	Used as a sedative and hypnotic.	Janca and Zentrich, 1996;
Asclepiadaceae	*Cynanchum vince-toxicum* Pers.	Roots	Decoction	Used as diuretic and laxative.	Mackù and Krejca, 1964
Boraginaceae	*Cynoglossum officinale* L.	Herb	Decoction, infusion	Externally applied in poultices on dermatitis, also as antirheumatic.	Janca and Zentrich, 1997
Lamiaceae	*Dalanum segetum* (Necker) Dost.	Herb	Decoction, infusion, tea	Used as expectorant.	Mackù and Krejca, 1964; Janca, 1995
Thymelaeaceae	*Daphne mezereum* L.	Bark	Extract, tincture, liniment	Externally applied as antirheumatic and on dermatitis.	Korbelár, 1981
Apiaceae	*Daucus carota* L.	Root	juice	As carmanetive, source of carote-noids; externallly applied on hemorrhoids	Bodlák and Zentrich, 1995; Korbelár, 1981; Janca and Zentrich, 1995
Brassicaceae	*Descurainia sophia* (L.) Webb. et Berth.	Herb	Tea, infusion	Used as a diuretic, externally applied on wounds.	Janca and Zentrich, 1998
Rutaceae	*Dictamnus albus* L.	Whole plants	Decoction, infusion, tincture	Internally applied as an adstringent, hemostatic, diuretic, digestive and antiphlogosis; externally applied on dermatosis.	Janca and Zentrich, 1997
Scrophularia-ceae	*Digitalis grandiflora* Mill.	Leaves	Ectract	Used as cardiotonic.	Mackù and Krejca, 1964
Scrophularia-ceae	*Digitalis lanata* Ehrh.	Leaves	Ectract	Used as cardiotonic.	Mackù and Krejca, 1964
Scrophularia-ceae	*Digitalis purpurea* L.	Leaves	Ectract	Used as cardiotonic.	Korbelár, 1981

Contd...

Table 3.1–*Contd...*

Family	*Botanical Name*	*Part Used*	*Method Use*	*Popular Use*	*References*
Dipsacaceae	*Dipsacus silvestris* Huds.	Whole plants	Tea	Tea used as a diuretic, spasmolysan; externally applied as an anti-eczematous	Bodlák *et al.*, 1995; Korbelár, 1981; Mackù and Krejca, 1964; Janca and Zentrich, 1996
Droseraceae	*Drosera rotundifolia* L.	Leaves	Infusion, tincture	Used as an antisclerotic, antibacterial and spasmolysan, also on hardening of the arteries and bronchitis.	Bodlák *et al.*, 1995; Korbelár, 1981; Mackù and Krejca, 1964; Krkavec and Hájková, 1962; Janca and Zentrich, 1998
Asteraceae	*Echinops sphaerocephalus* L.	Seeds	Crushed seeds, ethanol extract	Crushed seeds and extract used as parasympatholytic, for preventing heart diseases, for relieving high blood pressure	Bodlák *et al.*, 1995; Korbelár, 1981; Mackù and Krejca, 1964
Boraginaceae	*Echium vugare* L.	Herb	Infusion, decoction, tincture	Used as analgetic and antiphlogosis, externally applied as antirheumatic and on dermatitis.	Janca, 1995
Oenctheraceae	*Epilobium parviflorum* Schreb.	Leaves	Infusion	Used on pachyprostate.	Bodlák *et al.*, 1995; Korbelár, 1981; Mackù and Krejca, 1964; Janca and Zentrich, 1997
Apiaceae	*Eringium campestre* L.	Whole plants	Tea	Tea used as expectorant and for bronchitis cure, also as diuretic and antiphlogosis.	Janca and Zentrich, 1995
Apiaceae	*Eringium plantanum* L.	Whole plants	Tea	Tea used as expectorant and for bronchitis cure.	Bodlák and Zentrich, 1995; Bodlák *et al.*, 1995; Korbelár, 1981
Asteraceae	*Eupatorium cannabium* L.	Whole plants	Infusion, decoction, tincture	Drug taken as a hemostatic and tonic on blood pressure.	Janca and Zentrich, 1996
Scrophulariaceae	*Euphrasia rostkoviana* Hayne.	Herb	Decoction, wine	Internally applied on a catarrh, bronchitide and as digestive; externally applied on conjunctivitis (as eye-lavage).	Bodlák *et al.*, 1995; Korbelár, 1981; Mackù and Krejca, 1964; Krkavec and Hájková, 1962

Contd...

Table 3.1–*Contd...*

Family	*Botanical Name*	*Part Used*	*Method Use*	*Popular Use*	*References*
Ranunculaceae	*Ficaria verna* Huds.	Herb	Extract	Externally applied on dermatitis.	Mackù and Krejca, 1964
Asteraceae	*Filaginella uliginosa* Opiz.	Herb	Tea, decoction, oil	Used as a spasmolysan, antiphlogosis and hypotensive; externally applied on wounds and pardentosis in form oil.	Janca and Zentrich, 1996
Rosaceae	*Filipendula ulmaria* (L.) Maxim.	Leaves, flower	Decoction, infusion	Decoction or infusion used as an antirheumatic, antipyretic, diuretic, hypotensive and anti-laxative.	Bodlák *et al.*, 1995; Korbelár, 1981; Mackù and Krejca, 1964; Janca and Zentrich, 1997
Daucaceae	*Foeniculum vulgare* Mill.	Fruits	Tea, infusion, decoction	Used as a carminative, expectorant, lactagogue and spasmolysan.	Bodlák *et al.*, 1995; Korbelár, 1981 Mackù and Krejca, 1964
Rosaceae	*Fragaria vesca* L.	Leaves	Tea, infusion, decoction	Drug used on gallstone and liver diseases and as an antirheumatic, antiphlogosis; infusion taken over a long term is used as a blood depurative; externally applied on hemorrhoids and dentifrice.	Bodlák *et al.*, 1995; Korbelár, 1981; Mackù and Krejca, 1964; Krkavec and Hájková, 1962
Rhamnaceae	*Frangula alnus* Mill.	Bark	Infusion	Used as a laxative and anthemintic.	Bodlák *et al.*, 1995; Korbelár, 1981; Mackù and Krejca, 1964; Krkavec and Hájková, 1962
Oleaceae	*Fraxinus exelsior* L.	Leaves	Decoction, infusion	Used as diuretic, antirheumatic and antipyretic.	Krkavec and Hájková, 1962
Fumariaceae	*Fumaria officinalis* L.	Herb	Infusion	Infusion used on gallstone and liver diseases and for ureteritis.	Bodlák *et al.*, 1995; Korbelár, 1981; Mackù and Krejca, 1964; Krkavec and Hájková, 1962
Amaryllidaceae	*Galanthus nivalis* L.	Whole plants	Extract	Used as spasmolysan.	Korbelár, 1981
Fabaceae	*Galega officinalis* L.	Herb	Decoction, infusion	Used as a diuretic, antidiabetic and lactagogue.	Bodlák *et al.*, 1995; Korbelár, 1981; Krkavec and Hájková, 1962

Contd...

Table 3.1–*Contd...*

Family	*Botanical Name*	*Part Used*	*Method Use*	*Popular Use*	*References*
Asteraceae	*Galinsoga parviflora* Cav.	Herb	Tea, decoction	Used as an antiphlogosis, disinfectant, antipyretic and hypotensive; externally applied on dermatitis.	Janca and Zentrich, 1995
Rubiaceae	*Galium odoratum* (L.) Scop.	Herb	Decoction, infusion	Drug acts on trombosis, irritation and nervines, as a sedative, also as a diuretic.	Bodlák *et al.*, 1995; Korbelár, 1981, Mackù and Krejca, 1964; Krkavec and Hájková, 1962; Janca and Zentrich, 1995
Rubiaceae	*Galium verum* L.	Herb	Infusion	Drug acts as diuretic and spasmolysan.	Bodlák *et al.*, 1995; Korbelár, 1981; Mackù and Krejca, 1964
Fabaceae	*Genista germanica* L.	Flower heads	Tea, tincture	Tea used as diuretic, urinary bladder diseases; infusion for kidney, gall diseases and arthritis.	Bodlák and Zentrich, 1995; Korbelár, 1981
Fabaceae	*Genista tinctoria* L.	Herb	Infusion	Used as diuretic and cardio-stimulant.	Bodlák *et al.*, 1995; Korbelár, 1981; Mackù and Krejca, 1964
Gentianaceae	*Gentiana lutea* L.	Roots	Infusion, decoction, tincture, powder	Used as stomachic, anthelmintic, nerve-stimulant, spasmolysan and for cholechromeresis.	Bodlák *et al.*, 1995; Korbelár, 1981; Mackù and Krejca, 1964
Geraniaceae	*Geranium robertianum* L.	Herb	Decoction, infusion	Used as antilaxative and adstringent.	Mackù and Krejca, 1964
Rosaceae	*Geum rivale* L.	Rhizome	Decoction, infusion	Drug used as digestive, spasmolysan, anti-laxative and hemostatic, also as desinfection of buccal cavity; externally applied on hemorrhoids and dermatitis.	Mackù and Krejca, 1964
Rosaceae	*Geum urbanum* L.	Roots	Decoction, infusion	Drug used as digestive, spasmolysan, anti-laxative and hemostatic, also as desinfection of buccal cavity; externally applied on hemorrhoids and dermatitis.	Bodlák *et al.*, 1995; Korbelár, 1981; Mackù and Krejca, 1964; Krkavec and Hájková, 1962

Contd...

Table 3.1–*Contd...*

Family	*Botanical Name*	*Part Used*	*Method Use*	*Popular Use*	*References*
Lamiaceae	*Glechoma hederacea* L.	Herb	Decoction, infusion	Used for bronchitis, urinary-complaint, as an expectorant, digestive; externally applied on wounds, hemorrhoids and dermatitis.	Bodlák *et al.*, 1995; Korbelár, 1981; Mackù and Krejca, 1964; Krkavec and Hájková, 1962
Asteraceae	*Gnaphalium dioicum* L.	Flowers	Tea, decoction, infusion	Used as expectorant and antiphlogosis.	Mackù and Krejca, 1964
Silenaceae	*Gypsophilla paniculata* (L.) Fisch.	Rhizome	Tea, infusion, tincture	Used as an expectorant and an antiphlogosis.	Janca and Zentrich, 1998
Araliaceae	*Hedera helix* L.	Leaves	Decoction	Decoction used as cardiotonic and expectorant.	Korbelár, 1981
Asteraceae	*Helichrysum arenarium* (L.) Moench.	Flowers	Decoction, infusion, tea, extract	Used as spasmolysan, antiphlogosis and diuretic, also as drug acts on cholechromeresis.	Korbelár, 1981; Janca and Zentrich, 1996
Ranunculaceae	*Helleborus niger* L.	Roots	Extract	Used as cardiotonic, diuretic and laxative; externally applied for local anaesthesia.	Korbelár, 1981
Ranunculaceae	*Hepatica nobilis* Schreb.	Leaves	Decoction, infusion	Used for bronchitis, liver-complaint, and as an expectorant.	Bodlák *et al.*, 1995; Korbelár, 1981; Mackù and Krejca, 1964
Silenaceae	*Herniaria glabra* L.	Herb	Infusion	Used as diuretic.	Bodlák *et al.*, 1995; Korbelár, 1981; Mackù and Krejca, 1964
Silenaceae	*Herniaria hirsuta* L.	Herb	Infusion	Used as spasmolysan, antiphlogosis and diuretic.	Bodlák *et al.*, 1995; Korbelár, 1981
Fabaceae	*Hippocrepis comosa* L.	Herb	Infusion	Used as a tonic; externally applied on wounds.	Janca and Zentrich, 1998
Elaeagnaceae	*Hippophae rhamnoides* L.	Leaves, fruits	Juice of plants, decoction, oil from seeds	Tea, juice and decoction used as a stomachic, cytostatic agent, cholechromeresis as an antibiotic.	Bodlák *et al.*, 1995; Korbelár, 1981; Mackù and Krejca, 1964

Contd...

Table 3.1–*Contd...*

Family	*Botanical Name*	*Part Used*	*Method Use*	*Popular Use*	*References*
Cannabaceae	*Humulus lupulus* L.	Strobile	Decoction, infusion	Used as sedative, stomachic and anaphrodisia.	Bodlák *et al.*, 1995; Korbelár, 1981; Mackù and Krejca, 1964
Crassulaceae	*Hylotelephium maximum* (L.) Holub	Leaves	Decoction	Used as an adstringent, tonic, cardio-tonic and diuretic; externally applied on hemorrhoids and ulcers.	Janca and Zentrich, 1996
Solanaceae	*Hyoscyamus niger* L.	Whole plants	Decoction	Used as antiasthmatic, spasmolysan and analgesic; oil obtained from seeds is externally applied as antirheumatic and analgesic.	Bodlák *et al.*, 1995
Hypericaceae	*Hypericum perforatum* L.	Herb	Decoction, infusion, oil, tincture	Used as antiphlogosis, antibacterial, catalyze of blood circulation, liver and gall, also as antidepresive and stimulant, tonic; externally applied on frostbites, burns, hemorrhoids, dermatitis and against foot sweating.	Bodlák *et al.*, 1995; Korbelár, 1981; Mackù and Krejca, 1964; Krkavec and Hájková, 1962
Lamiaceae	*Hyssopus officinalis* L.	Herb	Infusion, oil	Internally applied as expectorant, digestive, diuretic, also used to preventive for erythroprosopalgia and unstrung; externally applied on wounds and dermatitis.	Bodlák *et al.*, 1995; orbelár, 1981; Mackù and Krejca, 1964
Daucaceae	*Imperatoria ostruthium* L.	Root	Decoction, infusion, tincture	Polychrest for alimentary tract and liver-complaint, gynopathy, bronchitis, as digestive, cytostatic agent, also as stimulant and tonic; externally applied on dermatosis, wounds and haemorrhoids.	Bodlák *et al.*, 1995; Korbelár, 1981; Mackù and Krejca, 1964
Asteraceae	*Inula helenium* L.	Root	Infusion, detoction, tincture, powder	Used as expectorant, stomachic, anthelmintic, antibacterial, antiphlo-gosis, cardio-sedative and analgesic.	Bodlák *et al.*, 1995; Korbelár, 1981; Mackù and Krejca, 1964; Janca and Zentrich, 1995

Contd...

Table 3.1–*Contd...*

Family	*Botanical Name*	*Part Used*	*Method Use*	*Popular Use*	*References*
Iridaceae	*Iris florentina* L.	Rhizome	Decoction, tea, infusion	Used as expectorant.	Bodlák *et al.*, 1995
Iridaceae	*Iris germanica* L.	Rhizome	Decoction, tea, infusion	Used as expectorant.	Bodlák *et al.*, 1995; Korbelár, 1981
Iridaceae	*Iris pallida* Lamarc	Rhizome	Decoction, tea, infusion	Used as expectorant.	Bodlák *et al.*, 1995
Juglandaceae	*Juglans regia* L.	Leaves, pericarp	Decoction, infusion, tincture	Drug applied on viscera catarrh, diabetes and as hemostatic; externally applied on dermatosis and hair-restorer.	Bodlák *et al.*, 1995; Korbelár, 1981; Mackù and Krejca, 1964; Janca and Zentrich, 1995
Dipsacaceae	*Knautia arvensis* L.	Herb	Infusion, decoction, tincture	Drug acts on metabolism as tonic, also as adstringent and diuretic.	Janca, 1995
Lamiaceae	*Lamium album* L.	Flower	Decoction, infusion, tincture	Drug acts on gynopathy (menoxenia, fluor albus, etc.), blood depurative; externally applied on varices, tumors and hemorrhoids.	Bodlák *et al.*, 1995; Korbelár, 1981; Mackù and Krejca, 1964
Apiaceae	*Laserpitium latifolium* L.	Roots, seeds	Infusion, tincture	Used as a diuretic, carminative and stomachic, also on atopomenorrhoea.	Janca and Zentrich, 1998
Lamiaceae	*Lavandula angustifolia* Mill.	Flower, herb	Decoction, tea, infusion, oil, tincture	Used as carminative, spasmolysan, antidepressant and sedative; externally applied as antirheumatic and on a gout and arthritis.	Bodlák *et al.*, 1995; Korbelár, 1981; Mackù and Krejca, 1964; Janca and Zentrich, 1995
Lamiaceae	*Leonurus cardiaca* L.	Herb	Decoction, infusion, tincture	Drug acts on cardiac neurosis, climacterium and nervine; also used as hypotensive, carminative and digestive.	Bodlák *et al.*, 1995; Korbelár, 1981; Mackù and Krejca, 1964; Krkavec and Hájková, 1962
Brassicaceae	*Lepidium ruderale* L.	Herb	Decoction, infusion	Used as a diuretic, antiphlogosis, tonic and digestive.	Janca and Zentrich, 1996;

Contd...

Table 3.1–*Contd...*

Family	*Botanical Name*	*Part Used*	*Method Use*	*Popular Use*	*References*
Daucaceae	*Levisticus officinale* Koch.	Roots	Infusion, detoction, tincture, powder	Drug used as carminative, diuretic and digestive; externally applied as anti-rheumatic and on a gout.	Bodlák *et al.*, 1995; Korbelár, 1981
Scrophulariaceae	*Linaria vulgaris* L.	Herb	Tea, decoction, liniment	Used as carminative, laxative and cardiotonic, also on ulcer disease; externally applied on dermatitis, venous ulceration and hemorrhoids.	Bodlák *et al.*, 1995; Korbelár, 1981; Mackù and Krejca, 1964; Janca and Zentrich, 1995
Linaceae	*Linium usitatissimum* L.	Seeds	Decoction, infusion, oil	Drug used as spasmolysan, expectorant and antiphlogosis; externally applied on exuberance prostate, herpes zoster and dermatitis.	Bodlák *et al.*, 1995; Korbelár, 1981
Boraginaceae	*Lithospermum officinale* L.	Fruits	Decoction, infusion, tincture	Used as diuretic.	Mackù and Krejca, 1964
Lobeliaceae	*Lobelia inflata* L.	Herb	Decoction, infusion, tincture, tea, powder	Used as expectorant, spasmolysan and antiasthmatic.	Mackù and Krejca, 1964
Loniceraceae	*Lonicera periclymenum* L.	Flower	Decoction, infusion	Used as a expectorant, anti-asthmatic, spasmolysan and analgetic.	Janca and Zentrich, 1998
Brassicaceae	*Lunaria rediviva* L.	Seeds	Infusion, tincture	Used as a sedative, antiepileptic and stimulant nervous.	Janca and Zentrich, 1995
Lamiaceae	*Lycopus europaeus* L.	Herb	Decoction	Used as sedative.	Korbelár, 1981
Silenaceae	*Lychnis flos-cuculi* L.	Herb	Tea, tincture	Used for bronchitis, on atopomenorrhoea and as an antiphlogosis.	Janca and Zentrich, 1998
Primulaceae	*Lysimachia nummularia* L.	Herb	Decoction, infusion	Infusion internally applied for bronchitis and phthisis, also as antirheumatic; decoction externally applied on wounds, dermatitis, hemorrhoids and as emollient.	Janca and Zentrich, 1997
Lythraceae	*Lythrum salicaria* L.	Whole plants	Tea, infusion, decoction	Used as antiphlogosis, also for medication of indigestion.	Janca, 1995

Contd...

Table 3.1–*Contd...*

Family	*Botanical Name*	*Part Used*	*Method Use*	*Popular Use*	*References*
Lamiaceae	*Majorana hortensis* Moench.	Herb	Tea, decoction, infusion, liniment	Drug used as expectorant, diuretic, stomachic, carminative, sedative and antiphlogosis; externally applied as antirheumatic.	Bodlák *et al.*, 1995; Korbelár, 1981; Mackù and Krejca, 1964; Janca and Zentrich, 1995
Malvaceae	*Malva silvestris* L.	Leaves, flower	Decoction, infusion	Drug used as expectorant and anti-phlogosis; externally applied on wounds and dermatitis.	Bodlák *et al.*, 1995; Korbelár, 1981; Mackù and Krejca, 1964; Krkavec and Hájková, 1962
Malvaceae	*Malva silvestris* L.	Leaves, flower	Decoction, infusion	Drug used as expectorant and anti-phlogosis; externally applied on wounds and dermatitis, also as gargle on tonsillitis.	Mackù and Krejca, 1964; Krkavec and Hájková, 1962
Euphorbiaceae	*Marcurialis annua* L	Herb	Infusion	Used as laxative and diuretic.	Janca and Zentrich, 1994
Lamiaceae	*Marrubium vulgare* L.	Herb	Decoction, infusion, syrup	Used as laxative, antiphlogosis and cardiac, also on gastritis, menoxenia, bronchitis; externally used as anti-phlogosis and on dermatitis.	Bodlák *et al.*, 1995; Korbelár, 1981; Mackù and Krejca, 1964
Fabaceae	*Medicago sativa* L.	Flower, seeds	Juice, powder	Used as an antirheumatic and tonic.	Janca and Zentrich, 1997
Fabaceae	*Melilotus officinalis* (L.) Lam.	Herb, flower	Decoction, infusion	Decoction or infusion used for bronchtis, urinary tract infection, climacteric hot flush and as hypo-tensive; externally applied on oncoides and lump, also as antirheumatic.	Bodlák *et al.*, 1995; Korbelár, 1981; Mackù and Krejca, 1964; Krkavec and Hájková, 1962
Lamiaceae	*Melissa officinalis* L.	Herb, leaves	Decoction, infusion, tincture, extract	Used as carminative, antidepressant and hypotensive; externally applied as antirheumatic, antiphlogosis and on dermatitis.	Bodlák *et al.*, 1995; Korbelár, 1981; Mackù and Krejca, 1964; Janca and Zentrich, 1995
Lamiaceae	*Mentha crispa* L.	Leaves	Infusion	Infusion used as spasmolysan, carminative, cholechromeresis and antibacterial.	Bodlák *et al.*, 1995; Korbelár, 1981; Mackù and Krejca, 1964; Janca and Zentrich, 1995

Contd...

Table 3.1–*Contd...*

Family	*Botanical Name*	*Part Used*	*Method Use*	*Popular Use*	*References*
amiaceae	*Mentha piperita* L.	Leaves, herb	Tea, infusion	Tea or infusion used as spasmolysan, carminative, cholechromeresis and antibacterial.	Bodlák *et al.*, 1995; Korbelár, 1981; Mackù and Krejca, 1964; Krkavec and Hájková, 1962; Janca and Zentrich, 1995
Menyanthaceae	*Menyanthes trifoliata* L.	Leaves	Decoction, infusion, tincture	Used as digestive tonic and taken as a blood and stomachic depurative.	Bodlák *et al.*, 1995; Korbelár, 1981; Mackù and Krejca, 1964; Krkavec and Hájková, 1962
Brassicaceae	*Nasturtium officinale* R.Br.	Herb	Decoction, infusion	Drug acts on cholechromeresis and as a tonic.	Korbelár, 1981
Lamiaceae	*Nepeta cataria* L.	Herb	Decoction, infusion, tea	Drug acts as stimulant nervous.	Mackù and Krejca, 1964; Krkavec and Hájková, 1962;
Nymphaeaceae	*Numphar lutea* (L.) Sm.	Rhizome	Extract	Used as sedative and stimulant nervous.	Korbelár, 1981
Nymphaeaceae	*Nymphaea alba* L.	Rhizome	Extract	Drug used as anaphrodisiac and cytostatic agent.	Korbelár, 1981; Janca and Zentrich, 1998
Lamiaceae	*Ocimum basilicum* L.	Herb	Tea, decoction, infusion	Internally applied as stomachic, carminative, expectorant, spasmolysan and for urinary tract infection, also as antidepressive and unstrung; externally applied as antiphlogosis and on dermatitis.	Bodlák *et al.*, 1995; Korbelár, 1981; Mackù and Krejca, 1964
Oenctheraceae	*Oenothera biennis* L.	Seeds	Oil	Used as a tonic, also treatment for constitutional dermatitis.	Janca and Zentrich, 1996
Fabaceae	*Ononis spinosa* L.	Roots	Tea, tincture	Tea and tincture used as diuretic	Bodlák and Zentrich, 1995; Korbelár, 1981; Mackù and Krejca, 1964; Krkavec and Hájková, 1962

Contd...

Table 3.1–*Contd...*

Family	*Botanical Name*	*Part Used*	*Method Use*	*Popular Use*	*References*
Asteraceae	*Onopordum acanthium* L.	Whole plants	Juice of plants, decoction	Juice or decoction of plants externally applied on dermatosis, ulcer disease; as an anti-inflammatory and as a carcinostatic.	Bodlák and Zentrich, 1995; Bodlák *et al.*, 1995; Korbelár, 1981; Mackù and Krejca, 1964; Janca and Zentrich, 1995
Lamiaceae	*Origanum vulgare* L.	Herb	Decoction, infusion, tincture, extract	Internally applied as stomachic, spasmolysan and for cholechromeresis, also as antidepresive, expectorant and antiphlogosis; externally applied as antiphlogosis and on dermatitis.	Bodlák *et al.*, 1995; Korbelár, 1981; Mackù and Krejca, 1964; Krkavec and Hájková, 1962
Oxalidaceae	*Oxalis acetosella* L.	Herb	Decoction, infusion	Used as diuretic.	Mackù and Krejca, 1964
Rosaceae	*Padus avium* Mill.	Fruits, bork, flower	Infusion, decoction, tincture	Drug from fruits used as an anti-laxative and sudorific; drug from leaves used on stomachal diseases; drug from flowers applied for bronchitis; and drug from bark used as an antirheumatic in form baths or poultices.	Janca and Zentrich, 1996
Paeoniaceae	*Paeonia officinalis* L.	Flower, roots	Extract	Used as spasmolysan and anticephalalgikum.	Bodlák *et al.*, 1995; Korbelár, 1981
Papaveraceae	*Papaver rhoeas* L.	Flower	Decoction, infusion	Used for bronchitis (for children) and as expectorant.	Bodlák *et al.*, 1995; Mackù and Krejca, 1964; Krkavec and Hájková, 1962; Janca and Zentrich, 1995
Asteraceae	*Petasites hybridus* (L.) P. Gaertn.	Roots	Decoction, infusion, tincture, extract	Used for bronchitis, tussis and pulmonitis and as spasmolysan; externally applied on dermatitis, hemorrhoids, bruises, wounds and sores.	Bodlák *et al.*, 1995; Korbelár, 1981; Mackù and Krejca, 1964; Krkavec and Hájková, 1962
Daucaceae	*Petroselinum hortense* Hoffm.	Roots, seeds	Decoction, infusion	Used as diuretic and digestive.	Bodlák *et al.*, 1995; Korbelár, 1981

Contd...

Table 3.1–*Contd...*

Family	*Botanical Name*	*Part Used*	*Method Use*	*Popular Use*	*References*
Apiaceae	*Phellandrium aquanticum* L.	Seeds, fruits	Infusion, tincture	Used as a diuretic, also as an anti-asthmatic.	Janca and Zentrich, 1996; Janca and Zentrich, 1998
Cichoriaceae	*Pilosella officinarium* Schultz.	Herb	Infusion, decoction, tincture	Internally applied as digestive and antiphlogosis, internally applied on dermatitis.	Janca, 1995
Daucaceae	*Pimpinella anisum* L.	Fruits	Tea, decoction, infusion, liniment	Used as carminative and expectorants.	Bodlák *et al.,* 1995; Korbelár, 1981; Mackù and Krejca, 1964
Daucaceae	*Pimpinella major* (L.) Huds.	Roots	Decoction, infusion, tincture	Used for bronchitis, tussis and pulmonitis and as diuretic, digestive; externally applied on dermatitis, bruises, wounds and sores.	Krkavec and Hájková, 1962
Daucaceae	*Pimpinella saxifraga* L.	Roots	Decoction, infusion, tincture	Used for bronchitis, tussis and pulmonitis and as digestive; externally applied on dermatitis, bruises, wounds and sores.	Bodlák *et al.,* 1995; Korbelár, 1981; Mackù and Krejca, 1964
Lentibulariaceae	*Pinguicula vulgaris* L.	Herb	Decoction, infusion, tea, extract	Used as an antisclerotic, antibacterial and spasmolysan, also on hardening of the arteries and bronchitis.	Bodlák *et al.,* 1995; Korbelár, 1981
Pirolaceae	*Pirola rotundifolia* L.	Leaves	Decoction	Used as antiphlogosis, antiseptic and diuretic.	Korbelár, 1981
Plantaginaceae	*Plantago lanceolata* L.	Leaves, seeds	Infusion, tea detoction, tincture, powder, syrup	Syrup, infusion, tea and/or decoction taken for bronchitis, and the common cold, also for indigestion; infusion, tincture and powder externally applied on turgidity and wounds.	Bodlák *et al.,* 1995; Mackù and Krejca, 1964; Krkavec and Hájková, 1962
Polemoniaceae	*Polemonium coerulum* L.	Roots	Decoction, infusion, tincture	Used as an adstringent, sedative and stimulant nervous.	Mackù and Krejca, 1964
Polygalaceae	*Polygala amara* L.	Whole plants	Decoction, infusion, tea	Used as a stomachic and expectorant.	Mackù and Krejca, 1964

Contd...

Table 3.1–*Contd...*

Family	*Botanical Name*	*Part Used*	*Method Use*	*Popular Use*	*References*
Liliaceae	*Polygonatum odoratum* (Mill.) Druce	Rhizome	Tea, tincture	Used as a diuretic and sedative; externally applied as an antirheumatic.	Janca and Zentrich, 1998
Polygonaceae	*Polygonum aviculare* L.	Herb	Decoction, liqueur	Used as diuretic and antirheumatic; externally applied on dermatitis and wounds.	Bodlák *et al.,* 1995; Korbelár, 1981; Mackù and Krejca, 1964; Krkavec and Hájková, 1962
Polygonaceae	*Polygonum bistoria* L.	Rhizome	Decoction, infusion, tea, extract	Used as adstringent, antiphlogosis and anti-laxative; externally applied as gargle on pharyngotonsillitis.	Korbelár, 1981
Polygonaceae	*Polygonum hydropiper* L.	Herb	Infusion	Internally used as hemostatic, diuretic, antirheumatic and expectorant; externally applied on dermatitis and arthritis.	Bodlák *et al.,* 1995; Korbelár, 1981; Mackù and Krejca, 1964; Janca and Zentrich, 1996
Salicaceae	*Populus nigra* L.	Browse	Decoction, infusion, tea, extract	Used as diuretic, antiphlogosis and antirheumatic; externally used as drug acts on hemorrhoids and burns.	Korbelár, 1981
Rosaceae	*Potentilla anserina* L.	Whole plants	Decoction, infusion	Used as adstringent, spasmolysan, antiphlogosis, anti-laxative and hemostatic.	Bodlák *et al.,* 1995; Korbelár, 1981; Mackù and Krejca, 1964
Rosaceae	*Potentilla argentea* L.	Herb	Infusion, decoction, tincture	Used as an adstringent, spasmolysan, antiphlogosis and hemostatic.	Janca and Zentrich, 1995
Rosaceae	*Potentilla erecta* (L.) Raeuschel	Roots	Decoction, tincture	Used as adstringent, anti-laxative and hemostatic.	Bodlák *et al.,* 1995; Korbelár, 1981; Mackù and Krejca, 1964
Primulaceae	*Primula elatior* (L) Hill.	Flower	Decoction, tincture	Used as expectorant, to treat influenza and for bronchitis.	Mackù and Krejca, 1964
Primulaceae	*Primula veris* L.	Flower, roots	Decoction, tincture, liqueur	Used as expectorant, to treat influenza and for bronchitis, also as a diuretic and antirheumatic.	Bodlák *et al.,* 1995; Korbelár, 1981; Mackù and Krejca, 1964

Contd...

Table 3.1–*Contd...*

Family	*Botanical Name*	*Part Used*	*Method Use*	*Popular Use*	*References*
Lamiaceae	*Prulella vulgaris* L.	Herb	Decoction, infusion, tea	Used as disinfecting, digestive and anti-laxative, externally applied as gargle on pharyngotonsillitis.	Krkavec and Hájková, 1962
Rosaceae	*Prunus spinosa* L.	Flowers, fruits	Tea; liqueur	Tea and liqueur used as an laxative, diaphoretic agent, diuretic stimulant and antispasmodic agent.	Bodlák and Zentrich, 1995; Bodlák *et al.*, 1995; Korbelár, 1981; Mackù and Krejca, 1964; Krkavec and Hájková, 1962; Janca and Zentrich, 1997
Lamiaceae	*Pulegium vulgare* Mill.	Herb	Infusion, decoction	Infusion used as spasmolysan, carminative, cholechromeresis and antibacterial.	Janca and Zentrich, 1995
Boraginaceae	*Pulmonaria officinalis* L.	Leaves	Infusion	Infusion used as expectorant and diuretic, also on bituminosis and bronchitis.	Bodlák *et al.*, 1995; Korbelár, 1981; Mackù and Krejca, 1964; Krkavec and Hájková, 1962; Janca and Zentrich, 1996
Asteraceae	*Pyrethrum parthenium* (L.) Smith	Herb	Infusion, decoction, tincture, tea	Used as an antipyretic, antiphlogosis and digestive, also drug from green leaves applied on neuralgic pains and rheumatic soreness.	Janca and Zentrich, 1996
Rosaceae	*Pyrus communis* L	Leaves, fruits	Decoction, infusion, tea, juice	Drug from leaves used as antiphlogosis and diuretic, juice from fruit used as diuretic, digestive, cardiotonic, also as drug acts on cholechromeresis.	Janca, 1995
Fagaceae	*Quercus petraea* L.	Bark, seeds	Decoction, infusion, tincture, extract, liniment	Bark decoction, infusion and liniment applied externally on frostbites, first degree burns, haemorrhoids, dermatitis and against foot sweating; seeds tincture or decoction gargled for sore throat, sinusitis and toothache.	Bodlák *et al.*, 1995; Korbelár, 1981; Mackù and Krejca, 1964

Contd...

Table 3.1–*Contd...*

Family	*Botanical Name*	*Part Used*	*Method Use*	*Popular Use*	*References*
Fagaceae	*Quercus robur* L.	Bark, seeds	Decoction, infusion, tincture, extract, liniment	Bark decoction, infusion and liniment applied externally on frostbites, first degree burns, haemorrhoids, dermatitis and against foot sweating; seeds tincture or decoction gargled for sore throat, sinusitis and toothache.	Bodlák *et al.*, 1995; Korbelár, 1981; Mackù and Krejca, 1964; Krkavec and Hájková, 1962
Rhamnaceae	*Rhamnus cathartica* L.	Fruits	Juice, macerate	Internally applied as laxative.	Bodlák and Zentrich, 1995; Bodlák *et al.*, 1995; Korbelár, 1981; Mackù and Krejca, 1964; Krkavec and Hájková, 1962
Crassulaceae	*Rhodiola rosea* L.	Rhizome	Tincture	Used as a tonic and and for prophylaxis, also as aphrodisiac.	Janca and Zentrich, 1996
Grossulariaceae	*Ribes nigrum* L.	Leaves, fruits	Tea, infusion, decoction for leaves, juice and macerate for fruits	Used to treat influenza and the common cold, also as antirheumatic.	Korbelár, 1981; Mackù and Krejca, 1964
Fabaceae	*Robinia pseudoacacia* L.	Flowers	Tea	Tea used as spasmolysan.	Bodlák and Zentrich, 1995; Bodlák *et al.*, 1995; Korbelár, 1981
Rosaceae	*Rosa canina* L.	Fruits	Tea, decoction, syrup, oil from seeds	Tea, decoction or oil used as diuretic, to spring fever and influenza, tonic.	Bodlák and Zentrich, 1995; Bodlák *et al.*, 1995; Korbelár, 1981; Mackù and Krejca, 1964; Krkavec and Hájková, 1962
Lamiaceae	*Rosmarinus officinalis* L.	Leaves	Decoction, infusion, tincture, extract, liniment	Drug used as carminative, digestive, antisclerotic and on brain-fag.	Bodlák *et al.*, 1995; Korbelár, 1981; Mackù and Krejca, 1964
Rubiaceae	*Rubia tinctorum* L.	Roots	Decoction, extract	Used as diuretic and antiphlogosis; externally applied on rhachitis and curing of ulcers.	Bodlák *et al.*, 1995; Korbelár, 1981; Janca and Zentrich, 1995

Contd...

Table 3.1–*Contd...*

Family	*Botanical Name*	*Part Used*	*Method Use*	*Popular Use*	*References*
Rosaceae	*Rubus fruticosus* L.	Leaves, fruits	Tea, decoction, juice, syrup	Syrup and juice of fruits used as an astringent for children; decoction or tea of leaves used as an astringent, also as a remedy for diabetes, as a vulnerary, anti-inflammatory, anti-diarrhoea and anti-tussis.	Bodlák and Zentrich, 1995; Bodlák *et al.*, 1995; Korbelár, 1981; Mackù and Krejca, 1964; Krkavec and Hájková, 1962; Janca and Zentrich, 1995
Rosaceae	*Rubus idaeus* L.	Leaves, fruits	Tea, decoction, juice, syrup	Syrup and juice of fruits used as an astringent for children; decoction or tea of leaves used as an astringent, also as a remedy for diabetes, as a vulnerary, anti-inflammatory, anti-diarrhoea and anti-tussis.	Bodlák and Zentrich, 1995; Bodlák *et al.*, 1995; Korbelár, 1981; Mackù and Krejca, 1964; Krkavec and Hájková, 1962; Janca and Zentrich, 1995
Rutaceae	*Ruta graveolens* L.	Herb	Infusion, macerate	Used as spasmolysan, utero-tonic, sedative, stomachic, antisclerotic and vaso-tonic.	Bodlák *et al.*, 1995; Korbelár, 1981; Mackù and Krejca, 1964
Salicaceae	*Salix alba* L.	Bark	Decoction	Internally applied as pyretic, analgetic, tonic (for nerves and vigilance), anti-rheumatic and anti-laxative; externally applied on cutaneous fungus and foot sweating.	Bodlák *et al.*, 1995; Korbelár, 1981; Mackù and Krejca, 1964
Lamiaceae	*Salvia officinalis* L.	Herb	Infusion, detoction, tincture, powder, syrup	Internally used as antiphlogosis, spasmolysan and antisudoral; externally applied as adstringent and antiphlogistic.	Bodlák *et al.*, 1995; Korbelár, 1981; Mackù and Krejca, 1964
Carpifoliaceae	*Sambucus ebulus* L.	Roots, leaves, fruits	Tea, infusion, decoction for leaves, juice, liqueur, syrup and macerate for fruits	Used as diaphoretic, laxative and antiphlogosis.	Mackù and Krejca, 1964

Contd...

Table 3.1–*Contd...*

Family	*Botanical Name*	*Part Used*	*Method Use*	*Popular Use*	*References*
Caprifoliaceae	*Sambucus nigra* L.	Flower, fruits	Tea, infusion, decoction for leaves, juice, liqueur, syrup and macerate for fruits	Flowerage used as diaphoretic, laxative and antiphlogosis; fruits used for neuralgia and artrodynia.	Bodlák *et al.*, 1995; Korbelár, 1981; Mackù and Krejca, 1964; Krkavec and Hájková, 1962
Rosaceae	*Sanguisorba officinalis* L.	Roots	Infusion	Infusion used as anti-laxative, antiseptic and pyretic; externally applied as gargles for a sore throat and a gingivitis.	Bodlák *et al.*, 1995; Korbelár, 1981; Mackù and Krejca, 1964; Krkavec and Hájková, 1962
Apiaceae	*Sanicula europaea* L.	Herb	Infusion	Polychrest for alimentary tract and liver-complaint; externally applied on dermatosis and haemorrhoids.	Bodlák *et al.*, 1995; Korbelár, 1981; Mackù and Krejca, 1964; Krkavec and Hájková, 1962
Caryophyllaceae	*Saponaria officinalis* L.	Roots	Macerate	Used as diaphoretic.	Bodlák *et al.*, 1995; Korbelár, 1981; Mackù and Krejca, 1964; Janca and Zentrich, 1995
Fabaceae	*Sarothamnus scoparius* Wimmer	Flower	Infusion	Used as antiphlogosis for medication of urinary tract diseases.	Janca, 1995
Lamiaceae	*Satureja hortensis* L.	Herb	Decoction, infusion	Decoction or infusion used for stomach catarrh and enteritis, also as anti-laxative, nervine, expectorant, hypotensive and anthelmintic.	Bodlák *et al.*, 1995; Korbelár, 1981; Mackù and Krejca, 1964
Saxifragaceae	*Saxifraga granulata* L.	Herb	Decoction, infusion, tincture	Used as diuretic and as drug acts on dissolving of gallbladder stones.	Mackù and Krejca, 1964; Krkavec and Hájková, 1962
Scrophulariaceae	*Scrophularia hodosa* L.	Rhizome	Tea, infusion, decoction	Used as diuretic and expectorant, externally applied on ulcers, sores, hemorrhoids, gathering wound and earaches.	Janca, 1995

Contd...

Table 3.1–*Contd...*

Family	*Botanical Name*	*Part Used*	*Method Use*	*Popular Use*	*References*
Crassulaceae	*Sedum acre* L.	Herb	Infusion	Used as hypotensive, externally applied on dermatitis, cutaneous fungus and haemorrhoids.	Korbelár, 1981
Crassulaceae	*Sempervivum tectorum* L.	Leaves	Juice	Juice from leaves applied on tubercles, wounds and burns.	Janca and Zentrich, 1995
Asteraceae	*Senecio jacobaea* L.	Herb	Tea, extract, tincture	Used as well as S. vulgaris.	Krkavec and Hájková, 1962; Janca and Zentrich, 1998
Asteraceae	*Senecio vulgaris* L.	Herb	Tea, extract, tincture	Used as drug acts on menstrual problems, also as diuretic and cardio-tonic; externally applied on dermatosis.	Krkavec and Hájková, 1962
Silenaceae	*Silene vulgaris* (Moench.) Garcke	Herb	Decoction	Internally applied as a diuretic; externally taken on wounds and dermatitis; also as an antirheumatic in form baths.	Janca and Zentrich, 1996
Brassicaceae	*Sinapis alba* L.	Seeds	Infusion	Used as stomachic and derivans, also as drug acts on cholechromeresis.	Janca, 1995
Solanaceae	*Solanum dulcamara* L.	Herb	Extracts, sirup, decoction	Internally used as antirheumatic and diuretic; decoction externally used on dermatitis.	Korbelár, 1981; Krkavec and Hájková, 1962; Janca and Zentrich, 1996
Asteraceae	*Solidago virgaurea* L.	Herb	Decoction, infusion	Drug used for urethrocystitis, coline-phritis, liver-complaint and as anti-rheumatic.	Bodlák *et al.*, 1995; Korbelár, 1981; Mackù and Krejca, 1964; Krkavec and Hájková, 1962
Rosaceae	*Sorbus aucuparia* L.	Fruits	Macerate	Used as diuretic and laxative.	Bodlák *et al.*, 1995; Korbelár, 1981; Mackù and Krejca, 1964; Krkavec and Hájková, 1962
Lamiaceae	*Stachys sylvatica* L.	Herb	Tea, infusion, decoction	Used as sedative, hemostatic, adstrin-gent and cardiotonic.	Janca and Zentrich, 1994
Dipsacaceae	*Succisa pratensis* Moench.	Herb, rhizome	Tea, infusion, decoction	Used as diuretic and tonic.	Janca and Zentrich, 1994

Contd...

Table 3.1–*Contd...*

Family	*Botanical Name*	*Part Used*	*Method Use*	*Popular Use*	*References*
Boraginaceae	*Symphytum officinale* L.	Roots	Decoction, infusion, tincture, extract, liniment	Internally used for relieving dermatosis, sarcitis, paradentosis, as hemostatic, antirheumatic; externally applied in poultices for eczema, erysipelas, also for medication of fractures, wounds and inflammations.	Bodlák *et al.*, 1995; Korbelár, 1981; Mackù and Krejca, 1964; Krkavec and Hájková, 1962
Asteraceae	*Tanacetum vulgare* L.	Leaves, flower	Decoction	Used as anthelmintic.	Bodlák *et al.*, 1995; Korbelár, 1981; Mackù and Krejca, 1964; Krkavec and Hájková, 1962; Janca and Zentrich, 1997
Asteraceae	*Taraxacum officinale* Web. Wiggers.	Whole plants	Decoction, infusion	Drug used as digestive, for cholechromeresis, diuretic and antidiabetic.	Bodlák *et al.*, 1995; Korbelár, 1981; Mackù and Krejca, 1964; Krkavec and Hájková, 1962
Lamiaceae	*Teucrium chamaedrys* L.	Herb	Decoction	Drug used as polychrest; drug acts on the function of livers, stomach, viscera, as antirheumatic, hemostatic and antibacterial.	Bodlák *et al.*, 1995; Korbelár, 1981; Mackù and Krejca, 1964
Ranunculaceae	*Thalictrum aquilegifolium* L.	Herb	Tincture	Externally applied as an antiphlogosis.	Janca and Zentrich, 1998
Lamiaceae	*Thymus serpyllum* L.	Herb	Decoction, infusion, tea, tincture, extract, liniment, oil	Used as spasmolysan, disinfection, diaphoretic, diuretic, expectorant, antipyretic, tonic and stomachic.	Bodlák *et al.*, 1995; Korbelár, 1981; Mackù and Krejca, 1964; Krkavec and Hájková, 1962
Lamiaceae	*Thymus vulgaris* L.	Herb	Decoction, infusion, tea, tincture	Used as spasmolysan, diaphoretic, diuretic, expectorant, antipyretic, tonic and stomachic.	Bodlák *et al.*, 1995; Korbelár, 1981; Mackù and Krejca, 1964; Janca and Zentrich, 1997
Tiliaceae	*Tilia cordata* Mill.	Flower	Decoction, infusion, tea, tincture	Used as spasmolysan, diaphoretic, diuretic, expectorant, antipyretic and stomachic.	Bodlák *et al.*, 1995; Korbelár, 1981; Mackù and Krejca, 1964; Krkavec and Hájková, 1962; Janca and Zentrich, 1995

Contd...

Table 3.1–*Contd...*

Family	*Botanical Name*	*Part Used*	*Method Use*	*Popular Use*	*References*
Tiliaceae	*Tilia platyphylla* Scop.	Flower	Decoction, infusion, tea, tincture	Used as spasmolysan, diaphoretic, diuretic, expectorant, antipyretic and stomachic.	Krkavec and Hájková, 1962
Euphorbiaceae	*Tithymalus amygdaloides* (L.) Hill.	Herb	Decoction	Used as a cytostatic agent and tonic; externally applied on wounds and dermatitis.	Janca and Zentrich, 1996
Cichoriaceae	*Tragopogon pratensis* L.	Rhizome	Decoction, tea	Used as stomachic, laxative, also on cholechromeresis.	Janca and Zentrich, 1998
Fabaceae	*Trifolium pratense* L.	Flower	Decoction	Decoction externally applied as desinfecting on wounds and dermatosis.	Bodlák *et al.*, 1995; Korbelár, 1981; Mackù and Krejca, 1964; Krkavec and Hájková, 1962
Asteraceae	*Tussilago farfara* L.	Flower, leaves	Tea, decoction, infusion, syrup	Tea, syrup and decoction used as antipyretic, expectorant, antitussive, also as drug acts on pharyngolaryngitis and rhinopneumonitis; externally applied as decoction on burns, dermatosis and as an antirheumatic.	Bodlák *et al.*, 1995; Korbelár, 1981; Mackù and Krejca, 1964
Ulmaceae	*Ulmus glabra* Huds.	Bark	Tea, infusion, decoction	Used as well as *U. leavis.*	Janca, 1995
Ulmaceae	*Ulmus laevis* Pallas	Bark	Tea, infusion, decoction	Used as digestive, carminative and laxative; externally applied on dermatitis.	Janca, 1995
Ulmaceae	*Ulmus minor* Mill.	Bark	Tea, infusion, decoction	Used as well as *U. leavis.*	Janca, 1995
Urticaceae	*Urtica dioica* L.	Whole plants	Tea, decoction, infusion, syrup	Internally used for relieving dermatosis, tonic, antirheumatic, hemostatic, antidiabetic, diuretic, to treat influenza and the common cold, infusion taken as blood depurative; externally applied in poultices for eczema, erysipelas, also for influenza.	Korbelár, 1981; Krkavec and Hájková, 1962, Janca, 1995

Contd...

Table 3.1–*Contd...*

Family	*Botanical Name*	*Part Used*	*Method Use*	*Popular Use*	*References*
Urticaceae	*Urtica urens* L.	Whole plants	Tea, decoction, infusion, syrup	Internally used for relieving dermatosis, tonic, antirheumatic, hemostatic, antidiabetic, diuretic, to treat influenza and the common cold, infusion taken as blood depurative; externally applied in poultices for eczema, erysipelas, also for influenza.	Bodlák *et al.*, 1995; Korbelár, 1981
Vacciniaceae	*Vaccinium myrtillus* L.	Leaves, fruits	Tea, infusion, decoction for leaves, juice, tea, liqueur and macerate for fruits	Used as adstringent, anti-laxative, disinfectant, cardiac, antidiabetic and diuretic.	Korbelár, 1981; Krkavec and Hájková, 1962
Vacciniaceae	*Vaccinium vitis-idaea* L.	Leaves, fruits	Tea, infusion, decoction for leaves, juice, tea, liqueur and macerate for fruits	Used as adstringent, anti-laxative, disinfectant and diuretic.	Bodlák *et al.*, 1995; Korbelár, 1981; Mackù and Krejca, 1964
Valerianaceae	*Valeriana officinalis* L.	Roots	Infusion, detoction, tincture, powder	Used as anticephalalgikum, sedative, spasmolysan and hypnotic.	Bodlák *et al.*, 1995; Korbelár, 1981; Mackù and Krejca, 1964; Krkavec and Hájková, 1962
Liliaceae	*Veratrum album* L.	Rhizome	Decoction, extract	Used as hypotensive, externally applied on local anaesthesia.	Korbelár, 1981
Scrophularia-ceae	*Verbascum densiflorum* Bertol.	Flower	Decoction, infusion, tea, tincture, extract, oil	Used as expectorant, spasmolysan, diuretic and antiinflammatory; externally applied on dermatosis and haemorrhoids.	Bodlák *et al.*, 1995; Korbelár, 1981; Mackù and Krejca, 1964
Scrophularia-ceae	*Verbascum phlomoides* L.	Flower	Decoction, infusion, tea, tincture, extract, oil	Used as expectorant, spasmolysan, diuretic and antiinflammatory; externally applied on dermatosis and haemorrhoids.	Bodlák *et al.*, 1995; Korbelár, 1981; Mackù and Krejca, 1964; Krkavec and Hájková, 1962

Contd...

Table 3.1–*Contd...*

Family	*Botanical Name*	*Part Used*	*Method Use*	*Popular Use*	*References*
Verbenaceae	*Verbena officinalis* L.	Herb	Decoction, infusion	Used as drug acts on bronchitis, expectorant, tonic, infusion gargled for sore throat and toothache; externally applied on dermatosis.	Bodlák *et al.*, 1995; Korbelár, 1981; Mackù and Krejca, 1964; Krkavec and Hájková, 1962
Scrophulariaceae	*Veronica officinalis* L.	Herb	Decoction, infusion, tincture	Used as antirheumatic and stomachic, externally applied on dermatosis.	Bodlák *et al.*, 1995; Korbelár, 1981; Mackù and Krejca, 1964; Krkavec and Hájková, 1962
Loniceraceae	*Viburnum opulus* L.	Fruts, bark	Tea, infusion, decoction	Used as spasmolysan for medication of mentrual diseases, externally and internally applied on dermatitis.	Janca, 1995
Apocynaceae	*Vinca minor* L.	Herb	Decoction, extract	Used as hemostatic, antiphlogosis and hypotensive.	Korbelár, 1981; Krkavec and Hájková, 1962
Violaceae	*Viola odorata* L.	Flower	Decoction, infusion, tincture, oil	Used as anemollient and a laxative, diuretic, tonic (on metabolism and vaso-active) and expectorant.	Bodlák *et al.*, 1995; Korbelár, 1981; Mackù and Krejca, 1964; Krkavec and Hájková, 1962; Janca, 1995
Violaceae	*Viola tricolor* L.	Herb	Decoction, infusion, tincture	Used as a diuretic, tonic (on metabolism and vaso-active) and expectorant.	Bodlák *et al.*, 1995; Korbelár, 1981; Mackù and Krejca, 1964; Krkavec and Hájková, 1962; Janca and Zentrich, 1995
Loranthaceae	*Viscum album* L.	Whole plants	Tincture, liniment	Tincture used as drug acts on conditioning of haematopiesis, as a diuretic and an antisclerotic; liniment on treatment for perniosis.	Bodlák *et al.*, 1995; Mackù and Krejca, 1964; Krkavec and Hájková, 1962

comprising 277 species in total and which comprise 54.8 per cent of all species of plants which have been utilized in folk medicine in the Czech Republic (Table 3.2). The majority of species are represented by the Asteraceae, Lamiaceae and Rosaceae family, with 37, 27, and 22 species respectively.

According to an ethno-botanical survey about the knowledge of plants in the Czech Republic, several species of plants are very popular in folk medicine and are very often utilized as ancillary teas and concoctions in the treatment of, in particular, colds, influenzas and illnesses of the upper and iower respiratory tract, and furthermore as a prevention against these above mentioned diseases. The most often utilized plants include: *Allium cepa, Chamomilla recutita, Crategus* sp., *Plantago* sp., *Prímula veris, Rosa canina, Sambucus nigra, Thymus* sp., *Tilia* sp., *Tusilago farfara* and *Urtica dioica*. Even though treatments for colds and illnesses of the respiratory tract are the most common and most used, there are also other species of plants against disease which are known among the people.

As an astringent and hemostatic, the most well known in the subconscious of the people are fresh leaves, respectively, the sap of the *Plantago* sp. leaves, a concoction from *Agrimonia eupatoria;* the most widespread medicine against piles are coverings and baths with the sap from the root of the *Daucus carota* carrot and a concoction from oak bark (*Quercus* sp.); most often utilized against problems of the limbs, including a treatment for rheumatism, are *Urtica* sp., *Sambucus nigra, Achillea* sp., *Betula pendula, Borago officinalis* and *Polygonum* sp. Most commonly ušed for common skin diseases and burns are *Chamomilla recutita, Chelidonium majus, Hypericum* sp., *Petasites hybridus, Urtica* sp. As an anthelmintics used *Allium sativum, Artemisia* sp., *Carum carvi, Cota tínctoria, Chenopodiumn ambrosioides, Tanacetum vulgare* and *Dryopteris filix-mas.* At least some of these plants are among the standard equipment of nearly every village household.

Fungi have a significant representation among the healing plants. Fungi range among the most popular dishes of Czech cuisine. The Czech people usually go to the woods to pick up some wild living fungi. Fungi collection happened to a Czech "national sporting event". Czech cuisine can offer various methods of preparation of different types of fungi, whereby 22 fungi species have different actions, such as astringent, hemostatic, antiaphrodisiac, tonic laxative, or anti-laxative, digestive, and also used for bronchitis, antiphlogosis and dermatitis (see Table 3.1).

A large number of plant species which have been utilized as folk medicine also grow in locations in Western and Southern Europe. If we compared the different utilization of these species of plants in folk medicine, we would come to the conclusion that the majority of these species have a very similar therapeutic utilization in traditional medicine throughout the Europe; nevertheless, with some species of plants, the therapeutic utilization diametrically differs *e.g., Achillea milefolium* has been utilized in the Czech Republic, Bulgaria and Italia as a diuretic and heamostatic; in Turkey this plant has been used as an anti-laxative which could correspond to its astringent and diuretic effects; however, in Spain, this plant has been utilized, among other purposes, as a laxative (Yesilada *et al.*, 1993, 1995; Pardo de Santayana *et al.*, 2005; Pavela, 2009).

Particularly interesting is the utilization of the onion (*Allium cepa*). In the Czech Republic, this vegetable has been utilized as a very effective medicine in the treatment of all types of colds with an irritating cough. In this situation, the onion is boiled and, after it is strained, it is mixed with honey and given several times a day. A medicine prepared in this way shows, very good expectorant effects especially in children. In other parts of Europe, this method of utilization has not yet been documented (Pierony, 2000; Pierony *at al.*, 2004; Camejo-Rodrigues *et al.*, 2003; Pavela, 2009).

Table 3.2: Most numerously represented families.

Families	*Number of Taxa*	*Per cent**
Apiaceae	8	2.8
Asteraceae	37	13.1
Brassicaceae	9	3.2
Daucaceae	11	3.9
Fagaceae	15	5.3
Lamiaceae	27	9.5
Linaceae	8	2.8
Rosaceae	22	7.8
Scrophulariaceae	9	3.2
Ranunculaceae	9	3.2
Total	277	54.8

* Percentage from all medicinal plants used in the traditional medicine of Czech Republic.

The use of medicinal plants by the Czech population has recently become a highly discussed issue among pharmacists, particularly with special regard to the harmonization of Czech legislation with European Union standards. The lack of relevant and scientific data on the prevalence and attitudes of the population in different regions of the country has been recognized as a hot issue as well, particularly in contrast to the longstanding tradition of medicinal plants collection and use (Novacek, 1998; Drabek and Zeidler, 2004; Prunerova 2006). Knotek *et al.* (2012) decided to carry-out a study to determine (i) which species of herbs are preferred by Czech adult respondents, (ii) what the most important predictors influencing herbal products utilization are and (iii) what are the information sources that influence Czech adult perception of herb utilization. This paper represents the first and unique study in Eastern Europe on the medicinal plants prevalence and use among adult pharmacy clients. The survey evidenced 143 plant species used for herbal products preparation, whereas local species were identified as the most popular ones, *i.e.*, *Mentha × piperita* (18.4 per cent of respondents), followed by *Melissa officinalis* (12.4 per cent), *Plantago lanceolata* (12.2 per cent), *Tilia cordata* (11.8 per cent) and *Matricaria recutita* (10.7 per cent).

In contrast to USA (Planta *et al.*, 2000), Czech adults more likely prefer literature as the major information source about the medicinal plants use and effects (Knotek *et al.*, 2012). Traditional knowledge on herbal medicine is very well documented in the

Czech literature. Books are traditionally very popular in CZ, especially among women, who prefer to use specialised literature to cure family members (Pavela, 2009; Knotek *et al.*, 2012).

Conclusion

European medicinal plants have helped us in the treatment of various diseases since the beginning of mankind. People have kept information about them, about their determination, collection and use, by means of oral traditions and, later and at present, by means of written statements. This review sums up basic information about medicinal plants used in Czech folk medicine, and it presents us with a complete survey of medicinal plants of an ethno botanical origin of the central European region.

The ethno-pharmaco-botanical review strengthens the firm belief that ethno botanical findings represent an important shared heritage, which should not be relegated to a narrow cultural and historical context but should be researched in order to provide new, useful and important knowledge. Today, in fact, many invaluable drugs have found their importance for the present as a result of this research. It is equally clear why a study aimed at discovering more efficient folk medicines must carefully compare every new biologically active compound with the potential healing effects of the plants.

Acknowledgements

This study was supported by grants of the Ministry of Education, Youth and Sports LH11133.

References

Bodlák, J., Severa F., and Vancura B. (1995). Priroda leci – bylinar na konci 20. stoleti. Granit, Pratur, Czech Republic.

Bodlák, J., and Zentrich J.A. (1995). Lécivá síla bodlákù. Fontána, Olomouc, Czech Republic.

Camejo-Rodrigues, J., Ascensão, L., Bonet, M.À., and Vallcs, J. (2003). An ethnobotanical study of medicinal and aromatical plants in the Natural Park of "Serra de São Mamede" (Portugal). *Journal of Etnopharmacology*, **89**: 199-209.

Drabek, P. (1972). The family pharmacy in the year 1595. *Ceskoslovenska Farmacie*, **21**: 197–198.

Drabek, P., and Zeidler, S.K. (2004). (1620–1689) on drugs. *Ceska a Slovenska Farmacie*, **53**: 47–50.

Dostál, J. (1989). Nová kvµtena CSSR, Vol. 1 and 2, Academia, Prague, Czech Republic

Janca, J., and Zentrich, J.A. (1994). Herbár lécivých rostlin. Vol. 1, Eminent, Prague, Czech Republic

Janca, J. (1995). Herbár lécivých rostlin. Vol. 2, Eminent, Prague, Czech Republic.

Janca, J., and Zentrich, J.A. (1995). Herbár léčivých rostlin. Vol. 3, Eminent, Prague, Czech Republic

Janca, J., and Zentrich, J.A. (1996). Herbár léčivých rostlin. Vol. 4. Eminent, Prague, Czech Republic

Janca, J., and Zentrich, J.A. (1997). Herbár léčivých rostlin. Vol. 5. Eminent, Prague, Czech Republic

Janca, J., and Zentrich, J.A. (1998). Herbár léčivých rostlin. Vol. 6., Eminent Prague, Czech Republic

Korbelár, E., and Endris, Z. (1958). Naše rostliny v lékarství. Státní Zdravotnické Nakladatelsví, Prague, Czech Republic

Korbelár, E. (1981). Naše rostliny v lékarství. Avicenium, Prague, Czech Republic

Knotek, K., Verner, V., Chaloupkova, P., and Kokoska, L. (2012). Prevalence and use of herbal products in the Czech Republic: Over-the-counter survey among adult pharmacies clients. *Complementary Therapies in Medicine,* **20**: 199–206

Krkavec, F., and Hájková, I. (1962). Léčivé rostliny Slezska. Krajské nakladatelství v Ostravµ, Ostrava, Czech Republic

Mackù, J., and Krejca, J. (1964). Atlas léčivých rostlin. Vydavatelstvo Slovenskej Akadémi Vied, Bratislava, Slovakia

Novacek, L. (1998). Development of the scientific basis of Czech pharmacy 1939–1945. *Ceska a Slovenska Farmacie,* **47**: 279–283.

Pardo de Santayana, M., Blanco, E., and Morales, R. (2005). Plants know as té in Spain: An etno-pharmaco-botanical rewiew. *Journal of Etnopharmacology,* **98**: 1-19.

Pavela, R. (2009). The plants used in folk medicine in Czech Republic. *In*: Recent Progress in Medicinal Plants, Vol. 25, Ed. By Singh, V. and Govil, J. Studium Press LLC, U.S.A., pp. 291-345.

Pierony, A. (2000). Medicinal plants and food medicines in the folk traditions of the upper Lucca Province, Italy. *Journal of Etnopharmacology,* **70**: 235-273.

Pierony, A., Quave, C.L., and Santoro, R.F. (2004). Folk pharmaceutical knowledge in the territory of the Dolomiti Lucane, inland southern Italy. *Journal of Etnopharmacology,* **95**: 373-384.

Planta, M., Gundersen, B., and Petitt, J.C. (2000). Prevalence of the use of HP in a low-income population. *Family Medicine,* **32**: 252-257.

Prunerova, M. (2006). Development of pharmacy in the CZ. *Casopis Lekaru Ceskych.,* **145**: 815–816.

Schultes, R.E. (1962). The role of the ethnobotanist in the search for new medicinal plants. *Lloydia* **25**: 257-266.

Yesilada, E., Honda, G., Sezik, E., Tabata, M., Goto, K., and Ikeshiro, Y. (1993). Traditional medicine in Turkey IV. Folk medicine in the Mediterranean subdivision. *Journal of Etnopharmacology*, **39**: 31-38.

Yesilada, E., Honda, G., Sezik, E., Tabata, M., Fujita, T., Tanaka, T., Takeda, Y., and Takaishi, Y. (1995). Traditional medicine in Turkey V. Folk medicine in the inner Taurus Mountains. *Journal of Ethnopharmacology*, **46**: 133-152.

Traditional and Folk Herbal Medicine: Recent Researches Vol. 2 (2014) *Pages* **115–132**
Editor: **V.K. Gupta**
Published by: **DAYA PUBLISHING HOUSE, NEW DELHI**

4

Preclinical Investigation of the Antineoplastic Activity of an Ayurvedic Rasayana Drug the Guduchi (*Tinospora cordifolia*) *in vivo*

Ganesh Chandra Jagetia[1*]

ABSTRACT

Cancer curing ability of an ayurvedic rasayana drug guduchi, Tinospora cordifolia (Willd.) Miers ex Hook. F. and Thoms. Family: Menispermaceae was evaluated in preclinical Ehrlich ascites tumor model. Guduchi was extracted in dichloromethane (TCE) and its antineoplastic activity was determined in the mice transplanted with Ehrlich ascites carcinoma (EAC). The EAC mice receiving 25, 30, 40, 50 and 100 mg/kg TCE showed a dose dependent elevation in tumor-free survival and a highest number of survivors was observed at a dose of 50 mg/kg TCE, where the average survival time (AST) and median survival time (MST) were approximately 56 and 55 days, respectively when compared with 19 days of saline treated controls. Administration of 50 mg/kg TCE resulted in 100 per cent long-term survivors (up to 90 days) as compared to none for saline treated controls. The effectiveness of TCE was studied during various stages of tumor development, where 50 mg/kg TCE was administered intraperitoneally after 1, 3, 6, 9, 12 or 15 days of tumor inoculation and these days have been arbitrarily designated as stage I, II, III, IV or V, respectively for reasons of clarity. The TCE was most effective when administered at stage I followed by II and III, where the number of long term survivors was approximately 33, 25 and 17 per cent, respectively. However, treatment of mice at stage IV and V was least effective as it did not increase long term survivors, despite an increase in AST and MST. Biochemical investigations revealed that the treatment of EAC mice with 50 mg/kg TCE caused a time dependent depletion in the glutathione

1 Department of Zoology, Mizoram University, Aizawl – 796 004, Mizoram, India
* *Corresponding author*: E-mail: gc.jagetia@gmail.com

activity up to 12 h post-treatment accompanied by a drastic elevation in lipid peroxidation. However, a maximum elevation in lipid peroxidation was observed at 6 h post-TCE treatment that declined gradually thereafter. TCE exerted cytotoxic effect on tumor cells by reducing the glutathione concentration and increasing lipid peroxidation simultaneously.

Keywords: Guduchi, *Ehrlich ascites carcinoma*, Mice, Survival, Glutathione, Lipid peroxidation.

Introduction

Cancer is one of the major killer diseases in the world and one in four deaths in United State is due to cancer. Despite several advances made in the treatment of cancer, the cancer mortality rates have not come down and remains at the level it were fifty years ago specifically for solid tumors (Jemal *et al.*, 2011). Chemotherapy is a major treatment modality for cancer, especially in the advanced stages and some of active molecules currently used in cancer treatment are derived from plants like *Catharanthus roseus, Podophyllum peltatum, Podophyllum emodii, Taxus brevifolia, Ochrosia elliptica* and *Campototheca acuminata*, have provided active principles which are in clinical use for controlling advanced stages of malignancies (Kinghorn and Balandrin, 1993) before their actual chemical synthesis. Most of these chemotherapeutic agents exhibit severe normal toxicity, resulting in undesirable side effects. Moreover, many of the active molecules sold for the treatment of cancer, are highly expensive, mutagenic, carcinogenic and teratogenic. Hence, there is a need to find alternative drugs, which are highly effective at non-toxic doses, inexpensive and accessible to common man. A need is therefore felt to search newer remedies, which are cheaper economically and do not have severe side effects of the pure compounds.

Medicines derived from plants have played a pivotal role in health care of ancient and modern cultures. Ayurveda, the Indian system of medicine mainly uses plant based drugs or formulations to treat various ailments including cancer. Recent surveys suggest that one in three Americans uses dietary supplements daily and the rate of usage is much higher in cancer patients, which may be up to 50 per cent of patients treated in cancer centers (Richardson *et al.*, 2000). The popularity for use of herbal medicines by general public is due to the belief that botanicals will provide some measure of benefit over and above modern allopathic medical approaches. They are also considered non-toxic or less toxic than the synthetic molecules. There is also the sense that taking supplements will allow some measure of self-choice in medical care (Eisenberg *et al.*, 1998).

Tinospora cordifolia (Willd.) Miers ex Hook. F. and Thoms is a large, glabrous, deciduous climbing shrub belonging to the family Menispermaceae (Anonymous, 1976; Nadkarni and Nadkarni, 1976). It is distributed throughout tropical Indian subcontinent and China, ascending to an altitude of 300 m. The plant is commonly known as Guduchi, Giloy, or Giloe, Amritha, which are Hindu mythological terms that refer to the heavenly elixir that have saved celestial beings from old age and kept them eternally young. Guduchi is widely used in veterinary folk medicine and Ayurvedic system of medicine for its general tonic, antiperiodic, anti-spasmodic, anti-inflammatory, antiarthritic, anti-allergic and anti-diabetic properties (Chopra *et al.*, 1956; Nadkarni and Nadkarni, 1976; Khosa and Prasad, 1971; Zhao *et al.*, 1991). The whole plant is used in Ayurvedic "Rasayanas" to improve the immune system and the body resistance against infections and root is known for its antistress, antileprotic and anti-malarial activities (Zhao *et al.*, 1991; Nayampalli, 1982). Guduchi has been reported to be active against throat cancer in man (Chauhan, 1995) and it has been reported to be non-toxic in acute toxicity studies, with almost no side effects (Anonymous, 1992; Jagetia *et al.*, 2002).

A variety of constituents have been isolated from *T. cordifolia* and their structures elucidated. They belong to different classes such as alkaloids, diterpenoid lactones, glycosides, steroids, sesquiterpenoid, phenolics, aliphatic compounds and polysaccharides. Leaves of *T. cordifolia* are rich in protein (11.2 per cent) and are fairly rich in calcium and phosphorus (Kumar *et al.*, 2000; Zhao *et al.*, 1991). Alkaloids like berberine, palmatine, tembetarine and magnoflorine have been isolated from the stem of *T. cordifolia*. The roots of *T. cordifolia* are also reported to contain other alkaloids like choline, tinosporin, isocolumbin, palmatine, tetrahydropalmatine and magnoflorine (Qudrat-I-Khuda *et al.*, 1964; Pachaly and Schneider, 1981; Bisset and Nwaiwu, 1983; Chi *et al.*, 1994; Padhya, 1986; Sarma *et al.*, 1998). Our preliminary studies on the stem extracts of *Tinospora cordifolia* have shown promising response in cultured human cancer cells, where various extracts of guduchi were found to reduce cell survival in a dose dependent manner. However, dichloromethane extract was found to be the most promising one and has been found to be non-toxic up to 1.2 g/ b. wt. (Jagetia *et al.*, 1998, 2002; Jagetia and Rao, 2006). Therefore, the present study was undertaken to investigate the antineoplastic activity of dichloromethane extract of *T. cordifolia* in mice transplanted with Ehrlich ascites carcinoma.

Materials and Methods

Collection and Extraction of Plant Material

The identification of the plant *Tinospora cordifolia* [Willd.) Miers ex Hook. F. and Thoms. (Family: Menispermaceae) was done by Prof. U. Rajagopal (a well known taxonomist of this area) Department of Botany, Mahatma Gandhi Memorial College, Udupi, India and the herbarium specimen (RB-TC 02) has been stored with us. The stems were collected in the month of March locally, shade dried and coarsely powdered with the help of a ball mill. The plant material (20 kg) was exhaustively extracted with 5 liters each of petroleum ether (60-80°C), chloroform and dichloromethane, respectively using a Soxhlet continuous extraction apparatus. The final dichloromethane extract (henceforth TCE) was concentrated *in vacuo* and dried under reduced pressure. An approximate yield of 1.2 per cent w/w was obtained.

Animal Care and Handling

The animal care and handling were done according to the guidelines set by the World Health Organization, Geneva, Switzerland, INSA (Indian National Science Academy, New Delhi, India) and the "Guide for the care and use of Laboratory Animals" (NIH publication #86-23, revised in 1985). Ten to twelve weeks old female Swiss albino mice weighing 30 to 36 g were selected from an inbred colony maintained under the controlled conditions of temperature (23 ± 2°C), humidity (50 ± 5 per cent) and light (14 and 10 h of light and dark, respectively). The animals had free access to the sterile food and water. Four animals were housed in a sterile polypropylene cage containing sterile paddy husk (procured locally) as bedding throughout the experiment. The study was approved by the Institutional Animal Ethical Committee of Kasturba Medical College, Manipal, India, where the studies were carried out.

Tumor Model

Ehrlich ascites carcinoma (EAC) procured from the Cancer Research Institute (ACTREC), Mumbai, India, was maintained and propagated by serial transplantation intraperitoneally in an aseptic environment. 10^6 viable EAC cells were injected intraperitoneally into each animal in an aseptic condition and the day of tumor inoculation was considered as day 0.

Preparation of Drug and Mode of Administration

The TCE was dissolved in 100 µl of ethanol and diluted with sterile physiological saline (SPS) containing 1 per cent carboxy methyl cellulose (CMC). Cyclophosphamide (CPA), which was used as a positive control was dissolved in sterile saline containing 1 per cent CMC. The SPS, CPA or TCE were administered intraperitoneally unless otherwise stated. All the drugs were freshly prepared immediately before use.

Selection of Optimum Dose

The dose of TCE was selected following the standard protocol recommended by the Drug Evaluation Branch, Drug Research and Development, NIH, USA (Geran *et al.*, 1972). Twenty-four hours after tumor inoculation, the animals were divided into the following groups:

SPS Control

The animals of this group received 0.3 to 0.36 ml of SPS containing 1 per cent CMC once daily, consecutively for 9 days.

CPA Control

This group of animals was injected with 25 mg/kg b. wt. of CPA (Geran *et al.*, 1972) once daily, consecutively for 9 days and served as the positive control.

TCE Group

The animals of this group were administered with 25, 30, 40, 50 and 100 mg/kg TCE once daily, consecutively for 9 days.

Stage Specific Evaluation

A separate experiment was carried out to evaluate the antineoplastic action of TCE during various stages of tumor development, where tumors were inoculated and allowed to grow for 1, 3, 6, 9, 12 and 15 days and for reasons of clarity these days have been arbitrarily designated as stage I, II, III, IV, V and VI, respectively. The animals in the above stages of development of tumor were divided into the following groups:

CPA Group

This group of animals received 25 mg/kg b. wt. of CPA, once daily for nine consecutive days at stage I, II, III, IV, V or VI of tumor development and served as concurrent positive control.

TCE Group

The animals of this group were administered with a single dose of 50 mg/kg TCE (the optimum dose) once daily, for nine consecutive days at stage I, II, III, IV, V or VI of tumor development.

The animals of all the experiments were monitored regularly for alteration in body weight, signs of toxicity and mortality. The weight of animals was recorded every third day up to 30 days after tumor inoculation in all the groups. A 33 per cent of drug related deaths or a weight loss of 5 g per mouse was considered as an index of toxicity (Geran *et al.*, 1972). The animal survival was monitored daily up to 120 days, since the survival of animals up to 120 days is roughly equivalent to 5 years survival in man (Nias, 1990). The tumor response was assessed on the basis of median survival time and tumor free survival. The MST and the AST were calculated from the animals dying within 120 days and those surviving 120 days were excluded from it. The MST and AST was calculated as follows

MST = First death + last death in the group/2

AST = Sum of animal death on different days/Number of animals

The increase in median life span (per cent IMLS) and increase in average life span (per cent IALS) were also calculated using the following formulae:

$$\text{IMLS} = \frac{\text{MST of treated mice} - \text{MST of control}}{\text{MST of control}} \times 100$$

$$\text{IALS} = \frac{\text{AST of treated mice} - \text{AST of control}}{\text{AST of control}} \times 100$$

Biochemical Analyses

A separate experiment was carried out to estimate glutathione and lipid peroxidation in the tumor cells. The animals were inoculated with tumor cells as described above and the tumor was allowed to grow for six days so as to get a reasonable volume for aspiration of cells. On seventh day, the tumor bearing animals were divided into the following groups:

SPS Control

The animals of this group received 0.3 to 0.36 ml of SPS containing 1 per cent CMC on day 7 post-tumor inoculation.

CPA Control

This group of animals was injected once with 25 mg/kg b. wt. of CPA on day 7 post-tumor inoculation.

TCE Group

The animals of this group were administered with a single injection of 50 mg/kg b. wt. of TCE on day 7 post-tumor inoculation.

Four animals from each group were sacrificed at 1.5, 3, 6, 9, 12, 18 or 24 h after the SPS, CPA or TCE administration. The tumor cells were aspirated in an aseptic condition and were washed with SPS thrice. The cells were counted under an inverted microscope (Labovert microscope, Ernst, Leitz GmbH, Wetzlar, Germany) and 1×10^6 cells were disrupted using an ultrasonicator (Virsonic, Virtis, NY, USA) and processed for the estimation of glutathione and lipid peroxidation assays as follows:

Glutathione (GSH)

GSH concentration was measured by the method of Moron *et al.* (1979). Briefly, proteins were precipitated by 25 per cent trichloroacetic acid (TCA), centrifuged and the supernatant was collected. The supernatant was mixed with 0.2 M sodium phosphate buffer pH 8.0 and 0.06 mM 5, 5'-dithiobis-(2)-nitrobenzoic acid (DTNB) and incubated for 10 minutes at room temperature. The absorbance of the sample/s was read against the blank at 412 nm in a UV-Visible double beam spectrophotometer (Shimadzu UV-260, Shimadzu Corp, Tokyo, Japan) and the GSH concentration was calculated from the standard curve.

Lipid Peroxidation (LPx)

LPx (TBARS) was estimated by the method of Beuege and Aust (1978). Briefly, the homogenate was mixed with TCA-TBA (thiobarbituric acid) -HCl and heated for 15 min in a boiling water bath. After centrifugation the absorbance was recorded at 535 nm using a UV-Visible double beam spectrophotometer. The lipid peroxidation has been expressed as thiobarbituric acid reactive substances (TBARS) in nmol per 10^6 cells. The concentration of LPx in the sample was determined against the standard curve of malonaldehyde.

Statistical Analysis

The statistical significance between the treatments was determined using the "Z" test for the survival studies (Abramowitz and Stegun, 1972), whereas student's '*t*' test was used for the biochemical estimations. A p value of <0.05 was considered statistically significant. All the data are expressed as mean ± SEM (standard error of the mean).

Results

The results are expressed in Figures 4.1 to 4.3 and Tables 4.1–4.2.

Selection of Optimum Dose

Transplanted EAC did not show spontaneous regression of tumor in transplanted mice of SPS group. The tumor bearing mice showed a constant weight gain and increase in the tumor volume due to cell multiplication and growth of EAC cells (Figure 4.1). The MST was found to be approximately 19 days, whereas the AST was 18 days for the SPS group (Table 4.1). Treatment of mice 24 h after EAC inoculation with 25 mg/kg CPA increased the MST and AST up to 29 and 28 days, respectively. This was accompanied by an increase in the IMLS and the IALS, which were found to be approximately 58 per cent and 56 per cent, respectively (Table 4.1).

Table 4.1: Effect of various doses of dichloromethane extract of guduchi (TCE) on Ehrlich ascites carcinoma in mice.

Treatment	*Survival*				*Percent Survival (day)*			
	MST	*IMLS*	*AST*	*IALS*	*30*	*60*	*90*	*120*
SPS	19		18	–	0	0	0	0
CPA 25	29	57. 9	28	55.8	16.7	0	0	0
TCE 25	21.5	13.1	20.5	14	0	0	0	0
TCE 30	26	36.8	25.7	43.3	0	0	0	0
TCE 40	35.5	86.8	34.9	94.3	91.7	66.7	0	0
TCE 50	55.5	192.1	56.4	213.3	100 [a]	100 [a]	100 [a]	33.3 [b]
TCE 100	26	36.8	25.7	43.3	0	0	0	0

a = $p < 0.0001$; b = $p < 0.05$ (when compared with SPS)

SPS: Sterile Physiological Saline; CPA: Cyclophosphamide; TCE: Dichloromethane extract of *Tinospora cordifolia.*

Treatment of tumor bearing mice with TCE resulted in a dose dependent increase in the MST and AST. The greatest effect was observed for the animals treated with 30, 40 or 50 mg/kg TCE, where the MST increased approximately up to 26, 35 and 53 days, respectively when compared with the SPS treated group (19 days). The AST was also elevated approximately up to 26, 35 and 56 days for 30, 40 and 50 mg/kg TCE, respectively, whereas the least effect was observed for 25 mg/kg TCE where MST and AST was found to be approximately 21and 20 days, respectively. This increase in AST and MST was non significant when compared with SPS group. The MST increased approximately by 6 and 26 days in 40 and 50 mg/kg treated animals, respectively than CPA treatment (Table 4.1). Of all the TCE doses tested, the highest anticancer activity was observed for 50 mg/kg, where 100 per cent healthy survivors were observed at the end of 110 days, whereas 33 per cent survivors were observed at the end of 120 days (Table 4.1). Treatment of EAC mice with 50 mg/kg TCE increased MST approximately by 34, 29, 20, 29 and 29 days when compared to 25, 30, 40 and 100 mg/kg of TCE and 25 mg/kg CPA respectively, whereas the AST elevated approximately by 36, 31, 21, 31 and 28 days, respectively. The treatment of mice 24 h after EAC inoculation with 25, 30, 40, 50 or 100 mg/kg TCE arrested the weight gain at all TCE doses indicating inhibition of tumor cell proliferation and growth (Figure 4.1). The administration of 100 mg/kg TCE was accompanied by toxic side effects like ruffling of hair, sluggishness and lacrimation in the recipients and none of the animals survived beyond day 6 post-tumor inoculation. Therefore, 50 mg/kg TCE was considered as an optimum dose, which resulted in the highest number of long-term survivors. It also did not induce any toxic effects in the form of debility, loss of body weight and death, hence further studies were carried out using this dose.

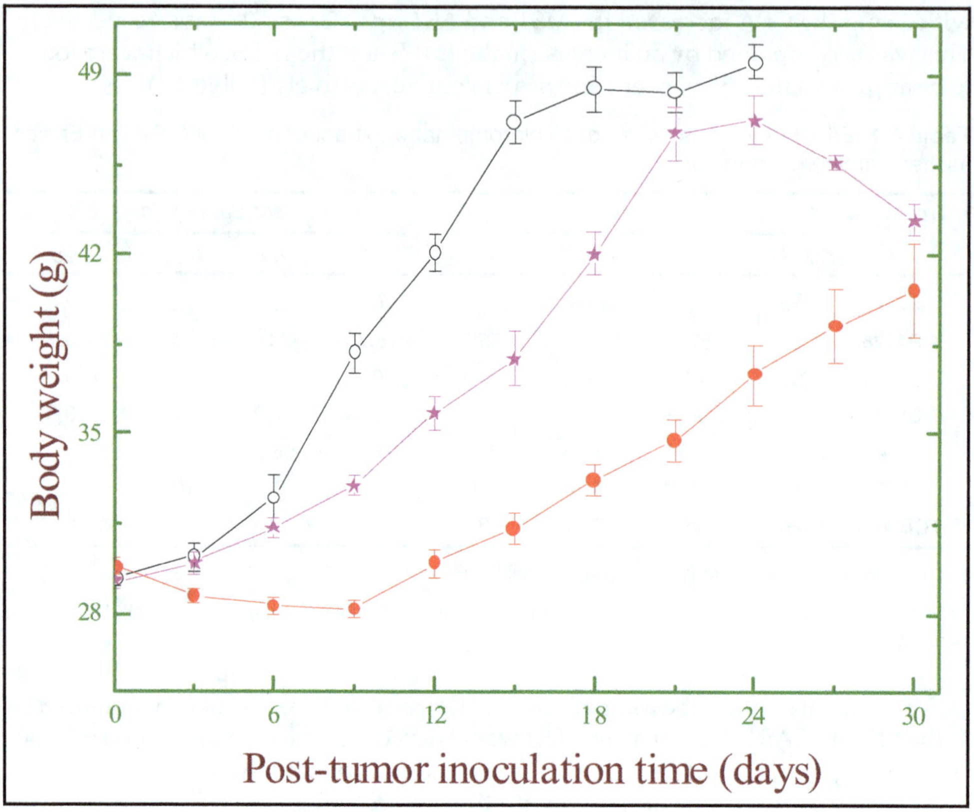

Figure 4.1: Effect of 50 mg/kg of dichloromethane extract of *Tinospora cordifolia* (TCE) in comparison with 25 mg/kg Cyclophosphamide (CPA) on the body weight of mice bearing Ehrlich ascites carcinoma. open circles: SPS alone, closed circles: CPA alone and Stars: TCE alone.

SPS: Sterile Physiological Saline; CPA: Cyclophosphamide; TCE: Dichloromethane extract of guduchi.

Stage Specific Evaluation

The evaluation of the anticancer activity of TCE was also carried out in tumor bearing animals at stage I, II, III, IV, V and VI, respectively by administering a single dose of 50 mg/kg TCE once daily or 25 mg/kg CPA for nine consecutive days, at stage I, II, III, IV, V or VI. The administration of 25 mg/kg CPA exerted a significant anticancer activity only when administered in the early stages of tumor development, which is indicated by the body weight changes (Table 4.2). None of the animals treated with CPA at various stages of tumor development survived beyond 36 days post-EAC inoculation (Table 4.2).

The survival of EAC mice declined in a stage specific manner. The MST for the tumor bearing animal that received CPA at stages I, II, III, IV, V and VI was approximately 29, 26, 24, 21, 19 and 18 days, respectively (Table 4.2), whereas AST was approximately 29, 26, 24, 21, 19 and 18 for stage I, II, III, IV, V and VI respectively. An IMLS, of 58, 39, 26 and 13 per cent was observed for stages I, II, III and IV respectively. The administration of CPA at the late stages (*i.e.* V and VI) proved ineffective in increasing the IMLS. The IALS decreased in a stage specific manner in the CPA group which was approximately 56, 47, 28, 17, 5 and 1 per cent, for stage I, II, III, IV, V and VI, respectively (Table 4.2).

Table 4.2: Effect of 50 mg/kg dichloromethane extract of guduchi (TCE) on the various stages of tumor regression in mice transplanted with Ehrlich ascites carcinoma.

Treatment	*Survival*				*Percent Survival (day)*			
	MST	*IMLS*	*AST*	*IALS*	*30*	*60*	*90*	*120*
SPS	19		18	–	0	0	0	0
TCE I	55.5	192.1	56.4	213.3	100[a]	50[b]	33.3[d]	33.3[d]
CPA 1	29	57.9	28	55.8	16.7	0	0	0
TCE II	48	152.6	42.7	137.2	91.7[a]	66.7[c]	25[d]	16.7
CPA II	26.5	39.5	26.3	46.6	8.3	0	0	0
TCE III	43.5	128.9	39	116.7	66.7[c]	33.3[d]	16.7	0
CPA III	24	26.3	23.1	28.5	0	0	0	0
TCE IV	35	84.2	30.6	70	25	0	0	0
CPA IV	21.5	13	21.1	17.4	0	0	0	0
TCE V	23.5	23.7	23.9	32.8	8.3	0	0	0
CPA V	19	–	18.9	5.3	0	0	0	0
TCE VI	20	5.3	19.7	9.9	0	0	0	0
CPA VI	18.5	–	18.2	1.6	0	0	0	0

a = $p < 0.0001$; b = $p < 0.001$; c = $p < 0.002$; d = $p < 0.05$ (when compared with SPS)

SPS: Sterile Physiological Saline; **CPA**: Cyclophosphamide; **TCE**: dichloromethane extract of *Tinospora cordifolia.*

Treatment of mice with 50 mg/kg TCE at stages I, II, III, IV, V or VI resulted in an increase in the MST approximately up to 55, 48, 43, 35, 23 and 20 days, whereas AST elevated to approximately 56, 43, 39, 31, 24 and 20 days, respectively, when compared with CPA treatment (Table 4.2). Despite an increase in IMLS and IALS after TCE treatment, a stage-specific decline was evident in IMLS and IALS and the lowest values were observed for stage VI. The IMLS of approximately 192, 153, 129, 84, 24 and 5 per cent and the IALS of approximately 213, 137, 117, 70, 33 and 10 per cent were recorded for stage I, II, III, IV, V and VI, respectively after 50 mg/kg TCE treatment (Table 4.2).

TCE was effective in arresting the weight gain in the animals due to remission of tumor, especially during the early stages of tumor growth. TCE was found to be efficient even in the mid stages of tumor progression in comparison with CPA. The use of TCE even during the late stages of tumor development proved effective in reducing the tumor burden and weight gain than that of CPA treatment (Figure 4.1).

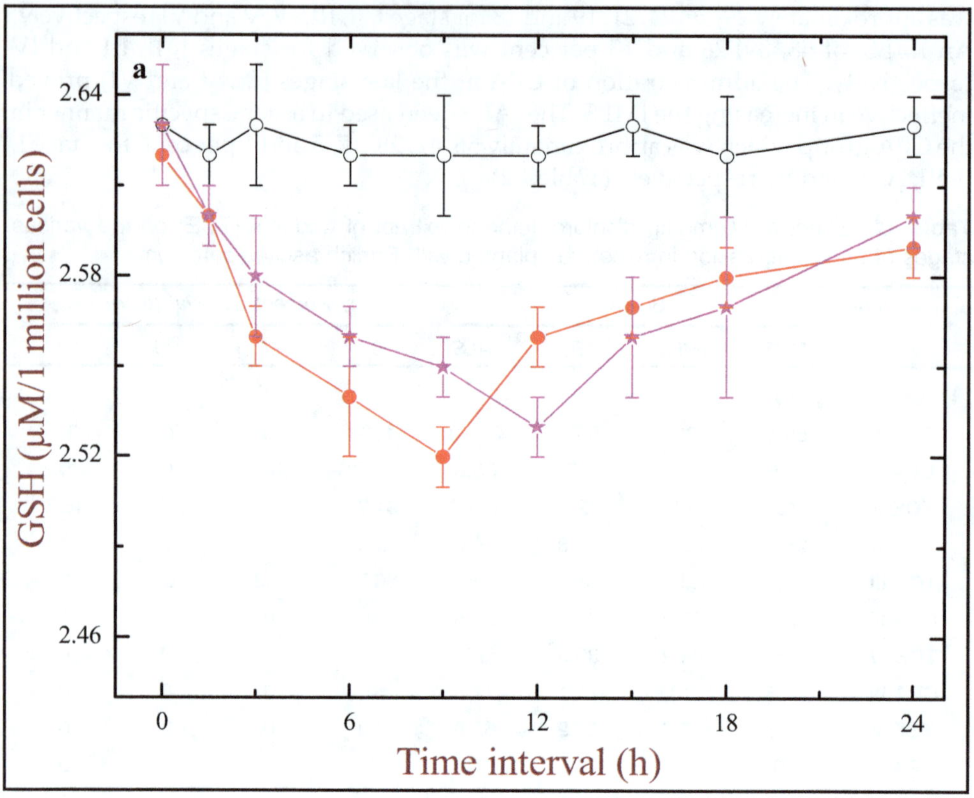

Figure 4.2: Effect of the treatment of 50 mg/kg of dichloromethane extract of Tinospora cordifolia (TCE) in comparison with cyclophosphamide (CPA) on the alteration in the Glutathione content Ehrlich ascites carcinoma. SPS (open circles); CPA (closed circles) and TCE (Stars).

SPS: Sterile Physiological Saline; CPA: Cyclophosphamide; TCE: dichloromethane extract of guduchi.

The administration of TCE was better than CPA as 33 and 17 per cent survivors ($p < 0.05$) were observed at the end of 120 days for stage I and III respectively, while no survivors were reported for CPA group. There was an increase of 26, 29, 31, 34, 36 and 37 days approximately in MST, while the AST elevated approximately by 28, 30, 33, 35, 37 and 38 days for stage I, II, III, IV, V and VI, respectively, when compared with the CPA group (Table 4.2).

Biochemical Analyses

Glutathione (GSH)

The GSH concentration of EAC cells, remained unaltered with time in EAC mice treated with SPS. Treatment of EAC mice with CPA treatment caused a time dependent decline in the GSH concentration in EAC cells until a nadir was reached at 9 h post-treatment. A further increase in the assay time caused a significant elevation at 12 h post-treatment, thereafter a steady state was attained (Figure 4.2). Similarly, TCE treatment showed a steady decline in GSH concentration in EAC cells and a nadir was reached at 12 h post-treatment. Thereafter the GSH concentration showed a gradual elevation up to 24 h. However, the concentration of GSH was 2.6 folds lower than the SPS group (Figure 4.2).

Lipid Peroxidation (LPx)

The baseline levels of lipid peroxidation remained unaltered with assay time in EAC mice treated with SPS. Treatment of EAC mice with CPA caused a time dependent increase in the LPx in EAC cells till a peak was reached at 12 h post-treatment. A further increase in the assay time caused a significant and gradual decline in the lipid peroxidation up to 24 h post-treatment (Figure 4.3). The pattern of elevation of LPx after TCE treatment was similar to that of CPA treatment except that the peak elevation was observed at 6 h post-treatment and LPx was higher in TCE group than that of CPA group (Figure 4.3).

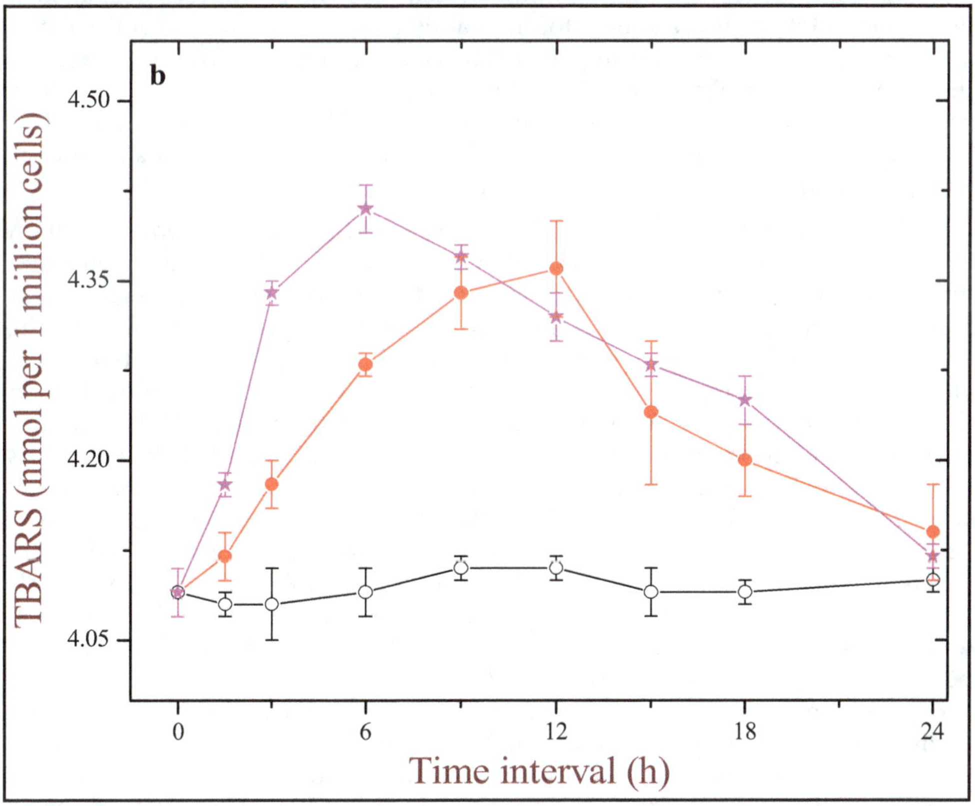

Figure 4.3: Effect of the treatment of 50 mg/kg of dichloromethane extract of Tinospora cordifolia (TCE) in comparison with cyclophosphamide (CPA) on the alteration in the Lipid peroxidation in mice bearing Ehrlich ascites carcinoma. SPS (open circles); CPA (closed circles) and TCE (Stars).

SPS: Sterile Physiological Saline; CPA: Cyclophosphamide; TCE: dichloromethane extract of guduchi.

Discussion

Humans have always relied on nature for survival since ancient times, which has been their main source of food, protection, clothing, transportation and medicines (Farnsworth *et al.*, 1985; Cragg *et al.*, 1997). Natural products have been regarded as important sources that could produce potential chemotherapeutic agents (Kim and Park, 2002). Plant-derived compounds, in particular have a special place in anticancer therapy and some of the new chemotherapeutic agents currently available for use in a clinical setting including paclitaxel, vincristine, podophyllotoxin and camptothecin have been derived initially from various palnts (Gerzon, 1980; Jardine, 1980; Kinghorn and Balandrin, 1993; Wall and Wani, 1993). Obviously natural products are extremely important as sources of medicinal agents. Although there are some new approaches to drug discovery, such as combinatorial chemistry and computer-based molecular modeling design, none of them can replace the importance of natural products in drug discovery and development. The whole plant of guduchi is used in Ayurveda as "Rasayanas" to improve the immune system and the body resistance against infections and its roots are known for their antistress, antileprotic and antimalarial activities (Zhao *et al.*, 1991; Nayampalli, 1982). It has been reported to be active against throat cancer in man (Chauhan, 1995). Guduchi has been reported to be non-toxic in acute toxicity studies, with almost no side effects. Therefore, it was desired to investigate the antineoplastic activity of *Tinospora cordifolia* in Swiss albino mice transplanted with Ehrlich ascites carcinoma.

The administration of various doses of TCE caused a dose dependent retardation in the tumor development, as is evident by the inhibition in body weight gain and increase in the life span (MST and AST). This indicates to the effectiveness of TCE in restricting tumor cell multiplication and growth. The reports regarding the antineoplastic action of dichloromethane extract of *T. cordifolia in vivo* are lacking. However, it has been reported to reduce clonogenic survival in cultured HeLa cells in a dose dependent manner earlier (Jagetia *et al.*, 2002; Jagetia and Rao, 2006). A similar effect has been reported earlier in mice treated with the extract of *Alstonia scholaris* (Jagetia and Baliga, 2003, 2004, 2005). Other Indian medicinal plants like *Ervatamia heyncana, Rubia cordifolia, Tylophora indica, Withania somnifera, Hygrophila spinosa* and *Podophyllum hexandrum* have been reported to possess antineoplastic activity in different tumor models *in vivo* and *in vitro* (Chitnis *et al.*, 1971, 1972; Adwankar *et al.*, 1980; Mazumdar *et al.*, 1997; Eisenberg *et al.*, 1998; Goel *et al.*, 1998). Of all the TCE doses tested, the highest anticancer activity was observed for 50 mg/kg, where 100 per cent healthy survivors were observed at the end of 110 days, while 33 per cent survivors were observed at the end of 120 days. In contrast, no survivors were reported for the SPS or CPA groups at the end of 120 days. Similarly, *Tylophora indica, Withania somnifera, Alstonia scholaris* and *Aegle marmelos* have been reported to exert an optimal activity only at a particular drug dose beyond which, it was either ineffective or toxic (Goel *et al.*, 1998; Jagetia and Baliga, 2003, 2004, 2005; Jagetia *et al.*, 2005).

Humans suffering from various neoplastic disorders come for clinical evaluation and treatment in the different stages of tumor development. Therefore, an effective anticancer agent should be able to kill tumor cells efficiently during any stage of the tumor. An effort has been made to screen the antineoplastic activity of 50 mg/kg TCE at various stages of tumor development with respect to the efficacy of 25 mg/kg of CPA. The results from the stage specific evaluation show that both TCE and CPA retarded the increase in the body weight gain in animals due to tumor development and increased the survival during the early stages effectively. This may be due to the effective killing of EAC cells during the early stages of tumor development by TCE, where the tumor burden is less when compared to late stages. TCE was effective even during mid and late stages of tumor development, while CPA treatment proved ineffective. This was evident by increased MST and AST during various stages of tumor development, when compared with CPA treatment. The effectiveness of TCE was better than CPA, where 33.3 and 16.7 per cent survivors were observed at the end of 120 days for stage I and III respectively, while no survivors were reported for CPA treatment. The studies of the anticancer activity of plants at different stages of tumor development are scanty. However, the extracts of *Alstonia scholaris* and *Aegle marmelos*, have been reported to inhibit EAC growth in various stages of tumor development (Jagetia and Baliga, 2003; Jagetia *et al.*, 2005).

T. cordifolia has been found to contain alkaloids, diterpenoid lactones, glycosides, steroids, sesquiterpenoid, phenolics, aliphatic compounds and polysaccharides. Alkaloids like berberine, palmatine, tembetarine and magnoflorine have been isolated from the stem of *T. cordifolia*. The roots of *T. cordifolia* are also reported to contain other alkaloids like berberine, choline, tinosporin, isocolumbin, palmatine, tetrahydropalmatine and magnoflorine (Qudrat-I-Khuda *et al.*, 1964; Pachaly and Schneider, 1981; Bisset and Nwaiwu, 1983; Chi *et al.*, 1994; Padhya, 1986; Sarma *et al.*, 1998). *Tinospora cordifolia* has also been found to contain an immunomodulator arabinoglycan (Chintalwar *et al.*, 1999). The anticancer activity exhibited by TCE can be attributed to the presence of berberine. Berberine has been shown to have an anticancer effect on mice bearing Ehrlich ascites carcinoma at a dose of 10 mg/kg b. wt. (Jagetia and Baliga, 2004). Berberine has been reported to reduce the *in vitro* growth of brain tumor cells, teratocarcinoma cells and HepG2 cells as well as also inhibition of the secretion of alpha-fetoprotein by HepG2 cells (Williamson, 2001). Even though berberine is one of the major alkaloids present in the stems of *T. cordifolia*, it can be inferred that there may be other important constituents in the dichloromethane extract, which might have contributed to the enhanced antineoplastic action of TCE than the positive control CPA.

The exact mechanism of action of TCE is not known. The antineoplastic activity of guduchi cannot be attributed to a single mechanism but several putative mechanisms may be operational simultaneously for effective tumor cell kill. TCE may have caused DNA damage in the EAC cells leading to the cytotoxic effect. This contention is supported by our earlier study where TCE has been found to increase micronuclei in a concentration dependent manner in HeLa cells (Jagetia *et al.*, 2002). Recently, it has been reported to increase molecular DNA damage (Jagetia and Rao, 2011). The reduced GSH contents and increased lipid peroxidation by TCE may also have contributed to the killing of EAC cells. Increased concentration of GSH in the tumor cells have been reported to make the tumor refractory to treatment, while depletion of glutathione has been reported to enhance the cell death and apoptosis of the tumor cells along with the loss of essential sulfhydryl groups that result in an alteration of the calcium homeostasis and eventually loss of cell viability (Axelson and Mannervik, 1983; Neal *et al.*, 2003). The lipid peroxidation is another important event related to cell death and has been reported to cause severe impairment of membrane function through increased membrane permeability and membrane protein oxidation and eventually cell death by damaging the cellular DNA (Bartsch and Nair, 2002; Marnett, 2002). The changes in the membrane fluidity coupled with the damage to DNA by TCE may have been responsible for the killing of tumor cells and prolongation of life span in EAC mice. The cytotoxic effect of TCE may be due to inhibition of topoisomerase II. TCE contains berberine that has been found to be a topoisomerase II inhibitor (Makhey *et al.*, 1995). This inhibiton of topoisomerase II would have not allowed any repair of the damaged DNA causing greater cell kill and effective tumor regression. Nuclear factor kappa B (NF-κB) consists of various dimeric complexes of Rel protein family and has been found to be constitutively expressed in several cancers (Sen, 2011). Guduchi may have inhibited the transactivation of NF-κB causing better tumor remission. Recently, berberine, an isoquinoline alkaloid and other phytochemicals present in the dichloromethane extract of guduchi have been reported to suppress the transactivation of NF-κB (Pandey *et al.*, 2008). Similarly, octacosanol present in the stem of guduchi has also been reported to suppress the transactivation of NF-κB (Thippeswamy *et al.*, 2008). Guduchi may have also suppressed the activation of COX-II and LOX-5 mRNA causing higher cell kill. Alcoholic extract of guduchi has been reported to inhibit their activities *in vitro* (Li *et al.*, 2004).

Conclusions

From our study it is clear that the TCE is a composite mixture of various alkaloids including berberine. The presence of other alkaloids may have a combinatorial effect on the antineoplastic action of TCE. It may have exerted its cytotoxic effects by reducing tumor glutathione concentration, increase lipid peroxidation, damaging DNA, inhibiting the transactivation of NF-κB, COX-II and LOX5 mRNAs. Encouraging results of pre-clinical and clinical trials of this extract and its other constituents could become a part of standard cancer treatment protocols and a powerful weapon for the treatment of cancer.

References

Abramowitz, M., and Stegun, I. A. (1972). Handbook of Mathematical Functions. Dover Publications, Inc. New York, Library of Congress Catalog card number 65-12253, pp. 925.

Adwankar, M. K., Chitnis, M. P., Khandalekar, D. D., and Bhadsavale, C. G. (1980). Anti-cancer activity of the extracts of *Rubia cordifolia* Linn. (NSC b668893). *Indian Journal of Experimental Biology*, **18(1):** 102.

Anonymous (1976) Wealth of India: Raw materials, Vol X. New Delhi: CSIR.

Anonymous (1992). Selected medicinal plants of India (A monograph of identity, safety), CHEMEXCIL Swami Prakashananda Ayurveda Research Center,

Bombay. Basic Pharmaceutical and Cosmetic Export Promotion Council. Bombay, India. pp. 319-322.

Axelson, K., and Mannervik, B. (1983). An essential role of cytosolic thioltransferase in protection of pyruvate kinase from rabbit liver against oxidative inactivation. *Federation of European Biochemical Societies Letters*, **152**: 114-118.

Bartsch, H., and Nair, J. (2002). Potential role of lipid peroxidation derived DNA damage in human colon carcinogenesis: studies on exocyclic base adducts as stable oxidative stress markers. *Cancer Detection and Prevention*, **26**: 308-312.

Bisset, N. G., and Nwaiwu, J. (1983). Quaternary alkaloids of *Tinospora species*. *Planta Medica*, **48**: 275-279.

Buege, J. A., and Aust, S. D. (1978). Microsomal lipid peroxidation. *Methods in Enzymolology*, **52**: 302-310.

Chauhan, K. (1995). Successful treatment of throat cancer with Ayurvedic drugs. *Suchitra Ayurved*, **47**: 840-842.

Chi, C. W., Chang, Y. F., Chao, T. W., Chiang, S. H., Eng, F. K. P., Liu, W. Y., and Liu, T. Y. (1994). Flowcytometric analysis of the berberine on the expression of glucocorticoid receptors in human hepatoma HepG2 cells. *Life Sciences*, **54**: 2099-2107.

Chintalwar, G., Jain, A., Sipahimalani, A., Banerji, A., Sumariwalla, P., Ramakrishnan, R., and Sainis, K. (1999). An immunologically active arabinogalactan from *Tinospora cordifolia*. *Phytochemistry*, **52(6):** 1089-1093.

Chitnis, M. P., Khandalekar, D. D., Adwankar, M. K., and Sahasrabudhe, M. B. (1971). Anticancer activity of the extracts of root, stem and leaf of *Ervatamia heyneana*. *Indian Journal Experimental Biology*, **9(2):** 268-270.

Chitnis, M. P., Khandalekar, D. D, Adwankar, M. K., and Sahasrabudhe, M. B. (1972). Anti-cancer activity of the extracts of stem and leaf of *Tylophora indica*. *Indian Journal of Medical Research*, **60(3):** 359-362.

Chopra, R. N., Nayar, S. L., and Chopra, I. C. (1956). Glossary of Indian Medicinal plants. New Delhi: Council for Scientific and Industrial Research, pp 21.

Cragg, G. M., and Newman, D. J. (1999). Discovery and development of antineoplastic agents from natural sources. *Cancer Investigation*, **17(2)**: 153-163.

Cragg, G. M., Newman, D. J., and Snader, K. M. (1997). Natural products in drug discovery and development. *Journal of Natural Products*, **60**: 52-60.

Eisenberg, D. M., Davis, R. B., Ettner, S. L., Appel, S., Wilkey, S., Van Rompay, M., and Kessler, R. C. (1998). Trends in alternative medicine use in the United States, 1990–1997. *Journal of American Medical Association*, **280**: 1569–1575.

Farnsworth, N. R., Akerele, O., Bingel, A. S., Soejarto, D. D., and Guo, Z. (1985). Medicinal plants in therapy. *World Health Organization Bulletin*, **63**: 965-981.

Geran, R. I., Greenberg, N. H., Mac Donald, M. M., Schumacher, A. M., and Abbott, B. J. (1972). Protocols for screening chemical agents and natural products against animal tumors and other biological systems. *Cancer Chemotherapy Reports Part, 3.* **2(3):** 25-27.

Gerzon, K. (1980). In Anticancer agents Based on Natural Product Models. Cassady, J. M., Douros, J. D. (eds.) Academic Press, pp. 271 – 317.

Goel, H. C., Prasad, J., Sharma, A., and Singh, B. (1998). Antitumour and radioprotective action of *Podophyllum hexandrum. Indian Journal of Experimental Biology*, **36(6):** 583-587.

Jagetia, G. C., and Baliga, M.S. (2003). Modulation of antineoplastic activity of cyclophosphamide by *Alstonia scholaris* in the Ehrlich ascites carcinoma-bearing mice. *Journal of Experimental Therapeutics and Oncology*, **3(5):** 272-82.

Jagetia, G. C., and Baliga, M. S. (2004). Effect of *Alstonia scholaris* in enhancing the anticancer activity of berberine in the Ehrlich ascites carcinoma bearing mice. *Journal of Medicinal Food*, **7(2):** 235-244.

Jagetia, G.C., and Baliga, M. S. (2005). The effect of seasonal variation on the antineoplastic activity of *Alstonia scholaris* R. Br. in HeLa cells. *Journal of Ethnopharmacology*, **96(1-2):** 37-42.

Jagetia, G. C., Nayak, V., and Vidyasagar, M. S. (1998). Evaluation of the antineoplastic activity of guduchi (*Tinospora cordifolia*) in cultured HeLa cells. *Cancer Letters*, **127(1-2):** 71-82.

Jagetia, G. C., Nayak, V., and Vidyasagar, M. S. (2002). Enhancement of radiation effect by Guduchi (*Tinospora cordifolia*) in HeLa cells. *Pharmaceutical Biology*, **40 (3)**: 179-188.

Jagetia, G.C., and Rao S.K. (2011). Assessment of radiation-induced DNA damage by comet assay in cultured HeLa cells treated with guduchi (*Tinospora cordifolia* Miers) before exposure to different doses of γ-radiation. *Research in Pharmaceutical Biotechnology*, **3(7)**: 93-103.

Jagetia, G.C., and Rao, S.K. (2006). Evaluation of cytotoxic effects of dichloromethane extract of guduchi (*Tinospora cordifolia* Miers Ex h\Hook f and Thoms) in cultured HeLa cells. *Evidence based Complimentary and Alternative Medicine*, **3(2):** 267-272.

Jagetia, G.C., Venkatesh P., and Baliga, M. S. (2005). *Aegle marmelos* (L.) Correa inhibits the proliferation of transplanted Ehrlich ascites carcinoma in mice. *Biological and Pharmaceutical Bulletin*, **28(1):** 58-64.

Jardine, I. (1980). In Anticancer agents Based on Natural Product Models. Cassady, J. M., Douros, J. D. (eds.) Academic Press, pp.319 – 351.

Jemal, A., Bray, F., Center, M.M., Ferlay, J., Ward, E., and Forman, D. (2011). Global cancer statistics. *CA Cancer Journal for Clinicians*, **61(2):** 69-90.

Khosa, R. L., and Prasad, S. (1971). Pharmacognostical studies on guduchi (*Tinospora cordifolia* Miers). *Journal of Research in Indian Medicine*, **6**: 261-9.

Kim, J., and Park, E. J. (2002). Cytotoxic anticancer candidates from natural resources. *Current Medicinal Chemical and Anti Cancer Agents*, **2(4):** 485-537.

Kinghorn, A. D., and Balandrin, M. F. (1993). Human medical agents from plants. American Chemical Society Symposium Series 534. Washington, DC, American Chemical Society, pp 80-95.

Kumar, S., Verma, N. S., Pande, D., and Srivastava, P. S. (2000). *In vitro* regeneration and screening of berberine in *Tinospora cordifolia*. *Journal of Medicinal and Aromatic Plant Sciences,* **22**: 61.

Li, R.W., Leach, D.N., Myers, S.P., Leach, G.J., Lin, G.D., Brushett, D.J., and Waterman, P.G. (2004). Anti-inflammatory activity, cytotoxicity and active compounds of *Tinospora smilacina* Benth. *Phytomedicine,* **18(1)**: 78-83.

Makhey, D., Gatto, B., Yu, C., Liu, A., Liu, L.F., and LaVoie, E.J. (1995). Protoberberine alkaloids and related compounds as dual inhibitors of mammalian topoisomerase I and II. *Medicinal Chemistry Reseach,* **5**: 1-12.

Marnett, L. J. (2002). Oxy radicals, lipid peroxidation and DNA damage. *Toxicology,* **27**: 219-222.

Mazumdar, U. K., Gupta, M., Maiti, S., and Mukherjee, D. (1997). Antitumor activity of *Hygrophila spinosa* on Ehrlich ascites carcinoma and sarcoma-180 induced mice. *Indian Journal of Experimental Biology,* **35(5):** 473-7.

Moron, M. S., Depierre, J. W., and Mannervik, B. (1979). Levels of glutathione, glutathione reductase and glutathione S-transferase activities in rat lung and liver. *Biochimica Biophysica Acta*, **582(1)**: 67-78.

Nadkarni, K. M., and Nadkarni, A. K. (1976). Indian Materia Medica, Vol 1. 3rd ed. Mumbai: M/S Popular Prakasan Pvt. Ltd.

Nayampalli, S., Ainapure, S. S., and Nadkarni, P. M. (1982). Study of antiallergic acid bronchodilator effects of *Tinospora cordifolia*. *Indian Journal of Pharmacology*, **14**: 64-66.

Neal, R., Matthews, R. H., Lutz, P., and Ercal, N. (2003). Antioxidant role of N-acetyl cysteine isomers following high dose irradiation. *Free Radical Biology and Medicine,* **34 (6)**: 689-695.

Nias, A. H. W. (1990). Radiation Biology. In Treatment of cancer. Sikora, K., Halnan, K. E. (eds) London. Chapman and Hall Medical, pp 53-75.

Pachaly, P., and Schneider, C. (1981). Alkaloids from *Tinospora cordifolia* Miers. *Archives of Pharmacology (Weinheim Ger),* **314**: 251-256.

Padhya, M. A. (1986). Biosynthesis of isoquinoline alkaloid berberine in tissue cultures of *Tinospora cordifolia*. *Indian Drugs,* **24**: 47-48.

Pandey, M.K., Sung, B., Kunnumakkara, A. B., Sethi, G., Chaturvedi, M.M., and Aggarwal, B.B. (2008). Berberine modifies cysteine 179 of IκBα kinase, suppresses nuclear factor-κB–regulated antiapoptotic gene products, and potentiates apoptosis. *Cancer Research,* **68(13):** 5370–5379.

Qudrat-I-Khuda, M., Khaleque, A., and Ray, N. (1964). *Tinospora cordifolia* L. constituents of the plant fresh from the field. *Scientific Res. (Dacca),* **1**: 177-83.

Richardson, M. A., Sanders, T., Palmer, J. L., Greisinger, A., and Singletary, S. E. (2000). Complementary/alternative medicine use in a comprehensive cancer center and the implications for oncology. *Journal of Clinical Oncology*, **18:** 2505–2514.

Sarma, D. N. K., Padma, P., and Khosa, R. L. (1998). Constituents of *Tinospora cordifolia* root. *Fitoterapia*, **69**: 541-2.

Sen, R. (2011). The origins of NF-κB. *Nature Immunology*, **12(8):** 686-688.

Thippeswamy, G., Sheela, M.L., and Salimath, B.P. (2008). Octacosanol isolated from *Tinospora cordifolia* downregulates VEGF gene expression by inhibiting nuclear translocation of NF-kB and its DNA binding activity. *European Journal of Pharmacology*, **588(2-3):** 141-150.

Wall, M. E., and Wani, M. C. (1993). In Human Medicinal Agents from Plants. Kinghorn, A. D., Balandrin, M. F. (eds). American Chemical Society Symposium Series 534, pp. 149 – 169.

Williamson, E. M. (2001). Synergy and other interactions in phytomedicines. *Phytomedicine*, **8(5):** 401-9.

Zhao, T.H., Wang, X., Rimando, A. M., and Che, C.T. (1991). Folkloric medicinal plants: *Tinospora sagittata* var. *cravaniana* and *Mahonia bealei*. *Planta Medica*, **57**: 505.

Traditional and Folk Herbal Medicine: *Pages* **133–151**
Recent Researches Vol. 2 (2014)
Editor: **V.K. Gupta**
Published by: **DAYA PUBLISHING HOUSE, NEW DELHI**

5

Centella asiatica: A Natural Remedy for the Treatment of Disease

Umakant Sahu[1], Rajpal Purabia[1], D. Nandini[3], Mayank Thakur[2] and Nagendra Singh Chauhan[1*]

ABSTRACT

Centella asiatica is valued as an ethnomedicines as well as in Ayurveda and Unani, the traditional Indian medicinal systems for thousands of years for various chronic diseases and disorders like asthma, skin disorders, ulcers, body aches, for improving memory, as a nervine tonic and in treatment of dropsy, elephantiasis, gastric catarrh, kidney troubles, leprosy, leucorrhoea and urethritis. Centella asiatica (L.) Urb. (syn.: Hydrocotyle asiatica) of the Apiaceae (Umbelliferae) family is a small herbaceous perennial plant, native to India, Bangladesh, Indonesia, Australia, China, Madagascar, the South Pacific and southern and middle Africa. The plant parts are rich in various bioactive compounds like asiaticoside, madecasoside, brahmic acid, Asiatic acids, Hydrocotylin, 3-glucosylquercetin and 3-glucosyl kaemferol. A Literature survey shows that C. asiatica is mainly known for its wound healing and memory enhancer properties. Various Research procedures were carried out regarding chemical constituents, pharmacological, and clinical uses. The present review is baseline information for future scientists one who are trying to develop newer drugs from natural resources are looking toward the Ayurveda, the Indian traditional system of medicine.

Keywords: Asiaticoside, Alzheimer's disease, Brahmi, *Centella asiatica*, Nervine tonic, Ethnomedicines.

1 Drugs Testing Laboratory Avam Anusandhan Kendra, Government Ayurvedic College Campus, Raipur – 492 010, Chhattisgarh, India

2 Department of Pharmaceutical Sciences, Dr. H.S. Gour University, Sagar, M.P., India

3 Sagar Institute of Pharmaceutical Sciences, Sagar, M.P., India

* *Corresponding author*: E-mail: chauhan.nagendra@gmail.com

Introduction

Since ancient times *C. asiatica* is used in India as traditional medicine, for various pathological diseases like leprosy's and in particularly for healing of wounds. In Ayurvedic system of medicine it is also recommended as a "brain tonic" in various mental disorders and in chronic diseases. *Centella asiatica* (L.) Urb. (syn.: *Hydrocotyle asiatica*) Of the Apiaceae (Umbelliferae) family is a perennial small herbaceous plant, native to India, China, Indonesia, Australia, South Pacific, Madagascar, Southern and Middle Africa preferably in damp swampy areas. This slender creeping plant with long internodes, rooting at nodes has long, prostrate, petioled leaves, filiform stems are reniform, oval or orbicular, deeply cordate. The small purple to white-green flowers are arranged in umbels arising from the axils of the leaves and fruit is ovoid, hard with strongly thickened pericarp (Kartnig *et al.*, 1998). There are many constituents in *Centella asiatica*, such as madecassoside (Figure 5.1a), Paeonoflorine (Figure 5.1b), madecassic acid (Figure 5.1c), asiatic acid (Figure 5.1d), asiaticoside (Figure 5.1e), castilliferol (Figure 5.1f), castillicetin (Figure 5.1g), terminolic acid, vanillic acid, succinic acid, asiaticoside-B, asiaticodiglycoside. Additionally, there are several pairs of isomers in *Centella asiatica* (Matsuda *et al.*, 2001; Sahu *et al.*, 1989). Madecassic acid and terminolic acid are a pair of structural isomers. Their structures are shown in Figure 5.1. It is listed officially in the Chinese Pharmacopoeia and used as an antipyretic and diuretic, in the treatment of heat stroke, diarrhoea, ulcerations, eczema, and traumatic diseases (Pedretti *et al.*, 2009). A preliminary approach has been developed for the identification of *C. asiatica* by TLC founded in the *European Pharmacopoeia* (2005), using drug powder as the starting material. Crude extract that contains isothankuniside and thankuniside showed antifertility action in mice (Duke *et al.*, 1985).

Taxonomical Classification

Class: Dicotyledenae
Sub-Class: Polypetalae
Series: Calyciflorae
Order: Umbellales
Family: Umbelliferae (Apiaceae)
Genus: *Centella*
Species: *asiatica*

It is well known that medicinal plants collected at different times and from different localities may considerably differ in their types and quantities of chemical components, therefore, resulting in different therapeutic efficacy. The differences in chemical components in *C. asiatica* will affect the quality of pharmaceutical products and the standardization of this herbal medicine, hence it is necessary to estimate variation in chemical components. The effective method is chromatographic finger printing analysis, which has been accepted as a strategy for quality assessment of herbal medicines by the WHO (2000), the FDA (2000), the EMA(2001), the Chinese State Food and Drug Administration SFDA (2000).

Figure 5.1: Phytoconstituents of *Centella asiatica*.

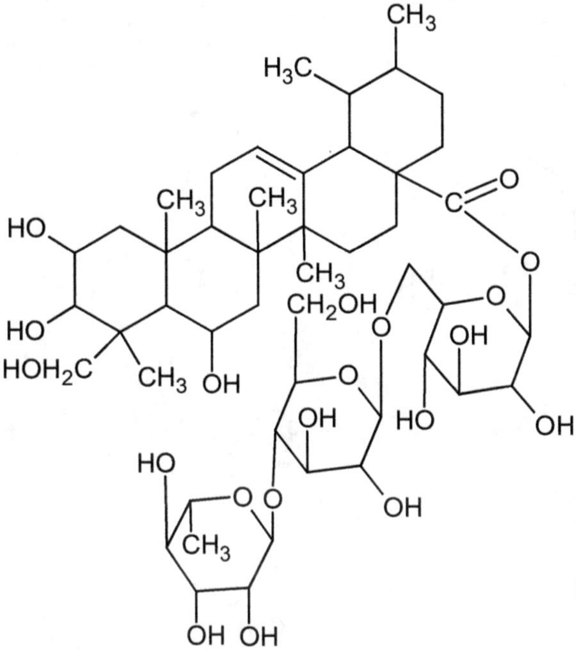

Medecassoside (a)

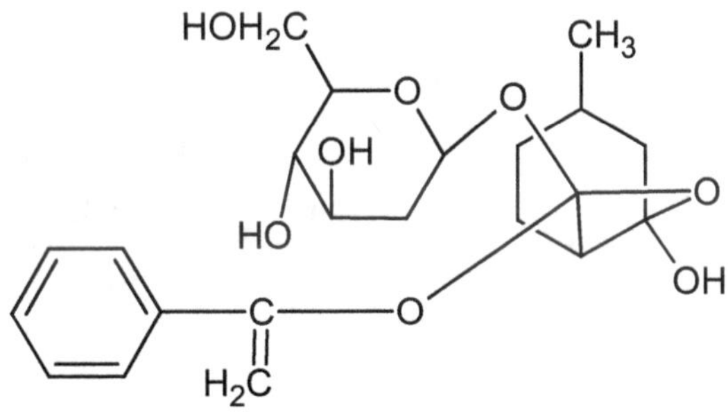

Paeonoflorin (b)

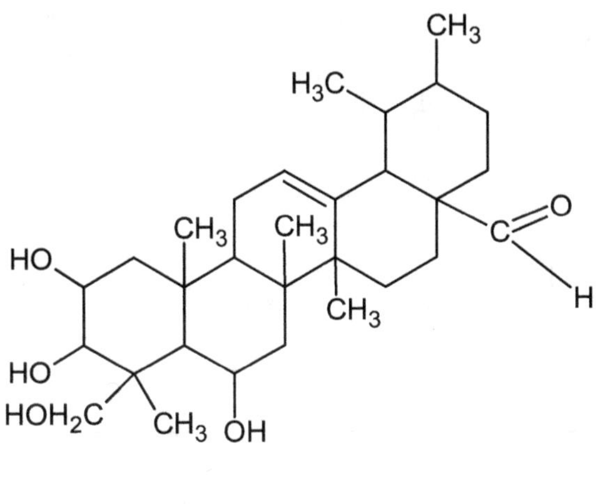

Medecassic acid (c)

Contd...

Figure 5.1–*Contd...*

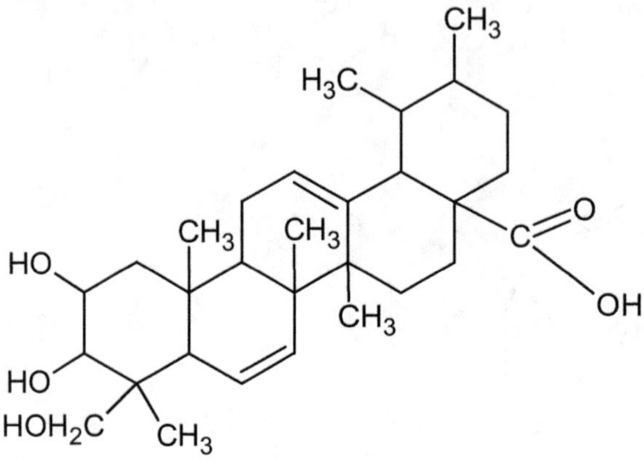

Asiatic Acid (d)

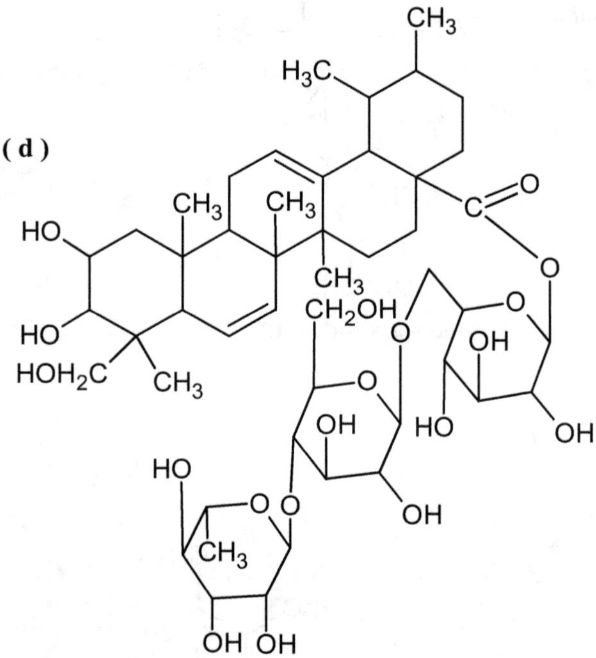

Asiaticoside (e)

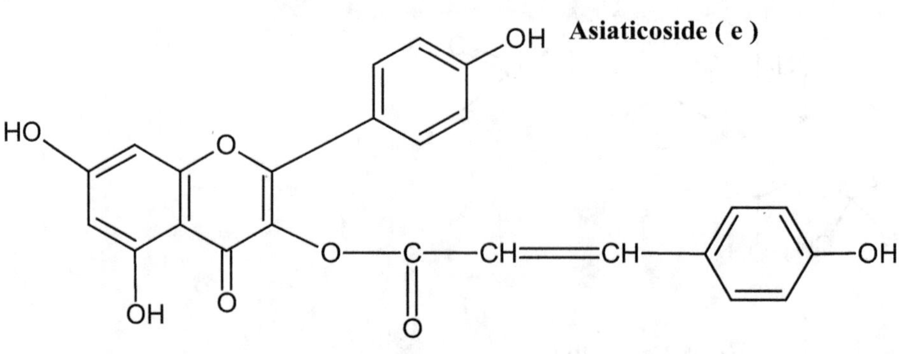

Castilliferol (f)

Contd...

Figure 5.1–*Contd...*

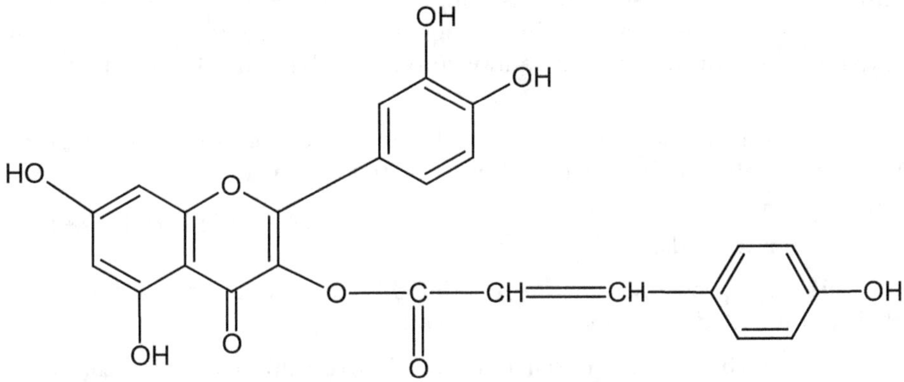

Castillicetin (g)

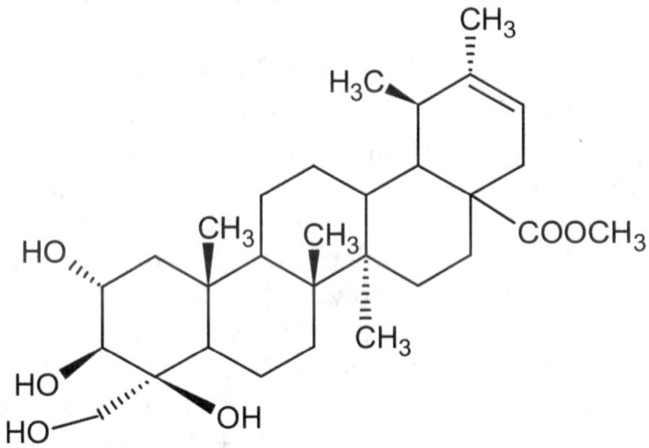

R = H (**h**)

R = rha(1-4)-glc-(1-6)-glc (**i**)

The Ethnopharmacological and Economical Values

Ethnomedicine are more precisely, study of folklore remedies. From last few decades in context to develop phytomedicine from traditional drug, vast numbers of researches are undergoing. In order of that *Centella asiatica* is drug having Socio-Economically important and traditional used for different purpose in different countries are choice in order of research. *Centella asiatica* is valued as an ethnomedicines as well as ISM drug in India and used for different ailments like urethritis, leucorrhoea, treatment of dropsy, elephantiasis, gastric catarrh, kidney troubles, leprosy skin disorders, ulcers and body aches for improving memory as a nervine tonic.

In Chinese traditional system this herb is used for dysentery, summer diarrhea, vomiting, jaundice, Hansen's disease (leprosy), nosebleeds, tonsillitis, fractures, measles, tuberculosis, urinary difficulties, endocrine tonic, scabies and as an adaptogen. Accounts of longevity and virility, Centella is also known as miracle of elixir.

In Bangladesh, whole plant is utilized to treat multiple ailments like dog bite, asthma, carminative, itching, leucorrhoea, malaria tumor and wounds.

In Nepal also this herb is used in rheumatism, indigestion, leprosy, poor memory, kill germs from wounds.

In Fiji for, eye problems, fractures, treating childhood tidal fevers, rib pain, swollen joints and unwanted pregnancy.

In Malaysia, the plant is used in hypertension, diuretic, urinary tract infections, diarrhoea lower blood pressure and decrease heart rate.

In Madagascar this herb traditionally utilized in leprosy, tuberculosis etc.

In Brazil for hair oil, used to prepare chutney elephantiasis and leprosy etc.

Phytochemical Studies

Preparations from *Centella asiatica* (*Hydrocotyle asiatica*) is listed in the Indian Pharmaceutical Codex, British Homoeopathic Pharmacopeia and Homoeopathisches Arzneibuch 1 (1978) (Germany) are well known for the treatment of various diseases, it shows wide range of pharmacological potential (Figure 5.2) wound-healing (Kieswetter *et al.*, 1964), asthma, leprosy, spasmolytic and tranquilizing effect (Ramaswamy *et al.*, 1970), ulcer-preventive (Huriez *et al.*, 1972), Antidepressive-sedative (Sakina and Dandiya *et al.*, 1990), Myocardial ischemia–reperfusion injury (Li *et al.*, 2007), Antiinflammatory (Tumen *et al.*, 2011), Antidepressant effects (Chen *et al.*, 2005). Asiaticoside (Figure 5.1e) has anxiolytict (Wijeweera *et al.*, 2006), Antioxidant (Meena *et al.*, 2012), antiulcer (Cheng *et al.*, 2004), Wound-healing property (Lee *et al.*, 2012) and Arthritis (Liu *et al.*, 2008) (Figure 5.2). *Centella asiatica* contains different phytoconstituents shows different activity such as isothankuniside and thankuniside showed antifertility action (Yunianto *et al.*, 2010), brahmoside and brahminoside, may be responsible for CNS and utero relaxant actions. The HPLC method described here for the quality controls of the triterpenes, which are responsible for the pharmacological activity, several methods, has been developed for the detection of phytoconstituent such as for the detection of madecassoside (Figure 5.1) in raw materials or pharmaceutical preparations is doe by HPLCUV (Zhang *et al.*, 2007) and HPLC (Zhang *et al.*, 2008) methods. Novel bioactive compound isolated from aerial part of the *C. asiatica* known as cadiyenol induces apoptosis of independence cell cycle of lymphoma cell and nitric oxide production in dose dependent manner in mouse of macrophages (Govindan *et al.*, 2007). Only a few analytical methods have been established for the quality control of *Centella asiatica* for instance, TLC-MS (Bonfill *et al.*, 2006), thin-layer chromatography (TLC) (Meng and Zeng *et al.*, 1988) and the prevailing HPLC-UV methods (Gunther and Wagner, 1996; Xiao *et al.*, 2000; Zeng *et al.*, 2000; Chen *et al.*, 2000; Schaneberg *et al.*, 2003; Rafamantananaa *et al.*, 2009). HPLC-UV method given by Schaneberg *et al.* (2003) has succeeded to give accurate quality profile of *Centella asiatica*, other method failed in baseline separation of madecassoside and its isomer asiaticoside (Figure 5.1e)–B. In addition, although the TLC-MS method works well for the identification of the triterpenoid compounds in *Centella asiatica*, the equipment is relatively expensive and may not be available in every laboratory. Moreover, the TLC method for quantification presents a challenge for precision and sensitivity in comparison with the HPLC method. *Centella asiatica* extracts and their structures were established by comparison of their spectral data (UV, IR, MS, ^{1}H- and ^{13}C-NMR) with the literature data (Sahu *et al.*, 1989; Matsuda *et al.*, 2001; Zhang *et al.*, 2005; He *et al.*, 2000). The purity of each compound was determined to be above 98 per cent by normalization of the peak areas detected by HPLC-ELSD. Structural isomers can be isolated by reversed-phase high performance liquid chromatography with high resolution on a C18 reversed-phase column with the addition of β-CD in the mobile phase. The effects of β-CD concentration and the pH of mobile phase on resolution were investigated. It was found that the resolution of the isomers increased with the increase in β-CD concentration when the mobile phase consisted of methanol-water (65: 35, v/v) at pH 4 (PAN *et al.*, 2007). Quantification of madecassoside and Asiaticoside (Figure 5.1e) using acetonitrile/water (29/17, v/v) in isocratic mode was proposed by Xingyi *et al.*, and for its isomer Zhang *et al.*, has proposed a method.

Two antioxidant flavonoids castilliferol (Figure 5.1f) and castillicetin (Figure 5.1g) also known isochlorogenic acid is isolated from the whole plant of *Centella asiatica* (Ravi *et al.*, 2008).

Pharmacological Activity

The *Centella asiatica* posses multifunctional pharmacological activity (Figure 5.2)

CNS Activity

As we know that *Centella asiatica* shows beneficial effect in various neurological disorder but recent study reveal that 70 per cent hydro-ethanolic extract of *Centella asiatica (CA)* gives potential activity on generalized anxiety disorder (GAD) in man. Hamilton's Brief Psychiatric Rating Scale (BPRS) was used to screen the subjects (Jana *et al.*, 2010). *Valeriana officinalis* and *centella asiatica* extract stimulated the glutamic acid decarboxylase (GAD) acivity(Awad *et al.*, 2007). Decline cognitive function is restore by in traditional medicine *Centell aasiatica* and assessed by using the computerized test battery (Wattanathorn *et al.*, 2007). Fresh leaf extract of *Centella asiatica* content the phytochemical constituent which is having stimulating properties to neuronal dendritic growth (Gadahadetal., 2008). The neuroprotection molecular mechanism of *Centella asiatica* is still remains to elucidate; study of Xu *et al.* (2008) revealing that ERK/RSK signaling pathway might mediate neuroprotection effect of *Centella asiatica* extract. Rasayana are set of rejuvenative measures to impart biological sustenance to the bodily tissues which are claimed to act as micronutrients. Some of these Rasayanas are organ and tissue specific. The popular Medhya Rasayanas are *Ashwagandha (Withania somnifera), Brahmi (Bacopa monnieri), Mandukaparni (Centella asiatica)* and *Sankhapuspi (Convolvulus pluricaulischois)*. Recent studies conducted on these traditional remedies on scientific parameters have shown promising results

(Singh *et al.*, 2008). Oxidative stress induced by 3-NPA completely prophylaxised by *Centella asiatica* (Shinomol *et al.*, 2008; Shinomol *et al.*, 2010). *Centella asiatica* can be used for enhancing neuronal dendrites in stress and other neurodegenerative and memory disorders. Rats treated with it showed a significant increase in dendritic branching points and dendritic length (intersections) along the length of dendrites of the amygdaloid neurons (Mohandas *et al.*, 2009). *C. asiatica* extracts can be used for effective management in treatment of epileptic seizures because shows anticonvulsant and neuroprotective activity (G V *et al.*, 2010). Asiatic acid having an inhibitory effect on acetyl cholinesterase (AChE) properties, no sedative effect on locomotor and a selective GABA (B) receptor agonist (Nasir *et al.*, 2011). *Bacopa monnieri* and *Centella asiatica* plants possess have nootropic activity and neuroprotective properties therapeutically it can be used for patients with memory loss (Shinomol *et al.*, 2011). *Centella asiatica* shows protective decrease in the superoxide dismutase and catalase activities in the liver, brain, kidneys and testes of rats and decrease in malondialdehyde levels (Sainath *et al.*, 2011). More than 16 plants are investigated along with *Centella asiatica* in order to discover new chemical active molecule to control and prevention of Alzheimer disease installation and progression. This can alter neuronal death and the development and the progression of Alzheimer's disease (da Rocha *et al.*, 2011). Asiaticoside (Figure 5.1e) a triterenoid saponin from *Centella asiatica* in parkinsonism induced model found to protect dopaminergic neuron by antagonizing MPTP induced neurotoxicity and to improve locomotor dysfunction (Xu *et al.*, 2011; Haleagrahara *et al.*, 2010).

Antidiabetic Activity

Lipogenic activity of *Centella asiatica* was reported and it improves the glucose metabolism and diabetic risk factors with varying degrees. Study proof that fifty-nine percent of *Centella asiatica* increased adipogenesis as did the standard troglitazone relative to the solvent controls. Both rho-iso alpha acids and A. nilotica reduced non fasting glucose in the db/db mouse model, *Centella* exhibits anti-inflammatory activity by inhibiting TNF alpha-stimulated free fatty acid release and attenuated TNF alpha inhibition of adiponectin secretion (Babish *et al.*, 2010). The inhibitory potential against key enzymes related to type 2 diabetes and hypertension was highest in hexane extract of selected Malaysian plants pagag, *Centella asiatcia* (Loh *et al.*, 2011). Combination of flavonoids desmind and troxerutin with *Centella asiatica* and Melilotus was administered orally to evaluate for 14 months for the treatment of diabetic cystoid muscular edema (CME) without macular thickening on the basis of best collected visual acuity, central retinal thickness at optical coherence tomography, retinal sensitivity and stability of fixation at microperimetry were did not show difference (Forte *et al.*, 2011). *Andrographis paniculata*, along with *Orthosiphon stamineus* and *Centella asiatica* interact with human cytochrome P450 2C19 (CYP2C19), a major hepatic CYP isoform involved in metabolism of many clinical drugs has not been investigated to date. Therefore should be careful when taken *Centella asiatica* and *Orthosiphon stamineus* components are with CYP2C19 substrates (such as omeprazole, barbiturates, proguanil, diazepam, and citalopram (Pan *et al.*, 2010).

Anti-inflammatory

Asiaticoside (Figure 5.1e) shows the inhibitory activity on collagen induced arthritis and reduced paw swelling and also exert inhibitory effect on COX-2, PGE2, TNF-alpha and IL-6 production (Li *et al.*, 2007; Liu *et al.*, 2008). Asiaticoside (Figure 5.1e) also shows activity on lipopolysaccharides induced lung injury and gives inhibitory effect on IL-6, TNF-alpha, IL-10 of broncho alveolar lavage fluid (Zhang *et al.*, 2008). *Centella asiatica* gives beneficial effect on inflammatory diseases and reduced NF-kappa B activation. Colitis was induced in rats by intra colonic instillation of dinitrobenzene sulphonic acid and DNBS administration, treatment with CA1 significantly reduced the appearance of diarrhoea and the loss of body weight (di Paola *et al.*, 2010). *Centella asiatica* contains Asiatic acid which leads to increasing the activities of CAT, SOD, and GPx in the liver probably related to the decrease in the level of MDA, iNOS, COX-2, and NF-κB in the edema paw (Huang *et al.*, 2011).

Wound Healing Activity

Because of it antioxidant activity CA extract also gives significant wound healing activity and in burn lower dose can repair probably mechanisms of action is effects of Asiaticoside (Figure 5.1e) on the levels of various cytokines (Kimura *et al.*, 2008). Madecassoside is triterpenesa active constituent of *Centella asiatica* shows activity on decreased nitric oxide levels and malondialdehyde content in the burn skin tissue. Its possible mechanism of action including antioxidative activity, collagen synthesis and angiogenesis (Liu *et al.*, 2008). Anti-aging activity described by Ayurveda also included *Centella asiatica* as anti-aging agent along with some rasayana plant (Datta *et al.*, 2010). Capsule preparation dosage form of *Centella asiatica* extract more effective in the wound healing promotion and suppress the scar in diabetic wound patients. Very few amount of *Centella asiatica* extract capsule group shows no \serious side effect and shorten the course of diabetic wound and it is safe for diabetic patients (Paocharoen *et al.*, 2010).

Antioxidant Activity

Centella asiatica extract in methanol and aqueous shows good antioxidant activity (free radical scavenging activity) against 1, 1-diphenly-2-picryl hydrazyl (Gupta *et al.*, 2009) and cytotoxic activities (Pittella *et al.*, 2009; Mustafa *et al.*, 2010). Significant percentage of beta-carotene and lutein help us to overcome from vitamin A deficiency and disorder (Chandrika *et al.*, 2010). *C. asiatica* is rich in flavones so use of herbal teas which may help to improve the illness condition (Ariffin *et al.*, 2011). Presence of significant amount of macadecassoside, Asiaticoside (Figure 5.1e), asiatic and madecassic acid (Figure 5.1c)(Figure 5.1c) in *Centella asiatica* are responsible for lipolytic activity, UV protection effect and antioxidant activity and it can be used in food and pharmaceutical industries (Hashim *et al.*, 2011)

Antimalarial Activity

Petroleum ether, acetone, hexane, ethyl acetate and methanol extract *Centella asiatica* of showed moderate larvicidal effects (Rahuman *et al.*, 2008). Hexane extracts of *Centella asiatica* show adulticidal, larvicidal effect moderate toxic effect and highest mortality on parasites on parasites was found in leaf hexane extract of *Centella asiatica*

(Bagavan *et al.*, 2009). Essential oils extracted by *Centella asiatica, Ipomoea cairica, Momordica charantia, Psidium guajava* and *Tridax procumbens* were evaluated for their topical repellency effects against malarial vector Anopheles in mosquito. Essential oils of *I. cairica, M. charantia* and *T. procumbens* are promising repellents activity at very less concentration (Rajkumar *et al.*, 2007).

Anticancer Activity

Methanolic extract of *Centella asiatica* induce apoptosis in different cancer cell lines MCF-7 cells most sensitive cell line for in vitro growth inhibitory activity (et al., 2008). Asiatic acid induced anticancer activity via caspase-9 which activated caspase-3 and poly (ADP-ribose) polymerase cleavage resulting in irreversible apoptotic death in the tumor cells (Tang *et al.*, 2009).

Skin Protective Activity

Aqueous extracts of *Centella asiatica* exhibits skin-protective activitiy of nanoencapsule extract compared to the unencapsulated crude extractswhich gives activity relativelyin large concentrations (Kwon *et al.*, 2011). Active compound madecassoside, extracted is shown significant effect on deep and superficial wrinkles, suppleness, firmness, roughness and skin hydration. Cosmetic cream formulations by phase inversion emulsification technique using ethanolic extracts leaves of *Centella asiatica* along with *Glycyrriza glabra, Curcuma longa* (roots), seeds of *Psorolea corlifolia, Cassia tora, Areca catechu, Punica granatum,* fruits of *Embelica officinale,* dried bark of *Cinnamon zeylanicum* and fresh gel of *Aloe vera.* Cream base formulations were assesd in terms of physicochemical parameters of formulations CAA1-CAA6, *i.e.*, pH, acid value, saponification value, viscosity, spreadability, layer thickness microbial count and skin sensitivity (Ahshawat *et al.*, 2008).

Antimicrobial Activity

Regenerated plants extracts of *Centella asiatica* inhibited bacterial growth against six bacterial strains using agar well diffusion method. Agar tube dilution method for antifungal assay showed inhibition of Mucor species, *Aspergillus fumigatus* and *Fusarium moliniformes* (Bibi *et al.*, 2011)

Hepatoprotective Activity

Asiaticoside (Figure 5.1e) is efficacious against Lipopolysaccharide/D-galactosamine -induced acute liver injury, significant protection as evidenced by caspase-3, alleviation of mortality, and improvement of liver pathological injury and decrease of elevated aminotransferases, hepatocytes apoptosis. Possible mechanism is related to inhibition of TNF-α and MAPKs (Zhang *et al.*, 2010). Liquid chromatography-tandem mass spectrometry (LC/MS-MS) analysis has done to measured inhibition of cytochrome P450 (CYP). The N-in-one cocktail is rapid and reliable methods that can be to monitor product variability prioritize products that require more detailed investigations (Savior *et al.*, 2010).AS inhibited CLP-induced the activation of mitogen-activated protein kinases (MAPKs) and nuclear factor-κB (NF-κB), the expression of cyclooxygenase-2 (COX-2) and inducible nitric oxide synthase (iNOS) protein in lung tissues, and the production of serum tumor necrosis

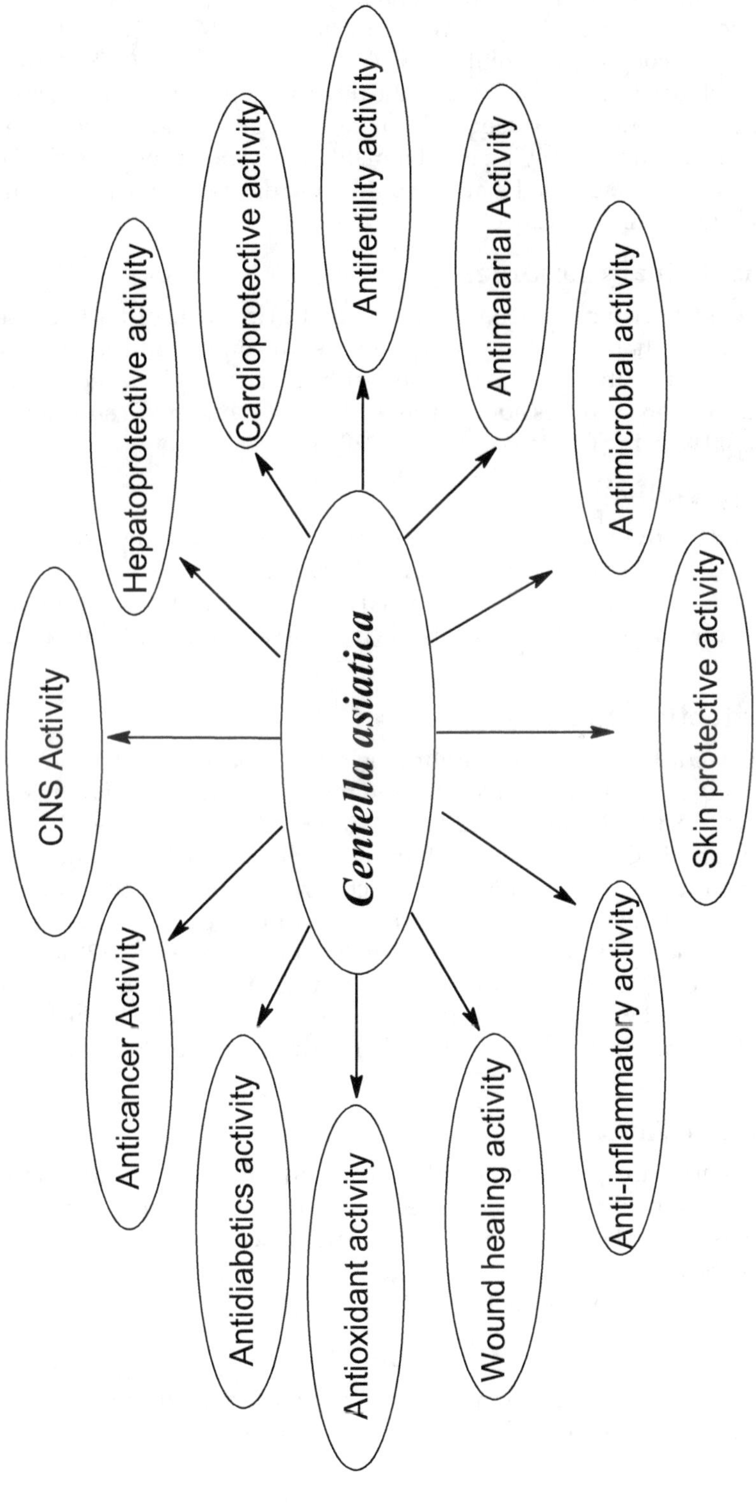

Figure 5.2: Pharmacological potentials of *Centella asiatica*.

factor (TNF-α) and interleukin-6 (IL-6). Probably protective effect of asiacoside due to related to up-regulation of PPAR-γ expression. Zhang LN *et al.*, 2010.Centella asiatica extracts and active constituents inhibited CYP2C9, CYP2D6 and CYP3A4 activities with varying potency with CYP2C9 being the most susceptible isoform to inhibition. Significant inhibition was observed for asiatic acid and CA ethanol and dichloromethane extracts, implying involvement of semi polar constituents from CA in the effect. This study suggested that CA could cause drug-herb interactions through CYP2C9 inhibition (Pan *et al.*, 2010).

Kidney and Urinary Disorder

Centella asiatica also been used in kidney and urinary disorder because of its antioxidant activity that protect the tubular epithelium and anti-fibrotic that induce apoptosis of pro-fibrotic myofibroblasts without adversely affecting tubular epithelium may slow progression of renal fibrosis, while toxic substances may exacerbate renal scarring (Wojcikowski *et al.*, 2008).

Anti-fertility Activity

Anti-fertility activity of *Centella asiatica* was investigated Sprague-Dawley adult male rats were divided into four groups control group,low dose group, medium dose and high dose groups. *Centella asiatica* showed some degeneration of spermatogenic cells and reduction of spermatozoa in the lumen of the seminiferous tubules (Yunianto *et al.*, 2010).

Cardio-protective Activity

Madecassoside has the protective effect on myocardial ischemia-reperfusion the protection ability possibly due to its anti-lipid peroxidation, anti-inflammation and anti-apoptosis function (Bian *et al.*, 2008). Asiaticoside (Figure 5.1e) is one of the compounds found in C. asiatica exert pharmacological potential on Phospholipase A (2) and LPS-induced Nitric Oxide (NO)which is responsible for abnormal activity in the central nervous system in some neuropsychiatric diseases. Due to C. sciatica's water extracts on the activity on Ca(2+)-independent Phospholipase A (2) and cytosolic Phospholipase A (2) complimentary Phospholipase A (2) may be useful to treat conditions associated with increased Phospholipase A (2)activity in the brain, such as epilepsy, stroke, multiple sclerosis and other neuropsychiatric disorders (Barbosa *et al.*, 2008).

Miscellaneous Activity

Centella asiatica are manly believed to be responsible for its wide therapeutic actions like wound healing, the herb is recommended for the treatment of various skin conditions such as leprosy, lupus, varicose ulcers, eczema, psoriasis, diarrhoea, fever, amenorrhea, diseases of the female genitourinary tract and also for relieving anxiety and improving cognition (Gohil *et al.*, 2010).

Conclusion

Centella asiatica is traditional medicinal plant widely used as folk medicine in some Asian country because of is tremendous range of activity, also has been confirmed by modern medical science in terms of pharmacodynamics investigations. Present review of *Centella asiatica* is one support to the researcher to develop a new remedy from the traditional ayurvedic drugs for the treatment of various life threatening disorder and diseases.

References

Ahshawat, M.S., Saraf, S., and Saraf, S. (2008). Preparation and characterization of herbal creams for improvement of skin viscoelastic properties. *International Journal of Cosmetic Science,* **30(3):** 183-93.

Ariffin, F., Heon, Chew, S., Bhupinder, K., Karim, A.A., and Huda, N.(2011). Antioxidant capacity and phenolic composition of fermented *Centella asiatica* herbal teas. *Journal of the Science of Food and Agriculture,* **91(15):** *2731-9*

Awad, R., Levac, D., Cybulska, P., Merali, Z., Trudeau, V.L., and Arnason, J.T.(2007). Effects of traditionally used anxiolytic botanicals on enzymes of the gamma-aminobutyric acid (GABA) system. *Canadian Journal of Physiology and Pharmacology,* **85(9):** 933-42.

Barbosa, R., Pittella, F., and Gattaz, W.F.(2008). *Centella asiatica* water extract inhibits iPLA2 and cPLA2 activities in rat cerebellum. *Phytomedicine,* **15(10):** 896-900.

Babykutty, S., Padikkala, J., Sathiadevan, P.P., Vijayakurup, V., Azis, T.K., Srinivas, P., and Gopala, S. (2008). Apoptosis induction of *Centell aasiatica* on human breast cancer cells. *African Journal of Traditional Complementary and Alternative Medicines,* **6(1)**: 9-16.

Babish, J.G., Pacioretty, L.M., Bland, J.S., Minich, D.M., Hu, J., and Tripp, M.L. (2010). Antidiabetic screening of commercial botanical products in 3T3-L1 adipocytes and db/db mice. *Journal of Medicinal Food,* **13(3):** 535-47

Bagavan, A., Kamaraj, C., Elango, G., Abduzzahir, A., and Abdul Rahuman, A.(2009). Adulticidal and larvicidal efficacy of some medicinal plant extracts against tick, fluke and mosquitoes. *Veterinary parasitology,* **166(3-4):** 286-92.

Bian, G.X., Li, G.G., Yang, Y., Liu, R.T., Ren, J.P., Wen, L.Q., Guo, S.M., and Lu, Q.J. (2008). Madecassoside reduces ischemia-reperfusion injury on regional ischemia induced heart infarction in rat. *Biological Pharmaceutical Bulletin,* **31(3):** 458-63.

Bibi, Y., Zia, M., Nisa, S., Habib, D., Waheed, A., and Chaudhary, F.M.(2011). Regeneration of *Centella asiatica* plants from non-embryogenic cell lines and evaluation of antibacterial and antifungal properties of regenerated calli and plants. *Journal of Biological Engineering,* **5(1):** 13.

Bonfill, M., Mangas, S., Cusidó, R.M., Osuna, L., Piñol, M.T., and Palazón, J. (2006). Identification of triterpenoid compounds of *Centella asiatica* by thin-layer chromatography and mass spectrometry. *Biomedical Chromatography,* **20**: 151.

Chandrika, U.G., Basnayake, B.M., Athukorala, I., Colombagama, P.W., and Goonetilleke, A.B. (2010). Carotenoid content and *in vitro* bioaccessibility of lutein in some leafy vegetables popular in Sri Lanka. *Journal of Nutritional Science and Vitaminology* (Tokyo), **6(3):** 203-7.

Chen, Y., Qin, L.P., Zheng, H.C., Jiang, Y., and Weng, Z.L. (2000). Determination of assiticoside and madecassoside in *Centella asiatica*(L.)Urban by RE-HPLC. *Chinese Traditional Patent Medicine*, **3**: 227.

Chen, Y., Han, T., Rui, Y., Yin, M., Qin, L., and Zheng, H. (2005). Effects of total triterpenes of *Centella asiatica* on the corticosterone levels in serum and contents of monoamine in depression rat brain. *Zhong Yao Cai*, **28(6):** 492-6.

da Rocha, M.D., Viegas, F.P., Campos, H.C., Nicastro, P.C., Fossaluzza, P.C., Fraga, C.A., Barreiro, E.J., and Viegas, C. Jr. (2011). The role of natural products in the discovery of new drug candidates for the treatment of neurodegenerative disorders II: Alzheimer's disease. *CNS Neurological Disorder Drug Targets*, **10(2):** 251-70.

Datta, H.S., and Paramesh, R. (2010). Trends in aging and skin care: Ayurvedic concepts. *Journal of Ayurveda and Integrative Medicine*, **1(2):** 110-3.

di Paola, R., Esposito, E., Mazzon, E., Caminiti, R., Toso, R.D., Pressi, G., and Cozzocrea, S. (2010). 3,5-Dicaffeoyl-4-malonylquinic acid reduced oxidative stress and inflammation in a experimental model of inflammatory bowel disease. *Free Radic Research*, **44(1):** 74-89.

Duke, J. (1985). Handbook of Medicinal Herbs. Boca Raton, FL CRC Press.

EMA (2001). Note for guidance on quality of herbal medicinal products. European Medicines Agency, London, p 6

European Pharmacopoeia (5th edn).*(2005)*.Vol. 2. Council of Europe: Strasbourg; 1236.

FDA (2000).Guidance for industry botanical drug products (draft guidance).FDA, Washington, DC, I, B, 2e-3e.

Forte, R., Cennamo, G., Finelli, M.L., Bonavolontà, P., de Crecchio, G., and Greco, G.M. (2011). Combination of flavonoids with *Centella asiatica* and Melilotus for diabetic cystoid macular edema without macular thickening. *Journal of Ocular Pharmacology and Therapeutics*, **27(2):** 109-13.

Gadahad, M.R., Rao, M., and Rao, G. (2008). Enhancement of hippocampal CA3 neuronal dendritic arborization by *Centella asiatica* (Linn) fresh leaf extracts treatment in adult rats. *Journal of Chinese Medical Association*, **71(1):** 6-13.

Gohil, K.J., Patel, J.A., and Gajjar, A.K. (2010). Pharmacological Review on *Centella asiatica*: A Potential Herbal Cure-all. *Indian Journal of Pharmaceutical Science*, **72(5):** *546-56*.

Govindan,G., Sambandhan,T.G., Govindan,M., sinskey,A., Vanessendelft,J., Adenan,I., and Rha, C.K. (2007). A Bioactive polyacetylene compound isolated *C. asiatica*. *Planta Medica*, **73:** 597-599.

Gupta, S., and Prakash, J. (2009).Studies on Indian green leafy vegetables for their antioxidant activity. *Plant Foods for Human Nutrition*, **64(1):** 39-45.

Gunther, B., and Wagner, H. (1996). Quantitative determination of triterpenes in extracts and phyto preparations of *Centella asiatica* (L.). *Urban Phytomedicine*, **3:** 59.

G.V., K. SP., V.L., and Rajendra W. (2010). The antiepileptic effect of *Centella asiatica* on the activities of Na/K, Mg and Ca-ATPases in rat brain during pentylenetetrazol-induced epilepsy. *Indian Journal of Pharmacology*, **42(2):** 82-6.

Haftek, M., Mac-Mary, S., Le Bitoux, M.A., Creidi, P., Seité, S., Rougier, A., and Humbert, P. (2008).Clinical, biometric and structural evaluation of the long-term effects of a topical treatment with ascorbic acid and madecassoside in photo aged human skin. *Experimental Dermatology*, **17(11):** 946-52.

Hashim, P., Sidek, H., Helan, M.H., Sabery, A., Palanisamy, U.D., and Ilham, M. (2011). Triterpene composition and bioactivities of *Centella asiatica*. *Molecules*, **16(2):** 1310-22

Haleagrahara, N., and Ponnusamy, K. (2010). Neuroprotective effect of *Centella asiatica* extract (CAE) on experimentally induced Parkinsonism in aged Sprague-Dawley rats. *Journal of Toxicological Sciences*, **35(1):** 41-7.

Huang, S.S., Chiu, C.S., Chen, H.J., Hou, W.C., Sheu, M.J, Lin, Y.C, Shie, P.H., and Huang, G.J.(2011). Antinociceptive activities and the mechanisms of anti-inflammation of asiatic acid in mice. *Evidence-based Complementary and Alternative Medicine*, **2011**: 895857.

Huriez, C.(1972). Action of the titrated extract of *Centella asiatica* in the cicatrization of leg ulcers. *Lille Medicale*, **17(Suppl. 3):** 574–579

Jana, U., Sur, T.K., Maity, L.N., Debnath, P.K., and Bhattacharyya, D. (2010). A clinical study on the management of generalized anxiety disorder with *Centella asiatica*. *Nepal Medical College Journal*, **12(1)**: 8-11.

Kartnig, T.(1998). Herbs, Spices and Medicinal Plants, L.E. Cracker, J.E. Simon (Eds.), Oryx Press, Arizona, USA, vol. 3: 145-173.

Kieswetter, H., Erfahrungs,berichttiber,Behandlung, von Wundenmit (1964). Asiaticosi (Madecassol). *Wiener Medizinische Wochenschrift*, 114: , 124-126.

Kimura, Y., Sumiyoshi, M., Samukawa, K., Satake, N., and Sakanaka, M. (2008). Facilitating action of Asiaticoside at low doses on burn wound repair and its mechanism. *European Journal of Pharmacology*, **584(2-3):** 415-23.

Kwon, M.C., Choi, W.Y., Seo, Y.C., Kim, J.S., Yoon, C.S., Lim, H.W., Kim, H.S., Ahn, J.H., and Lee, H.Y. (2012). Enhancement of the skin-protective activities of *Centella asiatica* L. urban by a nano-encapsulation process. *Journal of Biotechnology*, **157(1):** 100-6.

Lee, J.H., Kim, H.L., Lee, M.H., You, K.E., Kwon, B.J., Seo, H.J., and Park, J.C. (2012). Asiaticoside enhances normal human skin cell migration, attachment and growth *in vitro* wound healing model. *Phytomedicine*, **19(13):** 1223-7.

Li, G.G., Bian, G.X., Ren, J.P., Wen, L.Q., Zhang, M., and Lü, Q.J.(2007). Protective effect of madecassoside against reperfusion injury after regional ischemia in rabbit heart *in vivo*. *Yao Xue Xue bao.*, **42(5)**: 475-80.

Li, H.Z., Wan, J.Y., Zhang, L., Zhou, Q.X., Luo, F.L., and Zhang, Z. (2007). Inhibitiory action of asiaiticoside on collagen-induced arthritis in mice.*Yao Xue Xue Bao.*, **42(7):** 698-703.

Liu, M., Dai, Y., Yao, X., Li, Y., Luo, Y., Xia, Y., and Gong, Z. (2008). Anti-rheumatoid arthritis effect of madecassoside on type II collagen-induced arthritis in mice. *International Immunopharmacology*,**8 (11):** 1561-6

Liu, M., Dai, Y., Li,Y., Luo, Y., Huang, F., Gong, Z., and Meng, Q. (2008). Madecassoside isolated from *Centella asiatica* herbs facilitates burn wound healing in mice. *Planta Medica*, **74(8):** 809-15.

Loh, S.P., and Hadira, O. (2011). *In vitro* inhibitory potential of selected Malaysian plants against key enzymes involved in hyperglycemia and hypertension. *Malaysian Journal of Nutrition*, **17(1):** 77-86.

Matsuda, H., Morikawa, T., Ueda, H., and Yoshikawa, M. (2001). Medicinal foodstuffs. XXVII. Saponin constituents of Gotu Kola(2): structuresof new ursane- and oleanane-type triterpene oligoglycosides centella saponins B, C, and D from *Centella asiatica* cultivated in Sri Lanka. *Chemical Pharmacuetical Bulletin*, **49(10):** 1368.

Meena, H., Pandey, H.K., Pandey, P., Arya, M.C., and Ahmed, Z. (2012). Evaluation of antioxidant activity of two important memory enhancing medicinal plants *Baccopa monnieri* and *Centella asiatica*. *Indian Journal of Pharmacology*, **44(1):** 114-7.

Meng, Z., and Zeng.Y. (1988). Determination of Asiaticoside (Fig. 1e) contained in Sanjinpian. *Journal of China Pharmaceutical University*, **19**: 205.

Mohandas Rao, K.G., Muddanna,Rao, S., Gurumadhva, and Rao S. (2009). Enhancement of amygdaloid neuronal dendritic arborization by fresh leaf juice of *Centella asiatica* (Linn) during growth spurt period in rats. *Evidence-based Complementary and Alternative Medicine*, **6(2):** 203-10.

Mustafa, R.A., Abdul Hamid, A., Mohamed, S., and Bakar, F.A.(2010). Total phenolic compounds, flavonoids, and radical scavenging activity of 21 selected tropical plants. *Journal Food Sciences*, **75(1):** C28-35.

Nasir, M.N., Abdullah, J., Habsah, M., Ghani, R.I., and Rammes, G. (2011). Inhibitory effect of asiatic acid on acetylcholinesterase, excitatory post synaptic potential and locomotor activity. *Phytomedicine*, **19(3-4):** 311-6

Pan, Jian, Kai Guiqing, Yuan Chuanxun, Zhou Beibei, Jin Risheng, Yuan Yuan (2007). Separation and determination of madecassic acid in extracts of *Centella asiatica* using high performance liquid chromatography with β-cyclodextrin as mobile phase additive. *China Journal of Chromatography*, **25(3):** 316–318.

Pan, Y., Abd-Rashid, B.A., Ismail,Z., Ismail, R., Mak, J.W., Pook, P.C., Er, H.M., and Ong, C.E. (2011). *In vitro* modulatory effects of *Andrographis paniculata, Centella asiatica* and *Orthosiphon stamineus*on cytochrome P450 2C19 (CYP2C19). *Journal of Ethnopharmacology*, **133(2):** 881-7.

Pan Y, Abd-Rashid B.A, Ismail, Z., Ismail, R., Mak J.W., Pook, P.C., Er, H.M., and Ong, C.E. (2010). *In vitro* modulatory effects on three major human cytochrome P450 enzymes by multiple active constituents and extracts of *Centella asiatica*. *Journal of Ethnopharmacology*, **130(2):** 275-83.

Paocharoen, V. (2010). The efficacy and side effects of oral *Centella asiatica* extract for wound healing promotion in diabetic wound patients. *Journal of Medical Association of Thailand,* **93(Suppl 7):** S166-70.

Pittella, F., Dutra, R.C., Junio,r D.D., Lopes, M.T., and Barbosa N.R. (2009). Antioxidant and cytotoxic activities of *Centella asiatica* (L) Urb. *International Journal of Molecular Sciences,* **10(9):** 3713-21.

Pedretti, M. (1987). Erboristeria Domani 6, 45.

Rahuman, A.A., Venkatesan, P., and Gopalakrishnan, G. (2008). Mosquito larvicidal activity of oleic and linoleic acids isolated from *Citrullus colocynthis* (Linn.) Schrad. *Parasitology Research,* **103(6):** 1383-90.

Rajkumar, S., and Jebanesan, A. (2007). Repellent activity of selected plant essential oils against the malarial fever mosquito *Anopheles stephensi. Tropical Biomedicine.,* **24(2):** 71-5.

Ramaswamy, A. S., Periyasamy, S.M., and Basu, N. (1970). Pharmacological studies on *Centella asiatica*Linn. (Brahma manduki) (n. o. Urnbelliferae). *Indian Journal of Medical Research,* 1- 75.

Ravi Subban, A., Veerakumar, R., Manimaran, K., Hashim, M., and Indira, Balachandran.(2008). Two new flavonoids from *Centella asiatica* (Linn.). *Journal of Natural Medicine,* **62**: 369–373

Sahu, N.P., Roy, S.K., and Mahato, S.B. (1989). Spectroscopic determination of structures of triterpenoid trisaccharides from *Centella asiatica. Phytochemistry,* **28(10):** 28-52.

Sainath, S.B., Meena, R., Supriya, C.H., Reddy, K.P., and Reddy, P.S. (2011).Protective role of *Centella asiatica* on lead-induced oxidative stress and suppressed reproductive health in male rats. *Environmental Toxicology and Pharmacology,* **32(2):** 146-54.

Schaneberg, B.T., Mikell, J.R., Bedir, E., and Khan, I.A. (2003). An improved HPLC method for quantitative determination of six triterpenes in *Centella asiatica* extracts and commercial products. *Pharmazie,* **58**: 381.

Sakina, M. R., and Dandiya, P. C.(1990). Pycho-neuropharmacological profile of *Centella asiatica* extract. *Fitoterapia,* **61**: 291- 296.

Shinomol, G.K., and Muralidhara, H. (2008). Prophylactic neuroprotective property of *Centella asiatica* against 3-nitropropionic acid induced oxidative stress and mitochondrial dysfunctions in brain regions of prepubertal mice. *Neuro Toxicology,* **29(6):** 948-57

Shinomol, G.K., Ravikumar, and Muralidhara, H. (2010). Prophylaxis with *Centella asiatica* confers protection to prepubertal mice against 3-nitropropionic-acid-induced oxidative stress in brain. *Phytotherapy Research,* **24(6):** 885-92.

Shinomol, G.K., and Muralidhara,Bharath, M.M.(2011). Exploring the role of "Brahmi" (*Bocopa monnieri* and *Centella asiatica*) in brain function and therapy Recent Pat. *Endocrine, Metabolic and Immune Drug Discovery,* **5(1):** 33-49.

Singh, R.H., Narsimhamurthy, K., and Singh, G. (2008). Neuronutrient impact of Ayurvedic Rasayana therapy in brain aging. *Biogerontology*, **9(6):** 369-74.

Sevior, D.K., Hokkanen, J., Tolonen, A., Abass, K., Tursas, L., Pelkonen, O., and Ahokas, J.T. (2010). Rapid screening of commercially available herbal products for the inhibition of major human hepatic cytochrome P450 enzymes using the N-in-one cocktail. *Xenobiotica*, **40(4):** 245-54.

State Drug Administration of China (2000). Requirements for studying fingerprint of traditional Chinese medicine injection (Draft). *Chinese traditional patent medicine*, **22**: 671–675

Tang, X.L., Yang, X.Y., Jung, H.J., Kim, S.Y., Jung, S.Y., Choi, D.Y., Park, W.C., and Park, H. (2009). Asiatic acid induces colon cancer cell growth inhibition and apoptosis through mitochondrial death cascade. *Biological Pharmaceutical Bulletin*, **32(8):** 1399-405.

Tumen, I., Akkol, E.K., Süntar, I., and Kele°, H. (2011). Wound repair and anti-inflammatory potential of essential oils from cones of Pinaceae: preclinical experimental research in animal models. *Journal of Ethnopharmacology*, **137(3):** 1215-20.

Wattanathorn, J., Mator, L., Muchimapura, S., Tongun, T., Pasuriwong, O., Piyawatkul, N., Yimtae, K,Sripanidkulchai, B., and Singkhoraard, J. (2008). Positive modulation of cognition and mood in the healthy elderly volunteer following the administration of *Centella asiatica*. *Journal of Ethnopharmacology*, **116(2):** 325-32.

WHO (2000). General guidelines for methodologies on research and evaluation of traditional medicine.WHO, Geneva, p 1

Wijeweera, P., Arnason, J.T., Koszycki, D., and Merali, Z. (2006).Evaluation of anxiolytic properties of Gotukola—(*Centella asiatica*) extracts and Asiaticoside in rat behavioral models. *Phytomedicine*, **13(9-10):** 668-76.

Wojcikowski, K., Wohlmuth, H., Johnson, D.W., Rolfe, M., and Gobe, G. (2009). An *in vitro* investigation of herbs traditionally used for kidney and urinary system disorders: potential therapeutic and toxic effects. *Nephrology (Carlton)*, **14(1):** 70-9.

Xiao, J., Che, Z.T., and Bi, K.S. (2000). Pre-column derivatization HPLC method for the determination of asiaticoside in *Centella asiatica* and Sanjinpian. *Acta Pharmacologica Sinica*, **8:** 605.

Xu, C.L., Wang, Q.Z., Sun, L.M., Li, X.M., Deng, J.M., Li, L,F., Zhang, J., Xu, R., and Ma, S.P. (2012). Asiaticoside: Attenuation of neurotoxicity induced by MPTP in a rat model of Parkinsonism via maintaining redox balance and up-regulating the ratio of Bcl-2/Bax. *Pharmacology, Biochemistry and Behavior*, **100(3):** 413-8.

Xu, Y., Cao, Z., Khan, I., and Luo, Y. (2008). Gotu Kola (*Centella asiatica*) extract enhances phosphorylation of cyclic AMP response element binding protein in neuroblastoma cells expressing amyloid beta peptide. *Journal of Alzheimer's Disease*, **13(3):** 341-9.

Yunianto, I., Das, S., and Mat Noor, M. (2010). Antispermatogenic and antifertility effect of Pegaga (*Centella asiatica* L) on the testis of male Sprague-Dawley rats. *Clinical Terapeutica,* **161(3):** 235-9.

Yun, K.J., Kim, J.Y., Kim, J.B., Lee, K.W., Jeong, S.Y., Park, H.J., Jung, H.J., Cho, Y.W., Yun, K., and Lee, K.T. (2008). Inhibition of LPS-induced NO and PGE2 production by asiatic acid via NF-kappa B inactivation in RAW 264.7 macrophages: possible involvement of the IKK and MAPK pathways. *International Immunopharmacology,* **8(3):** 431-41.

Zeng, J.G., Song, Y.F., Hou, T.Z., and Xu, X. (2000). Determination of the triterpenes constituents in *Centella asiatica* by HPLC. *Primary Journal of Chinese Materia Medica,* **6:** 7.

Zhang, L., Li, H.Z., Gong, X., Luo, F.L., Wang, B., Hu, N., Wang, C.D., Zhang, Z., and Wan, J.Y. (2010). Protective effects of Asiaticoside on acute liver injury induced by lipopolysaccharide/D-galactosamine in mice. *Phytomedicine,* **17(10):** 811-9.

Zhang, L.L., Wang, H.S., and Yao, Q.Q.(2005). Chemical constituents from whole plant of *Centella asiatica*. *Chinese Traditional and Herbal Drugs,* **36(12):** 1761.

Zhang, L.N., Zheng, J.J., Zhang, L., Gong, X., Huang, H., Wang, C.D., Wang, B., Wu, M.J., Li, X.H., Sun, W.J., Liu, Y.J., and Wan, J.Y. (2011). Protective effects of asiaticoside on septic lung injury in mice. *Experimental and Toxicological Pathology,* **63(6)**: 519-25.

Zhang, Z., Qin, D.L., Wan, J.Y., Zhou, Q.X., Xiao, S.H., and Wu, K. (2008). Effects of asiaticoside on the balance of inflammatory factors of mouse's acute lung injury induced by LPS. *Zhongyaocai,* **31(4):** 547-9.

Traditional and Folk Herbal Medicine: Recent Researches Vol. 2 (2014) *Pages* **153–187**
Editor: **V.K. Gupta**
Published by: **DAYA PUBLISHING HOUSE, NEW DELHI**

6

Orthosiphon aristatus: A Review of Traditional Uses, Phytochemical Profile, and Pharmacological Properties

Tsun-Thai Chai[1,4,*], Fai-Chu Wong[1,4], Fazilah Abd Manan[2], Keng-Fei Ooh[1], and Nor Ismaliza Mohd Ismail[3,4]

ABSTRACT

Orthosiphon aristatus is a medicinal herb traditionally used as remedies for various ailments in Southeast Asia and in Europe. A fast-growing number of studies over the last decade have unravelled the phytochemical profiles of the herb as well as its pharmacological properties, which to different extends, have provided scientific basis to its traditional uses. Large numbers of phytochemical constituents have been isolated and identified from the herb, which include terpenes, flavonoids, caffeic acid derivatives and essential oils. Research to-date also revealed diverse bioactive properties, such as anti-inflammatory, antioxidant, antihypertensive, antimicrobial, anti-angiogenic, analgesic, hepatoprotective and hypoglycaemic properties in O. aristatus. In this review, an up-to-date overview and assessment of current knowledge of the traditional uses, phytochemistry and pharmacological properties of O. aristatus are presented.

Keywords: Herbal medicine, *Orthosiphon aristatus*, Pharmacology, Phytochemistry.

1 Department of Chemical Science, Faculty of Science, Universiti Tunku Abdul Rahman, Jalan Universiti, Bandar Barat 31900 Kampar, Malaysia.

2 Department of Biosciences and Health Sciences, Faculty of Biosciences and Medical Engineering, Universiti Teknologi Malaysia, 81310 Johor Bahru, Malaysia.

3 Department of Biological Science, Faculty of Science, Universiti Tunku Abdul Rahman, Jalan Universiti, Bandar Barat 31900 Kampar, Malaysia.

4 Centre for Biodiversity Research, Universiti Tunku Abdul Rahman, Malaysia.

* *Corresponding author*: E-mail: chaitt@utar.edu.my

Introduction

Orthosiphon aristatus (Blume) Miq. (Family Lamiaceae) is a native plant of tropical Asia (Malaysia, Singapore, Indonesia, Thailand, Cambodia, Vietnam, the Philippines, Papua New Guinea), temperate Asia (mainland China and Taiwan) as well as tropical Australia (Queensland). *O. aristatus* is referred to by different vernacular names, which include Java-tea (English), Thé de Java (France), kumis kucing (Indonesia, Malaysia), misai kucing (Malaysia) and ya nuat maeo (Thailand) (Germplasm Resources Information Network, 2009; Samy *et al.*, 2009; Wiart, 2006). Synonyms of the taxon include: *Clerodendranthus spicatus* (Thunb.) C. Y. Wu ex H. W. Li, *Clerodendrum spicatum* Thunb., *Ocimum aristatum* Blume, *Orthosiphon spicatus* (Thunb.) Bak., *Orthosiphon spiralis* (Linn.) Merrill, and *Orthosiphon stamineus* Benth. (Barnes *et al.*, 2007; Germplasm Resources Information Network, 2009; Khare, 2007). For convenience, we will refer to the plant as *Orthosiphon aristatus* throughout this review.

O. aristatus is a perennial herb that grows in the wild, along forest fringes, but also on wastelands and along roadsides (Samy *et al.*, 2009). Generally, the plant grows to a height of 75 cm. It has a quadrangular, purplish stem. Its leaves are simple, glabrous, lanceolate with serrated margins, and arranged in pairs. The flowers are usually light purple, borne along a maroon terminal raceme. The long protruding stamens give the flower its characteristic resemblance to cat's whiskers (Figure 6.1). The fruit of the plant is an oblong-ovoid nutlet of 1.5-2 mm in length (Samy *et al.*, 2009). In Malaysia, two varieties of *O. aristatus* can be found. The purple and white varieties can be distinguished by the colour of their corolla and calyx as well as their leaf characteristics (Chan and Loo, 2006).

O. aristatus is cultivated both as an ornamental plant and a medicinal herb in several tropical countries, including Malaysia and Indonesia (Handa *et al.*, 2006; Musa *et al.*, 2009; Wiart, 2006). In Malaysia, the plant is considered to have great potential for commercialisation (Jaganath and Ng, 2000). *O. aristatus* now appears in many commercial products in the forms of powdered herb, dried leaves, tea sachets, drinks, extracts, tablets and capsules.

Despite its widespread and long history of uses as an herbal medicine in different parts of the world, scientific research on the therapeutic potentials of *O. aristatus* only began to intensify over the last ten years or so. At present, the findings in the literature concerning the phytochemical profile and pharmacological properties of the plant have validated to a certain extent some of its purported therapeutic benefits. In this review, we endeavoured to compile an up-to date and comprehensive review of *O. aristatus* that encompasses its traditional uses, phytochemistry and pharmacology.

Traditional Uses

O. aristatus has been utilised in the folk or traditional medicine of Asian cultures for centuries. Besides its principal use as a diuretic, the herb is used as a remedy against renal and urinary disorders as well as various other diseases (Wiart, 2006).

In Malaysia, the aerial part of *O. aristatus* is used for controlling high blood pressure, rheumatic fever, gout, arthritis, and diabetes (Khatun *et al.*, 2011). Decoction of the plant is also consumed in Malaysia to eliminate bladder and kidney stones

Figure 6.1: (A) A flowering plant of *Orthosiphon aristatus*. (B) Long, protruding stamens of the flower which resemble cat's whiskers.

(Samy *et al.*, 2009). In Thailand, the plant is used as diuretic and anti-lithiatic agents (Premgamone *et al.*, 2001; Woottisin *et al.*, 2011) as well as treatment for dysuria (Ngamrojanavanich *et al.*, 2006). In Myanmar, the leaves of *O. aristatus* are used to treat diabetes, as well as urinary and renal diseases (Bwin and Gwan, 1967). In Vietnam, the plant is used as a diuretic, as well as treatment for influenza, eruptive fever, rheumatism, hepatitis, jaundice, and biliary lithiasis (Wiart, 2006). In Indonesia, the plant is used with a mixture of other herbs to treat jaundice, diabetes, gout, rheumatism, and arteriosclerosis (Samy *et al.*, 2009).

In the Philippines, the leaves of *O. aristatus* are used as diuretic and also as treatments for renal illness, arthritis, gallstones, and diabetes (Eusebio and Umali, 2004). In Taiwan, the herb is used for the treatment of urinary diseases, such as kidney inflammation, kidney stones, and dysuria (Hsu *et al.*, 2010). The Japanese, meanwhile, consume *O. aristatus* tea to facilitate body detoxification (Awale *et al.*, 2003b). In India, the leaves of *O. aristatus* are used as a diuretic in the treatment of nephrosis and severe oedema. Infusion of leaves is also used in the treatment of other kidney and bladder diseases as well as rheumatism and gout (Khare, 2007). In India, the herb is also used in combination with *Andrographis paniculata* Nees to treat diabetes (Wiart, 2006).

Table 6.1: Traditional uses of *O. aristatus*.

Health Problems/ Uses	*Parts of Plant Used/ Preparation*	*References*
Kidney and bladder stones	The whole plant	Samy *et al.*, 2009
Kidney stone	Mixture of *O. aristatus* leaves, *Phyllanthus urinaria* plant, *Stachytarpheta mutabilis* leaves, *Curcuma xanthorrhiza*, and young corn.	Samiyah, 2004
Kidney and bladder diseases, rheumatism, gout	Leaves	Khare, 2007
Oedema, eruptive fever, influenza, rheumatism, hepatitis, jaundice	Aerial part	Tezuka *et al.*, 2000
Diabetes	Leaves	Bwin and Gwan, 1967; Eusebio and Umali, 2004; Wiart, 2006
Hypertension	Leaves	Musa *et al.*, 2009
Body detoxification	Leaves	Awale *et al.*, 2003b
Diuretic agent	Leaves	Eusebio and Umali, 2004; Khare, 2007; Masuda *et al.*, 1992a
Antibacterial agent	The whole plant/leaves	Chen *et al.*, 1989; Samy *et al.*, 2009
Sexual energy booster	Flowers and leaves	Musa *et al.*, 2009
Healthy beverage (Java tea)	Leaves/tops of stems	Committee on Herbal Medicinal Products, 2011

The extract *of O. aristatus* is also a well-recognised traditional herbal medicinal product in a number of European countries, including Germany, France, Spain, and Poland. Europeans consume Java tea, an infusion of the dried leaves, for healthy urinary system, reduction of bacterial infection, and treatment of inflammatory diseases (Committee on Herbal Medicinal Products, 2011; Samy *et al.*, 2009).

Table 6.1 shows some traditional uses of *O. aristatus*. To date, investigations on the pharmacological properties of the plant have, to varying degrees, substantiated some of its traditional uses (see later sections).

Phytochemical Profiles

Over the years, enormous numbers of secondary metabolites from *O. aristatus* have been isolated and identified. As summarised in Table 6.2, the majority of these identified phytochemicals are of the terpene family. Among the identified terpenes, highly-oxygenated diterpenes represent the largest group, followed by triterpenes, sesquiterpenes and monoterpenes.

Table 6.2: Secondary metabolites identified in *O. aristatus.*

Components	*References*
Monoterpenes	
camphene	Hossain *et al.*, 2008
limonene	Hossain *et al.*, 2008
β-pinene	Hossain *et al.*, 2008
Sesquiterpenes	
α-humulene	Hossain *et al.*, 2008
β-bourbonene	Hossain *et al.*, 2008
β-caryophyllene	Hossain *et al.*, 2008
β-elemene	Hossain *et al.*, 2008
Diterpenes	
14-deoxo-14-O-acetylorthosiphol Y	Nguyen *et al.*, 2004
2-O-deacetylorthosiphol J	Awale *et al.*, 2003d
3-O-deacetylorthosiphol I	Awale *et al.*, 2003d
6-hydroxyorthosiphol B	Awale *et al.*, 2003d
7-O-deacetylorthosiphol B	Awale *et al.*, 2003d
neoorthosiphols A, B	Shibuya *et al.*, 1999
neoorthosiphonone A	Awale *et al.*, 2004
norstaminol A	Stampoulis *et al.*, 1999b
norstaminolactone A	Awale *et al.*, 2002b
norstaminols B, C	Awale *et al.*, 2002b
orthosiphols A and B	Masuda *et al.*, 1992b; Siddiqui and Ismail, 2011
orthosiphols F–I	Stampoulis *et al.*, 1999a

Contd...

Table 6.2–*Contd...*

Components	*References*
orthosiphols F–J	Tezuka *et al.*, 2000
orthosiphols O–Q	Awale *et al.*, 2002a
orthosiphols R-T	Awale *et al.*, 2002b
orthosiphonone C, D	Nguyen *et al.*, 2004
secoorthosiphol A-C	Awale *et al.*, 2002c
siphonols A–E	Awale *et al.*, 2003d
staminol A	Stampoulis *et al.*, 1999a
staminolactones A, B	Stampoulis *et al.*, 1999b; Tezuka *et al.*, 2000
staminols A, B	Tezuka *et al.*, 2000
staminols C, D	Nguyen *et al.*, 2004
Triterpenes	
betulinic acid	Banskota *et al.*, 2003; Hossain and Ismail, 2010
hydroxybetulinic acid	Hossain and Ismail, 2010
maslinic acid	Hossain and Ismail, 2010
oleanolic acid	Banskota *et al.*, 2003; Hossain and Ismail, 2010
ursolic acid	Banskota *et al.*, 2003; Hossain and Ismail, 2010
α-amyrin	Hossain and Ismail, 2010
β-amyrin	Hossain and Ismail, 2010
Flavonoids	
3'-hydroxy-5,6,7,4'-tetramethoxyflavone	Akowuah *et al.*, 2004b; Yam *et al.*, 2010
5, 4'-dihydroxy-6,7-dimethoxyflavone	Banskota *et al.*, 2003; Sumaryono *et al.*, 1991
5,6,7,3'-tetramethoxy-4'-hydroxy-8-C-prenylflavone	Hossain and Mizanur Rahman, 2011
5-hydroxy-6,7,3',4'-tetramethoxyflavone	Hossain and Mizanur Rahman, 2011
6-hydroxy-5,7,4'-trimethoxyflavone	Hossain and Mizanur Rahman, 2011
7,3',4'-tri-O-methylluteolin	Banskota *et al.*, 2003
eupatorin	Akowuah *et al.*, 2004b; Hossain and Mizanur Rahman, 2011; Olah *et al.*, 2003; Yam *et al.*, 2010
ladanein	Banskota *et al.*, 2003
salvigenin	Hossain and Mizanur Rahman, 2011
sinensetin	Akowuah *et al.*, 2004b; Hossain and Ismail, 2012; Olah *et al.*, 2003; Yam *et al.*, 2010
tetramethylscutellarein	Banskota *et al.*, 2003
Caffeic acid derivatives	
2,3-dicaffeoyltartaric acid	Banskota *et al.*, 2003; Sumaryono *et al.*, 1991
2-caffeoyltartaric acid	Banskota *et al.*, 2003; Sumaryono *et al.*, 1991
caffeic acid	Olah *et al.*, 2003

Contd...

Table 6.2–*Contd...*

Components	*References*
caftaric acid	Nuengchamnong *et al.*, 2011
cichoric acid	Olah *et al.*, 2003
danshensu	Nuengchamnong *et al.*, 2011
rosmarinic acid	Akowuah *et al.*, 2004b; Olah *et al.*, 2003
sagerinic acid	Nuengchamnong *et al.*, 2011
salvianolic acid B	Nuengchamnong *et al.*, 2011
Others	
1-octen-3-ol	Hossain *et al.*, 2008
aurantiamide acetate	Banskota *et al.*, 2003
methylripariochromene A	Guerin *et al.*, 1989
vomifoliol	Banskota *et al.*, 2003
β-sitosterol	Banskota *et al.*, 2003

In addition to members from the terpene family, other chemical constituents isolated from *O. aristatus* include, but are not limited to flavonoids, caffeic acid derivatives, methylripariochromene A, and ursolic acid. Based on a reported quantitative analysis using aqueous methanolic extract, sinensetin is the major flavonoid constituent of the plant (Banskota *et al.*, 2003; Sumaryono *et al.*, 1991). New entries are being added regularly to this expanding list of *O. aristatus*-derived phytochemicals. To illustrate the point, just last year, a new flavonoid compound (5,6,7,3′-tetramethoxy-4′-hydroxy-8-C-prenylflavone) was isolated and identified for the first time from this medicinal plant (Hossain and Mizanur Rahman, 2011).

Among the phytochemical compounds isolated from *O. aristatus*, some are present in abundant quantity. For instance, by using *O. aristatus* plants cultivated in Indonesia as starting material, isolation of methylripariochromene A was reported to have a 4 per cent yield relative to dry matter used (Guerin *et al.*, 1989). The structure of this compound had also been determined by X-ray crystallography (Guerin *et al.*, 1989). In contrast to methylripariochromene A, other chemical constituents are present in relatively minute quantities. For example, secoorthosiphols A-C, three of *O. aristatus*'s minor diterpene constituents, were found to have below 0.001 per cent yield from methanolic extracts. Among these three compounds, secoorthosiphols B had a mere 0.000061 per cent yield (Awale *et al.*, 2002c). With the advancement of techniques used in phytochemical isolation and identification, it is foreseeable that more and more novel chemical constituents will be reported for *O. aristatus* in the time to come.

Pharmacological Properties

Anti-inflammatory

Recent research using cells and animal models have provided evidence supporting the traditional use of *O. aristatus* as a remedy for inflammatory disorders. Eupatorin, sinensetin and ursolic acid were reported to be the key anti-inflammatory agents and the molecular basis of their actions were reported (Hsu *et al.*, 2010; Laavola *et al.*, 2012; Yam *et al.*, 2008; Yam *et al.*, 2010).

In vivo anti-inflammatory activity of *O. aristatus* was evidenced by the inhibition of the second phase of carrageenan-induced oedema in rats by methanolic extract and flavonoid-rich chloroform extract fraction the plant (Yam *et al.*, 2008; Yam *et al.*, 2010). The second phase of carrageenan-induced oedema involves prostaglandin synthesis and release. Hence, the anti-inflammatory action of the plant was proposed to be associated with inhibition of prostaglandin synthesis (Yam *et al.*, 2008; Yam *et al.*, 2010).

Recently, the molecular basis of the anti-inflammatory action of a flavonoid-rich chloroform extract fraction of *O. aristatus* leaves and its key bioactive compounds, eupatorin and sinensetin, has been elucidated (Laavola *et al.*, 2012). Eupatorin and sinensetin alleviated carrageenan-induced inflammation in mice. Furthermore, the two flavonoids attenuated the production of nitric oxide (NO), prostaglandins, and tumour necrosis factor a (TNF-α) in lipopolysaccharide-activated J774 murine macrophage cells (Laavola *et al.*, 2012). TNF-α is a key regulator of inflammatory response and plays a central role in the pathogenesis of inflammatory disorders (Bradley, 2008). The findings thus support the hypothesis that inhibition of the release or production of inflammatory mediators from activated inflammatory cells underlies the anti-inflammatory properties of eupatorin and sinensetin (Laavola *et al.*, 2012).

O. aristatus chloroform extract, its flavonoid-rich chloroform extract fraction, as well as eupatorin and sinensetin inhibited cyclooxygenase-2 (COX-2)-dependent prostaglandin synthesis and inducible NO synthase (iNOS)-dependent NO production possibly by down-regulating the expression of iNOS and COX-2 at the mRNA and/or protein levels (Laavola *et al.*, 2012). Eupatorin and sinensetin also down-regulated the activity of signal transducer and activator of transcription-1α (STAT1α), a transcription factor required for the activation of iNOS promoter, hence iNOS gene expression (Pautz *et al.*, 2010). Moreover, data collected also imply possible involvement of eupatorin and sinensetin as inhibitors of the iNOS and COX-2 enzymes (Laavola *et al.*, 2012). These findings suggest that anti-inflammatory action of eupatorin and sinensetin may be attributed to their ability to target the iNOS and COX-2 pathways both at the transcriptional and posttranscriptional levels.

Another study using lipopolysaccharide-stimulated RAW 264.7 murine macrophage cells as a model system (Hsu *et al.*, 2010) also reported evidence of anti-inflammatory activity in *O. aristatus*. In the study, ethanol extract of the plant inhibited the production of NO, intracellular reactive oxygen species and prostaglandin E_2 in the macrophage cells. Reduced mRNA and protein expression of iNOS and COX-2 in the cells provided a molecular basis for the mitigated prostaglandin synthesis. Ursolic acid was proposed to be responsible for the reduced production of NO and prostaglandin E_2 in the macrophage cells. Supporting this proposal is the concurrently high ursolic acid content and high NO inhibition effected by the ethanol extract compared to two other extracts (methanol and aqueous extracts) tested in the study (Hsu *et al.*, 2010). The inhibitory effect of ursolic acid isolated from *Plantago major* on COX-2-catalysed prostaglandin biosynthesis *in vitro* (Ringbom *et al.*, 1998) lends further support to this proposed role of ursolic acid in the *O. aristatus* extract.

The ability of ursolic acid to reduce production of NO and intracellular reactive oxygen species in macrophage cells (Hsu *et al.*, 2010) may be attributable to its antioxidant properties. The ability of ursolic acid to scavenge NO, hydroxyl radicals and superoxide anion radicals *in vitro* has been demonstrated (Ramachandran and Prasad, 2008). Thus, the reduction in NO accumulation in the macrophage cells (Hsu *et al.*, 2010) could be partly accounted for by NO scavenging, in addition to repressed iNOS expression. Furthermore, ursolic acid pretreatment protects human lymphocytes from UV-B-induced cell death, lipid peroxidation and DNA damage (Ramachandran and Prasad, 2008). This capacity of ursolic acid is of interest in the context of inflammation. In an inflammatory response, NO and reactive oxygen species can react to form peroxynitrite, a species capable of effecting tissue injury (Korhonen *et al.*, 2005). Hence, the potential role of ursolic acid in protecting cells or tissues against inflammation-associated oxidative injury merits further studies.

A number of compounds isolated from *O. aristatus* were reported to have NO scavenging activity (Awale *et al.*, 2003a; Awale *et al.*, 2003b; Awale *et al.*, 2003c; Awale *et al.*, 2003d; Awale *et al.*, 2004; Nguyen *et al.*, 2004). These compounds are of interest as potential anti-inflammatory agents considering the key role of NO as an inflammatory mediator (Korhonen *et al.*, 2005). Nevertheless, future characterisations of the *in vivo* effects of these compounds on animal models and/or components of the inflammatory pathways are necessary to validate their anti-inflammatory potential.

Diuretic

The diuretic effects of *O. aristatus* extracts have been investigated in several animal studies. These studies mostly examined the effects of oral administration of single extract doses. Furosemide and hydrochlorothiazide, two synthetic diuretics, were frequently used as positive controls.

Two studies analysed the diuretic activity of aqueous extracts of *O. aristatus*. Englert and Harnischfeger (1992) reported that oral administration of aqueous extract (750 mg/kg body weight) to rats did not enhance urinary output, but increased excretion of Na^+, K^+, and Cl^- ions. The saluretic effect was comparable to that of furosemide administered at 100 mg/kg body weight (Englert and Harnischfeger, 1992). By contrast, in a more recent study, oral doses of aqueous extract at lower concentrations (5 and 10 mg/kg body weight) induced dose-dependent diuresis over four hours following treatment. Excretion of Na^+ and Cl^- showed no discernible pattern of changes, whereas K^+ excretion was increased (Adam *et al.*, 2009). The diuretic effect of the extracts was less potent compared with furosemide and hydrochlorothiazide (10 mg/kg body weight). Based on the changes in urine and electrolyte excretion, it was proposed that the *O. aristatus* extracts probably had a different mechanism of action compared with the synthetic diuretics (Adam *et al.*, 2009).

Two other studies investigated the diuretic effects of hydroalcoholic extracts of *O. aristatus* in rats. Intraperitoneal administration of a hydroalcoholic extract of *O. aristatus* (50 mg/kg body weight) in rats significantly enhanced diuresis over the 2-24 hours following treatment (Beaux *et al.*, 1999). The diuretic effect was comparable to that induced by hydrochlorothiazide (10 mg/kg body weight). Nevertheless, while hydrochlorothiazide increased excretion of Na^+ and K^+, natriuretic and kaliuretic effects of the extract could not be affirmed. Olah *et al.* (2003) reported that 50 per cent and 70 per cent ethanolic extracts orally administered at 700 mg/kg body weight increased Na^+ and K^+ excretion. However, only 50 per cent ethanolic extract increased urine volume. HPLC analysis revealed the presence of caffeic acid derivatives (caffeic acid, cichoric acid, and rosmarinic acid) and flavonoids (sinensetin and eupatorin) in both ethanolic extracts. It was proposed that the higher abundance of these compounds in the 50 per cent ethanolic extract may explain its higher diuretic efficacy in comparison with the 70 per cent ethanolic extract (Olah *et al.*, 2003).

Only one study investigated the effects of repeated administrations of methanolic extracts of *O. aristatus* leaves in rats (Arafat *et al.*, 2008). In the study, single doses of 100 per cent and 50 per cent methanolic extracts (2 g/kg body weight) did not enhance urinary output following oral administration. By contrast, repeated administrations of both extracts (0.5 g/kg body weight) over seven days increased urinary output, with greater effectiveness exhibited by the 50 per cent methanolic extract. Moreover, urinary excretion of Na^+ and K^+ was only significantly increased when repeated oral administrations were performed with the 50 per cent methanolic extract. The mechanism of action of the 50 per cent methanolic extract has not been established. Nevertheless, the greater diuretic and saluretic effects of the 50 per cent methanolic extract may be associated with the presence of flavonoids and rosmarinic acid in the extract (Arafat *et al.*, 2008).

Arafat *et al.* (2008) found that the higher diuretic activity of 50 per cent methanolic extract in comparison with 100 per cent methanolic extract was also associated with higher (2.3-fold) potassium content in the former. The authors raised the possibility that the high potassium content of the extract may play a role in diuresis induction. However, earlier reports of the lack of diuresis induction by potassium aspartate or potassium chloride do not support this hypothesis (Beaux *et al.*, 1999; Englert and Harnischfeger, 1992). According to Galati *et al.* (2002), the potassium content of an extract is unlikely to be the key contributor to diuretic activity. At present, the biological relevance of the potassium content of *O. aristatus* extract in diuresis induction and the possibility of synergism between potassium and other bioactive compounds in the extract remains unclear.

Similar to *O. aristatus*, dandelion extract is a botanical diuretic used in the traditional medicine of several cultures (Clare *et al.*, 2009; Schütz *et al.*, 2006). Being a significant source of potassium, the dandelion leaf extract has the added advantage of compensating for the loss of K^+ from diuresis. Hence, the dandelion extract is devoid of side effects due to K^+ loss, which accompany some conventional diuretics (Clare *et al.*, 2009; Racz Kotilla *et al.*, 1974; Schütz *et al.*, 2006). Future research should investigate whether *O. aristatus*, also a rich source of potassium, offers a similar benefit. Capacity of *O. aristatus* to compensate for K^+ loss or mitigate hypokalaemia, if confirmed, may add more value to its use as a diuretic.

Little is known about the bioactive constituents responsible for the diuretic effects of *O. aristatus* extracts. Matsubara *et al.* (1999) reported a study in which methylripariochromene A isolated from the chloroform-soluble fraction of aqueous extract of *O. aristatus* was orally administered to rats at 100 mg/kg body weight. The treatment increased urinary output by 3-fold and was comparable in potency to hydrochlorothiazide (25 mg/kg body weight). The quantity of Na^+, K^+ and Cl^- ions excreted in the urine increased, although urinary concentration of these ions was unaltered. Hydrochlorothiazide, by contrast, increased the urinary concentration of Na^+ and Cl^-. Hence, the authors suggested that the mechanism of action underlying the diuretic activity of methylripariochromene A differs from that of hydrochlorothiazide (Matsubara *et al.*, 1999). Based on the findings of Matsubara *et al.* (1999), methylripariochromene A is likely one of the bioactive constituents responsible for diuresis induction. Nevertheless, possible involvement of other active compounds cannot be ruled out.

Adenosine A_1 receptor antagonists promote diuresis and natriuresis with low risks of renal dysfunction, a side effect of some conventional diuretics (Hocher, 2010). Recently, seven methoxy flavonoids acting as putative adenosine A_1 receptor antagonists were isolated from *O. aristatus*. The seven compounds are eupatorin, eupatoretin, sinensetin, pilloin, tetramethylscutellarein, 3′-hydroxy-4′,5,6,7-tetramethoxyflavone, and 5,6-dihydroxy-7,4′-dimethoxyflavone (Yuliana *et al.*, 2009). Information such as the *in vivo* efficacy, bioavailability and toxicity of these compounds is still unavailable and merits future investigations. The possibility that the diuretic activity of *O. aristatus* is mediated by adenosine A_1 receptor antagonists, if it can be confirmed, would make the plant an attractive option in diuretic therapy.

On the whole, multiple studies have demonstrated the diuretic activity of *O. aristatus*, justifying its traditional use as a diuretic agent. There remain gaps in our knowledge of the bioactive compounds which are responsible for the diuretic activity of the herb. Hence, future research should be directed towards unravelling the physiological and molecular basis of the diuretic effect of *O. aristatus* extracts and their active principles.

Anti-lithiatic

O. aristatus is traditionally used in the prevention and treatment of urinary stones (Hsu *et al.*, 2010; Samy *et al.*, 2009; Woottisin *et al.*, 2011). Diuretic activity is one of the mechanisms underlying the anti-lithiatic or stone-prevention effects of some medicinal herbs (Gürocak and Küpeli, 2006). The possibility that diuresis may account for the beneficial effects of *O. aristatus* in stone disease management has been proposed (Yuliana *et al.*, 2009). However, evidence of a direct or strong correlation between the diuretic effect of the herb and reduction in stone growth or formation is still unavailable.

About 80 per cent of all urinary stones are composed of calcium oxalate, whereas uric acid stones are the second most common (Knoll, 2010). The effects of *O. aristatus* in the management of both types of stones have been addressed by a limited number of animal and clinical studies.

In a prospective randomised clinical trial, Premgamone *et al.* (2001) investigated the effects of *O. aristatus* treatment on 48 patients with renal stones at least 10 mm in size. The patients were either given 250 mL of *O. aristatus* tea (made from a 2.5 g tea bag) twice daily, or treated with titrated sodium potassium citrate three times a day to maintain urinary pH between 6.2 and 6.8. Both treatments led to a time-dependent reduction in stone size over a period of 18 month. *O. aristatus* treatment resulted in a 28.6 ± 16 per cent reduction in stone size per year, which was not statistically different to the sodium potassium citrate treatment (33.8 ± 23.6 per cent). The chemical composition of the stones was unclear but was possibly mixed calculi containing both calcium oxalate and uric acid. Hence, stone size reduction in the patients may be attributable to increased excretion of calcium and uric acid (Premgamone *et al.*, 2001). Whether diuresis plays a role in the control of stone size in the patients is unclear.

Akanae *et al.* (2010) reported the beneficial effects of *O. aristatus* in a calcium oxalate stone-forming rat model. Daily doses of *O. aristatus* aqueous extract (200 mg/kg body weight) were administered by gastric intubations over two weeks. This was followed by the induction of hyperoxaluria and kidney calcium oxalate crystal deposition in the third week using ethylene glycol and vitamin D3. Pretreatment with the herbal doses alkalinised the urine, but had no effect on the urine volume of stone-forming rats. Notably, the number of crystal deposits in the renal tubules of the rats was reduced. The alleviation of oxidative stress in the kidney tissues was associated with increased protein expression of two antioxidant enzymes, superoxide dismutase and catalase (Akanae *et al.*, 2010). The findings of this study points to the importance of urine alkalinisation, but did not support the proposed role of diuresis in reducing stone formation. Importantly, *O. aristatus* may have the additional benefit of preventing renal tissue oxidative injury during stone formation by up-regulating the expression of antioxidant enzymes. More investigations are required to characterise the protective genes and proteins induced by *O. aristatus* and its active principles in relation to urolithiasis.

Interestingly, in contrast to the above study, Woottisin *et al.* (2011) found that *O. aristatus* did not reduce calcium crystal deposition in the kidneys of rats fed a glycolate diet to induce hyperoxaluria and kidney stones. Tablets, each containing 3.5 mg aqueous extract obtained from *O. aristatus* tea bag infusion, were orally administered daily over four weeks in the study. The reasons behind the discrepancy between the two studies (Akanae *et al.*, 2010; Woottisin *et al.*, 2011) which both involved induction of hyperoxaluria and calcium oxalate stone formation in rats is unclear. However, differences in the concentrations of extracts and their bioactive constituents may have contributed to the conflicting findings. Future research would allow better comparisons between studies if the bioactive chemical contents of the extracts used are characterised and reported.

The work of Arafat *et al.* (2008) hinted that *O. aristatus* may have an anti-lithiatic effect against uric acid stones. Their study demonstrated the hypouricaemic effect of single oral doses (0.5, 1, 2 g/kg body weight) of 50 per cent methanolic extract in rats at 6 hours after treatment. Nevertheless, the effect was mild compared with allopurinol, a xanthine oxidase inhibitor used to treat uric acid stones (Becker, 2007). Arafat *et al.* (2008) proposed that the antioxidant properties of the flavonoids, triterpenoids, and caffeic acid derivatives present in the extract may have attenuated xanthine oxidase-mediated uric acid formation. Consistent with the findings of Akanae *et al.* (2010), urine pH was also increased in response to the herbal treatment (Arafat *et al.*, 2008).

Risk factors involved in the pathogenesis of uric acid stones include hyperuricosuria, low urine volume, and persistently low urine pH (Ngo and Assimos, 2007). The study of Arafat *et al.* (2008) has shown the diuretic and urine alkalinisation effects of *O. aristatus*. Future studies should explore any direct correlation between *O. aristatus* treatment and mitigation of hyperuricosuria. Moreover, since urine pH is the key risk factor of uric acid stone formation (Ngo and Assimos, 2007), the ability of *O. aristatus* treatment to maintain elevated urine pH over a prolonged period should also be probed into.

It was suggested that *O. aristatus* may be used on a regular basis as a prophylactic and/or therapy, either on its own or in a combination therapy, to manage urinary stones (Akanae *et al.*, 2010; Premgamone *et al.*, 2001). However, more animal and clinical studies are required to fill the gaps in our knowledge of the effects of *O. aristatus* on stone passage, dissolution and prevention. At present, very little is known about the molecular mechanism of the herb in cases where anti-lithiatic effect was observed. The active principles involved also remains to be determined.

Antihypertensive

Studies investigating the antihypertensive effects of *O. aristatus* and its active compounds are scarce. The study of Matsubara *et al.* (1999) is at present the most comprehensive study which substantiates the antihypertensive effect of the plant and its use as a treatment for hypertension.

Matsubara *et al.* (1999) reported the antihypertensive effects of methylripariochromene A, a major constituent (2.3 per cent yield) isolated from water decoction of *O. aristatus* leaves. In an *in vivo* model, the compound was subcutaneously administered into conscious, stroke-prone, spontaneously hypertensive male rats at doses of 50 and 100 mg/kg body weight. The treatment continuously decreased the systolic blood pressure and heart rate of the rats over a period of 48 hours. In an *in vitro* model, methylripariochromene A suppressed agonists-induced contractions in endothelium-denuded rat thoracic aorta in order of high K^+ > *l*-phenylephrine > prostaglandin F2α. In an *ex vivo* model, methylripariochromene A reduced the contractile force of isolated bilateral guinea pig atria, without decreasing the beating rate. Furthermore, diuretic activity of the compound was confirmed in the same study. Matsubara *et al.* (1999) proposed that the antihypertensive effect of methylripariochromene A was mediated by its vasodilating action, decrease in cardiac output, and diuretic action.

In addition to methylripariochromene A, ten other compounds isolated from the water decoction of *O. aristatus* leaves also exhibited varying degrees of potency in suppressing K^+-induced contractions in endothelium-denuded rat thoracic aorta (Ohashi *et al.*, 2000). Based on IC_{50} values, the suppressive effects of orthosiphol A, orthosiphol B, orthosiphonone A, orthosiphonone B, and tertramethylscutellarein were more potent compared with methylripariochromene A (Ohashi *et al.*, 2000). Hence, a more comprehensive characterisation of the antihypertensive potential of these five compounds should be carried out.

A recent preliminary study reported antihypertensive effect of a 50 per cent methanolic extract orally administered to spontaneous hypertensive rats at doses of 250, 500, and 1000 mg/kg body weight (Azizan *et al.*, 2012). Following two weeks of treatment, systolic blood pressure of the rats decreased from 150 mm Hg to 114 mm Hg. Potency of the extract was comparable to irbesartan, administered at 20 mg/kg body weight in the positive control group. Antihypertensive effects of the extract were proposed to be mediated by diuresis and natriuresis (Azizan *et al.*, 2012), although no supporting data were presented.

There is only one clinical trial in the literature concerning the antihypertensive effect of *O. aristatus* extract. A combined nutraceutical containing alcoholic extract of *O. aristatus* leaves reduced the systolic and diastolic blood pressure as well as pulse pressure of hypertensive dyslipidaemic patients after eight weeks of treatment. The nutraceutical tested contained berberine, monakolin and policosanols, in addition to *O. aristatus* extract (Cicero *et al.*, 2012). One limitation of the study was that the antihypertensive activity of the constituents in the nutraceutical was not separately tested. Hence, the findings did not substantiate any convincing association between *O. aristatus* extract and antihypertensive effects in the subjects.

Antioxidant

Reactive oxygen species (ROS) and other free radicals promote tissue damage in living organisms and are implicated in diseases, such as arteriosclerosis, heart diseases and cancer, as well as in the aging process (Willcox *et al.*, 2004). The benefits and effectiveness of antioxidant therapies in the treatment of cancer and other diseases were discussed in several recent reviews (*e.g.* Di Domenico *et al.*, 2012; Small *et al.*, 2012; Tinkel *et al.*, 2012).

Over the last ten years, increasing levels of interests have been shown on the antioxidant potential of *O. aristatus*. The antioxidant activities of the plant have been demonstrated in multiple studies (*e.g.* Akowuah *et al.*, 2005; Hossain and Mizanur Rahman, 2011; Sahib *et al.*, 2009a; Yam *et al.*, 2010). A list of antioxidant parameters which have been evaluated in the extracts and/or solvent fractions of *O. aristatus* and with positive results is presented in Table 6.3.

DPPH scavenging activity was frequently used as an indicator of the antioxidant activity of *O. aristatus* extracts and/or solvent fractions. The DPPH scavenging assay IC_{50} values of *O. aristatus* extracts varied across studies, probably due to factors such as differences in extraction solvents used and the extraction procedures employed. In some studies, for example, the IC_{50} values of *O. aristatus* extracts (aqueous extract – 9.6 µg/mL; 95 per cent ethanolic extract – 21.4 µg/mL) were comparable to those of the reference compounds (butylated hydroxyl toluene – 21.1 µg/mL; ascorbic acid – 4.6 µg/mL) (Alshawsh *et al.*, 2012). In other studies, the IC_{50} values of *O. aristatus* were at least two-fold as high in comparison with the reference compounds. For example, Ho *et al.* (2010) reported that the IC_{50} values of the methanolic extracts of *O. aristatus* and their subfractions ranged between 147 and 386 µg/mL, whereas quercetin, butylated hydroxyl toluene and rosmarinic acid tested in the same assay yielded IC_{50} values of 11.4, 43.1 and 41.3 µg/mL, respectively.

Table 6.3: Antioxidant properties assessed in *O. aristatus*.

Antioxidant Parameters	*References*
2,2-diphenyl-1-picrylhydrazyl (DPPH) scavenging activity	Abdelwahab *et al.*, 2011; Akowuah *et al.*, 2005; Akowuah and Zhari, 2010; Akowuah *et al.*, 2004a; Alshawsh *et al.*, 2012; Chew *et al.*, 2011; Farhan *et al.*, 2012; Ho *et al.*, 2010; Matkowski, 2008; Nuengchamnong *et al.*, 2011; Olennikov and Tankhaeva, 2011; Yam *et al.*, 2007
Superoxide anion scavenging activity	Akowuah *et al.*, 2004a; Olennikov and Tankhaeva, 2011
Nitric oxide scavenging activity	Hsu *et al.*, 2010; Olennikov and Tankhaeva, 2011; Yam *et al.*, 2010
Trolox equivalent antioxidant capacity (TEAC)	Chew *et al.*, 2011; Hsu *et al.*, 2010; Yam *et al.*, 2007
Inhibition of Fe^{3+}-induced lipid peroxidation	Yam *et al.*, 2007
β-carotene bleaching assay	Abdelwahab *et al.*, 2011; Akowuah *et al.*, 2004b; Olennikov and Tankhaeva, 2011
Total phenolic content	Abdelwahab *et al.*, 2011; Akowuah *et al.*, 2004b; Hsu *et al.*, 2010; Matkowski, 2008
Total flavonoid content	Abdelwahab *et al.*, 2011; Chew *et al.*, 2011

As polyphenols are ubiquitous in the plant kingdom and represent an important source of antioxidants in plant products (Kondratyuk and Pezzuto, 2004), a number of studies have determined the total phenolic contents of *O. aristatus*. It was found that total phenolic contents of the plant ranged from 139.81 to 386 gallic acid equivalent per gram of extract (Abdelwahab *et al.*, 2011; Hsu *et al.*, 2010; Sahib *et al.*, 2009a).

Nitric oxide (NO) is an important signalling molecule that regulates a diverse range of physiological processes. However, excessive NO production may cause tissue damage in many organ systems (Vincent *et al.*, 2000). Excess NO may react with superoxide anion radicals to form the much more destructive peroxynitrite, which are responsible for most of the cytotoxic effects of NO (Pacher *et al.*, 2007). Hence, inhibition of NO accumulation is a beneficial therapeutic strategy. Notably, Yam *et al.* (2010) showed that a flavonoid-rich chloroform extract fraction of *O. aristatus* reduced the plasma level of NO in rats undergoing carrageenan-induced rat hind paw oedema. The potency of the extract fraction, orally administered at a dose of 1000 mg/kg body weight, was comparable to that of L-NG-nitroarginine methyl ester (L-NAME), a nitric oxide synthase inhibitor, given at a dose of 10 mg/kg. The same study also demonstrated the NO scavenging activity of the flavonoid-rich chloroform extract fraction *in vitro*. Consistent with this finding are the observations of inhibitory effects of *O. aristatus* on NO production in lipopolysaccharide-activated macrophage cell lines (Awale *et al.*, 2003b; Awale *et al.*, 2004; Hsu *et al.*, 2010). Hsu *et al.* (2010) reported that methanolic (50 μg/mL) and ethanolic (25 μg/mL) extracts of *O. aristatus* inhibited lipopolysaccharide-stimulated NO production in RAW 264.7 cells, while aqueous extract was ineffective. The ethanolic extract also inhibited intracellular production of reactive oxygen species in the same cells. These findings imply that the ethanolic extract may be able to reduce cytotoxicity associated with peroxynitrite formation at the cellular level.

Several studies employed HPLC analysis to detect and quantify potential antioxidant compounds in *O. aristatus* extracts. With the exception of ursolic acid (a triterpene compound), most studies focussed on analysing flavonoids and/or caffeic acid derivatives (Table 6.4).

Using a DPPH radical scavenging assay coupled with on-line HPLC–tandem mass spectrometry (HPLC-MS/MS), seven caffeic acid derivatives with potent DPPH scavenging activity were identified in the aqueous extract of *O. aristatus* leaves (Nuengchamnong *et al.*, 2011) (Table 6.4). Among the seven compounds, danshensu, caftaric acid, sagerinic acid and salvianolic acid B were reported for the first time in *O. aristatus*. The approach of Nuengchamnong *et al.* (2011), which quantitatively analysed the plant extract in the Multiple Reaction Monitoring (MRM) mode, is a powerful technique for rapid characterisation of potential antioxidant compounds in a complex, unfractionated plant extract.

The antioxidant properties of melanoidin, a phenolic polymer (4.4 kDa) isolated for the first time from the fermented leaves of *O. aristatus*, was recently reported (Olennikov and Tankhaeva, 2011). In comparison with rosmarinic acid and gallic acid, melanoidin exhibited only moderate scavenging activity against NO, DPPH, and 2,2'-azino-bis(3-ethylbenzothiazoline-6-sulphonic acid)(ABTS). Nevertheless, it had higher superoxide anion radical scavenging activity (IC_{50} = 43.5 µg/mL) than gallic acid (IC_{50} = 76.4 µg/mL). When compared with rosmarinic acid, melanoidin showed better chelating activity and higher antioxidant activity based on a beta-carotene bleaching assay.

Solvent of varying polarities have been used to extract antioxidants from *O. aristatus*. Akowuah *et al.* (2005) compared the effects of different extraction solvents (water, methanol, aqueous methanol (50 per cent), aqueous acetone (70 per cent), and chloroform) on antioxidant activity of the resulting extracts. The DPPH scavenging activities of the water, methanol, aqueous methanol, and aqueous acetone extracts were higher or comparable to those of pure compounds of butylated hydroxylanisole, quercetin, rosmarinic acids, sinensetin, eupatorin, and 3′-hydroxy-5,6,7,4′-tetramethoxyflavone. By contrast, chloroform extract was the extract exhibiting the lowest DPPH scavenging activity. Hence, polar solvents showed better antioxidant activity (Akowuah *et al.*, 2005). Supporting this finding is the study of Abdelwahab *et al.* (2011). The latter found that a crude methanolic extract of *O. aristatus* and its ethyl acetate, butanol and aqueous fractions had better antioxidant activity compared to the hexane and chloroform fractions. The variations in the antioxidant activities of different solvent extracts or fractions of *O. aristatus* is not surprising as the types of solvents chosen usually determine the major groups of active compounds that will be extracted. For example, Abdelwahab *et al.* (2011) found that hexane, chloroform and ethyl acetate fractions had higher flavonoid contents than fractions prepared with polar solvents such as aqueous methanol and butanol. Meanwhile, Akowuah *et al.* (2005) reported that rosmarinic acid was extracted effectively using water, methanol, 50 per cent methanol, and 70 per cent acetone, but was not extractable by using

Table 6.4: Potential radical scavenging compounds in *O. aristatus* analysed by HPLC.

Compounds	*Concentrations*	*Extracts*	*References*	*Remarks*
Eupatorin Sinensetin 3'-hydroxy-5,6,7,4'-tetra-methoxyflavone	5.05 per cent of extract dry mass 2.86 per cent of extract dry mass 1.01 per cent of extract dry mass	Chloroform extract fraction	Yam *et al.*, 2010	Potentially contributing to NO scavenging activity *in vivo* and *in vitro.*
Ursolic acid	1.8 per cent of extract dry mass 0.94 per cent of extract dry mass (calculated from data reported)	Ethanol Methanol	Hsu *et al.*, 2010	Potentially contributing to inhibiting lipopolysaccharide-activated NO production in RAW 264.7 cells.
Rosmarinic acid 3'-hydroxy-5,6,7,4'-tetra-methoxyflavone Eupatorin Sinensetin	5.1-29.9 per cent of leaf dry mass 0.05-0.69 per cent of leaf dry mass 0.34-3.37 per cent of leaf dry mass 0.22-1.76 per cent of leaf dry mass	Methanol	Akowuah *et al.*, 2004b	No clear evidence of antioxidant activity associated with the four compounds was presented.
Rosmarinic acid Sinensetin Eupatorin 3'-hydroxy-5,6,7,4'-tetra-methoxyflavone	0.978 per cent of leaf dry mass 0.433 per cent of leaf dry mass 0.146 per cent of leaf dry mass 0.017 per cent of leaf dry mass	50 per cent methanol Chloroform Chloroform Methanol	Akowuah *et al.*, 2005	DPPH scavenging activities of the extracts were compared with pure compounds of the identified phenolic antioxidants.
Rosmarinic acid Sinensetin	0.802 per cent of leaf dry mass 0.084 per cent of leaf dry mass	80 per cent methanol	Akowuah and Zhari, 2010	Potentially contributing to DPPH scavenging activity.
Rosmarinic acid Caffeic acid Danshensu Caftaric acid Sagerinic acid Salvianolic acid B Caffeic acid derivative	Note: Concentrations were only expressed in µg/mL in the report. Rosmarinic acid was apparently the major component among the seven compounds analysed.	Water	Nuengchamnong *et al.*, 2011	DPPH scavenging activity demonstrated.

chloroform. Methanol, ethanol and water were also found to vary in their effectiveness in extracting total phenolics from *O. aristatus* (Hsu *et al.*, 2010). Together, these findings suggest that there is no single solvent system that can extract all the antioxidants from *O. aristatus*.

Overall, findings in the literature have presented evidence of antioxidant activity of *O. aristatus* extracts and their subsequent fractions. At present, DPPH scavenging assay is frequently used to measure the antioxidant activity of *O. aristatus*. DPPH radicals do not occur in biological systems. In addition, the mode of action of phenolic antioxidants is often multifaceted and extends beyond merely free radical scavenging. Hence, despite the simplicity and convenience of the DPPH radical scavenging assay, its physiological relevance is modest (Dai and Mumper, 2010; Procházková *et al.*, 2011; Rathore *et al.*, 2011). Future studies should consider broadening their range of antioxidant assays to more physiological relevant ones, such as the superoxide anion radical and nitric oxide scavenging assays as well as lipid peroxidation inhibition assay. Assays of other properties which may indirectly contribute towards an antioxidant effect, such as chelating activity and inhibition of oxidases, should also be considered. More importantly, little is known about the efficacy of *O. aristatus* as an antioxidant agent *in vivo*. Future demonstrations of the antioxidant action of *O. aristatus* extracts and/or their subsequent solvent fractions on animal models are needed to provide more convincing evidence for the therapeutic efficacy of the plant as an antioxidant agent.

Anti-angiogenic

Angiogenesis is the process of forming new blood vessels from the pre-existing ones, regulated by various endogenous cytokines. Anti-angiogenic drugs, aimed at halting the growth of new blood vessels, are in various stages of testing or development for the treatment of cancer and chronic inflammatory diseases, such as rheumatoid arthritis (He *et al.*, 2012; Kerbel, 2000; Podar and Anderson, 2011; Schoettler and Brahn, 2009). Considering the current interests in developing novel anti-angiogenic agents for cancer treatment from phytochemicals (Sagar *et al.*, 2006a; Sagar *et al.*, 2006b), the anti-angiogenic potential of *O. aristatus* has attracted the attention of researchers.

A comparison of *O. aristatus* extracts prepared in petroleum ether, chloroform, methanol and water found the methanol extract to be the most potent anti-angiogenic agent in a rat aortic arch ring assay (Sahib *et al.*, 2009a). Administration of methanol extract (100 µg/mL) resulted in a 93 per cent inhibition in blood vessel growth. Such potency was comparable to the 100 per cent inhibition effected by the anti-angiogenic drug Suramin at the same concentration. It was suggested that the anti-angiogenic activity of *O. aristatus* could be attributed to its antioxidant activity (Sahib *et al.*, 2009a). However, this appears speculative as the study assessed the free radical scavenging activity of only the methanol extract, without comparing it with other extracts. A clear correlation between the relative levels of antioxidant activity and anti-angiogenic activity among the four extracts assessed was not presented. Furthermore, only two antioxidant parameters were investigated, free radical scavenging activity and total phenolic contents.

Molecular and cellular evidence for the anti-angiogenic activity of an ethanolic extract of *O. aristatus* leaves was reported by Ahamed *et al.* (2012). The extract inhibited migration and tube formation, two key stages of angiogenesis, in human umbilical vein endothelial cells. The extract also effectively reduced the concentration of vascular endothelial growth factor (VEGF), a key angiogenic factor, in the human colon tumour cells (*in vitro* model) and colon tumour tissue (*in vivo* model). Furthermore, Western blot study indicated that the extract inhibited the phosphorylation of VEGFR-2, a receptor of VEGF, in a dose-dependent manner. These results imply that *O. aristatus* may exert its anti-angiogenic effect by blocking the VEGF signalling pathway (Ahamed *et al.*, 2012).

Rosmarinic acid is one of the compounds present in an anti-angiogenic *O. aristatus* extract (Ahamed *et al.*, 2012). Huang and Zheng (2006) reported that rosmarinic acid (RA) inhibited all four important stages of angiogenesis, namely proliferation, migration, adhesion and tube formation, in human umbilical vein endothelial cells. In addition, RA significantly reduced the levels of reactive oxygen species (ROS) in the endothelial cells. ROS is an important mediator for angiogenesis. Therefore, the anti-angiogenic effect of RA is probably related to its antioxidative activity, which inhibits ROS-associated VEGF expression and interleukin-8 release.

The limited amount of data in the literature implies that the antioxidant activity of *O. aristatus* extracts may underlie their anti-angiogenic properties. More molecular and chemical evidence are required to substantiate this proposal. The potential role of rosmarinic acid in the anti-angiogenic action of *O. aristatus* also remains to be ascertained.

Anti-cancer

A survey of the literature revealed evidence of anti-cancer effects of *O. aristatus*. A series of diterpenes and flavonoids isolated from the herb were cytotoxic against cancer cell lines (Awale *et al.*, 2002a; Awale *et al.*, 2001; Awale *et al.*, 2002b; Stampoulis *et al.*, 1999b; Tezuka *et al.*, 2000). Many of these compounds, however, have not been fully characterised to establish their anti-cancer potency and mechanisms of action. Several recent studies explored the cellular and molecular bases of the antiproliferative and anti-tumour activity of *O. aristatus* extracts (Ahamed *et al.*, 2012; Doleckova *et al.*, 2012; Sahib *et al.*, 2009b; Salleh *et al.*, 2011). Such work has contributed to a better understanding of the mechanisms of action underlying the anti-cancer effects of the plant.

An ethyl acetate fraction derived from hot water extract of *O. aristatus* was found to inhibit the growth of human hepatocellular carcinoma cell line (HepG2) by inducing apoptosis (Salleh *et al.*, 2011). Signs of apoptosis, such as nuclear condensation and fragmentation as well as mitochondrial membrane dysfunction, were detected in the HepG2 cells. Changes in the expression of pro-apoptotic and anti-apoptotic Bcl-2 family proteins (Bax and Bcl-2), in addition to activation of caspase-3, caspase-8 and caspase-9, suggest that apoptosis in the cancer cells possibly involved the activation of mitochondria-mediated and death receptor-mediated pathways. Down-regulated cyclooxygenase-2 expression and inhibited translocation of NF-κB p65 from cytosol into nucleus may have also compromised the development and/or viability of the

HepG2 cells. Salleh *et al.* (2011) suggested that ethyl acetate fraction derived from the hot water extract of *O. aristatus* is a potential candidate for further development into a chemopreventive agent for human liver cancer. Rosmarinic acid and caffeic acid, major constituents detected in the ethyl acetate fraction, were suspected to have synergistically contributed to cancer cell apoptosis (Salleh *et al.*, 2011). The possibility of such synergy between the two compounds remains to be confirmed.

Doleckova *et al.* (2012) reported that eupatorin contributed significantly to the antiproliferative activity of a chloroform extract of *O. aristatus* leaves. Two lines of evidence support their proposition. Firstly, eupatorin reduced the viability of cancer cell lines to the same extent as the chloroform extract and with IC_{50} values in the micromolar range. Secondly, both eupatorin and chloroform extract induced cell cycle arrest at G2/*M phase* in the HeLa cancer cell. Eupatorin also induced cancer cell death. Notably, eupatorin was cytotoxic against a broad range of cancer cell lines, but less inhibitory towards normal cells. Visualisation of aberrant morphological changes in the microtubular system of HeLa cells and detection of molecular markers of apoptosis imply that eupatorin-induced cell death was mediated by mitotic catastrophe accompanied by apoptosis. In addition, it was proposed that the antiproliferative effect of eupatorin may be ascribed to its anti-angiogenic effect, potentially through the inhibition of vascular endothelial growth factor receptors (VEGFRs) (Doleckova *et al.*, 2012).

Convincing evidence for the anti-tumour effect of *O. aristatus* comes from the work of Ahamed *et al.* (2012). *In vivo* anti-tumour activity of 50 per cent ethanolic extract of *O. aristatus* was demonstrated against colorectal tumour in nude mice. Tumour growth was induced in the mice through subcutaneous injection of human colon tumour (HCT 116) cells. *O. aristatus* extract orally administered at doses of 100 and 200 mg/kg body weight over four weeks suppressed tumour growth by 47 per cent and 83 per cent, respectively. Histopathological examination of tumours excised from the mice revealed reduction in vasculature, increase in necrosis, disrupted tumour architecture, and reduction in the density of viable tumour cells. The extract also significantly reduced the level of vascular endothelial growth factor (VEGF) *in vitro* in tumour cells and *in vivo* in tumour tissue induced in the mice. Hence, the anti-tumour effects of *O. aristatus* ethanolic extract may be attributable to its anti-angiogenic properties, mediated by repression of the VEGF signalling pathway (Ahamed *et al.*, 2012). The same study detected a number of compounds, including rosmarinic acid, eupatorin, sinensetin, betulinic acid, and 3′-hydroxy-5,6,7,4′-tetramethoxyflavone, in the 50 per cent ethanolic extract and suggested possibility of their collective contribution to the anti-tumour activity of the extract. The chemical basis for the anti-tumour and anti-angiogenic effects observed in the study was less well-characterised and requires further investigations.

Sahib *et al.* (2009b) reported that *O. aristatus* methanolic extract synergistically enhanced the anti-proliferative effect of tamoxifen on oestrogen receptor-positive MCF-7 breast cancer cells. Tamoxifen is a drug used in endocrine-based chemoprevention of breast cancers (Bozovic-Spasojevic *et al.*, 2012). Tamoxifen (6.25 μg/mL), *O. aristatus* methanolic extract (25 μg/mL), and their combination inhibited the growth of breast cancer cells by 9.34 per cent, 36.86 per cent and 97.55 per cent, respectively (Sahib *et al.*, 2009b). The mechanism underlying this synergy remains unclear. Nonetheless, this preliminary finding implies potential use of *O. aristatus* methanolic extract and/or their bioactive compounds in combination with cancer-preventive agents, such as tamoxifen, in the management of hormone-responsive breast cancers.

Overall, evidence is accumulating which points to the potency of *O. aristatus* as a source of anti-cancer agents. Molecular and cellular studies suggest that the anti-cancer activities of the herb may be mediated by the induction of apoptosis (Doleckova *et al.*, 2012; Salleh *et al.*, 2011) and/or anti-angiogenic activity associated with repression of the VEGF signalling pathway (Ahamed *et al.*, 2012; Doleckova *et al.*, 2012). Future investigations should give more attention to *in vivo* confirmation of the anti-cancer potency of *O. aristatus* extract and its active compounds. In addition, the potential benefit of using the herb in combination with anti-cancer drugs is underexplored and deserves attention in future studies.

Antimicrobial

The antimicrobial activities of *O. aristatus* have been tested against a broad range of microorganisms, including both Gram-positive and Gram-negative bacteria, and fungal strains (Table 6.5). Disk-diffusion methods revealed that *O. aristatus* had minimal inhibition against most of the bacteria strains tested, frequently with inhibition zone less than 10 mm. The only bacteria for which inhibition zone larger than 10 mm was reported are Gram-positive *Bacillus subtilis* (11 mm) and *Staphylococcus aureus* (10.5 mm) (Alshawsh *et al.*, 2012; Jamal *et al.*, 2011). *O. aristatus* extract has also shown inhibitory activities against four out of five fungal strains tested (Hossain *et al.*, 2008) (Table 6.5).

Although the herb exhibited weak inhibitory activities (inhibition zone <10 mm) against most bacteria strains tested, it is worth pointing out that most of the available data were based primarily on disk-diffusion methods. With the exception of two Gram-positive bacteria (*S. aureus*, *Streptococcus agalactiae*)(Alshawsh *et al.*, 2012) and one Gram-negative bacteria (*Vibrio parahaemolyticus*)(Ho *et al.*, 2010), minimal inhibitory concentration (MIC) data are unavailable to allow a relatively reliable assessment of the antibacterial activities of the herb (Table 6.5).

The work of Alshawsh *et al.* (2012) found comparable antibacterial activities against Gram-positive *S. aureus* between *O. aristatus* extract and vancomycin. As vancomycin is an antibiotic known to inhibit Gram-positive bacterial cell wall synthesis, it is tempting to postulate that the herb may exert the same effect. This possibility deserves attention in future research.

Analgesic

Little has been investigated about the pain-relieving action of *O. aristatus*. The work of Yam *et al.* (2008) has provided early scientific evidence for the analgesic effect of the herb, justifying its traditional use for pain relief. A 50 per cent methanolic leaf extract, administered at 1000 mg/kg body weight, produced significant inhibition of the late phase of formalin-induced pains in rats. Furthermore, the extract exhibited dose-dependent analgesic activity in mice subjected to acetic acid-induced writhing test. The peripheral analgesic activity of *O. aristatus* extract was probably associated with repression of inflammatory pains. Sinensetin, eupatorin, 3′-hydroxy-5,6,7,4′-tetramethoxyflavone, and rosmarinic acid were suspected to be the active compounds responsible for the pain-relieving effect of the herb (Yam *et al.*, 2008).

Table 6.5: Reported activities of *O. aristatus* extracts against selected microorganisms.

Microorganism	*MIC (μg/ml)*	*Diameter of Zone of Inhibition (mm)*	*References*
Gram-positive bacteria			
Bacillus cereus		<8	Ho *et al.*, 2010
Bacillus subtilis		<8 11	Ho *et al.*, 2010 Jamal *et al.*, 2011
Listeria monocytogenes		<8	Ho *et al.*, 2010
Staphylococcus aureus	1560	10.5	Alshawsh *et al.*, 2012
Streptococcus agalactiae	3130	8.1	Alshawsh *et al.*, 2012
Streptococcus mutans (serotypes c and d)	7800-23400		Chen *et al.*, 1989
Gram-negative bacteria			
Escherichia coli		0 0	Ho *et al.*, 2010 Alshawsh *et al.*, 2012
Klebsiella pneumoniae		~8 0	Laikangbam *et al.*, 2009 Alshawsh *et al.*, 2012
Proteus mirabilis		~10	Laikangbam *et al.*, 2009
Pseudomonas stutzeri		~9	Laikangbam *et al.*, 2009
Salmonella enteritidis		<8	Ho *et al.*, 2010
Salmonella typhimurium		<8	Ho *et al.*, 2010
Vibrio parahaemolyticus	1560	8-10	Ho *et al.*, 2010
Fungi			
Botrytis cinerea HL 206	500	16.0	Hossain *et al.*, 2008
Colletotricum capsici HL 410	1000	13.5	Hossain *et al.*, 2008
Fusarium solani HL 115	500	14.5	Hossain *et al.*, 2008
Phytophthora capsici HL 97	Nd	Nd	Hossain *et al.*, 2008
Rhizoctonia solani HL 325	1000	13.5	Hossain *et al.*, 2008

Currently, increasing health awareness and safety concerns among the general public have made naturally-derived analgesic drugs preferable. Hence, the *in vivo* efficacy of *O. aristatus* as a natural pain-killing agent as well as its mechanism of action should be explored further in future research. Such research should pave the way for the development of the herb into natural analgesic drugs in future.

Anti-Pyretic

In Malaysia, *O. aristatus* is traditionally used to treat fever (Yam *et al.*, 2009). Fever, or pyrexia, is a physiological response characterised by an elevated body temperature beyond the normal range (Kozak *et al.*, 2000). Yam *et al.* (2009) reported the anti-pyretic action of a 50 per cent methanolic extract of *O. aristatus* in rats undergoing yeast-induced fever. Oral doses of the extract at 500 and 1000 mg/kg body weight significantly decreased the rectal temperature of the pyrexic rats. The anti-pyretic effect started as early as one hour after extract administration and persisted for four hours (Yam *et al.*, 2009).

Fever occurs when the thermoregulation process in the hypothalamus was interrupted, following an increase in the level of prostaglandin E_2 (PGE2) in the brain region (Jansky and Vybiral, 2004). Most anti-pyretic agents attenuate fever by inhibiting cyclooxygenase activity, hence reducing the level of PGE2 in the hypothalamus (Aronoff and Neilson, 2001; Flower and Vane, 1972; Simon, 1999). The anti-pyretic action of *O. aristatus* may also be mediated by the inhibition of prostaglandin synthesis and was potentially attributable to rosmarinic acid, the main phenolic compound in the extract (Yam *et al.*, 2009). The exact mechanism of action of the anti-pyretic activity of the herb and the potential role of rosmarinic acid requires further confirmation.

Hepatoprotective

Three studies using rat models subjected to chemical-induced liver injury have demonstrated the hepatoprotective effects of alcoholic extracts of *O. aristatus*. These studies have provided a scientific basis which justifies the use of the herb as a treatment for hepatitis.

Alshawsh *et al.* (2011) investigated the hepatoprotective activity of a 95 per cent ethanolic extract of *O. aristatus* against thioacetamide-induced hepatotoxicity in rats. Thioacetamide is hepatotoxic. During its metabolism, free radicals are produced which subsequently induce oxidative stress-mediated acute hepatitis and apoptosis of hepatocytes in the liver (Sun *et al.*, 2000). Alshawsh *et al.* (2011) found that thioacetamide treatment resulted in acute hepatocyte damage, as indicated by increased levels of liver function biomarkers (aspartate aminotransferase, alanine aminotransferase, alkaline phosphatase, bilirubin) and oxidative stress biomarker (malondialdehyde). Treatment with *O. aristatus* extract at 200 mg/kg body weight restored the levels of the biochemical markers to the near normal levels. Furthermore, liver sections of *O. aristatus*-treated rats showed only mild inflammation and necrosis of hepatocytes, in contrast to the severely disrupted cellular and tissue architecture in the hepatotoxic positive control group. These findings indicate that *O. aristatus* positively affected the recovery of liver structure in rats undergoing thioacetamide-induced liver cirrhosis (Alshawsh *et al.*, 2011).

The hepatoprotective activity of a methanolic extract of *O. aristatus* leaves was assessed in paracetamol-treated rats (Maheswari *et al.*, 2008). Levels of biochemical markers of hepatic damage, such as glutamate oxaloacetate transaminase, glutamate pyruvic transaminase, alkaline phosphatase, and lipid peroxides, were elevated in the paracetamol-treated rats. Treatment with *O. aristatus* extract (200 mg/kg) restored the altered levels of biochemical markers to the near normal profile. The liver sections of paracetamol-treated rats showed cloudy swelling and fatty degeneration of hepatocytes and necrosis of cells. Liver damage was also associated with increase in tissue lipid peroxidation and depletion in glutathione levels. Hence, Maheswari *et al.* (2008) proposed that the hepatoprotective effect of *O. aristatus* was due to the prevention of tissue glutathione depletion. However, this possibility remains to be confirmed.

The hepatoprotective effect of a hydromethanolic extract of *O. aristatus* has also been assessed in rats subjected to carbon-tetrachloride-induced liver toxicity (Yam *et al.*, 2007). The extract exhibited protective effect against necrotic injury in rat liver in a dose-dependent manner. In addition, increase of serum alanine transaminase and aspartate transaminase activities was inhibited, suggesting reduction in liver damage.

Biochemical and histopathological analyses on three different rat models have provided evidence of hepatoprotective activity of *O. aristatus* extracts. Nevertheless, the mechanism of action underlying the protective effects of the extracts is unclear. Alshawsh *et al.* (2011) and Yam *et al.* (2007) suggested that the hepatoprotective effect of *O. aristatus* may be attributed to the presence of the abundance of phenolic and flavonoids compounds in the extracts and their antioxidant properties. Future investigations should focus on uncovering the bioactive constituents responsible for the hepatoprotective effects of the herb.

Hypoglycaemic

Animal studies have provided evidence for the hypoglycaemic effect of *O. aristatus* extracts. Mariam *et al.* (1996) reported that an aqueous extract of the herb had produced a significant hypoglycaemic effect in normal and streptozotocin-induced diabetic rats over seven hours post-treatment. The efficacy of the extract, administered at 1000 mg/kg body weight in the diabetic rats, was similar to that of glibenclamide, an antidiabetic drug, administered at 10 mg/kg body weight. Furthermore, treatment with the extract (1000 mg/kg) lowered the blood glucose level of non-diabetic rats for about three hours following oral glucose loading (1500 mg/kg). Similarly, Sriplang *et al.* (2007) found that in both normal and streptozotocin-induced diabetic rats, aqueous extract of *O. aristatus*, at 500 mg/kg and higher, significantly decreased plasma glucose concentration following glucose loading. In diabetic rats, daily administration of the extract at 500 mg/kg body weight reduced plasma glucose levels after 7 and 14 days. Furthermore, experiments performed on perfused rat pancreas showed that the extract could potentiate glucose-induced insulin secretion when present at sufficient concentration (100 µg/mL). Therefore, Sriplang *et al.* (2007) suggested that the use of *O. aristatus* extract may be beneficial to patients with diabetes mellitus who have defective insulinotropic response to glucose. Further studies are needed to characterise the bioactive compounds of *O. aristatus* with regards to the mechanism of antihyperglycaemic actions, such as insulinotropic and glucose uptake activities.

A more recent study by Mohamed *et al.* (2011) investigated the antihyperglycaemic effect of *O. aristatus* leaves, which were extracted serially with solvents of increasing polarity (petroleum ether, chloroform, methanol, and water). Subcutaneous glucose tolerance test revealed that chloroform extract, given at 1g/kg body weight, significantly reduced the blood glucose level of glucose-loaded rats. The chloroform extract was proposed to have exerted antihyperglycaemic activity by improving glucose tolerance in the treated animals (Mohamed *et al.*, 2011). The chloroform extract was further fractionated and screening was continued with activity-guided fractionation in order to identify the active constituents responsible for antihyperglycaemic activity. A subfraction (C*f*2-B) was eventually obtained which clearly exhibited a blood glucose-lowering effect in fasting, normal rats after glucose loading. The C*f*2-B subfraction, however, did not exhibit a hypoglycaemic effect on blood glucose level in the streptozotocin-induced diabetic rats. These results imply that C*f*2-B functions similarly to metformin, which has a hypoglycaemic effect, but not an antihyperglycaemic effect in normal animal models. C*f*2-B possibly did not directly stimulate insulin secretion. Rather, it may have induced hypoglycaemia through an additional pancreatic mechanism, which potentially involves increased peripheral use of glucose by the tissues (Mohamed *et al.*, 2011). Phytochemical screening by thin layer chromatography revealed the presence of terpenoids and flavonoids, including sinensetin in the C*f*2-B subfraction. These compounds may have been, either individually or synergistically, responsible for the antihyperglycaemic effect of the C*f*2-B subfraction (Mohamed *et al.*, 2011). Potential synergism of these compounds and their functions remain to be ascertained.

At present, the hypoglycaemic potential of *O. aristatus* is underexplored in scientific research. Additional biochemical and pharmacological studies are necessary to identify the active principles responsible for the hypoglycaemic potential of *O. aristatus*, in addition to delineating the modes of action of the active compounds.

Conclusions

Pharmacological studies reported in the literature have provided justification to the uses of *O. aristatus* extracts in traditional medicine, *e.g.*, as anti-inflammatory, diuretic and anti-lithiatic agents. Notably, research to-date has yielded encouraging results for the potential of *O. aristatus* as a chemopreventive agent, although the plant is not traditionally used as an anti-cancer treatment. There are, however, knowledge gaps to be filled despite our current progress in understanding the therapeutic values of the plant. Among the diverse pharmacological properties of *O. aristatus*, some (*e.g.*, anti-inflammatory, diuretic, and anti-cancer) have been more extensively investigated than the others (*e.g.*, analgesic and anti-pyretic). Future research should devote more attention to the latter so that the medicinal potential of the herb can be more comprehensively characterised and exploited. Furthermore, our current understanding of the pharmacological activities of the herb and its active principles has been primarily built on *in vitro* models (*e.g.*, chemical assays and cancer cell lines) and laboratory animal studies. The promising pre-clinical therapeutic potential of the herb warrants more well-controlled human trials, with large sample sizes, to confirm the relevance of the pharmacological activities in humans. On the other hand, the elucidation of the mechanism of action of *O. aristatus* and its active principles is an area which calls for more investigations in future. Studies which assess the effects of *O. aristatus* and its active compounds on gene and protein expressions *in vivo* will shed light on the molecular mechanisms of the herb as a therapeutic agent. Meanwhile, a broad range of potentially bioactive compounds have been isolated and identified from the plant, many of which have not been fully characterised in terms of bioactivity. Considering the myriads of pharmacological activities of *O. aristatus*, phytochemicals derived from the plant may represent a potential source of natural compounds for the pharmaceutical industry.

Acknowledgments

Chai T.-T. would like to acknowledge the Universiti Tunku Abdul Rahman Research Fund (UTAR RF) for financing his research on the pharmacological properties of local flora. Ooh K.-F. is grateful to UTAR for providing a staff scholarship for his postgraduate research. Fazilah Abd Manan would like to acknowledge the Faculty of Biosciences and Medical Engineering, and Sustainability Research Alliance of Universiti Teknologi Malaysia for the teaching and research supports.

References

Abdelwahab, S.I., Mohan, S., Mohamed Elhassan, M., Al-Mekhlafi, N., Mariod, A.A., Abdul, A.B., Abdulla, M.A., and Alkharfy, K.M. (2011). Antiapoptotic and antioxidant properties of *Orthosiphon stamineus* benth (Cat's Whiskers): Intervention in the Bcl-2-mediated apoptotic pathway. *Evidence-based Complementary and Alternative Medicine*, **2011:** DOI: 10.1155/2011/156765.

Adam, Y., Somchit, M.N., Sulaiman, M.R., Nasaruddin, A.A., Zuraini, A., Bustamam, A.A., and Zakaria, Z.A. (2009). Diuretic properties of *Orthosiphon stamineus* Benth. *Journal of Ethnopharmacology*, **124**(1): 154-158.

Ahamed, M.B.K., Aisha, A.F.A., Nassar, Z.D., Siddiqui, J.M., Ismail, Z., Omari, S.M.S., Parish, C.R., and Majid, A.M.S.A. (2012). Cat's whiskers tea (*Orthosiphon stamineus*) extract inhibits growth of colon tumor in nude mice and angiogenesis in endothelial cells via suppressing VEGFR phosphorylation. *Nutrition and Cancer*, **64**(1): 89-99.

Akanae, W., Tsujihata, M., Yoshioka, I., Nonomura, N., and Okuyama, A. (2010). *Orthosiphon grandiflorum* has a protective effect in a calcium oxalate stone forming rat model. *Urological Research*, **38**(2): 89-96.

Akowuah, G.A., and Zhari, I. (2010). Effect of extraction temperature on stability of major polyphenols and antioxidant activity of *Orthosiphon stamineus* leaf. *Journal of Herbs, Spices and Medicinal Plants*, **16**(3-4): 160-166.

Akowuah, G.A., Ismail, Z., Norhayati, I., and Sadikun, A. (2005). The effects of different extraction solvents of varying polarities on polyphenols of *Orthosiphon stamineus* and evaluation of the free radical scavenging activity. *Food Chemistry*, **93:** 311-317.

Akowuah, G.A., Zhari, I., Norhayati, I., and Sadikun, A. (2004a). Radical scavenging activity of methanol leaf extracts of *Orthosiphon stamineus*. *Pharmaceutical Biology*, **42**(8): 629-635.

Akowuah, G.A., Zhari, I., Norhayati, I., Sadikun, A., and Khamsah, S.M. (2004b). Sinensetin, eupatorin, 3'-hydroxy-5, 6, 7, 4'-tetramethoxyflavone and rosmarinic acid contents and antioxidative effect of *Orthosiphon stamineus* from Malaysia. *Food Chemistry*, **87**(4): 559-566.

Alshawsh, M.A., Abdulla, M.A., Ismail, S., Amin, Z.A., Qader, S.W., Hadi, H.A., and Harmal, N.S. (2012). Free radical scavenging, antimicrobial and immunomodulatory activities of *Orthosiphon stamineus*. *Molecules*, **17**(5): 5385-5395.

Alshawsh, M.A., Abdulla, M.A., Ismail, S., and Amin, Z.A. (2011). Hepatoprotective effects of *Orthosiphon stamineus* extract on thioacetamide-induced liver cirrhosis in rats. *Evidence-based Complementary and Alternative Medicine*, **2011:** DOI: 10.1155/2011/103039.

Arafat, O.M., Tham, S.Y., Sadikun, A., Zhari, I., Haughton, P.J., and Asmawi, M.Z. (2008). Studies on diuretic and hypouricemic effects of *Orthosiphon stamineus* methanol extracts in rats. *Journal of Ethnopharmacology*, **118**(3): 354-360.

Aronoff, D.M., and Neilson, *E.G.* (2001). Antipyretics: mechanisms of action and clinical use in fever suppression. *The American Journal of Medicine*, **111**(4): 304-315.

Awale, S., Tezuka, Y., Banskota, A.H., Adnyana, I.K., and Kadota, S. (2003a). Nitric oxide inhibitory isopimarane-type diterpenes from *Orthosiphon stamineus* of Indonesia. *Journal of Natural Products*, **66**(2): 255-258.

Awale, S., Tezuka, Y., Banskota, A.H., and Kadota, S. (2003b). Inhibition of NO production by highly-oxygenated diterpenes of *Orthosiphon stamineus* and their structure-activity relationship. *Biological and Pharmaceutical Bulletin*, **26**(4): 468-473.

Awale, S., Tezuka, Y., Banskota, A.H., and Kadota, S. (2003c). Siphonols A-E: Novel nitric oxide inhibitors from *Orthosiphon stamineus* of Indonesia. *Bioorganic and Medicinal Chemistry Letters*, **13**(1): 31-35.

Awale, S., Tezuka, Y., Banskota, A.H., Ketut Adnyana, I., and Kadota, S. (2003d). Highly-oxygenated isopimarane-type diterpenes from *Orthosiphon stamineus* of Indonesia and their nitric oxide inhibitory activity. *Chemical and Pharmaceutical Bulletin*, **51**(3): 268-275.

Awale, S., Tezuka, Y., Banskota, A.H., Kouda, K., Kyaw, M.T., and Kadota, S. (2002a). Four highly oxygenated isopimarane-type diterpenes of *Orthosiphon stamineus*. *Planta Medica*, **68**(3): 286-288.

Awale, S., Tezuka, Y., Banskota, A.H., Kouda, K., Tun, K.M., and Kadota, S. (2001). Five novel highly oxygenated diterpenes of *Orthosiphon stamineus* from Myanmar. *Journal of Natural Products*, **64**(5): 592-596.

Awale, S., Tezuka, Y., Banskota, A.H., Shimoji, S., Taira, K., and Kadota, S. (2002b). Norstaminane- and isopimarane-type diterpenes of *Orthosiphon stamineus* from Okinawa. *Tetrahedron*, **58**(27): 5503-5512.

Awale, S., Tezuka, Y., Kobayashi, M., Ueda, J.Y., and Kadota, S. (2004). Neoorthosiphonone A; a nitric oxide (NO) inhibitory diterpene with new carbon skeleton from *Orthosiphon stamineus*. *Tetrahedron Letters*, **45**(7): 1359-1362.

Awale, S., Tezuka, Y., Shimoji, S., Taira, K., and Kadota, S. (2002c). Secoorthosiphols A–C: three highly oxygenated secoisopimarane-type diterpenes from *Orthosiphon stamineus*. *Tetrahedron Letters*, **43**(8): 1473-1475.

Azizan, N.A., Ahmad, R., Mohamed, K., Ahmad, M.Z., and Asmawi, Z. (2012). The *in vivo* antihypertensive effects of standardized methanol extracts of *Orthosiphon stamineus* on spontaneous hypertensive rats: A preliminary study. *African Journal of Pharmacy and Pharmacology*, **6**(6): 376–379.

Banskota, A.H., Tezuka, Y., Le Tran, Q., and Kadota, S. (2003). Chemical constituents and biological activities of Vietnamese medicinal plants. *Current Topics in Medicinal Chemistry*, **3**(2): 227-248.

Barnes, J., Anderson, L.A., and Phillipson, J.D. (2007). *Herbal Medicines*. Pharmaceutical Press, London.

Beaux, D., Fleurentin, J., and Mortier, F. (1999). Effect of extracts of *Orthosiphon stamineus* Benth, *Hieracium pilosella* L., *Sambucus nigra* L. and *Arctostaphylos uva-ursi* (L.) Spreng. in rats. *Phytotherapy Research*, **13**(3): 222-225.

Becker, G. (2007). Uric acid stones. *Nephrology*, **12**(SUPPL. 1): S21-S25.

Bozovic-Spasojevic, I., Azambuja, E., McCaskill-Stevens, W., Dinh, P., and Cardoso, F. (2012). Chemoprevention for breast cancer. *Cancer Treatment Reviews*, **38**(5): 329-339.

Bradley, J.R. (2008). TNF-mediated inflammatory disease. *Journal of Pathology*, **214**(2): 149-160.

Bwin, D.M., and Gwan, U.S. (1967). *Burmese Indigenous Medicinal Plants: 1. Plants with reputed Hypoglycaemic Action, Health and Myanmar Traditional Medicine*. Burma Medical Research Institute, Ministry of Health, Yangon, Myanmar: pp.126-128.

Chan, L.K., and Loo, P.S. (2006). Morphological similarities and differences between the two varieties of cat's whiskers (*Orthosiphon stamineus* Benth.) grown in Malaysia. *International Journal of Botany*, **2**(1): 1-6.

Chen, C.P., Lin, C.C., and Namba, T. (1989). Screening of Taiwanese crude drugs for antibacterial activity against *Streptococcus mutans*. *Journal of Ethnopharmacology*, **27**(3): 285-295.

Chew, K.K., Khoo, M.Z., Ng, S.Y., Thoo, Y.Y., Aida, W.M.W., and Ho, C.W. (2011). Effect of ethanol concentration, extraction time and extraction temperature on the recovery of phenolic compounds and antioxidant capacity of *Orthosiphon stamineus* extracts. *International Food Research Journal*, **18**(4): 1427-1435.

Cicero, A.F.G., De Sando, V., Izzo, R., Vasta, A., Trimarco, A., and Borghi, C. (2012). Effect of a combined nutraceutical containing *Orthosiphon stamineus* effect on blood pressure and metabolic syndrome components in hypertensive dyslipidaemic patients: A randomized clinical trial. *Complementary Therapies in Clinical Practice*, **18**(3): 190-194.

Clare, B., Conroy, R., and Spelman, K. (2009). The diuretic effect in human subjects of an extract of *Taraxacum officinale* folium over a single day. *Journal of Alternative and Complementary Medicine* **15**(8): 929-934.

Committee on Herbal Medicinal Products. (2011). Assessment report on *Orthosiphon stamineus* Benth., folium, London, United Kingdom.

Dai, J., and Mumper, R.J. (2010). Plant Phenolics: Extraction, analysis and their antioxidant and anticancer properties. *Molecules*, **15**(10): 7313-7352.

Di Domenico, F., Foppoli, C., Coccia, R., and Perluigi, M. (2012). Antioxidants in cervical cancer: Chemopreventive and chemotherapeutic effects of polyphenols. *Biochimica et Biophysica Acta–Molecular Basis of Disease*, **1822**(5): 737-747.

Doleckova, I., Rarova, L., Grúz, J., Vondrusova, M., Strnad, M., and Krystof, V. (2012). Antiproliferative and antiangiogenic effects of flavone eupatorin, an active constituent of chloroform extract of *Orthosiphon stamineus* leaves. *Fitoterapia*: DOI: 10.1016/j.fitote.2012.1006.1002.

Englert, J., and Harnischfeger, G. (1992). Diuretic action of aqueous *Orthosiphon* extract in rats. *Planta Medica*, **58**(3): 237-238.

Eusebio, J.E., and Umali, B.E. (2004). Inventory, documentation and status of medicinal plants research in the Philippines. *In*: Medicinal plants research in Asia. The framework and project workplans., Ed. By Batugal, P.A., Kanniah, J., Lee, S.Y., and Oliver, J.T., International Plant Genetic Resources Institute – Regional Office for Asia, the Pacific and Oceania (IPGRI-APO), Serdang, Malaysia pp. 147-183.

Farhan, M., Abdul Razak, S., Pin, K.Y., and Chuah, A.L. (2012). Antioxidant activity and phenolic content of different parts of *Orthosiphon stamineus* grown under different light intensities. *Journal of Tropical Forest Science*, **24**(2): 173-177.

Flower, R.J., and Vane, J.R. (1972). Inhibition of prostaglandin synthetase in brain explains the anti-pyretic activity of paracetamol (4-acetamidophenol). *Nature*, **240**(5381): 410-411.

Galati, E.M., Tripodo, M.M., Trovato, A., Miceli, N., and Monforte, M.T. (2002). Biological effect of *Opuntia ficus indica* (L.) Mill. (Cactaceae) waste matter: Note I: Diuretic activity. *Journal of Ethnopharmacology*, **79**(1): 17-21.

Germplasm Resources Information Network. (2009). Taxon: *Orthosiphon aristatus* (Blume) Miq.

Guerin, J.C., Reveillere, H.P., Ducrey, P., and Toupet, L. (1989). *Orthosiphon stamineus* as a potent source of methylripariochromene A. *Journal of Natural Products*, **52**(1): 171-173.

Gürocak, S., and Küpeli, B. (2006). Consumption of historical and current phytotherapeutic agents for urolithiasis: A critical review. *Journal of Urology*, **176**(2): 450-455.

Handa, S.S., Rakesh, D.D., and Vasisht, K. (2006). Compendium of Medicinal and Aromatic Plants–ASIA, ICS-UNIDO, Trieste, Italy.

He, K., Jin, K., Wang, H., and Teng, L. (2012). Anti-angiogenic therapy for colorectal cancer: On the way to getting better! *Hepato-Gastroenterology*, **59**(116): 1113-1117.

Ho, C.-H., Noryati, I., Sulaiman, S.-F., and Rosma, A. (2010). *In vitro* antibacterial and antioxidant activities of *Orthosiphon stamineus* Benth. extracts against food-borne bacteria. *Food Chemistry*, **122**(4): 1168-1172.

Hocher, B. (2010). Adenosine A1 receptor antagonists in clinical research and development. *Kidney International*, **78**(5): 438-445.

Hossain, M.A., and Ismail, Z. (2010). Isolation and characterization of triterpenes from the leaves of *Orthosiphon stamineus*. *Arabian Journal of Chemistry*: DOI: 10.1016/j.arabjc.2010.1010.1009.

Hossain, M.A., and Ismail, Z. (2012). Quantification and enrichment of sinensetin in the leaves of *Orthosiphon stamineus*. *Arabian Journal of Chemistry*: DOI: 10.1016/j.arabjc.2012.1002.1016.

Hossain, M.A., and Mizanur Rahman, S.M. (2011). Isolation and characterisation of flavonoids from the leaves of medicinal plant *Orthosiphon stamineus*. *Arabian Journal of Chemistry*: DOI: 10.1016/j.arabjc.2011.1006.1016.

Hossain, M.A., Ismail, Z., Rahman, A., and Kang, S.C. (2008). Chemical composition and anti-fungal properties of the essential oils and crude extracts of *Orthosiphon stamineus* Benth. *Industrial Crops and Products*, **27**(3): 328-334.

Hsu, C.L., Hong, B.O.H., Shan, Y.U., and Yen, G.C. (2010). Antioxidant and anti-inflammatory effects of *Orthosiphon aristatus* and its bioactive compounds. *Journal of Agricultural and Food Chemistry*, **58**(4): 2150-2156.

Huang, S.S., and Zheng, R.L. (2006). Rosmarinic acid inhibits angiogenesis and its mechanism of action *in vitro*. *Cancer Letter*, **239**: 271-280.

Jaganath, I.B., and Ng, L.T. (2000). *Herbs–The Green Pharmacy of Malaysia*. Vinpress Sdn. Bhd and Malaysian Agricultural Research and Development Institute, Kuala Lumpur, Malaysia.

Jamal, P., Karim, I.A., Abdullah, E., Raus, R.A., and Hashim, Y.Z. (2011). Phytochemical screening for antibacterial activity of potential Malaysian medicinal plants. *African Journal of Biotechnology*, **10**(81): 18795-18799.

Jansky, L., and Vybiral, S. (2004). Thermal homeostasis in systemic inflammation: modulation of neuronal mechanisms. *Frontiers in Bioscience*, **9**: 3068-3084.

Kerbel, R.S. (2000). Tumor angiogenesis: Past, present and the near future. *Carcinogenesis*, **21**: 505-515.

Khare, C. (2007). *Indian Medicinal Plants–An Illustrated Dictionary*. Springer, New York.

Khatun, M.A., Harun-Or-Rashid, M., and Rahmatullah, M. (2011). Scientific validation of eight medicinal plants used in traditional medicinal systems of Malaysia: a review. *American-Eurasian Journal of Sustainable Agriculture*, **5**(1): 67-75.

Knoll, T. (2010). Epidemiology, pathogenesis, and pathophysiology of urolithiasis. *European Urology Supplements*, **9**(12): 802-806.

Kondratyuk, T., and Pezzuto, J. (2004). Natural product polyphenols of relevance to human health. *Pharmaceutical Biology*, **42**(Supplement 1): 46-63.

Korhonen, R., Lahti, A., Kankaanranta, H., and Moilanen, E. (2005). Nitric oxide production and signaling in inflammation. *Current Drug Targets: Inflammation and Allergy*, **4**(4): 471-479.

Kozak, W., Kluger, M.J., Tesfaigzi, J., Kozak, A., Mayfield, K.P., Wachulec, M., and Dokladny, K. (2000). Molecular mechanisms of fever and endogenous antipyresis. *Annals of the New York Academy of Sciences*, **917:** 121-134.

Laavola, M., Nieminen, R., Yam, M., Sadikun, A., Asmawi, M., Basir, R., Welling, J., Vapaatalo, H., Korhonen, R., and Moilanen, E. (2012). Flavonoids eupatorin and sinensetin present in *Orthosiphon stamineus* leaves inhibit inflammatory gene expression and STAT1 activation. *Planta Medica*, **78**(8): 779-786.

Laikangbam, R., Devi, M.D., and Singh, S.R. (2009). Anti-bacterial efficacy of elite medicinal plants on urolithiasis inducing flora. *Journal of Food, Agriculture and Environment*, **7**(2): 40-45.

Maheswari, C., Maryammal, R., and Venkatanarayanan, R. (2008). Hepatoprotective activity of "*Orthosiphon stamineus*" on liver damage caused by paracetamol in rats. *Jordan Journal of Biological Sciences*, **1**(3): 105-108.

Mariam, A., Asmawi, M.Z., and Sadikun, A. (1996). Hypoglycaemic activity of the aqueous extract of *Orthosiphon stamineus*. *Fitoterapia*, **67**(5): 465-468.

Masuda, T., Masuda, K., and Nakatani, N. (1992a). Orthosiphol A, a highly oxygenated diterpene from the leaves of *Orthosiphon stamineus*. *Tetrahedron Letters*, **33**(7): 945-946.

Masuda, T., Masuda, K., Shiragami, S., Jitoe, A., and Nakatani, N. (1992b). Orthosiphol A and B, novel diterpenoid inhibitors of TPA (12-O-tetradecanoylphorbol-13-acetate)-induced inflammation, from *Orthosiphon stamineus*. *Tetrahedron*, **48**(33): 6787-6792.

Matkowski, A. (2008). Antioxidant activity of extracts and different solvent fractions of *Glechoma hederacea* L. and *Orthosiphon stamineus* (Benth.) Kudo. *Advances in Clinical and Experimental Medicine*, **17**(6): 615-624.

Matsubara, T., Bohgaki, T., Watarai, M., Suzuki, H., Ohashi, K., and Shibuya, H. (1999). Antihypertensive actions of methylripariochromene A from *Orthosiphon aristatus*, an Indonesian traditional medicinal plant. *Biological and Pharmaceutical Bulletin*, **22**(10): 1083-1088.

Mohamed, E.A.H., Mohamed, A.J., Asmawi, M.Z., Sadikum, A., Ebrika, O.S., and Yam, M.F. (2011). Antihyperglycemic effect of *Orthosiphon stamineus* Benth leaves extract and its bioassay-guided fractions. *Molecules*, **16:** 3787-3801.

Musa, Y., Azimah, A.K., and Zaharah, H. (2009). *Tumbuhan Ubatan Popular Malaysia*. Malaysian Agricultural Research and Development Institute, Serdang, Malaysia.

Ngamrojanavanich, N., Manakit, S., Pornpakakul, S., and Petsom, A. (2006). Inhibitory effects of selected Thai medicinal plants on Na^+,K^+-ATPase. *Fitoterapia*, **77**(6): 481-483.

Ngo, T.C., and Assimos, D.G. (2007). Uric acid nephrolithiasis: Recent progress and future directions. *Reviews in Urology*, **9:** 17-27.

Nguyen, M.T.T., Awale, S., Tezuka, Y., Chien-Hsiung, C., and Kadota, S. (2004). Staminane- and isopimarane-type diterpenes from *Orthosiphon stamineus* of Taiwan and their nitric oxide inhibitory activity. *Journal of Natural Products*, **67**(4): 654-658.

Nuengchamnong, N., Krittasilp, K., and Ingkaninan, K. (2011). Characterisation of phenolic antioxidants in aqueous extract of *Orthosiphon grandiflorus* tea by LC-ESI-MS/MS coupled to DPPH assay. *Food Chemistry*, **127**(3): 1287-1293.

Ohashi, K., Bohgaki, T., Matsubara, T., and Shibuya, H. (2000). Indonesian medicinal plants. XXIII. Chemical structures of two new migrated pimarane-type diterpenes, neoorthosiphols A and B, and suppressive effects on rat thoracic aorta of chemical constituents isolated from the leaves of *Orthosiphon aristatus* (Lamiaceae). *Chemical and Pharmaceutical Bulletin*, **48**(3): 433-435.

Olah, N.-K., Radu, L., Mogo°an, C., Hanganu, D., and Gocan, S. (2003). Phytochemical and pharmacological studies on *Orthosiphon stamineus* Benth. (Lamiaceae) hydroalcoholic extracts. *Journal of Pharmaceutical and Biomedical Analysis*, **33**(1): 117-123.

Olennikov, D.N., and Tankhaeva, L.M. (2011). Physicochemical characteristics and antioxidant activity of melanoidin pigment from the fermented leaves of *Orthosiphon stamineus*. *Brazilian Journal of Pharmacognosy*, **22**(2): 284-290.

Pacher, P., Beckman, J.S., and Liaudet, L. (2007). Nitric oxide and peroxynitrite in health and disease. *Physiological Reviews*, **87**(1): 315-424.

Pautz, A., Art, J., Hahn, S., Nowag, S., Voss, C., and Kleinert, H. (2010). Regulation of the expression of inducible nitric oxide synthase. *Nitric Oxide–Biology and Chemistry*, **23**(2): 75-93.

Podar, K., and Anderson, K.C. (2011). Emerging therapies targeting tumor vasculature in multiple myeloma and other hematologic and solid malignancies. *Current Cancer Drug Targets*, **11**(9): 1005-1024.

Premgamone, A., Sriboonlue, P., Disatapornjaroen, W., Maskasem, S., Sinsupan, N., and Apinives, C. (2001). A long-term study on the efficacy of a herbal plant, *Orthosiphon grandiflorus*, and sodium potassium citrate in renal calculi treatment. *Southeast Asian Journal of Tropical Medicine and Public Health*, **32**(3): 654-660.

Procházková, D., Boušová, I., and Wilhelmová, N. (2011). Antioxidant and prooxidant properties of flavonoids. *Fitoterapia*, **82**(4): 513-523.

Racz Kotilla, E., Racz, G., and Solomon, A. (1974). The action of *Taraxacum officinale* extracts on the body weight and diuresis of laboratory animals. *Planta Medica*, **26**(3): 212-217.

Ramachandran, S., and Prasad, N.R. (2008). Effect of ursolic acid, a triterpenoid antioxidant, on ultraviolet-B radiation-induced cytotoxicity, lipid peroxidation and DNA damage in human lymphocytes. *Chemico-Biological Interactions*, **176**(2-3): 99-107.

Rathore, G.S., Suthar, M., Pareek, A., and Gupta, R.N. (2011). Nutritional antioxidants: A battle for better health. *Journal of Natural Pharmaceuticals*, **2**(1): 2-14.

Ringbom, T., Segura, L., Noreen, Y., Perera, P., and Bohlin, L. (1998). Ursolic acid from *Plantago major*, a selective inhibitor of cyclooxygenase-2 catalyzed prostaglandin biosynthesis. *Journal of Natural Products*, **61**(10): 1212-1215.

Sagar, S.M., Yance, D., and Wong, R.K. (2006a). Natural health products that inhibit angiogenesis: A potential source for investigational new agents to treat cancer–Part 1. *Current Oncology*, **13**(1).

Sagar, S.M., Yance, D., and Wong, R.K. (2006b). Natural health products that inhibit angiogenesis: A potential source for investigational new agents to treat cancer–Part 2. *Current Oncology*, **13**(3).

Sahib, H.B., Aisha, A.F., Yam, M.F., Asmawi, M.Z., Ismail, Z., Salhimi, S.M., Othman, N.H., and Abdul Majid, A.M.S. (2009a). Anti-angiogenic and anti oxidant properties of *Orthosiphon stamineus* Benth. methanolic leaves extract. *International Journal of Pharmacology*, **5**(2): 162-167.

Sahib, H.B., Ismail, Z., Othman, N.H., and Abdul Majid, A.M.S. (2009b). *Orthosiphon stamineus* Benth. Methanolic extract enhances the anti-proliferative effects of tamoxifen on human hormone dependent breast cancer. *International Journal of Pharmacology*, **5**(4): 273-276.

Salleh, S.A., Rajab, N.F., Abdullah, N.R., Ismail, Z., Mouatt, P., Dowell, A., and Muhamad, S. (2011). *In vitro* chemopreventive activity of an ethyl acetate fraction derived from hot water extract of *Orthosiphon stamineus* in HepG2 cells. *Journal of Medicinal Plants Research*, **5**(10): 1892-1899.

Samiyah, M.N. (2004). *'Herba dan ulam-ulaman: Misai kucing: penawar buah pinggang yang mudah dan murah'*, *Agromedia*, **15**: 68-69.

Samy, J., Sugumaran, M., and Lee, K.L.W. (2009). *Herbs of Malaysia: An Introduction to the Medicinal, Culinary, Aromatic and Cosmetic Use of Herbs*. Marshall Cavendish (Malaysia) Sdn Bhd, Shah Alam, Malaysia: pp.176-177.

Schoettler, N., and Brahn, E. (2009). Angiogenesis inhibitors for the treatment of chronic autoimmune inflammatory arthritis. *Current Opinion in Investigational Drugs*, **10**(5): 425-433.

Schütz, K., Carle, R., and Schieber, A. (2006). Taraxacum–A review on its phytochemical and pharmacological profile. *Journal of Ethnopharmacology*, **107**(3): 313-323.

Shibuya, H., Bohgaki, T., and Ohashi, K. (1999). Two novel migrated pimarane-type diterpenes, neoorthosiphols A and B, from the leaves of *Orthosiphon aristatus* (Lamiaceae). *Chemical and Pharmaceutical Bulletin*, **47**(6): 911-912.

Siddiqui, M.J.A., and Ismail, Z. (2011). Simultaneous analysis of bioactive markers from *Orthosiphon stamineus* benth leaves extracts by reverse phase high performance liquid chromatography. *Tropical Journal of Pharmaceutical Research*, **10**(1): 97-103.

Simon, L.S. (1999). Role and regulation of cyclooxygenase-2 during inflammation. *The American Journal of Medicine*, **106**(5): 37S-42S.

Small, D.M., Coombes, J.S., Bennett, N., Johnson, D.W., and Gobe, G.C. (2012). Oxidative stress, anti-oxidant therapies and chronic kidney disease. *Nephrology*, **17**(4): 311-321.

Sriplang, K., Adisakwattana, S., Rungsipipat, A., and Yibchok-anun, S. (2007). Effects of *Orthosiphon stamineus* aqueous extract on plasma glucose concentration and lipid profile in normal and streptozotocin-induced diabetic rats. *Journal of Ethnopharmacology*, **109**(3): 510-514.

Stampoulis, P., Tezuka, Y., Banskota, A.H., Kim Qui, T., Saiki, I., and Kadola, S. (1999a). Staminol A, a novel diterpene from *Orthosiphon stamineus*. *Tetrahedron Letters*, **40**(22): 4239-4242.

Stampoulis, P., Tezuka, Y., Banskota, A.H., Tran, K.Q., Saiki, I., and Kadota, S. (1999b). Staminolactones A and B and norstaminol A: Three highly oxygenated staminane-type diterpenes from *Orthosiphon stamineus*. *Organic Letters*, **1**(9): 1367-1370.

Sumaryono, W., Proksch, P., Wray, V., Witte, L., and Hartmann, T. (1991). Qualitative and quantitative analysis of the phenolic constituents from *Orthosiphon aristatus*. *Planta Medica*, **57**(2): 176-180.

Sun, F., Hayami, S., Ogiri, Y., Haruna, S., Tanaka, K., Yamada, Y., Tokumaru, S., and Kojo, S. (2000). Evaluation of oxidative stress based on lipid hydroperoxide, vitamin C and vitamin E during apoptosis and necrosis caused by thioacetamide in rat liver. *Biochimica et Biophysica Acta (BBA)–Molecular Basis of Disease*, **1500**(2): 181-185.

Tezuka, Y., Stampoulis, P., Banskota, A.H., Awale, S., Tran, K.Q., Saiki, I., and Kadota, S. (2000). Constituents of the Vietnamese medicinal plant *Orthosiphon stamineus*. *Chemical and Pharmaceutical Bulletin*, **48**(11): 1711-1719.

Tinkel, J., Hassanain, H., and Khouri, S.J. (2012). Cardiovascular antioxidant therapy: A review of supplements, pharmacotherapies, and mechanisms. *Cardiology in Review*, **20**(2): 77-83.

Vincent, J.L., Zhang, H., Szabo, C., and Preiser, J.C. (2000). Effects of nitric oxide in septic shock. *American Journal of Respiratory and Critical Care Medicine*, **161**(6): 1781-1785.

Wiart, C. (2006). *Medicinal Plants of the Asia-Pacific: Drugs from the Future?* World Scientific Publishing Co. Pte. Ltd., Singapore.

Willcox, J., Sarah, A., and Catignani, G. (2004). Antioxidants and prevention of chronic disease. *Critical Reviews in Food Science and Nutrition.*, **44:** 275-295.

Woottisin, S., Hossain, R.Z., Yachantha, C., Sriboonlue, P., Ogawa, Y., and Saito, S. (2011). Effects of *Orthosiphon grandiflorus, Hibiscus sabdariffa* and *Phyllanthus amarus* extracts on risk factors for urinary calcium oxalate stones in rats. *Journal of Urology*, **185**(1): 323-328.

Yam, M.F., Ang, L.F., Basir, R., Salman, I.M., Ameer, O.Z., and Asmawi, M.Z. (2009). Evaluation of the anti-pyretic potential of *Orthosiphon stamineus* Benth standardized extract. *Inflammopharmacology*, **17**(1): 50-54.

Yam, M.F., Asmawi, M.Z., and Basir, R. (2008). An investigation of the anti-inflammatory and analgesic effects of *Orthosiphon stamineus* leaf extract. *Journal of Medicinal Food,* **11**(2)**:** 362-368.

Yam, M.F., Basir, R., Asmawi, M.Z., and Ismail, Z. (2007). Antioxidant and hepatoprotective effects of *Orthosiphon stamineus* Benth. standardized extract. *The American Journal of Chinese Medicine,* **35**(1)**:** 115-126.

Yam, M.F., Lim, V., Salman, I.M., Ameer, O.Z., Ang, L.F., Rosidah, N., Abdulkarim, M.F., Abdullah, G.Z., Basir, R., Sadikun, A., and Asmawi, M.Z. (2010). HPLC and anti-inflammatory studies of the flavonoid rich chloroform extract fraction of *Orthosiphon stamineus* leaves. *Molecules,* **15**(6)**:** 4452-4466.

Yuliana, N.D., Khatib, A., Link-Struensee, A.M.R., Ijzerman, A.P., Rungkat-Zakaria, F., Young, H.C., and Verpoorte, R. (2009). Adenosine A1 receptor binding activity of methoxy flavonoids from *Orthosiphon stamineus*. *Planta Medica,* **75**(2)**:** 132-136.

Traditional and Folk Herbal Medicine: Recent Researches Vol. 2 (2014) *Pages* **189–219**
Editor: **V.K. Gupta**
Published by: **DAYA PUBLISHING HOUSE, NEW DELHI**

7

Standardization, Pharmacological Evaluation and Safety Studies of Sitopaladi Churna, an Ayurvedic Herbal Formulation

Rayees Sheikh[1], Gurdarshan Singh[1*],
Surjeet Singh[1], Naresh Satti[1], and Vijay Gupta[1]

ABSTRACT

The herbal formulation Sitopaladi Churna, which is a mixture of 4 extracts such as Bambusa arundinacea, Piper longum, Elletaria cardamum and Cinnamomum zeylanicum, was investigated for its anti-tussive and anti-histaminic activity. Anti-tussive activity was evaluated both by mechanically induced and chemical induced coughing in guinea pigs whereas, the anti-histamic activity was evaluated on isolated rat peritoneal mast cells. Sitopaladi Churna (SC-1) significantly reduced the coughing in both mechanically and chemically induced coughing models. It also reduced the histamine release from rat peritoneal mast cells significantly. The results furnished by SC-1 for both the activities were in dose/concentration dependent manner. The present review also reports the standardization of Sitopaladi Churna with piperine as reference compound which is a major constituent of Piper longum. Additionally the acute and sub-acute toxicity was done on laboratory animals as per the OECD Guidelines to identify undesirable pharmacodynamic properties of the SC-1 that may have relevance to its human safety.

***Keywords*:** Sitopaladi Churna, Anti-tussive, Anti-histaminic, Histamine, Piperine, OECD, Pharmacodynamic.

---

1 CSIR-Indian Institute of Integrative Medicine, Canal Road, Jammu – 180 001 (J&K State), India

* *Corresponding author*: E-mail: singh_gd@iiim.ac.in

Introduction

Medicinal plants are widely distributed throughout the world especially in tropical countries. It is estimated that about 25 per cent of all modern medicines are directly or indirectly derived from higher plants (WHO, 2005, 2002a,b, 1999a,b, 1998a,b, 1990, 1981, 1979; De Smet, 1995; Duke and Martinez, 1994; Majno, 1975; Ackerknecht, 1973). Thus, herbal medicine has led to the discovery of a number of new drugs, and non-drug substances.

Ayurveda is the most ancient health care system and is practiced widely in India, Sri lanka and other countries. Numerous drugs have entered the international pharmacopoeia through the study of ethno pharmacology and traditional medicines. Traditional medicines are used by about 80 per cent of the world's population. These only used for primary health care, not just in rural areas of developing countries, but also in developed countries, where modern medicines are predominantly used (Kamboj, 2000; Bodekar, 1994). Plants are most important source of herbal medicines and play important role in enhancing the longevity and quality of life in gaining prominence throughout the world and still people do not rely on herbal medicines due to lack of scientific validation, quality control and safety of medicinal products which are contaminated with biological and environmental pollutants and adulterants.

Through out human history people have relied on natural products and plants in particular to promote and maintain good health and to fight sickness, pain and disease. Now-a-day's modern (allopathic) medicine has in most parts of the world, replaced traditional medical practices to the benefit of individual and public health. The effective western style medicine is neither accessible nor affordable to hundreds of millions of people worldwide. This is more true than in India, in spite of the rapid and widespread diffusion of western medical practices throughout the country during the past 150 years. The majority of the developing countries rely on traditional medicines, many a times in the form of home remedies for its health care needs.

Total global herbal market is of size 62.0 billion dollars. In this, India's contribution is only one billion dollars. The table in annexure 5 explains the position of India in the global market. European union is the biggest market with the share 45 per cent of total herbal market. North America accounts for 11 per cent, Japan 16 per cent, Asean countries 19 per cent and rest of European Union 4.1 per cent. Countries like Japan and China have successfully marketed their traditional medicines abroad. Their alternative therapies are well-accepted in Europe and US. Product like Ginseng – the famed aphrodisiac from China is having the same property as of Ashwagandha-an ayurvedic medicine, yet it accounts for over US $ 800 M of international market as compared to all our herbs put together (which is less than US $ I M).

When compared to the Chinese and the Japanese level of penetration in the global market India is not at all figuring anywhere. The reason is simple, lack of quality control and scientific evaluation of Indian products. Although there are many formulations available in the market but neither they are scientifically evaluated nor authenticated or quality controlled.

But there are also positive signals for us in the global market. India has 16 Agro-climatic zones, 10 vegetative zones, 15 biotic provinces, 426 biomes, 45000 different plant species and 15000 medicinal plants that include 7000 ayurveda, 700 in Unani medicine, 600 in Siddha medicine and 30 in modern medicine. This makes India one among 12 mega biodiverse countries of the world, which despite having only 2.5 per cent total land area, accounting for over 8 per cent of the recorded species of the world. The forecast is that the global market for herbal products is expected to be $5 trillion by 2050. Herbal remedies would become increasingly important especially in developing countries. India, with its biodiversity has a tremendous potential and advantage in this emerging area.

Herbal medicines are very important since, in common with conventional medicines, they contain biologically active substances that may produce non-trivial side effects when taken in excessive amounts and on the other hand, have no therapeutic value, when taken in low doses (Garg and Bhutani, 2008).

Lack of standardization of polyherbal drug formulation is a serious problem in validating efficacy and maintaining quality control for manufacture of traditional herbal drugs. As commercialization of the herbal medicine has happened, assurance of safety, quality and efficacy of medicinal plants and herbal products have become an important issue. The herbal raw material is prone to a lot of variations due to several factors, the important ones being the identity of the plants and seasonal variations (which has a bearing on the time of collection), the ecotypic, genotypic and chemotypic variations, drying and storage conditions and the presence of xenobiotic (Dixit *et al.,* 2008). Standardization as defined by American Herbal Product Association: "Standardization refers to the body of information and control necessary to produce material of reasonable consistency. This is achieved by minimizing the inherent variation of natural product composition through quality assurance practices applied to agricultural and manufacturing processes (Waldesch *et al.,* 2003).

The Indian herbal industry is growing in a tremendous rate. More number of herbal products have arrived in the market. The safety and efficacy of herbal products are dependent upon the standardization of these herbal drugs. The traditional approach towards standardization is insufficient for current herbal market and hence there is need for more advanced techniques for standardization. There are basically two techniques used for standardization and these are chromatographic fingerprinting and DNA fingerprinting. The chromatographic fingerprinting is based on the chromatographic separation and identification of marker compound from other constituents. For this purpose TLC, HPTLC, HPLC, LC-MS, LC-NMR, GC-MS, GC-FID and SFC methods are used. The other method used is DNA fingerprinting. As the DNA fingerprint of genome remain the same irrespective of the plant part used while the photochemical content will varies with the plant part used, physiology and environment, hence this is the well established and highly precious method for standardization of herbal drug.

Toxicology or safety data and stability of herbal drugs for all practical purposes is unknown or not available. Several researchers also reported that most herbal medicines exhibit organ specific toxicity. There has been minimal research to assess possible systemic toxicity that might be associated due to chronic administration of herbal products (Miller, 1998)

Sitopaladi Churna (SC-1) is a polyherbal formulation containing the extracts of *Bambusa arundinacea, Piper longum, Elettaria cardamum* and *Cinnamomum zeylanicum.* Its traditional uses in folk medicine are multiple, and some of its therapeutic effects including antitussive, analgesic, antipyretic (Ayurvedacharya *et al.*, 1951; Pattanayak *et al.*, 2010) and anti-histaminic (Ahirwar *et al.*, 2008) have been confirmed by various experimental studies.

Sitopaladi Churna, manufactured by a reputed Indian drug company, Indian Medicinal Pharmaceutical Corporation Limited (*IMPCL*), Mohan Nainital, Uttranchal was taken up for investigation for "standardization, pharmacological evaluation and safety studies" under the collaborative research project between, Indian Institute of Integrative Medicine, Jammu; Indian Medicinal Pharmaceutical Corporation Limited, Nainital and Department of Science and Technology, Govt., of India, New Delhi.

Materials and Methods

Drug and Chemicals

Raw materials and formulation (Sitopaladi Churana), were supplied by IMPCL (Indian Medicines Pharmaceutical Corporation Ltd., Mohan (Nainital), Uttranchal, India. Biochemical kits were purchased from M/s Bayer Diagnostics, India and all other chemicals were purchased from Sigma Chemicals, Bangalore, India.

Standardization

Chemical Standardization of Formulation, SC-1 by HPLC

Standardization of Sitopaladi Churna, a poly herbal formulation was carried out on the basis of piperine as a reference compound. This compound is the major constituent of *Piper longum* and that of the formulation (SC-1) also. Different batches of SC-1 were taken for extraction and these were standardized by HPLC. SC-1 (1 g) was extracted with chloroform in a Soxhlet's extraction apparatus for 8 hr. Chloroform was evaporated to get residue and it was dissolved in 25 ml of HPLC grade methanol. The sample was filtered through Millipore filter (0.45mm) and 10 ml of the sample was injected in to the HPLC system.

Chromatographic analysis of formulation SC-1 was carried out on Merck 5µm RP 18e reverse-phase column (4.6 x 250 mm). The mobile phase consisted of an isocratic mixture of water methanol (7: 3) and it was delivered at a flow rate of 1.0 ml/min, the column temperature was set at 30°C whereas the detection was carried out at 254 nm while using the Photo diode Array (PDA) detector.

Determination of Essential Oils by GC and GC/MS

Formulation (100 g) was taken in a round bottom flask fitted with a Clevenger's apparatus. Water (700 ml) was added to the formulation and the mixture was hydro distilled for 3 hr. The distillate was trapped in hexane to dissolve oils. Hexane layer was removed and dried over anhydrous sodium sulphate. Removal of hexane yielded an oil which was analyzed by GC and GC/MS

Stability Study of Sitopaladi Churna (SC-1)

Stability studies of the formulation SC-1 were carried out on the basis of two parameters *i.e.* temperature and relative humidity. The stability study was done on the basis of single marker, piperine which was assayed on the basis of High Pressure Liquid Chromatography (HPLC). The shelf-life of the formulation was expressed and the time required for 10 per cent reduction in the potency of the formulation was taken as the basis of the shelf-life calculations.

The 100 g of powder formulation was kept in high density plastic containers. The samples/formulation was stored at two different conditions namely, ambient conditions, (25°C ± 2°C/60 per cent ± 5 per cent R.H.) and accelerated conditions, (40°C ± 2°C/75 per cent ± 5 per cent R.H). The samples stored at ambient conditions were drawn for testing at pre-defined time intervals *viz.* 0, 1, 2, 3, 4, 6, 7, 8, 9 and 12 months whereas samples of accelerated conditions were drawn at 0, 1, 2, 3, 4, 5 and 6 months interval from the date of storage. The chromatographic analysis was performed using Thermofinnagan HPLC system, which consisted of a PDA detector (UV6000LP), pump (P4000 Quat. Gradient), On-line degasser (SCM1000), autosampler (AS3000) and column oven. The data acquisition and processing was performed using Chromquest 4.0 software.

The optimized chromatographic conditions for the estimation of piperine were mobile phase which consisting of a mixture of methanol (Rankem) and HPLC grade water (Millipore) (70: 30 v/v), flow rate 1.0 ml/min, RP-18, 5 µm (Merck) column was used for analysis. The column oven temperature was set at 40°C to provide efficiency to separation whereas detection wavelength was set at 254 nm, total analysis run time was 30 min.

Animals

Healthy Wistar rats (male and female), Swiss mice (Female) and Guinea pigs (male), were kept in experimental room. The room was well-ventilated (10 air changes per hour) with 100 per cent fresh air. A 12-hr light/dark photoperiod was maintained. Room temperature and relative humidity was set to be maintained between 22 ± 2°C and 40 -80 per cent respectively according to CPCSEA guidelines

Acute Toxicity

The acute toxicity of the formulation, Sitopaladi Churna (SC-1), was evaluated in mice using the fixed dose procedure (OECD Guidelines No. 420, pages 1/14- 14/14 (2001). The test substance (SC-1) at a single dose of 2000 mg/kg body weight was administered to five animals orally by gavage. The animals were observed for toxic symptoms continuously for the first 4 hr. after dosing. Finally, the number of deaths were noted after 24 hr. And these animals were then maintained for further 13 days with observations made daily.

Sub-acute Toxicity Study

Wistar albino rats of either sex were segregated into 12 groups and 10 animals (5 male and 5 female) were used at each dose level. An additional satellite group of 10 animals (5 per sex) in the control and in the top (highest dose) group for observation, of reversibility or persistence of any toxic effects after the treatment period was also used. At least three dose levels and a concurrent control was used. SC-1 was administered for 28 days at doses of 250,500 and 1000 mg/kg body weight respectively. Control groups received the vehicle only. Toxic manifestations, mortality, feed and water intake were monitored daily and body weight changes were recorded every 7 days till the end of the study.

Body Weight, Feed and Water Intake

All animals were observed twice daily to see any abnormal clinical signs or observable toxicity. Body weights were recorded weekly (Day 0, 7, 14, 27 and 28) using a sensitive balance and food and water consumption was also measured daily.

Relative Organs Weight

At the end of the experiment, all the animals were euthanized by CO_2 inhalation. Different organs like, the brain, heart, lungs, liver, stomach, spleen, kidneys and gonads were carefully dissected out and weighed in grams (absolute organ weight).

Clinical Laboratory Investigations

Urine Analysis

On 27th day, 24 hr. urine samples from all rats were collected by placing the animals in the metabolic cages with free access to aqua-guard drinking water. Urinalysis was conducted for volume, specific gravity, pH, protein, glucose, occult blood, and ketone, using Multistix 10 SG (Bayer Diagnostics, Baroda, India). Urinary sediments (pus/epithelial cells, casts and crystals of calcium oxalate) were examined microscopically.

Haematological Analysis

On day 28th, all animals were fasted overnight, and blood collected from the retro-orbital plexus in two types of eppendrof tubes: one with 5 per cent EDTA for immediate analysis of haematological parameters and others to separate serum for biochemical estimations.

Hematological analysis was performed using an automated haematology analyzer (Humacount; Human, Weisbaden, Germany), including RBC and WBC and differential leucocyte count, haemoglobin (Hb), platelet (Plt), Hematocrit (Ht).

Clinical Biochemistry

Glucose, triglyceride, cholesterol, total protein, total bilirubin, alkaline phosphatase, serum glutamate oxaloacetate transaminase (SGOT), serum glutamate pyruvate transaminase (SGPT), blood urea, blood urea nitrogen (BUN), uric acid and creatinine were assayed from the serum by enzymatic methods using ready made diagnostic kits (Bayer Diagnostic) with biochemistry analyzer (CHEM-7, Erba, India).

Pathology

Gross Necropsy and Organ Weight

All animals were subjected to full, detailed gross necropsy which includes careful examination of the external surface of the body, all orifices, and the cranial, thoracic and abdominal cavities and their contents. Animals were then sacrificed by ether anesthesia. The liver, kidneys, adrenals, testes, ovaries, spleen, brain and heart were dissected out, washed and transferred to an ice-cold saline solution to wash out any blood strain. The organs were weighed and portions of these organs were fixed in 10 per cent formalin for histopathological examinations.

Histopathology

Full histopathology was carried out on the preserved organs and tissues of all animals in the control and high dose group (OECD, Guideline, 407).

Pharmacological Evaluation of the SC-1

Antitussive Activity of SC-1

The clinical use of Sitopaladi Churna is primarily for cough and cold. The formulation was evaluated in the experimental animals for the cough and cold.

Study 1: Mechanically Induced Coughing in Guinea Pigs

Male Guinea pigs (weighing 350-400 g), five for each group, were anaesthetized with 25 per cent (w/v) of urethane (5 ml/kg, i.p.) and tracheomized. The tracheal cannula was connected to a force transducer (Gross polygraph, model 7D). Coughing was induced by stimulating the walls of the trachea with a polyethylene tube of small diameter according to (Dhuley,1999). Stimuli were performed at 10 min intervals. The basal value was the sum of number of coughing outbursts in 1 hour.

Sitopaladi Churna, was given orally at doses of 125, 250 and 500 mg/kg. Antitussive activity was estimated by comparing the total number of coughs for 1 hr. before and after drug treatment in comparison to standard drug, codeine phosphate.

Study 2: Citric Acid Induced Coughing in Guinea Pigs

Male Guinea pigs (weighing 300-400 g), five for each group were exposed 5 per cent citric acid at a pressure of about 500 mm Hg through a nebulizer (Buxco, USA) containing 5-6 ml of the irritant solution. The animals were selected by the number of the outbursts of coughing observed during an aerosol exposure 24 h before the test; animals with more than 30 or fewer than ten outbursts were not used for the study. Test material was given orally 1 hr. prior to the initiation of the test. Anti-tussive activity was calculated by comparing the number of coughs prior to and after treatment according to (Tanaka and Maruyama, 2005).

Study 3: Anti-histaminic Activity

Preparation of Rat Peritoneal Mast Cells (RPMC)

RPMC were isolated from rat peritoneal fluid and purified as per the procedures described by (Shin and Lee, 2003) without any modification. Mast cell preparations were about 95 per cent pure as assessed by toluidine blue staining. Viability of cell was >97 per cent as determined by trypan blue exclusion test.

Determination of Histamine Release from RPMCs

Purified mast cells ($2x10^6$) were resuspended in a buffer containing 10 mM HEPES, 130 mM NaCl, 5 mM KCl, 1.0 mM $MgCl_2$, 1.4 mM $CaCl_2$, 5.6 mM glucose). Pre-incubation of the mast cells (10^6 cells/ml) was done in presence of test materials for 10 min at 37 °C. Compound 48/80 (10µg/ml) was added and cells were further incubated for 20 min. The reaction was stopped by cooling the tubes in ice. Cells were centrifuged (400×g for 5 min) and histamine was measured in supernatants as described by (Shore *et al.*, 1959) using excitation and emission wavelengths of 353 and 438 respectively.

Results

Chemical Standardization of Formulation by HPLC

An HPLC-derived fingerprint profile of formulation SC-1 is shown in Figures 7.1 and 7.2. Chemical standardization of SC-1 was carried out on the basis of piperine as reference compound. SC-1 was found to contain 0.067 per cent of reference compound piperine (Table 7.1).

Table 7.1: Quantitative analysis of SC-1.

	Per cent of Piperine (per100 g of formulation)				
	Expt. 1	*Expt. 2*	*Expt. 3*	*Mean*	*SD*
SC-1	0.068	0.065	0.068	0.067	0.00173

Determination of Essential Oils by GC and GCMS

The major and identified peaks of Sitopaladi Churna by GLC and GC/MS are shown in Tables 7.2 and 7.3; Figures 7.3 and 7.4 respectively. The extractive value of essential oil from SC-1 was found to be 0.0871 g/100 g of the formulation.

Stability Study

Shelf life of SC-1 at accelerated conditions, (40°C ± 2°C/75 per cent ± 5 per cent R.H) and ambient conditions, (25°C ± 2°C/60 per cent ± 5 per cent R.H.) according to HPLC assay was found to be 9 and 18 months respectively (Tables 7.4 and 7.5; Figures 7.5 and 7.6).

Acute Toxicity Study

After oral administration up to the dose of 2000 mg/kg body weight in mice, no mortality or any signs of behavioral changes or toxicity was observed. Hence, the LD_{50} of SC-1 is more than 2000 mg/kg p.o. (Table 7.6).

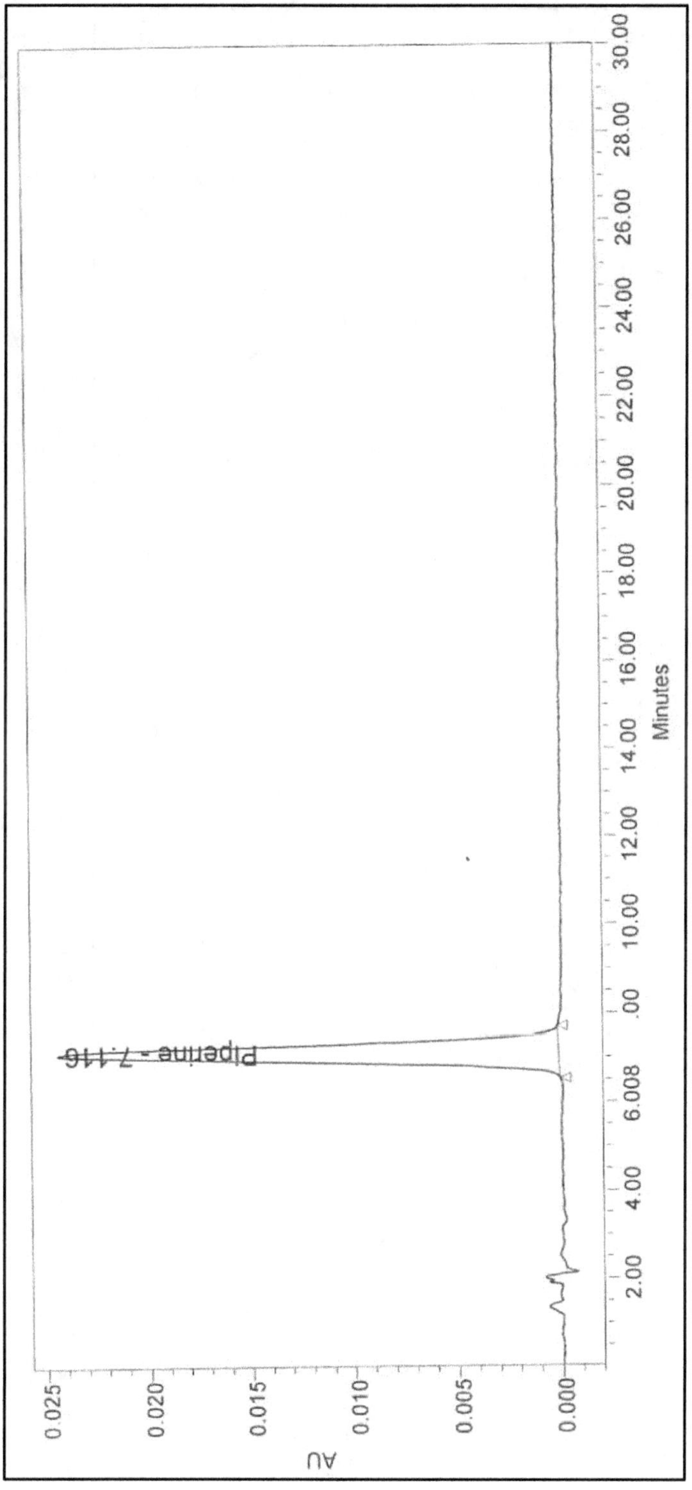

Figure 7.1: HPLC chromatogram (254 nm) of reference compound (Piperine).

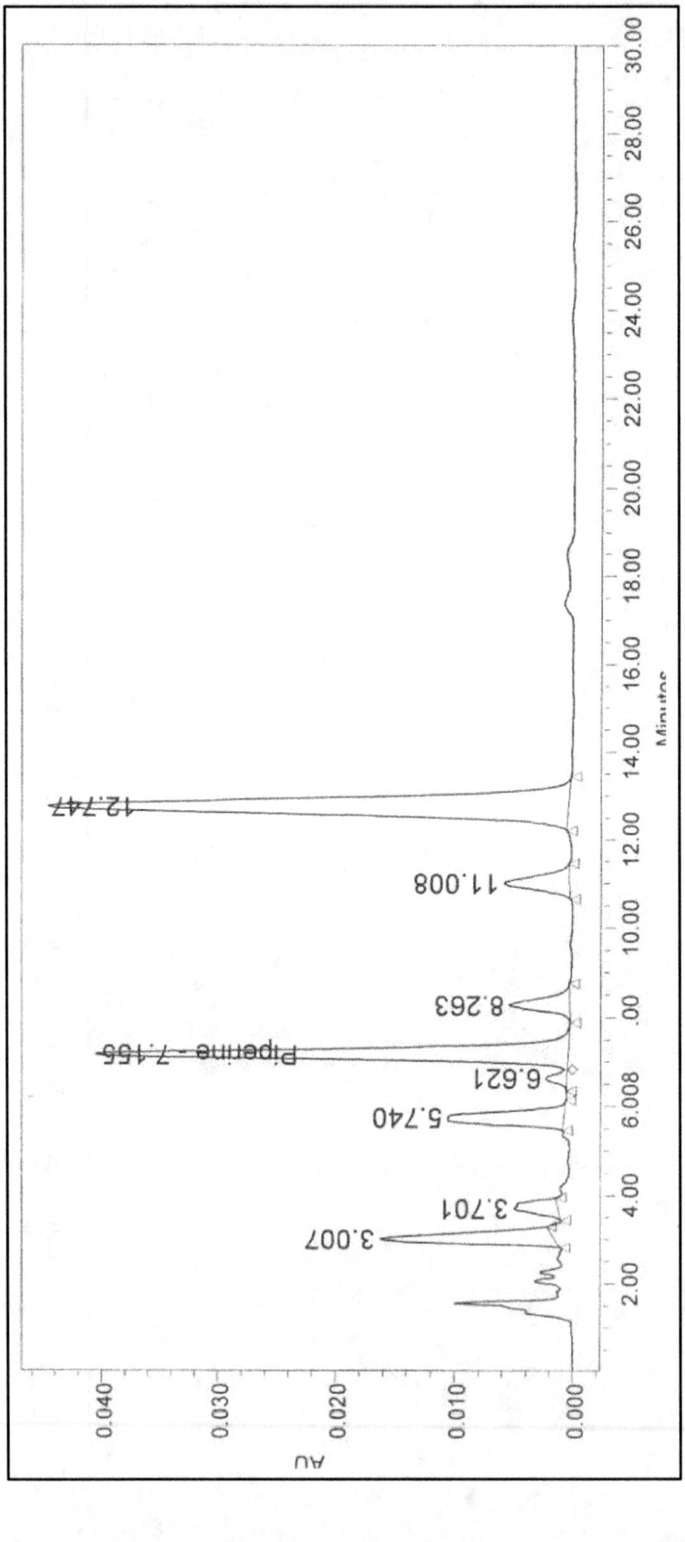

Figure 7.2: HPLC chromatogram (254nm) of chloroform extract of SC-1.

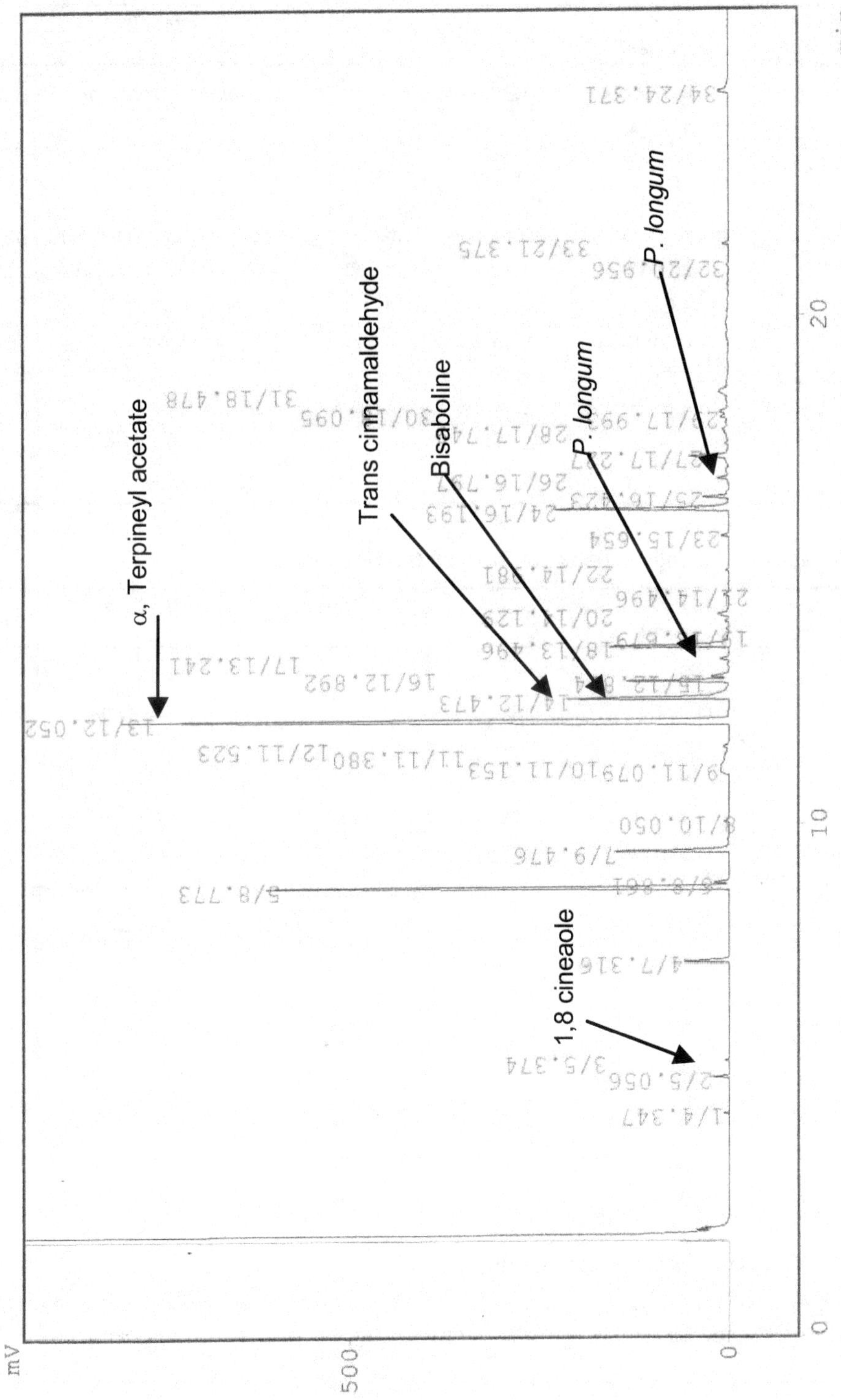

Figure 7.3: GLC fingerprint of SC-1 essential oil. Conditions: Inj. Vol., 0.2 µl; GLC, Shimadzu; Model, GC-17A; Column, BP-10; Inj. Temp.–240°C; Det Temp. 260°C; Prog-90c/2m@7c.m to 220 c/5m

Table 7.2: Major and identified peaks of SC-1 by GLC analysis.

Retention Time (min)	*Area (per cent)*	*Compound*	*Ingredient*
5.374	0.0795	1,8-Cinole	*Elletaria cardamomum*
7.316	1.6514	–	–
8.773	20.4032	–	–
9.476	5.0286	–	–
12.052	35.3413	α-terpineyle acetate	*Elletaria cardamomum*
12.47	8.0035	Trans	*Cinnamomum zeylanicum*
12.814	4.0880	Cinnamaldehyde	*Piper longum*
13.241	0.3408	Bisaboline	*Piper longum*
13.496	5.7638	–	–
16.193	8.2467	–	–
16.797	0.3570	–	–
17.227	2.1135	–	*Piper longum*

Table 7.3: Major and identified peaks of SC-1 by GC/MS analysis.

Retention Time (min)	*Mass (per cent)*	*Total*	*Marker*	*Ingredient*
2.93	136	0.2	–	*Elletaria*
3.57	154	0.5	–	*Elletaria*
4.60	136	1.6	–	*Elletaria*
5.57	152	16.5	–	–
6.32	136	5.7	α-Terpineol	*Elletaria*
7.42	139	3.6	–	–
8.02	131	4.7	Trans-cinnamaldehyde	*C. zeylanicum*
9.47	181	33.0	α-Terpineol acetate	*Elletaria*
9.60	164	3.9	–	–
10.67	204	3.2	Bisaboline β	*P. longum*
12.47	206	1.4	–	–
13.27	204	9.2	–	–
13.65	187	2.4	–	–
19.47	223	5.4	–	–

Sub-acute Toxicity Study

Effect of SC-1 on General Behaviour

No pre-terminal death was recorded for any group including the control and any dose up to the maximum of 1000 mg/kg for 28 days during the period of treatment.

Table 7.4: Stability studies of Sitopladi Churna under accelerated conditions, 40ºC ± 2ºC/ 75 per cent ± 5 per cent R.H.

Time (Months)	*Assay (per cent)*	*Log Assay*	*K (Degradation Rate constant)*
0 month	100.000	2.000	0.0122059
1 month	100.000	2.000	
1.5 month	98.188	1.9920	
3 month	98.776	1.9946	
4 month	95.672	1.9807	
5 month	96.500	1.9845	
6 month	91.903	1.9633	

Shelf-life at accelerated conditions, 40ºC ± 2ºC/75 per cent ± 5 per cent R.H = 9 months.

Table 7.5: Stability studies of Sitopladi Churna under ambient conditions, 25ºC ± 2ºC/60 per cent ± 5 per cent R.H.

Time (Months)	*Assay (per cent)*	*Log Assay*	*K (Degradation Rate constant)*
0 month	100.000	2.000	0.0059878
1 month	100.772	2.000	
1.5 month	100.000	2.000	
3 month	99.217	1.9966	
4 month	96.527	1.9846	
5 month	97.001	1.9867	
6 month	94.307	1.9745	
7 month	97.760	1.9901	
9 month	94.832	1.9769	
12 month	93.750	1.9719	

Shelf-life at ambient conditions, 25ºC + 2ºC/60 per cent ± 5 per cent R.H. – 18 months.

Table 7.6: Acute oral toxicity study of SC-1 in Swiss mice.

Weight (g) Mouse	*Dose (mg/kg)*	*Toxicity Symptoms*	*Mortality*	*Necropsy after Terminal Sacrifice*
23	2000	Nil	No	NAD
24	2000	Nil	No	NAD
22	2000	Nil	No	NAD
22	2000	Nil	No	NAD
24	2000	Nil	No	NAD

Mouse No.1: Used for sighting study; **Mice No. 2-5** used for main study.

NAD: No abnormalities detected.

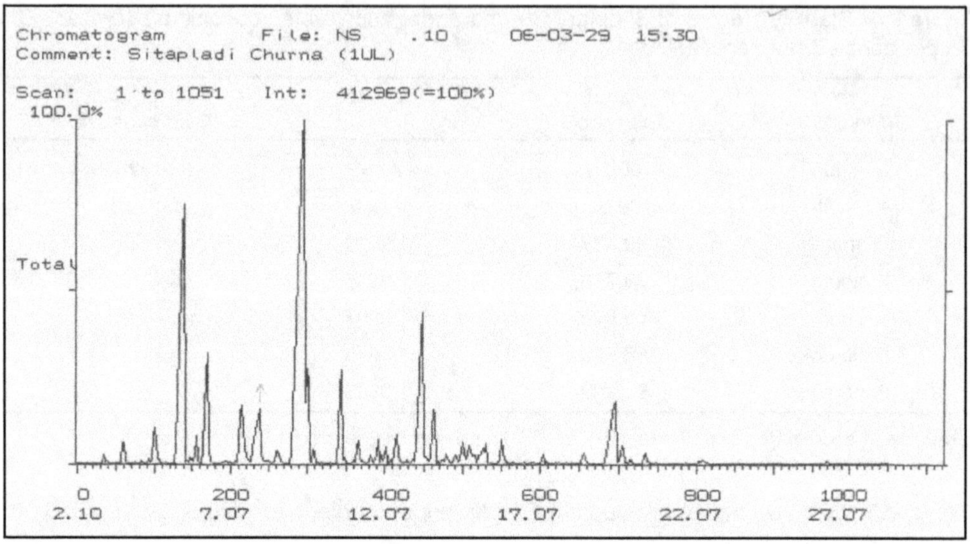

Figure 7.4: Total Ion Chromatogram (TIC) of SC-1 oil under GC-MS.

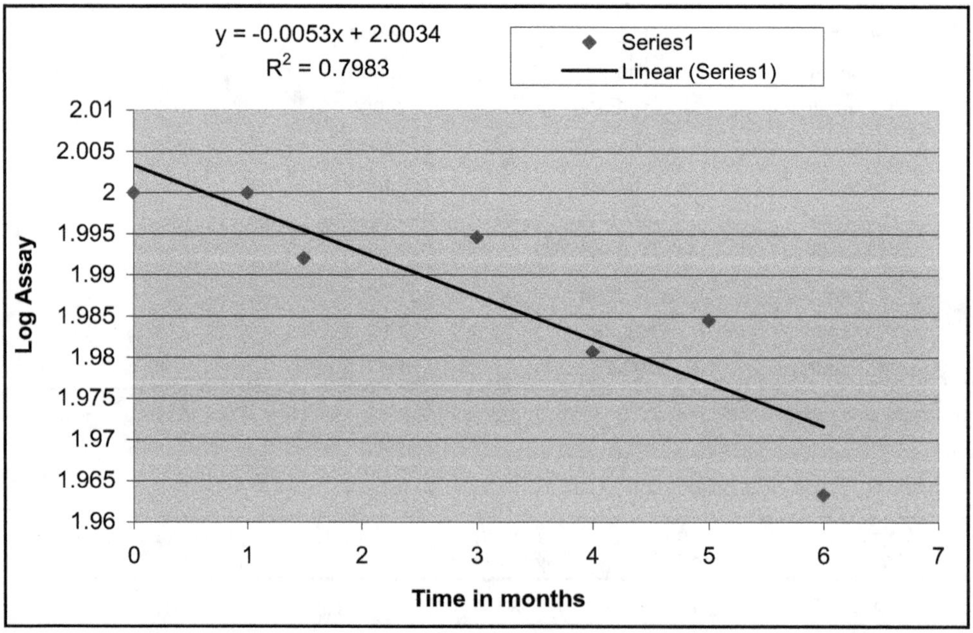

Figure 7.5: Stability Data of Sitopaladi Churna (SC-1) at accelerated conditions, 40ºC ± 2°C/75 per cent ± 5 per cent R.H).

Effect of SC-1 on Feed and Water Intake

Daily feed and water in-take by the individual groups of rats was observed for 28 days. The results of weekly mean ± S.E. values of the feed and water-in take

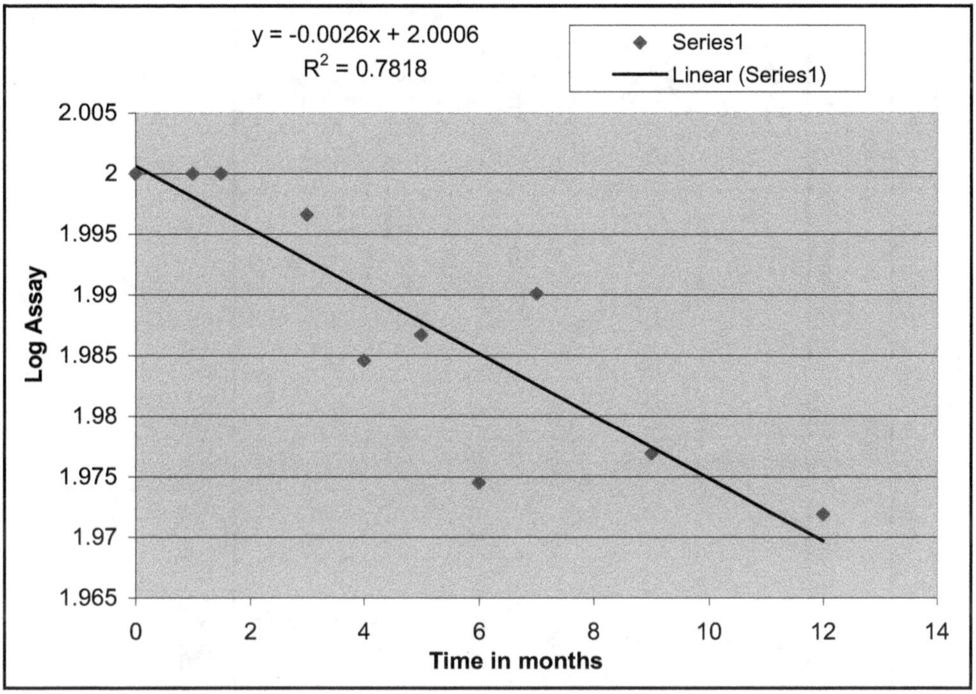

Figure 7.6: Stability data of Sitopaladi Churna at ambient conditions, 25°C ± 2°C/60 per cent ± 5 per cent R.H).

revealed (data not shown) that none of the group of animals has shown any change in the in-take values of both food and water indicating that the test drug has shown no adverse effect.

Effect of SC-1 on Body Weights

No significant changes in the body weights were observed in any of the treatment groups(male and female) in comparison to the control as shown in Figures 7.7 and 7.8.

Effect of SC-1 on Urine Parameters

No abnormality was observed in urine of treated animals, male and female. No significant variation was found in any of the parameters when compared with the control (Tables 7.7 and 7.8).

Effect of SC-1 on Haematological Parameters

The effect of SC-1 on the haematological parameters is shown in Table 7.9. The results revealed that all the values were found within the normal limits in all groups including drug treated and control animals.

Effect of SC-1 on Biochemical Parameters

The biochemical tests, which were performed on serum include, glucose, urea, creatinine, SGOT, SGPT, cholesterol, total protein, bilirubin, uric acid and triglycerides.

Table 7.7: Urine analysis of male rats treated with SC-1 for 28 days.

	Analysis	*Treatment (mg/kg; p.o)*					
		Control	*250*	*500*	*1000*	*Satellite*	
						Control	*1000*
a.	24 hour urine output per group (ml)	12 ml	13 ml	14 ml	12 ml	14 ml	15 ml
b.	Macroscopic examination						
	Colour	Pale yellow	Pale yellow	Pale yellow	Pale yellow	Pale yellow	Pale yellow
	Specific gravity	1.011	1.010	1.014	1.012	1.013	1.015
	pH	6.7	6.7	7.1	6.8	6.9	6.6
c.	Chemical examination						
	Glucose	Absent	Absent	Absent	Absent	Absent	Absent
	Albumin	Traces	Traces	Traces	Traces	Traces	Traces
	Ketones	Negative	Negative	Negative	Negative	Negative	Negative
d.	Microscopic examination						
	WBC/Puss cells	0-2	0-1	0-1	0-1	0-2	1-2
	RBC	Nil	Nil	Nil	Nil	Nil	Nil
	Epithelial cells	0-1	0-2	1-2	0-1	2-3	0-2
	Casts	Nil	Nil	Nil	Nil	Nil	Nil
	Crystals of calcium oxalate	Nil	Nil	Nil	Nil	Nil	Nil

Values are mean ± S.E of Wistar rat(n=5/sex/dose), having different doses of SC-1 (pooled data). No statistical difference observed between the SC-1 treated groups and control males and females using Student-Newman-Keul's multiple comparison test.

Table 7.8: Urine analysis of Female rats treated with SC-1 for 28 days.

	Analysis	*Treatment (mg/kg; p.o)*					
		Control	*250*	*500*	*1000*	*Satellite*	
						Control	*1000*
a.	24 hour urine output per group (ml)	12 ml	14 ml	14 ml	12 ml	10 ml	15 ml
b.	Macroscopic examination						
	Colour	Pale yellow	Pale yellow	Pale yellow	Pale yellow	Pale yellow	Pale yellow
	Specific gravity	1.015	1.013	1.01	1.014	1.015	1.016
	pH	6.6	6.5	6.8	6.5	6.8	6.4
c.	Chemical examination						
	Glucose	Absent	Absent	Absent	Absent	Absent	Absent
	Albumin	Traces	Traces	Traces	Traces	Traces	Traces
	Ketones	Negative	Negative	Negative	Negative	Negative	Negative
d.	Microscopic examination						
	WBC/Puss cells	0-1	0-2	0-1	0-1	0-2	1-2
	RBC	Nil	Nil	Nil	Nil	Nil	Nil
	Epithelial cells	0-1	0-2	1-2	0-2	1-2	0-2
	Casts	Nil	Nil	Nil	Nil	Nil	Nil
	Crystals of calcium oxalate	Nil	Nil	Nil	Nil	Nil	Nil

Values are mean ± S.E of Wistar rat(n=5/sex/dose), having different doses of SC-1 (pooled data). No statistical difference observed between the SC-1 treated groups and control males and females using Student-Newman-Keul's multiple comparison test.

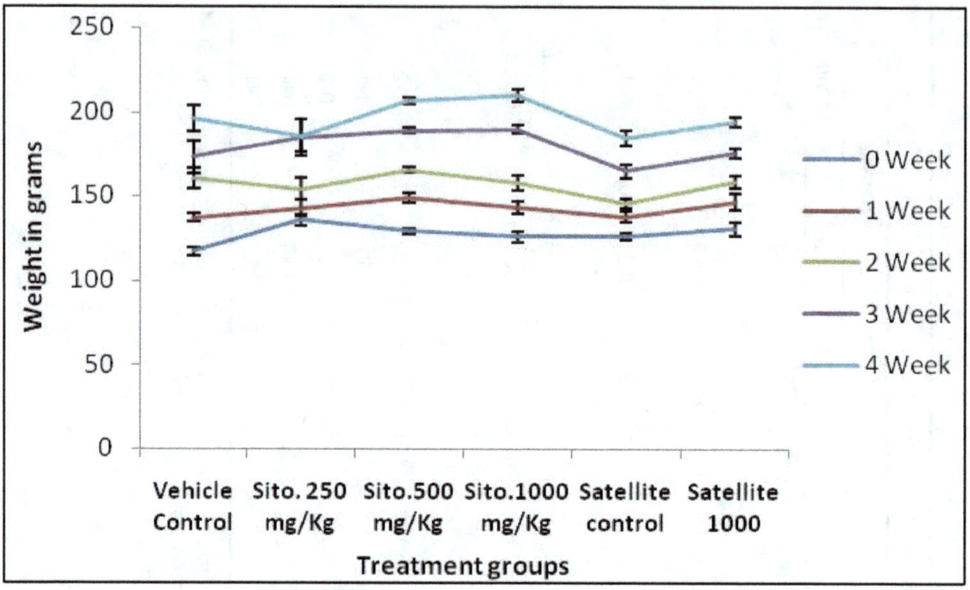

Figure 7.7: Mean weekly body weight of male rats treated with SC-1 for 28 days.

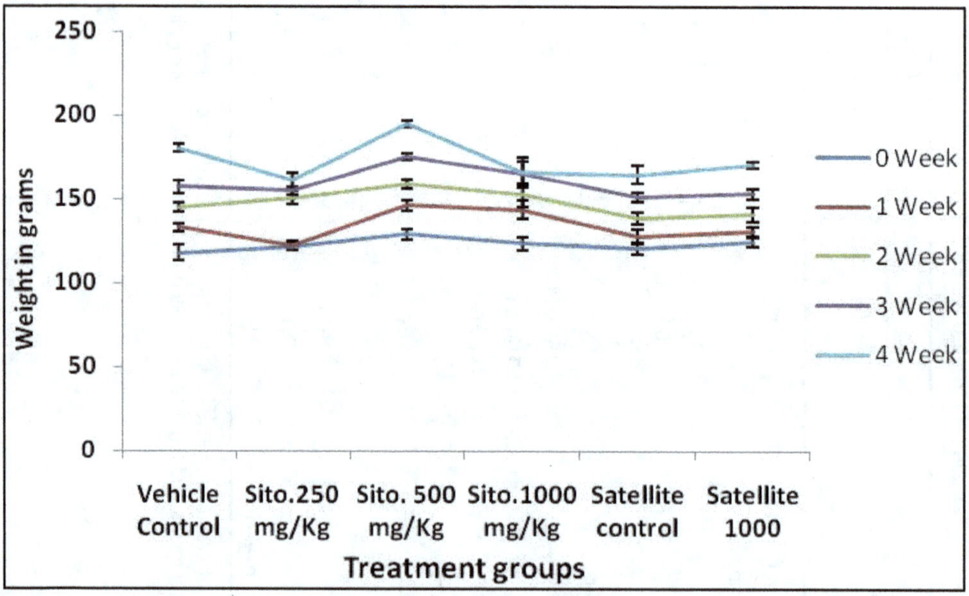

Figure 7.8: Mean weekly body weight of female rats treated with SC-1 for 28 days.

The results of the study from all the treated and control groups (Table 7.10; Figures 7.9 and 7.10), showed that all the biochemical parameters were within the normal limits except for the triglycerides, uric acid and SGPT. A significant increase in triglycerides, SGPT and uric acid ($p < 0.05$) was found in all the three doses of treatments including satellite groups in both males and females when compared to

Table 7.9: Mean hematological values of male and female rats treated with SC-1 for 28 days.

Sex (Treatment) (mg/kg p.o.)	*WBC (10^9/L)*	*LYM (per cent)*	*GRA (per cent)*	*RBC (10^{12}/L)*	*HGB (g/dL)*	*HCT (per cent)*	*PLT (10^9/L)*	*CT (min.)*	*BT (min.)*
Male									
Control	11.3 ± 1.3	74.34 ± 2.5	21.71 ± 3.4	9.06 ± 0.5	14.85 ± 0.9	49.86 ± 3.2	636.4 ± 3.2	2.17 ± 0.06	8.45 ± 0.33
250	19.38 ± 2.5	73.57 ± 3.84	20.78 ± 2.5	8.35 ± 0.48	13.65 ± 0.6	48.96 ± 2.4	570.4 ± 3.8	1.9 ± 0.22	6.74 ± 0.16
500	16.68 ± 1.2	71.97 ± 3.6	22.42 ± 2.2	7.98 ± 0.4	12.37 ± 0.8	44.91 ± 3.0	910.9 ± 67.0	1.8 ± 0.23	8.69 ± 0.23
1000	18.79 ± 3.9	70.82 ± 3.7	24.9 ± 4.4	9.36 ± 0.5	14.78 ± 0.8	52.79 ± 2.7	863.7 ± 86	2.1 ± 0.31	7.02 ± .18
Satellite									
Control	14.69 ± 1.9	70.0 ± 4.2	24.78 ± 2.6	6.61 ± 0.17	10.52 ± 0.32	37.1 ± 0.9	665.9 ± 24.0	1.3 ± 0.1	7.83 ± .31
1000	16.31 ± 2.6	70.1 ± 4.0	25.25 ± 2.5	6.24 ± 0.53	10.2 ± 0.65	36.4 ± 1.8	741.8 ± 70.7	1.04 ± 0.02	8.6 ± 0.5
Female									
Control	11.9 ± 0.99	72.2 ± 5.4	24.16 ± 3.8	8.35 ± 0.18	13.76 ± 0.3	50.29 ± 1.1	684.3 45.7	1.76 ± 0.37	8.48 ± .29
250	12.33 ± 2.3	71.77 ± 5.0	22.19 ± 4.9	6.87 ± 0.2	11.07 ± 0.1	40.97 ± 0.7	672.0 ± 55.8	2.24 ± 0.34	7.38 0.20
500	10.72 ± 3.2	74.12 ± 3.8	23.61 ± 3.2	7.23 ± 0.5	11.25 ± 0.7	41.64 ± 2.4	610.1 ± 16.5	1.6 ± 0.1	8.36 0.26
1000	11.58 ± 2.8	73.78 ± 2.1	21.87 ± 2.6	7.69 ± 0.4	11.04 ± 0.7	45.44 ± 2.5	606.2 ± 73.9	2.1 ± 0.2	8.90 0.16
Satellite									
Control	11.26 ± 1.9	61.71 ± 3.8	33.3 ± 3.0	4.5 ± 0.16	10.14 ± 0.4	47.52 ± 0.96	529.3 ± 28.4	1.06 ± 0.03	9.15 ± 0.51
1000	11.23 ± 0.3	66.08 ± 4.2	27.28 ± 3.4	5.8 ± 0.4	9.58 ± 0.6	45.73 ± 2.2	547.0 ± 15.2	1.1 ± 0.03	8.77 ± 0.88

Hematological values(Mean ± S.E.), of Wistar rats (n = 5/sex/dose), having different doses of SC-1. No statistical difference observed between the SC-1 treated groups and control males and females using Student-Newman-Keul's multiple comparison test. WBC, white blood cell; LYM., lymphocyte; GRA., granulocyte; RBC, red blood cell; HCT, hematocrit; PLT, platelet, CT, clotting time; BT, bleeding time.

Table 7.10: Mean blood biochemical parameters of rats treated with SC-1for 28 days sub-acute oral toxicity study.

Treatment (mg/kg p.o.)	*Control*	*250*	*500*	*1000*	*Satellite*	
					Control	*1000*
Male						
Glu. (mg/dL)	99.00 ± 6.31	74.40 ± 3.33	71.20 ± 4.02	71.60 ± 4.81	88.21 ± 5.20	75.33 ± 6.11
Trigl. (mg/dL)	53.40 ± 5.17	101.6 ± 1.72*	102.0 ± 2.12*	101.8 ± 8.31*	66.50 ± 4.61	98.22 ± 4.52*
Chol. (mg/dL)	60.20 ± 9.89	54.6 ± 2.73	56.8 ± 2.82	56.6 ± 3.96	58.25 ± 3.28	55.74 ± 4.15
TP (g/dL)	7.40 ± 0.39	7.90 ± 0.07	7.50 ± 0.17	7.58 ± 0.20	7.56 ± 0.41	8.15 ± 0.29
Bilu (mg/dL)	0.34 ± 0.02	0.44 ± 0.02	0.44 ± 0.02	0.48 ± 0.02	0.31 ± 0.02	0.40 ± 0.04
SGOT(IU/L)	168.20 ± 6.70	169.8 ± 6.98	171.6 ± 7.44	175.8 ± 8.42	158.41 ± 5.65	153.8 ± 4.85
SGPT(IU/L)	62.60 ± 7.96	138.4 ± 2.73*	139.0 ± 3.04*	137.8 ± 3.52*	74.84 ± 6.99	129.7 ± 5.42*
BUN (mg/dL)	46.6 ± 1.78	41.4 ± 2.38	39.6 ± 1.72	39.6 ± 1.83	53.51 ± 2.04	42.62 ± 3.08
Creat. (mg/dL)	1.12 ± 0.03	1.20 ± 0.07	1.18 ± 0.05	1.18 ± 0.05	1.05 ± 0.02	1.18 ± 0.04
U Acid (mg/dL)	1.20 ± 0.05	1.66 ± 0.08*	1.72 ± 0.07*	1.66 ± 0.08*	1.39 ± 0.07	1.54 ± 0.05*
Female						
Glu. (mg/dL)	93.60 ± 4.74	79.6 ± 5.50	76.40 ± 3.91	74.80 ± 3.56	82.35 ± 6.01	71.63 ± 5.31
Trigl. (mg/dL)	58.00 ± 4.46	116.2 ± 8.33*	119.6 ± 18.69*	130.4 ± 11.90*	61.55 ± 3.67	108.26 ± 6.27*
Chol. (mg/dL)	63.00 ± 2.07	72.2 ± 3.73	73.2 ± 4.51	69.6 ± 7.91	68.28 ± 4.17	65.42 ± 5.87
TP (g/dL)	6.90 ± 0.35	7.90 ± 0.07	7.76 ± 0.13	7.62 ± 0.18	7.12 ± 0.45	7.89 ± 0.39
Bilu (mg/dL)	0.40 ± 0.03	0.48 ± 0.03	0.42 ± 0.03	0.58 ± 0.03	0.39 ± 0.04	0.45 ± 0.07
SGOT(IU/L)	148.2 ± 8.43	172.2 ± 6.40*	175.8 ± 6.15*	179.8 ± 8.51*	139.38 ± 4.63	161.8 ± 4.98*
SGPT(IU/L)	66.8 ± 6.78	148.2 ± 4.91*	146.2 ± 5.30*	143.6 ± 10.65*	64.54 ± 5.89	138.5 ± 6.10*
BUN (mg/dL)	49.20 ± 1.53	37.6 ± 1.16	37.2 ± 1.07	40.4 ± 1.69	50.21 ± 2.89	44.25 ± 5.01
Creat. (mg/dL)	1.04 ± 0.04	1.06 ± 0.06	1.08 ± 0.04	1.22 ± 0.02	1.16 ± 10.03	1.23 ± 0.04
U.Acid (mg/dL)	1.00 ± 0.05	1.50 ± 0.10	1.52 ± 0.11	1.76 ± 0.14	1.30 ± 0.04	1.43 ± 0.04

Values are mean ± SD (n = 5/sex/dose); Glu= Glucose;Trigl= Triglycerides;Chol= Cholesterol; TP= Total protein; Bilu= Bilirubin; BUN= Blood Urea Nitrogen; Creat= Createnine; U. Acid= Uric Acid

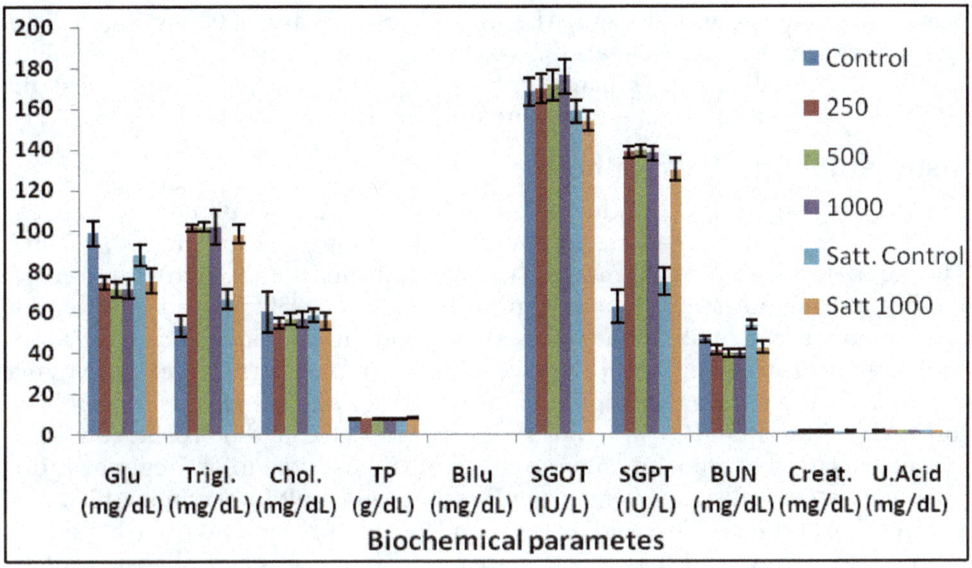

Figure 7.9: Mean blood biochemical parameters of male rats treated with SC-1 for 28 days sub- acute oral toxicity study.

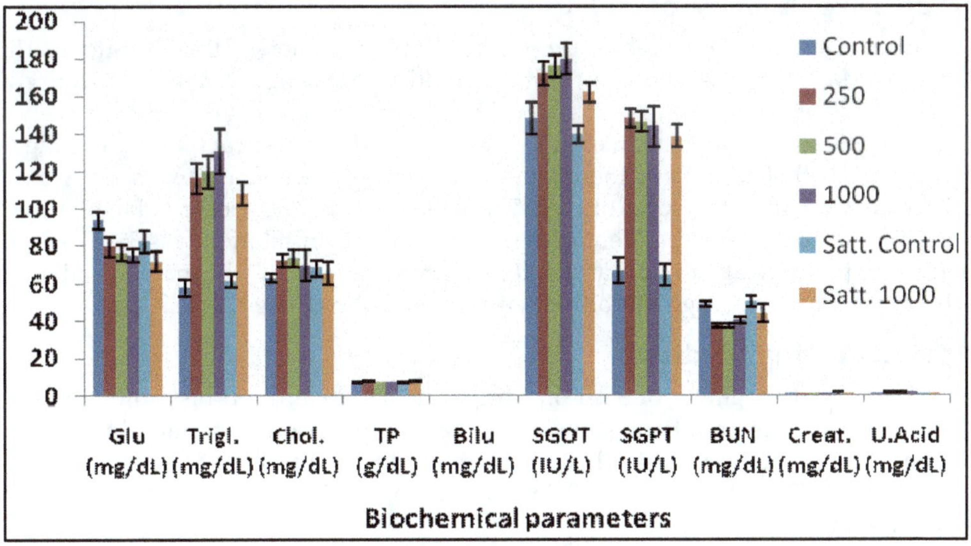

Figure 7.10: Mean Blood biochemical parameters of female rats treated with SC-1 for 28 days sub-acute oral toxicity study.

the control group. SGOT was also increased dose dependently in female treated rats as compared to control animals.

Gross Necropsy and Organ Weights

The macroscopic examination of the vital organs of control, treatment (250, 500 and 1000 mg/kg) and satellite groups (1000 mg/kg) did not reveal any significant changes except liver and kidneys. The liver of 250, 500 and 1000 mg/kg of *SC-1* administered male and female rats showed treatment related enlargement, mottling to yellowish discoloration. Similarly, organs weight of all animals (both sexes) did not show any significant changes as compared to vehicle control group (Table 7.11).

Histopathological Examination

The vital organs (trachea, lungs, brain, heart, spleen, lymph nodes, stomach, intestine, urinary bladder and testes/ovary) of the control, treatment (250, 500 and 1000 mg/kg) and satellite (1000 mg/kg) group rats treated with SC-1 in either sex did not show any significant histopathological changes. But, 250, 500 and 1000 mg/kg dose group rats both male and female showed mild diffuse vacuolar degeneration, focal mononuclear cell infiltration and coagulative necrosis in hepatic parenchyma in both male and female rats (Figures 7.11–7.13). Kidneys of three rats from 500mg dose and four rats from 1000 mg/kg dose group revealed congestion of blood vessels and glomerular capillaries, haemorrhages, diffuse vacuolar degeneration, mononuclear cell infiltration in intertubular spaces and multifocal tubular dilatation (Figures 7.14 and 7.15). Two mid dose group (500 mg/kg) rats and three high dose group male rat revealed loss of a few spermatogonial cells and oligospermia in somniferous tubules of testes (Figures 7.16 and 7.17).

Antitussive Activity of SC-1

Data of mechanical coughing depicted in Table 7.12 showed that the Sitopaladi Churna at different doses showed significant reduction in cough frequency. It showed 30, 44.44 and 66.66 per cent inhibition at 125, 250 and 500mg/kg dose respectively, whereas codeine phosphate protected coughing up to 80 per cent at the dose of 5 mg/kg (Table 7.12). The data of the chemical coughing in guinea pigs showed that Sitopaladi Churna at the doses of 125, 250 and 500 mg/kg and codeine phosphate at 5 mg/kg caused dose-dependent cough inhibition. The inhibition was 38.23, 47.42 and 65.21 per cent at 125, 250 and 500 mg/kg of Sitopaladi Churna respectively whereas 80 per cent cough inhibition with codeine phosphate (Table 7.13).

Anti-histaminic Activity

The inhibitory effects of SC-1 on compound 48/80 induced histamine release from RPMCs on albino rats are shown in Table 7.14. SC-1 dose dependently showed a significant inhibition rate in histamine release at 200 μg/ml. SC-1 showed an inhibition of 72.43 per cent ($p < 0.05$) compared to 82.67 per cent by 15 μg/ml Fexofenadine ($p < 0.05$).

Discussion

Sitopaladi Churna (SC-1), a polyherbal formulation has been used as an ayurvedic medicine in India since long for maintaining healthy respiratory system including cough and cold. However, a literature survey has indicated that the toxicity of this formulation has not been evaluated so far. The evaluation of toxicity is essential, if the extract or formulation has to be used for human applications for the chronic treatment of diseases. Toxicological study also helps in determination of safe dose

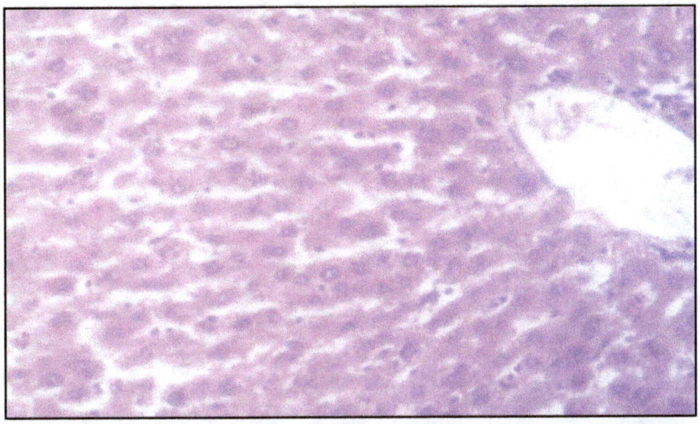

Figure 7.11: Liver of a control rat showing normal hepatic parenchyma (H and E sectionX400).

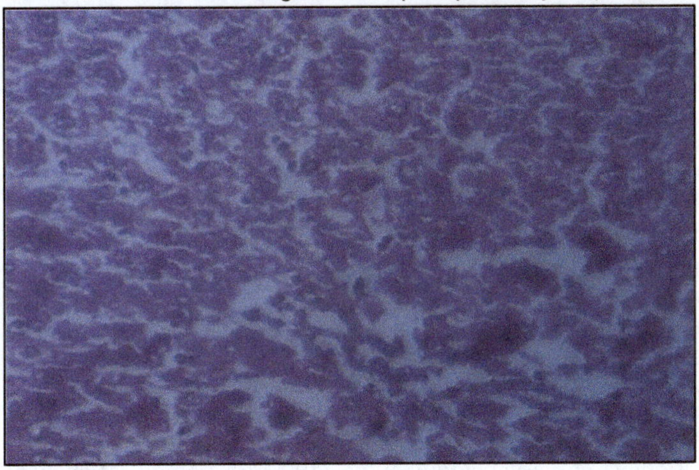

Figure 7.12: Liver of a 1000 mg of Sitopaladi Churna rat showing diffuse vacuolar degeneration hepatic parenchyma (H and E sectionX400).

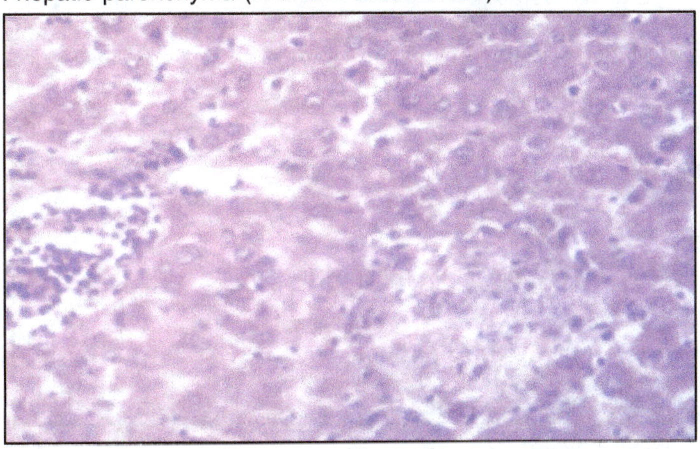

Figure 7.13: Liver of a Sitopaladi Churna (1000mg) treated rat showing focal mononuclear cell infiltration and coagulative necrosis (H and E sectionX400).

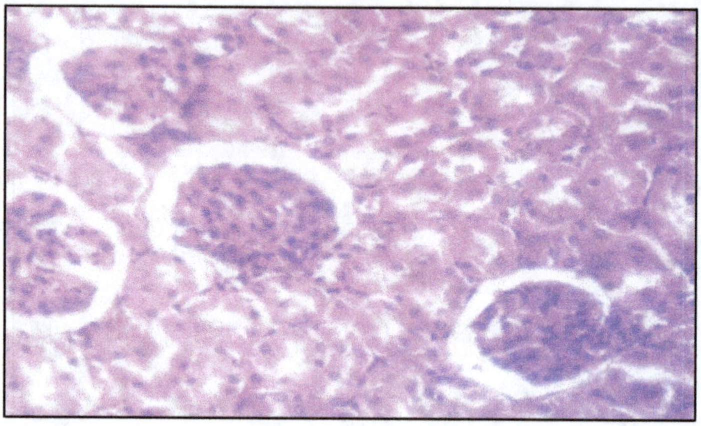

Figure 7.14: Kidneys of a control rat showing normal glomeruli and renal tubules (H and E sectionX400).

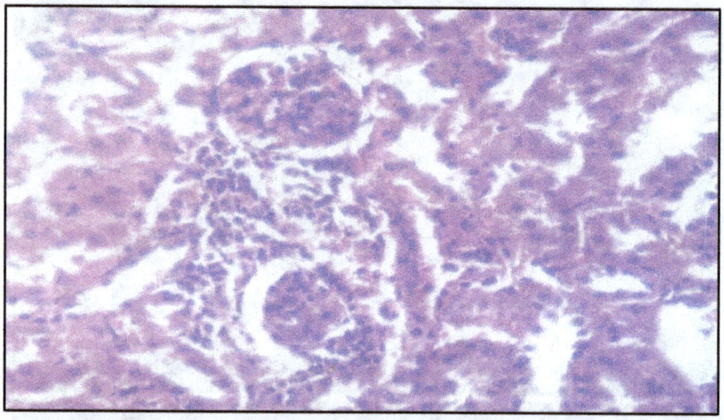

Figure 7.15: Kidneys of a Sitopaladi Churna (1000mg) treated rat showing glomerular congestion, focal mononuclear cell infiltration in intertubular spaces and mild tubular dilatation (H and E sectionX400).

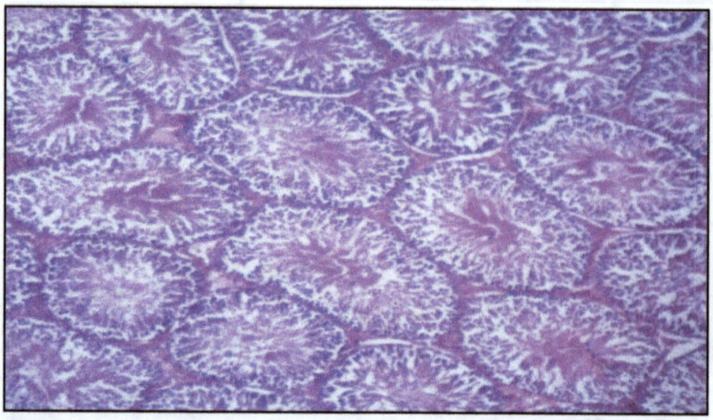

Figure 7.16: Testes of a control rat showing normal somniferous tubules with spermatozoa in the lumen (H and E sectionX400).

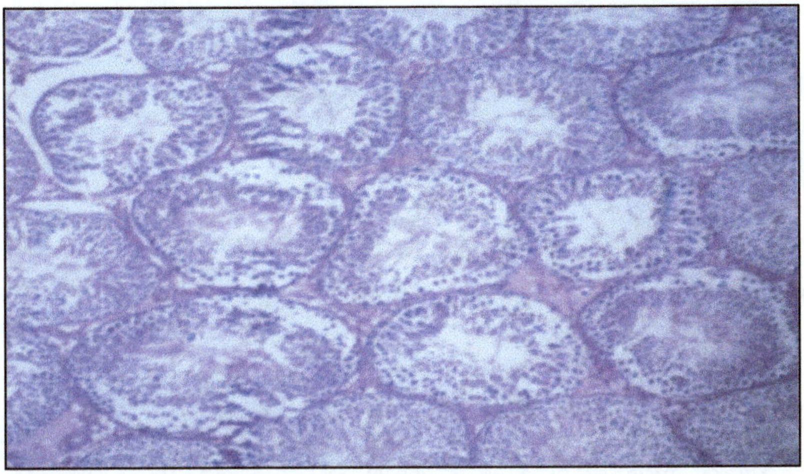

Figure 7.17: Testes of a high dose group rat (1000 mg Sitopaladi Churna) showing loss of a few spermatogonial cells in the seminiferous tubules with oligospermia in the lumen (H and E sectionX400).

and to ascertain the margin of safety evaluation. Therefore, the present study was carried out to evaluate the acute and sub-acute toxicity in laboratory animals according to (OECD Guidelines Nos. 420 and 407) respectively.

In the acute toxicity study, mice administered with formulation, SC-1 at the dose of 2,000 mg/kg did not show any signs of toxicity in the entire period of 14 days of observation. No mortality or any signs of behavioral changes or toxicity was observed. Moreover, gross and pathological examinations of the internal organs revealed no pathological abnormality. These results suggest that the SC-1 is non-toxic after an acute exposure to the dose of 2,000 mg/kg. In general, the acute treatment with the SC-1 did not show any index of toxicity as compared to the control group from the beginning until the 14th day after the treatment for the surviving animals.

In repeated dose study in Wistar rats (28 days oral sub-acute toxicity study), treatment with SC-1 did not affect the water and food consumption (data not shown) of thie animals. The body weight of the animals treated with SC-1 once a day during 28 days (did not show any decrease in the body weight exposed to different doses of SC-1 in comparison to the control animals. The internal organ weights were not significantly changed in treated animals as compared with those of the control. Urine analysis showed no abnormal findings. The qualitative analysis of urine samples from different groups (treated as well as control), showed no significant changes in specific gravity, pH, Albumin, WBC/Pus cells, Epithelial cells. During the experimental period, there were no treatment-related effects on the hematological parameters evaluated. The results showed that all the values were found within the normal limits in treated as well as in control animals.

Rats administered with SC-1 showed slight changes in some biochemical parameters. All the biochemical values were found to be present within the reference limits (Liang *et al.*, 1999), except for the increase in the uric acid, SGPT and triglyceride

Table 7.11: Mean organ weight (g) of rats treated with SC-1 for 28 days sub-acute toxicity.

Treatment (mg/kg p.o.)	*Brain*	*Heart*	*Lung*	*Liver*	*Spleen*	*Kidney*	*Testis*
Male							
Control	1.48 ± 0.06	0.72 ± 0.06	1.39 ± 0.09	8.32 ± 0.41	0.68 ± 0.06	0.80 ± 0.05	1.30 ± 0.032
250	1.52 ± 0.03	0.67 ± 0.03	1.50 ± 0.05	6.50 ± 0.04	0.89 ± 0.03	0.70 ± 0.08	1.44 ± 0.175
500	1.55 ± 2.23	0.66 ± 0.29	1.44 ± 0.06	6.95 ± 0.20	0.78 ± 0.06	0.72 ± 0.04	1.35 ± 0.03
1000	1.60 ± 0.03	0.68 ± 0.03	1.41 ± 0.06	7.51 ± 0.32	0.79 ± 0.07	0.74 ± 0.01	1.35 ± 0.03S
Satellite							
Control	1.53 ± 0.07	0.75 ± 0.07	1.36 ± 0.08	7.58 ± 0.03	0.72 ± 0.06	0.77 ± 0.06	1.33 ± 0.04
1000	1.59 ± 0.05	0.69 ± 0.05	1.43 ± 0.06	8.02 ± 0.50	0.72 ± 0.05	0.79 ± 0.04	1.40 ± 0.05
Female							
Control	1.40 ± 0.06	0.62 ± 0.07	1.29 ± 0.08	8.02 ± 0.43	0.63 ± 0.05	0.75 ± 0.06	-
250	1.55 ± 0.02	0.57 ± 0.03	1.72 ± 0.17	5.70 ± 0.29	0.72 ± 0.03	0.53 ± 0.08	-
500	1.51 ± 0.01	0.63 ± 0.01	1.38 ± 0.03	6.68 ± 0.10	0.75 ± 0.02	0.64 ± 0.06	-
1000	1.60 ± 0.02	0.57 ± 0.07	1.17 ± 0.12	6.85 ± 0.23	0.84 ± 0.03	0.66 ± 0.03	-
Satellite							
Control	1.42 ± 0.05	0.65 ± 0.06	1.26 ± 0.06	7.92 ± 0.02	0.65 ± 0.06	0.71 ± 0.04	-
1000	1.57 ± 0.04	0.61 ± 0.05	1.23 ± 0.07	7.55 ± 0.40	0.82 ± 0.06	0.76 ± 0.03	-

levels both in male and female animals. SGOT levels also showed an increase in female treated rats as compared to control animals. However, these changes were not significant enough that could lead to the development of the biochemical shifts and further to toxicity and need further detailed investigation.

Table 7.12: Effect of SC-1 on coughing induced in Guinea pigs by mechanical stimulation of the trachea.

Treatment	*Dose (mg/kg)*	*No. of Coughs*		*Per cent of Iinhibition*
		Before	*After*	
Control	–	9	10	–
	125	10	7	30
SC-1	250	9	5	44.44
	500	9	3	66.66
Codeine Phosphate	5	10	2	80

N= 5.

Table 7.13: Effect of SC-1 on citric acid induced coughing in guinea pigs.

Treatment	*Dose (mg/kg)*	*Frequency of Coughs*		*Per cent of Iinhibition*
		Before	*After*	
SC-1	125	20.4 ± 2.4	12.6 ± 2.1	38.23 ± 2.2
	250	17.5 ± 1.9	9.2 ± 1.0	47.42 ± 3.1
	500	18.4 ± 1.4	6.4 ± 0.7	65.21 ± 3.3
Codeine Phosphate	5	18.2 ± 1.5	2.6 ± 0.51	85.71 ± 3.3

N= 5.

Table1 7.4: Effect of SC-1 on compound 48/80 induced histamine release from RPMC.

Treatment	*Dose (μg/ml)*	*Per cent Inhibition Release*
	12.5	40.93 ± 0.98*
	25	50.81 ± 0.77*
SC-1	50	56.85 ± 1.13*
	100	63.93 ± 0.69*
	200	72.43 ± 1.99*
Fexofenadine	15	82.67 ± 3.49*

Data are presented as mean ± SE (5 independent experiments).

RPMC were treated with test materials as described in Materials and Methods section. * $p< 0.05$ vs. control representing spontaneous release of histamine from RPMC. (Student's t-test).

There were no gross pathological alteration was found in the internal organs between male and female or between drug treated or control groups. SC-1 at the doses of 250-1000 mg/kg. p.o. did not produce organ swelling, atrophy or hypertrophy between male and female rats. The organ-to-body weight-ratio is one of the fundamental judgement to diagnose whether the organs were exposed to injury or not. The data showed that, there were no significant changes of organ-to-body weight ratios in any of the groups, treated as well as control.

The gross and histopathological examination of liver and kidneys showed toxicity related changes which could be supported with data of biochemical changes such as increased levels of triglycerides, SGPT and uric acid. The testes of 500 mg/kg and 1000 mg/kg treated male rats showed testicular toxicity which needs further studies.

Coughing is a normal physiological response to irritation of the laryngo-tracheo-bronchial system caused by mechanical or chemical stimulation. It may be painful and fatiguing and require suppression by anti-tussive drugs. In animals, coughing has been elicited by mechanical (Tedeschi *et al.*, 1959) or chemical irritation (Miyagoshi *et al.*, 1986) and by electrical stimulation (Cavanagh *et al.*, 1976) of tracheal mucosa or by nerve stimulation (Pickering and James, 1979). Of all these methods, chemical and mechanical stimulation are more similar to the physiological event and also to the experimental models generally used in man (Dhuley, 1999).

In the present study, the anti-tussive activity of formulation, SC-1 has been compared with that of codeine phosphate against coughing induced in two different models, mechanical induced and citric acid induced coughing in guinea pigs. A third model, anti-histaminic activity using rat peritoneal mast cells was also used as histamine plays an important role in sneezing and coughing. SC-1 administered orally has shown marked anti-tussive activity, comparable to that of codeine phosphate against mechanical and citric acid induced coughing. SC-1 at the doses of 125, 250 and 500 significantly suppress the coughing dose dependently in both the models. SC-1 also showed significant inhibition of compound 48/80 induced histamine release from RPMC. SC-1 produced concentration dependent inhibitory effect. Maximum inhibitory effect (72.43 per cent) was recorded in presence of 200 µg of SC-1. Fexofenadine, produced a maximum of 82.67 per cent inhibitory effect at a concentration of 15 µg (concentration dependent effect of fexofenadine is not shown).

Conclusion

In conclusion, SC-1 can be considered safe, as it did not cause any mortality or showed adverse changes in general behaviour, growth rate, feed and water consumption, body weight and different vital organ to body weight ratios. The various haematological and biochemical parameters evaluated after 28 days of drug treatment showed no abnormalities and values were in the normal physiological limits and well comparable with vehicle treated control rats. It also did not cause abnormalities in necropsy and histopathological findings. Keeping these results into consideration, the NOAEL (No-observed-adverse-effect level) in case of 28-day sub-acute oral toxicity was found to be 1000 mg/kg p.o. b.wt.

The study revealed that SC-1 is devoid of any toxic risk and gave useful scientific knowledge and support to launch this formulation in the global herbal market.

Acknowledgements

The authors highly acknowledge Indian Medicinal Pharmaceutical Corporation Limited (IMPCL), Uttarakhand, India for providing us the SC-1 samples. The authors also thank Department of Science & Technology, Govt. of India for financial support.

References

Ackerknecht, E.H. (1973). Therapeutics: from the Primitives to the Twentieth Century. Hafner Press, New York.

Ahirwar, B., Ahirwar, D., and Ram, A. (2008). Antihistaminic effect of Sitopaladi churna extract. *J Res Pharm and Tech.*, **1-2**: 89-92.

Ayurvedacharya, K.S.A., Ratnawali, B., and Saasthan, C.S. (1951). Publisher of Indian Sanskrit Literature and Distributors Varanashi. p.242.

Bodekar, G.C. (1994). Traditional health care and public policy: recent trend. A paper presented at International Conference Canada and edited by Anwar I. in Traditional Health System and Public Policy, international Development Research Center, Ottawa, Canada.

Cavanagh, R.L., Gylys, J.A., and Bierwagen, M.E.(1976). Antitussive properties of Butorphanol. *Arch. Int. Pharmacodyn.*, **220**: 258–268

De Smet, P.A.G.M. (1995). Health risks of herbal remedies. *Drug Safety*. **13**: 81-93.

Dhuley, Jayant N. (1999). Antitussive effect of *Adhatoda vasica* extract on mechanical or chemical stimulation-induced coughing in animals. *J Ethnopharmacol.*, **67(3):** 361-5.

Dixit,V.K., and Yadav, N.P. (2008). Recent approaches in herbal drug standardization. *Integr Biol.*, **2 (3)**: 195-203.

Duke, J.A., and Martinez, R.V. (1994). Handbook of Ethnobotanicals (Peru). CRC Press, Boca Raton, FL.

Garg, S., and Bhutani, K.K. (2008). Chromatographic analysis of Kutajarista–an ayurvedic polyherbal formulation. *Phytochem Anal.*, **19(4)**: 323-8.

Kamboj, V.P. (2000). Herbal medicines. *Current Science*, **78 (1):** 35–39

Liang, C.T., Chang, M.H., Hong, C.C., and Huang, K.J. (1999). Establishing the blood chemistry references values for 658 SPF rats and mice. *J. Chin. Soc. Vet. Sci.*, **25**: 55–68.

Majno, G.M. (1975). Healing Hand: Man and Wound in the Ancient World. Harvard University Press, Cambridge, MA.

Miyagoshi, M., Amagaya, S., and Ogihara, Y. (1986). Antitussive effects of 1-ephedrine, amygdalin and makyokansokito (Chinese traditional medicine) using a cough model induced by sulphur dioxide gas in mice. *Planter Med.*, **52**: 275 –278

Miller, L.G. (1998). Herbal medicines. Selected clinical considerations focusing on known or potential drug-herb interactions. *Arch Intern Med.*, **158:** 2200–2211

Organization for Economic Cooperation and Development (OECD), (1995). Repeated dose 28-day oral toxicity study in rodents. In: OECD Guideline for the Testing of Chemicals. Section 4: Health Effects, No. 407, Adopted by the Council on 27th July 1995, 8 pp.

OECD, 2001.OECD Guidelines No. 420, pages 1/14- 14/14 (2001). The Organisation of Economic Co-operation and 680 Development. The OECD Guideline for Testing of Chemical: 681420 Acute Oral Toxicity. OECD, Paris, pp. 1–14.

Pattanayak, P., Panda, S.K, Dash, S., and Mishra, S.K. (2010). Study of anti-tussive activity of Sitopaladi churna a polyherbal formulation. *Int J Pharm Sci.*, **4(2):** 65-67.

Pickering, R.W., and James, G.W.L. (1979). The antitussive activity of a novel compound RU 20201. *Arzneim.-Forsch. Drug Res.*, **29**: 287–289

Shin, T.Y., and Lee, J.K. (2003). Effect of *Phlomis umbrosa* root on mast cell-dependent immediate-type allergic reactions by anal therapy. *Immunopharmacol Immunotoxicol.*, **25(1):** 73-85

Shore, P.A., Burkhalter, A., and Cohn, V.H. (1959). A method for fluorometric assay of histamine in tissues. *J Pharmacol Exp Ther.*, **127:** 182-186

Tanak, M., and Maruyama, K. (2005). Mechanisms of capsaicin- and citric-acid-induced cough reflexes in Guinea Pigs. *J Pharmacol Sci.*, **99**: 77 – 82.

Tedeschi, D.H., Tedeschi, R.E., and Fellows, E. J. (1959). The effect of tryptanine on the central nervous system, including a pharmacological procedure for the evaluation of iproniazid-like drugs. *J. of Pharmacol. Exp. Ther.*, **126**: 223-232

Waldesch, F.G., Konigswinter, B.S., and Remagen, H.B. (2003). Herbal medicinal products- Scientific and regulatory basis for development quality assurance and marketing authorization, published by Medpharm Stuttagart and CRC press, Washington DC, : 37-52.

WHO. (1979). The International Pharmacopeia, General Methods of Analysis, 3rd edn. World Health Organization, Geneva. 1.

WHO, (1981). The International Pharmacopeia, Quality Specifications, 3rd edn.World Health Organization, Geneva. 2.

WHO, (1990). The Use of Essential Drugs. Eighth report of the WHO Expert committee. World Health Organization, Geneva.

WHO, (1998a). Quality Control Methods for Medicinal Plant Materials, World Health Organization, Geneva

WHO, (1998b). Guidelines for the Appropriate Use of Herbal Medicines. WHO Regional Publications, Western Pacific Series. WHO Regional office for the Western Pacific, Manila. 3.

WHO, (1999a). Quality Control Methods for Medicinal Plant Materials. World Health Organization, Geneva.

WHO, (1999b). WHO Monographs on Selected Medicinal Plants, World Health Organization, Geneva. 1.

WHO, (2002a). WHO Monographs on Selected Medicinal Plants, World Health Organization, Geneva. 2.

WHO, (2002b). Ancient Remedies, New Disease. World Health Organization, Geneva.

WHO, (2005). WHO Global Atlas of Traditional, Complementary and Alternative Medicine. World Health Organization, Geneva. 1 and 2.

Traditional and Folk Herbal Medicine: Recent Researches Vol. 2 (2014) *Pages* **221–247**
Editor: **V.K. Gupta**
Published by: **DAYA PUBLISHING HOUSE, NEW DELHI**

8

In vitro Antimicrobial and Antioxidant Potency of Two Nutraceutical Plants of Cucurbitaceae

Mital J. Kaneria[1], Kalpna D. Rakholiya[1] and Sumitra V. Chanda[1*]

ABSTRACT

The use of natural antimicrobial compounds, especially extracted from plants, as a food preservative is nowadays widely prevalent. The present study was designed to evaluate in vitro antimicrobial and antioxidant potential of aerial part of Lagenaria siceraria (Molina) Standl. and Momordica charantia L. belonging to the family Cucurbitaceae. Extraction was done sequentially in soxhlet apparatus using different solvents like petroleum ether, toluene, ethyl acetate, acetone, water of increasing polarity. The antimicrobial activity was assessed by agar well diffusion method against fifteen pathogenic microorganisms, which included five Gram positive bacteria, five Gram negative bacteria and five fungi. Antioxidant property was measured by DPPH free radical scavenging assay and superoxide anion radical scavenging assay. Total phenol and flavonoid content was measured by Folin-Ciocalteu reagent method and aluminium chloride colorimetric method respectively. Significant antimicrobial activity was observed in M. charantia. Bacteria were more susceptible than fungi, while Gram negative bacteria were more susceptible than Gram positive bacteria. Remarkable antioxidant activity and highest total phenol content was in acetone extract of L. siceraria. The study supports the folkloric uses of L. siceraria and M. charantia.

Keywords Cucurbitaceae, *Lagenaria siceraria*, *Momordica charantia*, Antimicrobial activity, Antioxidant activity, Nutraceutical plants.

---

1 Phytochemical, Pharmacological and Microbiological Laboratory, Department of Biosciences, Saurashtra University, Rajkot, Gujarat, India

* *Corresponding author*: E-mail: svchanda@gmail.com

Introduction

Nutraceutical can be defined as, "a food or a part of food that provides medical or health benefits, including the prevention and/or treatment of a disease". Nutraceuticals, in broad sense, are food or part of food playing a significant role in modifying and maintaining normal physiological functions that sustain healthy human beings (Das *et al.*, 2012). Many natural phytochemicals extracted from plants have gained a lot of attention since they exhibit important biological activities due to their potent antioxidant and antimicrobial properties, least side effects and economically viable. Food habits and other lifestyle changes in tropical countries have led to serious problems of micronutrient deficiency and other nutritionally related health problems. Nutraceutical refers to food components which have a medicinal or beneficial effect on human health. Such foods are also called functional foods. Nutraceuticals or functional foods provide an opportunity to improve the health of people, reduce health care costs and support economic development in rural communities. Nutraceuticals are nothing but phytochemicals or secondary metabolites of the plants which are present in herbs, fruits and vegetable which are consumed daily as part of our diet (Dembitsky *et al.*, 2011). Increased vegetable consumption is correlated with reduced risks of cardiovascular disease, stroke, arthritis, inflammatory bowel diseases, and some cancers (Dauchet *et al.*, 2010; Griep, *et al.*, 2011). Plants have provided mankind with useful drugs for centuries and still remain one of the best reservoirs of traditional medicine and can be prolific resources for new drugs.

Microorganisms have developed resistance to many of the existing antibiotics and antimicrobials due to extensive use and it has become an everlasting problem in clinical treatment of infectious diseases and has become a global health issue (Mazumdar *et al.*, 2010; Chanda *et al.*, 2011c). On the other side, the number of new antibiotics being brought to the market has shown a precipitous decline with only a few new antibacterial agents approved for clinical use (Moellering Jr., 2011). In addition, food industries face the problem of food getting contaminated with some microbes which leads to food spoilage, deteriorate the quality of food products and reduces the shelf life (Lv *et al.*, 2011); further consumption of such contaminated food cause many infections. Microorganisms are also responsible for the outbreaks of food borne diseases (Tauxe, 1997).

Systemic fungal infections have drastically increased over the past three decades due to the rising immunocompromised population as a result of transplantation, cancer chemotherapy, steroid therapy and, in particular, HIV infection (AIDS). Plants and their secondary metabolites are a promising source to provide structurally diverse bioactive compounds as potentially therapeutic agents, including antimicrobials. Many attempts have been made to control microbial growth and to reduce the incidence of food poisoning and spoilage with synthetic chemicals but they are associated with adverse effects like hypersensitivity, allergic reaction and immunity suppression (Cakir *et al.*, 2005). Synthetic drugs are not only expensive and inadequate for the treatment of microbial diseases but are also often adulterated and have some side effects. Therefore, there is a need for alternative drugs for the treatment of infectious diseases. Consumers demand food products with fewer synthetic additives but expect increased safety, quality and shelf- life (Roller, 2003). This demand has led to renewed interest in the use of natural antimicrobials to preserve foods. Medicinal herbs represent a rich source from which novel antibacterial and antifungal chemotherapeutic agents may be obtained (Chanda and Rakholiya, 2011).

A free radical is a molecule containing one or more unpaired electrons in the outermost shell. Some examples are the reactive oxygen species (ROS) and reactive nitrogen species (RNS) that are generated through endogenous processes such as respiration, metabolism and phagocytosis under normal physiological conditions. They are also produced by exogenous systems such as exposure to radiation, toxic chemicals, organic solvents, pesticides, cigarette smoking and alcohol consumption and by eating oxidized polyunsaturated fats (Kaneria and Chanda, 2013). The excessive amounts of ROS and RNS are the most common dangerous species among the free radicals and can cause many cell disorders through their action on lipids, amino acids, double bonds of polyunsaturated fatty acids and nitrogenous bases of DNA (Satpathy *et al.*, 2011; Salgado *et al.*, 2012). Consequently, they contribute to the pathogenesis of various oxidative stress-related degenerative diseases and disorders including cancer, stroke, cardiovascular, Alzheimer's and Parkinson's diseases (Nguimbou *et al.*, 2012). Oxidative stress is a process where the physiological balance between prooxidants and antioxidants is disrupted in favour of the former, leading to potential damage to the organism (Dar *et al.*, 2011).

The excessive amount of free radicals can be scavenged by the protective role of natural and synthetic antioxidant agents. The protective mechanism generally functions at several different levels within the cells in human body by inhibiting the formation of free radical species, intercepting radical-chained reactions, converting existing free radicals into less harmful molecules and repairing oxidative damage (Sulaiman *et al.*, 2011). Antioxidants are substances that can inhibit or diminish oxidizable substrate action. The synthetic antioxidants such as butylated hydroxytoluene (BHT), butylated hydroxyanisole (BHA), propyl gallate (PG), and tertiary butylhydroquinone (TBHQ) have been suspected to cause or promote adverse health effects and toxic to human beings (Lagouri *et al.*, 2011; Mbaebie *et al.*, 2012; Adamez *et al.*, 2012). Unlike synthetic drugs, bioactive natural products have beneficial effect on the whole organism and cause no unwanted side effects. Fruits and vegetables provide many essential nutrients and also contain several other phytochemicals which have been suggested to be responsible for health benefits due to their antioxidant properties and other positive effects (Yahia *et al.*, 2011).

Plant-derived ingredients possessing antioxidant and antimicrobial properties have the advantage of being readily accepted by consumers, as they are considered to be natural. In this context, medicinal plants are being viewed as the most potential candidates, particularly on account of their proven antioxidant activity, lesser side effects, and relatively lower cost as compared to synthetic ones (Bhatt *et al.*, 2012). Nowadays there is consumer demand for natural food additives and additionally, the burden of proof of safety may be less rigorous than that required for synthetic antioxidants (Pukalskas *et al.*, 2012).

Phenolic compounds which are secondary metabolites in plants are one of the most widely occurring groups of phytochemicals that exhibit antiallergenic, antimicrobial, antioxidant, antiartherogenic, antithrombotic, antiinflammatory, vasodilatory and cardioprotective effects (Su *et al.*, 2009; Khan *et al.*, 2012). Antimicrobial activities of phenolic compounds may involve multiple modes of action. The type of the microorganism and its cell wall structure are thought to play an important role (Shan *et al.*, 2007). Phenolic compounds can denature enzymes (Furneri *et al.*, 2002) but they can also bind to substrates such as minerals, vitamins and carbohydrates making them unavailable for microorganisms (Shahidi and Naczk, 2004). Furthermore, phenols can be absorbed to the cell wall, resulting in a disruption of the membrane structure and function (Hugo and Bloomfield, 1971). This has fostered research in the screening of two Cucurbitaceae plants *viz. Lagenaria siceraria* (Molina) Standl. and *Momordica charantia* L. for their antioxidant and antimicrobial potential. Hence, the objectives of this work were to evaluate antimicrobial and antioxidant potential of different solvent extracts of *L. siceraria* and *M. charantia*. The botanical description and reported uses of both the plants is given below.

1. *Lagenaria siceraria* (Molina) Standl.

- **Vernacular name:** Dudhi
- **Family:** Cucurbitaceae
- **Part used:** Aerial parts
- **Voucher specimen number:** PSN328
- **Action/uses:** *Lagenaria siceraria* is official in Ayurvedic Pharmacopoeia of India, and having composition of variety of essential phytoconstituents, so that *L. siceraria* fruits are traditionally used for its cardio protective, cardio tonic, general tonic, diuretic, and aphrodisiac, antidote to certain poisons and scorpion stings, alternative purgative and cooling effects. It cures pain, ulcers, and fever and is used for pectoral-cough, asthma and other bronchial disorders. Extracts of the plant have shown antibiotic activity.

- **Commercial utility:** Cultivated for its fruits and seeds. Fruits used as vegetable.
- **Reported activities:**
    - Antihyperlipidemic effect (Ghule *et al.*, 2009);
    - Cytotoxic activity (Ghosh *et al.*, 2009);

- Anthelmintic activity (Smita *et al.*, 2009);
- Antioxidant activity (Jadhav *et al.*, 2010; Rakholiya *et al.*, 2011);
- Cardioprotective effect (Mali and Bodhankar, 2010);
- Antimicrobial activity (Chanda *et al.*, 2010b).

2. *Momordica charantia* L.

- ☆ **Vernacular name:** Karela
- ☆ **Family:** Cucurbitaceae
- ☆ **Part used:** Aerial parts
- ☆ **Voucher specimen number:** PSN333
- ☆ **Action/uses:** Roots: Bitter, Acrid, astringent, ophthalamic; Leaves: Bitter, anthelmintic, antipyretic, emetic and purgative; Fruits: Bitter, acrid, thermogenic, depurative, vulnerary, stimulant, purgative, antidiabetic, carminative, digestive, stomachic, anti-inflammatory, fat loss, fever (malarial), galactagogue, gout, hydrophobia, hyperglycemia, jaundice, kidney (stone), laxative, leprosy, leucorrhea, liver, piles, pneumonia, febrifuge and tonic. Its hypoglycemic activity has been reported in pulps, seeds and leaves *in vivo*. Bitter gourd fruit juice can stimulate glucose uptake of skeletal muscle cells.
- ☆ **Commercial utility:** Cultivated for its fruits. Fruits used as vegetable as well as in ayurvedic, antidiabetic preparations, anticancerous, antiviral, antibacterial, digestive stimulant, hypoglycemic activity.
- ☆ **Reported activities:**
  - Immunosuppressive activities (Leung *et al.*, 1987);
  - Antimutagenic activity (Guevara *et al.*, 1990);
  - Anti-HIV and antitumor activities (Lee-Huang *et al.*, 1995);
  - Antispermatogenic and androgenic activities (Naseem *et al.*, 1998);
  - Antimalarial activity (Munoz *et al.*, 2000);
  - Antiulcer activity (Yesilada *et al.*, 1999; Gurbuz *et al.*, 2000; Alam *et al.*, 2009);
  - Larvicidal efficacy (Prabakar and Jebanesan, 2004);
  - Toxicity and antidiabetic activity (Batran *et al.*, 2006);
  - Antifeedant activity (Bing *et al.*, 2008);

- Antioxidant activity (Parekh and Chanda, 2007b; Kubola and Siriamornpun, 2008; Wu and Ng, 2008; Liu *et al.*, 2010; Rakholiya *et al.*, 2011);
- Antiprotozoal and cytotoxic activities (Mesia *et al.*, 2008);
- Antidiabetic activity (Kumar *et al.*, 2008; Tan *et al.*, 2008; Yuan *et al.*, 2008; Kumar *et al.*, 2009; Nivitabishekam *et al.*, 2009; Jarald *et al.*, 2009; Gunjan *et al.*, 2010; Chang *et al.*, 2011);
- Antimicrobial activity (Braca *et al.*, 2008; Nair and Chanda, 2008; Parekh and Chanda, 2008; Chanda *et al.*, 2010b);
- Antiinflammatory effect (Lii *et al.*, 2009);
- Antileishmanial activity (Gupta *et al.*, 2010);
- Neuroprotective effect (Malik *et al.*, 2011).

Materials and Methods

Chemicals

2, 2-diphenyl-1-picrylhydrazyl (DPPH), Nitroblue tetrazolium (NBT), Nicotinamide Adenine Dinucleotide reduced (NADH), Phenazine methosulphate (PMS), gallic acid, ascorbic acid, quercetin, Folin-Ciocalteu's reagent, aluminium chloride, potassium acetate, Tris-HCl, Dimethyl sulfoxide (DMSO), nutrient broth, Sabouraud dextrose broth, Muller Hinton No. 2, Sabouraud dextrose agar, agar powder were obtained from Hi-Media, Mumbai, India; petroleum ether, toluene, acetone, ethyl acetate, methanol, etc. were obtained from Merck, India.

Plant Material

Aerial part of *Lagenaria siceraria* (Molina) Standl. and *Momordica charantia* L. belonging to the family Cucurbitaceae were collected in the month of August, 2008, from Jam-Jodhpur, Jamnagar, Gujarat, India. The plant was compared with voucher specimen deposited at Department of Biosciences, Saurashtra University, Rajkot, Gujarat, India. Voucher specimen No. of *L. siceraria* and *M. charantia* were PSN328 and PSN333 respectively. The aerial parts were separated, washed thoroughly with tap water, shade dried, homogenized to fine powder and stored in airtight bottles.

Sequential Extraction Method

The dried powder was extracted sequentially (Wiart *et al.*, 2004) by soxhlet apparatus (Lin *et al.*, 1999), using different solvents depending upon their polarities like petroleum ether, toluene, ethyl acetate, acetone and water (Flow Chart 8.1). The extracts were concentrated and freed of solvent under reduced pressure, using a rotary evaporator. The dried crude concentrated extracts were weighed to calculate the extractive yield and stored in a refrigerator (4°C) in air tight bottles.

Antimicrobial Assay

Microorganisms Tested

The microorganisms studied were obtained from National Chemical Laboratory (NCL), Pune, India. They consisted of five Gram positive bacteria [*Bacillus megaterium*

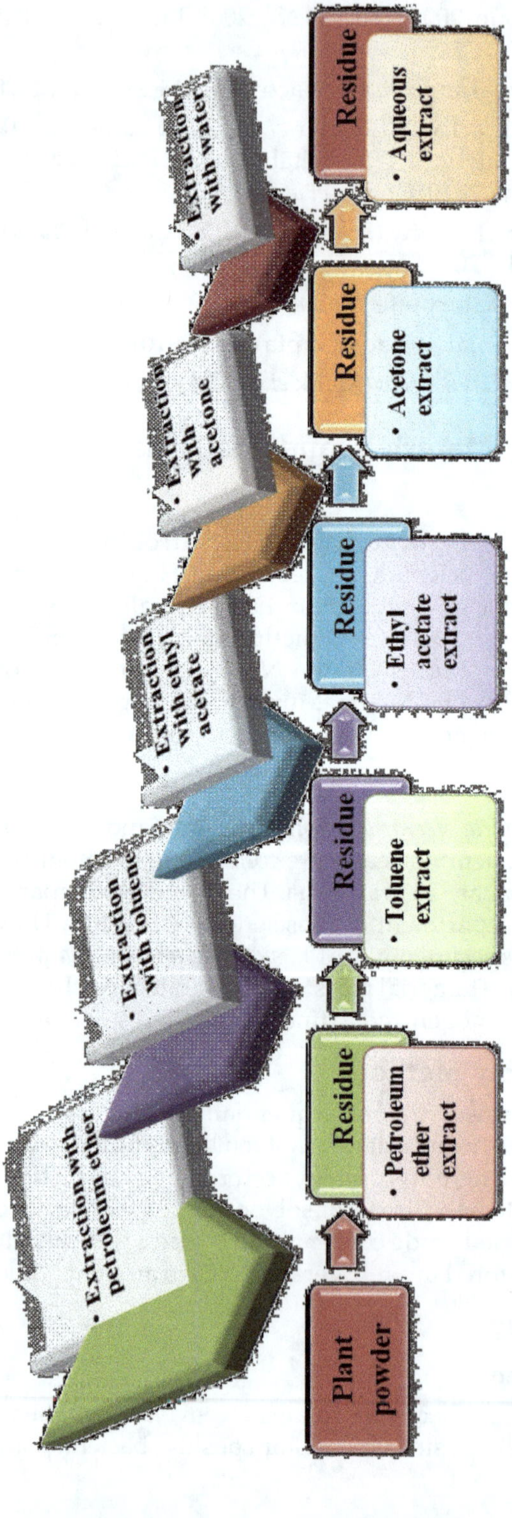

Flow Chart 8.1: Sequential extraction in soxhlet apparatus.

ATCC9885 (*BM*), *Bacillus subtilis* ATCC6633 (*BS*), *Corynebacterium rubrum* ATCC14898 (*CR*), *Staphylococcus aureus* ATCC25923 (*SA*), *Staphylococcus epidermidis* ATCC12228 (*SE*)], five Gram negative bacteria [*Citrobacter freundii* ATCC10787 (*CF*), *Enterobacter aerogenes* ATCC13048 (*EA*), *Klebsiella pneumoniae* NCIM2719 (*KP*), *Proteus mirabilis* NCIM2241 (*PM*), *Salmonella typhimurium* ATCC23564 (*ST*)] and five fungi [*Candida albicans* ATCC2091 (*CA*), *Candida glabrata* NCIM3448 (*CG*), *Cryptococcus leuteolus* ATCC32044 (*CL*), *Candida neoformans* NCIM3542 (*CN*), *Candida tropicalis* ATCC4563 (*CT*)]. The bacteria and fungi were maintained on nutrient agar and MGYP medium (Hi Media, India) respectively, at 4°C and sub-cultured before use. The microorganisms studied are clinically important ones causing several infections, food borne diseases, spoilages, skin infection and it is essential to overcome them through some active therapeutic agents.

Determination of Antimicrobial Assay

In vitro antimicrobial activity of the different solvent extracts was studied against pathogenic microbial strains by the agar well diffusion method (Perez *et al.*, 1990; Chanda *et al.*, 2010b). Muller Hinton No. 2/Sabouraud dextrose agar (Hi-media) was used for the antibacterial and antifungal susceptibility test respectively. The different solvent extracts were diluted in 100 per cent DMSO to give a concentration of 20 mg ml^{-1}. The Muller Hinton agar/Sabouraud dextrose agar was melted and cooled to 48-50°C and a standardized inoculum (1.5×10^8 CFU ml^{-1}, 0.5 McFarland) was then added aseptically to the molten agar and poured into sterile Petri dishes; wells (8.5 mm) were prepared in the seeded agar plates. The test compound (100 µl) was introduced into the well. The plates were incubated over night at 37°C and 28°C for 24 h and 48 h respectively, for bacteria and fungi. DMSO was used as negative control. The microbial growth was determined by measuring the diameter of the zone of inhibition and the mean values are presented with ± S.E.M.

Quantitative Phytochemical Analysis

Determination of Total Phenol Content

The amount of total phenol content (TPC), in different solvent extracts was determined by Folin-Ciocalteu's reagent method (Mc Donald *et al.*, 2001). The extract (0.5 ml) and 0.1 ml Folin-Ciocalteu's reagent (0.5 N) were mixed, and the mixture was incubated at room temperature for 15 min. Then, 2.5 ml saturated sodium carbonate solution was added and further incubated for 30 min at room temperature, and the absorbance was measured at 760 nm using a UV–VIS Spectrophotometer (Shimadzu, Japan) against a blank sample. Total phenol content is expressed in terms of gallic acid equivalent (mg g^{-1} of extracted compounds).

Determination of Total Flavonoid Content

The amount of flavonoid content (TFC) in different solvent extracts was determination by aluminium chloride colorimetric method (Chang *et al.*, 2002). The reaction mixture (3.0 ml), consisted of 1.0 ml sample (1 mg ml^{-1}), 1.0 ml methanol, 0.5 ml $AlCl_3$ (1.2 per cent) and 0.5 ml potassium acetate (120 mM) and was incubated at room temperature for 30 min. The absorbance of all samples was measured at 415 nm using a UV–VIS Spectrophotometer (Shimadzu, Japan), against a blank sample. The calibration curve was made by preparing a quercetin (5 to 60 μg ml^{-1}) solution in methanol. The flavonoid content is expressed in terms of quercetin equivalent (mg g^{-1} of extracted compounds).

Antioxidant Assays

DPPH Free Radical Scavenging Activity

The free radical scavenging activity of different solvent extracts was measured by using DPPH by the modified method described by Mc Cune and Johns (2002). The reaction mixture 3.0 ml consisting of 1.0 ml methanol, 1.0 ml DPPH (0.3 mM) and 1.0 ml of solvent extracts of different concentrations diluted by methanol, was incubated for 10 min, in dark, after which the absorbance was measured at 517 nm using digital spectrophotometer (Systronic, India), against a blank sample. Ascorbic acid (2 to 16 μg ml^{-1}) was used as positive control (Liu *et al.*, 2011). The percentage inhibition was determined by comparing the results of the test and the control. Percentage of inhibition was calculated using the formula:

Per cent Inhibition = [1–(A/B)] × 100

Where, B = absorbance of blank (DPPH plus methanol), A = absorbance of sample (DPPH, methanol plus sample)

Superoxide Anion Radical Scavenging Activity

The superoxide anion radical scavenging activity of different solvent extracts was measured following the method described by Robak and Gryglewski (1988). Superoxide radicals are generated by oxidation of NADH and assayed by the reduction of NBT. The reaction mixture 3.0 ml consisted of 1.0 ml of the solvent extracts of different concentrations diluted by distilled water, 0.5 ml Tris HCl buffer (16 mM, pH 8), 0.5 ml NBT (0.3 mM), 0.5 ml NADH (0.93 mM) and 0.5 ml PMS (0.12 mM). The superoxide radical generating reaction was started by the addition of PMS solution to the mixture. The reaction mixture was incubated at 25°C for 5 min and then the absorbance was measured at 560 nm using digital spectrophotometer (Systronic, India), against a blank sample. Gallic acid (50 to 225 μg ml^{-1}) was used as a positive control (Robak and Gryglewski, 1988). Percentage of inhibition was calculated as described above.

Results

Extractive Yield

The dry powder of both the plants was extracted by sequential extraction method. The extractive yield of both the plants in different solvent is given in Figure 8.1. The yield of soluble substances, expressed as mg g^{-1} dry weight of aerial parts of *L. siceraria* and *M. charantia*. The extractive yield was varied among the solvent used of both the plants. Organic extracts showed very less extractive yield as compared to aqueous extract. Amongst all the different solvent extracts, maximum yield was in aqueous extract of *L. siceraria*.

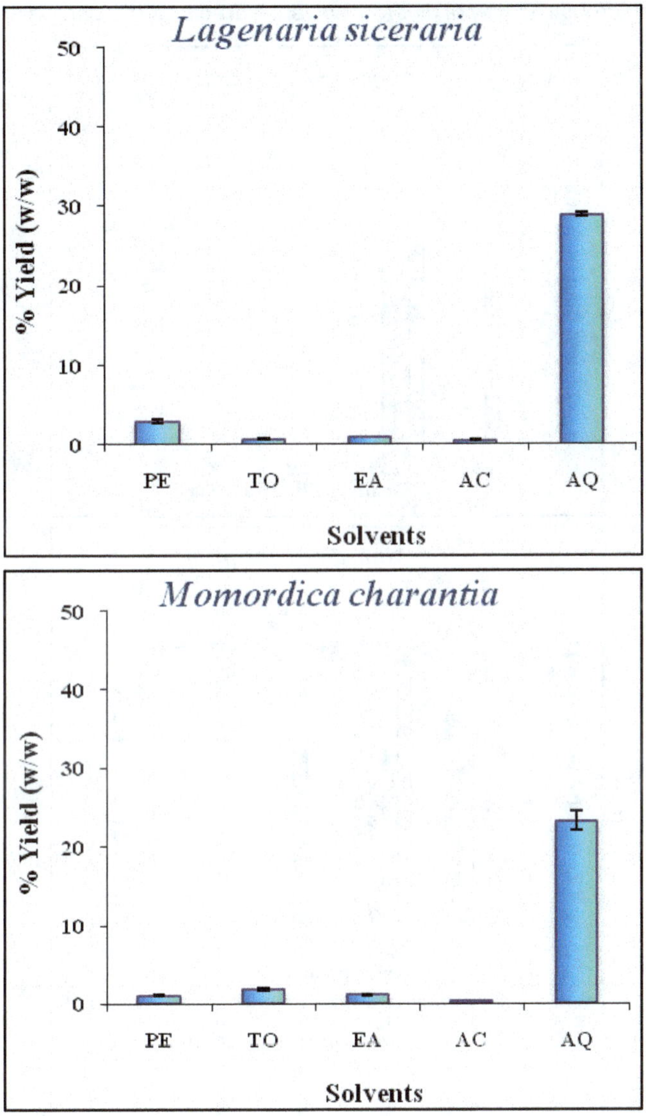

Figure 8.1: Percentage yield of different solvent extracts of *L. siceraria* and *M. charantia.*

Antimicrobial Activity

The antimicrobial activity of different solvent extracts of aerial parts of *L. siceraria* and *M. charantia* is shown in Figure 8.2. *B. megaterium* was susceptible to only toluene extract of *L. siceraria*. *B. subtilis* was susceptible to all the solvent extracts except acetone and aqueous extracts of *L. siceraria*. *K. pneumoniae* was susceptible to all different solvent extracts and highest activity was shown by acetone extract of *L. siceraria*. *P. mirabilis* was susceptible to all the solvent extracts except acetone extract of *L. siceraria*. The aqueous extract of *L. siceraria* showed activity against *C. albicans*

Figure 8.2: Antimicrobial activity of different solvent extracts of *L. siceraria* and *M. charantia*.

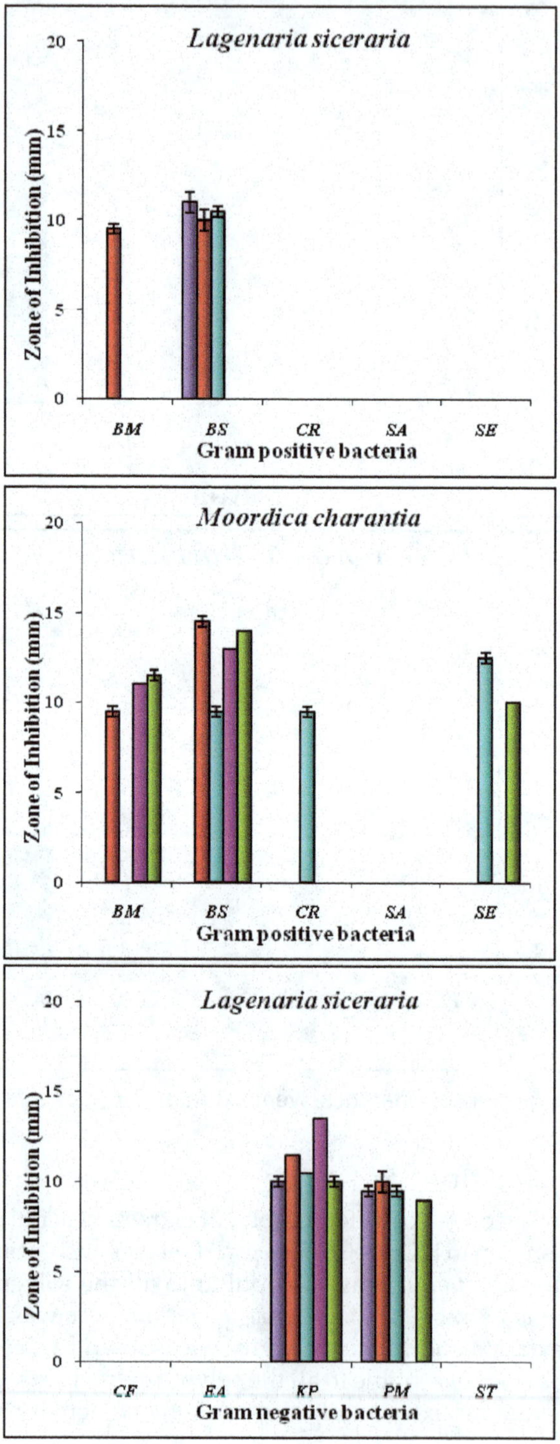

Contd...

Figure 8.2–*Contd...*

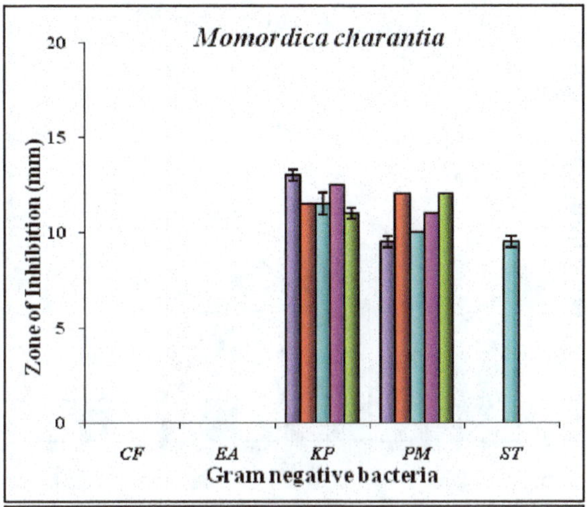

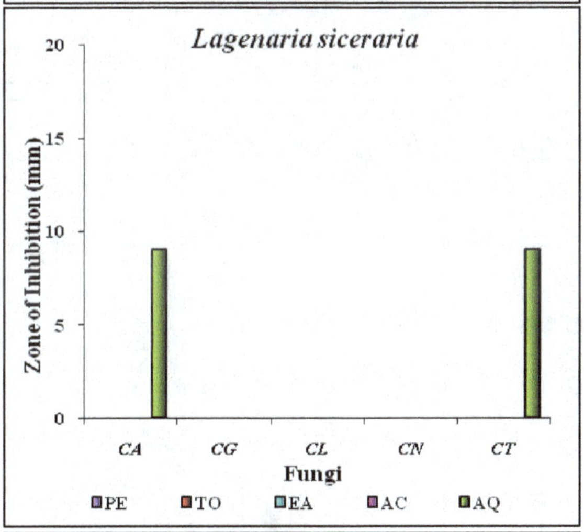

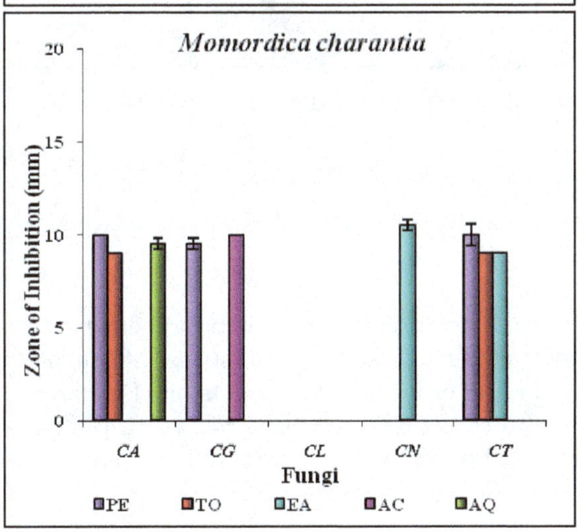

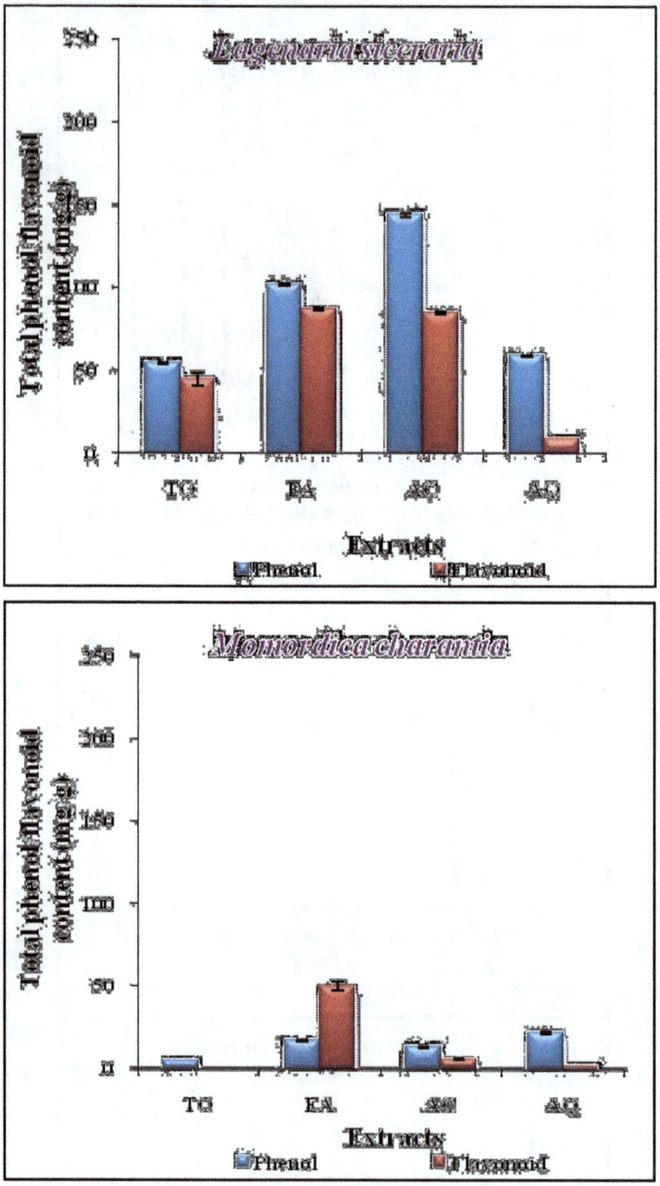

Figure 8.3: Total phenol and flavonoid content of different solvent extracts of *L. siceraria* and *M. charantia.*

and *C. tropicalis*. *B. subtilis* was susceptible to all the solvent extracts except petroleum ether extract of *M. charantia*. *B. megaterium* was susceptible to all the solvent extracts except petroleum ether and ethyl acetate extracts of *M. charantia*. The ethyl acetate extract of *M. charantia* showed activity against *C. rubrum*. The ethyl acetate and aqueous extracts of *M. charantia* showed activity against *S. epidermidis*. *K. pneumoniae* and

P. mirabilis were susceptible to each one of the solvent extracts of *M. charantia*. Petroleum ether extract of *M. charantia* showed highest activity against *K. pneumoniae*. *S. typhimurium* was susceptible to only ethyl acetate extract of *M. charantia*. *C. glabrata* was susceptible to petroleum ether and acetone extracts, while *C. neoformans* was susceptible to ethyl acetate extract of *M. charantia*.

Total Phenol and Flavonoid Contents

Total phenol content and flavonoid content of both the plants are shown in Figure 8.3. In both the plants in all the extracts, total phenol content was considerably more than flavonoid content except in ethyl acetate extract of *M. charantia*. All the solvent extracts of *L. siceraria* showed more total phenol and flavonoid contents than *M. charantia*. Highest amount of total phenol content was in acetone extract of *L. siceraria*.

Antioxidant Activities

The DPPH free radical scavenging activity of both the plants is shown in Figure 8.4 and Table 8.1. Four different solvent extracts of the two plants were evaluated for their DPPH free radical scavenging activity. Out of 8 extracts investigated, 4 extracts showed IC_{50} value of more than 1000 μg ml^{-1} (Table 8.1) while the remaining 4 extracts showed varied levels of DPPH free radical scavenging activity (Table 8.1). IC_{50} values ranged from 71 to 660 μg ml^{-1} (Table 8.1). Ascorbic acid was used as a standard and its IC_{50} value was 11.4 μg ml^{-1} (Table 8.1). Amongst all the extracts, the lowest IC_{50} value was of acetone extract of *L. siceraria* (71 μg ml^{-1}) and the highest IC_{50} value was of ethyl acetate extract of *M. charantia* (660 μg ml^{-1}). The acetone extract of *L. siceraria* showed the best DPPH free radical scavenging activity.

Table 8.1: IC_{50} values of DPPH free radical (DPPH) and superoxide anion radical (SO) scavenging activity in different solvent extracts of *L. siceraria* and *M. charantia*.

Assays	*Plant Name*	*IC_{50} Values (μg/ml)*			
		Toluene	*Ethyl Acetate*	*Acetone*	*Aqueous*
DPPH	*Lagenaria siceraria*	>1000	**107.5**	**71**	**212.5**
	Momordica charantia	>1000	**660**	>1000	>1000
	Standard (Ascorbic acid)		**11.4**		
SO	*Lagenaria siceraria*	>1000	>1000	**660**	**855**
	Momordica charantia	>1000	>1000	>1000	>1000
	Standard (Gallic acid)		**185**		

The superoxide anion radical scavenging activity of both the plants is shown in Figure 8.5 and Table 8.1. Four different extracts of the two plants were evaluated for their superoxide anion radical scavenging activity. Out of 8 extracts investigated, 6 extracts showed IC_{50} value of more than 1000 μg ml^{-1} (Table 8.1) while the remaining 2 showed varied levels of superoxide anion radical scavenging activity (Table 8.1). Gallic acid was used as a standard and its IC_{50} value was 185 μg ml^{-1} (Table 8.1). The IC_{50} values of acetone extract and aqueous extract were 660 and 855 μg ml^{-1}

Figure 8.4: DPPH free radical scavenging activity of different solvent extract of *L. siceraria* and *M. charantia.*

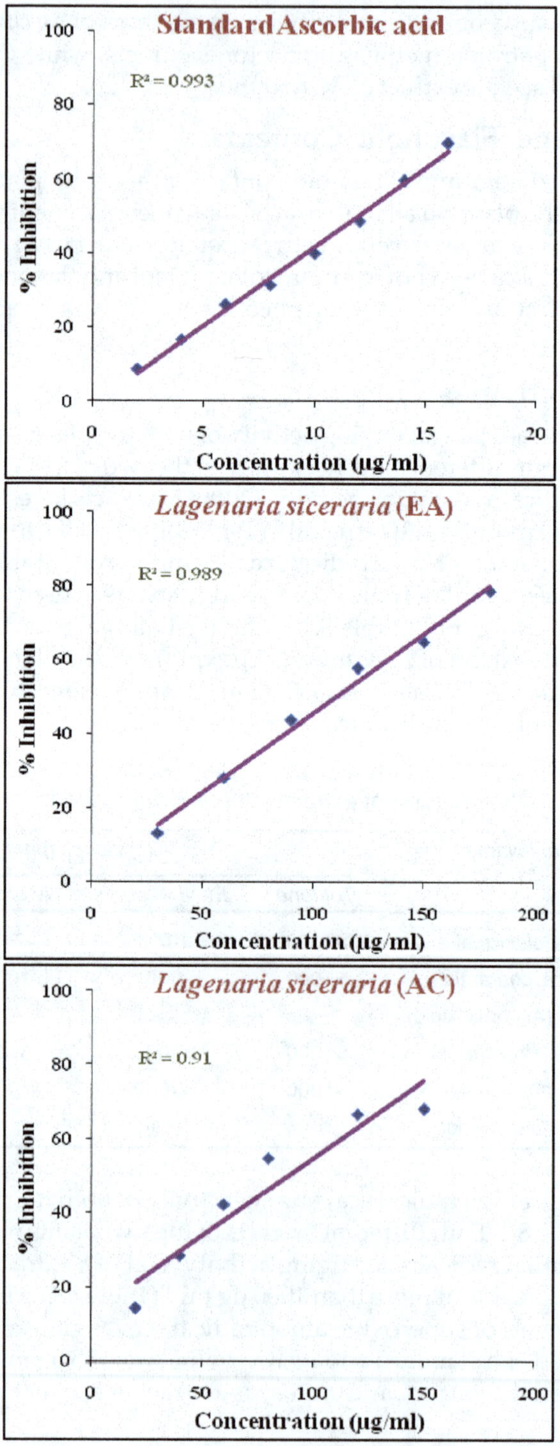

Contd...

Figure 8.4–*Contd...*

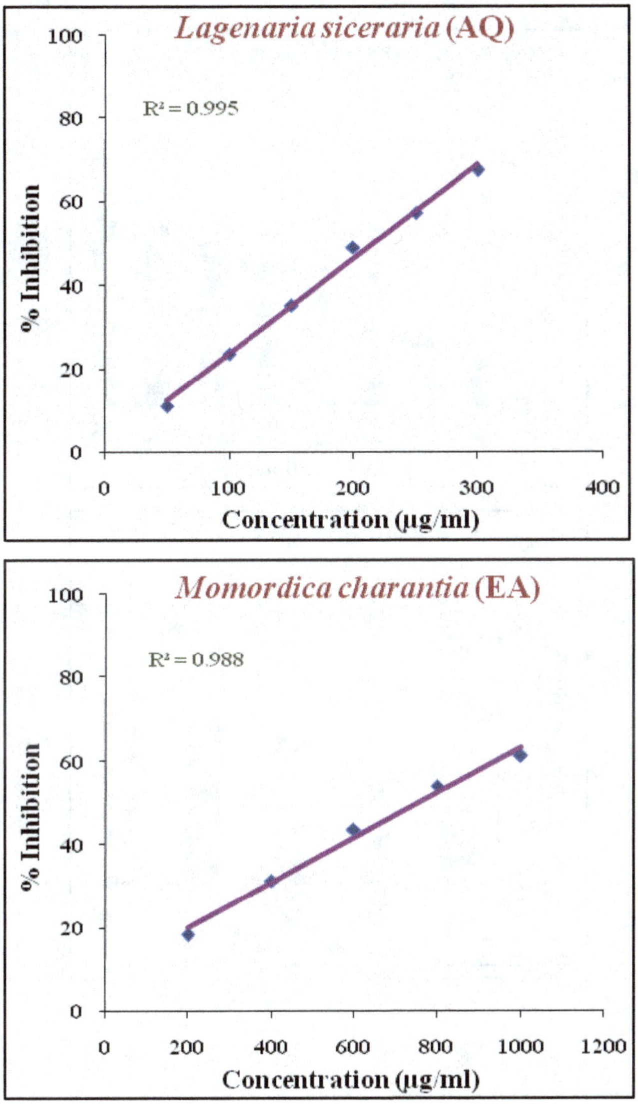

respectively. The acetone extract of *L. siceraria* showed the best superoxide anion radical scavenging activity.

Discussion

The extractive yield depends on solvents, time and temperature of extraction as well as the chemical nature of sample. Under the same time and temperature conditions, the solvent used and the chemical property of sample are the most important factors (Shimada *et al.*, 1992; Rakholiya *et al.*, 2011; Kaneria *et al.*, 2012b). In the present study,

Figure 8.5: Superoxide anion radical scavenging activity of different solvent extracts of *L. siceraria*.

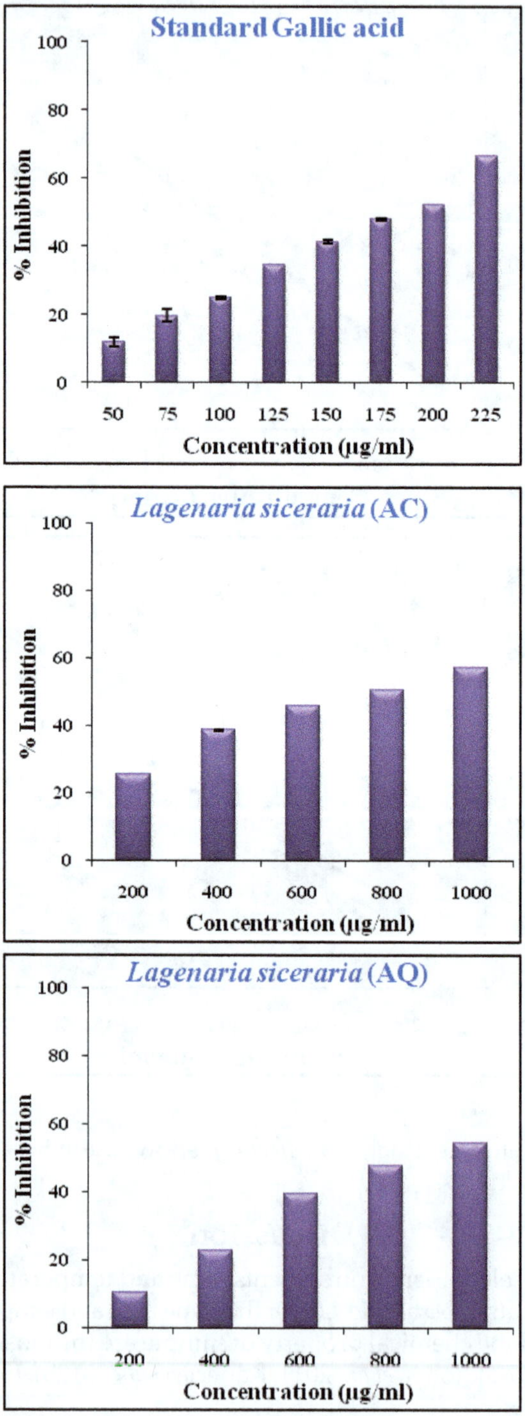

the extractive yield was maximum in aqueous extracts of both the plants. This may be because the phytoconstituents present in these plants are extracted better in water. Similar results were also reported by (Itagaki *et al.*, 2009; Chanda *et al.*, 2011a; Kaneria *et al.*, 2012a).

All the solvent extracts of *M. charantia* and *L. siceraria* showed antibacterial activity more towards Gram negative bacteria than Gram positive bacteria; fungi were more resistant than bacteria. It is generally believed that Gram positive bacteria are more susceptible to plant extracts than Gram negative bacteria because the cell wall of Gram positive bacteria is made up of a single membrane which is considerably more accessible to permeation of plant extracts. On the other hand the cell wall of Gram negative bacteria is lipid layered which becomes an effective barrier for the plant extracts (Kiran *et al.*, 2008; Kaneria *et al.*, 2009; Bajpai *et al.*, 2010; Sandri *et al.*, 2007; Chanda *et al.*, 2010b). But in the present work, Gram negative bacteria were more susceptible than Gram positive bacteria, which is a contradiction to the general belief. These are very promising results because comparatively less plant extracts show activity against Gram negative bacteria. Akinsulire *et al.* (2007), Doughari *et al.* (2007) and Chanda *et al.* (2010a) also reported plant extracts showing antibacterial activity against Gram negative bacteria.

Further when antibacterial and antifungal activity is compared, bacteria were more susceptible than fungi. Resistance of fungi to plant extracts can be explained by chitinous structure of the cell wall, which does not allow easy penetration of bioactive substances (Ogundare *et al.*, 2006). Various works also reported similar results (Stefanovic *et al.*, 2009; Marijana *et al.*, 2010; Chanda and Kaneria, 2011; Chanda *et al.*, 2012). The results obtained in our study are promising because different solvent extracts of *M. charantia* could be important in future formulations for treatment of multi resistant Gram-negative pathogens, because of their production of extended-spectrum β-lactamase, Enterobacteriaceae, responsible for serious infections in community and hospital patients (Slama, 2008). The most considerable antibacterial activity was obtained against *K. pneumoniae* and *P. mirabilis*.

Phenolics are secondary plant metabolites that are present in all plants and plant products and are known to act as antioxidants not only because of their ability to donate hydrogen/electrons to free radicals but also because they are stable radical intermediates (Maillard *et al.*, 1996). Thus, measurement of polyphenols and antioxidant activity in herbs has become an important tool to understand the reactive values of plant species (Amarowicz *et al.*, 2010).

Several mechanisms have been proposed to be involved in the antioxidant activity such as hydrogen donation, termination of free radical mediated chain reaction, prevention of hydrogen abstraction, chelation of catalytic ions, and elimination of peroxides (Gordon, 1990). Antioxidant activity is a dependent system and the characteristic of a particular system can influence the outcome of the analysis. Hence, a single assay would not be representative of the antioxidant potential of plant extracts. In the present study, we employed two different models of antioxidant assays which could provide a more reliable approach to assess the antioxidant potential of different solvent extracts of *L. siceraria* and *M. charantia*. DPPH and SO assays are simple, rapid methods commonly used to assess antioxidant activities *in vitro*. DPPH radical is a stable organic free radical with absorption maximum at 517 nm. It loses this absorption when accepting an electron or a free radical species, which results in a visually noticeable discoloration from purple to yellow. Concentration of sample at which the inhibition percentage reaches 50 per cent is its IC_{50} value. IC_{50} value is negatively related to the antioxidant activity, as it expresses the amount of antioxidant needed to decrease its radical concentrate by 50 per cent. The lower IC_{50} value indicates higher antioxidant activity of the sample tested. The superoxide anion radical, included in free radical species, is a harmful factor that can induce ageing and destroy cell membranes. It can be generated by oxidative stress. It is also considered to play an important role in the peroxidation of lipid. Therefore, studying the superoxide anion radical scavenging effect was an important aspect of this study. Our present results confirmed that *L. siceraria* indeed had remarkable antioxidant activity.

Conclusions

In the present work, significant antimicrobial activity was observed in *M. charantia*. Bacteria were more susceptible than fungi, while Gram negative bacteria were more susceptible than Gram positive bacteria. Remarkable antioxidant activity and highest total phenol content was in acetone extract of *L. siceraria*. It appears that these plants could be a potential source of antimicrobial and antioxidant agents. The extraction of active compounds from different solvent extracts of these plants, may lead to their use as food preservatives as well as in pharmaceutical and natural therapies for the treatment of infectious diseases. Nevertheless, further research is required to isolate and characterize the active compounds that are responsible for the antimicrobial and antioxidant activities.

Acknowledgements

The authors thank Prof. S.P. Singh, Head, Department of Biosciences, Saurashtra University, Rajkot, Gujarat, India for providing excellent research facilities and Mr. V. Chanda for editing and correcting the English language. The authors, Mr. Mital Kaneria and Ms. Kalpna Rakholiya are also thankful to University Grants Commission, New Delhi, India for providing financial support.

References

Adamez, J.D., Samino, E.G., Sanchez, E.V., and Gonzalez-Gomez, D. (2012). *In vitro* estimation of the antibacterial activity and antioxidant capacity of aqueous extracts from grape-seeds (*Vitis vinifera* L.). *Food Control*, **24:** 136-141.

Alam, S., Asad, M., Asdaq, S.M.B., and Prasad, V.S. (2009). Antiulcer activity of methanolic extract of *Momordica charantia* L. in rats. *Journal of Ethnopharmacology*, **123:** 464-469.

Akinsulire, O.R., Aibinu, *I.E.*, Adenipekun, T., Adelowotan, T., and Odugbemi, T. (2007). *In vitro* antimicrobial activity of crude extracts from plants *Bryophyllum pinnatum* and *Kalanchoe crenata*. *African Journal of Traditional Complementary and Alternative Medicine*, **4:** 338 – 344.

Amarowicz, R., Estrella, I., Hernandez, T., Robredo, S., Troszynska, A., Kosinska, A., and Pegg, R.B. (2010). Free radical-scavenging capacity, antioxidant activity, and phenolic composition of green lentil (*Lens culinaris*). *Food Chemistry*, **121:** 705-711.

Bajpai, V.K., Na, M., and Kang, S.C. (2010). The role of bioactive substances in controlling foodborne pathogens derived from *Metasequoia glyptostroboides* Miki ex Hu. *Food and Chemical Toxicology*, **48:** 1945-1949.

Batran, S.A.S., El-Gengaihi, S.E., and Shabrawya, O.A. (2006). Some toxicological studies of *Momordica charantia* L. on albino rats in normal and alloxan diabetic rats. *Journal of Ethnopharmacology*, **108:** 236-242.

Bhatt, I.D., Dauthal, P., Rawat, S., Gaira, K.S., Jugran, A., Rawal, R.S., and Dhar, U. (2012). Characterization of essential oil composition, phenolic content, and antioxidant properties in wild and planted individuals of *Valeriana jatamansi* Jones. *Scientia Horticulturae*, **136:** 61-68.

Bing, L., Guo-cai, W., Ji, Y., Mao-xin, Z., and Guang-wen, L. (2008). Antifeedant activity and active ingredients against *Plutella xylostella* from *Momordica charantia* leaves. *Agricultural Sciences in China*, **7:** 1466-1473.

Bors, W., Michel, C., and Stettmaier, K. (2001). Structure–activity relationships governing antioxidant capacities of plant polyphenols. *Methods in Enzymology*, **335:** 166-180.

Braca, A., Siciliano, T., D'Arrigo, M., and Germano, M.P. (2008). Chemical composition and antimicrobial activity of *Momordica charantia* seed essential oil. *Fitoterapia*, **79:** 123-125.

Cakir, A., Kordali, S., Kilic, H., and Kaya, E. (2005). Antifungal properties of essential oil and crude extracts of *Hypericum linarioides* Bosse. *Biochemical Systematics and Ecology*, **33:** 245-256.

Chanda, S., and Kaneria, M. (2011). Indian nutraceutical plant leaves as a potential source of natural antimicrobial agents. *In:* Science Against Microbial Pathogens: Communicating Current Research and Technological Advances, 3rd Series, Ed. By Mendez – Vilas, A. Formatex, Spain, **2:** 1251-1259.

Chanda, S., and Rakholiya, K. (2011). Combination therapy: Synergism between natural plant extracts and antibiotics against infectious diseases. *In:* Science Against Microbial Pathogens: Communicating Current Research and Technological Advances, 3rd Series, Ed. By Mendez – Vilas, A. Formatex, Spain, **1:** 520-529.

Chanda, S.V., and Kaneria M.J. (2012). Optimization of conditions for the extraction of antioxidants from leaves of *Syzygium cumini* L. using different solvents. *Food Analytical Methods*, **5:** 332-338.

Chanda, S., Dudhatra, S., and Kaneria, M. (2010a). Antioxidative and antibacterial effects of seeds and fruit rind of nutraceutical plants belonging to the Fabaceae family. *Food and Function*, **1:** 308-315.

Chanda, S., Baravalia, Y., Kaneria, M., and Rakholiya, K. (2010b). Fruit and vegetable peels – strong natural source of antimicrobics. *In:* Current Research, Technology and Education Topics in Applied Microbiology and Microbial Biotechnology, 2nd Series, Ed. By Mendez – Vilas, A. Formatex, Spain, pp. 444-450.

Chanda, S., Dave, R., and Kaneria, M. (2011a). *In vitro* antioxidant property of some Indian medicinal plants. *Research Journal of Medicinal Plant,* **5:** 169-179.

Chanda, S., Kaneria, M., and Nair, R. (2011b). Antibacterial activity of *Psoralea corylifolia* L. seed and aerial parts with various extraction methods. *Research Journal of Microbiology,* **6:** 124-131.

Chanda, S., Rakholiya, K., and Nair, R. (2011c). Antimicrobial activity of *Terminalia catappa* L. leaf extracts against some clinically important pathogenic microbial strains. *Chinese Medicine,* **2:** 171-177.

Chanda, S., Kaneria, M., and Baravalia, Y. (2012). Antioxidant and antimicrobial properties of various polar solvent extracts of stem and leaves of four *Cassia* species. *African Journal of Biotechnology,* **11:** 2490-2503.

Chang, C.C., Yang, M.H., Wen, H.M., and Chern, J.C. (2002). Estimation of total flavonoid content in propolis by two complementary colorimetric methods. *Journal of Food and Drug Analysis,* **10:** 178-182.

Chang, C.I., Tseng, H.I., Liao, Y.W., Yen, C.H., Chen, T.M., Lin, C.C., and Cheng, H.L. (2011). *In vivo* and *in vitro* studies to identify the hypoglycaemic constituents of *Momordica charantia* wild variant WB24. *Food Chemistry,* **125:** 521-528.

Dar, M.Y., Shah, W.A., Rather, M.A., Qurishi, Y., and Hamid, A. (2011). Chemical composition, *in vitro* cytotoxic and antioxidant activities of the essential oil and major constituents of *Cymbopogon jawarancusa* (Kashmir). *Food Chemistry,* **129:** 1606-1611.

Das, L., Bhaumik, E., Raychaudhuri, U., and Chakraborty, R. (2012). Role of nutraceuticals in human health. *Journal of Food Science and Technology,* **49:** 173-183.

Dauchet, L., Montaye, M., Ruidavets, J. B., Arveiler, D., Kee, F., Bingham, A., Ferrieres, J., Haas, B., Evans, A., Ducimeticre, P., Amouyel, P., and Dallongeville, J. (2010). Association between the frequency of fruit and vegetable consumption and cardiovascular disease in male smokers and non-smokers. *European Journal of Clinical Nutrition,* **64:** 578–586.

Dembitsky, V.M., Poovarodom, S., Leontowicz, H., Leontowicz, M., Vearasilp, S., Trakhtenberg, S., and Gorinstein, S. (2011). The multiple nutrition properties of some exotic fruits: Biological activity and active metabolites. *Food Research International,* **44:** 1671-1701.

Doughari, J.H., Elmahmood, A.M., and Manzara, S. (2007). Studies on the antibacterial activity of root extracts of *Carica papaya* L. *African Journal of Microbiology Research,* **1:** 37-41.

Furneri, P.M., Marino, A., Saija, A., Uccella, N., and Bisignano, G. (2002). *In vitro* antimycoplasmal activity of oleuropein. *International Journal of Antimicrobial Agents*, **20:** 293-296.

Ghosh, K., Chandra, K., Ojha, A.K., Sarkar, S., and Islam, S.S. (2009). Structural identification and cytotoxic activity of a polysaccharide from the fruits of *Lagenaria siceraria* (Lau). *Carbohydrate Research*, **344:** 693-698.

Ghule, B.V., Ghante, M.H., Saoji, A.N., and Yeole, P.G. (2009). Antihyperlipidemic effect of the methanolic extract from *Lagenaria siceraria* Stand. fruit in hyperlipidemic rats. *Journal of Ethnopharmacology*, **124:** 333-337.

Gordon, M.F. (1990). The mechanism of antioxidant action *in vitro*. *In:* Food antioxidants. Ed. By Hudson, B.J.F. Elsevier Applied Science, London, pp. 1-18.

Griep, L.O., Verschuren, W.M., Kromhout, D., Ocke, M.C., and Geleijnse, J.M. (2011). Raw and processed fruit and vegetable consumption and 10-year stroke incidence in a populationbased cohort study in The Netherlands. *European Journal of Clinical Nutrition*, **65:** 791–799.

Guevara, A.P., Lim-Sylianco, C., Dayrit, F., and Finch, P. (1990). Antimutagens from *Momordica charantia*. *Mutation Research*, **230:** 121-126.

Gunjan, M., Jha, A.K., Jana, G.K., Ishra, U., and Verma, P. (2010). An antidiabetic study of *Momordica charantia* fruit. *Journal of Pharmacy Research*, **3:** 1223-1224.

Gupta, S., Raychaudhuri, B., Banerjee, S., Das, B., Mukhopadhaya, S., and Datta, S.C. (2010). Momordicatin purified from fruits of *Momordica charantia* is effective to act as a potent antileishmania agent. *Parasitology International*, **59:** 192-197.

Gurbuz, I., Akyuz, C., Yesilada, E., and Sener, B. (2000). Anti-ulcerogenic effect of *Momordica charantia* L. fruits on various ulcer models in rats. *Journal of Ethnopharmacology*, **71:** 77-82.

Hugo, W. B., and Bloomfield, S.F. (1971). Studies on the mode of action of the phenolic antibacterial agent fentichlor against *Staphylococcus aureus and Escherichia coli*. III. The effect of fentichlor on the metabolic activities of *Staphylococcus aureus* and *Escherichia coli*. *Journal of Applied Bacteriology*, **34:** 579-591.

Itagaki, S., Kurokava, T., Nakata, C., Saito, Y., Oikawa, S., Kobayashi, T., Hirano, T., and Iseki, K. (2009). *In vitro* and *in vivo* antioxidant properties of ferulic acid: a comparative study with other natural oxidation inhibitors. *Food Chemistry*, **114:** 466-471.

Jadhav, T.A., Joshi, Y.M., and Kadam, V.J. (2010). *In vitro* antioxidant activity of ethanolic extract of the leaves of *Lagenaria siceraria* (Molina) Standl. *Journal of Pharmacy Research* **3:** 257-260.

Jarald, E., Bangar, O., Edwin, S., Ahmad, S., and Jamalludin, S. (2009). Antidiabetic activity of few Indian medicinal plants vs their combination in alloxan induced diabetic rats. *Journal of Pharmacy Research*, **2:** 1760-1763.

Kaneria, M., Baravalia, Y., Vaghasiya, Y., and Chanda, S. (2009). Determination of antibacterial and antioxidant potential of some medicinal plants from Saurashtra region, India. *Indian Journal of Pharmaceutical Sciences*, **71:** 406-412.

Kaneria, M.J., and Chanda, S.V. (2013). The effect of sequential fractionation technique on the various efficacies of pomegranate (*Punica granatum* L.). *Food Analytical Methods,* DOI 10.1007/s12161-012-9412-8; **6**: 164-175.

Kaneria, M.J., Bapodara, M.B., and Chanda, S.V. (2012a). Effect of extraction techniques and solvents on antioxidant activity of pomegranate (*Punica granatum* L.) leaf and stem. *Food Analytical Methods,* **5:** 396-404.

Kaneria, M., Kanani, B., and Chanda, S. (2012b). Assessment of effect of hydroalcoholic and decoction methods on extraction of antioxidants from selected Indian medicinal plants. *Asian Pacific Journal of Tropical Biomedicine,* **2:** 195-202.

Khan, R.A., Khan, M.R., Sahreen, S., and Ahmed, M. (2012). Evaluation of phenolic contents and antioxidant activity of various solvent extracts of *Sonchus asper* (L.) Hill. *Chemistry Central Journal,* **6:** 12-18.

Kiran, S.R., Devi, P.S., and Reddy, K.J. (2008). Evaluation of *in vitro* antimicrobial activity leaf and stem essential oils of *Chloroxylon swietenia* Dc. *World Journal of Microbiology and Biotechnology,* **24:** 1909-1914.

Kubola, J., and Siriamornpun, S. (2008). Phenolic contents and antioxidant activities of bitter gourd (*Momordica charantia* L.) leaf, stem and fruit fraction extracts *in vitro*. *Food Chemistry,* **110:** 881-890.

Kumar, G.S., Shetty, A.K., and Salimath, P.V. (2008). Modulatory effect of bitter gourd (*Momordica charantia* Linn.) on alterations in kidney heparan sulfate in streptozotocin-induced diabetic rats. *Journal of Ethnopharmacology,* **115:** 276-283.

Kumar, R., Balaji, S., Uma, T.S., and Sehgal, P.K. (2009). Fruit extracts of *Momordica charantia* potentiate glucose uptake and up-regulate Glut-4, PPARγ and PI3K. *Journal of Ethnopharmacology,* **126:** 533-537.

Lagouri, V., Guldas, M., and Gurbuz, O. (2011). *In vitro* antioxidant/free radical scavenging and antibacterial properties of endemic oregano and thyme extracts from Greece. *Food Science and Biotechnology,* **20:** 1487-1493.

Lee-Huang, S., Huang, P.L., Chen, H.C., Huang, P.L., Bourinbaiar, A., Huang, H.I., and Kung, H.F. (1995). Anti-HIV and anti-tumor activities of recombinant MAP30 from bitter melon. *Gene,* **161:** 151-156.

Leung, S.O., Yeung, H.W., and Leung, K.N. (1987). The immunosuppressive activities of two abortifacient proteins isolated from the seeds of bitter melon (*Momordica charantia*). *Immunopharmacology,* **13:** 159-171.

Lii, C.K., Chen, H.W., Yun, W.T., and Liu, K.L. (2009). Suppressive effects of wild bitter gourd (*Momordica charantia* Linn. var. abbreviate ser.) fruit extracts on inflammatory responses in RAW264.7 macrophages. *Journal of Ethnopharmacology,* **122:** 227-233.

Lin, J., Opoku, A.R., Geheeb-Keller, M., Hutchings, A.D., Terblanche, S.E., Jager, A.K., and Van-Staden, J. (1999). Preliminary screening of some traditional Zulu medicinal plants for anti-inflammatory and antimicrobial activities. *Journal of Ethnopharmacology,* **68:** 267-274.

Liu, C.H., Yen, M.H., Tsang, S.F., Gan, K.H., Hsu, H.Y., and Lin, C.N. (2010). Antioxidant triterpenoids from the stems of *Momordica charantia*. *Food Chemistry,* **118:** 751-756.

Liu, J., Wang, C., Wang, Z., Zhang, C., Lu, S., and Liu, J. (2011). The antioxidant and free-radical scavenging activities of extract and fractions from corn silk (*Zea mays* L.) and related flavone glycosides. *Food Chemistry,* **126:** 261-269.

Lv, F., Liang, H., Yuan, Q., and Li, C. (2011). *In vitro* antimicrobial effects and mechanism of action of selected plant essential oil combinations against four food-related microorganisms. *Food Research International,* **9:** 3057-3064.

Maillard, M.N., Soum, M.H., Boivia, P., and Berset, C. (1996). Antioxidant activity of barley and malt: relationship with phenolic content. *Lebensmittel-Wissenschaft Und Technologie,* **29:** 238-244.

Mali, V.R., and Bodhankar, S.L. (2010). Cardioprotective effect of *Lagenaria siceraria* (LS) fruit powder in isoprenalin-induced cardiotoxicity in rats. *European Journal of Integrative Medicine,* **2:** 143-149.

Malik, Z.A., Singh, M., and Sharma, P.L. (2011). Neuroprotective effect of *Momordica charantia* in global cerebral ischemia and reperfusion induced neuronal damage in diabetic mice. *Journal of Ethnopharmacology,* **133:** 729-734.

Marijana, K., Branislav, R., and Slobodan, S. (2010). Antimicrobial activity of the lichen *Lecanora frustulosa* and *Parmeliopsis hyperopta* and their divaricatic acid and zeorin constituents. *African Journal of Microbiology Research,* **4:** 885-890.

Mazumdar, K., Kumar, K.A., and Dutta, N.K. (2010). Potential role of the cardiovascular non-antibiotic (helper compound) amlodipine in the treatment of microbial infections: scope and hope for the future. *International Journal of Antimicrobial Agents,* **36:** 295-302.

Mbaebie, B.O., Edeoga, H.O., and Afolayan, A.J. (2012). Phytochemical analysis and antioxidants activities of aqueous stem bark extract of *Schotia latifolia* Jacq. *Asian Pacific Journal of Tropical Biomedicine,* **2:** 118-124.

Mc Cune, L.M., and Johns, T. (2002). Antioxidant activity in medicinal plants associated with the symptoms of diabetes mellitus used by the indigenous peoples of the North American boreal forest. *Journal of Ethnopharmacology,* **82:** 197-205.

Mc Donald, S., Prenzler, P.D., Autolovich, M., and Robards, K. (2001). Phenolic content and antioxidant activity of olive extracts. *Food Chemistry,* **73:** 73-84.

Mesia, G.K., Tona, G.L., Nanga, T.H., Cimanga, R.K., Apers, S., Cos, P., Maes, L., Pieters, L., and Vlietinck, A.J. (2008). Antiprotozoal and cytotoxic screening of 45 plant extracts from Democratic Republic of Congo. *Journal of Ethnopharmacology,* **115:** 409-415.

Moellering Jr. R.C. (2011). Discovering new antimicrobial agents. *International Journal of Antimicrobial Agents,* **37:** 2-9.

Munoz, V., Sauvain, M., Bourdy, G., Callapa, J., Rojas, I., Vargas, L., Tae, A., and Deharo, E. (2000). The search for natural bioactive compounds through a multidisciplinary approach in Bolivia. Part II. Antimalarial activity of some plants used by Mosetene Indians. *Journal of Ethnopharmacology,* **69:** 139-155.

Nair, R., and Chanda, S. (2008). Antimicrobial activity of *Terminalia catappa, Manilkara zapota* and *Piper betel* leaf extracts. *Indian Journal of Pharmaceutical Sciences,* **70:** 390-393.

Naseem, M.Z., Patil, S.R., Patil, S.R., Ravindra, and Patil, S.B. (1998). Antispermatogenic and androgenic activities of *Momordica charantia* (Karela) in albino rats. *Journal of Ethnopharmacology,* **61:** 9-16.

Nguimbou, R.M., Njintang, N.Y., Makhlouf, H., Gaiani, C., Scher, J., and Mbofung, C.M.F. (2012). Effect of cross-section differences and drying temperature on the physicochemical, functional and antioxidant properties of giant taro flour. *Food Bioprocess Technology,* DOI 10.1007/s11947-012-0846-1.

Nivitabishekam, S.N., Asad, M., and Prasad, V.S. (2009). Pharmacodynamic interaction of *Momordica charantia* with rosiglitazone in rats. *Chemico-Biological Interactions,* **177:** 247-253.

Ogundare, A.O., Adetuyi, F.C., and Akinyosoye, F.A. (2006). Antimicrobial activities of *Vernonia tenoreana. African Journal of Biotechnology,* **5:** 1663-1668.

Parekh, J. and Chanda, S. (2007b). *In vitro* screening of antibacterial activity of aqueous and alcoholic extracts of various Indian plant species against selected pathogens from Enterobacteriaceae. *African Journal of Microbiology Research,* **1:** 92-99.

Parekh, J., and Chanda, S. (2008). Antibacterial activity of aqueous and alcoholic extracts of 34 Indian medicinal plants against some *Staphylococcus* species. *Turkish Journal of Biology,* **32:** 63-71.

Perez, C., Paul, M., and Bazerque, P. (1990). An antibiotic assay by the agar well diffusion method. *Acta Biologiae et Medicine Experimentalis,* **15:** 113-115.

Prabakar, K., and Jebanesan, A. (2004). Larvicidal efficacy of some Cucurbitacious plant leaf extracts against *Culex quinquefasciatus* (Say). *Bioresource Technology,* **95**: 113-114.

Pukalskas, A., Venskutonis, P.R., Salido, S., de Waard, P., and Beek, T.A. (2012). Isolation, identification and activity of natural antioxidants from horehound (*Marrubium vulgare* L.) cultivated in Lithuania. *Food Chemistry,* **130:** 695-701.

Rakholiya, K., Kaneria, M., and Chanda, S. (2011). Vegetable and fruit peels as a novel source of antioxidants. *Journal of Medicinal Plants Research,* **5:** 63-71.

Robak, J. and Gryglewski, R.J. (1988). Flavonoids are scavengers of superoxide anions. *Biochemical Pharmacology,* **37:** 837-841.

Roller, S. (2003). Natural antimicrobials for the minimal processing of foods. Woodhead Publishing Limited, Cambridge, pp. 320.

Salgado, J.M., Ferreira, T.R.B., Biazotto, F.O., and Dias, C.T.S. (2012). Increased antioxidant content in juice enriched with dried extract of pomegranate (*Punica granatum*) Peel. *Plant Foods for Human Nutrition,* **67:** 39-43.

Sandri, I.G., Zacaria, J., Fracaro, F., Delamare, A.P.L., and Echeverrigaray, S. (2007). Antimicrobial activity of the essential oils of Brazilian species of the genus *Cunila* against food borne pathogens and spoiling bacteria. *Food Chemistry,* **103:** 823-828.

Satpathy, G., Tyagi, Y., and Gupta, R. (2011). Preliminary evaluation of nutraceutical and therapeutic potential of raw *Spondias pinnata* K., an exotic fruit of India. *Food Research International,* **44:** 2076-2087.

Shahidi, F., and Naczk, M. (2004). Phenolics in food and nutraceuticals. Boca Raton, FL: CRC Press. pp352-355.

Shan, B., Cai, Yi, Z., Brooks, J.D., and Corke, H. (2007). The *in vitro* antibacterial activity of dietary spice and medicinal herb extracts. *International Journal of Food Microbiology,* **117:** 112-119.

Shimada, K., Fulikawa, K., Yahara, K., and Nakamura, T. (1992). Antioxidative properties of Xanthem autoxidation of soyabean oil in cyclodextrin emulsion. *Journal of Agricultural and Food Chemistry,* **40:** 945-948.

Slama, T.G. (2008). Gram-negative antibiotic resistance: there is a price to pay. *Critical Care,* **12:** S4 doi: 10.1186/cc6820

Smita, T., Rashmi, T., Farooque, P.M., and Daud, P.S. (2009). *In vitro* anthelmintic activity of seed extract of *Lagenaria siceraria* (Molina.) Standley fruit. *Journal of Pharmacy Research,* **2:** 1194-1195.

Stefanovic, O., Stenojevic, D., and Comic, L. (2009). Inhibitory effect of *Torilis anthriscus* on growth of microorganisms. *Central European Journal of Biology,* **4:** 493-498.

Su, X., Wang, Z., and Liu, J. (2009). *In vitro* and *in vivo* antioxidant activity of *Pinus koraiensis* seed extract containing phenolic compounds. *Food Chemistry,* **117:** 681-686.

Sulaiman, S.F., Yusoff, N.A., Eldeen, I.M., Seow, E.M., Sajak, A.A., Supriatno and Ooi, K.L. (2011). Correlation between total phenolic and mineral contents with antioxidant activity of eight Malaysian bananas (*Musa* sp.). *Journal of Food Composition and Analysis,* **24**: 1-10.

Tan, M.J., Ye, J.M., Turner, N., Hohnen-Behrens, C., Ke, C.Q., Tang, C.P., Chen, T., Weiss, H.C., Gesing, E.R., Rowland, A., James, D.E., and Ye, Y. (2008). Antidiabetic activities of triterpenoids isolated from bitter melon associated with activation of the AMPK Pathway. *Chemistry and Biology,* **15:** 263-273.

Tauxe, R.V. (1997). Emerging foodborne diseases: an evolving public health challenge. *Dairy, Food Environmental Sanitation,* **17:** 788-795.

Wiart, C., Hannah, A., Yassim, M., Hamimah, H., and Sulaiman, M. (2004). Antimicrobial activity of *Acalypha siamensis* Oliv. ex Gage. *Journal of Ethnopharmacology,* **95:** 285-286.

Wu, S.J., and Ng, L.T. (2008). Antioxidant and free radical scavenging activities of wild bitter melon (*Momordica charantia* Linn. var. abbreviata Ser.) in Taiwan. *LWT Food Science and Technology,* **41:** 323-330.

Yahia, E.M., Gutierrez-Orozco, F., and Leon, C.A. (2011). Phytochemical and antioxidant characterization of mamey (*Pouteria sapota* Jacq. H.E. Moore and Stearn) fruit. *Food Research International,* **44:** 2175-2181.

Yesilada, E., Gurbuz, I., and Shibata, H. (1999). Screening of Turkish anti-ulcerogenic folk remedies for anti-*Helicobacter pylori* activity. *Journal of Ethnopharmacology,* **66:** 289-293.

Yuan, X., Gu, X., and Tang, (2008). Purification and characterisation of a hypoglycemic peptide from *Momordica charantia* L. Var. abbreviata Ser. *Food Chemistry,* **111:** 415-420.

Traditional and Folk Herbal Medicine: Recent Researches Vol. 2 (2014)
Editor: **V.K. Gupta**
Published by: **DAYA PUBLISHING HOUSE, NEW DELHI**

Pages **249–272**

9

Applying the Cultural Salience Index to Medicinal Plants Used by the Baixa dos Quelés *Quilombola* Community in the Jeremoabo Municipality in Bahia, Brazil

V.S.Almeida[1]* and F.P.S. de F. Bandeira[1]

ABSTRACT

A large part of the Latin American population does not have access to allopathic medicine and one alternative to this is the use of medicinal plants. In Brazil there are a number of traditional communities that use medicinal plants to cure diseases. One of these traditional groups is the quilombola people (descended from African peoples who rebelled against the past colonial system, seeking refuge in locations which had their own social and political organization). This work aims to identify those medicinal plants of greatest cultural salience for the quilombola community and the factors that determine their patterns of use. In order to do this, we used freelists and semi-structured interviews with twelve local specialists and then applied the cultural salience index. Seventy-one plant species were cited, distributed across 52 genera and 28 families. Fourteen disease systems were identified, based on the WHO (World Health Organization) disease classification system. Diseases of the respiratory (21.23 per cent) and digestive (20.54 per cent) systems were most often cited as used in medicinal plant treatment. In descending order, the five species of greatest cultural salience value were: Gochnatia oligocephala (Gardner) Cabrera (Asteraceae) (0.52); Cordia curassavica (Jacq.) Roem. and Schult (Boraginaceae) (0.37); Schinus terebinthifolius Raddi (Anacardiaceae) (0.27); Mimosa

1 State University of Feira de Santana (Universidade Estadual de Feira de Santana), Dept. Biological Sciences, Post-graduate Programme in Botany, Av. Universitária s/n, 44031-460, Feira-de-Santana, BA, Brazil.

* *Corresponding author*: E-mail: vansoal@yahoo.com.br

tenuiflora (Willd.) Poir (Mimosaceae) (0.25); and Lippia thymoides Mart. and Schauer (Verbenaceae) (0.23). The G. oligocephala species was most often recommended in local therapeutics for gastritis, while Cordia curassavica was recommended for combating influenza symptoms. The knowledge that these traditional communities have regarding medicinal plants may be considered in public health policies aimed at meeting the needs of these populations, thereby valuing their culture and disseminating their knowledge.

Keywords: Caatinga, Diseases, Ethnobotany, Salience index, Freelists, Medicinal plants, Quilombolas.

Introduction

Humanity has always used plants to cure diseases of the body and soul. Many traditional or autochthonous societies possess a vast natural pharmacopeia, derived in large part from the plant resources found in the natural environment the population occupies or cultivated in anthropically altered environments (Amorozo, 2002a). Secondary compounds in plants play an important ecological role in inhibiting the germination and growth of other plants and they attract insects and other animals for pollination, seed dispersal and chemical defence against insects and herbivory (Harborne, 1993). Many of these compounds are bioactive and may serve as medicines for humans (Kinghorn, 1994).

During the 20th century, until the 1970s, and principally following the 2nd World War, the discovery of antibiotics and the ever-increasing growth of medicines based on synthetic drugs meant that naturally occurring drugs were, to some extent, abandoned and there was some scepticism about them. However due, amongst other factors, to the growing price of medication and the side effects of synthetic drugs, research into drugs of plant origin has once again been revived (Silva, 2002).

It is estimated that 50 per cent of the Latin American population has little or no access to medication (Jorquera, 1993), which leads to a demand for traditional alternatives to treat health problems. Data revealed by the World Health Organization (WHO), demonstrates that 80 per cent of the population of developing countries use traditional practices in primary care, 85 per cent of this total use medicinal plants or preparations, (Brazilian Ministry of Health, 2001) and, at global level, WHO also recognizes and recommends herbal medicine for prophylactic, curative, palliative or diagnostic purposes.

Things are no different in Brazil where, due to its precarious economic, social and sanitary circumstances and the difficulty in accessing health centres, the use of plants to cure diseases occurs very frequently and for certain ethnic groups often represents the only therapeutic resource. Another reason for using plants is the excellent cost-benefit relationship, in other words, low cost and efficiency in the prevention and treatment of diseases (Brazilian Ministry of Health, 2001).

Over the last decade, ethnobotanical investigations have broken down barriers and moved away from the exclusivity of the early studies of ethnobotany, which had a merely descriptive focus on the use of medicinal plants, towards the development of a quantitative approach. This research has increased over the last decades, introducing quantitative indices to evaluate the different cultural significances of plants in different human cultures (Camou-Guerrero *et al.*, 2008).

The number of studies about medicinal plants conducted in Brazil that use indices to evaluate the importance of plant species has increased: Vieira *et al.* (2008), Silva and Peixoto (2009), Gomes (2010) and Almeida *et al.* (2006) have all sought to clarify how these communities share and view the plant world.

In the Northeast, particularly in Bahia, the only research in the field of ethnobotany regarding medicinal plants using quantitative methods and based in quilombola communities are the studies conducted by Gomes (2010) in the Casinhas community in Jeremoabo and by Almeida and Bandeira (2010), who used the Lawrence index to evaluate the local value attached to the use of plants in the Baixa dos Quelés and Casinhas communities, situated in the Jeremoabo municipality in Bahia. It is therefore necessary to extend studies in one of the states in the Northeast with a high concentration of both quilombola communities and of biodiversity.

This research, therefore, aims to identify those medicinal plants of greatest cultural salience, as well as the factors that determine their patterns of use by the Baixa dos Quelés quilombola community in the municipality of Jeremoabo, Bahia, Brazil.

Materials and Methods

Study Area

The quilombola community of Baixa dos Quelés (09°58′58.9″S; 38°26′17.0″W) belongs to the municipality of Jeremoabo and is approximately 24 km from the municipal centre. Jeremoabo is situated in the northeast region of the state of Bahia, 371 km from Salvador, and borders on the following municipalities: Canudos, Macururé, Pedro Alexandre, Santa Brígida, Paulo Afonso, Rodelas, Novo Triunfo, Coronel João de Sá, Pedro Alexandre and Sítio do Quinto.

The municipality of Jeremoabo belongs to the semi-arid region, has a dry climate and an average annual temperature of 24°C, with a maximum of 28°C and a minimum of 20.2°C. The annual average rainfall is 654 mm, with a maximum of 2,273 and a minimum of 276 mm. The rainy season is concentrated from May to July (SEBRAE, 1995). The vegetation is not homogenous, with ecotonal areas between Cerrado-Caatinga-seasonal forest; Cerrado-Caatinga; arboreal Caatinga without palm trees; dense arboreal Caatinga without palm trees; and open arboreal Caatinga with palm trees (CEI, 1994).

Study Community

Baixa dos Quelés has 35 families and 120 inhabitants. Most of its population is composed of small-scale rural farmers, who carry out subsistence agriculture to support their families, mainly of *Manihot esculenta Crantz* (cassava), *Phaseolus vulgaris* L. (beans), *Zea mays* L. (maize). Each family has its own growing area and fields. To supplement family income, some families raise animals, such as cattle, hens and pigs in units some distance away from their houses, while others use their gardens. There is also a group of small-scale beekeepers who use the forest for honey production. The community does not have a health centre or receive medical visits; when the populaton needs health care, they use services located in Jeremoabo. Water supply to their houses is through pipes; most houses have a septic tank and when rubbish is not burnt in the garden it is disposed of on wasteland around the settlement.

Data Collection

Data was collected between September 2010 and January 2011. Twelve members of the community were interviewed: 8 men and 4 women. The snow-ball technique was used to identify subjects; this consisted in identifying local specialists (people recognized by the community as having an excellent knowledge of plants) through residents' recommendations. Thus, the first specialist recommended the next one and so on, in succession (Albuquerque and Lucena, 2004). The analysis of body systems and the therapeutic indications cited by the specialists were adapted according to the World Health Organization's "International Statistical Classification of Diseases and Related Health Problems (ICD-10)".

We initially used free listing, which consists of identifying, through a key question, those flora elements with medicinal potential most used by the local population. The question for the interviewees was: What "sticks from the forest" do you most often use as medicine? According to Quinlan (2005), freelists enable the collection of rapid and easy data, have been frequently used in ethnobotanical investigations of medicinal plants and serve to: identify the items in a emic category or cultural domain; recognize that in listing items people tend to place them in order of familiarity, in other words, of those they know best; indicate who knows most about a subject, since those who know the most make the most citations; and highlight the terms most often mentioned in the location. After the listing we conducted a semi-structured interview with questions about age, schooling and economic activity, as well as about plant unit, form of administration, part used, habitat, and collection location for every plant species cited in the freelists.

We calculated the salience index from the data collected through freelists. This index assesses to what degree a plant is culturally important for a social group, thus, the most important plants are those used by a large number of people for the same category of use, and those of low cultural importance are those used by a smaller number of people (Trotter and Logan, 1986). The "salience index" was calculated for men and women for each plant species using Visual Anthropac-Freelists 4.0 (Borgatti, 1992).

In line with Quinlan (2005), in order to calculate the salience index for each species, we took the freelist of each participant and gave it a value according to the order in which the species was cited, for example: the species in first place received the maximum value for the total number of species cited in the list and this procedure was followed with the other species, according to the cited position (inverted rank). We then divided this by the total number of species cited in the list. To calculate the index's "total salience" we totalled the values for each species (obtained from each informant) and divided each one by the total number of informants.

The botanical material mentioned was collected with the help of local informants, deposited in the Herbarium of the State University of Feira de Santana (*Herbário da Universidade Estadual de Feira de Santana*: HUEFS) and identified through support from specialists or comparisons with exsiccates. We adopted the APG III (2009) system for family classification.

Results and Discussion

Cultural and Socio-Economic Characterization of the Baixa Dos Quelés Quilombola Community

The oldest residents in the communities of Baixa dos Quelés and Baixa da Lagoa (another quilombola community 1.5 km from Baixa dos Quelés, which many residents consider part of Baixa dos Quelés), tell that there was a king called Rui Barbosa (a land owner and enslaver), who owned extensive lands in the region and used slave labour to plant sugarcane. At that time, the enslaver Rui Barbosa ordered the construction of slave quarters to prevent the escape of many of the Black people who were rebels who had been fleeing to Caldeirão da Negra, a nearby location, these days known as Raso. Many of them were captured by the slave hunter. Fortunately, at the end of slavery, the slave quarters were never completed and only some stone walls in Baixa da Lagoa. Regarding the construction of this stone wall were left, one of the residents says: "......The wall has some large stones, I don't know how people managed to pick up those huge stones, the building isn't totally finished, but there are two high walls, which allow us to see, don't they, the construction of this wall and that slavery existed there". (A.L.J.).

Within the community, we observed the self-recognition of quilombola identity and a sense of wanting to preserve the culture and social values are things that we observed within the community. For example, not everyone belongs to the Baixa dos Quelés village. Receiving new residents occurs by consensus and through the authorization of village residents, as one of them said:

"......A family that the people called old Lourenço, who lived in Baixa da Lagoa, a Black family, right, who lived in the region people called Baixa da Lagoa, they called it the Lagoa dos Negros because only coloured people lived there, very dark-skinned people, Black people. But then it went like this, right, someone who's black-skinned marries someone who's fairer-skinned and then you get this mixed quality, but they don't stop being descendents from this region; their origin is still quilombola" (A.L.J.).

"......Let's say that the community comes together and sees if it can donate this plot of land to that person, because suddenly someone can come from outside, who we think may not be viable for the community, who could damage it, yes, really." (A.L.J.)

Of the twelve interviewees, 66.6 per cent were born in the Baixa dos Quelés community, 8.3 per cent had lived there for 10 to 20 years and 25.1 per cent for more than 20 years. Of these, 25 per cent of the women and 87.5 per cent of the men were born in the area. Due to a lack of work prospects, some residents had left the area, but had returned because they had not been able to find better living conditions. The age range varied from 37 – 92 years old, with most being in the 40 to 50 age range (41.66 per cent). About 75 per cent of the women and 50 per cent of the men were older than fifty.

According to the interviewees, knowledge about medicinal plants is transmitted by the oldest residents, normally by parents. However they know of the existence of an expert in the community, someone who better understands plants, who is also the transmitter of knowledge about the use of medicinal plants to the whole community.

Most of the interviewees have 2 or 3 children and all of these attend between 1st and 4th grade of primary school at the school situated in the village. Homes have piped water but no electricity, which is only available at the school. Access to health centres is difficult, so that when not making use of allopathic remedies many people use plants to cure their diseases or they combine the two medical practices.

Medicinal Plants: Knowledge, Use and Handling

Seventy-one species of plants were cited, distributed across 52 genera and 28 families (Table 9.1). The plant families most represented were: Leguminosae (sensu Engler) (35.71 per cent), Euphorbiaceae (28.57 per cent), Solanaceae (10.71 per cent) and Anacardiaceae (10.71 per cent) (Figure 9.1). Albuquerque (2001) found 60 species of medicinal plants in Recife; Junior *et al.* (2005) surveyed 60 species of medicinal use in urban communities in the municipality of Itaberaba, BA; Morais *et al.* (2005), working with indigenous Indians in Tapebas, CE, recorded 63 medicinal plants; Albuquerque and Andrade (2002) found 48 species of medicinal plants cited by rural farmers in the municipality of Alagoinhas, PE; while Gomes (2010) and Franco and Barros (2006), working in quilombola communities in Bahia and Piauí, surveyed 87 and 82 species of medicinal plants, respectively. Gomes *et al.* (2008), making an inventory of medicinal plants in open-air markets in the semi-arid regions of the states of Bahia and Pernambuco, found a higher number of species from the leguminosae family. A similar fact was recorded by Loiola *et al.* (2010), when investigating the potential use of the leguminosae family in four rural locations in Rio Grande do Norte. These same authors cited the leguminosae family as an important group of plants, a significant source of natural resources, with economic, social and ecological potential, particularly for inhabitants of the semi-arid region.

Certain diseases related to quilombolas such as: casting out evil, dispelling negative energy, inflammation of the uterus, women's health problems, swelling (caused by a fall), congestion (stroke), swelling at the bottom of the stomach, broken bones, painful urination, healing wounds or injuries, numbness in the foot, boils, (Table 9.1) were converted into the WHO classification system (2000). We found fourteen categories of diseases, with certain modifications: Endocrine, nutritional and metabolic diseases; Diseases of the nervous systems; Diseases of the eye and adnexa; Diseases of the circulatory system; Diseases of the respiratory system; Diseases of the digestive system; Diseases of the skin and subcutaneous tissue; Diseases of the musculoskeletal system and connective tissue; Diseases of the genitourinary system; General symptoms and signs; Injury; Poisoning and certain other consequences of external causes, Neoplastic Diseases; and "Cultural diseases". Cultural diseases are those which refer to manifestations interpreted by the community as diseases but which do not have a cause based on science (Pinto *et al.*, 2006).

Diseases of the respiratory (21.23 per cent) and digestive (20.54 per cent) systems were most often indicated for medicinal plant treatment, followed by cultural (9.58

Table 9.1: Plant species cited, collection location, part used and therapeutic indications made by subjects from the Baixa dos Quelés village, Jeremoabo, Bahia.

Vernacular Name (Portuguese)	*Species/Family*	*Habitat*	*Collec. Locat.*	*Part Used*	*Mode of Preparation*	*Ethnopharamcological Indication*
Acançu	*Periandra mediterranea* (Vell.) Taub./Papilionaceae	Subshrub	P	Root	Syrup or tea	Influenza
Alecrim/alecrinho	*Lippia thymoides* Mart. and Schauer/Verbenaceae	Subshrub	F/G	Leaf, stalk	Tea, bath, Smoker	Something evil in the body, dispel negative energy, cast out evil
Algaroba	*Prosopis juliflora* Sw. D.C./ Mimosaceae	Tree	G	Leaf	Syrup	Influenza
Algodão-criolo	*Gossypium* sp./Malvaceae	Shrub	G	Leaf	Juice, tea	Headache, oral thrush
Alfavaca	*Monniera trifolia* Loefl./Rutaceae	Herb	G	Seed, leaf	Juice	Inflammation of the eye
Araçá	*Psidium guajava* L.	Shrub	G	Leaf	Tea	Stomach ache
Araçá-branco	*Mitranthes gardneriana* O. Berg./Myrtaceae	Tree	F	Leaf	Tea	Dysentery
Aroeira	*Schinus terebinthifolius* Raddi./Anacardiaceae	Tree	F	Bark, leaf, seed	Tea, water, lick, bath, hydro-alcoholic extract (sugarcane rum or white wine)	Inflammation of the uterus, swelling at the bottom of the stomach, wound healing, influenza
Barbatimâo/ babatenã	*Abarema cochliacarpos* (B.A. Gomes) Barneby and J.W. Grimes/Mimosaceae	Tree	F	Bark, leaf	Water, tea, bath	Inflammation of the uterus
Barbatimâo/ babatenã branco	*Miconia ferruginata* DC./ Melastomataceae	Shrub	F	Bark	Roasted	Itching, burns, injuries
Babosa	*Aloe vera* (L.) Burm/Liliaceae	Subshrub	G	Leaf	Juice	Pains in legs
Bandola grande	*Pisonia laxa* Netto./ Nyctaginaceae	Tree	F/Fi	Shaving (bark)	Water/hydro-alcoholic extract	Swelling caused by a fall, pain in back and kidneys

Contd...

Table 9.1–*Contd...*

Vernacular Name (Portuguese)	*Species/Family*	*Habitat*	*Collec. Locat.*	*Part Used*	*Mode of Preparation*	*Ethnopharamcological Indication*
Boldo	*Plectranthus barbatus* (Andr.) Benth/Labiaceae	Subshrub	G	Leaf	Tea	Stomach ache, food poisoning
Brauna	*Schinopsis brasiliensis* Engl./ Anacardiaceae	Tree	F	Bark	Tea, roasted	Wound healing
Caatinga-de-cheiro	*Cordia curassavica* (Jacq.) Roem. & Schult./Boraginaceae	Subshrub	F/G	Leaf, root, flower	Tea, bath/lick	Digestion, food poisoning, influenza, stomach ache, diarrhoea, fever, cast out evil
Cajueiro branco	*Anacardium occidentale* L./ Anacardiaceae	Tree	Fi	Bark	Tea, bath	Inflammation of the uterus
Cambabá/ guinezinho	*Petiveria alliacea* L./ Phytolaccaceae	Shrub	G	Root, leaf	Tea, bath, hydro-alcoholic extract	Influenza. headache
Camboatá	*Cupania oblongifolia* Mart./ Sapindaceae	Tree	F/Fi/G	Bark	Water/bark/tea	Snake bite, dysentery, stomach ache, weakness
Candeia branca	*Miconia albicans* (Sw.)/ Melastomataceae	Shrub	F	Leaf/stalk	Tea/bath	Evil spirits
Candeia preta	*Gochnatia oligocephala* (Gardner) Cabrera/Asteraceae	Tree	F/Fi	Leaf/bark	Bath, tea/water, leaf chewing (direct use)	Gas, gastritis, sore throat, evil spirits, stomach ache
Capim-santo	*Cymbopogon citratus* (DC.) Stapf/Poaceae	Herb	G/Fi	Leaf	Tea, bath	Influenza
Catingueira/ catinga de porco	*Poincianella microphylla* (Mart. ex G. Don) L.P.Queiroz/ Caesalpiniaceae	Shrub	F	Flower, stalk, bark	Tea, bath	Food poisoning
Catuaba	Sp 1	Tree	F	Bark	Tea, lick	Lack of air
Cedro	*Cedrela fissilis* Vell./Meliaceae	Tree	F	Bark, leaf	Bath, smoker	Headache, fever

Contd...

Table 9.1–*Contd...*

Vernacular Name (Portuguese)	*Species/Family*	*Habitat*	*Collec. Locat.*	*Part Used*	*Mode of Preparation*	*Ethnopharamcological Indication*
Cidreira	*Lippia alba* (Mill.) N.E. Br./ Verbenaceae	Herb	G	Leaf	Tea	Food poisoning, stomach ache
Cipó de grajaú	*Piriadacus erubescens* (DC.) Pichon/Bignoniaceae	Liana	Fi	Leaf	Tea	Influenza
Corona	*Cestrum axillares* Vell./ Solanaceae	Subshrub	G	Leaf	Bath	Evil spirits
Chacuretá	*Swartzia apetala* Raddi./ Papilionaceae	Tree	P	Bark	Tea, water, hydro-alcoholic extract/syrup	Pain in back and kidneys
Crista de galo	*Heliotropium elongatum* Willd. ex Cham. Boraginaceae	Subshrub	G	Root/leaf	Bath, tea	Stroke, headache
Endio	Sp 2	Herb	G	Seed	Tea	Colic, stomach ache
Erva doce	*Pimpinella anisum* L./Apiaceae	Herb	M	Seed	Tea	Colic
Eucalipto	*Eucalyptus* sp./Myrtaceae	Tree	G	Leaf	Tea	Influenza
Genipapo	*Tocoyena formosa* (Cham. and Schltdl.) K. Schum./Rubiaceae	Tree	F	Leaf	Poultice	Numbness in foot
Gergelim	*Sesamum indicum* L./pedaliaceae	Subshrub	G	Seed	Maceration	Congestion (stroke)
Girassol	*Helianthus annuus* L./Asteraceae	Subshrub	G	Seed		Congestion (stroke)
Graviola	*Annona muricata* L./Annonaceae	Tree	G	Leaf, bark,	Water/tea seed	High blood pressure, prostate, tiredness
Guela de camaleão	Sp3	Tree	F	Bark	Poultice	Broken bones
Imburana/ umburana de cheiro	*Commiphora leptophloeos* (Mart.) J.B. Gillett/Burseraceae	Tree	F	Bark	Tea, syrup, water	Influenza, bad circulation

Contd...

Table 9.1–*Contd...*

Vernacular Name (Portuguese)	*Species/Family*	*Habitat*	*Collec. Locat.*	*Part Used*	*Mode of Preparation*	*Ethnopharamcological Indication*
Hortelã	*Plectrantus amboinicus* (Lour) Spreng./Lamiaceae	Herb	M	Leaf	Tea/juice	Influenza, stomach ache
Jatobá	*Hymenaea courbaril* L./ Caesalpiniaceae	Tree	F	Bark	Smoking, syrup, bath, water, tea, hydro-alcoholic extract	Influenza, respiratory problems, catarrh, back pain, vitamin deficiency
Jatobá	*Hymenaea courbaril* L./ Caesalpiniaceae	Tree	F	Bark	Smoking, syrup, bath, water, tea, hydro-alcoholic extract	Influenza, respiratory problems, catarrh, back pain, vitamin deficiency
Jiquirizeiro	*Piptadenia viridiflora* Kunth./ Mimosaceae	Tree	Fi/F	Root/bark	Water/hydro-alcoholic extract	Bad digestion, food poisoning
Jurema-preta	*Mimosa tenuiflora* (Willd) Poir./ Mimosaceae	Tree	F/Fi	Bark	Water, bath, tea, syrup, roasted, hydro-alcoholic extract	Toothache, wound healing, antibiotics, influenza
Jurubeba	*Solanum paniculatum* L/ Solanaceae.	Shrub	G	Fruit	Lick, tea	Influenza
Malva-de-jegue	*Sida cordifolia* Forsk./Malvaceae	Subshrub	G	Root	Tea	Inflammation of the uterus
Mamona	*Ricinus communis* L./ Euphorbiaceae	Shrub	G	Seed	Direct use (oil extract)	Food poisoning
Manacá	*Brunfelsia uniflora* (Pohl) D. Don/Solanaceae	Shrub	F/Fi	Bark/root	Tea	Prostate, pain in legs, rheumatism
Mangaba	*Hancornia speciosa* Gomes/ Apocynaceae	Tree	F/P	Latex	Direct use	Swelling caused by a fall, back pain
Mangueira	*Mangifera indica* L./ Anacardiaceae	Tree	G	Leaf	Tea	Influenza

Contd...

Table 9.1–*Contd...*

Vernacular Name (Portuguese)	*Species/Family*	*Habitat*	*Collec. Locat.*	*Part Used*	*Mode of Preparation*	*Ethnopharamcological Indication*
Marmeleiro	*Croton sonderianus* Mull. Arg./ Euphorbiaceae	Subshrub	F	Bark	Water	Uterus, gastritis
Moleque-duro	*Croton* sp./Euphorbiaceae	Shrub	G	Leaf	Bath	Cast out evil
Mororó	*Bauhinia subclavata* Benth. in Mart./Caesalpiniaceae	Shrub	F	Bark	Tea, syrup	Influenza, back pain
Mostarda	Sp4	Shrub	G	Seed	Roasted/infusion	Headache, congestion (stroke)
Pau d arco roxo	*Tabebuia impetiginosa* Mart./ Bignoniaceae	Tree	F	Bark	Water/hydro-alcoholic extract	Back pain, pains in the body
Pau-ferro	*Libidibia ferrea* (Mart. Ex Tul.) L. P. Queiroz/Caesalpiniaceae	Tree	Fi	Bark	Tea, water, roasted	Influenza
Pau-pra-tudo	*Zanthoxylum riedelianum* Engl./ Rutaceae	Shrub	F/Fi	Leaf, bark, root	Tea, bath, hydro-alcoholic extract	Cancer
Pereiro	*Aspidosperma pyrifolium* Mart./ Apocynaceae	Tree	F	Bark	Maceration	Smack on the eye
Pinha	*Annona squamosa* L./ Annonaceae	Shrub	G	Leaf	Tea	´Bad' digestion
Pinhão-manso	*Jatropha* sp./Euphorbiaceae	Shrub	G	Seed, bark	Tea, roasted	Congestion, prayers, stomach ache
Pinhão-roxo	*Jatropha gossypiifolia* L./ Euphorbiaceae	Shrub	G	Leaf	Bath, prayer	Cast out evil, evil spirits, congestion, headache
Pulga de batata	Sp5	Climber	G	Root	Tea/syrup	Headache, influenza
Quebra-façāo	*Croton* sp./Euphorbiaceae	Shrub	Fi/F	Bark	Bath, water, tea/ hydro-alcoholic extract	Influenza, cast out evil, dysentery, Spinal pain, Swelling caused by a fall

Contd...

Table 9.1–*Contd...*

Vernacular Name (Portuguese)	*Species/Family*	*Habitat*	*Collec. Locat.*	*Part Used*	*Mode of Preparation*	*Ethnopharamcological Indication*
Quina-quina	*Chiococca brachiata* Ruiz and Pav./Rubiaceae	Shrub	F	Bark	Tea	Back pain, cough
Quixabeira	*Sideroxylum obtusifolium* (Roem. Et. Sch.)/Sapotaceae	Tree	F	Bark	Tea, water, syrup, hydro-alcoholic extract	Swelling caused by a fall, influenza, back pain, inflamed kidney
Sete dor	Sp6	Creeper	G	Flower/leaf	Tea	Stomach ache
Sucupira/sipipira	*Bowdichia virgilioides* Kunth./ Papilionaceae	Tree	F	Bark, seed	Water, hydro-alcoholic extract	Back pain
Umbu de ameixa	*Ximenia americana* L./ Olacaceae	Tree	F	Bark	Tea, water, bath	Inflamed throat, inflammation of the female genital tract, tooth ache
Unha de gato	*Piptadenia stipulacea* (Benth.) Ducke/Leg. Mimosaceae	Creeper	F	Bark	Hydro-alcoholic extract	Broken bones
Vassorrinha de botão	Sp7	Subshrub	G	Leaf/stalk	Juice	Broken bones
Vassorrinha de Nossa Senhora	*Scoparia dulcis* L./ Scrophulariaceae	Subshrub	G	Stalk/leaf	Bath, juice	Prayers, evil eye
Velandinho	*Croton pulegioides* Mull. Arg./ Euphorbiaceae	Subshrub	F	Root, leaf	Tea	Sinusitis
Velame/Velande	*Croton heliotropiifolius* Kunth/ Euphorbiaceae	Subshrub	F	Leaf	Tea, syrup	Influenza

F: Forest; G: Garden; P: "Platter" or Plain; Fi: Field, M: Market.

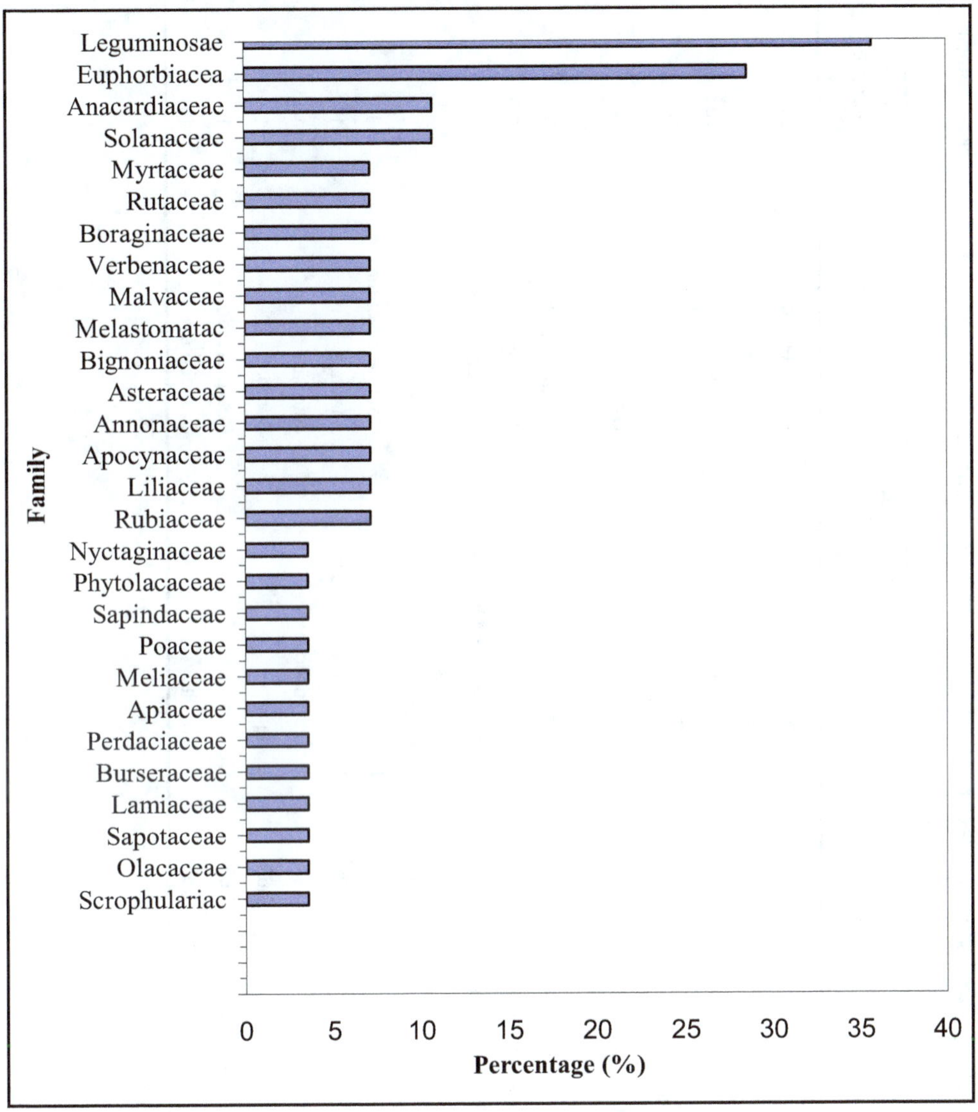

Figure 9.1: Percentage of plant families cited by specialists from the Baixa dos Quelés community, Jeremoabo, BA.

per cent), genitourinary (8.90 per cent) and musculoskeletal (8.90 per cent) systems (Table 9.2). Similar results have been found in quilombola communities in the semi-arid region by Monteles and Pinheiro (2007), Vieira *et al.* (2008), Gomes (2010) and Franco and Barros (2006). The symptoms most often mentioned in these categories were, respectively: influenza, "stomach ache", "cast out evil", inflammation of the uterus/vagina and back pain (Table 9.2). Influenza, stomach ache, tapeworms and sore throat were the diseases of greatest incidence in this community, as listed by residents. Poor sanitary conditions and lack of community infrastructure contributed to the emergence of these illnesses.

Table 9.2: Percentage of citations within each disease category, most cited symptoms within Each Category and most cited plant by category.

Category	*Citations (per cent)*	*Most Cited Symptom within each category*	*Most Cited Plant by Category*	*No. of plants Recorded per Category*
Diseases of the respiratory system	21.23	Influenza	Catinga-de-cheiro	21
Diseases of the digestive system	20.54	Gastritis	Candeia	18
"Cultural diseases"	9.58	"Cast out evil"	Alecrim	10
Diseases of the genitourinary system	8.90	Inflammation of the uterus/vagina	Aroeira	10
Diseases of the musculoskeletal system and the connective tissue	8.90	Back pain	Mangaba	11
Diseases of the circulatory system	8.21	Stroke	Pinhão-manso	10
Diseases of the skin and the subcutaneous tissue	7.53	Injuries	Jurema-preta	7
Injuries, poisoning and certain other consequences of external causes	5.47	Swelling caused by a fall	Mangaba	6
Diseases of the nervous system	3.42	Head ache	Cedro	5
General symptoms and signs	2.05	Fever	Catinga-de-cheiro	3
Diseases of the eye and adnexa	1.36	Inflammation of the eye	Alfavaca	1
Endocrine, nutritional and metabolic diseases.	0.68	Vitamin deficiency	Jatobá	1
Others	0.68	Dengue	Cambabá	1
Neoplasm	0.68	Cancer	Pau-pra-tudo	1
Poisoning	0.68	Snake bite	Camboatá	1

The *Gochnatia oligocephala* species (Gardner) Cabrera (**candeia**[1]) was most often recommended in local therapeutics to treat gastritis, *Cordia curassavica* (Jacq.) Roem. and Schult (**catinga-de-cheiro**) to combat influenza symptoms and *Schinus terebinthifolius* Raddi (**aroeira**) for the treatment of inflammation of the uterus/vagina (Table 9.2). In the work conducted by Patzlaff (2007), this latter plant was related to the same therapeutic property.

Ten species were mentioned in the treatment of "cultural diseases", such as curing "the evil eye", "casting out evil" "removing something evil from the body" and "evil spirits", these are administered through a bath to dispel negative energy, through smoking or through prayer. *Cestrum axillares* (**corona**) and *Lippia thymoides* Mart. and Schauer (**alecrim)** are used together in the smoking process. *Lippia thymoides* is the species most often mentioned within the cultural diseases category and is used to ward off spirits through bathing. *Petiveria alliacea* (**cambabá**) was not cited as a plant related to prayer rituals in this work, although Almeida and Bandeira (2010), working in the Casinhas quilombola community, recorded this species as a plant associated with prayer rituals and blessings. Camargo (2007) mentions this species as linked to the African mystical-religious universe in Brazil. During slavery, this plant, also known as "tames the master", was prepared as a "magical" remedy and used by the slaves to debilitate the enslavers, in order to defend Black women from harassment by their enslavers.

Croton sp. (**quebra-facão**) is the most versatile species and is indicated for the treatment of five body systems: digestive; respiratory; "cultural"; musculoskeletal; and injuries, poisoning and certain other consequences of external causes. *Cordia curassavica* (**catinga-de-cheiro**) and *Sideroxylum obtusifolium* (Roem. Et. Sch.) (**quixabeira**), serve to treat four body systems. Nine species are recommended for three body systems: *Gochnatia oligocephala* (**candeia**), *Schinus terebinthifolius* (**aroeira**) *Pisonia laxa* Netto (**bandola**), *Cupania oblongifolia* Mart. (**camboatá**), *Hymenaea courbaril* L. (**jatobá**), *Mimosa tenuiflora* (Willd) Poir. (**jurema-preta**), *Jatropha* sp (**pinhão-manso**), *Jatropha gossypiifolia* L. (**pinhão-roxo**), and *Ximenia americana* L.(**umbu de ameixa**); and the vast majority of the species, 61.42 per cent, were recommended to treat a single system. The *Ximenia americana* (**umbu de ameixa**) and *Libidibia ferrea* (Mart. Ex. Tul.) L. P. Queiroz (**pau ferro**) species presented with the same indications for medicinal use (anti-inflammatory and influenza treatment respectively) in this research as they did in studies conducted by Vieira *et al.* (2008) in a quilombola community in Piauí.

The part of the plant most often used by the local ethno-medical system to treat diseases is the leaf (37.64 per cent), followed by "shavings or scrapings" (in other words, the bark) (32.94 per cent), the root (11.76 per cent), the seeds (11.76 per cent), the stalk (5.88 per cent), the flowers (3.52 per cent), the fruit (1.17 per cent) and the latex (1.17 per cent). Chemical compounds are distributed across different parts of plants, with the vast majority concentrated in the leaves. Using the leaves as remedies

---

1 Translator's note: plant names throughout the text are given as follows: Latin names in italics and common Portuguese names in bold.

may be seen as a handling strategy, in collecting a part which does not compromise plant development (Moreira and Guarim-Neto, 2009). However, in many species in the semi-arid region leaves are not available throughout the year, which leads the local population to search for other parts of the plant, such as the bark (Albuquerque and Andrade 2002). In this research we recorded a significant percentage of bark use, a part which, when not collected appropriately or exploited rationally, may compromise the maintenance of species populations (Gomes *et al.*, 2008). In order to collect this bark, some people remove it from the bottom upwards, since they believe that the "evil", the disease, exits this way: "it exits in this direction and if it's done in the opposite one, it's no good as medicine" (M.S.B.). Other people, however, state that they remove the bark in the opposite direction, from the top downwards. They also collect it in small quantities, so as not to exhaust their resources; at least, they demonstrate this concern in the interview discussions, as one resident asserts: "We only take off a small piece to make the tea, to do what's necessary, and the rest we leave there, yes, not for the person to stock up on it, not like that. We just take a little, just the amount that we really need, right." (A.L.J.). However, in order to make an assessment of the sustainability of this handling, one would need to conduct a study of the impact of extraction on the species whose bark is used. At the same time, encouraging cultivation of these species in gardens may help to minimize species vulnerability.

The most utilized mode of preparation is tea, representing 41.86 per cent of cases; however, other modes were also cited: baths (13.17 per cent), hydro-alcoholic extract (12.40 per cent), syrups and licks (7.75 per cent), "water" (7.75 per cent), juice (5.42 per cent), roasting (4.65 per cent), direct use (fresh) (4.65 per cent), poultice (1.55 per cent) and smoking (0.77 per cent). The term "water" here refers to a mode of preparation in which part of the plant is placed in cold water for a certain length of time (Lorenzi and Matos, 2008), while the term hydro-alcoholic is when the plant is placed in an alcoholic solution, such as sugarcane rum or white wine. Plants are administered alone or in association with others, as happens with the so-called "Forest Syrup", which combines several plants: *Bauhinia subclavata* Benth. *in* Mart (**mororó**), *Commiphora leptophloeos* (Mart.) J.B. Gillett (**imburana**), *Hymenaea courbaril* L. (**jatobá**), *Schinus terebinthifolius* Raddi (**aroeira**), *Croton heliotropiifolius* Kunth (**velame**), *Sideroxylum obtusifolium* (Roem. Et. Sch.) (**quixabeira**), *Prosopis juliflora* Sw. D.C (**algaroba**), *Mimosa tenuiflora* (Willd) Poir. (**jurema-preta**) and *Libidibia ferrea* (Mart. Ex Tul.) L. P. Queiroz (**pau-ferro**). Similar observations were made by Abreu (2000) and Franco and Barros (2006). Other plant parts are placed in extractors with a base of white wine and onion, as is the case with *Croton sonderianus* (**marmeleiro**); whereas for *Bowdichia virgilioides* (**sucupira**) the base is white wine alone and for *Petiveria alliacea* L. (**cangabá**) and *Croton* sp (**quebra- facão**) it is sugarcane rum.

Of the total seventy-one species cited, the vast majority are trees (40.6 per cent) and shrubs (24.6 per cent) (Figure 9.2). These medicinal plants are collected from gardens, forests, fields, "platters"/plains and markets (Figure 9.3). For the quilombolas, "platters" or plains are places at some distance from the village, the flat tops of mountains, where small plants grow. Gardens and forests are the environments most frequently exploited by the residents. Although the vast majority of native plants

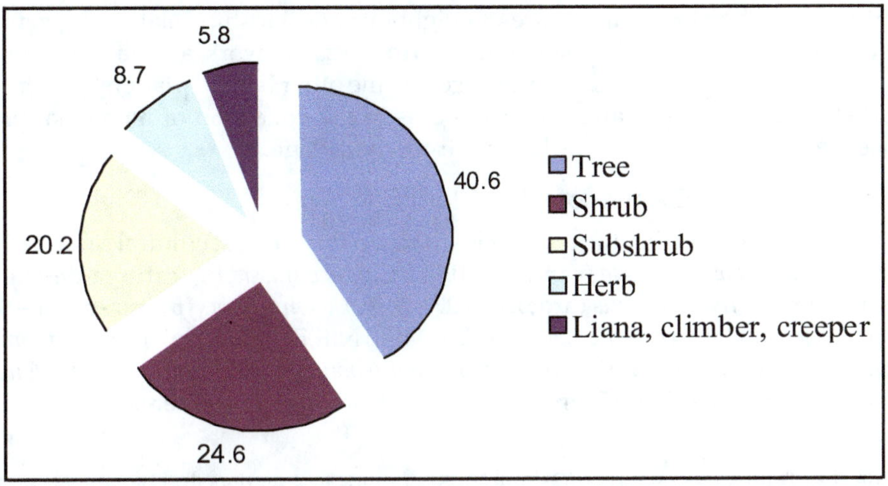

Figure 9.2: Habitat of medicinal plant species cited by subjects from the Baixa dos Quelés community, Jeremoabo, BA.

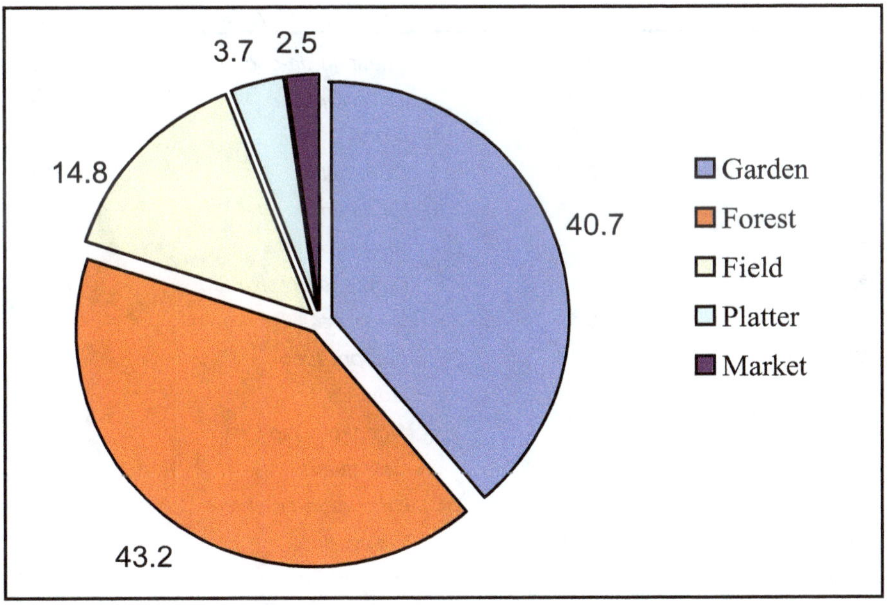

Figure 9.3: Collection environments for medicinal plants cited by specialists from the Baixa dos Quelés community, Jeremoabo, BA.

are collected in the forest, we confirmed that some are cultivated in gardens, such as, *Poincianella microphylla* (Mart.ex G. Don) L. P. Queiroz (**catingueira**), *Bauhinia subclavata* (**mororó**) and *Scoparia dulcis* L (**vassourinha**), alongside introduced plants, such as, *Chenopodium ambrisioides* L. **(mastruz)**, *Monniera trifolia* Loefl (**alfafaca**) and *Plectrantus amboinicus* (Lour) Spreng (**hortelã**). According to Amorozo (2002b), these gardens all represent a germplasm bank, where plants of local interest are disseminated, and contribute to the close-knit nature of neighbour and kinship relationships, in that they provide elements which circulate across the social network, as well as information about their employment and significance. On the other hand, "platters"/plains and markets are less sought after environments in the collection of medicinal plants, since they are situated at some distance from the village.

The Cultural Salience of the Plants

In descending order, the ten species that had the greatest cultural salience value were: *Gochnatia oligocephala* (**candeia** – 0.52), *Cordia curassavica* (**catinga-de-cheiro** – 0.38), *Schinus terebinthifolius* (**aroeira**–0.27), *Mimosa tenuiflora* (**Jurema-preta** – 0.25), *Lippia thymoides* (**alecrim**–0.24), *Lippia alba* (Mill.) N.E.Br. (**cidreira** – 0.19), *Anacardium ocidentale* L. (**cajueiro** – 0.17), *Sideroxylum obtusifolium* (**quixabeira** – 0.16), *Cupania oblongifolia* (**camboatá** – 0.15) and *Miconia ferruginata* DC. (**barbatemão** – 0.15). (Table 9.3).

Table 9.3: Total salience index (Quilan 2005) of medicinal plants cited by specialists from the Baixa dos Quelés community, Jeremoabo, Bahia.

Vernacular name (Portuguese)	*Salience Index*	*Vernacular Name (Portuguese)*	*Salience Index*
Candeia	0.52	*Caatinga-de-cheiro*	0.38
Aroeira	0.27	*Jurema-preta*	0.25
Alecrim	0.24	*Cidreira*	0.19
Cajueiro-branco	0.17	*Quixabeira*	0.16
Camboatá	0.15	*Barbatemão*	0.15
Pinhão-manso	0.15	*Jatobá*	0.15
Pinhão-roxo	0.14	*Umbu-de-ameixa*	0.14
Quebra-facão	0.14	*Mangaba*	0.14
Bandola	0.13	*Pau-pra-tudo*	0.12
Cambabá	0.11	*Graviola*	0.10
Mororó	0.09	*Algodão-criolo*	0.09
Sete-dor	0.09	*Catingueira*	0.09
Manacá	0.09	*Umburana-de-cheiro*	0.08
Araçá-branco	0.08	*Vassourinha*	0.08
Sucupira	0.08	*Malva-de-jegue*	0.08
Cedro	0.07	*Girassol*	0.07
Braúna	0.07	*Genipapo*	0.07
Mostarda	0.07	*Crista-de-galo*	0.07
Barbatenã-branco	0.07	*Pau-ferro*	0.06
Moleque-duro	0.06	*Gergelim*	0.06
Mangueira	0.06	*Capim-santo*	0.06
Cipó-de-grajaú	0.06	*Guela-de-camaleão*	0.05

Contd...

Table 9.3–*Contd...*

Vernacular name (Portuguese)	Salience Index	Vernacular Name (Portuguese)	Salience Index
Mameleiro	0.05	*Acançu*	0.05
Unha-de-gato	0.05	*Corona*	0.04
Endio	0.04	*Chacuretá*	0.04
Pereiro	0.03	*Boldo*	0.03
Candeia-branca	0.03	*Pulga-de-batata*	0.03
Erva-doce	0.03	*Catuaba*	0.03
Jurubeba	0.03	*Mamona*	0.02
Eucalipto	0.02	*Velame*	0.02
Pinha	0.02	*Babosa*	0.02
Jequirizeiro	0.02	*Quina-quina*	0.01
Hortelã	0.01	*Vassourinha-de-botão*	0.01
Pau-d'arco-roxo	0.01	*Velandinho*	0.01
Alfavaca	0.01	*Algaroba*	0.01
Araçá	0.01		

G. oligocephala and *C. curassavica* were the only species that received more than 50 per cent citation frequency (75 and 58.3 per cent). *G. oligocephala* was also the species of greatest cultural value in the work conducted by Almeida and Bandeira (2010) in the Baixa dos Quelés community, taking into account various use categories. In the community studied, *G. oligocephala* or "**candeia**" is principally used and indicated for the treatment of gastritis. As well as being medicinal, "**candeia**" is also used for fuel, construction and rituals and is recognized as a melliferous plant.

In our study, of the ten species of greatest cultural salience, four (*Gochnatia oligocephala, Schinus terebinthifolius, Lippia alba* and *Anacardium ocidentale*) were considered species of greatest cultural salience in the studies conducted by Gomes (2010) in the quilombola community of Casinhas, located approximately 50 km from the community studied here. Approximately thirty-eight percent (37.93 per cent) of the medicinal plants inventoried by that author were recorded in this study.

Of the ten species of greatest cultural salience, all had the highest citation frequency, except for *Anacardium ocidentale* (**cajueiro)** and *Sideroxylum obtusifolium* (Roem. Et. Sch.) (**quixabeira)**, which had an order of frequency below species such as *Cupania oblongifolia* Mart. (**camboatá**), *Pisonia laxa* Netto (**bandola**,) *Jatropha gossypiifolia* L. (**pinhão-roxo)** and *Croton* sp. (**quebra-facão**). This difference is due to the citation order in which the plant is mentioned, which therefore influenced the final value of the cultural salience index.

Figure 9.4 presents the salience index for the species cited by both the men and the women. In descending order: *Gochnatia oligocephala* (**candeia** – 0.46), *Schinus terebinthifolius* (**aroeira** – 0.41), *Mimosa tenuiflora* (**jurema-preta** – 0.36), *Sideroxylum*

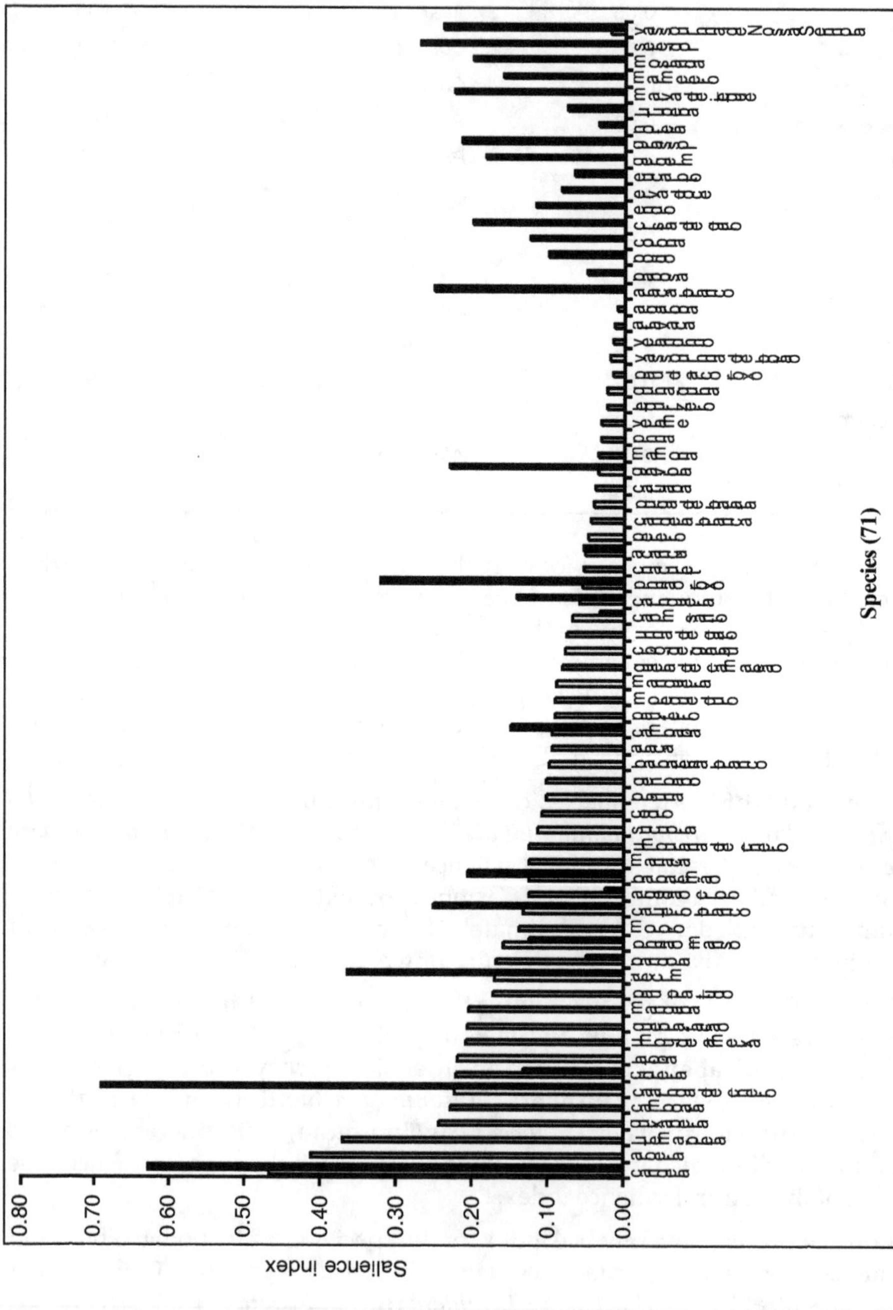

Figure 9.4: Salience index for medicinal plants as cited by men (white) and women (black) from the community of Baixa dos Quelés, Jeremoabo, BA.

obtusifolium (**quixabeira** – 0.24), *Cupania oblongifolia* (**camboatá** – 0.23), *Cordia curassavica* (**catinga-de-cheiro** – 0.22), *Lippia alba* (**cidreira** – 0.22), *Hymenaea courbaril* (**jatobá** – 0.21), *Ximenia americana* (**umbu-de-ameixa** – 0.20) and *Croton* sp (**quebra-facão** – 0.20) were the species of greatest cultural salience mentioned by the men.

Similarly, *Cordia curassavica* (**catinga-de-cheiro** – 0.68), *Gochnatia oligocephala* (**candeia** – 0.62), *Lippia thymoides* (**alecrim**–0.14), *Jatropha gossypiifolia* (**pinhão-roxo** – 0.32), Sp6 (**sete-dor** – 0.26), *Mitranthes gardneriana* (**araçá-branco** 0.25), *Anacardium occidentale* (**cajueiro-branco**- 0.31), *Scoparia dulcis* (**vassourinha de Nossa-Senhora**– 0.23), *Annona muricata* (**graviola** – 0.23), *Sida cordifolia* (**malva-de-jegue** – 0.22) were those species most often mentioned by the women. *Schinus terebinthifolius, Anacardium occidentale, Miconia ferruginata* and *Sida cordifolia* have therapeutic properties recognized by the community in combating diseases linked to the female reproductive system. The preparation of these plants generally occurs through "seated bathing" and is frequently associated with more than one plant. The women mentioned fewer medicinal plants than the men (36.8 per cent and 63.2 per cent, respectively), although many fewer women (4) were interviewed than men (8). Different results were found by Camou-Guerrero *et al.* (2008) when studying knowledge and value use of plants between men (25) and women (34) in the Rarámuri community in Mexico, when examining the category of medicinal plants. For this category, the women cited 54 per cent and the men 46 per cent.

Some authors have ascertained to different aspects related to men and women in the cultural domain of use of plants (Vieira *et al.*, 2008; Camou-Guerrero, 2008; Amorozo, 1996; Borba and Macedo, 2006; and Colaço, 2006). These differences of knowledge and perception are explained as a consequence of the gender-based division of labour found in traditional societies (Camou-Guerrero *et al.*, 2008). When Almeida and Bandeira (2010) studied the cultural significance of plants in quilombola communities, they did not find significant differences, although they were able to observe that differences exist in the acquisition of knowledge about plant resources, according to men and women's production activities.

The data presented in this study reveals that the Baixa dos Quelés community depends directly on local plant resources to cure infirmities of body and soul. Native species were those that had the greatest cultural salience, while, according to residents, some of those species cultivated in the gardens and fields are not readily available in the location, as is the case with *Cedrela fissilis* (**cedro**), *Commiphora leptophloeos* (**umburana de cheiro**), *Croton* sp. (**quebra-facão**) and *Sideroxylum obtusifolium* (**quixabeira**); these thus merit attention in programmes aimed at managing and conserving the Caatinga. The knowledge that these traditional communities have regarding medicinal plants may be considered in public health policies aimed at meeting the needs of these populations, thereby valuing their culture and disseminating their knowledge.

Acknowledgements

To the post-graduate course in Botany at the State University of Feira de Santana, BA, to which this research study is affiliated. To the Bahia State Research Foundation (*Fundação de Amparo a Pesquisa do Estado da Bahia*: FAPESB) for granting the Doctoral scholarship. To the local specialists from the Baixa dos Quelés community, for their help and kindness in contributing to this research.

References

Abreu, J.R. (2000). Diversidade de recursos vegetais do cerrado utilizados pelos quilombolas Mimbó (Amarante, Piauí, Brasil). *Dissertação de Mestrado-Universidade Federal de Pernambuco*, Recife, p. 68

Albuquerque, U.P., and Lucena, R.F.P. (2004). Métodos e técnicas de pesquisa etnobotânica. Recife: Ed. Livro Rápido/NUPEEA, 189p.

Albuquerque, U.P.de., and Andrade, L.H.C. (2002). Conhecimento botânico tradicional e conservação em uma área de Caatinga no estado de Pernambuco, Nordeste do Brasil. *Acta Botânica Brasílica*, **16**(3): 273-285.

Albuquerque, U.P.de. (2001). The use of medicinal plants by the cultural descendants of African people in Brazil. *Acta Farmaceutica Bonaerense*, **20**(2): 139-144.

Almeida, C.F.C.B.R., Amorim, E.L.C., Albuquerque, U.P., and Maia, M.B.S. (2006). Medicinal plants popularly used in the Xingó region–a semi-arid location in Northeastern Brazil. *Journal of Ethnobiology and Ethnomedicine*, **2**(15): 1-9.

Almeida, V.S., and Bandeira, F.P.S.F. (2010). O significado cultural do uso de plantas da caatinga pelos quilombolas do Raso da Catarina, município de Jeremoabo, Bahia, Brasil. *Rodriguésia*, **61**(2): 195-209.

Amorozo, M.C.M. (2002a). Agricultura Tradicional, Espaços de Resistência e o Prazer de Plantas. In: Albuquerque, U.P. et al. (orgs.), *Atualidades em Etnobiologia e Etnoecologia*. Recife: Sociedade Brasileira de Etnobiologia e Etnoecologia, pp. 123-131.

Amorozo. M.C.M. (2002b). Diversidade de plantas medicinais em Santo Antônio do Leverger, MT, Brasil. *Acta Botânica Brasílica*, **16**(2) : 189-203.

Amorozo, M.C.M. (1996). A abordagem etnobotânica na pesquisa de plantas medicinais. *In*: Di Stasi L. C. (org.). Plantas medicinais: arte e ciência–Um guia de estudo interdisciplinar. São Paulo, Editora da Universidade Estadual Paulista, pp. 47-68.

APG III (2009). An update of the Angiosperm Phylogeny Group classi cation for the orders and families of owering plants: APG III. *Botanical Journal of the Linnean Society*, **161**: 105-121.

Borba, A.M., and Macedo, M. (2006). Plantas medicinais usadas para saúde bucal pela comunidade do bairro Santa Cruz, Chapada dos Guimarães, Mato Grosso, Brasil. *Acta Botanica Brasílica*, **20**(4): 771–782.

Borgatti, S.P. (1992). ANTHROPAC 4,0 Methods Guide Natick, MA: Analytic Technologies. Brummitt, R. K. and Powell, C. E. (Eds). 1992. Authors of Plant Names: a list of authors of scientific names of plants, with recommended standard forms of their names, inclunding abbreviations. Royal Botanic Garden Kew, London, p. 732.

BMH – Brazilian Ministry of Health (2001). Secretaria de Políticas de Saúde. Proposta de política nacional de plantas medicinais e medicamentos fitoterápicos. Brasília.

Camargo, M.T.L.A. (2007). Contribuição etnofarmacobotânica ao estudo de Petiveria alliacea L.–Phytolacaceae ("amansa-senhor") e a atividade hipoglicemiante relacionada a transtornos mentais. *Rodriguésia*, **23**(1): 21-27.

Camou-Guerrero, A., Reyes-García A.V., Martínez-Ramos, M., and Casas, A. (2008). Knowledge and use value of plant species in a Rarámuri community: a gender perspective for conservation. *Human Ecology*, **(36)**: 259-272.

CEI (1994). Centro de Estatística e Informacões-básicas dos municípios baianos: região nordeste. *Salvador*, *11*: 35-37.

Colaço, M.O. (2006). Significado cultural do uso e manejo de plantas da caatinga pelos índios Pankararé do Raso da Catarina/BA. Dissertação. Universidade de Feira de Santana. Feira de Santana, 94p.

Franco, E.A.P., and Barros, R.F.M. (2006). Uso e diversidade de plantas medicinais no Quilombo Olho D'água dos Pires, Esperantina, Piauí. *Revista Brasileira de Plantas Medicinais*, (8): 78-88.

Gomes, E.C.S., Vilar, F.C.R., Perz, J.O., Vilar, R.C., Freire, J.L.O., Lima, A.N., and Dias, T. J. (2008). Plantas da caatinga de uso terapêutico: Levantamento etnobotânico. *Engenharia Ambiental*, **5**: 74–85.

Gomes, T.B. (2010). Uso e diversidade de plantas medicinais em uma comunidade quilombola Casinhas no Raso da Catarina- BA. Monografia. Universidade Estadual da Bahia.

Jorquera, C.S. (1993). Utilizacion industrial de plantas medicinales. Workshop presented in Unido in Latin america, Panajachel, Guatemala.

Júnior, J.P.A., Yala -O Suna, A.T.J., Queiroz, S.R.O. D., and Rios, A.P. (2005). Levantamento etnobotânico e etnofarmacológico de plantas medicinais do município de Itaberaba -BA para cultivo e preservação. Sitientibus Série Ciências Biológicas, **5(1):** 39-44.

Kinghorn, A.D. (1994). The discovery of drugs from higher plants in: Gullo, V. P. (ed.), *The discovery of natural products with therapeutic potential*. Butterworth-Heinemann, Boston, M.A, pp. 81-108.

Loiola, M.I.B., Paterno, C.G.B., Diniz, J.A., Calado, J.F., and Oliveira, A.C.P. (2010). Leguminosas e seu potencial de uso em comunidades rurais de São Miguel do Gostoso – RN. Revista Caatinga, **23(3)**: 59-70.

Lorenzi, H., and Matos, F.J.A. (2008). Plantas medicinais no Brasil: Nativas e exóticas. 2ª ed. Nova Odessa: Instituto Plantarum de Estudos da Flora LTDA, 544 p.

Monteles, R., and Pinheiro, C.U.B. (2007). Plantas medicinais em um quilombo maranhense: uma perspectiva etnobotânica. *Revista de Biologia e Ciências da Terra*, **7**(2): 38-48.

Morais, S.M.de., Dantas, J.D.P., Silva, A.R.A.da., and Magalhães, E.F. (2005). Plantas medicinais usadas pelos índios Tapebas do Ceará. *Revista Brasileira de Farmacognosia*. **15**(2): 169-177.

Moreira, D.L., and Neto-Guarim, G. (2009). Usos múltiplos de plantas do cerrado: um estudo etnobotânico na comunidade sítio Pindura, Rosário Oeste Mato Grosso Brasil. *Polibotânica*, **27**: 159-190.

Patzlaff, R.G. (2007). Estudo etnobotânico de plantas de uso medicinal e místico na comunidade da Capoeira Grande, Pedra de Guaratiba. Dissertação de mestrado, Instituto de Pesquisas Jardim Botânico, Rio de Janeiro. 124p.

Pinto, E.de P.P., Amorozo, M.C.deM., and Furlan, A. (2006). Conhecimento popular sobre plantas medicinais em comunidades rurais de mata atlântica – Itacaré, BA, Brasil· *Acta Botânica Brasílica*, **20(4)**: 751-762.

Quinlan, M. (2005). Considerations for collecting Freelists in the Field: Examples from Ethnobotany. *Field Methods*, **17**: 1-16.

SEBRAE (1995). Secretaria da Indústria Comércio e Turismo/Serviço de Apoio as Micro e Pequenas Empresas da Bahia. Diagnóstico de municípios: Micro-área de Paulo Afonso e Jeremoabo. SEBRAE, Salvador, p. 111.

Silva, I.M., and Peixoto, A.L. (2009). O abajurú (*Chrysobalanus icaco* L. e *Eugenia rotundifolia* Casar.) comercializado na cidade do Rio de Janeiro, Brasil. *Revista Brasileira de Farmacognosia*. **19(1B)**: 325-332.

Silva, R.B.L. (2002). A etnobotânica de plantas medicinais da comunidade quilombola de Curiaú, Macapá-AP, Brasil. Dissertação de Mestrado em Agronomia – Universidade Federal Rural da Amazônia, Belém, p. 172.

Trotter, R.T., and Logan, M.N. (1996). Informant Consensus: a New Approach for Identifying Potentially Effectice Medicinal Plants. In: Etkin, N. L. (Ed.). Indigenous Medicine and Diet: biohevioral approaches. Redgrave Bedford Hills, New York, p. 91-112.

Vieira, F.J., Santos, L.G.P., Barros, R.F.M., and Araújo, J.L.L. (2008). Quilombola of Macacos community, São Miguel do Tapuio City, Piauí State: History, use and conservation of plant resourses. *Funtional Ecosystems and Communities*, **2**: 81-87.

WHO (2000). World Health Organization. Classificação estatística internacional de doenças e problemas relacionados à saúde. Editora da Universidade de São Paulo, São Paulo.

Traditional and Folk Herbal Medicine: Recent Researches Vol. 2 (2014) *Pages* **273–357**
Editor: **V.K. Gupta**
Published by: **DAYA PUBLISHING HOUSE, NEW DELHI**

10

Fabaceae Family Medicinal Plants Used in Folk Medicinal Practices of Bangladesh for Treatment of Cancer

Rownak Jahan[1], Taufiq Rahman[2] and Mohammed Rahmatullah[1*]

ABSTRACT

Bangladesh has a number of traditional medicinal systems like Ayurveda, Unani, homeopathy, and folk medicine, which has been practiced in this region for centuries. All these traditional medicinal systems claim to cure cancer, a disease affecting millions of people worldwide, and against which allopathic treatments through chemotherapy, radiotherapy or surgery can alleviate the disease at the primary stages but even then with severe side-effects. It is thus always of interest to find novel drugs, which can prove efficacious in the treatment of the disease with lesser or no adverse reactions on the human body. Plants have always been a source of effective anti-cancer drugs, of which some examples may be found in vincristine, vinblastine, and taxol. Folk medicinal system of Bangladesh relies primarily on medicinal plants for treatment of cancer, and most of these plants are largely unexplored in terms of relevant scientific research. In ethnomedicinal surveys conducted over the last several years, it was observed that folk medicinal practitioners (Kavirajes) use a total of 164 plant species for treatment of cancer. Among them, the Fabaceae family contributed the majority of species with 16 plant species. A perusal of scientific literature suggests that the Fabaceae family plant species can be potential sources of novel anti-cancer drugs. Some instances of new anti-cancer

1 Faculty of Life Sciences, University of Development Alternative, House No. 78, Road No. 11A (new), Dhanmondi R/A, Dhaka-1205, Bangladesh

2 Department of Pharmacology, University of Cambridge, Tennis Court Road, CB2 1PD, Cambridge, UK

* *Corresponding author*: E-mail: rahamatm@hotmail.com

compounds isolated from the Fabaceae family plants can be found in examples of milletinol, isobavachalcone, and xanthoxyletin, suggesting that other Fabaceae family plants can also yield novel anti-cancer compounds. In fact, activities against a number of cancer cell lines has been described for Fabaceae family species like Retama monosperma (l.) Boiss., Tephrosia purpurea (L.) Pers., and Indigofera cassioides Rottler ex DC., to name only a few. This chapter shall focus on the phytochemical constituents, anti-cancer bioactivity, and therapeutic potential of the 16 Fabaceae family anti-cancer plants used by Kavirajes of Bangladesh. These 16 plant species include Abrus precatorius, Acacia arabica, Acacia catechu, Acacia nilotica, Bauhinia variegata, Cassia angustifolia, Cassia fistula, Cassia sophera, Desmodium gangeticum, Erythrina variegata, Indigofera tinctoria, Lens culinaris, Mimosa pudica, Samanea saman, Sesbania sesban, and Vicia faba.

Keywords: Anti-cancer, Fabaceae, Folk medicine, Bangladesh.

Introduction

Cancer is not only one of the major public health problem in many parts of the world, but also a disease that is dreaded by the world population because of the difficulties in curing this disease, particularly in the advanced stages. Even in the United States, it has been projected that a total of 1,638,910 new cancer cases and 577,190 deaths from cancer will occur during 2012 (Siegel *et al.*, 2012). Cancer has been defined as diseases in which abnormal cells divide without control and are able to invade other tissues. Cancer cells can spread to other parts of the body through the blood and lymph systems. There are more than 100 different types of cancer, and each cancer is named for the organ in which it starts, *e.g.*, cancer that occurs in the lungs first is known as lung cancer. The National Cancer Institute of USA has categorized different cancers into several broad categories as shown below.

- ☆ **Carcinoma**–cancer that begins in the skin or in tissues that line or cover internal organs.
- ☆ **Sarcoma**–cancer that begins in bone, cartilage, fat, muscle, blood vessels, or other connective or supportive tissue.
- ☆ **Leukaemia**–cancer that starts in blood-forming tissue such as the bone marrow and causes large numbers of abnormal blood cells to be produced and enter the blood.
- ☆ **Lymphoma and myeloma**–cancers that begin in the cells of the immune system.
- ☆ **Central nervous system cancers**–cancers that begin in the tissues of the brain and spinal cord.

Essentially, under normal body conditions, cells of various tissues proliferate in a controlled manner, *i.e.*, they only undergo division and proliferation when necessary (for example to replace dead or damaged cells, or during the growth phase of any living being, or due to signal transduced by a growth factor or a particular stimuli). Under normal cell growth, cells stop dividing when it reaches a certain point when new cells are no longer required. When cells undergo abnormal proliferation, they gain the ability of uncontrolled growth. This uncontrolled growth can be due to genetic mutations, exposure to chemicals, or a variety of other factors including family history, lifestyle and food habits. Under these conditions, a particular organ or tissue of the body may experience abnormal proliferation of cells in that organ, thus creating a lump or tumor in the initial phase. The lump can grow and remain in the same place or can spread through the blood and lymph systems to other parts of the body and continue their abnormal proliferation in that part also. When this happens, this condition is called metastasis and is more difficult to cure or incurable than cancer in one particular organ of the body. Virtually all cancers, including cancers of the blood and the lymphatic system (leukemia, multiple myeloma, and lymphoma), can form metastatic tumors. Although rare, the metastasis of blood and lymphatic system cancers to the lungs, heart, central nervous system, and other tissues has been reported. However, even some cancers which are limited to a particular organ, like pancreatic cancer has no known cure in allopathic medicine.

At present, the various allopathic methods for treatment of cancer include chemotherapy, (where the disease is treated with one or a combination of anti-cancer drugs), radiotherapy (where radiation is directed to cancerous areas of the body to destroy cancer cells), and surgery (like mastectomy, where one or both breasts are removed partially or completely if the person has breast cancer). However, all cancer treatments have serious side-effects which may include from nausea and vomiting to serious weakening of the body and the body's defenses, leading to rise of opportunistic infections. For instance, common chemotherapy agents act by killing cells that divide rapidly, which is one of the main properties of most cancer cells. This means that chemotherapy also harms normal cells that divide rapidly under normal circumstances. Examples of these cells are cells in the bone marrow, digestive tract, and hair follicles. Damage to these cells lead to most common side-effects of chemotherapy: myelosuppression (decreased production of blood cells, hence also immunosuppression), mucositis (inflammation of the lining of the digestive tract), and alopecia (hair loss). Possible short and long-term side-effects of radiotherapy may include tiredness, skin reactions, loss of appetite, hair loss, infertility, lymphoedema, shortness of breath, and difficulty in swallowing. Surgery can lead to permanent amputation of an organ (like a limb) with resultant loss of functions of that organ along with emotional disturbances that may accompany loss of an organ.

Because of the adverse side-effects of allopathic methods of cancer treatment, plus the fact that not all cancers, or cancers that have passed a particular stage cannot be cured through allopathic methods, attention has recently re-focused on alternative methods of therapy. Complementary or alternative methods of therapy have existed for centuries (possibly from the earliest stages of human evolution) before the advent of modern or allopathic medicine, and co-exist even today, side-by-side with allopathic treatment. Examples of such alternative therapy exist in almost all countries. In the Indian sub-continent (of which Bangladesh is a part), various forms of alternative therapy have existed for hundreds of years. The four most common forms of such therapy in Bangladesh are Ayurveda, Unani, homeopathy, and folk medicine.

Folk Medicinal Practice in Bangladesh

Of the four major traditional medicinal practices of Bangladesh, folk medicinal practice is the most common as the practitioners do not have to go through any regulatory procedures to establish their practice. Ayurvedic, Unani and homeopathy practitioners have to undergo a period of training in a Government-approved institution with a Government approved course curricula, and pass through regularly conducted examinations and obtain certificates, before a Government approved regulatory body grants them license to practice. As such, the procedure is similar to that undergone by allopathic physicians. Following obtaining of licenses, an Ayurvedic or a homeopathic practitioner gets the title of Doctor, while a Unani practitioner gets the title of Hekim or Hakim. A few years previously, an Ayurvedic practitioner would get the title of Kaviraj, but in the last few years this has changed.

Folk medicinal practitioners, on the other hand do not have to undergo any formal training in any Government approved institution, nor do they have their own course curricula or formulary. Practically, anybody can indulge in folk medicinal practices. These practitioners mainly rely on medicinal plants for treatment, which is quite similar to Ayurveda or Unani, but differs from these two traditional practices in simplicity. Folk medicinal practitioners, otherwise known as Kavirajes or Vaidyas use simple decoctions, paste, juice, or other simple preparations of medicinal plants in their treatment, which preparations they make with their own hands at their homes, and with plants collected by them. This is quite unlike Ayurveda or Unani or even homeopathy, where formulations can be quite complex and involve a multitude of preparation processes like fermentation, distillation, dilution or calcinations. Ayurvedic and Unani medicines are not usually prepared by the practitioners themselves; rather there are pharmaceutical companies, which are specialized in preparations of Ayurvedic or Unani formulations and which are inspected on a regular basis by special regulatory bodies as to the uniformity of formulations from batch to batch, and conformity to Ayurvedic and Unani pharmacopoeias.

Folk medicinal practitioners (hereinafter referred to as Kavirajes) are the most numerous among all the other type of medical practitioners in Bangladesh, including allopathic doctors. They mostly practice in the villages and at least one or two can be found in each of the 86,000 villages of the country. They also practice in the towns and cities. Their major clientele consists of the rural population and the urban poor, who either lack access or affordability to visit allopathic doctors, clinics or hospitals. As such, Kavirajes form the first tier of providing primary health-care in Bangladesh. More often than not, a Kavirajes practice is a family practice and passed on to successive generations, where a son or a daughter may pick up the knowledge of medicinal plants from his or her parents. However, occasionally Kavirajes have apprentices, who stay with the Kaviraj for a number of years till the apprentice believes that he or she has picked up enough knowledge on medicinal plants to be able to practice alone.

As mentioned earlier, anybody can take up the Kaviraj profession without undergoing any formal training. As such, the question arises as to how and where did the Kaviraj acquire his medicinal plant knowledge in the first place? There are no clear cut answers. In the last few years, we had been conducting extensive ethnomedicinal surveys among the various Kavirajes of Bangladesh in practically all districts of the country. Our survey has also included tribal medicinal practitioners (TMPs), for tribal medicine can be considered also as folk medicine, and in reality the practicing mode of a TMP does not appreciably differ from a Kaviraj in the sense that both utilizes simple preparations of a plant or plant part to cure ailments. A number of such surveys has been published, a few of which have been referenced in this chapter (Nawaz *et al.*, 2009; Rahmatullah *et al.*, 2009a-c; Hasan *et al.*, 2010; Hossan *et al.*, 2010; Mollik *et al.*, 2010a,b; Rahmatullah *et al.*, 2010a-g; Haque *et al.*, 2011; Jahan *et al.*, 2011; Rahmatullah *et al.*, 2011a,b; Sarker *et al.*, 2011; Rahmatullah *et al.*, 2012a-d). Various Kavirajes have given different answers to the above query as to their source of medicinal plant knowledge. Some has mentioned that Kaviraj practice has existed in his or her family for generations and they have acquired it from their father or mother or a close relative; others have obtained such knowledge from their 'guru' with whom they have apprenticed for several years; others have obtained such knowledge from reading books (mainly Ayurvedic books written for the lay person or written in a very comprehensible manner); yet others have obtained plants, and diseases cured by the plants in dreams; some have experimented with plants on domestic animals and later tried it on human beings or have seen animals consuming such and such plant during distress (mostly for gastrointestinal disorders, aches, fever, and rheumatism); a few have obtained such knowledge from hearsay; and still others use medicinal plants, which they believe resemble the ailment treated to a certain extent, for example, red flowers of *Delonix regia* (Bojer) Raf. or *Hibiscus rosa sinensis* L. are administered orally for treatment of excessive bleeding during menstruation. Since blood has a red color, it is believed by the Kavirajes that administration of red flowers can compensate for the blood loss (kind of compensation for the redness lost by the body). Two things have resulted from all these variations in how a Kaviraj selects plants for ailment treatment. The first is that even two Kavirajes practicing in even the same village very rarely uses the same plant for treatment of a common ailment. This leads to a selection of a wide diversity of plants, even though all plants may be growing in the same region. This fact has been noticed in our over 200 ethnomedicinal surveys conducted among the Kavirajes of Bangladesh. The second fact, as obtained from our surveys, is that while the vast majority of the Kavirajes really have a quite extensive knowledge on the medicinal properties of at least their local flora, a few indulge in nothing but quackery.

Kavirajes have neither at their disposal nor use modern diagnostic procedures to detect cancer. So naturally the question arises as to how do the Kavirajes detect cancer? First of all, our interviews with hundreds of Kavirajes show that cancer and diabetes are two terms that Kavirajes have picked up from allopathic doctors and the general public. More frequently, a patient whose cancer has been determined by allopathic doctors as terminal cancer resorts to Kavirajes as their last hope. So Kavirajes learn that a patient has cancer after it has been diagnosed by allopathic doctors. However, many Kavirajes detect cancer by the following symptoms: presence of a visible lump or swelling, which does not go away following treatment by the Kaviraj but increases in size with time, presence of multiple illnesses, rapid wasting away of a patient's body and health, and finally any ailment not recognized by the Kaviraj. A number of Kavirajes (particularly those who have read Ayurvedic treatises) detect cancer according to the symptoms as described in classical Ayurvedic texts. Charaka and Sushruta Samhitas, two classical Ayurvedic texts, have described cancer as inflammatory or non-inflammatory swelling and have mentioned them as either 'Granthi' (minor neoplasm) or 'Arbuda' (major neoplasm). Ayurvedic classification of neoplasm depends on the status of the three body-control systems, namely 'Vata' (air or the nervous system), 'Pitta' (fire or the venous system), and 'Kapha' (water or the arterial system). One or two of the three body-control systems may be out of control in benign neoplasm, while all three body-control systems get out of control in major neoplasm (Balachandran and Govindarajan, 2005). From this perspective, Ayurveda has classified cancer into three major groups. Group 1 consists of diseases that can be classified as clear malignancies and which includes arbuda and granthi; examples of such diseases are 'mamsarbuda' (melanoma), 'raktarbuda' (leukaemia), and 'mukharbuda' (oral cancer). Group 2 diseases can consist of diseases like incurable ulcers ('tridosaj gulmas'), which means abdominal tumors like carcinomas of the stomach and liver or lymphomas. Group 3 diseases are diseases which can lead to cancer like 'visarpa' (erysipelas), 'asadhya kamala' (incurable jaundice) and 'nadi vrans' (sinusitis) (Prasad, 1987; Singh, 2002).

It may seem strange to an outsider not familiar with the practices of a Kaviraj as to how a Kaviraj can detect and treat cancer without a proper diagnosis as per modern clinical procedures. However, our ethnomedicinal surveys have shown that Kavirajes do treat cancer with medicinal plants, and considering the number of cancer patients who visit the Kavirajes, it is a realistic view to assume that the Kavirajes must be having some success in their treatment, however limited that success can be. It is also strange that most Kavirajes treat cancer without diagnosing which organ of the body that the cancer has affected. Cancer is treated as a whole, and no distinctions are made, say as if the cancer is a small cell lung cancer or a chloangiocarcinoma. Nevertheless, as will be discussed in the rest of this chapter, most of the medicinal plants used by the Kavirajes of Bangladesh for treatment of cancer (irrespective of organ) has been shown in a multitude of scientific reports to have anti-cancer properties, which exist in either the whole plant or any of the phytochemicals present within the plant.

Fabaceae Family Plants Used by Kavirajes for Treatment of Cancer in Bangladesh

In our completed over 200 ethnomedicinal surveys thus far among the Kavirajes of the mainstream Bengali-speaking population of the country and among various tribes, it was observed that a total of 164 plant species were used by Kavirajes and TMPs for treatment of cancer. This review shall focus on only the Fabaceae family plants, of which 16 species were used for cancer treatment. As a group, the Fabaceae family formed the largest group among the various families of plants used for treatment of cancer within Bangladesh. Fabaceae family contains a number of plants, which have found world-wide uses in various traditional medicinal systems for treatment of cancer. Fabaceae family plant species, which are used outside Bangladesh for cancer treatment shall be reviewed to some extent in the Discussion part of this chapter. This part shall cover the scientific validation and potential therapeutic uses (based on their reported phytochemical constituents) of the Fabaceae family plants used in Bangladesh. A list of these plants is shown in Table 10.1. The 16 plant species were *Abrus precatorius, Acacia arabica, Acacia catechu, Acacia nilotica, Bauhinia variegata, Cassia angustifolia, Cassia fistula, Cassia sophera, Desmodium gangeticum, Erythrina variegata, Indigofera tinctoria, Lens culinaris, Mimosa pudica, Samanea saman, Sesbania sesban,* and *Vicia faba.*

Abrus precatorius

Phytochemical Constituents

From whole plants, the oleanene-type triterpenoids abruslactone, methyl abrusgenate and abrusgenic acid has been reported (Chiang *et al.*, 1983); also reported are two anti-tumor proteins – abrin A and abrin B (Lin *et al.*, 1978a); a flavone, abrectorin (Bhardwaj *et al.*, 1982); a tetraoxygenated flavone, and premtorin (6,7,4'-trimethoxy-3'-hydroxyflavone) (Bhardwaj *et al.*, 1988; 1989) (Figure 10.1).

Seeds are known to contain abrin and hypaphorine (Ahmad and Rahman, 1965); abrin and ricin (Lin and Tung, 1972); abrins A and C (Wei *et al.*, 1974); 5β-cholanic acid (Mandava *et al.*, 1974a); polysaccharide containing D-galacturonic acid, D-galactose, L-arabinose, L-rhamnose, and D-xylose (Akhtar *et al.*, 1972); an indole plant growth inhibitor – N,N-dimethyl-L-tryptophan (Mandava *et al.*, 1974b); flavonoids – abrectorin (I: R = R1 = H) and desmethoxycentaureidin-7-O-rutinoside (Bhardwaj *et al.*, 1980); abrin A (Herrmann and Behnke, 1981); abrins A-D (Lin *et al.*, 1981); alkaloids – trigonelline, choline, hypaphorine, and abrine (Karawya *et al.*, 1980); anthocyanins – xyloglucosyldelphinidin and (p-coumarylgalloyl)-glucosyldelphinidin (Karawya *et al.*, 1981a); 7,9,15-trimethylheneicosane, n-hentriacontane, pentacosanoic acid, α-amyrin, β-sitosterol, and campestrol (Bhaumik, 1987); abrusin (8-C-glucosylscutellarein 6,7-dimethyl ether, I) and its 2''-O-apioside (Markham *et al.*, 1988); a sapogenol – abrisapogenol J [3β-22β-dihydroxy-11-oxoolean-13(18)-ene], sophoradiol and its 22-O-acetate, hederagenin Me ester, tri-Me tryptophan dipolar ion, kaikasaponin III Me ester, abrine and its 2''-O-apioside (Kinjo *et al.*, 1991); saponins and C-glycosyl flavones – 3-O-[β-D-glucurunopyranosyl-(1→2)-β-D-glucopyranosyl]hederagenin (abrus-saponin I), 3-O-[β-D-glucuronopyranosyl-(1→2)-β-D-glucopyranosyl]oleanolic acid 28-β-D-glucopyranosyl ester (abrus-saponin II), 3-O-[β-D-glucuronopyranosyl-(1→2)-β-D-glucopyranosyl]oleanolic acid, 6-C-β-D-glucopyranosyl-4',5-dihydroxy-7,8-dimethoxyflavone (precatorin I), 6-C-[β-D-apiofuranosyl-(1→2)-β-D-glucopyranosyl]-4',5-dihydroxy-7,8-dimethoxyflavone (precatorin II), 6-C-[β-D-apiofuranosyl-(1→2)-β-D-glucopyranosyl]-4',5-dihydroxy-7-methoxyflavone (precatorin III), along with kaikasaponin I, 3-O-[β-D-xylopyranosyl-(1→2)-β-D-galactopyranosyl-(1→2)-β-D-glucuronopyranosyl] sophoradiol, kaikasaponin III, kaikasaponin III methyl ester, phaseoside IV, abrusin, abrusin 2''-O-β-D-apioside, 4',5-dihydroxy-6,7-dimethoxyflavone, 4',5,7-trihydroxy-6-methoxyflavone, abrine, and N,N,N-trimethyl tryptophan (Ma *et al.*, 1998); a flavonol glycoside, 7,3',5'-trimethoxy-4'-hydroxy flavones-3-O-β-D-galactosyl-(1→4)-α-L-xyloside (Yadava and Reddy, 2002); gallic acid and ellagic acid (Bagul *et al.*, 2005);

Table 10.1: Fabaceae family plants used in folk medicines of Bangladesh for treatment of cancer

Botanical Name	*Family*	*Local Name*	*Parts Used*	*Formulation and Mode of Administration*
Abrus precatorius L., syn. *Abrus abrus* (L.) W. Wight., *Glycine abrus* L. English: Crab's eye, Jequerity, Precatory bean, Rosary pea	Fabaceae	Josthi modhu, Laal koonch, Kuch kori	Root	Decoction of roots is orally administered.
Acacia arabica (Lam.) Willd. var. *indica* Benth., syn. *Acacia nilotica* (L.) Delile var. *indica* (Benth.) A.F. Hill, *Acacia nilotica* (L.) Willd. ex Del. subsp. *indica* (Benth) Brenan English: Indian Gum Arabic tree	Fabaceae	Babla	Gum	Gum is orally administered following mixing with water.
Acacia catechu (L. f.) Willd., syn. *Acacia catechu* (L. f.) Willd. var. *catechuoides* (Roxb.) Prain, *Acacia catechuoides* (Roxb.) Benth., *Acacia sundra* (Roxb.) Bedd., *Acacia wallichiana* DC., *Mimosa catechu* L. f., *Mimosa catechuoides* Roxb. English: Black catechu, Black cutch, Catechu, Catechu tree, Cutch tree, Khair tree (India), Wadalee gum tree	Fabaceae	Khoyer	Whole plant, flower	Decoction of whole plant is orally administered for tumors.Leaves of *Catharanthus roseus*, stems of *Eleusine coracana* and *Vitex negundo*, leaves of *Clerodendrum viscosum*, *Justicia adhatoda*, and *Myristica fragrans*, flowers of *Acacia catechu*, and fruits of *Morus alba* are macerated together and made into pills. The pills are taken daily till cure. (**Garo** tribe)
Acacia nilotica (L.) Delile syn. *Acacia arabica* (Lam.) Willd., *Acacia scorpioides* W. Wight, *Acacia vera* Willd., *Mimosa arabica* Lam., *Mimosa nilotica* L., *Mimosa scorpioides* L. English: Gun Arabic tree	Fabaceae	Babla	Leaf, gum, seed	A decoction of leaves, gum and seeds in water is orally administered.
Bauhinia variegata L. syn. *Bauhinia candida* Ait., *Bauhinia purpurea* auct. English: Butterfly tree, Mountain ebony, Orchid tree, Poor man's orchid	Fabaceae	Kanchon	Whole plant	A decoction of whole plant is orally administered.
Cassia angustifolia Vahl., syn. *Senna alexandrina* Mill., *Senna angustifolia* (Vahl) Batka, *Cassia senna* L., *Cassia acutifolia* Del. English: Senna, Arabian senna	Fabaceae	Sonamukhi	Leaf, root, fruit	A decoction of leaves, roots and fruits is orally administered.

Contd...

Table 10.1–*Contd...*

Botanical Name	*Family*	*Local Name*	*Parts Used*	*Formulation and Mode of Administration*
Cassia fistula L., syn. *Bactyrilobium fistula* Willd., *Cassia bonplandiana* DC., *Cassia excels* H.B. K., *Cassia fistuloides* Colladon, *Cassia rhombifolia* Roxb., *Cathartocarpus excelsus* G.Don, *Cathartocarpus fistula* Pers., *Cathartocarpus fistuloides* (Colladon) G.Don, *Cathartocarpus rhombifolius* G.Don English: Cassia Stick Tree, Golden rain, Golden shower, Indian Laburnum, Pudding-pipe tree, Purging Cassia	Fabaceae	Bandor lathi, shonalu	Whole plant, fruit	Decoction of whole plant is orally administered. Fruits are orally taken (**Garo** tribe).
Cassia sophera L., syn. Senna sophera (L.) Roxb. *Cassia canca* Cav., *Cassia esculenta* Roxb., *Cassia frutescens* Miller, *Cassia geminiflora* Schrank, *Cassia linearis* Michaux, *Cassia linneata* Michaux, *Cassia occidentalis* L. var. *glabra* DC., *Cassia occidentalis* L. var. *sophera* (L.)Kuntze, *Cassia patula* Aiton, *Cassia proboscidea* Pollard, *Cassia sophera* L. var. *ligustrinoides* Benth., *Cassia torosa* Cav., *Chamae fistula sophera* G.Don, *Ditremexa sophera* (L.) Britton and Wilson English: Senna, Senna sophera, African senna	Fabaceae	Helchi, thonthoni	Whole plant	Decoction of whole plant is orally administered.
Desmodium gangeticum (L.) DC., syn. *Desmodium gangeticum* (L.) DC. Var. *Maculatum* (L.) Baker, *Hedysarum gangeticum* L., *Hedysarum maculatum* L., *Desmodium natalitum* Sond., *Desmodium polygonoides* Baker English: Tick tree	Fabaceae	Shalpani	Whole plant	Decoction of whole plant is orally administered.
Erythrina variegata L. syn. *Erythrina indica* Lam., *Erythrina orientalis* (L.) Merr. English: Tiger's claw	Fabaceae	Madar, phalte madar	Seed, leaf, stem, bark, fruit	Decoction of seeds, leaves and stems, alternately barks or fruits is orally administered.

Contd...

Table 10.1–*Contd...*

Botanical Name	*Family*	*Local Name*	*Parts Used*	*Formulation and Mode of Administration*
Indigofera tinctoria L., syn. *Indigofera anil* L. var. *orthocarpa* DC, *Indigofera argentea* L., non Burm.f., *Indigofera. Bergii* Vatke, *Indigofera cinerascens* DC, *Indigofera houer* Forssk., *Indigofera indica* Lam., *Indigofera sumatrana* Gaertn., *Indigofera ornithopodioides* Schum., *Indigofera orthocarpa* (DC.) Berg. English: Black henna, Commercial indigo, Common indigo, Dyer's indigo, East Asian indigo plant, Frank indigo, Indian indigo, Indian maddar, Indigo, True indigo	Fabaceae	Neel, baita gach	Leaf, seed, fruit	Decoction of leaves, seeds and fruits are orally administered.
Lens culinaris Medik., syn. *Cicer lens* Willd., *Ervum lens* L., *Lens esculenta* Moench English: Lentil	Fabaceae	Moshur dal	Leaf, bark, seed	Decoction of leaves, barks and seeds is orally administered.
Mimosa pudica L., syn. *Mimosa hispidula* Kunth English: Shame plant	Fabaceae	Lojjaboti	Leaf, root, flower	Decoction of leaves, roots and flowers is orally administered.
Samanea saman (Jacq.) Merr., syn. *Albizia saman* (Jacq.) F. Muell., *Pithecellobium saman* (Jacq.) Benth. English: Raintree	Fabaceae	Koroi	Leaf, bark, seed	Decoction of leaves, barks and seeds is orally administered.
Sesbania sesban (L.) Merr., syn. *Aeschynomene aegyptiaca* (Pers.) Steud., *Aeschynomene elongata* Salisb., *Aeschynomene sesban* L., *Dolichos aeschynome-sesban* Forssk., *Sesbania punctata* DC., *Sesbania tchadica* A. Chev., *Emerus sesban* (L.) Kuntze, *Sesban aegyptiaca* Poiret, *Sesbania aegyptiaca* Poiret, *Sesbania confaloniana* (Chiov.) Chiov., *Sesbania pubescens* sensu auct English: Common sesban, Egyptian rattle pod, River bean	Fabaceae	Joyonti, dhonche	Leaf, bark, flower, seed	Decoction of leaves, barks, flowers and seeds is administered orally.
Vicia faba L. syn. *Faba vulgaris* Moench, *Faba sativa* Bernh., *Vicia esculenta* Salisb. English: Fava bean, Horse bean	Fabaceae	Borboti	Leaf, fruit, seed	Tumor, cancer. Decoction of leaves, fruits and seeds is orally administered.

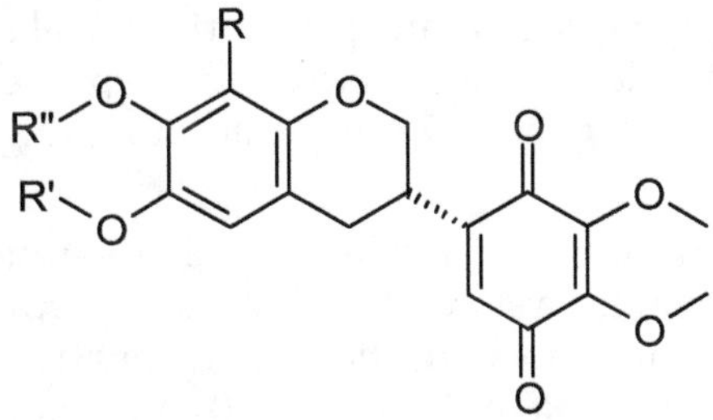

R = R''= -CH_3, R' = -H (abruquinone A)
R = R' = R''= -CH_3 (abruquinone B)
R = R''= -CH_3 ,R'= -H (abruquinone C)
R = R'= -CH_3, R'' = -H (abruquinone D)

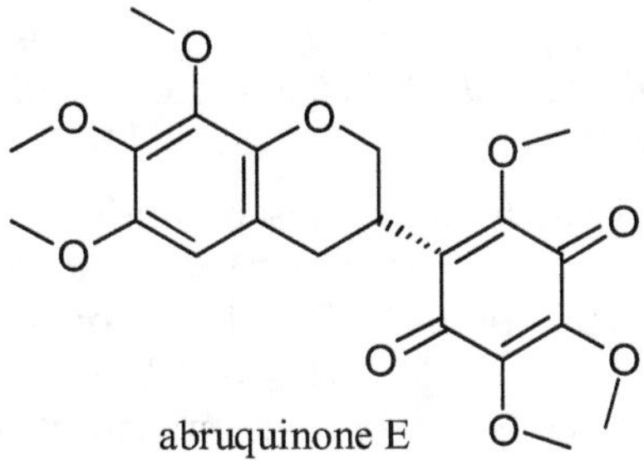

abruquinone E

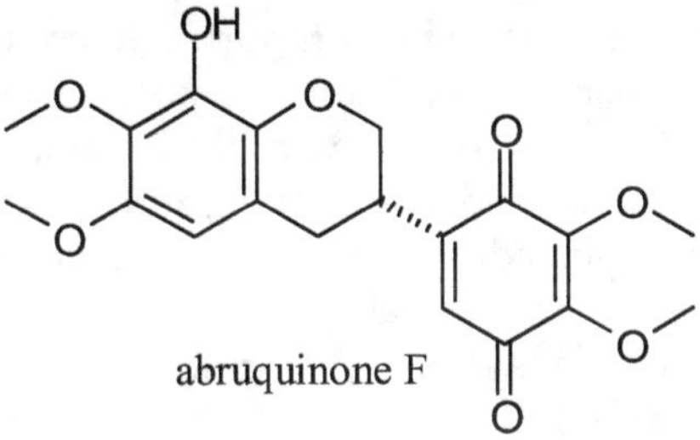

abruquinone F

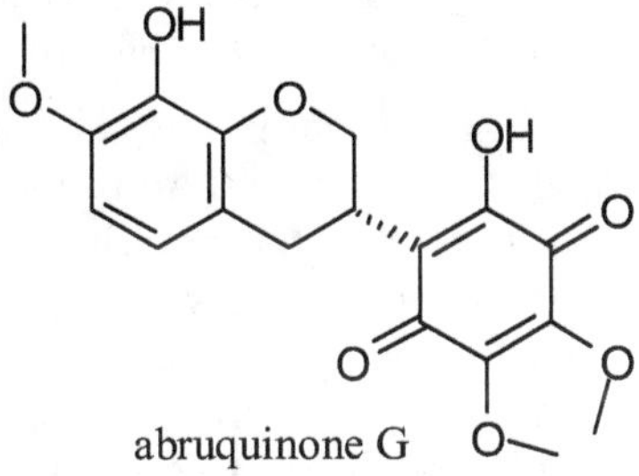

abruquinone G

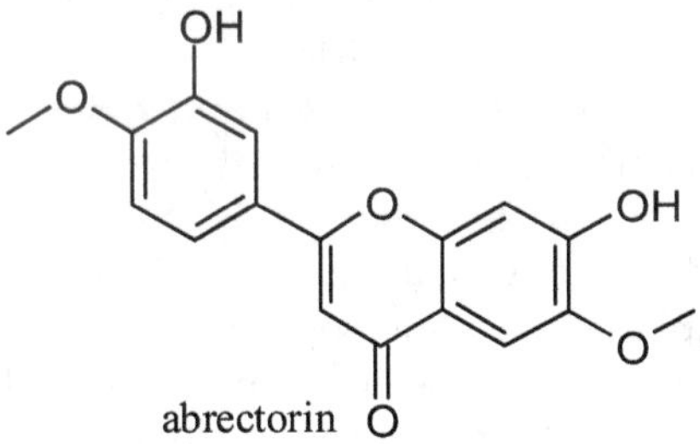

abrectorin

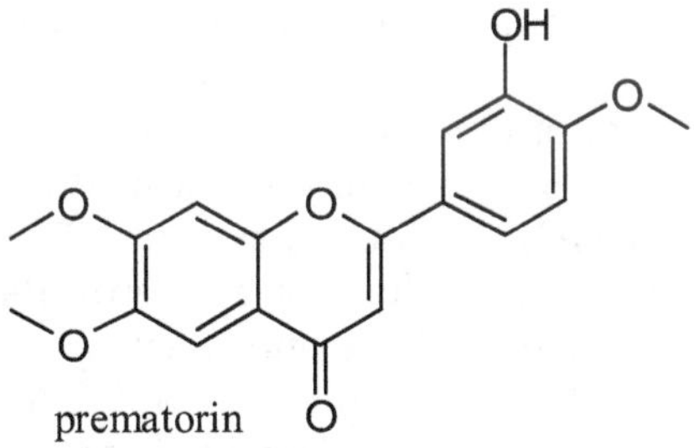

prematorin

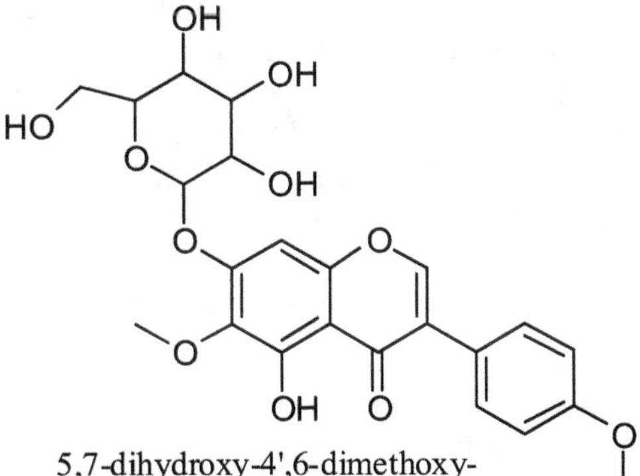

5,7-dihydroxy-4',6-dimethoxy-isoflavone 7-O-β-D-galactopyranoside

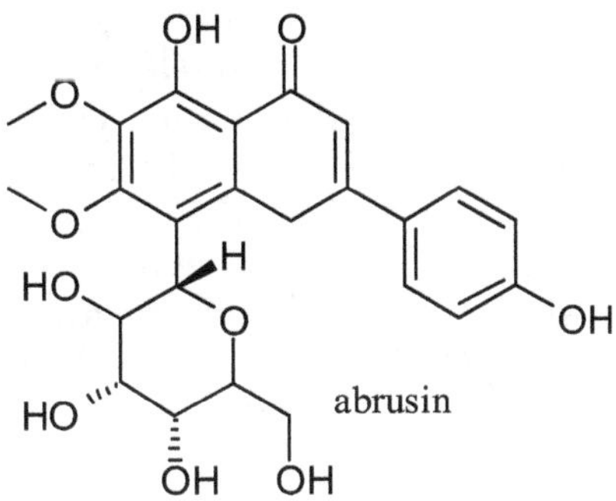

abrusin

Figure 10.1: Some flavonoids present in *Abrus precatorius*.

alkaloids – N-methyltryptophan (abrine), hypaphorine, and precatorine (Sivakumar and Alagesaboopathi, 2007); phenols – syringic acid, caffeic acid, and (-)-epicatechin (Vadivel *et al.*, 2011); and gallic acid and glycyrrhizin (Verma *et al.*, 2011) (Figures 10.2–10.5).

From roots, a polysaccharide containing glucose, xylose, and uronic acid residues has been isolated, which on fractionation yielded a starch-type glucan and an acidic xylan (Haq *et al.*, 1973). Also reported are abruquinones (isoflavanquinones I) (R = H, R1 = Me; R = OMe, R1 = Me, H) (Lupi *et al.*, 1979); steroids – abricin (I) and abridin (II) as well as cycloartenol, α-amyrin, campesterol, cholesterol, stigmasterol, β-sitosterol, and squalene (Siddiqui *et al.*, 1978); a triterpenoid sapogenin, abruslactone A (Chang *et al.*, 1982); isoflavanquinones – abruquinones A, B, D, E, and F (Kuo *et al.*, 1995); and an isoflavone, [5,7-dihydroxy-4′,6-dimethoxyisoflavone 7-O-β-D-galactopyranoside] (Saxena and Sharma, 1999) (Figure 10.2).

Leaves and aerial parts (*i.e.*, leaves and stems) have been reported to contain glycyrrhizin (Akinloye and Adalumo, 1981); abruslactone A (oleanene-type triterpene) (Chang *et al.*, 1982); flavonoids – vitexin (5,7,4-trihydroxyflavan glycoside), 7,4-dihydroxyflavonol diglycoside (liquiritigenin 7-mono- and diglycoside), and taxifolin-3-glucoside (El-Gengaihi *et al.*, 1988); triterpene glycosides–abrusosides A-D and the cycloartane-type aglycon, abrusogenin (the structures of abrusosides A-D were determined to be, respectively, 3-O-β-D-glucopyranosyl, 3-O-β-D-glucopyranosyl-(1→2)-β-D-6-methylglucuronopyranosyl+ ++, 3-O-β-D-glucopyranosyl-(1→2)-β-D-glucopyranosyl, and 3-O-β-D-glucopyranosyl-(1→2)-β-D-glucuronopyranosyl derivatives of abrusogenin) (Choi *et al.*, 1989); abrusoside E, 3β-O-[β-D-glucuronopyranosyl-(1→2)-β-D-glucopyranosyl](20S, 22S)-3β,22-dihydroxy-9,19-cyclolanost-24-en-26, 29-dioic acid δ-lactone (Kennelly *et al.*, 1996); triterpenoids – (20S,22S)-3β,22-dihydroxycucurbita-5(10),24-diene-26,29-dioic acid δ-lactone, 3-O-[6′-methyl-β-D-glucuronopyranosyl]-3β,22β-dihydroxyolean-12-en-29-oic acid Me ester, 3-O-β-D-glucuronopyranosylsophoradiol Me ester, and sophoradiol (Kim *et al.*, 2002); triterpenoid saponins – olean-12-ene-3β,23,28-trihydroxy-3-O-{[β-D-glucopyranosyl-(1→4)-β-D-glucopyranosyl-(13]-[β-D-glucopyranosyl-(12)]-β-D-fucopyranoside}, and olean-12-ene-3β,16β,23,28-tetrahydroxy-3-O-{[β-D-glucopyranosyl-(1→4)-β-D-glucopyranosyl-(1→3)]-[β-D-glucopyranosyl-(1→2)]-β-D-fucopyranoside} (Anam, 2003); the isoflavanquinones, abruquinone B and abruquinone G (Limmatvapirat *et al.*, 2004); a triterpenoid saponin, 3-O-β-D-glucopyranosyl-(1→2)-β-D-glucopyranosyl subprogenin D along with the triterpenoids – subprogenin D, abrusgenic acid, triptotriterpenic acid B, abruslactone A, abrusogenin, and abrusoside C (Xiao *et al.*, 2012); and triterpenoid glycosides, namely, 3β,24-dihydroxyurs-12-en-28-oic acid-24-O-β-D-xylopyranodide and 3β,24-dihydroxyurs-12-en-28-oic acid-3-O-β-D-xylopyranoside (Gupta and Chhabra, 2012) (Figures 10.2–10.4).

Carbohydrates reported from leaves, stems, roots and seeds of the plant include galactose, arabinose and xylose (Karawya *et al.*, 1981b).

Anti-cancer Activity

Phytoagglutinins from the plant inhibited the growth of Yoshida sarcoma cells (Tomita *et al.*, 1972). Abrus agglutinin, a hetero-tetrameric galactose specific lectin,

Figure 10.2: Some triterpenoids present in *Abrus precatorius*.

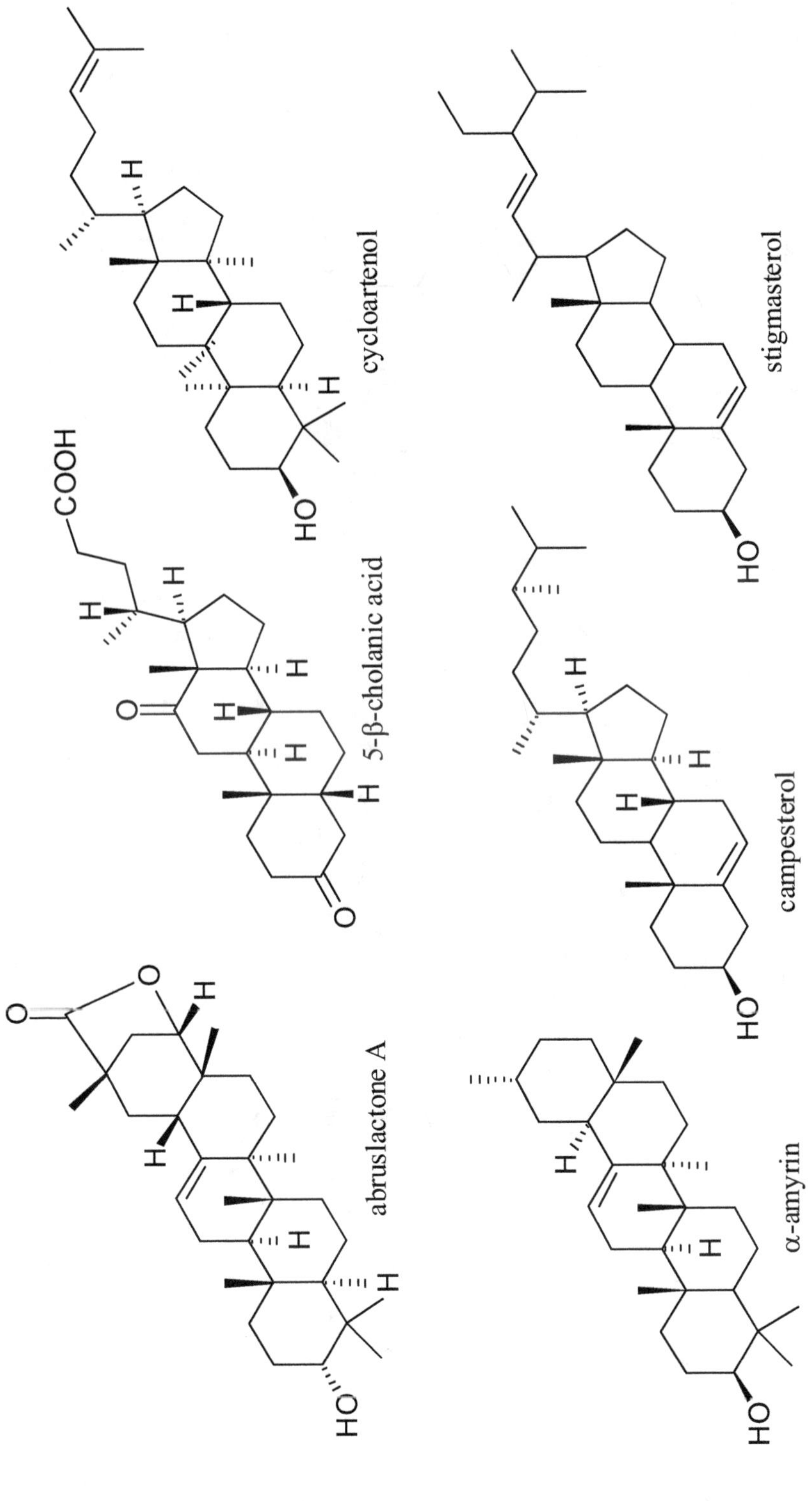

Contd...

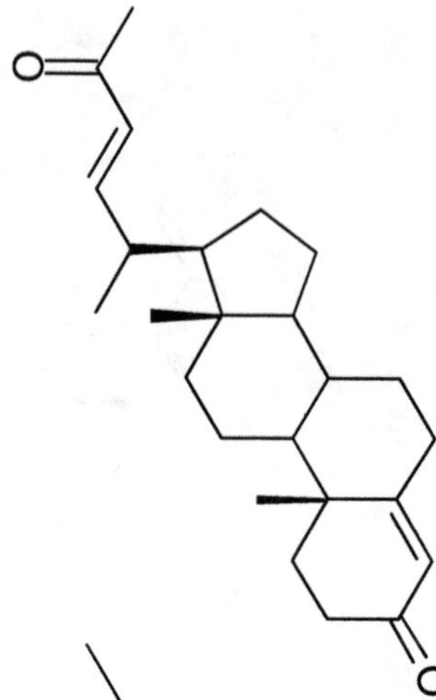

abridin

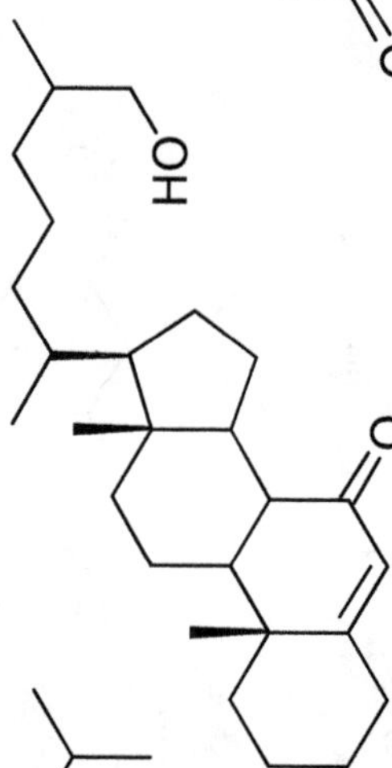

abricin

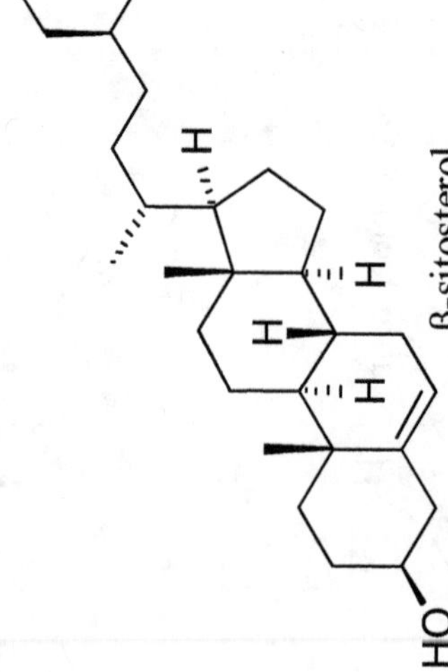

β-sitosterol

Figure 10.2–*Contd...*

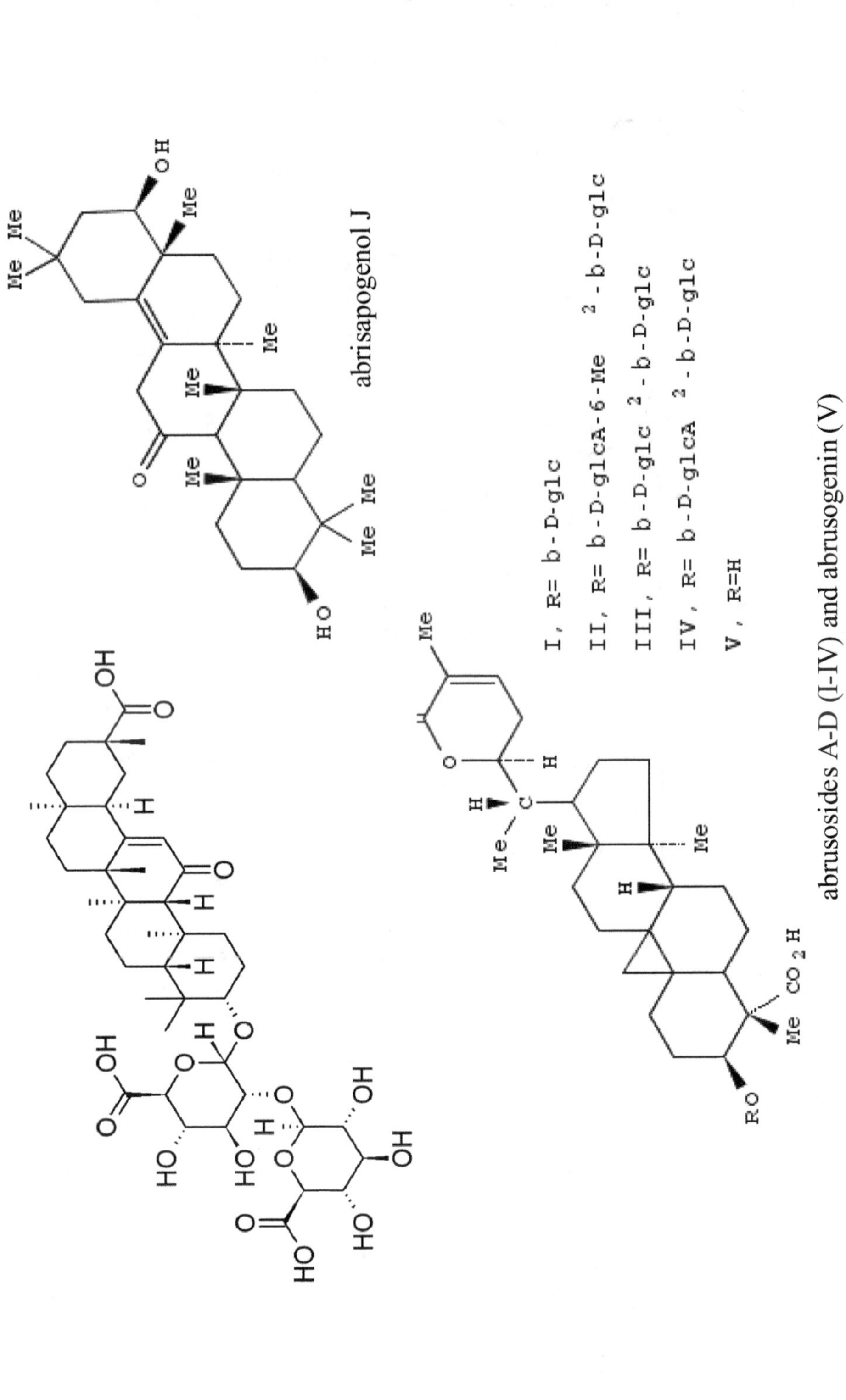

Figure 10.3: Some triterpenoids present in *Abrus precatorius*.

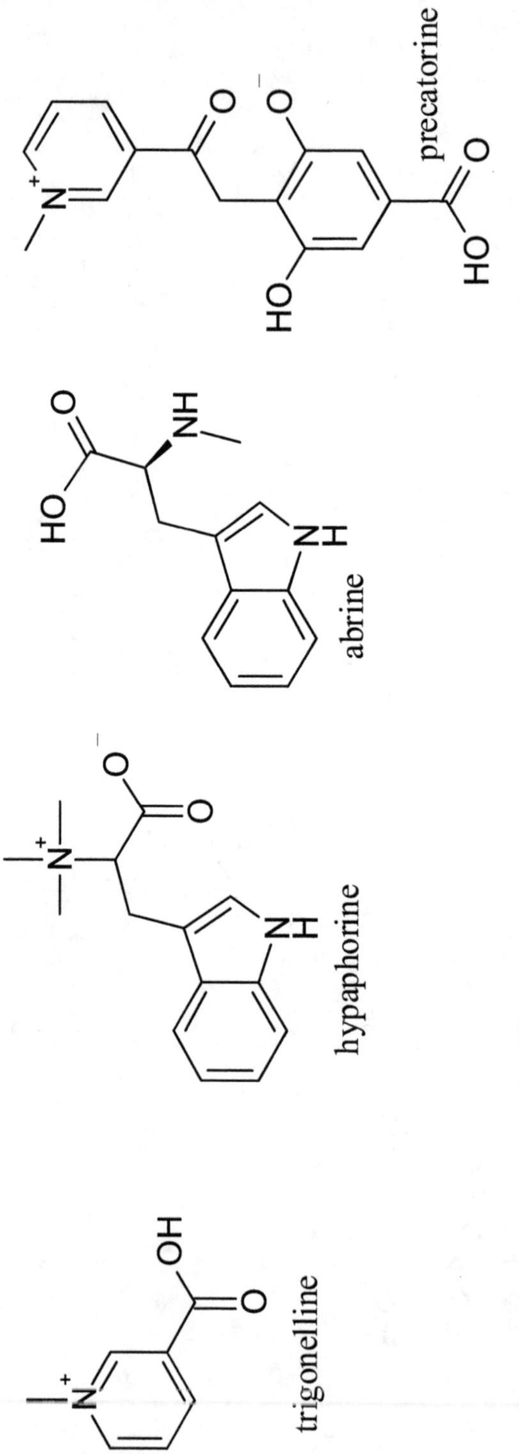

Figure 10.4: Some alkaloids present in *Abrus precatorius*.

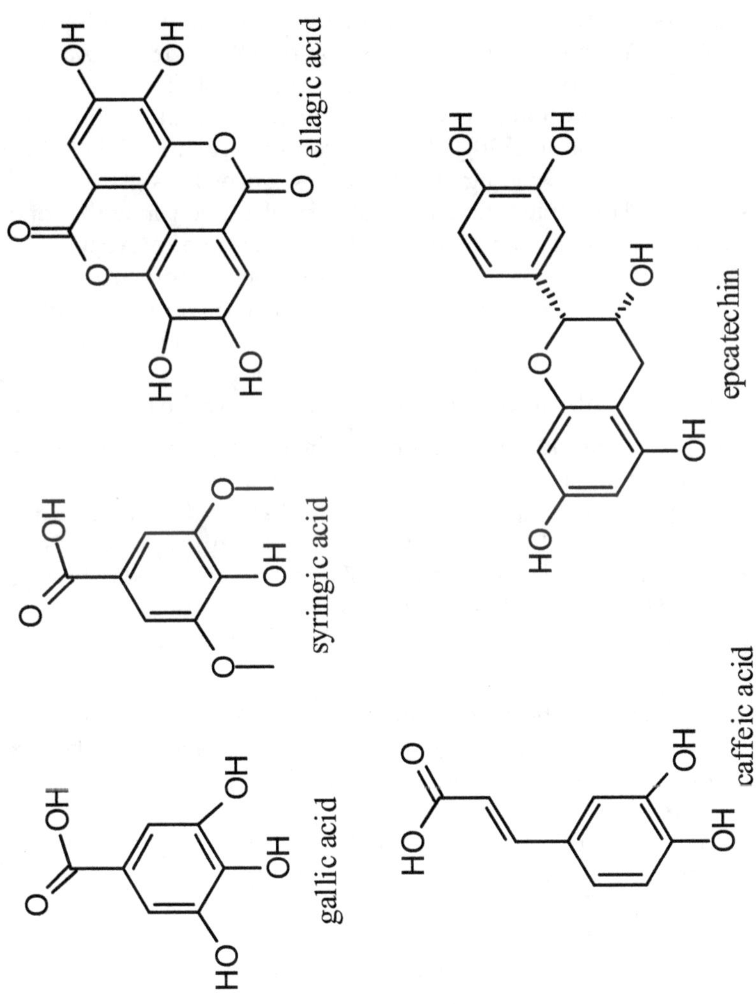

Figure 10.5: Some phenolics present in *Abrus precatorius*.

isolated from the seeds showed anti-tumor activity against Dalton's lymphoma ascites tumorogenic model in mice, in both native and heat denatured condition by stimulating the innate immune system (Ghosh and Maiti, 2007). Abrin, a toxic lectin isolated from seeds of the plant, exhibited anti-tumor activity in mice when used against Dalton's Lymphoma Ascites (DLA) and Ehrlich's Ascites Carcinoma (EAC) cells (Ramnath *et al.*, 2002). Abrin has been shown to induce apoptosis in murine DLA cells by stimulating the expression of pro-apoptotic caspase-3, and at the same time blocking the expression of the anti-apoptotic gene Bcl-2 (Ramnath *et al.*, 2006; Ramnath *et al.*, 2009). Abrin reportedly also inhibited Ehrlich ascites tumor development (Lin and Tung, 1972). Abrin A reportedly agglutinated both Sarcoma 180 cells and Ehrlich ascites tumor cells, while abrin B could inhibit the growth of Ehrlich ascites tumor cells (Lin *et al.*, 1978b). Immunotoxins prepared by conjugation with monoclonal antibodies (MAbs SF25, XF8, and AF20) to the surface antigen of human hepatocellular carcinoma cell line has been tested against pancreas cancer cells (Sato *et al.*, 1993). Immunotoxins, prepared by conjugation an abrin variant, monensin with monoclonal antibodies to malignant melanoma (MAb 9.2.27), to sarcoma (MAb TP-3) and to lymphoma (MAb HD 39) enhanced the specific cytotoxic activity of the immunotoxins (Godal, 1993). Abrin-derived peptide fraction has been shown to have potent anti-tumor activity against DLA model in tumor-bearing mice and HeLa cells (Bhutia *et al.*, 2009a,b). Thus while abrin is known to be highly toxic (Verma *et al.*, 2011), it can have potential uses as an anti-cancer agent through utilizing the toxic properties.

Among other phytochemical constituents of the plant, zinc salt of glycyrrhizin (zinc glycyrrhizate), as well as glycyrrhizic acid has been shown to ameliorate the genotoxicity and nephrotoxicity induced by the anti-cancer drug cisplatin in mice (Xu *et al.*, 1993; Arjumand and Sultana, 2011). Glycyrrhizin reportedly inhibited transplanted mandibular gland fibro-sarcoma of mice (Yu *et al.*, 2004). Glycyrrhizic acid has been reported to be a potential terpenoid along with ursolic acid and oleanolic acid as one of the promising anti-cancer therapeutic drug of the future (Kuttan *et al.*, 2011).

Gallic and ellagic acid has been reported in seeds of *A. precatorius*. Gallic-based indanone derivatives have been suggested as potential anti-cancer agents (Saxena *et al.*, 2008). The compound has been found to induce G2/M phase cell cycle arrest via regulating 14-3-3β release from Cdc25C and Chk2 activation in human bladder transitional carcinoma cells (Ou *et al.*, 2010). It also reportedly induced apoptosis in A375.S2 human melanoma cells through both caspase-dependent and –independent pathways (Lo *et al.*, 2010). The anti-cancer property of gallic acid has been shown in A549, a human lung adenocarcinoma cell line (Maurya *et al.*, 2011). It has been shown to suppress the migration and invasion of PC-3 human prostate cancer cells via inhibition of matrix metalloproteinase-2 and -9 signaling pathways (Liu *et al.*, 2011). The compound also inhibited migration and invasion in human osteosarcoma U-2 OS cells through suppressing the matrix metalloproteinase-2/-9, protein kinase B (PKB) and protein kinase C (PKC) signaling pathways (Liao *et al.*, 2012). Autoxidation of gallic acid has been observed to induce reactive oxygen species (ROS)-dependent death in human prostate cancer LNCaP cells (Russell *et al.*, 2012). The compound has further been found to induce the apoptosis of human osteosarcoma cells *in vitro* and *in vivo* via the regulation of mitogen-activated protein kinase pathways (Liang *et al.*, 2012).

Ellagic acid is another potential candidate for treatment of various forms of cancer. The compound is known to inhibit DNA topoisomerases *in vitro* (Constantinou *et al.*, 1995). In combination with quercetin and resveratrol, ellagic acid reportedly induced apoptosis and caused transient cell cycle arrest in human leukemia (MOLT-4) cells (Mertens-Talcott and Percival, 2005). Ellagic acid rich fractions from muscadine grapes (*Vitis rotundifolia*) have been shown to induce cell death in Caco-2 human colon carcinoma cells (Mertens-Talcott *et al.*, 2006). A chitosan-ellagic acid film has been developed as a local drug delivery system to induce apoptotic death of human melanoma cells (Kim *et al.*, 2009). Extract of *Punica granatum* fruit peel containing ellagic acid as the most abundant phenol has been shown to induce apoptosis in breast cancer (MCF-7) cells through increase of the pro-apoptotic gene, Bax and the decrease of the anti-apoptotic gene, Bcl-2 (Dikmen *et al.*, 2011). Modulation of 7,12-dimethylbenz[a]anthracene (DMBA)-induced hamster buccal pouch carcinomas has been reported with a combination of ellagic acid with chlorophyllin (Vidya Priyadarsini *et al.*, 2012).

Various phytosterols like campesterol, β-sitosterol and stigmasterol are present in *A. precatorius*. A protective role of β-sitosterol and the other phytosterols has been reported for colon, prostate and breast cancer (Awad *et al.*, 2000; Awad and Fink, 2000). The Korean fermented vegetable food, kimchi and one of its component, β-sitosterol, has been reported to reduce oncogenic H-ras(v12)-induced DNA synthesis (Park *et al.*, 2003), suggesting that they have both preventive and curative effects in cancer. The compound, isolated from *Poncirus trifoliate* (L.) Raf. has been shown to inhibit colon cancer cell growth but did not demonstrate any cytostatic activity against non-cancerous COS-1 fibroblast cells (Jayaprakasha *et al.*, 2007). Growth inhibition of the human colon cancer cell line (COLO 320 DM) has also been reported for this compound, isolated from *Asclepias curassavica* L.; also the number of aberrant crypt and crypt multiplicity was found to be reduced on administration of this compound in 1,2-dimethylhydrazine (DMH)-initiated rats (Baskar *et al.*, 2010). Extract of *Typhonium blumei* Nicolson and Sivad demonstrated inhibitory effects against the cancer cells, namely A549 lung cancer cells, LNCaP prostate cancer cells and MCF-7 breast cancer cells; the active ingredients were identified as campesterol and β-sitosterol (Hsu *et al.*, 2011).

Acacia arabica

Phytochemical Constituents

Reported phytochemicals include saponins and α-spinasterol in whole plant, proanthocyanidins and other phenolics in leaves, and catechin gallate derivatives in bark (quoted in Parmar *et al.*, 2010). Leaves have also been reported to contain 1-dotriacontanol, methyl 3,4-dihydroxy-5-methoxy benzoate, methyl gallate, 3,5,7,2′,3′,4′-hexahydroxy flavan, and 3′-formyl-3,5,7,2′,4′-pentahydroxy flavan (Bimla and Kalidhar, 2005) (Figure 10.6).

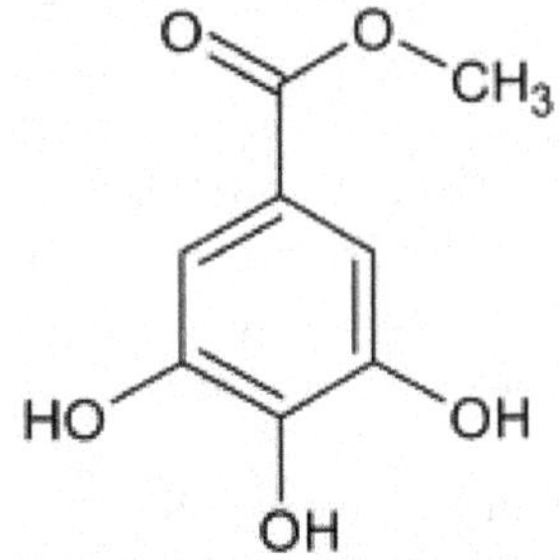

Figure 10.6: Methyl gallate present in *A. arabica*.

Anti-cancer Activity

The anti-cancer activity of gallic acid and derivatives has been discussed above.

Acacia catechu

Phytochemical Constituents

From aqueous extract of the plant, two new phenolic compounds, 5-hydroxy-2-[2-(4-hydroxyphenyl)acetyl]-3-methoxylbenzoic acid and (2S,3S)-3,7,8,3′,4′-pentahydroxyflavane have been isolated along with rhamnetin, 4-hydroxyphenyl ethanol, 3,3′,5,5′,7-pentahydroxyflavane, and fisetinidol (Li *et al.*, 2011). Other phytochemicals reported for whole plant include catechin, epicatechin, epicatechin-3-O-gallate, epigallocatechin-3-O-gallate, caffeine, flavanol dimmers, and flavonol glycosides (Shen *et al.*, 2006). Twelve compounds, namely, 4-hydroxybenzoic acid, kaempferol, quercetin, 3,4′,7-trihydroxyl-3′,5′-dimethoxyflavone, catechin, epicatechin, afzelechin, epiafzelechin, mesquitol, ophioglonin, aromadendrin, and phenol has been reported from whole plant (Li *et al.*, 2010). Catechin has also been shown in oil (Dubey *et al.*, 2009) and heartwood (Hye *et al.*, 2009). Quercetin has been reported in bark (Roy *et al.*, 2012) (Figure 10.7).

Crude extract of leaves were observed to contain carbohydrates, steroids, alkaloids, glycosides, tannins, saponins, flavones and phenolic compounds (Gulzar *et al.*, 2010). A new flavones glycoside – 5,7,3′,4′-tetrahydroxy-3-methoxyflavone-7-O-β-D-galactopyranosyl-(1→4)-O-β-D-glucopyranoside was isolated from stem (Yadava and Sodhi, 2002).

Anti-cancer Activity

Catechins, epicatechin and galloyl derivatives of catechin all have known anti-cancer activities.

In one formulation of cancer treatment used by the Garo tribe of Bangladesh, leaves of *Catharanthus roseus* (L.) G. Don., stems of *Eleusine coracana* (L.) Gaertn. and *Vitex negundo* L., leaves of *Clerodendrum viscosum* Vent., *Justicia adhatoda* L., and *Myristica fragrans* Houtt., flowers of *Acacia catechu*, and fruits of *Morus alba* L. are macerated together and made into pills. The pills are taken daily till cure. A number of anti-cancer alkaloids are present in *C. roseus* and some of which, like vinblastine and vincristine has been in use for nearly four decades as an anti-cancer drug (van Der Heijden *et al.*, 2004). Vincristine is an effective agent for treatment of acute lymphocytic cell leukemia, Hodgkin diseases and non-Hodgkin disease (Lu *et al.*, 2003). Chrysoplenetin and chrysoplenol, isolated from *V. negundo* has been shown to be effective against human pancreatic cancer cell line, PANC-1 (Awale *et al.*, 2011). Chrysoplenetin has also been found to be active against a panel of 25 cell lines belonging to lung, breast, CNS, colon, melanoma, ovarian, prostate and stomach cancer in the same study. Vitexins, which are lignin compounds and have been isolated from *V. negundo* exhibited inhibitory activity against breast, prostate and ovarian cancer cells and induced apoptosis with cleavage of PARP, up-regulation of Bax and down-regulation of Bcl-2 (Zhou *et al.*, 2009). Essential oil from the seeds of *M. fragrans* and an isolated component, myristicin showed cytotoxic activity against a colon cancer cell line, Caco-2 cells (Piras *et al.*, 2012). Phenolic-containing organic

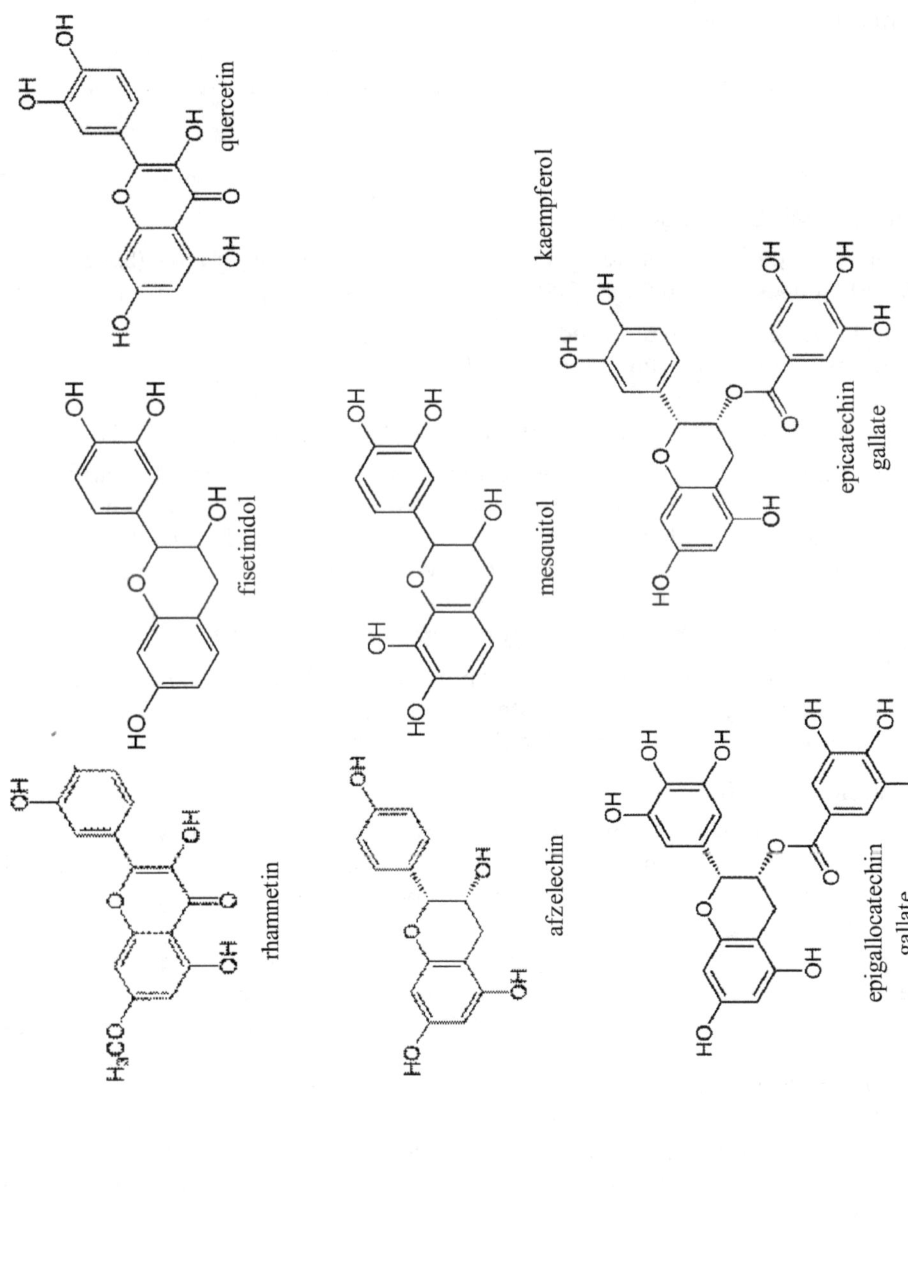

Figure 10.7: Some flavonoids and polyphenols present in *Acacia catechu.*

extracts of leaves of *M. alba* inhibited human hepatoma HepG2 hepatoma cells through G2/M phase cell cycle arrest, induction of apoptosis, and inhibition of topoisomerase IIα activity (Naowaratwattana *et al.*, 2010); the extracts contained rutin, isoquercetin, various derivatives of kaempferol and quercetin glycosides, chlorogenic acid, and caffeoylquinic derivatives. Geranylated flavanones isolated from roots of *M. alba* also showed cytotoxicity against human cancer cell lines including THP-1 monocytic leukemia cell line (Smejkal *et al.*, 2010). Anthocyanins, cyanidine 3-rutinoside and cyanidine 3-glucoside, isolated from the same plant exhibited an inhibitory effect on the migration and invasion of a human lung cancer cell line (A549 human lung carcinoma) (Chen *et al.*, 2006).

Acacia nilotica

Phytochemical Constituents

From whole plant, reported phytochemicals include kaempferol (Singh *et al.*, 2008), and gallic acid (Kaur *et al.*, 2005).

Ethyl gallate has been reported from leaves (Kalaivani *et al.*, 2011a). γ-Sitosterol has also been reported from leaves (Sundarraj *et al.*, 2012). From aerial parts an androstene steroid, 3β,acetoxy-17β-hydroxy-androst-5-ene has been isolated (Chaubal *et al.*, 2003), and D-pinitol – a bioactive carbohydrate (Chaubal *et al.*, 2005).

From the bark, a polyphenol, (+)-catechin-5-galloyl ester, gallic acid, methyl ester of gallic acid, naringenin, and (+)-catechin (Khalid *et al.*, 1989); and a cassane diterpene, niloticane (Eldeen *et al.*, 2010) have been reported (Figure 10.8). Gallic, protocatechuic, caffeic and ellagic acids along with quercetin have also been reported from bark (Singh *et al.*, 2009). Peltogynoids – acanilol A and acanilol B and the triterpene lupenone has been isolated from stem bark (Ahmadu *et al.*, 2010).

From pods (fruits), phytochemicals reported include gallocatechin 5-O-gallate, methyl gallate, gallic acid, catechin, catechin 5-O-gallate, 1-O-galloyl-β-D-glucose, 1,6-di-O-galloyl-β-D-glucose, and digallic acid (Salem *et al.*, 2011) (Figure 10.8).

Anti-cancer Activity

The anti-carcinogenic activity of bark extract has been shown by inhibition of development of pre-neoplastic lesions in response to DMBA; notably, a similar effect was observed with bark extract of another species of the *Acacia* genera, *Acacia auriculiformis* A. Cunn. ex Benth. (Kaur *et al.*, 2002). Bark extract demonstrated chemopreventive potential by preventing lipid peroxidation and promoting the enzymatic and non-enzymatic anti-oxidant system during hepatocellular carcinoma induced by N-nitrosodiethylamine (NDEA) followed by weekly subcutaneous injections of carbon tetrachloride (Singh *et al.*, 2009). Aqueous extracts of gum, flower and leaf also showed anti-cancer and anti-mutagenic properties against DMBA-induced skin papillomageensis in Swiss albino mice (Meena *et al.*, 2006). Phenolic compounds isolated from the plant demonstrated anti-uveal melanoma activity as well as inhibitory activity against cutaneous melanoma, ovarian cancer and glioblastoma (Salem *et al.*, 2011).

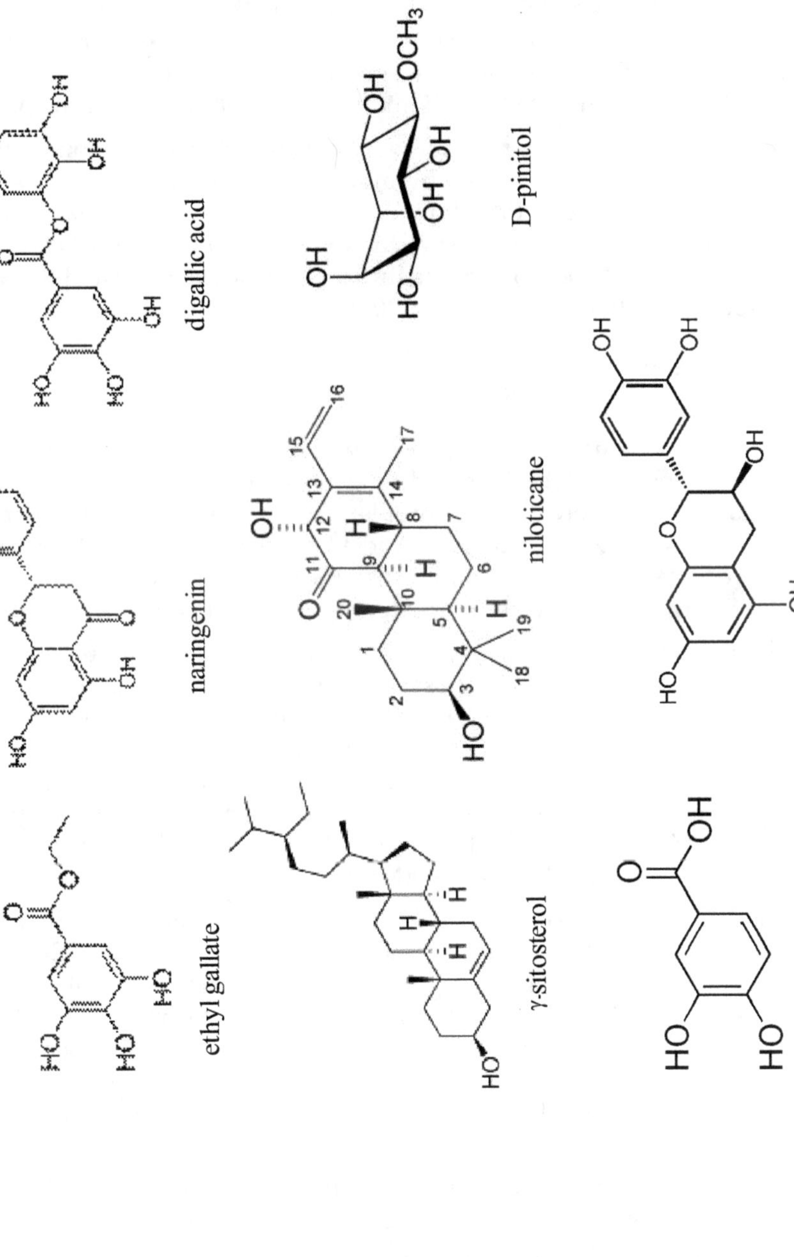

Figure 10.8: Some constituents of *Acacia nilotica*.

Ethyl gallate, isolated from leaves showed cytotoxic activity with IC_{50} value less than 100 microg/ml for Vero cell lines, and 72 microg/ml for HeLa cell lines (Kalaivani *et al.*, 2011a). Ethanol extract of leaves showed cytotoxicity against Vero and HeLa cell lines with IC_{50} values of 53.6 microg/ml and 28.9 microg.ml, respectively (Kalaivani *et al.*, 2011b).

Leaf extract and γ-sitosterol obtained from leaves showed growth inhibition, cell cycle arrest and apoptosis in human breast cancer and lung cancer cell lines, MCF-7 and A549. The IC_{50} values of extract on MCF-7 and A549 cancer cells were, respectively, 493.3 ± 15.2 and 696.6 ± 11.5 microg/ml (Sundarraj *et al.*, 2012).

From a related species, *Acacia victoriae* (Bentham), triterpenoid saponins – avicins have been shown to induce apoptosis by mitochondrial perturbation (caspase 3 activation and total cleavage of PARP) in Jurkat human T cell line (Haridas *et al.*, 2001). Avicins induced cell cycle arrest at the G1 phase in human MDA-MB-453 breast cancer cell line and apoptosis of the Jurkat (T-cell leukaemia) and the MDA-MB-453 breast cancer cell line (Mujoo *et al.*, 2001).

Bauhinia variegata

Phytochemical Constituents

Although the specific nature of components was not specified, it has been reported that whole plants contain glycosides and flavonoids (Rasul *et al.*, 1989). Various flavonoids isolated from different parts of the plant include quercetin, rutin, quercitrin, apigenin, and apigenin 7-O-glucoside (Abd-El-Wahab *et al.*, 1987). A flavone glycoside, 5-hydroxy-7,3′,4′,5′-tetramethoxyflavone 5-O-β-D-xylopyranosyl-(1→2)-α-L-rhamnopyranoside has also been reported (Yadava and Reddy, 2001).

Leaf phytochemical components include a triterpene saponin, named as 23-hydroxy-3α-[O-α-L-1C4-rhamnopyranosyl-(1″→4′)-O-α-L-4C1-arabinopyranosyl-oxy]olean-12-en-28-oic acid O-α-L-1C4-rhamnopyranosyl-(1′″→4′″)-O-β-D-4C1-glucopyranosyl-(1′″→6′″)-O-β-D-4C1-glucopyranosyl ester, flavonoids and cinnamic acid derivatives (Mohamed *et al.*, 2009); a flavonoid aglycone, naringenin (Arafa, 2001); tripernoids and a fatty ester of tripernoid, namely, α-amyrin caprylate, lupeol, nor-α-amyrin, and 3β-28-dihydroxyl olean-12-enyl-palmitate (Saha *et al.*, 2011); and two flavonoids, rutin and kaempferol (Modh *et al.*, 2011).

From the stems, a flavanone glycoside has been isolated and identified as 5,7-dihydroxyflavanone-4′-O-β-L-rhamnopyranosyl-β-D-glucopyranoside (Gupta *et al.*, 1979); other components reported include β-sitosterol, lupeol, and 5,7-dimethoxyflavanone 4′-O-β-L-rhamnopyranosyl-β-D-glucopyranoside (Gupta *et al.*, 1980); a flavonol glycoside, kaempferol-3-glucoside (Gupta and Chauhan, 1984); and from the stem barks, a polyphenolic aromatic glycoside – 2,4,8,9,10-pentahydroxy-3,7-dimethoxyanthracene-6-O-β-L-rhamnopyranoside (Bodakhe *et al.*, 2010) (Figure 10.9).

Roots have been reported to contain an anti-inflammatory flavonol glycoside – 5,7,3′,4′-tetrahydroxy-3-methoxy-7-O-α-L-rhamnopyranosyl(1→3)-O-β-D-galactopyranoside (Yadava and Reddy, 2003).

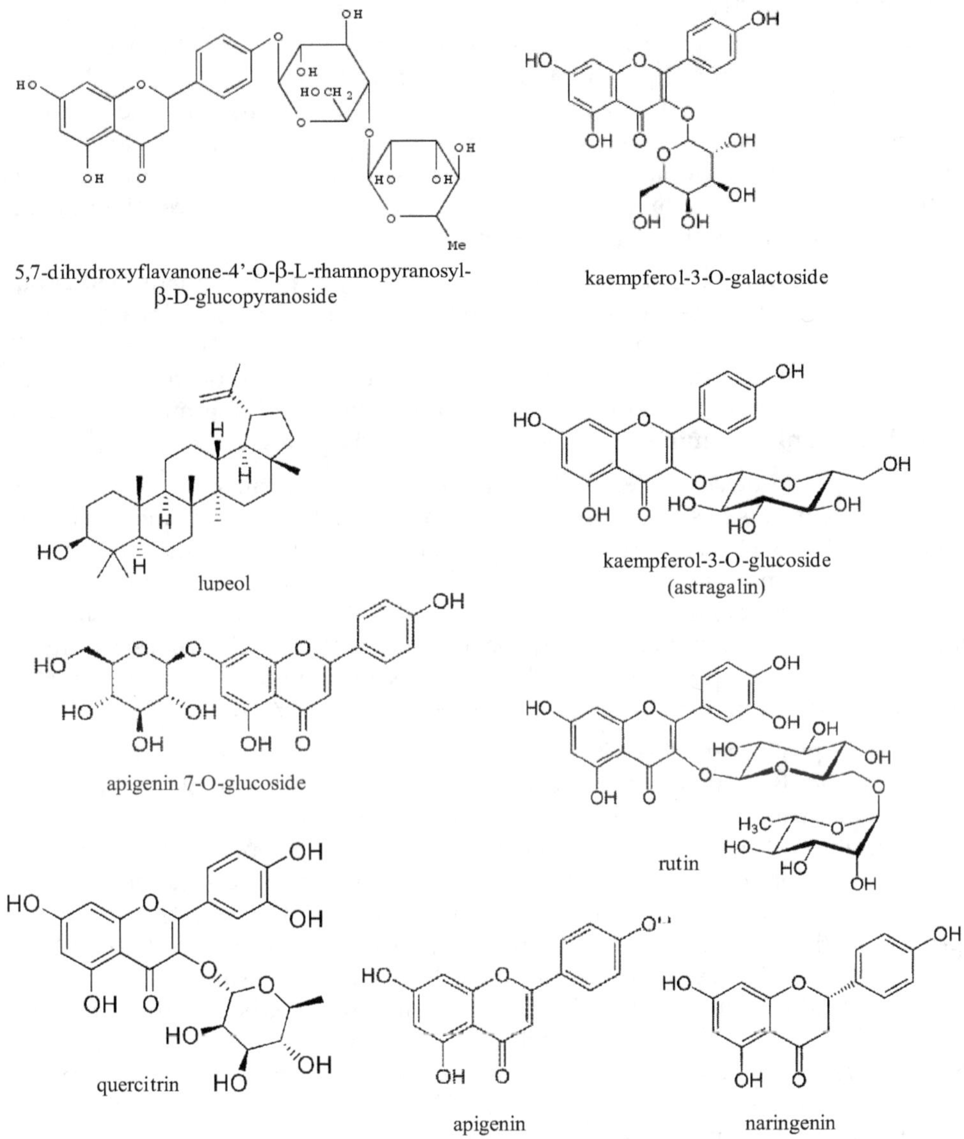

Figure 10.9: Some constituents of *Bauhinia variegata*.

Phytochemical components present in flower include kaempferol-3-galactoside and kaempferol-3-rhamnoglucoside (Rahman and Begum, 1966) (Figure 10.9).

Phytochemicals reported from pods (fruits) of the plant include 16-Me methyltetratriacontanoate, 7-hydroxynonadecan-9-one, 10-acetoxydotriacontane, 9-hydroxypentatriacontan-7-one, lupeol, and β-sitosterol (Singh *et al.*, 2011).

Anti-cancer Activity

The anti-tumor activity of ethanol extract of whole plant has been evaluated against Dalton's Ascitic Lymphoma (DAL) in Swiss albino mice, where extract administration led to a significant enhancement of mean survival time in tumor bearing mice (Rajkapoor *et al.*, 2003). Ethanol extract of whole plant led to effective suppression of liver tumor induced by N-nitrosodiethylamine (DEN); the extract was further found to be cytotoxic against human epithelial larynx cancer (HEp2) and human breast cancer (HBL-100) cells (Rajkapoor *et al.*,

2006). In skin papilloma model in Swiss albino mice, whole plant methanolic extract administration led to significant prevention with delayed appearance and reduction in the cumulative number of papillomas induced by DMBA and croton oil; in anti-mutagenicity studies, the extract administration significantly prevented micronucleus formation and chromosomal aberrations in bone marrow cells of cyclophosphamide-administered mice (Agrawal and Pandey, 2009).

The nephroprotective effect of ethanolic extract of stem against cisplatin-induced nephropathy in rats has been reported (Saumya *et al.*, 2011). A flavanone isolated from stem demonstrated cytotoxicity against human tumor cell lines representig leukemia, non-small cell lung, colon, central nervous system, melanoma, ovarian, renal, prostate and breast cancers (Rajkapoor *et al.*, 2009). Extract of flower has been shown to have chemopreventive potential on 2-stage skin carcinogenesis, induced by a single topical application of DMBA followed one week later by repeated application of croton oil in Swiss albino mice (Pandey and Agrawal, 2010).

Various phytochemicals with anti-cancer activity like rutin, kaempferol, lupeol, quercetin, and apigenin are present within the plant. Rutin is known to have both anti-cancer properties as well as can ameliorate the toxicity associated with administration of existing anti-cancer drugs. It has been reported that rutin can ameliorate cyclophosphamide (anti-cancer alkylating agent) induced reproductive toxicity in male rats (Abarikwu *et al.*, 2012). The modulating effects of rutin and quercetin on the mutagenic anti-cancer drug, mitomycin C has been reported (Undeðer *et al.*, 2004). Rutin has been shown to induce cell cycle arrest and apoptosis in murine leukemia WEHI-3 cells *in vitro* and *in vivo*; the compound has been further shown to inhibit human leukemia HL-60 cells implanted into mice (Lin *et al.*, 2012). Anti-proliferative effects have been observed with ethanol extract of the plant *Lactuca indica* L. (Asteraceae) in human leukemic HL-60 cell lines; bio-active phenolic compounds responsible for this anti-proliferative effect has been attributed to quercetin, caffeic acid, rutin and chlorogenic acid with quercetin having the highest inhibitory activity (Chen *et al.*, 2007a).

Experimental and in silico analysis has demonstrated the efficiency and possible use of kaempferol and quercetin as lead molecules for designing novel anti-tumor drugs (Gacche *et al.*, 2011). *Moringa oleifera* Lam. leaf extract reportedly demonstrated anti-proliferative and apoptotic effects in human tumor (KB) cell line; the active ingredients were found to be quercetin and kaempferol (Sreelatha *et al.*, 2011a). Organic extracts of *Morus alba* (mulberry) leaf reportedly caused inhibition of HepG2 hepatoma cells (a cell line related to hepatocellular carcinoma) through G2/M phase cell cycle arrest, induction of apoptosis, and inhibition of topoisomerase IIα activity; the various extracts were found to contain rutin, isoquercetin, and various derivatives of kaempferol and quercetin glycosides, as well as chlorogenic acid and caffeoylquinic acid derivatives (Naowaratwattana *et al.*, 2010). Kaempferol and quercetin (present in *Gingko biloba* L. extract) reportedly induced apoptosis in various oral cancer cell lines (SCC-1483, SCC-25 and SCC-QLL1), showed cleavage of poly (ADP-ribose) polymerase (PARP). Induction of apoptosis was caspase-3-dependent (Kang *et al.*, 2010). Other studies have also demonstrated the synergistic anti-proliferative action of kaempferol and quercetin in cultured human cancer cell lines like human gut cancer cells (HuTu-80 and Caco-2) and breast cancer cells (PMC42) (Ackland *et al.*, 2005). In chronic myelogenous leukemia cell line K562 and promyelocitic human leukemia U937, kaempferol induced apoptosis through a mechanism involving decrease of B-cell lymphoma-2 (Bcl-2, an apoptosis regulator protein) and increasing expression of Bax protein (a pro-apoptotic protein). There was also significant activation of caspases 3 and 9 along with PARP cleavage. Kaempferol treatment further increased the expression and mitochondrial localization of the NAD-dependent deacetylase SIRT3 (sirtuin3), inhibited phosphoinositide-3 kinase (PI3K) and induced dephosphorylation of Akt (also known as protein kinase B) at Ser473 and Thr308, leading to overall inactivation of Akt and activation of the mitochondrial phase of the apoptotic program (Marfe *et al.*, 2009).

Kaempferol-7-O-β-D-glucoside (KG) isolated from *Smilax china* L. rhizome reportedly displayed marked anti-cancer activity in several cancer cell lines, of which Hela human cervix carcinoma cells were the most sensitive. The compound induced G2/M cell cycle arrest in a p53-independent manner, the growth arrest being correlated with cyclin B1 and Cdk1 decrease. It was further observed that there was an increase in apoptosis, inhibition of nuclear factor-κB (NF-κB) nuclear translocation, up-regulation of Bax and down-regulation of Bcl-2 in HeLa cells (Xu *et al.*, 2008). KG has also been reported to induce apoptosis in cancer lines A375 and HL60 and arrest cell cycle at G1 phase (Li *et al.*, 2007). Two kaempferol coumaroyl glycosides (platanoside and tiliroside (isolated from the methanolic extract of *Platanus orientalis* L. buds exhibited *in vitro* cytotoxicity against a panel of human leukaemic cell lines (Dimas *et al.*, 2000).

Quercetin, a dietary antioxidant, present also in anumber of fruits and vegetables, is increasingly considered to be a promising anti-cancer agent. The anti-cancer activity of this compound has variously been attributed to cell cycle regulation, interaction with type II estrogen binding sites, and tyrosine kinase inhibition (Lamson and Brignall, 2000). For instance, cell growth and metastasis-associated properties in A431 cells overexpressing epidermal growth factor receptor was found to be inhibited by quercetin, which inhibited the tyrosine kinase activity of the receptor (Huang *et al.*, 1999). Quercetin furthermore has been shown to inhibit protein, DNA and RNA synthesis in MCF7 human breast cancer cells (Rodgers and Grant, 1998).

A study found that exposure of gastric cancer cells AGS and MKN28 to quercetin resulted in pronounced pro-apoptotic effect (Wang *et al.*, 2011a). The chemopreventive action of quercetin along with epigallocatechin-3-O-gallate and curcumin has been reviewed (Jagtap *et al.*, 2009). Quercetin also has been shown to induce apoptosis in human hepatocellular carcinoma (HCC) cell lines and enhanced the apoptotic effect of the chemotherapeutic agent, paclitaxel, in HA22T/VGH cells. The levels of apoptotic sub-G0/G1, reactive oxygen species and annexin V were increased prior to cell death, suggesting that reactive oxygen species is involved in quercetin-mediated apoptosis in human hepatocelluar carcinoma cells (Chang *et al.*, 2006). Quercetin also potentiated the ffect of the anti-cancer drug adriamycin (ADR) in a multidrug-resistant MCF-7 human breast cancer cell line (Scambia *et al.*, 1994). This suggests that quercetin can act both alone or could be used in combination therapy with existing anti-cancer drugs. A binary system containing quercetin-cyclodextrin has been found to decrease B16F10 melanoma growth and impaired tumor vascularization in BDF1 mice (Kale *et al.*, 2006).

The effects of lupeol has been studied in normal human melanocytes, and human metastatic (451Lu) and non-metastatic (WM35) cells, as well as athymic mice bearing 451Lu cell-originated tumors. While lupeol was found to have only a marginal effect on normal human melanocyte cells, it significantly decreased the viability of 451Lu and WM35 melanoma cells. Also, lupeol significantly reduced 451Lu tumor growth in athymic nude mice. The anti-cancer activity of lupeol has been attributed to several factors including (a) G1-S phase cell cycle arrest and apoptosis, (b) down-regulation of Bcl2 and up-regulation of Bax, (c) activation of caspase-3 and induction of poly(ADP)ribose polymerase cleavage, (d) decreased expression of cyclin D1, cyclin D2, and cdk2, and (e) increased expressed of p21 protein (Saleem *et al.*, 2008). In the human prostate cancer cell (PC-3), lupeol inhibited cell proliferation and induced apoptosis through a mechanism involving inhibition of expression of cyclin B, cdc25C and plk1, up-regulation of Bax, caspases 3 and 9, and apaf1 genes and down-regulation of the anti-apoptotic Bcl-2 gene (Prasad *et al.*, 2008a). The apoptogenic activity of lupeol has also been reported in human prostate cancer cells (lymph node carcinoma of the prostate, LNCaP) (Prasad *et al.*, 2008b). Lupeol has been shown to suppress overexpression of cellular FLICE-like inhibitory protein, which is reported to confer chemoresistance in pancreatic cancer (PaC) cells (Murtaza *et al.*, 2009). Overall, triterpenes like lupeol, ursolic and oleanolic acids, and betulinic acid, although they have not undergone extensive clinical trilas, are already being suggested as the next generation of anti-cancer drugs (Laszczyk, 2009; Petronelli *et al.*, 2009).

Apigenin has been reported to inhibit expression of vascular endothelial growth factor (VEGF) and angiogenesis in human A549 lung cancer cell. VEGF transcriptional activation was inhibited through the hypoxia-inducible factor 1 (HIF-1) binding site leading to specific decrease of HIF-1α. Apigenin further inhibited AKT and p70S6K1 activation. *In vivo*, apigenin significantly inhibited tumor growth in nude mice by having an inhibitory effect on angiogenesis (Liu *et al.*, 2005). The compound demonstrated inhibitory action on prostate (DU-145) and breast (MDA-MB-231) cancer cells, which was mediated through estrogen receptor, ER-β. Pro-apoptotic activity in these two cell lines was mediated through caspase 3 activation (Mak *et al.*, 2006). It has been reported that apigenin can function as both an antiestrogen and a protein kinase inhibitor with activity against breast cancer cells (MCF7) with acquired resistance to tamoxifen (OHT) or fulvestrant; the inhibitory effect on cell growth is mediated through both ER-α-dependent and –independent mechanisms (Long *et al.*,

2008). The inhibitory effect of apigenin on MDA-MB-453 human breast cancer cells has been further studied. It was shown that apigenin significantly inhibited the proliferation of the cancer cells in dose- and time-dependent manner with IC_{50} values of 59.44 and 35.15 microM at 24 and 72h, respectively. Induction of apoptosis was observed, and cytochrome c was released in cells exposed to apigenin at 72h. Subsequently, caspase 9, which plays a role in mitochondria-mediated apoptosiswas cleaved. Apigenin also activated caspase 3, a downstream caspase to caspase 9. Caspase 3 activation was accompanied by cleavages of caspases 6, 7 and 8. The authors have hypothesized that apigenin-induced apoptosis in the breast cancer cell line occurs through the activation of both intrinsic and extrinsic apoptotic pathways (Choi and Kim, 2009a). In ErbB2 overexpressing human breast cancer MDA-MB-453 cells, it has been observed that the cells were resistant to 5-fluorouracil, but combined treatment with apigenin significantly reduced the resistance. Cellular proliferation was significantly inhibited with the combination of the two compounds leading to apoptosis and drastic reduction of ErbB2 expression. The apoptotic effect has been postulated to occur through the down-regulation of ErbB2 expression and Akt signaling (Choi and Kim, 2009b). In 12-O-tetradecanoylphorbol-13-acetate (TPA)-induced human breast adenocarcinoma MCF-7 cells, apigenin time- and dose-dependently down regulated the matrix metalloproteinase (MMP)-9 enzymatic activities and its mRNA expression. Furthermore, there was also down-regulation of the nuclear factor (NF-κB) subunit p65 and activator protein (AP)-1 subunit c-Fos protein expressions in nucleus, and transcriptional activity of NF-κB and AP-1. The results strongly suggest that apigenin possess goog anti-metastatic potential through the inhibition of MMP-9 enzyme via NF-κB/AP-1 signaling (Liao *et al.*, 2012).

Proteasome inhibition is considered to be an important approach in cancer therapy. Apigenin has been shown to have proteasome-inhibitory activity in the highly metastatic human breast MDA-MB-231 cells and in its xenografts. Furthermore, apigenin induced apoptotic cell death in the cancer cells and in the tumors (Chen *et al.*, 2007).

Apigenin-induced apoptosis of leukemia cells has been shown to be mediated by a bimodal and differentially regulated residue-specific phosphorylation of heat-shock protein (hsp)-27, the protein kinases involved being primarily p38 and PKCδ (Gonzalez-Mejia *et al.*, 2010). Apigenin has been shown to block proliferation through cell cycle arrest in G2/M phase for myeloid HL60 and G0/G1 phase for erythroid TF1 cells. In both leukemia cell lines, the major target of apigenin was the JAK/STAT signaling pathway. However, in HL60 cells, apigenin also inhibited PI3K/PKB pathway and induced caspase-dependent apoptosis. In contrast, apoptosis was not detected in TF1 cells; rather an initiation of autophagy was noticed, which in turn led to a reduced susceptibility toward a commonly used cancer therapeutic agent, vincristine. The results suggest that patients should observe proper dietary precautions regarding foods containing apigenin, as it may interfere with their cancer treatment with existing drugs (Ruela-de-Sousa *et al.*, 2010). On the other hand, apigenin has been reported to sensitize HeLa cells to the anti-cancer drug paclitaxel-induced apoptosis through suppressing superoxide dismutase (SOD) activity, leading to accumulation of reactive oxygen species (ROS) and cleavage of caspase 2 (Xu *et al.*, 2011).

In prostate cancer cells (PC-3 and 22Rv1), *in vitro* apigenin-mediated apoptotic response has been shown to be due to inhibition of class I histone deacetylases (HDACs), specifically HDAC1 and HDAC3 at the protein and expression level. This inhibition resulted in histone H3 and H4 acetylation, as well as localized hyperacetylation of histone H3 on the p21/waf1 promoter. A corresponding increase in p21/waf1 and Bax protein and mRNA expression was observed, resulting in cell cycle arrest and apoptosis in both cell lines. In PC-3 xenografts in athymic nude mice, oral administration of apigenin led to reductions in tumor growth, reduced HDAC activity, and reduced protein expressions of HDAC1 and HDAC3, and an increase in p21/waf1 expression (Pandey *et al.*, 2011).

Cassia angustifolia

Phytochemical Constituents

From whole plants, reported constituents include glycosides belonging to the anthraquinone family, namely sennosides A-D (dimeric products of aloe emodin and/or rhein) along with tinnevellin glucoside (Franz, 1993) (Figure 10.10).

From leaves and pods (fruits), reported constituents include the naphthalene glycoside, tinnevellin glucoside (Lemli *et al.*, 1981); anthraquinone derivatives (Singh and Rao, 1982); sennoside G (Tanaka *et al.*, 1982); aloe-emodin dianthrone diglucoside (Nakajima *et al.*, 1985); an anthraquinone glycoside, emodin 8-O-sophoroside (Kinjo *et al.*, 1994); kaempferol 3-O-β-glucoside and isorhamnetin 3-O-β-glucoside (Singh *et al.*, 1995); seventeen different 1,8-dihydroxyanthranoids (anthraquinones and their bianthranyls) (Metzger and Reif, 1996); a flavonol glucoside, kaempferol-3-O-[(6‴-O-trans-sinapoyl)-β-D-glucopyranosyl(1→6]-β-D-glucopyranoside and apigenin-6,8-di-C-glycoside (Wu *et al.*, 2009) (Figure 10.10).

From the seeds, reported constituents include two anthraquinone derivatives, 1-hydroxy-3,6,7,8-tetramethoxy-2-isopropylanthraquinone and 1,5,7-trihydroxy-8-methoxy-3-methylanthraquinone (Gupta *et al.*, 1998); 5,7,4′-trihydroxy-6,8,3′,6′-tetramethoxy-2,3-dihydroflavonol and 1-hydroxy-6,8-dimethoxy-3-methylanthraquinone (Gupta *et al.*, 2002); a glycoside, 2-acetyl-3-methyl-8-methoxy-1,4-naphthoquinone 6-O-β-D-glucopyranoside (Wang *et al.*, 2007); and an oleanene-type triterpenoid glycoside, 3-O-{β-D-glucuronopyranosyl-(1→4)-[β-D-galactopyranosyl-(1→2)]-β-D-xylopyranosyl-(1→3)-β-D-glucopyranosyl}-2,16α-dihydroxy-4,20-bis(hydroxymethyl)olean-12-ene-28-oic acid (Khan and Srivastava, 2009).

Anti-cancer Activity

None has been reported thus far for the whole plant or its extracts. However, a number of the phytochemicals present in the plant having components like kaempferol, emodin (although emodin has not been directly reported from *Cassia angustifolia* yet, other *Cassia* species contain emodin making it highly likely for this plant to contain emodin also) and aloe-emodin in there are known to demonstrate anti-cancer activities. The anti-cancer activity of kaempferol has been discussed earlier. Aloe-emodin has been found effective against a wide variety of cancer cells. It appears to have selective activity against neuroectodermal tumors (Pecere *et al.*, 2000). In

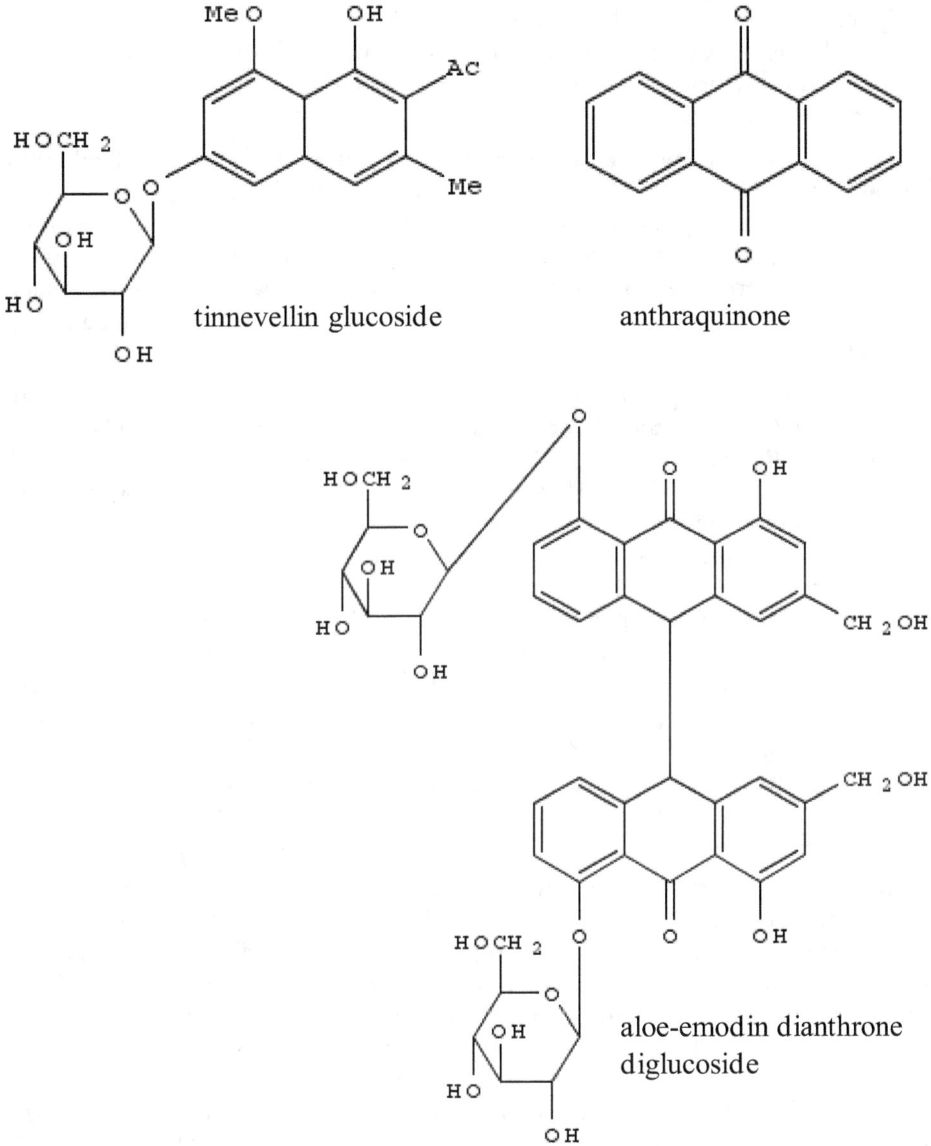

Figure 10.10: Some constituents of *Cassia angustifolia*.

human hepatoma cells, aloe-emodin has been found to induce apoptosis through p53- and p21-dependent apoptotic pathways (Kuo *et al.*, 2002; Shieh *et al.*, 2004). The compound has been found to inhibit prostate cancer cell growth through down-regulation of androgen receptor (Cha *et al.*, 2005). In T24 human bladder cancer cells, aloe-emodin reportedly induced apoptosis through the p53-dependent apoptotic pathway (Lin *et al.*, 2006). Anti-cancer effects of aloe-emodin have also been observed in hepatocarcinoma HepG2 cells (Lu *et al.*, 2007). The compound induced apoptosis in human gastric carcinoma cells (Chen *et al.*, 2007). The anti-cancer effect of aloe-emodin on cervical cancer cells was observed to involve G2/M cell cycle arrest and induction of differentiation (Guo *et al.*, 2007). Aloe-emodin demonstrated anti-tumor properties via induction of transglutaminase 2 activity in B16-F10 melanoma cells (Tabolacci *et al.*, 2010).

In human multiple myeloma, emodin showed cytotoxic activity through inhibition of Janus-activated kinase 2 (JAK 2) (Muto *et al.*, 2007). Emodin reportedly induced apoptosis of human tongue squamous cancer SCC-4 cells through production of reactive oxygen species and mitochondria-dependent pathways (Lin *et al.*, 2009). It has further been reported that in human tongue squamous cancer SCC-4 cells, aloe-emodin induced cell death through S-phase arrest and caspase-dependent pathways (Chiu *et al.*, 2009).

Emodin, besides acting alone, can also potentiate the cytotoxicity of other anti-cancer drugs. For instance, the compound has been found to enhance gefitinib-induced cytotoxicity via Rad51 down-regulation and ERK1/2 inactivation (Chen *et al.*, 2009). Suppression of ERCC1 and Rad51 expression through ERK 1/2 inactivation is also essential in emodin-mediated cytotoxicity in human non-small cell lung cancer cells (Ko *et al.*, 2010a). Emodin has been shown to enhance cisplatin-induced cytotoxicity via down-regulation of ERCC1 and inactivation of ERK 1/2 (Ko *et al.*, 2010b). Emodin has also been shown to enhance sensitivity of gall bladder cancer cells to platinum drugs via glutathione depletion and MRP1 downregulation Wang *et al.*, 2010). In gall bladder cancer cells, emodin reportedly potentiated the anti-cancer effect of cisplatin through the generation of reactive oxygen species and the inhibition of surviving expression (Wang *et al.*, 2011). Emodin also has been observed to abrogate cisplatin-induced nephrotoxicity in rats (Ali *et al.*, 2011). The protective effects of emodin against cisplatin-induced oxidative stress have also been observed in cultured human kidney (HEK 293) cells (Waly *et al.*, 2012).

Emodin induced apoptotic death in murine myelomonocytic leukemia WEHI-3 cells *in vitro* and enhanced phagocytosis in leukemia mice *in vivo* (Chang *et al.*, 2011). The anti-tumor and apoptosis-promoting properties of emodin against pancreatic cancer in mice have been attributed to inhibition of Akt activation (Wei *et al.*, 2011). Aloe-emodin has been found to be an anti-proliferative and differentiating agent on human U937 monoblastic leukemia cells (Tabolacci *et al.*, 2011). The anti-tumor effects of emodin have been shown *in vitro* and *in vivo* on LS1034 human colon cancer cells, where emodin caused apoptotic cell death (Ma *et al.*, 2012). Emodin has further been found to regulate apoptosis in human liver cancer cells (Yu *et al.*, 2012).

Cassia fistula

Phytochemical Constituents

Reported components in the whole plant include kaempferol and a mixture of anthraquinones like chrysophanol, rhein and physcion (Mahesh *et al.*, 1984); free and combined forms of chrysophanol, emodin and physcion (Khan *et al.*, 2005); and sennosides A and B and kaempferol (Bhope *et al.*, 2010) (Figure 10.11).

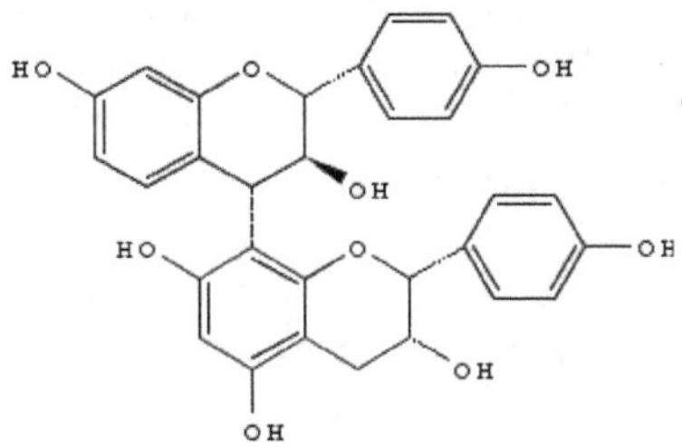

a dimeric proanthocyanidin

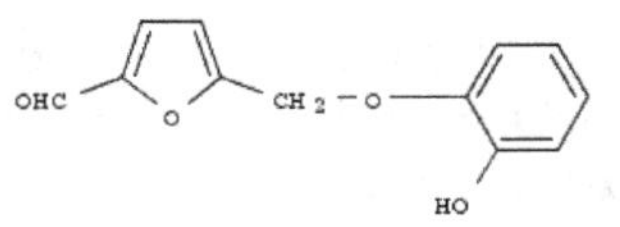

I

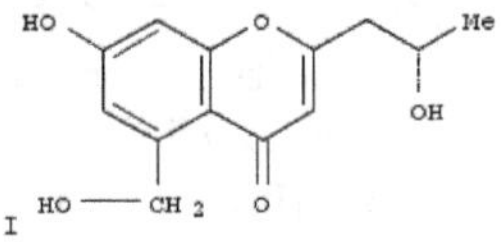

II

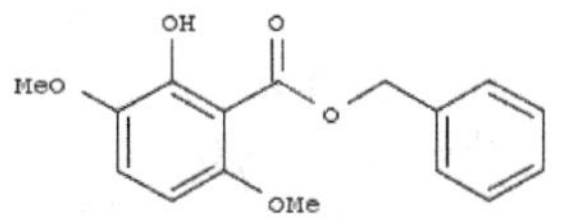

III

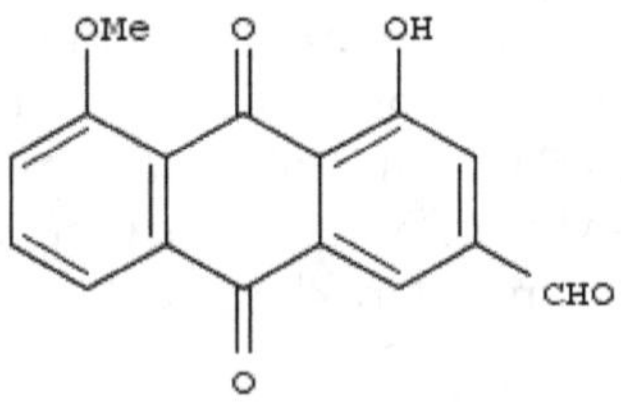

3-formyl-1-hydroxy-8-methoxy-anthraquinone

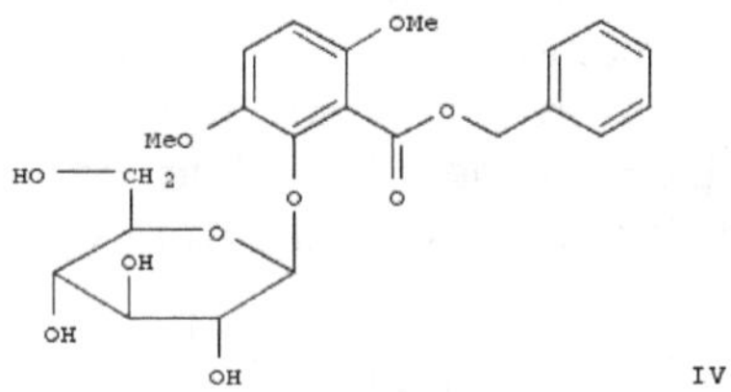

IV

5-(2-hydroxyphenoxymethyl)furfural (I), (2'S)-7-hydroxy-5-hydroxymethyl-2-(2'-hydroxypropyl)chromone (II), benzyl 2-hydroxy-3,6-dimethoxybenzoate (III), benzyl 2□-O-D-glucopyranosyl-3,6-dimethoxynenzoate (IV)

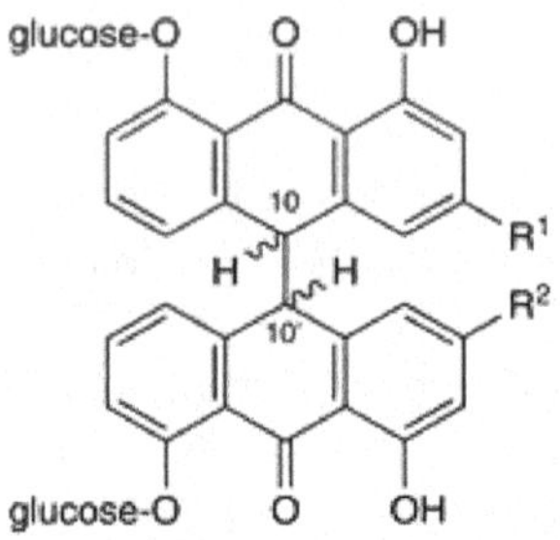

	R^1	R^2	Isomeric form
sennoside A	COOH	COOH	(+), *trans*
sennoside B	COOH	COOH	*meso, cis*

Figure 10.11: Some constituents of *Cassia fistula.*

Phytochemicals reported from leaves include 4′-methoxykaempferol, 4′-methoxyquercetin, p-coumaric acid, and 3′,4′,7-trimethoxyquercetin (Abraham *et al.*, 1988); sennoside B (Dutta and De, 2002); cis-heptacosanyl-5-hydroxypentadec-2-enoate and octacosan-5,8-diol (Singh *et al.*, 2005); (-)-epicatechin (Nagore *et al.*, 2010); essential oil from leaf contain phytol (Tzakou *et al.*, 2007).

Phytochemicals reported from the bark include fistucacidin, an optically inactive leucoanthocyanidin with the structure of 3,4,7,8,4′-pentahydroxyflavan (Rao and Venkateswarlu, 1965; Murty *et al.*, 1967). Other reported components from bark include lupeol, β-sitosterol and hexacosanol (Sen and Shukla, 1968); rhein (Khaleque and

Haroon, 1970). From the stem bark, two flavonol glycosides and one xanthone glycoside has been reported, their structures being, respectively, 5,7,3′,4′-tetrahydroxy-6,8-dimethoxyflavone-3-O-α-arabinopyranoside, 5,7,4′-trihydroxy-6,8,3′-trimethoxyflavone-3-O-α-L-rhamnosyl (1→2)-O-β-D-glucopyranoside, and 1,8-dihydroxy-3,7-dimethoxyxanthone-4-O-α-L-rhamnosyl (1→2)-O-β-D-glucopyranoside (Gupta *et al.*, 1989). Other components reported from the bark include eight long-chain hydrocarbons, 1-hexacosanol, 1-octacosanol, palmitic acid, stearic acid, oleic acid, linoleic acid, heptacosyl eicosanate, glyceryl-1-tetraeicosanoate; three sterols, β-sitosterol, stigmasterol, β-sitosteryl-3-O-D-glucopyranoside; one triterpene, lupeol; eight anthraquinones, chrysophanol, emodin, physcion, citreorosein, rhein, rhein Me ester, ziganein, 1,4,5-trihydroxyanthraquinone; two coumarins, isoscopoletin, scopoletin; two chromones, 2,5-dimethyl-7-hydroxychromone, 2,5-dimethyl-7-methoxychromone; three aromatic compounds, isovanillic acid, vanillic acid, and 2,4-dihydroxybenzaldehyde (Lee *et al.*, 2001); from the stem bark, an anthraquinone, namely, 1,8-dihydroxy-6-methoxy-3-Me- anthraquinone (Ledwani and Singh, 2006); and catechin from stem bark (Pitchai and Manikkam, 2011).

Fistucacidin has also been reported from sapwood but with the structure of 5,4′-dihydroxyflavan-3,4-diol (Reddy, 1968). Other components reported from sapwood include (-)-epiafzelechin, (+)-catechin, kaempferol, dihydrokaempferol, 1,8-dihydroxy-3-methylanthraquinone, and a dimeric proanthocyanidin (Patil and Deshpande, 1982).

Phytochemicals reported from heartwood include fistucacidin and two anthraquinones – barbaloin and rhein (Rao and Venkateswarlu, 1965; Murty *et al.*, 1967).

Phytochemical components reported in flowers include ceryl alcohol and its monoacetate, kaempferol and tetraacetate of kaempferol, rhein and its diacetyl derivative, rhamnose, an aglycone fistulin, and a bianthraquinone present as a glycoside (Kumar *et al.*, 1966); gibberellic acid (Sircar *et al.*, 1970); kaempferol, a proanthocyanidin (leucopelargonidin tetramer having a free glycol unit), leucopelargonidin trimer with no free glycol unit, higher polymers of propelargonodin, and rhein as glycoside (Narayanan and Seshadri, 1972); stigmasterol, sitosterol and 28-isofucosterol (Ghosh *et al.*, 1982); essential oil from flowers contain (E)-nerolidol and 2-hexadecanone (Tzakou *et al.*, 2007); rhein (Duraipandiyan *et al.*, 2011); and fistulic acid (Anbu *et al.*, 2011) (Figure 10.11).

Phytochemicals from pods (fruits) include lignoceric acid, n-triacontanol, n-triacontane-1,30-diol, n-triacontyl lignocerate and delignocerate of n-triacontane-1,30-diol (Agrawal *et al.*, 1970); fistulic acid (Agrawal *et al.*, 1972); 5-nonatetracontanone and 2-hentriacontanone (Misra *et al.*, 1996); a diterpene – 3β-17-norpimar-8(9)-en-15-one (Misra *et al.*, 1997); an anthraquinone, 3-formyl-1-hydroxy-8-methoxy-anthraquinone (Rani and Kalidhar, 1998); sennoside B (Dutta and De, 2002); anthocyanins – cyaniding-3-O-galactoside and petunidin-3-O-glucoside and flavan-3-ols–catechin and epicatechin (Jain *et al.*, 2011); and fistulic acid (Anbu *et al.*, 2011) (Figure 10.11).

A flavone glycoside from defatted seeds, namely, 5,3′,4′,-trihydroxy-6-methoxy-7-O-α-L-rhamnopyranosyl-(1→2)-O-β-D-galactopyranoside (Yadava and Verma, 2003); malvalic acid and sterculic acid from seed oil (Daulatabad *et al.*, 1987); linoleic acid, oleic acid, stearic acid, palmitic acid, caprylic acid, and myristic acid from seed oil (Sayeed *et al.*, 1999); 5-(2-hydroxyphenoxymethyl)furfural (I), (2′S)-7-hydroxy-5-hydroxymethyl-2-(2′-hydroxypropyl)chromone (II), benzyl 2-hydroxy-3,6-dimethoxybenzoate (III), benzyl 2β-O-D-glucopyranosyl-3,6-dimethoxynenzoate (IV), 5-hydroxymethylfurfural, (2′S)-7-hydroxy-2-(2′-hydroxypropyl)-5-methylchromone, and two oxyanthraquinones – chrysophanol and chrysophanein from seeds (Kuo *et al.*, 2002); roseanone from seeds (Jothy *et al.*, 2011); from seeds, benzyl 2-hydroxy-3,6-dimethoxybenzoate and its dimer dibenzyl 2,2′-dihydroxy-3,6,3″,6″-tetramethoxy-biphenyl-1,1′-dicarboxylate (Sartorelli *et al.*, 2012) (Figure 10.11).

Reported phytochemicals from roots include betulinic acid, 7-methylphyscion and β-sitosterol (Vaishnava *et al.*, 1993); a flavonoid glycoside, rhamnetin 3-O-gentiobioside (Vaishnav and Gupta, 1996); 1,8-dihydroxy-6-methoxy-3,7-dimethyl anthraquinone (Das and Jogi, 1998); a flavonol diglycoside, rhamnetin 3-O-β-D-glucopyranosyl-(1→6)-β-D-glucopyranoside (Vaishnav and Jain, 2004); 7-methylphysicon and 3β-hydroxy-Lup-20(29)en-oic acid (Vaishnav *et al.*, 2006). Chrysophanol and physcion were isolated from non-differentiating callus cultures (Ahuja *et al.*, 1988).

Anti-cancer Activity

Methanolic extract of seeds has been tested on Ehrlich Ascites Carcinoma (EAC) and on the life span of tumor-bearing mice. Treatment with the extract increased life span, decreased tumor volume and viable tumor cell count in tumor bearing hosts (Gupta *et al.*, 2000). Notably, seeds contain chrysophanol. Chrysophanic acid has been shown to block proliferation of epidermal growth factor receptor (EGFR)-overexpressing SNU-C5 human colon cancer cells through inhibition of EGF-induced receptor phosphorylation, and suppression of activation of downstream signaling molecules such as Akt and ERK (Lee *et al.*, 2011).

The chemopreventive efficacy of bark extract has been shown in DMBA-induced hamster buccal pouch carcinogenesis (Vasudevan *et al.*, 2008). Bark is known to contain a number of compounds like lupeol, β-sitosterol, and emodin, the anti-cancer properties of which have been discussed before. Catechin, another compound reported in bark may also mediate the anti-cancer effect. Synergistic anti-cancer activity with induction of apoptosis has been demonstrated with catechin and curcumin in human colon adenocarcinoma HCT15, HCT 116, and human larynx carcinoma HepG2 cell lines (Manikandan *et al.*, 2012). Catechin hydrate has been shown to suppress MCF-7 breast cancer cells through TP53/caspase-mediated apoptosis by increasing the expression of the pro-apoptotic genes such as caspase-3, -8, and -9 and TP53 (Alshatwi, 2010). Catechin hydrate reportedly also inhibited proliferation and mediated apoptosis of SiHa human cervical cancer cells by inducing several fold increases in caspases 3, 8 and 9 (Al-Hazzani and Alshatwi, 2011).

Anti-proliferative activity of aqueous extract of whole plant on HeLa cancer cell line has been reported (Manosroi *et al.*, 2012). Notably, anti-cancer compounds like kaempferol, emodin and chrysophanol have been reported for whole plant.

Cassia sophera

Phytochemical Constituents

A novel anthraquinone, sopheranin (1,3,6,8-tetrahydroxy 2-methyl 7-vinyl anthraquinone) along with β-sitosterol, chrysophanol, physcion and emodin has been reported from heartwood (Malhotra and Misra, 1982). A new cycloartane triterpene glycoside, named cyclosophoside A (29-O-β-D-glucopyranoside of 1α,3β,24χ,25-tetrahydroxy-9,19-cycloartan-29-oic acid) has been isolated from seeds (Zhao *et al.*, 2007). Essential oil from seeds have been reported to contain palmitic acid, linoleic acid, elaidic acid, stearic acid, 5-isopropyl-6-methyl-3-heptyne-2,5-diol, undecyl lauric acid, oleic acid, arachidic acid, and 3α,7β-dihydroxy-5β,6β-epoxycholestane (Mostafa *et al.*, 2007). A new flavonol-8-C-glycoside has been reported from leaves (Tiwari and Bajpai, 1980).

Anti-cancer Activity

The anti-cancer activities of β-sitosterol and emodin have been discussed before.

Desmodium gangeticum

Phytochemical Constituents

From whole plant, reported phytochemicals include aminoglucosyl glycerolipid, trans-5-hexadecenoic acid, 1-tritriacontanol, 1-heptadecanol, β-sitosterol, β-amyrone, gangetin, glycosphingolipid, 5-methoxy N,N-dimethyl tryptamine, 8-C-prenyl-5,7,5′-trimethoxy 3′,4′-methylene dioxy flavones, salicylic acid, 5-O-methylgenistein-7-O-β-D-glucopyranoside, 3,4-dihydroxy benzoic acid, kaempferol-7-O-β-D-glucopyranoside, 5-methoxy N,N-dimethyl tryptamine N_b-oxide, β-sitosterol-3-O-β-D-glucopyranoside, rutin, quercetin-7-O-β-D-glucopyranoside, and uridine triacetate (Mishra *et al.*, 2005) (Figure 10.12).

Phytochemicals present in roots include gangetial (Varaprasad *et al.*, 2009). Various parts of the plant, including the roots, also reportedly contain alkaloids such as tryptamines, phenethylamines and their N-oxides; pterocarpans such as gangetin, gangetinin, desmodin and desmocarpin; phospholipids, sterols, and flavones glycosides, 4′,5,7-trihydroxy-8-prenylflavone-4′-O-α-L-rhamnopyranosyl-(1→6)-β-D-glucopyranoside, and 8-C-prenyl-5,7,5′-trimethoxy-3′,4′-methylenedioxy flavones (quoted from Rastogi *et al.*, 2011).

Aerial parts of the plant reportedly contain sterols like N,N-dimethyltryptamine, 5-methoxy-N,N-dimethyltryptamine and their oxides and derivatives (Behari and Varshney, 1986). Indole-3-alkyl amines and phenylethylamines have also been reported from aerial parts (Ghosal and Bhattacharya, 1972) (Figure 10.12).

Anti-cancer Activity

The anti-cancer activites of rutin, β-sitosterol, quercetin and their derivatives have been discussed earlier.

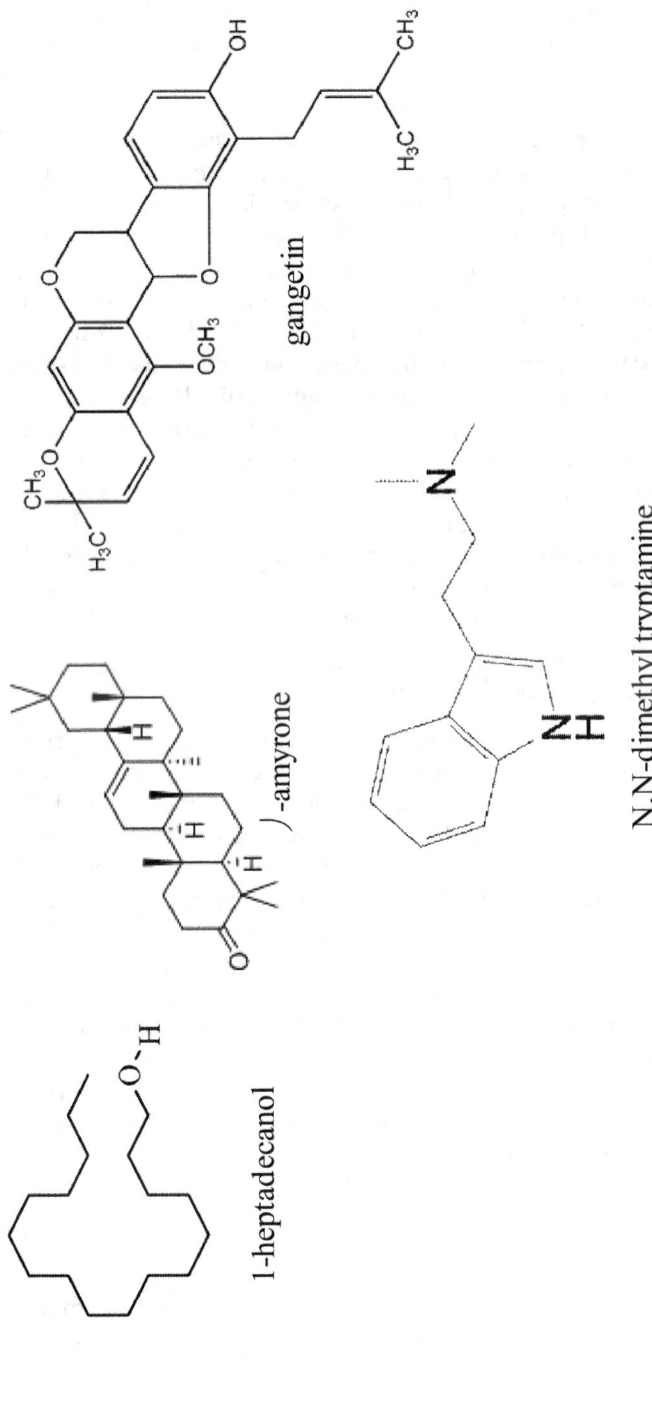

Figure 10.12: Some constituents of *Cassia sophera*.

Erythrina variegata

Phytochemical Constituents

From various parts of the plant reported alkaloids include (+)-3-demethoxyerythratidinone, (+)-erythraline, (+)-erythrinine, (+)-erysonine, (+)-erysotine, (+)-erysodine, (+)-erysovine, (+)-11-hydroxy-epi-erythratidine, (+)-epi-erythratidine, (+)-erysodienone, (+)-erysotrine, (+)-erysopitine, (+)-11-β-hydroxyerysotrine[(+)-erythrartine (quoted in Kumar *et al.*, 2010) (Figure 10.13). Other reported compounds from various parts of the plant are eight spiroamine alkaloids along with three carboxylated indole-3-alkylmamines, namely hypaphorine, its methyl ester, and N,N-dimethyltryptophan (Ghosal *et al.*, 1972). Pyranoisoflavones, robustone and 4-O-methylalpinum isoflavone has been reported from the plant along with the isoflavonoids epilupeol, 6-hydroxygenistein, and 3β,28-dihydroxyolean-12-ene (Rahman *et al.*, 2007). Orientanol B (9-hydroxy-3-methoxy-2-γ,γ-dimethylallylpterocarpan), erycristagallin (3,9-dihydroxy-2,10-di(γ,γ-dimethylallyl)-6a,11a-dehydropterocarpan), cristacarpin, sigmoidin K, erystagallin A (3,6a-dihydroxy-9-methoxy-2,10-di(γ,γ-dimethylallyl)pterocarpan), and 2-(γ,γ-dimethylallyl)-6a-hydroxyphaseollidin have also been reported (Sato *et al.*, 2002). Bidwillon B has been reported from whole plant (Sato *et al.*, 2004). Other compounds reported from the plant include decarbomethoxyerymelanthine, stachydrine, sterols, and the isoflavones, namely, scandenone, 4′,5,7-trihydroxy-8-prenylisoflavone, and 4′,5,7-trihydroxy-8-methyl isoflavone (Rahman *et al.*, 2010). Xanthoxyletin has also been reported from the plant (Rasul *et al.*, 2011a).

Alkaloids reported from leaves include scoulerine, (+)-coreximine, I-reticuline, and erybidine (quoted in Kumar *et al.*, 2010).

From bark, reported compounds include erysovine and stachydrine, as well as erythrabyssin, octacosyl ferulate, alkyl ferulates and alkyl phenolates (Singh *et al.*, 1975). Isoflavonoids reported from bark include erythrinins A, B and C, osajin and alpinum isoflavone together with the styrene, oxyresveratrol and dihydrostilbene dihydroxyresveratrol (Chawla *et al.*, 1988). The flavonoids, abyssinone V (4′-hydroxy-3′,5′-diprenylisoflavonone), erycricstagallin (3,9-dihydroxy-2,10-diprenylpterocarp-6a-ene), and 4-hydroxy-6,3,5-triphenylisoflavonone (Hedge *et al.*, 1997). From stem bark, three new isoflavones have been reported: 5,4′-dihydroxy-8-(3,3-dimethylallyl)-2″ethoxyisopropylfurano[4,5: 6,7] isoflavone, 5,7,4′-trihydroxy-6-(3,3-dimethylallyloxiranylmethyl) isoflavone, and 5,4′-dihydroxy-8-(3,3-dimethylallyl)-2″-hydroxymethyl-2″methylpyranol [5,6: 6,7] isoflavone, along with a new isoflavanone – 5,4′-dihydroxy-2′-methoxy-8-(3,3-dimethylallyl)-2″,2″-dimethylpyranol [5,6: 6,7] isoflavanone and seven other compounds, namely, euchrenone b 10, isoerysenegalensein E, wighteone, labumetin, lupiwighteone, erythrodiol, and oleanolic acid (Xiaoli *et al.*, 2006) (Figure 10.13).

From flowers, reported compounds include isoquinoline (erythritol) and isococculinine alkaloids (Chawla and Sharma, 1993).

From roots, reported constituents include two diphenylpropan-1,2 diols– cryvarinols A and B, three isoflavonoids – eryvarins M, N and O, two

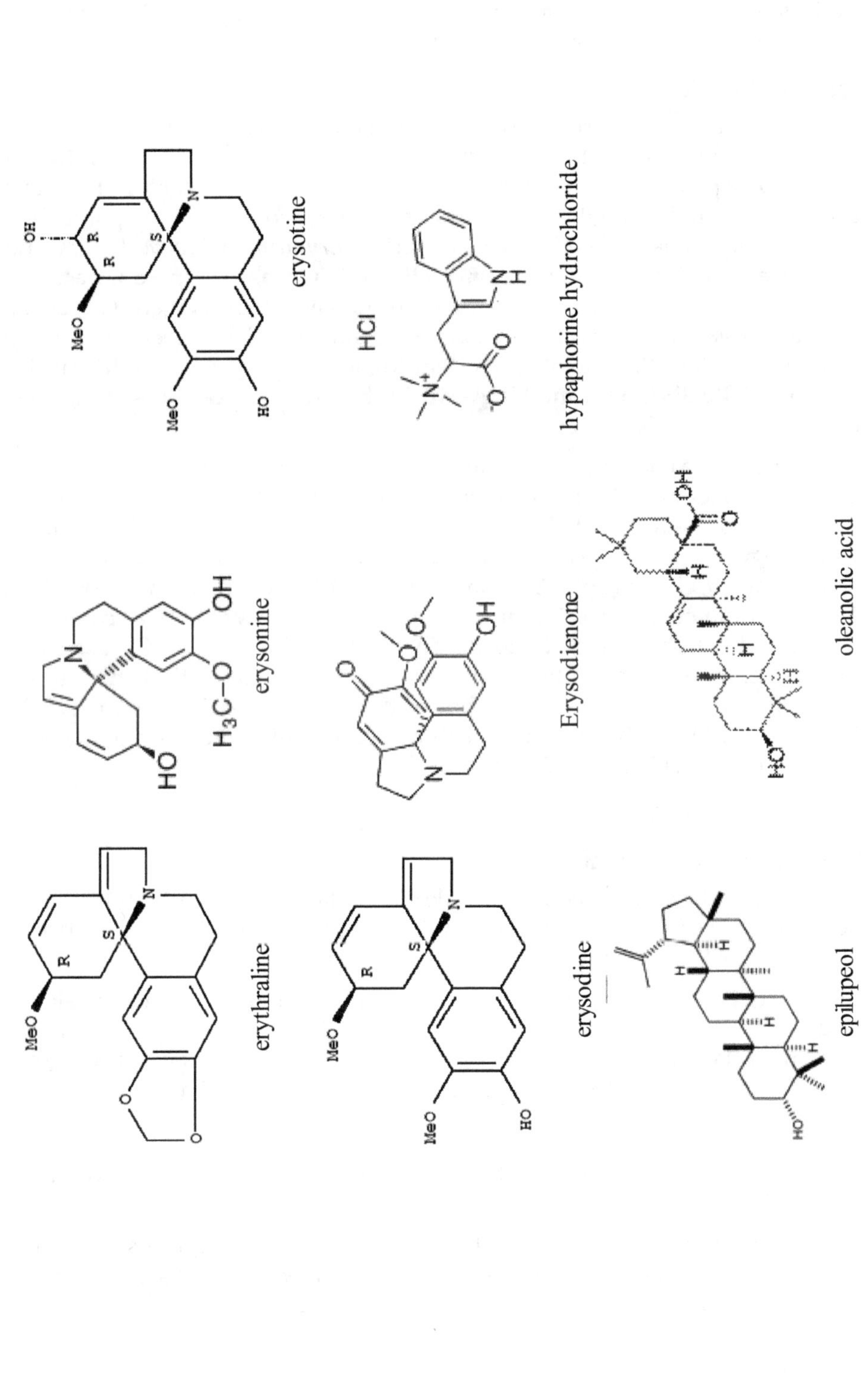

Figure 10.13: Some constituents of *Erythrina variegata*.

2-arylbenzofurans – eryvarins P and Q, and a 3-aryl-2,3-dihydrobenzofuran – eryvarin R, as well as eryvarins A and B, and eryvarins F and G (Tanaka *et al.*, 2000; Tanaka *et al.*, 2002; Tanaka *et al.*, 2003; Tanaka *et al.*, 2004).

Anti-cancer Activity

Xanthoxyletin, a coumarin present in the plant has been shown to induce S-phase arrest and apoptosis in human gastric adenocarcinoma SGC-7901 cells (Rasul *et al.*, 2011a). The plant is rich in pterocarpans. Although the pterocarpans from this plant are yet to be studied for their anti-cancer activities, neorautenol and phaseollin, two other pterocarpans, isolated from stem bark of *Erythrina addisoniae* Hutch. and Dalziel showed toxicity in H4IIE hepatoma cells. Both increased caspase 3/7 activity and nuclear fragmentation as markers for apoptotic cell death (Wätjen *et al.*, 2007). Pterocarpans isolated from *Erythrina abyssinica* Lam. ex DC. has been reported to inhibit growth of MCF7, tamoxifen-resistant MCF7 (MCF7/TAMR), adriamycin-resistant MCF7 (MCF7/ADR) and MDA-MB-231 breast cancer cell lines (Nguyen *et al.*, 2009).

Oleanolic acid, a constituent reported from the plant has been shown to exhibit moderate cytotoxicity on HT-29 colon cancer cells (Juan *et al.*, 2008). The compound also induced apoptosis in HuH7 human hepatocellular carcinoma cells through a mechanism involving release of mitochondrial cytochrome c into the cytosol, activation of caspases 3 and 9, cleavage of PARP, suppression of NF-κB and modulation of the mRNA expression of X-linked inhibitor of apoptotic protein (XIAP) as compared with untreated cells (Shyu *et al.*, 2010). The methanol extract of *Clerodendrum infortunatum* L. reportedly exhibited anti-cancer activity against Ehrlich's Ascites Carcinoma in mice; oleanolic acid was one of the major ingredients in the extract (Sannigrahi *et al.*, 2012). Tht anti-cancer activity of this compound has recently been reviewed (Pollier and Goossens, 2012).

Indigofera tinctoria

Phytochemical Constituents

Indirubin has been reported from whole plant (Han, 1994) (Figure 10.14). Other compounds reported in whole plants include gallic acid, quercitrin and myricetin (Bakasso *et al.*, 2008) (Figure 10.14).

From leaf extract, various phytochemicals reported include flavonoids, alkaloids, glycosides, and terpenoids (Verma and Suresh, 2002). From aerial parts, furano-flavones and a flavonol glycoside has been reported (Narender *et al.*, 2006). From aerial parts, trans-tetracos-15-enoic acid has been also isolated (Singh *et al.*, 2006).

Anti-cancer Activity

Indirubin has been shown useful for the treatment of chronic myelocytic leukaemia (Han, 1994; Steriti, 2002).

Thioredoxin reductase (TrxR) and thioredoxin (Trx) are overexpressed in many aggressive tumors and the tumor cells appear to be dependent on Trx system. Myricetin has been found to have strong inhibitory effects on mammalian Trxs with an IC50

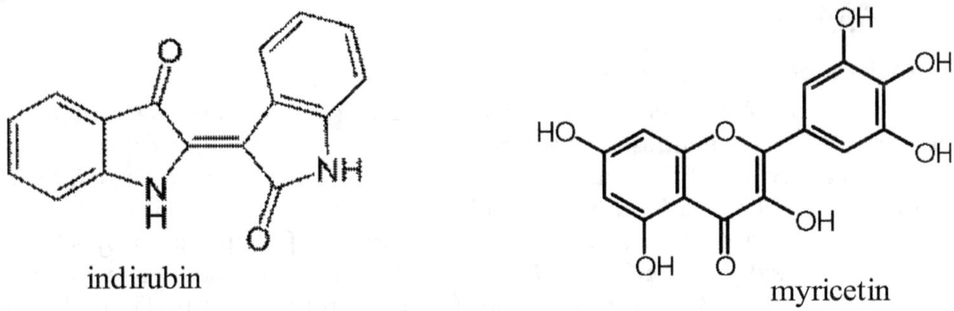

Figure 10.14: Some constituents of *Indigofera tinctoria*.

value of 0.62 micromol/L. In addition, myricetin inhibited the growth of A549 cells (adenocarcinomic human alveolar basal epithelial cells). A similar effect was observed with quercetin (Lu *et al.*, 2006). Myricetin has been shown to suppress invasion and migration of human lung adenocarcinoma A549 cells through inhibition of matrix metalloproteinase-2 (MMP-2) and urokinase-plasminogen activator (u-PA). Moreover, an inhibitory effect was also seen of the ERK signaling pathway, and inhibition of activation of NF-κB, c-Fos and c-Jun. The compound, therefore, can play a beneficial role in cancer metastasis (Shih *et al.*, 2009). Myricetin has been shown to inhibit TPA or epidermal growth factor (EGF)-induced JB6 P+ mouse epidermal cell transformation through inhibition of MEK kinase activity and suppressed TPA or EGF-induced phosphorylation of ERK or p90 ribosomal S6 kinase. The results suggest that myricetin may play an active role in inhibiting tumor development and neoplastic cell transformation (Lee *et al.*, 2007). Myricetin has also been shown to suppress UVB-induced skin cancer. In mouse skin epidermal JB6 P+ cells, myricetin suppressed UVB-induced cyclooxygenase-2 (COX-2) expression, and UVB-induced activation of activator protein-1 (AP-1) and nuclear factor-κB (NF-κB). Myricetin also inhibited Fyn kinase, and subsequently attenuated UVB-induced phosphorylation of mitogen-activated protein kinases (MAPK or ERK). *In vivo* experiments showed that pre-treatment with myricetin significantly suppressed UVB-induced skin tumor incidence (Jung *et al.*, 2008). Myricetin reportedly demonstrated moderate activity by inhibiting Wnt signaling pathway and so promoting cell survival in human biliary tract cancer cells (Wachter *et al.*, 2012).

Other species belonging to the Indigofera genera has been reported to have anti-cancer activity. Methanolic extract of leaves of *Indigofera cassioides* Rottler ex DC. has been shown to demonstrate *in vitro* cytotoxicity against HeLa, HEp-2, HEpG-2, MCF-7, HT-29, Vero and NIH 3T3 cells and *in vivo* cytotoxicity against EAC and DLA tumor bearing mice but was not cytotoxic against normal cells (Kumar *et al.*, 2011a). Aqueous extract of *Indigofera aspalathoides* Vahl ex DC. showed protective action against 20-methylcholanthrene (20-MCA) induced fibrosarcoma in male albino rats (Kumar *et al.*, 2011b). Aqueous extract of *Indigofera suffruticosa* Mill has been suggested for use as an alternative anti-cancer agent because of its low toxicity and high effectiveness in inhibiting growth of solid tumors (Viera *et al.*, 2007).

Lens culinaris

Phytochemical Constituents

From whole plant, reported phytochemicals include ferulic acid, kaempferol-3-rutonosyl-7-O-α-L-rhamnoside, kaempferol-3-rutinosyl-7-O-β-glucoside, and p-coumaric acid (Duke, 1992); furnaoacetylenes wyerone and wyerone epoxide, as well as the pterocarpan variabilin (Robeson, 1978); 4-chloro-1H-indole-3-N-methylacetamide, itaconic acid, arbutin, gentisic acid 5-O-β-d-apiofuranosyl-(1→2)-β-d-xylopyranoside, and (6S,7Z,9R)-9-hydroxymegastigma-4,7-dien-3-one-9-O-β-d-apiofuranosyl-(1→2)-β-d-glucopyranoside (Tsopmo and Muir, 2010) (Figure 10.15).

From seeds, the following phytochemicals have been reported: esculin, folacin, gadoleic acid, linoleic acid, linolenic acid, myristic acid, oleic acid, oxalic acid, palmitic acid, palmitoleic acid, paragalactoaraban, and stearic acid (Duke, 1992); 9-hydroperoxy-10,12-octadecadienoic acid (Gallasch and Spiteller, 2000); from seed coats – (+)-catechin-3-glucose, (+)-catechin, and (-)-epicatechin, various dimer, trimer and tetramer proanthocyanidins constituting of catechin, gallocatechin, and catechin gallate units, procyanidins and prodelphinidins from pentamers to nonamers (Dueñas *et al.*, 2003); from seeds, soyasaponins Bb and βg, kaempferol 3-O-β-glucopyranosyl(1→2)-O-[α-rhamnopyranosyl(1→6)]-β-galactopyranoside-7-O-α-rhamnopyranoside and kaempferol 3-O-β-glucopyranosyl(1→2)-O-[α-rhamnopyranosyl(1→6)]-β-glucopyranoside-7-O-α-rhamnopyranoside (Taylor *et al.*, 2007); and furosine (Rodríguez *et al.*, 2007).

Anti-cancer Activity

The seeds of plants grown in the Canadian prairies are rich in selenium, which is an essential micronutrient needed for healthy well-being and protection against cancer (Thavarajah *et al.*, 2007). The chemopreventive effect of raw and cooked seeds has been demonstrated against azoxymethane-induced aberrant crypt foci in rats (Faris *et al.*, 2009). The anti-cancer effect of arbutin against A375 human malignant melanoma cells has been reported (Nawarak *et al.*, 2009).

Mimosa pudica

Phytochemical Constituents

Phytochemicals reported for whole plants include crocetin, mimosine and norepinephrine (Duke, 1992; Tiwari and Spenser, 1965) (Figure 10.16).

Leaves have been reported to contain crocetin dimethyl ether and β-carotene (Duke, 1992); aerial parts contain C-glycosylflavones (Englert *et al.*, 1994); compared to whole plant, stem, and seed, leaves contain significantly higher amounts of 5,7,3′,4′-tetrahydroxy-6-C-[β-D-apiose-(1→4)]-β-D-glycopyranosyl flavone, isorientin, orientin, isovitexin and vitexin (Zhang *et al.*, 2011) (Figure 10.16).

Seeds contain D-glucuronic acid, linoleic acid, linolenic acid, oleic acid, palmitic acid, sitosterol, and stearic acid (Duke, 1992).

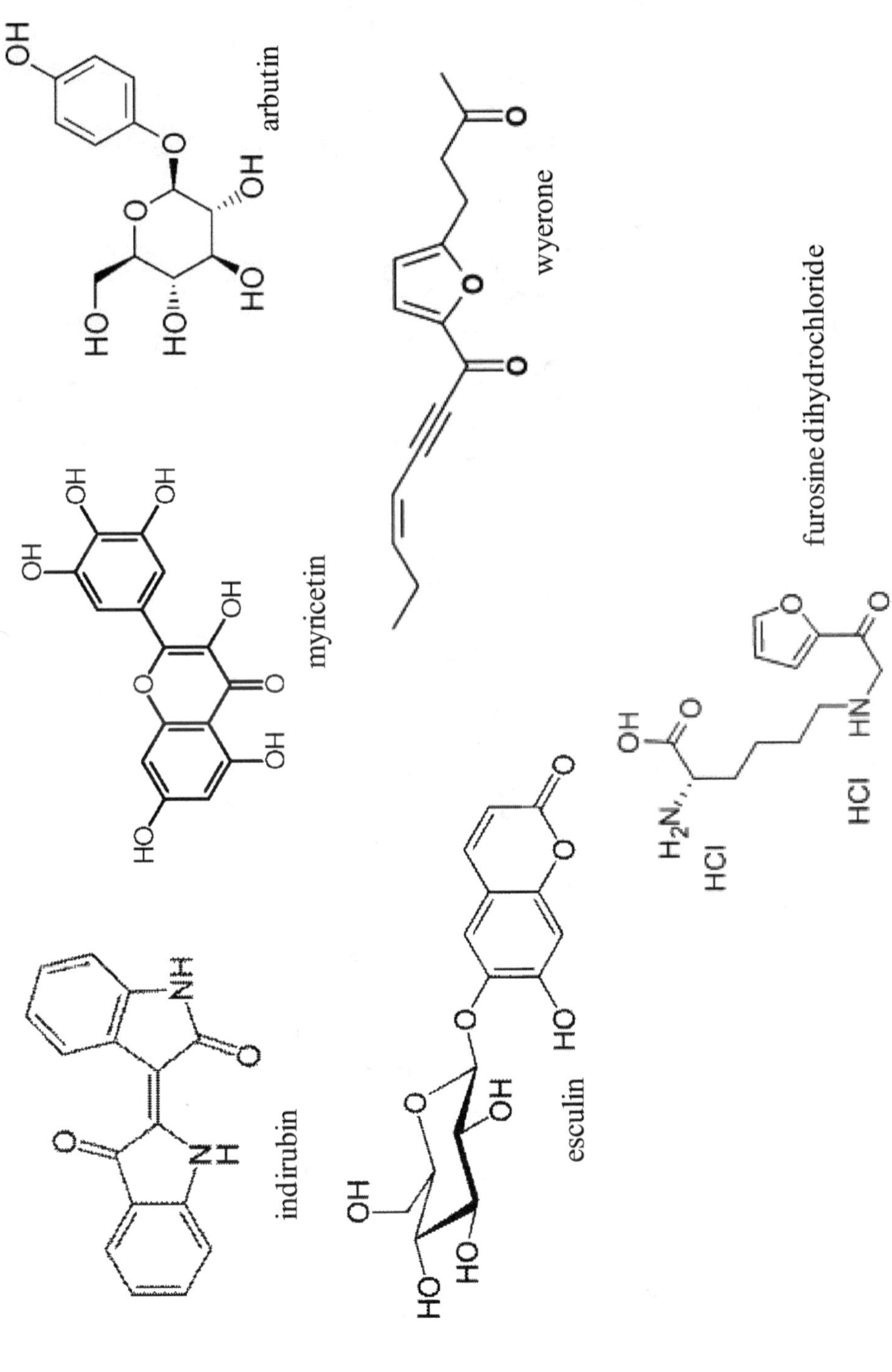

Figure 10.15: Some constituents of *Lens culinaris*.

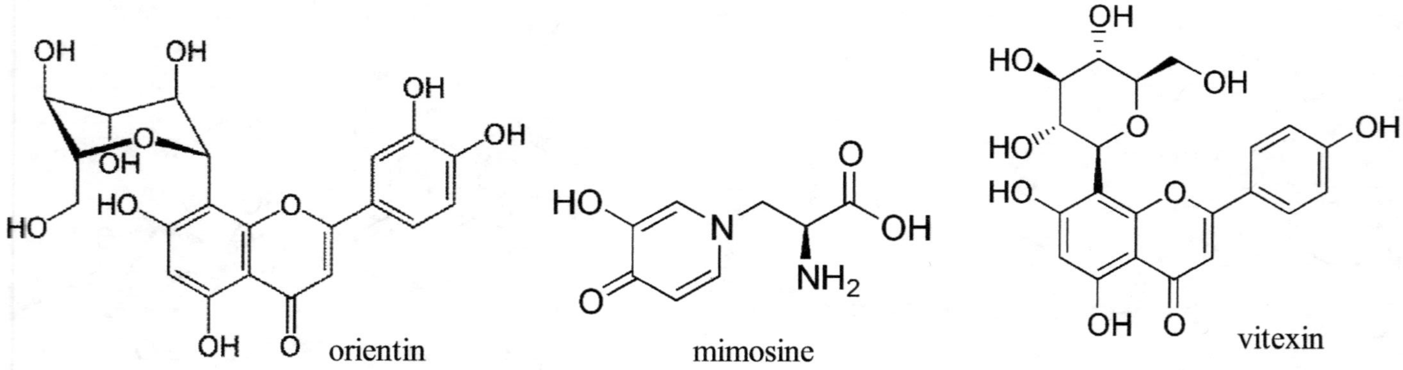

Figure 10.16: Some constituents of *Mimosa pudica*.

Anti-cancer Activity

Vitexin has been reported to have cytotoxic effect on breast, prostate and ovarian cancer cells with induction of apoptosis. Also noted was cleavage of PARP, up-regulation of Bax and down-regulation of Bcl-2 (Zhou *et al.*, 2009). Mimosine has been shown to block cell cycle progression and suppress proliferation of human lung cancer cells both *in vitro* and *in vivo*. Mimosine inhibited cyclin D1 and induced p21WAF1 expression in human H226 lung cancer cells grown in nude mice and induced apoptosis to suppress tumor growth (Chang *et al.*, 2000).

Samanea saman

Phytochemical Constituents

Phytochemicals reported from the whole plant include the alkaloid, pithecolobine (Wiesner *et al.*, 1968); serotonin and norepinephrine (Applewhite, 1973); saponins (samanin A and samanin B), samanin A being a glycoside of acacic acid containing glucose, arabinose, xylose and rhamnose attached to C-3, C-16, and C-21 hydroxyl groups, and samanin B being an acacic acid derivative with the above sugar moieties attached to one or both of the –OH groups at C-3 and C-16 (Varshney, 1976; Varshney *et al.*, 1978); lupeol, α-spinasterol, echinocystic acid, and acacic acid lactone (Dan *et al.*, 1982) (Figure 10.17).

Leaves are known to contain phenols, glycosides, terpenoids, alkaloids, saponins, flavanoids and tannins though their identities were not elucidated (Shanmugakumaran *et al.*, 2011).

Reported phytochemicals from the bark include octacosanoic acid, β-D-glucoside of α-spinasterol, α-spinasterol, lupeol, hexacosanol, lupenone, and α-spinasterone (Nigam *et al.*, 1971); a saponin, samanin E (glycoside of acacic acid containing galactose, glucose, xylose, fucose and rhamnose in the molar ratio 2: 6: 2: 1: 2) (Varshney *et al.*, 1985) (Figure 10.17).

A saponin, saponin C – glycoside of acacic acid (3β,16β,21β-trihydroxyolean-12-en-28-oic acid) with glucose, arabinose, xylose, fucose, and rhamnose as sugars in the molar ratio of (6: 1: 2: 3: 4) which are present on all the 3 hydroxyl groups and also on the carboxyl group has been reported from the wood of the plant (Varshney and Jain, 1977; Varshney *et al.*, 1985).

The seeds reportedly contain a saponin, namely samanin A (glycoside of acacic acid containing D-glucose, L-arabinose, D-xylose and L-rhamnose in a molar ratio of 3: 4: 2: 6) (Varshney and Vyas, 1976).

Palmitic acid and 1,8-cineole has been reported from volatile oils of fruits (Ogunwande *et al.*, 2006). Tannins have been reported from the fruits and whose components have been describes as cyaniding, catechin, epicachin, anthocyanin monoglycones, delphinidin and malidin (Ukoha *et al.*, 2011).

Anti-cancer Activity

α-Spinasterol octadecanonate and α-spinasterol-3-O-β-D-glucopyranoside, isolated from dry root of *Momordica dioica* reportedly showed anti-cancer activity in

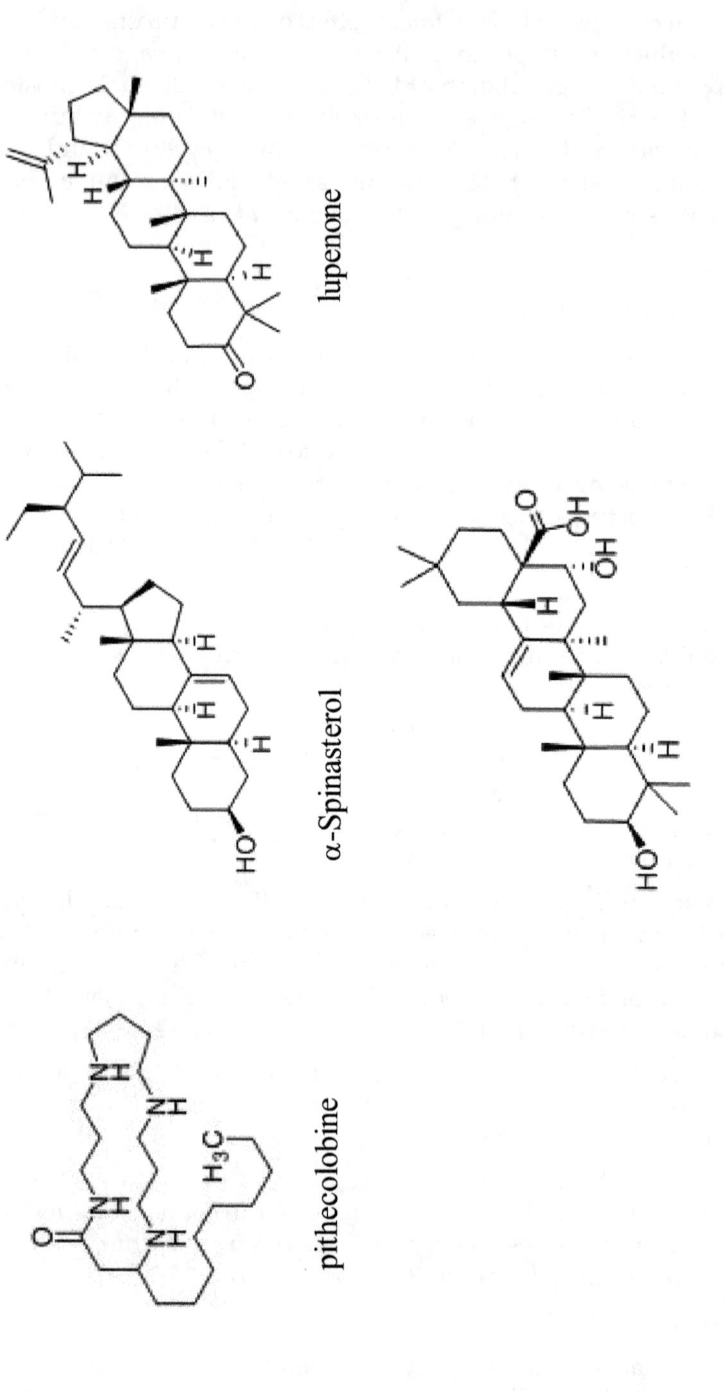

Figure 10.17: Some constituents of *Samanea saman*.

pharmacological testing on cancer cell, L1210 (Luo *et al.*, 1998). Echinocystic acid has been shown to induce apoptosis in human HepG2 cells, as characterized by DNA fragmentation, activation of caspases 3,8 and 9, and PARP cleavage. The compound further induced the truncation of Bid protein and reduction of Bcl-2 protein. Jun kinase (JNK, c-Jun NH_2-terminal kinase) and p38 kinases were activated indicating that these two kinases-mediated mitochondrial pathways might be involved in the observed apoptosis of the cancer cell line (Tong *et al.*, 2004).

Sesbania sesban

Phytochemical Constituents

From whole plants, sterols, saponins, flavonoids, sesbanine, glycosides and kaempferol has been reported (Chatterjee and Pakrashi, 1992). Reported constituents from whole plant also include a kaempferol trisaccharide (el-Sayed, 1991), and oleanolic acid 3β-D-glucuronide (Das *et al.*, 2011) (Figure 10.18).

Anti-cancer Activity

A kaempferol trisaccharide, isolated from the plant has been reported as an anti-tumor promoter (el-Sayed, 1991). Ethanol extract of a plant from the same genus, *Sesbania grandiflora* demonstrated anti-cancer activity against Ehrlich Ascites carcinoma in Swiss albino mice (Sreelatha *et al.*, 2011b). A protein fraction from the same plant also showed anti-proliferative activity and induced apoptosis in Dalton's lymphoma ascites (DLA) and colon cancer cells (SW-480) (Laladhas *et al.*, 2010).

Vicia faba

Phytochemical Constituents

4-Chloro-6-methoxyindole has been reported from beans (fruits) (Yang *et al.*, 1984; Tiedink *et al.*, 1991). Beans are reported to further contain phytic acid, phenolic compounds and phytase (Oomah *et al.*, 2011), as well as a kaempferol glycoside (Tselepi *et al.*, 2011) (Figure 10.19).

Two amide-linked conjugates of jasmonic acid, N-[(3R,7R)-(-)-jasmonoyl]-(S)-dopa and N-[(3R,7R)-(-)jasmonoyl]-dopamine along with N-[(3R,7R)-(-)-jasmonoyl]-(S)-tyrosine has been reported from flowers (Kramell *et al.*, 2005) (Figure 10.19).

Anti-cancer Activity

Administration of beans in the diet of rats led to significant suppression in the formation of malignant epithelial tumors (cribriform carcinoma) in rats induced with DMBA (el-Aaser *et al.*, 1994). Incidences of hepatomas and bladder papillomas developed in mice as a result of administration of dibutylnitrosamine (DBN) precursors (dibutylamine and nitrite) were also reportedly reduced significantly by administration of beans (Zakhary *et al.*, 1994). A 15-kDa Bowman-Birk type trypsin inhibitor (VFTI-G1), isolated from seeds of beans reportedly showed specific anti-proliferative activity toward HepG2 hepatoma cells by inducing chromatin condensation and cell apoptosis (Fang *et al.*, 2011). A kaempferol glycoside, isolated from beans was found to inhibit the growth of human cancer cell lines MCF7, HeLa and HepG2 through inhibition of topoisomerases I and II (Tselepi *et al.*, 2011).

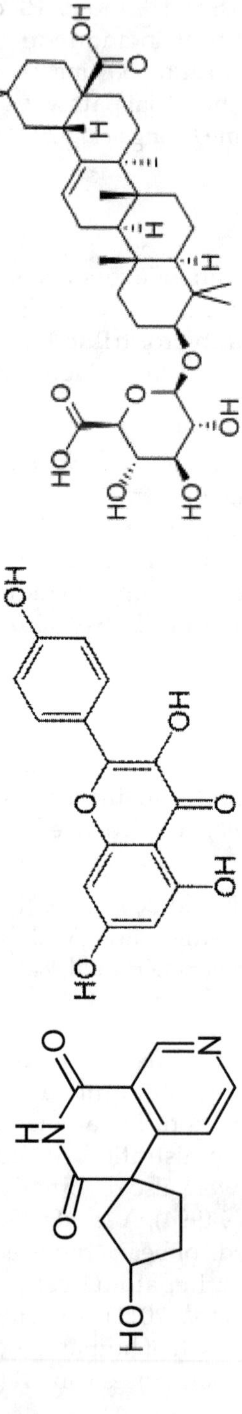

oleanolic acid 3β-D-glucuronide

kaempferol

(+) sesbanine

Figure 10.18: Some constituents of *Sesbania sesban*.

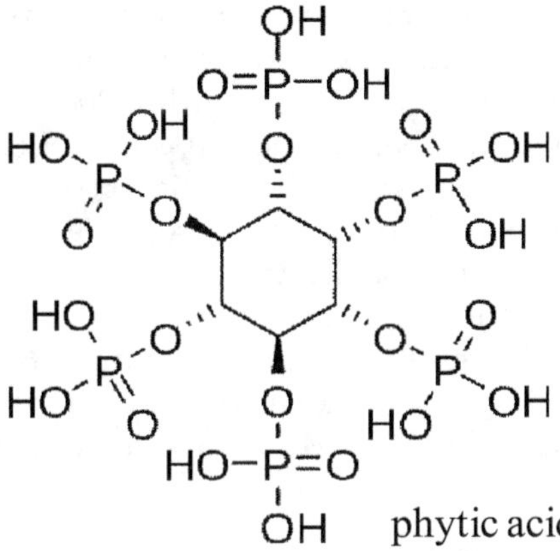

phytic acid

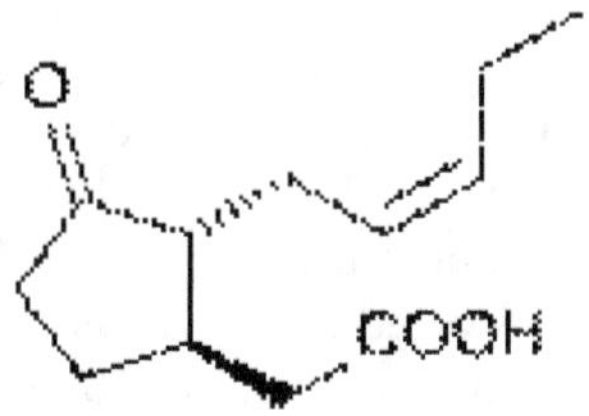

jasmonic acid

Figure 10.19: Some constituents of *Vicia faba.*

Phytic acid (inositol hexaphosphate, IP6) has been shown to inhibit human rhabdomyosarcoma *in vitro* and *in vivo* (Vucenik *et al.*, 1998). Phytic acid showed a dose-dependent cytotoxic effect in human leukemic cell lines and fresh chronic myelogenous leukaemia (CML) progenitor cells through extensive down-modulation of genes involved in transcription and cell cycle regulation and up-regulation of cell cycle inhibitors (Deliliers *et al.*, 2002). Phytic acid blocked HeLa tumor cell growth by activating apoptotic mechanisms (cytochrome c release, activation of caspases 3 and 9), as well as by inhibiting the Akt/NFκB-mediated cell survival pathway (Ferry *et al.*, 2002).

Phytic acid was found to inhibit growth of advanced prostate carcinoma DU145 cells through both blocking cell cycle progress and turning on the apoptotic machinery of the cell. Growth inhibition of cells was both time- and dose-dependent. There was an increased expression of cyclin-dependent kinase inhibitors (CDKIs), Cip1/p21 and Kip1/p27. Kinase activities associated with CDK2, 4 and 6 and cyclins E and D1 were inhibited. There was increased binding of Kip1/p27 and Cip1/p21 with cyclin D1 and E. Phytic acid increased hypophosphorylation of retinoblastoma (Rb)-related proteins, pRb/p107 and pRb2/p130. At higher doses and longer treatment times, phytic acid caused a marked increase in apoptosis which was accompanied by increased levels of cleaved PARP and active caspase 3 (Singh *et al.*, 2003). In fact, in this type of prostate cancer cells, p21/Cip1 and p27/Kip 1 seem to be essential molecular targets of phytic acid to demonstrate its anti-tumor efficacy (Roy *et al.*, 2009).

The compound was shown to suppress DMBA-induced mouse skin tumor development through modulation of p53 and Bcl-2 expressions and activation of caspases (Gupta *et al.*, 2003; Singh and Gupta, 2008). Administration of IP6 in drinking water led to decreases in tumor incidences and tumor multiplicity in UVB-induced skin tumors in SKH1 hairless mice (Kolappaswamy *et al.*, 2009).

The anti-cancer mechanism of action by IP6 in K-562 leukaemia cells has been shown to be associated with the modulation of multiple genes involved in immunity,

Wnt and IGF pathways, as well as PI3 kinase signaling, and apoptosis (Bozsik *et al.*, 2007). The anti-cancer effect of phytic acid (inositol hexaphosphate, IP6) seems to be enhanced in the presence of inositol (Vucenik and Shamsuddin, 2003). Several mechanisms have been postulated to be working behind the anti-cancer effect of the two compounds, namely, reduction in cell proliferation, differentiation of malignant cells (cytostatic effect), enhanced immunity and anti-oxidant properties, the latter two contributing to tumor cell destruction. Additionally, the compounds appear to enhance the anti-cancer effect of conventional chemotherapy, control cancer metastases, and improve quality of life (Vucenik and Shamsuddin, 2006). Novel irinotecan-loaded liposome using phytic acid has been shown to have high therapeutic efficiency in mouse colon 26 and human COLO 320DM tumor xenografts in mice (Hattori *et al.*, 2009).

Discussion

It is of interest to note that all 16 of the Fabaceae family plants used by the Kavirajes and TMPs of Bangladesh for treatment of cancer has been described in a number of scientific studies to have anti-cancer properties, arising from the presence of anti-cancer phytochemicals of various types. The available scientific reports validate the use of these Fabaceae family plants in the treatment of cancer. While cancer is not diagnosed by the Kavirajes or TMPs through modern diagnostic procedures, nevertheless, it appears that they have a general idea of cancer, irrespective of its origin, and treat cancer with plants that have been shown to have anti-cancer phytochemicals. It is also noteworthy that we observed only one tribe, namely the Garo tribe, who treats cancer. Many other tribes, where we conducted ethnomedicinal surveys do not have any idea of cancer. On the other hand, the Kavirajes of the mainstream population (*i.e.* the folk medicinal practitioners), who are in contact with allopathic doctors and have picked up some allopathic terminology, are the medicinal practitioners who treat cancer. What is surprising is that even while these Kavirajes do not use modern diagnostic procedures for detection of cancer and treats more or less all cancer types with the same plant(s), they have succeded in selecting plants with anti-cancer properties, as demonstrated throughout this Chapter.

Some of the plants used by the Kavirajes and the TMPs for treatment of cancer like *Abrus precatorius* or *Erythrina variegata* have been well studied as to their phytochemical constituents, among which are present a number of chemicals, which are effective against different types of cancer cell lines. Other plants like *Indigofera tinctoria*, *Mimosa pudica*, *Samanea saman* or *Vicia faba* needs more research on elucidation of their phytochemical constituents. Irrespective of the plant being well studied phytochemically or not, all plants used by the Kavirajes had at least one phytochemical, which has been shown in scientific studies to have anti-cancer properties, and some plants have a number of phytochemicals with anti-cancer properties, as reported in the scientific literature. The major conclusion is that modern scientific research is validating the use of the plants by the Kavirajes for treatment of cancer. It is thus of the utmost importance that all plants used by the Kavirajes for treatment of cancer be subjected to more thorough scientific analysis, for that could well lead to discovery of new and more efficacious anti-cancer drugs with minimal side-effects.

It is being increasingly recognized that plants can be potential sources of a new generation of anti-cancer drugs. Three of the currently used important anti-cancer compounds, namely, vincristine, vinblastine and taxol are all plant-derived. Gallic acid, ellagic acid, glycyrrhizin, quercetin, rutin, catechin, and epi-catechin, to name only a few, have already demonstrated potential to undergo clinical trials in humans as effective anti-cancer agents. These compounds are present in one or more of the Fabaceae family plants used by the Kaviarjes for treatment of cancer. In fact, two major classes of compounds present in the Fabaceae family plants described above are polyphenols (flavones, flavanoids, anthocyanins, anthocyanidins) and saponins. Both groups of compounds are at present being isolated from various plant sources and investigated seriously for their anti-cancer potentials, as discussed below.

A number of studies have indicated the anti-cancer effects of flavones, flavanoids, anthocyanins/anthocyanidins and saponins, isolated from various plants. Only a few instances will be cited here because the literature in this area is vast and goes beyond the scope of this Chapter. The anti-proliferative and anti-angiogenic effect of the flavone, eupatorin, isolated from leaves of *Orthosiphon stamineus* Benth. has been reported (Dolecková *et al.*, 2012). Gossypin (from *Hibiscus vitifolius* L.) has been shown to induce G2/M phase cell cycle arrest in human malignant glioma U251 cells (Shi *et al.*, 2012). The anti-cancer effects of the flavonoid diosmetin on cell cycle progression and proliferation of MDA-MB 468 breast cancer cells has been attributed to activation of CYP1 (Androutsopoulos *et al.*, 2009). Flavopiridol, a compound isolated from an indigenous plant of India, *Dysoxylum binectariferum* (Roxb.) Hook. f. ex Hiern. has reportedly selective anti-tumor activity both *in vitro* and *in vivo* for prostate carcinoma cells (Drees *et al.*, 1997).

The flavonoids diosmetin [from fruit peel of *Citrus limon* (L.) Burm.] and luteolin (from *Daucus carota* L.) have been shown to exert synergistic cytostatic effects in human hepatoma HepG2 cells (Androutsopoulos and Spandidos, 2012). Silibinin, isolated from *Silybum marianum* (L.) Gaertn. inhibited renal cell carcinoma progression via suppression of the EGFR/MMP-9 signaling pathway (Liang *et al.*, 2012). Luteolin, which is present in celery, green pepper, thyme, perilla, chamomile tea, carrots, olive oil, peppermint, rosemary, navel oranges, and oregano, has been shown to inhibit Aurora B kinase (a protein that functions in the attachment of the mitotic spindle to the centromere) activity and block proliferation of cancer cells (Xie *et al.*, 2012). Induction of mitochondria-mediated apoptosis in human gastric adenocarcinoma SGC-7901 cells by kuraridin and Nor-kurarinone isolated from *Sophora flavescens* Aiton has been reported (Rasul *et al.*, 2011b).

Other flavonoid compounds with reported anti-cancer effects include alpinetin (isolated from *Alpinia katsumadai* Hayata) with antiproliferative effect in BxPC-3 pancreatic cancer cells (Du *et al.*, 2012); didymin from *Monarda didyma* L. with apoptosis-inducing effect in neuroblastoma through inhibiting N-Myc and up-regulating RKIP (Singhal *et al.*, 2012); baicalein from *Terminalia arjuna* (Roxb.) Wight and Arn. bark with inhibitory action against migration and invasive properties of human hepatoma cells (Chiu *et al.*, 2011); icariin from *Epimedium grandiflorum* C. Morren with apoptosis-inducing activity in mouse MLTC-10 Leydig tumor cells mediated through activation of the mitochondrial pathway and down-regulation of the expression of piwil4 (Wang *et al.*, 2011c); and amentoflavone from *Araucaria angustifolia* (Bertol.) Kuntze with apoptotic inducing activity in human cervical cancer cells mediated through suppressing E7 expression, cell cycle arrest at sub-G1 phase and involvement of mitochondrial pathways (Lee *et al.*, 2011b).

Anthocyanins from black rice and Chinese blueberry (*Vaccinium uliginosum* L.) have been shown to inhibit, respectively, breast cancer cells *in vitro* and *in vivo*, and DLD-1 and COLO205 colorectal adenocarcinoma cells (Hui *et al.*, 2010; Zu *et al.*, 2010).

Like the polyphenolic compounds discussed above, saponins from various plants also have reported anti-cancer activities. Saponins isolated from the *Asparagus* genera of plants have been shown to induce apoptosis in human hepatoma cell line HepG2 through mitochondrial-mediated pathway (Ji *et al.*, 2012). Total saponins of *Tupistra chinensis* Baker reportedly induced apoptosis in A549 cells (Huang *et al.*, 2012). Steroidal saponins from *Dioscorea zingiberensis* Wright had an apoptosis-inducing effect in cancer cells (Tong *et al.*, 2012). Both *in vitro* and *in vivo*, anti-tumor activity of saponins from *Pulsatilla chinensis* (Bunge) Regel has been seen in human liver tumor 7402 cells (Xu *et al.*, 2012). Tubeimoside I [a triterpenoid saponin extracted from the tubers of *Bolbostemma paniculatum* (Maxim.) Franquet] demonstrated cytotoxicity in human choriocarcinoma JEG-3 cells through induction of cytochrome c release and apoptosis induction via the mitochondrial-related signaling pathway (Huang *et al.*, 2011). Platycodin D, isolated from roots of *Platycodon grandiflorum* (Jacq.) A. DC. has been shown to induce apoptosis in MCF-7 human breast cancer cells (Yu and Kim, 2010). Macranthoside B is a hederagenin saponin extracted from *Lonicera macaranthoides* Hand. Mazzet, which has shown anti-tumor activities *in vitro* and *in vivo* (Wang *et al.*, 2009). Steroid saponins from *Paris polyphylla* var. *yunnanensis* Franch has been shown to demonstrate *in vitro* and *in vivo* anti-cancer activity (Yan *et al.*, 2009). The triterpenoid saponins, ardisiacrispin (A + B), isolated from *Ardisia crenata* Sims showed pro-apoptotic and microtubule-disassembly effects on human hepatoma Bel-7402 cells (Li *et al.*, 2008).

In conclusion, Fabaceae family plants have an important role to play in the discovery and development of novel anti-cancer drugs. The list of Fabaceae family plant species is growing with growing scientific research on anti-cancer phytochemicals present in such plants. Besides the plants used in Bangladesh, a partial list of other Fabaceae family plants with reported anti-cancer phytochemical constituents include *Sophora flavescens* Aiton, *Sutherlandia frutescens* (L.) R.Br., *Retama monosperma* L., *Tephrosia purpurea* (L.) Pers., *Indigofera cassioides* Rottl. ex DC., *Millettia leucantha* Kurz., *Leucaenia leucocephala* (Lam.) de Wit, *Sesbania grandiflora* (L.) Pers., *Genista sessilifolia* DC., *Genista tinctoria* L., *Cajanus cajan* (L.) Millsp., *Gleditsia sinensis* Lam., *Milletia reticulata* Benth., *Erythrina abyssinica* Lam. ex DC., *Cassia tora* L., *Pterogyne nitens* Tul., *Psoralea corylifolia* L., *Senna italica* Mill., *Glycyrrhiza glabra* L., *Phaseolus vulgaris* L., *Erythrina addisoniae* Hutch., *Pterodon pubescens* Benth., *Pterocarpus santalinus* L.f., and *Copaifera multijuga* Hayne (Rasul *et al.*, 2011; Vorster *et al.*, 2012; Merghoub *et al.*, 2011; Gulecha and Sivakuma, 2011; Kumar *et al.*, 2011; Rayanil *et al.*, 2011; Haggag *et al.*, 2011; Sreelatha *et al.*, 2011; Rigano *et al.*, 2010; Luo *et al.*, 2010; Lee *et al.*, 2010; Fang *et al.*, 2010; Nguyen *et al.*, 2009; Rejiya *et al.*, 2009; Regasini *et al.*, 2009; Wang *et al.*, 2011d; Masoko *et al.*, 2009; Asi and Hosseinzadeh, 2008; Aparacio-Fernández *et al.*, 2008; Wätjen *et al.*, 2007; Vieira *et al.*, 2008; Kwon *et al.*, 2006; Lima *et al.*, 2003).

References

Abarikwu, S.O., Otuechere, C.A., Ekor, M., Monwuba, K., and Osobu, D. (2012). Rutin ameliorates cyclophosphamide-induced reproductive toxicity in male rats. *Toxicology International*, **19**: 207-214.

Abd-El-Wahab, S.M., Wassel, G.M., Ammar, N.M., and Hanna, T. (1987). Flavonoid constituents in the different organs of selected *Bauhinia* species and their effect on blood glucose. *Herba Hungarica*, **26**: 27-39.

Abraham, K.J., Daniel, M., and Sabnis, S.D. (1988). Phytoalexins of *Cassia fistula* Linn. and *Morinda tomentosa* Heyne. *National Academy Science Letters (India)*, **11**: 101-102.

Ackland, M.L., van de Waarsenburg, S., and Jones, R. (2005). Synergistic antiproliferative action of the flavonols quercetin and kaempferol in cultured human cancer cell lines. *In Vivo*, **19**: 69-76.

Agrawal, G.D., Rizvi, S.A.I., Gupta, P.C., and Tewari, J.D. (1970). Chemical examination of *Cassia fistula* pods. *Proceedings of the National Academy of Sciences, India, Section A: Physical Sciences*, **40**: 338-340.

Agrawal, G.D., Rizvi, S.A.I., Gupta, P.C., and Tewari, J.D. (1972). Structure of fistulic acid, a new coloring matter from the pods of *Cassia fistula*. *Planta Medica*, **21**: 150-155.

Agrawal, R.C., and Pandey, S. (2009). Evaluation of anticarcinogenic and antimutagenic potential of *Bauhinia variegata* extract in Swiss albino mice. *Asian Pacific Journal of Cancer Prevention*, **10**: 913-916.

Ahmad, K., and Rahman, A.F.M. (1965). P-sterone, a keto steroid from *Abrus precatorius*. *Pakistan Journal of Biological and Agricultural Sciences*, **8**: 218.

Ahmadu, A., Abdulkarim, A., Grougnet, R., Myrianthopoulos, V., Tillequin, F., Magiatis, P., and Skaltsounis, Alexios-Leandros (2010). Two new peltogynoids from *Acacia nilotica* Delile with kinase inhibitory activity. *Planta Medica*, **76**: 458-460.

Ahuja, A., Parshad, R., and Kaushik, J.P. (1988). Anthraquinones from callus cultures of *Cassia fistula*. *Fitoterapia*, **59**: 496-497.

Akhtar, N., Hasan, S.S., and Mahboobuddin (1972). Polysaccharide components of *Abrus precatorius*. *Science and Industry (Karachi)*, **9**: 204-206.

Akinloye, B.A., and Adalumo, L.A. (1981). *Abrus precatorius* leaves – a source of glycyrrhizin. *Nigerian Journal of Pharmacy*, **12**: 405.

Al-Hazzani, A.A., and Alshatwi, A.A. (2011). Catechin hydrate inhibits proliferation and mediates apoptosis of SiHa human cervical cancer cells. *Food and Chemical Toxicology*, **49**: 3281-3286.

Ali, B.H., Al-Salam, S., Al Husseini, I.S., Al-Lawati, I., Waly, M., Yasin, J., Fahim, M., and Nemmar, A. (2011). Abrogation of cisplatin-induced nephrotoxicity by emodin in rats. *Fundamental and Clinical Pharmacology*, (in press).

Alshatwi, A.A. (2010). Catechin hydrate suppresses MCF-7 proliferation through TP53.caspase-mediated apoptosis. *Journal of Experimental and Clinical Cancer Research*, **29**: 167.

Anam, E.M. (2003). Triterpenoid saponin from *Abrus precatorius* (Fabaceae). *Indian Journal of Chemistry, Section B: Organic Chemistry Including Medicinal Chemistry*, **42B**: 386-389.

Anbu, S.P., Arokiaraj, A., and Vijayarajan, D. (2011). Isolation, spectral and optical analysis of fistulic acid – a phytochemical constituent of *Cassia fistula* Linn. *Indian Journal of Science and Technology*, **4**: 422-424.

Androutsopoulos. V.P., Mahale, S., Arroo, R.R., and Potter, G. (2009). Anticancer effects of the flavonoid diosmetin on cell cycle progression and proliferation of MDA-MB 468 breast cancer cells due to CYP1 activation. *Oncology Reports*, **21**: 1525-1528.

Androutsopoulos, V.P., and Spandidos, D.A. (2012). The flavonoids diosmetin and luteolin exert synergistic cytostatic effects in human hepatoma HepG2 cells via CYP1A-catalyzed metabolism, activation of JNK and ERK and P53/P21 up-regulation. *Journal of Nutritional Biochemistry*, (in press).

Aparicio-Fernández, X., Reynoso-Camacho, R., Castaño-Tostado, E., García-Gasca, T., González de Mejia, E., Guzmán-Maldonado, S.H., Elizondo, G., Yousef, G.G., Lila, M.A., and Loarca-Pina, G. (2008). Antiradical capacity and induction of apoptosis on HeLa cells by a *Phaseolus vulgaris* extract. *Plant Foods for Human Nutrition*, **63**: 35-40.

Applewhite, P.B. (1973). Serotonin and norepinephrine in plant tissues. *Phytochemistry*, **12**: 191-192.

Arafa, R. (2001). Naringenin as an antidermatophyte flavonoid aglycone from *Bauhinia variegata*. *Al-Azhar Journal of Microbiology*, **51**: 17-23.

Arjumand, W., and Sultana, S. (2011). Glycyrrhizic acid: a phytochemical with a protective role against cisplatin-induced genotoxicity and nephrotoxicity. *Life Sciences*, **89**: 422-429.

Asi, M.N., and Hosseinzadeh, H. (2008). Review of pharmacological effects of *Glycyrrhiza* sp. and its bioactive compounds. *Phytotherapy Research*, **22**: 709-724.

Awad, A.B., and Fink, C.S. (2000). Phytosterols as anticancer dietary components: evidence and mechanism of action. *Journal of Nutrition*, **130**: 2127-2130.

Awad, A.B., Chan, K.C., Downie, A.C., and Fink, C.S. (2000). Peanuts as a source of β-sitosterol, a sterol with anticancer properties. *Nutrition and Cancer*, **36**: 238-241.

Awale, S., Linn, T.Z., Li, F., Tezuka, Y., Myint, A., Tomida, A., Yamori, T., Esumi, H., and Kadota, S. (2011). Identification of chrysoplenetin from *Vitex negundo* as a potential cytotoxic agent against PANC-1 and a panel of 39 human cancer cell lines (JFCR-39). *Phytotherapy Research*, **25**: 1770-1775.

Bagul, M., Srinivasa, H., Padh, H., and Rajani, M. (2005). A rapid densitometric method for simultaneous quantification of gallic acid and ellagic acid in herbal raw materials using HPTLC. *Journal of Separation Science*, **28**: 581-584.

Bakasso, S., Lamien-Meda, A., Lamien, C.E., Kiendrebeogo, M., Millogo, J., Ouedraogo, A.G., and Nacoulma, O.G. (2008). Polyphenol contents and antioxidant activities of five *Indigofera* species (Fabaceae) from Burkina Faso. *Pakistan Journal of Biological Sciences*, **11**: 1429-1435.

Balachandran, P., and Govindrajan, R. (2005). Cancer – an ayurvedic perspective. *Pharmacological Research*, **51**: 19-30.

Baskar, A.A., Ignacimuthu, S., Paulraj, G.M., and Al Numair, K.S. (2010). Chemopreventive potential of β-sitosterol in experimental colon cancer model – an *in vitro* and *in vivo* study. *BioMed Central Complementary and Alternative Medicine*, **10**: 24.

Behari, M., and Varshney, A. (1986). Sterols from *Desmodium* species. *Indian Drugs*, **23**: 434-435.

Bhardwaj, D.K., Bisht, M.S., and Mehta, C.K. (1980). Flavonoids from *Abrus precatorius*. *Phytochemistry*, **19**: 2040-2041.

Bhardwaj, D.K., Gupta, A.K., and Sharma, A.K. (1982). Syntheses of abrectorin and its isomers. *Indian Journal of Chemistry, Section B: Organic Chemistry Including Medicinal Chemistry*, **21B**: 98-100.

Bhardwaj, D.K., Gupta, A.K., Jain, A., and Shrawat, V.K. (1988). Synthesis of premtorin and its isomers. *Indian Journal of Chemistry, Section B: Organic Chemistry Including Medicinal Chemistry*, **27B**: 261-263.

Bhardwaj, D.K., Gupta, A.K., Jain, A., Shrawat, V.K., and Sharma, A. (1989). Constitution of premtorin: synthetic studies. *Proceedings of the Indian National Science Academy, Part A: Physical Sciences*, **55**: 50-53.

Bhaumik, H.L. (1987). Hydrocarbons, fatty acid, triterpenoid and sterols in the seeds of *Abrus precatorius*. *Science and Culture*, **53**: 23-24.

Bhope, S.G., Kuber, V.V., and Nagore, D.H. (2010). Validated HPTLC method for simultaneous quantification of sennoside A, sennoside B, and kaempferol in *Cassia fistula* Linn. *Acta Chromatographica*, **22**: 481-489.

Bhutia, S.K., Mallick, S.K., Maiti, S., and Maiti, T.K. (2009a). Inhibitory effect of Abrus abrin-derived peptide fraction against dalton's lymphoma ascites model. *Phytomedicine*, **16**: 377-385.

Bhutia, S.K., Mallick, S.K., Maiti, S., Mishra, D., and Maiti, T.K. (2009b). Abrus abrin derived peptides induce apoptosis by targeting mitochondria in HeLa cells. *Cell Biology International*, **33**: 720-727.

Bimla, M., and Kalidhar, S.B. (2005). Phytochemical study of the leaves of *Acacia arabica* (Lamk.). *National Journal of Plant Improvement*, **7**: 93-94.

Bodakhe, S.H., Ram, A., Bodakhe, K.S., and Pandey, D.P. (2010). New polyphenolic aromatic glycoside from *Bauhinia variegata* L. stem bark. *Asian Journal of Chemistry*, **22**: 3549-3553.

Bozsik, A., Kökény, S., and Olah, E. (2007). Molecular mechanisms for the antitumor activity of inositol hexakisphosphate (IP6). *Cancer Genomics and Proteomics*, **4**: 43-51.

Cha, T.L., Qiu, L., Chen, C.T., Wen, Y. and Hung, M.C. (2005). Emodin down-regulates androgen receptor and inhibits prostate cancer cell growth. *Cancer Research*, **65**: 2287-2295.

Chang, H.M., Chiang, T.C., and Mak, T.C.W. (1982). Isolation and structure elucidation of abruslactone A: a new oleanene-type triterpene from the roots and vines of *Abrus precatorius* L. *Journal of the Chemical Society, Chemical Communications*, **20**: 1197-1198.

Chang, H.C., Weng, C.F., Yen, M.H., Chuang, L.Y., and Hung, W.C. (2000). Modulation of cell cycle regulatory protein expression and suppression of tumor growth by mimosine in nude mice. *International Journal of Oncology*, **17**: 659-665.

Chang, Y.F., Chi, C.W., and Wang, J.J. (2006). Reactive oxygen species production is involved in quercetin-induced apoptosis in human hepatoma cells. *Nutrition and Cancer*, **55**: 201-209.

Chang, Y.C., Lai, T.Y., Yu, C.S., Chen, H.Y., Yang, J.S., Chueh, F.S., Lu, C.C., Chiang, J.H., Huang, W.W., Ma, C.Y., and Chung, J.G. (2011). Emodin induces apoptotic death in murine myelomonocytic leukaemia WEHI-3 cells *in vitro* and enhances phagocytosis in leukemia mice *in vivo*. *Evidence-based Complementary and Alternative Medicine*, **2011**: 523596.

Chatterjee, A., and Pakrashi, S. (1992). *In*: The Treatise of Indian Medicinal Plants. Deep Publication, New Delhi, vol. 2, p121.

Chaubal, R., Mujumdar, A.M., Puranik Vedavati, G., Deshpande, V.H., and Deshpande, N.R. (2003). Isolation and X-ray study of an anti-inflammatory active androstene steroid from *Acacia nilotica*. *Planta Medica*, **69**: 287-288.

Chaubal, R., Pawar Pushpa, V., Hebbalkar, G.D., Tungikar, V.B., Puranik Vedavati, G., Deshpande, V.H., and Deshpande, N.R. (2005). Larvicidal activity of *Acacia nilotica* extracts and isolation of D-pinitol – a bioactive carbohydrate. *Chemistry and Biodiversity*, **2**: 684-688.

Chawla, A.S., Krishnan, T.R., Jackson, A.H., and Scalabrin, D.A. 91988). Alkaloidal constituents of *Erythrina variegata* bark. *Planta Medica*, **54**: 526-528.

Chawla, H.M., and Sharma, S.K. (1993). Erythritol, a new isoquinoline alkaloid from *Erythrina variegata* flower. *Fitoterapia*, **64**: 15-17.

Chen, P.N., Chu, S.C., Chiou, H.L., Kuo, W.H., Chiang, C.L., and Hsieh, Y.S. (2006). Mulberry anthocyanins, cyaniding 3-rutinoside and cyaniding 3-glucoside, exhibited an inhibitory effect on the migration and invasion of a human lung cancer cell line. *Cancer Letters*, **235**: 248-259.

Chen, Y.H., Chen, H.Y., Hsu, C.L., and Yen, G.C. (2007a). Induction of apoptosis by the *Lactuca indica* L. in human leukemia cell line and its active components. *Journal of Agricultural and Food Chemistry*, **55**: 1743-1749.

Chen, D., Landia-Piwowar, K.R., Chen, M.S., and Dou, Q.P. (2007b). Inhibition of proteasome activity by the dietary flavonoid apigenin is associated with growth inhibition in cultured breast cancer cells and xenografts. *Breast Cancer Research*, **9**: R80.

Chen, S.H., Lin, K.Y., Chang, C.C., Fang, C.L., and Lin, C.P. (2007c). Aloe-emodin-induced apoptosis in human gastric carcinoma cells. *Food and Chemical Toxicology*, **45**: 2296-2303.

Chen, R.S., Jhan, J.Y., Su, Y.J., Lee, W.T., Cheng, C.M., Ciou, S.C., Lin, S.T., Chuang, S.M., Ko, J.C., and Lin, Y.W. (2009). Emodin enhances gefitinib-induced cytotoxicity via rad51 down regulation and ERK 1/2 inactivation. *Experimental Cell Research*, **315**: 2658-2672.

Chiang, T.C., Chang, H.M., and Mak, T.C. (1983). New oleanene-type triterpenes from *Abrus precatorius* and X-ray crystal structure of abrusgenic acid-methanol 1: 1 solvate. *Planta Medica*, **49**: 165-169.

Chiu, T.H., Lai, W.W., Hsia, T.C., Yang, J.S., Lai, T.Y., Wu, P.P., Ma, C.Y., Yeh, C.C., Ho, C.C., Lu, H.F., Wood, W.G., and Chung, J.G. (2009). Aloe-emodin induces cell death through S-phase arrest and caspase-dependent pathways in human tongue squamous cancer SCC-4 cells. *Anticancer Research*, **29**: 4503-4511.

Chiu, Y.W., Lin, T.H., Huang, W.S., Teng, C.Y., Liou, Y.S., Kuo, W.H., Lin, W.L., Huang, H.I., Tung, J.N., Huang, C.Y., Liu, J.Y., Wang, W.H., Hwang, J.M., and Kuo, H.C. (2011). Baicalein inhibits the migration and invasive properties of human hepatoma cells. *Toxicology and Applied Pharmacology*, **255**: 316-326.

Choi, E.J., and Kim, G.H. (2009a). Apigenin induces apoptosis through a mitochondria/caspase-pathway in human breast cancer MDA-MB-453 cells. *Journal of Clinical Biochemistry and Nutrition*, **44**: 260-265.

Choi, E.J., and Kim, G.H. (2009b). 5-Fluorouracil combined with apigenin enhances anticancer activity through induction of apoptosis in human breast cancer MDA-MB-453 cells. *Oncology Reports*, **22**: 1533-1537.

Choi, Y.H., Hussain, R.A., Pezzuto, J.M., Kinghorn, A.D., and Morton, J.F. (1989). Abrusosides A-D, four novel sweet-tasting triterpene glycosides from the leaves of *Abrus precatorius*. *Journal of Natural Products*, **52**: 1118-1127.

Constantinou, A., Stoner, G.D., Mehta, R., Rao, K., Runyan, C., and Moon, R. (1995). The dietary anticancer agent ellagic acid is a potent inhibitor of DNA topoisomerases *in vitro*. *Nutrition and Cancer*, **23**: 121-130.

Dan, M.S., Dan, S.S., and Mukhopadhayay, P. (1982). Chemical examination of three indigenous plants. *Journal of the Indian Chemical Society*, **59**: 419-420.

Das, C., and Jogi, S.R. (1998). 1,8-Dihydroxy-6-methoxy-3,7-dimethyl anthraquinone from root of *Cassia fistula*. *Oriental Journal of Chemistry*, **14**: 145-146.

Das, N., Chandran, P., and Chakraborty, S. (2011). Potent spermicidal effect of oleanolic acid 3-β-D-glucuronide, an active principle isolated from the plant *Sesbania sesban* Merrill. *Contraception*, **83**: 167-175.

Daulatabad, C.D., Hosamani, K.M., and Alagawadi, K.R. (1987). Epoxy and cyclopropenoid fatty acids in *Cassia* seed oils. *Journal of the Oil Technologists' Association of India (Mumbai, India)*, **19**: 41-42.

Deliliers, G.L., Servida, F., Fracchiolla, N.S., Ricci, C., Borsotti, C., Colombo, G., and Soligo, D. (2002). Effect of inositol hexaphosphate (IP(6)) on human normal and leukaemic haematopoietic cells. *British Journal of Haematology*, **117**: 577-587.

Dikmen, M., Ozturk, N., and Ozturk, Y. (2011). The antioxidant potency of *Punica granatum* L. fruit peel reduces cell proliferation and induces apoptosis on breast cancer. *Journal of Medicinal Food*, **14**: 1638-1646.

Dimas, K., Demetzos, C., Mitaku, S., Marselos, M., Tzavaras, T., and Kokkinopoulos, D. (2000). Cytotoxic activity of kaempferol glycosides against human leukaemic cell lines *in vitro*. *Pharmacological Research*, **41**: 83-86.

Dolecková, I., Rárová, L., Grúz, J., Vondrusová, M., Strnad, M., and Kryštof, V. (2012). Antiproliferative and antiangiogenic effects of flavones eupatorin, an active constituent of chloroform extract of *Orthosiphon stamineus* leaves. *Fitoterapia*, **83**: 1000-1007.

Drees, M., Dengler, W.A., Roth, T., Labonte, H., Mayo, J., Malspeis, L., Grever, M., Sausville, E.A., and Fiebig, H.H. (1997). Flavopiridol (L86-8275): selective antitumor activity *in vitro* and activity *in vivo* for prostate carcinoma cells. *Clinical Cancer Research*, **3**: 273-279.

Du, J., Tang, B., Wang, J., Sui, H., Jin, X., Wang, L., and Wang, Z. (2012). Antiproliferative effect of alpinetin in BxPC-3 pancreatic cancer cells. *International Journal of Molecular Medicine*, **29**: 607-612.

Dubey, N., Dubey, N., Mehta, R., and Saluja, A. (2009). Estimation of catechin in Ayurvedic oil formulations containing *Acacia catechu*. *Journal of AOAC International*, **92**: 1021-1026.

Dueñas, M., Sun, B., Hernández, T., Estrella, I., and Spranger, M.I. (2003). Proanthocyanidin composition in the seed coat of lentils (*Lens culinaris* L.). *Journal of Agricultural and Food Chemistry*, **51**: 7999-8004.

Duke, James A. (1992). *In*: Handbook of phytochemical constituents of GRAS herbs and other economic plants. Boca Raton, FL. CRC Press.

Duraipandiyan, V., Ignacimuthu, S., and Paulraj, M.G. (2011). Antifeedant and larvicidal activities of Rhein isolated from the flowers of *Cassia fistula* L. *Saudi Journal of Biological Sciences*, **18**: 129-133.

Dutta, A., and De, B. (2002). *Cassia fistula*: a source of sennoside B (chromosome morphology and sennoside content). *Acta Horticulturae* **576** (Proceedings of the International Conference on Medicinal and Aromatic Plants, 2001): 45-48.

el-Aaser, A.A., Zakhary, N.I., el-Guindy, S.M., hafiez, A.R., halawa, F., and Mokhtar, N. (1994). Effect of soybean, *Vicia faba*, and vitamin C on the carcinogenicity of DMBA. *Nutrition and Cancer*, **22**: 195-200.

Eldeen, I.M., Van Heerden, F.R., and Van Staden, J. (2010). *In vitro* biological activities of niloticane, a new bioactive cassane diterpene from the bark of *Acacia nilotica* subsp. *kraussiana*. *Journal of Ethnopharmacology*, **128**: 555-560.

El-Gengaihi, S., Karawya, M.S., Wassel, G., and Ibrahim, N. (1988). Investigation of flavonoids of *Abrus precatorius* L. *Herba Hungarica*, **27**: 27-33.

el-Sayed (1991). A rare kaempferol trisaccharide anti-tumor promoter from *Sesbania sesban*. *Pharmazie*, **46**: 679-680.

Englert, J., Jiang, Y., Cabalion, P., Oulad-Ali, A., and Anton, R. (1994). C-Glycosylflavones from aerial parts of *Mimosa pudica*. *Planta Medica*, **60**: 194.

Fang, S.C., Hsu, C.L., Lin, H.T., and Yen, G.C. (2010). Anticancer effects of flavonoid derivatives isolated from *Millettia reticulata* Benth in SK-Hep-1 human hepatocellular carcinoma cells. *Journal of Agricultural and Food Chemistry*, **58**: 814-820.

Fang, E.F., Hassanien, A.A., Wong, J.H., Bah, C.S., Soliman, S.S., and Ng, T.B. (2011). Isolation of a new trypsin inhibitor from the Faba bean (*Vicia faba* cv. Giza 843) with potential medicinal applications. *Protein and Peptide Letters*, **18**: 64-72.

Faris, M.A., Takruri, H.R., Shomaf, M.S., and Bustanji, Y.K. (2009). Chemopreventive effect of raw and cooked lentils (*Lens culinaris* L.) and soybeans (*Glycine max*) against azoxymethane-induced aberrant crypt foci. *Nutrition Research*, **29**: 355-362.

Ferry, S., Matsuda, M., Yoshida, H., and Hirata, M. (2002). Inositol hexakisphosphate blocks tumor cell growth by activating apoptotic machinery as well as by inhibiting the Akt/NFκB-mediated cell survival pathway. *Carcinogenesis*, **23**: 2031-2041.

Franz, G. (1993). The senna drug and its chemistry. *Pharmacology*, **47**: 12-16.

Gacche, R.N., Shegokar, H.D., Gond, D.S., Yang, Z., and Jadhav, A.D. (2011). Evaluation of selected flavonoids as antiangiogenic, anticancer, and radical scavenging agents: an experimental and in silico analysis. *Cell Biochemistry and Biophysics*, **61**: 651-663.

Gallasch, B.A., and Spiteller, G. (2000). Synthesis of 9,12-dioxo-10(Z)-dodecenoic acid, a new fatty acid metabolite derived from 9-hydroperoxy-10,12-octadecadienoic acid in lentil seed (Lens culinaris Medik.). *Lipids*, **35**: 953-960.

Ghosal, S., and Bhattacharya, S.K. (1972). *Desmodium* alkaloids. Pt-II chemical and pharmacological evaluation of *Desmodium gangeticum*. *Planta Medica*, **22**: 434-440.

Ghosal, S., Dutta, S.K., and Bhattacharya, S.K. (2010). *Erythrina* – chemical and pharmacological evaluation II. Alkaloids of *Erythrina variegata* L. *Journal of Pharmaceutical Sciences*, **61**: 1274-1277.

Ghosh, P., Thakur, S., Itoh, T., and Matsumoto, T. (1982). Sterols from flowers of *Cassia siamea* Lam., *Cassia sophera* Linn. and *Cassia fistula* Linn. *Indian Journal of Chemistry, Section B: Organic Chemistry Including Medicinal Chemistry*, **21B**: 796-797.

Ghosh, D., and Maiti, T.K. (2007). Immunomodulatory and anti-tumor activities of native and heat denatured Abrus agglutinin. *Immunobiology*, **212**: 589-599.

Godal, A. (1993). Abrin variants and immunotoxins. *PCT International Application*, WO 9300367 A1 19930107.

Gonzalez-Mejia, M.E., Voss, O.H., Murnan, E.J., and Doseff, A.I. (2010). Apigenin-induced apoptosis of leukemia cells is mediated by a bimodal and differentially regulated residue-specific phosphorylation of heat-shock protein-27. *Cell Death and Disease*, **1**: e64.

Gulecha, V., and Sivakuma, T. (2011). Anticancer activity of *Tephrosia purpurea* and *Ficus religiosa* using MCF 7 cell lines. *Asian Pacific Journal of Tropical Medicine*, **4**: 526-529.

Gulzar, A., Manjul, P.S., Anita, S., Upendra, K., and Yatendra, K. (2010). Preliminary phytochemical and antimicrobial screening of leaves extract of *Acacia catechu* Willd. *Journal of Pharmacy Research*, **3**: 2583-2584.

Guo, J.M., Xiao, B.X., Liu, Q., Zhang, S., Liu, D.H., and Gong, Z.H. (2007). Anti-cancer effect of aloe-emodin on cervical cancer cells involves G2/M arrest and induction of differentiation. *Acta Pharmacologica Sinica*, **28**: 1991-1995.

Gupta, A.K., Vidyapati, T.J., and Chauhan, J.S. (1979). 5,7-dihydroxyflavanone-4'-O-b-L-rhamnopyranosyl-β-D-glucopyranoside from the stem of *Bauhinia variegata*. *Indian Journal of Chemistry, Section B: Organic Chemistry Including Medicinal Chemistry*, **18B**: 85-86.

Gupta, A.K., Vidyapati, T.J., and Chauhan, J.S. (1980). Chemical examination of the stem of *Bauhinia variegata*. *Planta Medica*, **38**: 174-176.

Gupta, A.K., and Chauhan, J.S. (1984). Constituents from the stem of *Bauhinia variegata*. *National Academy Science Letters (India)*, **7**: 15-16.

Gupta, V., Agrawal, A., and Tiwari, H.P. (1989). Isolation and characterization of two flavonols and a xanthone glycoside from the stem bark of *Cassia fistula* Linn. *Indian Journal of Chemistry, Section B: Organic Chemistry Including Medicinal Chemistry*, **28B**: 282-284.

Gupta, A., Siddiqui, I.R., Singh, J., and Sharma, J.P. (1998). Two new anthraquinone derivatives from the seeds of *Cassia angustifolia*. *Indian Journal of Chemistry, Section B: Organic Chemistry Including Medicinal Chemistry*, **37B**: 615-617.

Gupta, M., Mazumder, U.K., Rath, N., and Mukhopadhyay, D.K. (2000). Antitumor activity of methanolic extract of *Cassia fistula* L. seed against Ehrlich ascites carcinoma. *Journal of Ethnopharmacology*, **72**: 151-156.

Gupta, A., Singh, J., and Sharma, J.P. (2002). Anthraquinone and dihydroflavonol from *Cassia angustifolia*. *Journal of the Indian Chemical Society*, **79**: 289-290.

Gupta, K.P., Singh, J., and Bharathi, R. (2003). Suppression of DMBA-induced mouse skin tumor development by inositol hexaphosphate and its mode of action. *Nutrition and Cancer*, **46**: 66-72.

Gupta, S.P., and Chhabra, B.S. (2012). New triterpenoid glycosides from *Abrus precatorius*. *Journal of the Indian Chemical Society*, **89**: 675-677.

Haggag, *E.G.*, Kamal, A.M., Abdelhady, M.I., El-Sayed, M.M., El-Wakil, E.A., and Abd-El-hamed, S.S. (2011). Antioxidant and cytotoxic activity of polyphenolic compounds isolated from the leaves of *Leucenia leucocephala*. *Pharmaceutical Biology*, **49**: 1103-1113.

Han, R. (1994). Highlight on the studies of anticancer drugs derived from plants in China. *Stem Cells*, **12**: 53-63.

Haq, Q.N., Awal, A., and Kiamuddin, M. (1973). Polysaccharides from the roots of *Abrus precatorius*. *Bangladesh Journal of Scientific and Industrial Research*, **8**: 47-52.

Haque, M.A., Shaha, M.K., Ahmed, S.U., Akter, R., Rahman, H., Chakravotry, S., Imran, A.H.M.N., Islam, M.T., Das, R.C., and Rahmatullah, M. (2011). Use of inorganic substances in folk medicinal formulations: a case study of a folk medicinal practitioner in Tangail district, Bangladesh. *American Eurasian Journal of Sustainable Agriculture*, **5**: 415-423.

Haridas, V., Higuchi, M., Jayatilake, G.S., Bailey, D., Mujoo, K., Blake, M.E., Arntzen, C.J., and Gutterman, J.U. (2001). Avicins: triterpenoid saponins from *Acacia victoriae* (Bentham) induce apoptosis by mitochondrial perturbation. *Proceedings of the National Academy of Sciences USA*, **98**: 5821-5826.

Hasan, M.M., Annay, M.E.A., Sintaha, M., Khaleque, H.N., Noor, F.A., Nahar, A., Seraj, S., Jahan, R., Chowdhury, M.H., and Rahmatullah, M. (2010). A survey of medicinal plant usage by folk medicinal practitioners in seven villages of Ishwardi Upazilla, Pabna district, Bangladesh. *American Eurasian Journal of Sustainable Agriculture*, **4**: 326-333.

Hattori, Y., Shi, L., Ding, W., Koga, K., Kawano, K., Hakoshima, M., and Maitani, Y. (2009). Novel irinotecan-loaded liposome using phytic acid with high therapeutic efficacy for colon tumors. *Journal of Controlled Release*, **136**: 30-37.

Hedge, V.R., Dai, P., Patel, M.J., Puar, M.S., Das, P., Pai, J., Bryant, R., and Cox, P.A. (1997). Phospholipase A2 inhibitors from an *Erythrina* species from Samoa. *Journal of Natural Products*, **60**: 537-539.

Herrmann, M.S., and Behnke, D.W. (1981). A characterization of abrin A from the seeds of the *Abrus precatorius* plant. *Biochimica et Biophysica Acta, Protein Structure*, **667**: 397-410.

Hossan, M.S., Hanif, A., Agarwala, B., Sarwar, M.S., Karim, M., Rahman, M.T., Jahan, R., and Rahmatullah, M. (2010). Traditional use of medicinal plants in Bangladesh to treat urinary tract infections and sexually transmitted diseases. *Ethnobotany Research and Applications*, **8**: 61-74.

Hsu, H.F., Huang, K.H., Lu, K.J., Chiou, S.J., Yen, J.H., Chang, C.C., and Houng, J.Y. (2011). *Typhonium blumei* extract inhibits proliferation of human lung adenocarcinoma A549 cells via induction of cell cycle arrest and apoptosis. *Journal of Ethnopharmacology*, **135**: 492-500.

Huang, Y.T., Hwang, J.J., Lee, P.P., Ke, F.C., Huang, J.H., Huang, C.J., Kandaswami, C., Middleton, E. Jr., and Lee, M.T. (1999). Effects of luteolin and quercetin, inhibitors of tyrosine kinase, on cell growth and metastasis-associated properties in A431 cells overexpressing epidermal growth factor receptor. *British Journal of Pharmacology*, **128**: 999-1010.

Huang, P., Yu, C., Liu, X.Q., Ding, Y.B., Wang, Y.X., and He, J.L. (2011). Cytotoxicity of tubeimoside I in human choriocarcinoma JEG-3 cells by induction of cytochrome c release and apoptosis via the mitochondrial-related signaling pathway. *International Journal of Molecular Medicine*, **28**: 579-587.

Huang, W., Zhang, H., Zou, K., Chen, J., Li, X., Liu, C., and Huang, N. (2012). Total saponins of *Tupistra chinensis* induces apoptosis in A549 cells. *Neoplasma*, (in press).

Hui, C., Bin, Y., Xiaoping, Y., Long, Y., Chunye, C., Mantian, M., and Wenhua, L. (2010). Anticancer activities of an anthocyanin-rich extract from black rice against breast cancer cells *in vitro* and *in vivo*. *Nutrition and Cancer*, **62**: 1128-1136.

Hye, M.A., Taher, M.A., Ali, M.Y., Ali, M.U., and zaman, S. (2009). Isolation of (+)-catechin from *Acacia catechu* (Cutch tree) by a convenient method. *Journal of Scientific Research*, **1**: 300-305.

Jagtap, S., Meganathan, K., Wagh, V., Winkler, J., Hescheler, J., and Sachinidis, A. (2009). Chemopreventive mechanism of the natural compounds, epigallocatechin-3-O-gallate, quercetin and curcumin against cancer and cardiovascular diseases. *Current Medicinal Chemistry*, **16**: 1451-1462.

Jahan, F.I., Hasan, M.R.U., Jahan, R., Seraj, S., Chowdhury, A.R., Islam, M.T., Khatun, Z., and Rahmatullah, M. (2011). A Comparison of Medicinal Plant Usage by Folk Medicinal Practitioners of two Adjoining Villages in Lalmonirhat district, Bangladesh. *American Eurasian Journal of Sustainable Agriculture*, **5**: 46-66.

Jain, R., Chaurasia, S., Saxena, R.C., and Jain, D.K. (2011). Identification of flavonoids in the fruit of *Cassia fistula* by high performance liquid chromatography-electrospray mass spectrophotometry. *Asian Journal of Chemistry*, **23**: 2109-2111.

Jayaprakasha, G.K., Mandadi, K.K., Poulose, S.M., Jadegoud, Y., Nagana Gowda, G.A., and Patil, B.S. (2007). Inhibition of colon cancer cell growth and antioxidant activity of bioactive compounds from *Poncirus trifoliata* (L.) Raf. *Bioorganic and Medicinal Chemistry*, **15**: 4923-4932.

Ji, Y., Ji, C., Yue, L., and Xu, H. (2012). Saponins isolated from *Asparagus* induce apoptosis in human hepatoma cell line HepG2 through a mitochondrial-mediated pathway. *Current Oncology*, **19 (Supplement 2)**: eS1-eS9.

Jothy, S.L., Zakaria, Z., Chen, Y., Lau, Y.L., Latha, L.Y., Shin, L.N., and Sasidharan, S. (2011). Bioassay-directed isolation of active compounds with antiyeast activity from a *Cassia fistula* seed extract. *Molecules*, **16**: 7583-7592.

Juan, M.E., Planas, J.M., Ruiz-Gutierrez, V., Daniel, H., and Wenzel, U. (2008). Antiproliferative and apoptosis-inducing effects of maslinic and oleanolic acids, two pentacyclic triterpenes from olives, on HT-29 colon cancer cells. *British Journal of Nutrition*, **100**: 36-43.

Jung, S.K., Lee, K.W., Byun, S., Kang, N.J., Lim, S.H., Heo, Y.S., Bode, A.M., Bowden, G.T., Lee, H.J., and Dong, Z. (2008). Myricetin suppresses UVB-induced skin cancer by targeting Fyn. *Cancer Research*, **68**: 6021-6029.

Kalaivani, T., Rajasekaran, C., and Mathew, L. (2011a). Free radical scavenging, cytotoxic, and hemolytic activities of an active antioxidant compound ethyl gallate from leaves of *Acacia nilotica* (L.) Wild. Ex. Delile subsp. *indica* (Benth.) Brenan. *Journal of Food Science*, **76**: T144-T149.

Kalaivani, T., Rajasekaran, C., Suthindhiran, K., and Mathew, L. (2011). Free radical scavenging, cytotoxic and hemolytic activities from leaves of *Acacia nilotica* (L.) Wild. ex. Delile subsp. *indica* (Benth.) Brenan. *Evidence-based Complementary and Alternative Medicine*, **2011**: 274741.

Kale, R., Saraf, M., Juvekar, A., and Tayade, P. (2006). Decreased b16f10 melanoma growth and impaired tumour vascularization in BDF1 mice with quercetin-cyclodextrin binary system. *The Journal of Pharmacy and Pharmacology*, **58**: 1351-1358.

Kang, J.W., Kim, J.H., Song, K., Kim, S.H., Yoon, J.H., and Kim, K.S. (2010). Kaempferol and quercetin, components of *Gingko biloba* extract (EGb 761), induce caspase-3-dependent apoptosis in oral cavity cancer cells. *Phytotherapy Research*, **24 Suppl 1**: S77-S82.

Karawya, M.S., El-Gengaihi, S., Wassel, G., and Ibrahim, N. (1980). Phytochemical studies of *Abrus precatorius* alkaloids. *Herba Hungarica*, **19**: 21-25.

Karawya, M.S., El-Gengaihi, S., Wassel, G., and Ibrahim, N.A. (1981a). Anthocyanins from the seeds of *Abrus precatorius*. *Fitoterapia*, **52**: 175-177.

Karawya, M.S., El-Gengaihi, S., Wassel, G., and Ibrahim, N.A. (1981b). Carbohydrates of *Abrus precatorius*. *Fitoterapia*, **52**: 179-181.

Kaur, K., Arora, S., Hawthorne, M.E., Kaur, S., Kumar, S., and Mehta, R.G. (2002). A correlative study on antimutagenic and chemopreventive activity of *Acacia auriculiformis* A. Cunn. and *Acacia nilotica* (L.) Willd. Ex Del. *Drug and Chemical Toxicology*, **25**: 39-64.

Kaur, K., Michael, H., Arora, S., Harkonen, P., and Kumar, S. (2005). *In vitro* bioactivity-guided fractionation and characterization of polyphenolic inhibitory fractions from *Acacia nilotica* (L.) Willd. ex Del. *Journal of Ethnopharmacology*, **99**: 353-360.

Kennelly, E.J., Cai, L., Kim, N.-C., and Kinghorn, A.D. (1996). Potential sweetening agents of plant origin. Part 31. Abrusoside E, a further sweet-tasting cycloartane glycoside from the leaves of *Abrus precatorius*. *Phytochemistry*, **41**: 1381-1383.

Khaleque, A., and Haroon, S.N. (1970). Glycosides and aglycones. VIII. Constituents of *Cassia fistula*. *Scientific Researches (Dacca)*, **7**: 63-66.

Khalid, S.A., Yagi, S.M., Khristova, P., and Duddeck, H. (1989). (+)-Catechin-5-galloyl ester as a novel natural polyphenol from the bark of *Acacia nilotica* of Sudanese origin. *Planta Medica*, **55**: 556-558.

Khan, M., Shingare, M.S., Siddiqui, A.R., Angadi, S.S., Maske, P.V., Yahya, Mohd., and Siddiqui, M.A. (2005). TLC-colourimetric estimation of free and combined forms of chrysophanol, emodin, and physcione in some *Cassia* species. *Indian Journal of Pharmaceutical Sciences*, **67**: 767-770.

Khan, N.A., and Srivastava, A. (2009). Antifungal activity of bioactive triterpenoid saponin from the seeds of *Cassia angustifolia*. *Natural Product Research*, **23**: 1128-1133.

Kim, N.-C., Kim, D.S.H.L., and Kinghorn, A.D. (2002). New triterpenoids from the leaves of *Abrus precatorius*. *Natural Product Letters*, **16**: 261-266.

Kim, S., Liu, Y., Gaber, M.W., Bumgardner, J.D., Haggard, W.O., and Yang, Y. (2009). Development of chitosan-ellagic acid films as a local drug delivery system to induce apoptotic death of human melanoma cells. *Journal of Biomedical Materials Research. Part B, Applied Biomaterials*, **90**: 145-155.

Kinjo, J., Matsumoto, K., Inoue, M., Takeshita, T., and Nohara, T. (1991). Studies on leguminous plants. Part XIX. A new sapogenol and other constituents in abri semen, the seeds of *Abrus precatorius* L. *Chemical and Pharmaceutical Bulletin*, **39**: 116-119.

Kinjo, J., Ikeda, T., Watanabe, K., and Nohara, T. (1994). An anthraquinone glycoside from *Cassia angustifolia* leaves. *Phytochemistry*, **37**: 1685-1687.

Ko, J.C., Su, Y.J., Lin, S.T., Jhan, J.Y., Ciou, S.C., Cheng, C.M., and Lin, Y.W. (2010a). Suppression of ERCC1 and Rad51 expression through ERK 1/2 inactivation is essential in emodin-mediated cytotoxicity in human non-small cell lung cancer cells. *Biochemical Pharmacology*, **79**: 655-664.

Ko, J.C., Su, Y.J., Lin, S.T., Jhan, J.Y., Ciou, S.C., Cheng, C.M., Chiu, Y.F., Kuo, Y.H., Tsai, M.S., and Lin, Y.W. (2010b). Emodin enhances cisplatin-induced cytotoxicity via down-regulation of ERCC1 and inactivation of ERK 1/2. *Lung Cancer*, **69**: 155-164.

Kolappaswamy, K., Williams, K.A., Benazzi, C., sarli, G., McLeod, C.G.Jr., Vucenik, I., and DeTolla, L.J. (2009). Effect of inositol hexaphosphate on the development of UVB-induced skin tumors in SKH1 hairless mice. *Comparative Medicine*, **59**: 147-152.

Kramell, R., Schmidt, J., Herrmann, G., and Schliemann, W. (2005). N-(jasmonoyl)tyrosine-derived compounds from flowers of broad beans (*Vicia faba*). *Journal of Natural Products*, **68**: 1345-1349.

Kumar, A., Pande, C.S., and Kaul, R.K. (1966). Chemical examination of *Cassia fistula* flowers. *Indian Journal of Chemistry*, **4**: 460.

Kumar, A., Lingadurai, S., Jain, A., and Barman, N.R. (2010). *Erythrina variegata* Linn: A review on morphology, phytochemistry, and pharmacological aspects. *Pharmacognosy Review*, **4**: 147-152.

Kumar, R.S., Rajkapoor, B., and Perumal, P. (2011a). *In vitro* and *in vivo* anticancer activity of *Indigofera cassioides* Rottl. Ex. DC. *Asian Pacific Journal of Tropical Medicine*, **4**: 379-385.

Kumar, S.S., Karrunakaran, C.M., Rao, M.R., and Balasubramanian, M.P. (2011b). Inhibitory effects of *Indigofera aspalathoides* on 20-methylcholanthrene-induced chemical carcinogenesis in rats. *Journal of Carcinogenesis*, **10**: 1.

Kuo, S.C., Chen, S.C., Chen, L.H., Wu, J.B., Wang, J.P., and Teng, C.M. (1995). Potent antiplatelet, anti-inflammatoey and antiallergic isoflavanquinones from the roots of *Abrus precatorius*. *Planta Medica*, **61**: 307-312.

Kuo, P.L., Lin, T.C., and Lin, C.C. (2002). The antiproliferative activity of aloe-emodin is through p53-dependent and p21-dependent apoptotic pathway in human hepatoma cell lines. *Life Sciences*, **71**: 1879-1892.

Kuo, Y.H., Lee, P.H., and Wein, Y.S. (2002). Four new compounds from the seeds of *Cassia fistula*. *Journal of Natural Products*, **65**: 1165-1167.

Kuttan, G., Pratheeshkumar, P., Manu, K.A., and Kuttan, R. (2011). Inhibition of tumor progression by naturally occurring terpenoids. *Pharmaceutical Biology*, **49**: 995-1007.

Kwon, H.J., Hong, Y.K., Kim, K.H., Han, C.H., Cho, S.H., Choi, J.S., and Kim, B.W. (2006). Methanolic extract of *Pterocarpus santalinus* induces apoptosis in HeLa cells. *Journal of Ethnopharmacology*, **105**: 229-234.

Laladhas, K.P., Cheriyan, V.T., Puliappadamba, V.T., Bava, S.V., Unnithan, R.G., Vijayammal, P.L., and Anto, R.J. (2010). A novel protein fraction from *Sesbania grandiflora* shows potential anticancer and chemopreventive efficacy, *in vitro* and *in vivo*. *Journal of Cellular and Molecular Medicine*, **14**: 636-646.

Lamson, D.W., and Brignall, M.S. (2000). Antioxidants and cancer, part 3: quercetin. *Alternative Medicine Review*, **5**: 196-208.

Laszczyk, M.N. (2009). Pentacyclic triterpenes of the lupine, oleanane and ursane group as tools in cancer therapy. *Planta Medica*, **75**: 1549-1560.

Ledwani, L., and Singh, M. (2006). Isolation and characterization of anthraquinones from the stem bark of *Cassia* species. *Journal of the Indian Chemical Society*, **83**: 383-385.

Lee, C.K., Lee, P.H., and Kuo, Y.H. (2001). The chemical constituents from the aril of *Cassia fistula* L. *Journal of the Chinese Chemical Society (Taipei, Taiwan)*, **48**: 1053-1058.

Lee, K.W., Kang, N.J., Rogozin, E.A., Kim, H.G., Cho, Y.Y., Bode, A.M., Lee, H.J., Surh, Y.J., Bowden, G.T., and Dong, Z. (2007). Myricetin is a novel natural inhibitor of neoplastic cell transformation and MEK1. *Carcinogenesis*, **28**: 1918-1927.

Lee, S.J., Park, K., Ha, S.D., Kim, W.J., and Moon, S.K. (2010). *Gleditsia sinensis* thorn extract inhibits human colon cancer cells: the role of ERK1/2, G2/M-phase cell cycle arrest and p53 expression. *Phytotherapy Research*, **24**: 1870-1876.

Lee, M.S., Cha, E.Y., Sul, J.Y., Song, I.S., and Kim, J.Y. (2011a). Chrysophanic acid blocks proliferation of colon cancer cells by inhibiting EGFR/mTOR pathway. *Phytotherapy Research*, **25**: 833-837.

Lee, S., Kim, H., Kang, J.W., Kim, J.H., Lee, D.H., Kim, M.S., Yang, Y., Woo, E.R., Kim, Y.M., Hong, J., and Yoon, D.Y. (2011b). The biflavonoid amnetoflavone induces apoptosis via suppressing E7 expression, cell cycle arrest at sub-G_1 phase, and mitochondria-emanated intrinsic pathways in human cervical cancer cells. *Journal of Medicinal Food*, **14**: 808-816.

Lemli, J., Toppet, S., Cuveele, J., and Janssen, G. (1981). Naphthalene glycosides in *Cassia senna* and *Cassia angustifolia*. Studies in the field of drugs containing anthracene derivatives. XXXII. *Planta Medica*, **43**: 11-17.

Li, Y.L., Gan, G.P., Zhang, H.Z., Wu, H.Z., Li, C.L., Huang, Y.P., Liu, Y.W., and Liu, J.W. (2007). A flavonoid glycoside isolated from *Smilax china* L. rhizome *in vitro* anticancer effects on human cancer cell lines. *Journal of Ethnopharmacology*, **113**: 115-124.

Li, M., Wei, S.Y., Xu, B., Guo, W., Liu, D.L., Cui, J.R., and Yao, X.S. (2008). Pro-apoptotic and microtubule-disassembly effects of ardisiacrispin (A+B), triterpenoid saponins from *Ardisia crenata* on human hepatoma Bel-7402 cells. *Journal of Asian Natural Products Research*, **10**: 739-746.

Li, X., Wang, H., Liu, C., and Chen, R. (2010). Chemical constituents of *Acacia catechu*. *Zhongguo Zhong Yao Za Zhi*, **35**: 1425-1427.

Liang, C.Z., Zhang, X., Li, H., Tao, Y.Q., Tao, L.J., Yang, Z.R., Zhou, X.P., Shi, Z.L., and Tao, H.M. (2012). *Cancer Biotherapy and Radiopharmaceuticals* (in press).

Liao, Y.F., Rao, Y.K., and Tzeng, Y.M. (2012). Aqueous extract of *Anisomeles indica* and its purified compound exerts anti-metastatic activity through inhibition of NF-κB/AP-1-dependent MMP-9 activation in human breast cancer MCF-7 cells. *Food and Chemical Toxicology*, **50**: 2930-2936.

Liao, C.L., Lai, K.C., Huang, A.C., Yang, J.S., Lin, J.J., Wu, S.H., Gibson, W.W., Lin, J.G., and Chung, J.G. Gallic acid inhibits migration and invasion in human osteosarcoma U-2 OS cells through suppressing the matrix metalloproteinase-2/-9, protein kinase B (PKB) and PKC signaling pathways. *Food and Chemical Toxicology*, **50**: 1734-1740.

Lima, S.R., Junior, V.F., Christo, H.B., Pinto, A.C., and Fernandes, P.D. (2003). *In vivo* and *in vitro* studies on the anticancer activity of *Copaifera multijuga* Hayne and its fractions. *Phytotherapy Research*, **17**: 1048-1053.

Limmatvapirat, C., Sirisopanaporn, S., and Kittakoop, P. (2004). Antitubercular and antiplasmodial constituents of *Abrus precatorius*. *Planta Medica*, **70**: 276-278.

Li, X.C., Liu, C., Yang, L.X., and Chen, R.Y. (2011). Phenolic compounds from the aqueous extract of *Acacia catechu*. *Journal of Asian Natural Product Research*, **13**: 826-830.

Liang, L., Li, L., Zeng, J., Gao, Y., Chen, Y.L., Wang, Z.Q., Wang, X.Y., Chang, L.S., and He, D. (2012). Inhibitory effect of silibinin on EGFR signal-induced renal cell carcinoma progression via suppression of the EGFR/MMP-9 signaling pathway. *Oncology Reports*, **28**: 999-1005.

Lin, J.G., Chen, G.W., Li, T.M., Chouh, S.T., Tan, T.W., and Chung, J.G. (2006). Aloe-emodin induces apoptosis in T24 human bladder cancer cells through the p53 dependent apoptotic pathway. *The Journal of Urology*, **175**: 343-347.

Lin, J.P., Yang, J.S., Lin, J.J., Lai, K.C., Lu, H.F., Ma, C.Y., Sai-Chuen, W.R., Wu, K.C., Chueh, F.S., Gibson Wood, W., and Chung, J.G. (2012). Rutin inhibits heman leukemia tumor growth in a murine xenograft model *in vivo*. *Environmental Toxicology*, **27**: 480-484.

Lin, J.-Y., and Tung, T.-C. (1972). Phytotoxic proteins. Abrin and ricin. *Journal of the Chinese Biochemical Society*, **1**: 1P-20P.

Lin, J.Y., Lee, T.C., and Tung, T.A. (1978a). Isolation of antitumor proteins abrin-a and abrin-b from *Abrus precatorius*. *In*: Practical Application of Biochemistry in Economically Developed Countries, Proceedings of FAOB Symposium., 1st, pp128-140.

Lin, J.Y., Lee, T.C., and Tung, T.C. (1978). Isolation of antitumor proteins abrin-A and abrin-B from *Abrus precatorius*. *International Journal of Peptide and Protein Research*, **12**: 311-317.

Lin, J.Y., Lee, T.C., Hu, S.T., and Tung, T.C. (1981). Isolation of four isotoxic proteins and one agglutinin from jequiriti bean (*Abrus precatorius*). *Toxicon*, **19**: 41-51.

Lin, S.Y., Lai, W.W., Ho, C.C., Yu, F.S., Chen, G.W., Yang, J.S., Liu, K.C., Lin, M.L., Wu, P.P., Fan, M.J., and Chung, J.G. (2009). Emodin induces apoptosis of human tongue squamous cancer SCC-4 cells through reactive oxygen species and mitochondria-dependent pathways. *Anticancer Research*, **29**: 327-335.

Liu, K.C., Huang, A.C., Wu, P.P., Lin, H.Y., Chueh, F.S., Yang, J.S., Lu, C.C., Chiang, J.H., Meng, M., and Chung, J.G. (2011). Gallic acid suppresses the migration and invasion of PC-3 human prostate cancer cells via inhibition of matrix metalloproteinase-2 and -9 signaling pathways. *Oncology Reports*, **26**: 177-184.

Liu, L.Z., Fang, J., Zhou, Q., Hu, X., Shi, X., and Jiang, B.H. (2005). Apigenin inhibits expression of vascular endothelial growth factor and angiogenesis in human lung cancer cells: implication of chemoprevention of lung cancer. *Molecular Pharmacology*, **68**: 635-643.

Lo, C., Lai, T.Y., Yang, J.H., Yang, J.S., Ma, Y.S., Weng, S.W., Chen, Y.Y., Lin, J.G., and Chung, J.G. (2010). Gallic acid induces apoptosis in A375.S2 human melanoma cells through caspase-dependent and –independent pathways. *International Journal of Oncology*, **37**: 377-385.

Long, X., Fan, M., Bigsby, R.M., and Nephew, K.P. (2008). Apigenin inhibits antiestrogen-resistant breast cancer cell growth through estrogen receptor-α-dependent and estrogen receptor-α-independent mechanisms. *Molecular Cancer Therapeutics*, **7**: 2096-2108.

Lu, Y., Hou, S.X., and Chen, T. (2003). Advances in the study of vincristine: and anticancer ingredient from *Catharanthus roseus*. *Zhongguo Zhong Yao Za Zhi*, **28**: 1006-1009.

Lu, J., Papp, L.V., Fang, J., Rodriguez-Nieto, S., Zhivotovsky, B., and Holmgren, A. (2006). Inhibition of mammalian thioredoxin reductase by some flavonoids: implications for myricetin and quercetin anticancer activity. *Cancer Research*, **66**: 4410-4418.

Lu, G.D., Shen, H.M., Ong, C.N., and Chung, M.C. (2007). Anticancer effects of aloe-emodin on HepG2 cells: Cellular and proteomic studies. *Proteomics. Clinical Applications*, **1**: 410-419.

Luo, L., Li, Z., Zhang, Y., and Huang, R. (1998). Triterpenes and steroidal compounds from *Momordica dioica*. *Yao Xue Xue Bao*, **33**: 839-842.

Luo, M., Liu, X., Zu, Y., Fu, Y., Zhang, S., Yao, L., and Efferth, T. (2010). Cajanol, a novel anticancer agent from Pigeonpea [*Cajanus cajan* (L.) Millsp.] roots, induces apoptosis in human breast cancer cells through a ROS-mediated mitochondrial pathway. *Chemico-biological Interactions*, **188**: 151-160.

Lupi, A., Delle Monache, F., Marini-Bettolo, G.B., Costa, D.L.B., and D'Albuquerque, I.L. (1979). Abruquinones: new natural isoflavaquinones. *Gazzetta Chimica Italiana*, **109**: 9-12.

Ma, C.-M., Nakamura, N., and Hattori, M. (1998). Saponins and C-glycosyl flavones from the seeds of *Abrus precatorius*. *Chemical and Pharmaceutical Bulletin*, **46**: 982-987.

Ma, Y.S., Weng, S.W., Lin, M.W., Lu, C.C., Chiang, J.H., Yang, J.S., Lai, K.C., Lin, J.P., Tang, N.Y., Lin, J.G., and Chung, J.G. 92012). Antitumor effects of emodin on LS1034 human colon cancer cells *in vitro* and *in vivo*: roles of apoptotic cell death and LS1034 tumor xenografts. *Food and Chemical Toxicology*, **50**: 1271-1278.

Mahesh, V.K., Sharma, R., Singh, R.S., and Upadhya, S.K. (1984). Anthraquinones and kaempferol from *Cassia* species section *Fistula*. *Journal of Natural Products*, **47**: 733.

Mak, P., Leung, Y.K., Tang, W.Y., harwood, C., and Ho, S.M. (2006). Apigenin suppresses cancer cell growth through ERβ. *Neoplasia*, **8**: 896-904.

Malhotra, S., and Misra, K. (1982). A new anthraquinone from *Cassia sophera* heartwood. *Planta Medica*, **46**: 247-249.

Mandava, N., Anderson, J.D., Dutky, S.R., and Thompson, M.J. (1974a). Novel occurrence of 5β-cholanic acid in plants. Isolation from jequirity bean seeds (*Abrus precatorius*). *Steroids*, **23**: 357-361.

Mandava, N., Anderson, J.D., and Dutky, S.R. (1974). Indole plant-growth inhibitor from *Abrus precatorius* seeds. *Phytochemistry*, **13**: 2853-2856.

Manikandan, R., Beulaja, M., Arulvasu, C., Sellamuthu, S., Dinesh, D., Prabhu, D., Babu, G., Vaseeharan, B., and Prabhu, N.M. (2012). Synergistic anticancer activity of curcumin and catechin: an *in vitro* study using human cancer cell lines. *Microscopy Research and Technique*, **75**: 112-116.

Manosroi, J., Boonpisuttinant, K., manosroi, W., and Manosroi, A. (2012). Anti-proliferative activities on HeLa cancer cell line of Thai medicinal plant recipes selected from MANOSROI II database. *Journal of Ethnopharmacology*, **142**: 422-431.

Marfe, G., Tafani, M., Indelicato, M., Sinibaldi-Salimei, P., Reali, V., Pucci, B., Fini, M., and Russo, M.A. (2009). Kaempferol induces apoptosis in two different cell lines via Akt inactivation, Bax and SIRT3 activation, and mitochondrial dysfunction. *Journal of Cellular Biochemistry*, **106**: 643-650.

Markham, K.R., Wallace, J.W., Babu, Y.N., Murty, V.K., and Rao, M.G. (1988). 8-C-Glucosylscutellarein 6,7-dimethyl ether and its 2''-O-apioside from *Abrus precatorius*. *Phytochemistry*, **28**: 299-301.

Masoko, P., Gololo, S.S., Mokgotho, M.P., Eloff, J.N., Howard, R.I., and Mampuru, L.J. (2009). Evaluation of the antioxidant, antibacterial, and antiproliferative activities of the acetone extract of the roots of *Senna italica* (Fabaceae). *African Journal of Traditional, Complementary and Alternative Medicines*, **7**: 138-148.

Maurya, D.K., Nandakumar, N., and Devasagayam, T.P. (2011). Anticancer property of gallic acid in A549, a human lung adenocarcinoma cell line, and possible mechanisms. *Journal of Clinical Biochemistry and Nutrition*, **48**: 85-90.

Meena, P.D., Kaushik, P., Shukla, S., Soni, A.K., Kumar, M., and Kumar, A. (2006). Anticancer and antimutagenic properties of *Acacia nilotica* (Linn.) on 7,12-dimethylbenz(a)anthracene-induced skin papillomagenesis in Swiss albino mice. *Asian Pacific Journal of Cancer Prevention*, **7**: 627-632.

Merghoub, N., Benbacer, L., El Btaouri, H., Ait Benhassou, H., Terryn, C., Attaleb, M., Madoulet, C., Benjouad, A., El Mzibri, M., Morjani, H., and Amzazi, S. (2011). *In vitro* antiproliferative effect and induction of apoptosis by *Retama monosperma* L. extract in human cervical cancer cells. *Cellular and Molecular Biology (Noisy-le-grand)*, **57 Suppl**: OL1581-1591.

Mertens-Talcott, S.U., and Percival, S.S. (2005). Ellagic acid and quercetin interact synergistically with resveratrol in the induction of apoptosis and cause transient cell cycle arrest in human leukaemia cells. *Cancer Letters*, **218**: 141-151.

Mertens-Talcott, S.U., Lee, J.H., Percival, S.S., and Talcott, S.T. (2006). Induction of cell death in Caco-2 human colon carcinoma cells by ellagic acid rich fractions from muscadine grapes (*Vitis rotundifolia*). *Journal of Agricultural and Food Chemistry*, **54**: 5336-5343.

Metzger, W., and Reif, K. (1996). Determination of 1,8-dihydroxyanthranoids in Senna. *Journal of Chromatography, A*, **740**: 133-138.

Mishra, P.K., Singh, N., Ahmad, G., Dube, A., and Maurya, R. (2005). Glycolipids and other constituents from *Desmodium gangeticum* with antileishmanial and immunomodulatory activities. *Bioorganic and Medicinal Chemistry Letters*, **15**: 4543-4546.

Misra, T.N., Singh, R.S., Pandey, H.S., and Pandey, R.P. (1996). Chemical constituents of hexane fraction of *Cassia fistula* roots. *Fitoterapia*, **67**: 173-174.

Misra, T.N., Singh, R.S., Pandey, H.S., and Singh, B.K. (1997). A new diterpene from *Cassia fistula* pods. *Fitoterapia*, **68**: 375-376.

Modh, K.M., Parmar, P.T., Panigrahi, B., Anand, I.S., and Patel, C.N. (2011). Pharmacognostical and phytochemical evaluation of leaves of *Bauhinia variegata* Linn. *Pharmacognosy Journal*, **3**: 45-49.

Mohamed, M.A., Mammoud, M.R., and hayen, H. (2009). Evaluation of antinociceptive and anti-inflammatory activities of a new triterpene saponin from *Bauhinia variegata* leaves. *Zeitscrift für Naturforschung. C, Journal of Biosciences*, **64**: 798-808.

Mollik, M.A.H., Hossan, M.S., Paul, A.K., Rahman, M.T., Jahan, R., and Rahmatullah, M. (2010a). A comparative analysis of medicinal plants used by folk medicinal healers in three districts of Bangladesh and inquiry as to mode of selection of medicinal plants. *Ethnobotany Research and Applications*, **8**: 195-218.

Mollik, M.A.H., Hassan, A.I., Paul, T.K., Sintaha, M., Khaleque, H.N., Noor, F.A., Nahar, A., Seraj, S., Jahan, R., Chowdhury, M.H., and Rahmatullah, M. (2010b). A survey of medicinal plant usage by folk medicinal practitioners in two villages by the Rupsha River in Bagerhat district, Bangladesh. *American Eurasian Journal of Sustainable Agriculture*, **4**: 349-356.

Mostafa, M., Ahmed, M., Jahan, I.A., and Choudhury, J.U. (2007). Composition of oil from the seeds of *Cassia sophera* Linn. *Bangladesh Journal of Scientific and Industrial Research*, **42**: 75-78.

Mujoo, K., Haridas, V., Hoffmann, J.J., Wächter, G.A., Hutter, L.K., Lu, Y., Blake, M.E., Jayatilake, G.S., Bailey, D., Mills, G.B., and Gutterman, J.U. (2001). Triterpenoid saponins from *Acacia victoriae* (Bentham) decrease tumor cell proliferation and induce apoptosis. *Cancer Research*, **61**: 5486-5490.

Murtaza, I., Saleem, M., Adhami, V.M., Hafeez, B.B., and Mukhtar, H. (2009). Suppression of cFLIP by lupeol, a dietary triterpene, is sufficient to overcome resistance to TRAIL-mediated apoptosis in chemoresistant human pancreatic cancer cells. *Cancer Research*, **69**: 1156-1165.

Murty, V.K., Rao, T.V.P., and Venkateswarlu, V. (1967). Chemical examination of *Cassia fistula*. *Tetrahedron*, **23**: 515-518.

Muto, A., Hori, M., Sasaki, Y., Saitoh, A., Yasuda, I., Maekawa, T., Uchida, T., Asakura, K., Nakazato, T., Kaneda, T., Kizaki, M., Ikeda, Y., and Yoshida, T. (2007). Emodin has a cytotoxic activity against human multiple myeloma as a Janus-activated kinase 2 inhibitor. *Molecular Cancer Therapeutics*, **6**: 987-994.

Nagore, D.H., Ghosh, V.K., Patil, M.J., and Wahile, A.M. (2010). Validated HPTLC method for quantification of epicatechin in extracts of leaves of *Cassia fistula* Linn. *Acta Chromatographica*, **22**: 259-265.

Nakajima, K., Yamauchi, K., and Kuwano, S. (1985). Isolation of a new aloe-emodin dianthrone diglucoside from senna and its potentiating effect on the purgative activity of sennoside A in mice. *Journal of Pharmacy and Pharmacology*, **37**: 703-706.

Naowaratwattana, W., De-Eknamkul, W., and De Mejia, *E.G.* (2010). Phenolic-containing organic extracts of mulberry (*Morus alba* L.) leaves inhibit HepG2 hepatoma cells through G2/M phase arrest, induction of apoptosis, and inhibition of topoisomerase IIα activity. *Journal of Medicinal Food*, **13**: 1045-1056.

Narayanan, V., and Seshadri, T.R. (1972). Proanthocyanidins of *Cassia fistula*. *Indian Journal of Chemistry*, **10**: 379-381.

Narender, T., Khaliq, T., Puri, A., and Chander, R. (2006). Antidyslipidemic activity of furano-flavonoids isolated from *Indigofera tinctoria*. *Bioorganic and Medicinal Chemistry Letters*, **16**: 3411-3414.

Nawarak, J., Huang-Liu, R., Kao, S.H., Liao, H.H., Sinchaikul, S., Chen, S.T., and Cheng, S.L. (2009). Proteomics analysis of A375 human malignant melanoma cells in response to arbutin treatment. *Biochim Biophys Acta*, **1794**: 159-167.

Nawaz, A.H.M.M., Hossain, M., Karim, M., Khan, M., Jahan, R., and Rahmatullah, M. (2009). An ethnobotanical survey of Rajshahi district in Rajshahi division, Bangladesh. *American Eurasian Journal of Sustainable Agriculture*, **3**: 143-150.

Nguyen, P.H., Le, T.V., Thuong, P.T., Dao, T.T., Ndinteh, D.T., Mbafor, J.T., Kang, K.W., and Oh, W.K. (2009). Cytotoxic and PTB1B inhibitory activities from *Erythrina abyssinica*. *Bioorganic and Medicinal Chemistry Letters*, **19**: 6745-6749.

Nigam, S.K., Misra, G., and Mitra, C.R. (1971). Constituents of *Samanea saman* bark. *Phytochemistry*, **10**: 1954-1955.

Ogunwande, I.A., Walker, T.M., Setzer, W.N., and Essien, E. (2006). Volatile constituents from *Samanea saman* (Jacq.) Merr. Fabaceae. *African Journal of Biotechnology*, **5**: 1890-1893.

Oomah, B.D., Luc, G., Leprelle, C., Drover, J.C., Harrison, J.E., and Olson, M. (2011). Phenolics, phytic acid, and phytase in Canadian-grown low-tannin faba bean (*Vicia faba* L.) genotypes. *Journal of Agricultural and Food Chemistry*, **59**: 3763-3771.

Ou, T.T., Wang, C.J., Lee, Y.S., Wu, C.H., and Lee, H.J. (2010). Gallic acid induces G2/M phase cell cycle arrest via regulating 14-3-3β release from Cdc25C and Chk2 activation in human bladder transitional carcinoma cells. *Molecular Nutrition and Food Research*, **54**: 1781-1790.

Pandey, M., Kaur, P., Shukla, S., Abbas, A., Fu, P., and Gupta, S. (2011). Plant flavones apigenin inhibits HDAC and remodels chromatin to induce growth arrest and apoptosis in human prostate cancer cells: *In vitro* and *in vivo* study. *Molecular Carcinogenesis*, (in press).

Pandey, S., and Agrawal, R.C. (2010). Chemopreventive potential of *Bauhinia variegata* flower extract against DMBA-induced skin papillomagenesis in mice. *Pharmacologyonline*, **1**: 39-46.

Park, K.Y., Cho, E.J., Rhee, S.H., Jung, K.O., Yi, S.J., and Jhun, B.H. (2003). Kimchi and an active component, β-sitosterol, reduce oncogenic H-Ras(v12)-induced DNA synthesis. *Journal of Medicinal Food*, **6**: 151-156.

Parmar, K.A., Patel, A.N., Prajapati, S.N., and Patel, R.I. (2010). Anti viral in HEL cell, HeLa cell cultures, antibacterial and antioxidant activity of *Acacia arabica* seeds extracts by the use of DPPH free radical method. *Journal of Chemical and Pharmaceutical Research*, **2**: 324-332.

Patil, A.D., and Deshpande, V.H. (1982). A new dimeric proanthocyanidin from *Cassia fistula* sapwood. *Indian Journal of Chemistry, Section B: Organic Chemistry Including Medicinal Chemistry*, **21B**: 626-628.

Pecere, T., Gazzola, M.V., Mucignat, C., Parolin, C., Vecchia, F.D., Cavaggioni, A., Basso, G., Diaspro, A., Salvato, B., Carli, M., and Palù, G. (2000). Aloe-emodin is a new type of anticancer agent with a selective activity against neuroectodermal tumors. *Cancer Research*, **60**: 2800-2804.

Petronelli, A., Pannitteri, G., and Testa, U. (2009). Triterpenoids as new promising anticancer drugs. *Anticancer Drugs*, **20**: 880-892.

Piras, A., Rosa, A., Marongiu, B., Atzeri, A., Dessi, M.A., Falconieri, D., and Porcedda, S. (2012). Extraction and separation of volatile and fixed oils from seeds of *Myristica fragrans* by supercritical CO2: chemical composition and cytotoxic activity on Caco-2 cancer cells. *Journal of Food Science*, **77**: C448-C453.

Pitchai, D., and Manikkam, R. (2011). Hypoglycemic and insulin mimetic impact of catechin isolated from *Cassia fistula*: a substantiate in silico approach through docking analysis. *Medicinal Chemistry Research*, 1-13.

Pollier, J., and Goossens, A. (2012). Oleanolic acid. *Phytochemistry*, **77**: 10-15.

Prasad, G.C. (1987). Studies on cancer in Ayurveda and its management. *Journal of the Royal Asiatic Society of Great Britain and Ireland*, **3**: 147-167.

Prasad, S., Nigam, N., Kalra, N., and Shukla, Y. (2008a). Regulation of signaling pathways involved in lupeol induced inhibition of proliferation and induction of apoptosis in human prostate cancer cells. *Molecular Carcinogenesis*, **47**: 916-924.

Prasad, S., Kalra, N., and Shukla, Y. (2008). Induction of apoptosis by lupeol and mango extract in mouse prostate and LNCaP cells. *Nutrition and Cancer*, **60**: 120-130.

Rahman, W., and Begum, S.J. (1966). Flower pigments: flavonoids from the white flowers of *Bauhinia variegata*. *Naturwissenschaften*, **53**: 385.

Rahman, M.Z., Sultana, S.J., Faruqee, C.F., Ferdous, F., Rahman, M.S., Islam, M.S., Rashid, M.A. (2007). Phytochemical and biological investigation of *Erythrina variegata*. *Saudi Pharmaceutical Journal*, **15**: 140-145.

Rahman, M.Z., Rahman, M.S., Kaisar, A., Hossain, A., and Rashid, M.A. (2010). Bioactive isoflavones from *Erythrina variegata* L. *Turkish Journal of Pharmaceutical Sciences*, **7**: 21-28.

Rahmatullah, M., Ferdausi, D., Mollik, M.A.H., Azam, M.N.K., Rahman, M.T., and Jahan, R. (2009a). Ethnomedicinal Survey of Bheramara Area in Kushtia District, Bangladesh. *American Eurasian Journal of Sustainable Agriculture*, **3**: 534-541.

Rahmatullah, M., Noman, A., Hossan, M.S., Rashid, M.H., Rahman, T., Chowdhury, M.H., and Jahan, R. (2009b). A survey of medicinal plants in two areas of Dinajpur district, Bangladesh including plants which can be used as functional foods. *American Eurasian Journal of Sustainable Agriculture*, **3**: 862-876.

Rahmatullah, M., Das, A.K., Mollik, M.A.H., Jahan, R., Khan, M., Rahman, T., and Chowdhury, M.H. (2009c). An ethnomedicinal survey of Dhamrai sub-district in Dhaka district, Bangladesh. *American Eurasian Journal of Sustainable Agriculture*, **3**: 881-888.

Rahmatullah, M., Ferdausi, D., Mollik, M.A.H., Jahan, R., Chowdhury, M.H., and Haque, W.M. (2010a). A survey of medicinal plants used by Kavirajes of Chalna area, Khulna district, Bangladesh. *African Journal of Traditional, Complementary and Alternative Medicines*, **7**: 91-97.

Rahmatullah, M., Khatun, M.A., Morshed, N., Neogi, P.K., Khan, S.U.A., Hossan, M.S., Mahal, M.J., and Jahan, R. (2010b). A randomized survey of medicinal plants used by folk medicinal healers of Sylhet Division, Bangladesh. *Advances in Natural and Applied Sciences*, **4**: 52-62.

Rahmatullah, M., Kabir, A.A.B.T., Rahman, M.M., Hossan, M.S., Khatun, Z., Khatun, M.A., and Jahan, R. (2010c). Ethnomedicinal practices among a minority group of Christians residing in Mirzapur village of Dinajpur district, Bangladesh. *Advances in Natural and Applied Sciences*, **4**: 45-51.

Rahmatullah, M., Momen, M.A., Rahman, M.M., Nasrin, D., Hossain, M.S., Khatun, Z., Jahan, F.I., Khatun, M.A., and Jahan, R. (2010d). A randomized survey of medicinal plants used by folk medicinal practitioners in Daudkandi sub-district of Comilla district, Bangladesh. *Advances in Natural and Applied Sciences*, **4**: 99-104.

Rahmatullah, M., Mollik, M.A.H., Ahmed, M.N., Bhuiyan, M.Z.A., Hossain, M.M., Azam, M.N.K., Seraj, S., Chowdhury, M.H., Jamal, F., Ahsan, S., and Jahan, R. (2010e). A survey of medicinal plants used by folk medicinal practitioners in two villages of Tangail district, Bangladesh. *American Eurasian Journal of Sustainable Agriculture*, **4**: 357-362.

Rahmatullah, M., Mollik, M.A.H., Islam, M.K., Islam, M.R., Jahan, F.I., Khatun, Z., Seraj, S., Chowdhury, M.H., Islam, F., Miajee, Z.U.M., and Jahan, R. (2010f). A survey of medicinal and functional food plants used by the folk medicinal practitioners of three villages in Sreepur Upazilla, Magura district, Bangladesh. *American Eurasian Journal of Sustainable Agriculture*, **4**: 363-373.

Rahmatullah, M., Jahan, R., Khatun, M.A., Jahan, F.I., Azad, A.K., Bashar, A.B.M., Miajee, Z.U.M., Ahsan, S., Nahar, N., Ahmad, I., and Chowdhury, M.H. (2010g). A pharmacological evaluation of medicinal plants used by folk medicinal practitioners of Station Purbo Para Village of Jamalpur Sadar Upazila in

Jamalpur district, Bangladesh. *American Eurasian Journal of Sustainable Agriculture*, **4**: 170-195.

Rahmatullah, M., Ishika, T., Rahman, M., Swarna, A., Khan, T., Monalisa M.N., Seraj, S., Mou, S.M., Mahal, M.J., and Biswas, K.R. (2011a). Plants prescribed for both preventive and therapeutic purposes by the traditional healers of the Bede community residing by the Turag River, Dhaka district. *American Eurasian Journal of Sustainable Agriculture*, **5**: 325-331.

Rahmatullah, M., Azam, M.N.K., Rahman, M.M., Seraj, S., Mahal, M.J., Mou, S.M., Nasrin, D., Khatun, Z., Islam, F., and Chowdhury, M.H. (2011b). A survey of medicinal plants used by Garo and non-Garo traditional medicinal practitioners in two villages of Tangail district, Bangladesh. *American Eurasian Journal of Sustainable Agriculture*, **5**: 350-357.

Rahmatullah, M., and Biswas, K.R. (2012a). Traditional medicinal practices of a Sardar healer of the Sardar (Dhangor) community of Bangladesh. *Journal of Alternative and Complementary Medicine*, **18**: 10-19.

Rahmatullah, M., Hasan, A., Parvin, W., Moniruzzaman, M., Khatun, A., Khatun, Z., Jahan, F.I., and Jahan, R. (2012b). Medicinal plants and formulations used by the Soren clan of the Santal tribe in Rajshahi district, Bangladesh for treatment of various ailments. *African Journal of Traditional, Complementary and Alternative Medicines*, **9**: 342-349.

Rahmatullah, M., Khatun, Z., Hasan, A., Parvin, W., Moniruzzaman, M., Khatun, A, Mahal, M.J., Bhuiyan, M.S.A., Mou, S.M., and Jahan, R. (2012c). Survey and scientific evaluation of medicinal plants used by the Pahan and Teli tribal communities of Natore district, Bangladesh. *African Journal of Traditional, Complementary and Alternative Medicines*, **9**: 366-373.

Rahmatullah, M., Azam, M.N.K., Khatun, Z., Seraj, S., Islam, F., Rahman, M.A., Jahan, S., Aziz, M.S., and Jahan, R. (2012d). Medicinal plants used for treatment of diabetes by the Marakh sect of the Garo tribe living in Mymensingh district, Bangladesh. *African Journal of Traditional, Complementary and Alternative Medicines*, **9**: 380-385.

Rajkapoor, B., Jayakar, B., and Murugesh, N. (2003). Antitumour activity of *Bauhinia variegata* on Dalton's ascetic lymphoma. *Journal of Ethnopharmacology*, **89**: 107-109.

Rajkapoor, B., Jayakar, B., Murugesh, N., and Sakthisekaran, D. (2006). Chemoprevention and cytotoxic effect of *Bauhinia variegata* against N-nitrosodiethylamine induced liver tumors and human cancer cell lines. *Journal of Ethnopharmacology*, **104**: 407-409.

Rajkapoor, B., Murugesh, N., and Krishna, D.R. (2009). Cytotoxic activity of a flavanone from the stem of *Bauhinia variegata* Linn. *Natural Product Research*, **23**: 1384-1389.

Ramnath, V., Kuttan, G., and Kuttan, R. (2002). Antitumor effect of abrin on transplanted tumors in mice. *Indian Journal of Physiology and Pharmacology*, **46**: 69-77.

Ramnath, V., Kuttan, G., and Kuttan, R. (2006). Induction of pro-apoptotic caspase-3 expression in Dalton's Lymphoma Ascites (DLA) cells by abrin. *Amala Research Bulletin*, **26**: 191-197.

Ramnath, V., Rekha, P.S., Kuttan, G., and Kuttan, R. (2009). Regulation of Caspase-3 and Bcl-2 expression in Dalton's Lymphoma Ascites cells by abrin. *Evidence-based Complementary and Alternative Medicine*, **6**: 233-238.

Rani, M., and Kalidhar, S.B. (1998). A new anthraquinone derivative from *Cassia fistula* Linn. pods. *Indian Journal of Chemistry, Section B: Organic Chemistry Including Medicinal Chemistry*, **37B**: 1314-1315.

Rao, T.V.P., and Venkateswarlu, V. (1965). Fistucacidin from the bark and heartwood of *Cassia fistula*. *Bulletin of the National Institute of Sciences of India*, **31**: 28-33.

Rastogi, S., Pandey, M.M., and Rawat, A.K.S. (2011). An ethnomedicinal, phytochemical and pharmacological profile of *Desmodium gangeticum* (L.) DC. and *Desmodium adscendens* (Sw.) DC. *Journal of Ethnopharmacology*, **136**: 283-296.

Rasul, N., Saleem, B., and Nawaz, R. (1989). Preliminary phytochemical screening of four common plants of family Caesalpiniaceae. *Pakistan Journal of Pharmaceutical Sciences*, **2**: 55-57.

Rasul, A., Khan, M., Yu, B., Ma, T., and Yang, H. (2011a). Xanthoxyletin, a coumarin induces S phase arrest and apoptosis in human gastric adenocarcinoma SGC-7901 cells. *Asian Pacific Journal of Cancer Prevention*, **12**: 1219-1223.

Rasul, A., Yu, B., Yang, L.F., Ali, M., Khan, M., Ma, T., and Yang, H. (2011b). Induction of mitochondria-mediated apoptosis in human gastric adenocarcinoma SGC-7901 cells by kuraridin and Nor-kurarinone isolated from *Sophora flavescens*. *Asian Pacific Journal of Cancer Prevention*, **12**: 2499-2504.

Rayanil, K.O., Bunchornmaspan, P., and Tuntiwachwuttikul, P. (2011). A new phenolic compound with anticancer activity from the wood of *Millettia leucantha*. *Archives of Pharmacal Research*, **34**: 881-886.

Reddy, K.K. (1968). Isolation of a novel type of leucoanthocyanidin, fistucacidin (5,4'-dihydroxyflavan-3,4-diol) from *Cassia fistula* sapwood. *Leather Science (Madras)*, **15**: 171-172.

Regasini, L.O., Castro-Gamboa, I., Silva, D.H., Furlan, M., Barreiro, E.J., Ferreira, P.M., Pessoa, C., Lotufo, L.V., de Moraes, M.O., Young, M.C., and Bolzani Vda, S. (2009). Cytotoxic guanidine alkaloids from Pterogyne nitens. *Journal of Natural Products*, **72**: 473-476.

Rejiya, C.S., Cibin, T.R. and Abraham, A. (2009). Leaves of *Cassia tora* as a novel cancer therapeutic – an in vitro study. *Toxicology In Vitro*, **23**: 1034-1038.

Rigano, D., Russo, A., Formisano, C., Cardile, V., and Senatore. F. (2010). Antiproliferative and cytotoxic effects on malignant melanoma cells of essential oils from the aerial parts of *Genista sessilifolia* and *G. tinctoria*. *Natural Product Communications*, **5**: 1127-1132.

Robeson, D.J. (1978). Furanoacetylene and isoflavonoid phytoalexins in *Lens culinaris*. *Phytochemistry*, **17**: 807-808.

Rodgers, E.H., and Grant, M.H. (1998). The effect of the flavonoids, quercetin, myricetin and epicatechin on the growth and enzyme activities of MCF7 human breast cancer cells. *Chemico-Biological Intercations*, **116**: 213-228.

Rodríguez, C., Frías, J., Vidal-Valverde, C., and Hernández, A. (2007). Total chemically available (free and intrachain) lysine and furosine in pea, bean, and lentil sprouts. *Journal of Agricultural and Food Chemistry*, **55**: 10275-10280.

Roy, S., Gu, M., Ramasamy, K., Singh, R.P., Agarwal, C., Siriwardana, S., Sclafani, R.A., and Agarwal, R. (2009). P21/Cip1 and p27/Kip1 are essential molecular targets of inositol hexaphosphate for its antitumor efficacy against prostate cancer. *Cancer Research*, **69**: 1166-1173.

Roy, A., Lakshmi, T., and Geetha, R.V. (2012). Estimation of quercetin in *Acacia catechu* ethanolic bark extract by HPLC method. *International Journal of PharmTech Research*, **4**: 501-505.

Ruela-de-Sousa, R.R., Fuhler, G.M., Blom, N., Ferreira, C.V., Aoyama, H., and Peppelenbosch, M.P. (2010). Cytotoxicity of apigenin on leukemia cell lines: implications for prevention and therapy. *Cell Death and Disease*, **1**: e19.

Russell, L.H.Jr., Mazzio, E., Badisa, R.B., Zhu, Z.P., Agharahimi, M., Oriaku, E.T., and Goodman, C.B. (2012). Autoxidation of gallic acid induces ROS-dependent death in human prostate cancer LNCaP cells. *Anticancer Research*, **32**: 1595-1602.

Saha, S., Subrahmanyam, E.V.S., Kodangala, C., and Shastry, S.C. (2011). Isolation and characterization of triterpenoids and fatty acid ester of triterpenoid from leaves of *Bauhinia variegata*. *Pharma Chemica*, **3**: 28-37.

Saleem, M., Maddodi, N., Abu Zaid, M., Khan, N., bin hafeez, B., Asim, M., Suh, Y., Yun, J.M., Setaluri, V., and Mukhtar, H. (2008). Lupeol inhibits growth of highly aggressive human metastatic melanoma cells *in vitro* and *in vivo* by inducing apoptosis. *Clinical Cancer Research*, **14**: 2119-2127.

Salem, M.M., Davidorf, F.H., and Abdel-Rahman, M.H. (2011). *In vitro* anti-uveal melanoma activity of phenolic compounds from the Egyptian medicinal plant *Acacia nilotica*. *Fitoterapia*, **82**: 1279-1284.

Sannigrahi, S., Mazumder, U.K., Pal, D., and Mishra, S.L. (2012). Terpenoids of methanol extract of *Clerodendrum infortunatum* exhibit anticancer activity against Ehrlich's ascites carcinoma (EAC) in mice. *Pharmaceutical Biology*, **50**: 304-309.

Sarker, S., Seraj, S., Sattar, M.M., Haq, W.M., Chowdhury, M.H., Ahmad, I., Jahan, R., Jamal, F., and Rahmatullah, M. (2011). Medicinal plants used by folk medicinal practitioners of six villages in Thakurgaon district, Bangladesh. *American Eurasian Journal of Sustainable Agriculture*, **5**: 332-343.

Sartorelli, P., Lago, J.H.G., Cunha, R.L.O.R., Kitamura, R.O.S., and Young, M.C.M. (2012). A new minor dimeric ester from seeds of *Cassia fistula* L. (Leguminosae). *Natural Product Research*, **26**: 36-41.

Sato, N., Uenishi, N., and Yamazaki, S. (1993). *European Patent Application*, EP 526888 A2 19930210.

Sato, M., Tanaka, H., Fujiwara, S., Hirata, M., Yamaguchi, R., Etoh, H., and Tokuda, C. (2002). Antibacterial property of isoflavonoids isolated from *Erythrina variegata* against cariogenic oral bacteria. *Phytomedicine*, **9**: 427-433.

Sato, M., Tanaka, H., Yamaguchi, R., Kato, K., and Etoh, H. (2004). Synergistic effects of mupirocin and an isoflavanone isolated from *Erythrina variegata* on growth and recovery of methicillin-resistant *Staphylococcus aureus*. *International Journal of Antimicrobial Agents*, **24**: 241-246.

Saumya, P.R., Mishra, S., Sahoo, S., and Panda, P.K. (2011). Nephroprotective effect of *Bauhinia variegata* (Linn.) whole stem extract against cisplatin-induced nephropathy in rats. *Indian Journal of Pharmacology*, **43**: 200-202.

Saxena, H.O., Faridi, U., Srivastava, S., Kumar, J.K., Darokar, M.P., Luqman, S., Chanotiya, C.S., Krishna, V., Negi, A.S., and Khanuja, S.P. (2008). Gallic acid-based indanone derivatives as anticancer agents. *Bioorganic and Medicinal Chemistry Letters*, **18**: 3914-3918.

Saxena, V.K., and Sharma, D.N. (1999). A new isoflavone from the roots of *Abrus precatorius*. *Fitoterapia*, **70**: 328-329.

Sayeed, M.A., Ali, M.A., Khan, G.R.M.A.M., and Rahman, M.S. (1999). Studies on the characterization and glyceride composition of *Cassia fistula* seed oil. *Bangladesh Journal of Scientific and Industrial Research*, **34**: 144-148.

Scambia, G., Ranelletti, F.O., Panici, P.B., De Vincenzo, R., Bonanno, G., Ferrandina, G., Piantelli, M., Bussa, S., Rumi, C., Cianfriglia, M., and Mancuso, S. (1994). Quercetin potentiates the effect of adriamycin in a multidrug-resistant MCF-7 human breast-cancer cell line: P-glycoprotein as a possible target. *Cancer Chemotherapy and Pharmacology*, **34**: 459-464.

Sen, A.B., and Shukla, Y.N. (1968). Chemical examination of *Cassia fistula*. *Journal of the Indian Chemical Society*, **45**: 744.

Shanmugakumaran, S.D., Satheesh Kumar, G., Layambhica, V., and Natraj, K. (2011). Pharmacognostical, phytochemical and antimycobacterial perspectives of the leaf extracts of *Samanea saman* (Jacq.) Merr. *International Journal of Pharmacy and Technology*, **3**: 3202-3217.

Shen, D., Wu, Q., Wang, M., Yang, Y., Lavoie, E.J., and Simon, J.E. (2006). Determination of the predominant catechins in *Acacia catechu* by liquid chromatography/electrospray ionization-mass spectrometry. *Journal of Agricultural and Food Chemistry*, **54**: 3219-3224.

Shi, L., Chen, J., Wang, Y.Y., Sun, G., Liu, J.N., Zhang, J.X., Yan, W., Qian, C.F., Liu, N., Fu, Z., You, Y.P., and Zeng, Y. Gossypin induces G2/M arrest in human malignant glioma U251 cells by the activation of Chk1/Cdc25C pathway. *Cellular and Molecular Neurobiology*, **32**: 289-296.

Shieh, D.E., Chen, Y.Y., Yen, M.H., Chiang, L.C., and Lin, C.C. (2004). Emodin-induced apoptosis through p53-dependent pathway in human hepatoma cells. *Life Sciences*, **74**: 2279-2290.

Shih, Y.W., Wu, P.F., Lee, Y.C., Shi, M.D., and Chiang, T.A. (2009). Myricetin suppresses invasion and migration of human lung adenocarcinoma A549 cells; possible mediation by blocking the ERK signaling pathway. *Journal of Agricultural and Food Chemistry*, **57**: 3490-3499.

Shyu, M.H., Kao, T.C., and Yen, G.C. (2010). Oleanolic acid and ursolic acid induce apoptosis in HuH7 human hepatocellular carcinoma cells through a mitochondrial-dependent pathway and downregulation of XIAP. *Journal of Agricultural and Food Chemistry*, **58**: 6110-6118.

Siddiqui, S., Siddiqui, B.S., and Naim, Z. (1978). Studies in the steroidal constituents of the seeds of *Abrus precatorius* Linn. (scarlet variety). *Pakistan Journal of Scientific and Industrial Research*, **21**: 158-161.

Siegel, R., Naishadham, D., and Jemal, A. (2012). Cancer Statistics, 2012. *CA: A Cancer Journal for Clinicians*, **62**: 10-29.

Singh, H., Chawla, A.S., Jindal, A.K., Conner, A.H., and Rowe, J.W. (1975). Investigation of *Erythrina* spp. VII. Chemical constituents of *Erythrina variegata* Var. *orientalis* bark. *Lloydia*, **38**: 97-100.

Singh, P., and Rao, M.M. (1982). Crude drug and anthraquinone yield of discarded parts of the senna plant (*Cassia angustifolia* Vahl). *Current Science*, **51**: 146.

Singh, R.M. (2002). An assessment of ayurvedic concept of cancer and a new paradigm of anticancer treatment in Ayurveda. *Journal of Alternative and Complementary Medicine*, **8**: 609-614.

Singh, R.P., Agarwal, C., and Agarwal, R. (2003). Inositol hexaphosphate inhibits growth, and induces G1 arrest and apoptotic death of prostate carcinoma DU145 cells: modulation of CDKI-CDK-cyclin and pRb-related protein-E2F complexes. *Carcinogenesis*, **24**: 555-563.

Singh, R.S., Singh, H., Pandey, H.S., Pandey, R.P., and Singh, S. (2005). Two new aliphatic compounds from *Cassia fistula* L. *Indian Journal of Chemistry, Section B: Organic Chemistry Including Medicinal Chemistry*, **44B**: 2372-2374.

Singh, B., Chandan, B.K., Sharma, N., Bhardwaj, V., Satti, N.K., Gupta, V.N., Gupta, B.D., Suri, K.A., and Suri, O.P. (2006). Isolation, structure elucidation and in vivo hepatoprotective potential of trans-tetracos-15-enoic acid from *Indigofera tinctoria* Linn. *Phytotherapy Research*, **20**: 831-839.

Singh, J., and Gupta, K.P. (2008). Inositol hexaphosphate induces apoptosis by coordinative modulation of P53, Bcl-2 and sequential activation of caspases in 7,12 dimethylbenz[a]anthracene exposed mouse epidermis. *Journal of Environmental Pathology, Toxicology and Oncology*, **27**: 209-217.

Singh, R., Singh, B., Singh, S., Kumar, N., Kumar, S., and Arora, S. (2008). Anti-free radical activities of kaempferol isolated from *Acacia nilotica* (L.) Willd. Ex Del. *Toxicology In Vitro*, **22**: 1965-1970.

Singh, B.N., Singh, B.R., Sarma, B.K., and Singh, H.B. (2009). Potential chemoprevention of N-nitrosodiethylamine-induced hepatocarcinogenesis by polyphenolics from *Acacia nilotica* bark. *Chemico-Biological Interactions*, **181**: 20-28.

Singh, R.S., Sinha, R., Pandey, H.S., and Rao, C.V. (2011). Four new aliphatic compounds from *Bauhinia variegata* pods. *Journal of Medicinal and Aromatic Plant Sciences*, **33**: 144-149.

Singhal, J., Nagaprashantha, L.D., Vatsyayan, R., Ashtosh, A.S., and Singhal, S.S. (2012). Didymin induces apoptosis by inhibiting N-Myc and upregulating RKIP in neuroblastoma. *Cancer Preventive Research (Philadelphia, PA.)*, **5**: 473-483.

Sircar, P.K., Dey, B., Sanyal, T., Ganguly, S.N., and Sircar, S.M. (1970). Gibberellic acid in the floral parts of *Cassia fistula*. *Phytochemistry*, **9**: 735-736.

Sivakumar, R., and Alagesaboopathi, C. (2007). Phytochemical studies on the red and white forms of *Abrus precatorius* L. *Advances in Plant Sciences*, **20**: 413-416.

Smejkal, K., Svacinová, J., Slapetová, T., Scneiderová, K., Dall'acqua, S., Innocenti, G., Závalová, V., Kollár, P., Chudik, S., Marek, R., Julínek, O, Urbanová, M., kartal, M., Csöllei, M., and Dolezal, K. (2010). Cytotoxic activities of several geranyl-substituted flavanones. *Journal of Natural Products*, **73**: 568-572.

Sreelatha, S., Jeyachitra, A., and Padma, P.R. (2011a). Antiproliferation and induction of apoptosis by *Moringa oleifera* leaf extract on human cancer cells. *Food and Chemical Toxicology*, **49**: 1270-1275.

Sreelatha, S., Padma, P.R., and Umasankari, E. (2011b). Evaluation of anticancer activity of ethanol extract of *Sesbania grandiflora* (Agati Sesban) against Ehrlich ascites carcinoma in Swiss albino mice. *Journal of Ethnopharmacology*, **134**: 984-987.

Steriti, R. (2002). Nutritional support for chronic myelogenous and other leukemias: a review of the scientific literature. *Alternative Medicine Review*, **7**: 404-409.

Sundarraj, S., Thangam, R., Sreevani, V., Kaveri, K., Gunasekaran, P., Achiraman, S., and Kannan, S. (2012). γ-Sitosterol from *Acacia nilotica* L. induces G2/M cell cycle arrest and apoptosis through c-Myc suppression in MCF-7 and A549 cells. *Journal of Ethnopharmacology*, **141**: 803-809.

Tabolacci, C., Lentini, A., Mattioli, P., Provenzano, B., Oliverio, S., Carlomosti, F., and Beninati, S. (2010). Antitumor properties of aloe-emodin and induction of transglutaminase 2 activity in B16-F10 melanoma cells. *Life Sciences*, **87**: 316-324.

Tabolacci, C., Oliverio, S., Lentini, A., Rossi, S., Galbiati, A., Montesano, C., Mattioli, P., Provenzano, B., Facchiano, F., and Beninati, S. (2011). Aloe-emodin as antiproliferative and differentiating agent on human U937 monoblastic leukemia cells. *Life Sciences*, **89**: 812-820.

Tanaka, H., Murata, R., Yoshida, A., and Hayashi, S. (1982). Analytical studies on the active constituents in crude drugs. V. The structure of sennoside G, a new glucoside from senna. *Chemical and Pharmaceutical Bulletin*, **30**: 1550-1556.

Tanaka, H., Etoh, H., Shimizu, H., Makita, T., and Tateishi, Y. (2000). Two new isoflavonoids from *Erythrina variegata*. *Planta Medica*, **66**: 578-579.

Tanaka, H., *Hirata*, M., *Etoh*, H., Watanabe, N., Shimizu, H., Ahmad, M., Terada, Y., and Fukai, T. (2002). Two diphenylpropan-1,2-diol syringates from the roots of *Erythrina variegata*. *Journal of Natural Products*, **65**: 1933-1935.

Tanaka, H., Hirata, M., Etoh, H., Shimizu, H., Sako, M., Murata, J., Murata, H., Darnaedi, D., and Fukai, T. (2003). Eryvarins F and G, two 3-phenoxychromones from the roots of *Erythrina variegata*. *Phytochemistry*, **62**: 1243-1246.

Tanaka, H., Hirata, M., Etoh, H., Sako, M., Sato, M., Murata, J., Murata, H., Darnaedi, D., and Fukai, T. (2004). Six new constituents from the roots of *Erythrina variegata*. *Chemistry and Biodiversity*, **1**: 1101-1108.

Taylor, W.G., Fields, P.G., and Sutherland, D.H. (2007). Fractionation of lentil seeds (*Lens culinaris* Medik.) for insecticidal and flavonol tetraglycoside components. *Journal of Agricultural and Food Chemistry*, **55**: 5491-5498.

Thavarajah, D., Vandenberg, A., George, G.N., and Pickering, I.J. (2007). Chemical form of selenium in naturally selenium-rich lentils (*Lens culinaris* L.) from Saskatchewan. *Journal of Agricultural and Food Chemistry*, **55**: 7337-7341.

Tiedink, H.G., de Haan, L.H., Jongen, W.M., and Koeman, J.H. (1991). *In-vitro* testing and the carcinogenic potential of several nitrosated indole compounds. *Cell Biology and Toxicology*, **7**: 371-386.

Tiwari, H.P., and Spenser, I.D. (1965). Precursors of mimosine in *Mimosa pudica*. *Canadian Journal of Biochemistry*, **43**: 1687-1691.

Tiwari, R.D., and Bajpai, M. (1980). A new flavonol-8-C-glycoside from the leaves of *Cassia sophera*. *Phytochemistry*, **5**: 437-438.

Tomita, M., Kurokawa, T., Onozaki, K., Osawa, T., Sakurai, Y., and Ukita, T. (1972). Surface structure of murine ascites tumors. II. Difference in cytotoxicity of various phytoagglutinins toward Yoshida sarcoma cells *in vitro*. *International Journal of Cancer*, **10**: 602-606.

Tong, X., Lin, S., Fujii, M., and Hou, D.X. (2004). Molecular mechanisms of echinocystic acid-induced apoptosis in HepG2 cells. *Biochemical and Biophysical Research Communications*, **321**: 539-546.

Tong, Q.Y., He, Y., Zhao, Q.B., Qing, Y., Huang, W., and Wu, X.H. (2012). Cytotoxicity and apoptosis-inducing effect of steroidal saponins from *Dioscorea zingiberensis* Wright against cancer cells. *Steroids*, (in press).

Tselepi, M., Papachristou, E., Emmanouilidi, A., Angelis, A., Aligiannis, N., Skaltsounis, A.L., Kouretas, D., and Liadaki, K. (2011). Catalytic inhibition of eukaryotic topoisomerases I and II by flavonol glycosides extracted from *Vicia faba* and *Lotus edulis*. *Journal of Natural Products*, **74**: 2362-2370.

Tsopmo, A., and Muir, A.D. (2010). Chemical profiling of lentil (*Lens culinaris* Medik.) cultivars and isolation of compounds. *Journal of Agricultural and Food Chemistry*, **58**: 8715-8721.

Tzakou, O., Loukis, A., and Said, A. (2007). Essential oil from the flowers and leaves of *Cassia fistula* L. *Journal of Essential Oil Research*, **19**: 360-361.

Ukoha, P.O., Cemaluk, E.A.C., Nnamdi, O.L., and Madus, E.P. (2011). Tannins and other phytochemical of the *Samanea saman* pods and their antimicrobial activities. *African Journal of Pure and Applied Chemistry*, **5**: 237-244.

Undeðer, U., Aydin, S., Ba°aran, A.A., and Ba°aran, N. (2004). The modulating effects of quercetin and rutin on the mitomycin C induced DNA damage. *Toxicology Letters*, **151**: 143-149.

Vadivel, V., Nandety, A., and Biesalski, H.K. (2011). Antioxidant, free radical scavenging and type II diabetes-related enzyme inhibition properties of traditionally processed Jequirity bean (*Abrus precatorius* L). *International Journal of Food Science and Technology*, **46**: 2505-2512.

Vaishnav, M.M., and Gupta, K.R. (1996). Rhamnetin 3-O-gentiobioside from *Cassia fistula* roots. *Fitoterapia*, **67**: 78-79.

Vaishnav, M.M., and Jain, P. (2004). A flavonol diglycoside from *Cassia fistula*. *Journal of the Institution of Chemists (India)*, **76**: 115-117.

Vaishnav, M.M., Sahu, D., and Jain, J.K. (2006). Phytochemical and pharmacological investigation of *Cassia fistula* root. *Journal of the Institution of Chemists (India)*, **78**: 173-175.

Vaishnava, M.M., Tripathi, A.K., and Gupta, K.R. (1993). Constituents of *Cassia fistula* roots. *Fitoterapia*, **64**: 93.

van Der Heijden, R., Jacobs, D.I., Snoeijer, W., Hallard, D., and Verpoorte, R. (2004). The *Catharanthus* alkaloids: pharmacology and biotechnology. *Current Medicinal Chemistry*, **11**: 607-628.

Varaprasad, M.V., Balakrishna, K., Sukumar, E., and Patra, A. (2009). Gangetial, a new pterocarpan from the roots of *Desmodium gangeticum*. *Journal of the Indian Chemical Society*, **86**: 654-656.

Varshney, I.P. (1976). Glycosides and carbohydrates from the members of the family Leguminoseae. *Research Journal – Science – University of Indore*, **4**: 13-22.

Varshney, I.P. and Vyas, P. (1976). Samanin A, a new saponin from the seeds of *Pithecolobium saman* Benth. *Indian Journal of Chemistry, Section B: Organic Chemistry Including Medicinal Chemistry*, **14B**: 814-815.

Varshney, I.P., and Jain, D.C. (1977). Study of Samanin C – a new saponin from *Pithecillobium saman* wood. *Indian Journal of Pharmacy*, **39**: 80-81.

Varshney, I.P., Vyas, P., Srivastava, H.C., and Singh, P.P. (1978). Study of saponins (samanin-A and samanin-B) from *Pithecolobium saman* Benth. *Indian Journal of Chemistry, Section B: Organic Chemistry Including Medicinal Chemistry*, **16B**: 166-167.

Varshney, I.P., Jain, D.C., Beg, M.F.A., and Srivastava, H.C. (1985). Samanin-C and samanin-E new saponins from *Pithecolobium saman*. *Fitoterapia*, **56**: 281-283.

Vasudevan, K., Manoharan, S., Alias, L.M., Balakrishnan, S., Vellaichamy, L., and Gitanjali, M. (2008). Evaluation of chemopreventive efficacy of *Cassia fistula* in 7,12-dimethyl benz(a) anthracene (DMBA) induced oral carcinogenesis. *International Journal of Chemical Sciences*, **6**: 1341-1354.

Verma, S.M., and Suresh, K.B. (2002). Phytochemical investigations of *Indigofera tinctoria* Linn leaves. *Ancient Science of Life*, **21**: 235-239.

Verma, D., Tiwari, S.S., Srivastava, S., and Rawat, A.K.S. (2011). Pharmacognostical evaluation and phytochemical standardization of *Abrus precatorius* L. seeds. *Natural Product Sciences*, **17**: 51-57.

Vidya Priyadarsini, R., Kumar, N., Khan, I., Thiyagarajan, P., Kondaiah, P., and Nagini, S. (2012). Gene suppression signature of DMBA-induced hamster buccal pouch carcinomas: modulation by chlorophyllin and ellagic acid. *Public Library of Science One*, **7**: e34628.

Vieira, J.R., de Souza, I.A., do Nascimento, S.C., and Leite, S.P. (2007). *Indigofera suffruticosa*: an alternative anticancer therapy. *Evidence-based Complementary and Alternative Medicine*, **4**: 355-359.

Vieira, C.R., Marques, M.F., Soares, P.R., Matuda, L., de Oliveira, C.M., Kato, L., da Silva, C.C., and Guillo, L.A. (2008). Antiproliferative activity of *Pterodon pubescens* Benth. seed soil and its active principle on human melanoma cells. *Phytomedicine*, **15**: 528-532.

Vorster, C., Stander, A., and Joubert, A. (2012). Differential signaling involved in *Sutherlandia frutescens*-induced cell death in MCF-7 and MCF-12A cells. *Journal of Ethnopharmacology*, **140**: 123-130.

Vucenik, I., Kalebic, T., Tantivejkul, K., and Shamsuddin, A.M. (1998). Novel anticancer function of inositol hexaphosphate: inhibition of human rhabdomyosarcoma *in vitro* and *in vivo*. *Anticancer Research*, **18**: 1377-1384.

Vucenik, I., and Shamsuddin, A.M. (2003). Cancer inhibition by inositol hexaphosphate (IP6) and inositol: from laboratory to clinic. *Journal of Nutrition*, **133**: 3778s-3784s.

Vucenik, I., and Shamsuddin, A.M. (2006). Protection against cancer by dietary IP6 and inositol. *Nutrition and Cancer*, **55**: 109-125.

Wachter, J., Neureiter, D., Alinger, B., Pichler, M., Fuereder, J., Oberdanner, C., Di Fazio, P., Ocker, M., Berr, F., and Kiesslich, T. (2012). Influence of five potential anticancer drugs on wnt pathway and cell survival in human biliary tract cancer cells. *International Journal of Biological Sciences*, **8**: 15-29.

Waly, M.I., Ali, B.H., Al-Lawati, I., and Nemmar, A. (2012). Protective effects of emodin against cisplatin-induced oxidative stress in cultured human kidney (HEK 293) cells. *Journal of Applied Toxicology*, (in press).

Wang, Z.J., Wu, Q.P., Tang, L.Y., Fu, M.H., He, Y., Gong, Q.F., and Huang, L.Q. (2007). Two new glycosides from the genus of *Cassia*. *Chinese Chemical Letters*, **18**: 1218-1220.

Wang, J., Zhao, X.Z., Qi, Q., Tao, L., Zhao, Q., Mu, R., Gu, H.Y., Wang, M., Feng, X., and Guo, Q.L. (2009). Macranthoside B, a hederagenin saponin extracted from *Lonicera macranthoides* and its anti-tumor acyivities *in vitro* and *in vivo*. *Food and Chemical Toxicology*, **47**: 1716-1721.

Wang, W., Sun, Y.P., Huang, X.Z., He, M., Chen, Y.Y., Shi, G.Y., Li, H., Yi, J., and Wang, J. (2010). Emodin enhances sensitivity of gallbladder cancer cells to platinum drugs via glutathione depletion and MRP1 downregulation. *Biochemical Pharmacology*, **79**: 1134-1140.

Wang, K., Liu, R., Li, J., Mao, J., Lei, Y., Wu, J., Zeng, J., Zhang, T., Wu, H., Chen, L., Huang, C., and Wei, Y. (2011a). Quercetin induces protective autophagy in gastric cancer cells: involvement of Akt-mTOR- and hypoxia-induced factor 1α-mediated signaling. *Autophagy*, **7**: 966-978.

Wang, W., Sun, Y., Li, X., Li, H., Chen, Y., Tian, Y., Yi, J., and Wang, J. (2011b). Emodin potentiates the anticancer effect of cisplatin on gallbladder cancer cells through the generation of reactive oxygen species and the inhibition of surviving expression. *Oncology Reports*, **26**: 1143-1148.

Wang, Q., Hao, J., Pu, J., Zhao, L., Lü, Z., Hu, J., Yu, Q., Wang, Y., Xie, Y., and Li, G. (2011c). Icariin induces apoptosis in mouse MLTC-10 Leydig tumor cells through activation of the mitochondrial pathway and down-regulation of the expression of piwil4. *International Journal of Oncology*, **39**: 973-980.

Wang, Y., Hong, C., Zhou, C., Xu, D., and Qu, H.B. (2011d). Screening antitumor compounds psoralen and isopsoralen from *Psoralea corylifolia* L. seeds. Evidence-based Complementary and Alternative Medicine, **2011**: 363052.

Wätjen, W., Kulawik, A., Suckow-Schnitker, A.K., Chovolou, Y., Rohrig, R., Ruhl, S., Kampkötter, A., Addae-Kyereme, J., Wright, C.W., and Passreiter, C.M. (2007). Pterocarpans phaseollin and neorautenol isolated from *Erythrina addisoniae* induce apoptotic cell death accompanied by inhibition of ERK phosphorylation. *Toxicology*, **242**: 71-79.

Wei, C.H., Hartman, F.C., Pfuderer, P., and Yang, W.-K. (1974). Purification and characterization of two major toxic proteins from seeds of *Abrus precatorius*. *Journal of Biological Chemistry*, **249**: 3061-3067.

Wei, W.T., Chen, H., Ni, Z.L., Liu, H.B., Tong, H.F., Fan, L., Liu, A., Qiu, M.X., Liu, D.L., Guo, H.C., Wang, Z.H., and Lin, S.Z. (2011). Antitumor and apoptosis-promoting properties of emodin, an anthraquinone derivative from *Rheum officinale* Baill, against pancreatic cancer in mice via inhibition of Akt activation. *International Journal of Oncology*, **39**: 1381-1390.

Wiesner, K., MacDonald, D.M., Bankiewicz, C., and Orr, D.E. (1968). Structure of pithecolobine. II. *Canadian Journal of Chemistry*, **46**: 1881-1886.

Wu, Q.P., Wang, Z.J., Tang, L.Y., Fu, M.H., and He, Y. (2009). A new flavonoid glucoside from *Cassia angustifolia*. *Chinese Chemical Letters*, **20**: 320-321.

Xiao, Z.-H., Wang, F.-Z., Sun, A.-J., Li, C.-R., Huang, C.-G., and Zhang, S. (2012). A new triterpenoid saponin from *Abrus precatorius* Linn. *Molecules*, **17**: 295-302.

Xiaoli, L., Naili, W., Sau, W.M., Chen, A.S., and Xinsheng, Y. (2006). Four isoflavonoids from the stem bark of *Erythrina variegata*. *Chemical and Pharmaceutical Bulletin (Tokyo)*, **54**: 570-573.

Xie, F., Lang, Q., Zhou, M., Zhang, H., Zhang, Z., Zhang, Y., Wan, B., Huang, Q., and Yu, L. (2012). The dietary flavonoid luteolin inhibits Aurora B kinase activity and blocks proliferation of cancer cells. *European Journal of Pharmaceutical Sciences*, **46**: 388-396.

Xu, Z.L., Guo, J.H., Song, S.T., Li, M.J., and Wu, D.Z. (1993). The effect of zinc glycyrrhizate on toxicity and anticancer activity of cisplatin in mice. *Yao Xue Xue Bao*, **28**: 567-571.

Xu, W., Liu, J., Li, C., Wu, H.Z., and Liu, Y.W. (2008). Kaempferol-7-O-β-D-glucoside (KG) isolated from *Smilax china* L. rhizome induces G2/M phase arrest and apoptosis on HeLa cells in a p53-independent manner. *Cancer Letters*, **264**: 229-240.

Xu, Y., Xin, Y., Diao, Y., Lu, C., Fu, J., Luo, L., and Yin, Z. (2011). Synergistic effects of apigenin and paclitaxel on apoptosis of cancer cells. *Public Library of Science One*, **6**: e29169.

Xu, Q.M., Shu, Z., He, W.J., Chen, L.Y., Yang, S.L., Yang, G., Liu, Y.L., and Li, X.R. (2012). Antitumor activity of *Pulsatilla chinensis* (Bunge) Regel saponins in human liver tumor 7402 cells *in vitro* and *in vivo*. *Phytomedicine*, **19**: 293-300.

Yadava, R.N., and Reddy, V.M.S. (2001). A new flavones glycoside, 5-hydroxy-7,3′,4′,5′-tetramethoxyflavone 5-O-β-D-xylopyranosyl-(1→2)-α-L-rhamnopyranoside from *Bauhinia variegata* Linn. *Journal of Asian Natural Products Research*, **3**: 341-346.

Yadava, R.N., and Reddy, V.M.S. (2002). A new biologically active flavonol glycoside from the seeds of *Abrus precatorius* Linn. *Journal of Asian Natural Products Research*, **4**: 103-107.

Yadava, R.N., and Reddy, V.M.S. (2003). Anti-inflammatory activity of a novel flavonol glycoside from the *Bauhinia variegata* Linn. *Natural Product Research*, **17**: 165-169.

Yadava, R.N., and Verma, V. (2003). A new biologically active flavones glycoside from the seeds of *Cassia fistula* (Linn.) *Journal of Asian Natural Products Research*, **5**: 57-61.

Yadava, R.N., and Sodhi, S. (2002). A new flavone glycoside: 5,7,3′,4′-tetrahydroxy-3-methoxyflavone-7-O-β-D-galactopyranosyl-(1→4)-O-β-D-glucopyranoside from the stem of *Acacia catechu* Willd. *Journal of Asian Natural Product Research*, **4**: 11-15.

Yan, L.L., Zhang, Y.J., Gao, W.Y., Man, S.L., and Wang, Y. (2009). *In vitro* and *in vivo* anticancer activity of steroid saponins of *Paris polyphylla* var. *yunnanensis*. *Experimental Oncology*, **31**: 27-32.

Yang, D., Tannenbaum, S.R., Büchi, G., and Lee, G.C. (1984). 40Chloro-6-methoxyindole is the precursor of a potent mutagen (4-chloro-6-methoxy-2-

hydroxy-1-nitroso-indolin-3-one oxime) that forms during nitrosation of the fava bean (*Vicia faba*). *Carcinogenesis*, **5**: 1219-1224.

Yu, H., Ge, S.F., and Wang, Y.G. (2004). Inhibition effect of glycyrrhizin on transplanted mandibular gland fibrosarcoma of mice. *Hua Xi Kou Qiang Yi Xue Za Zhi*, **22**: 452-455.

Yu, J.S., and Kim, A.K. (2010). Platycodin D induces apoptosis in MCF-7 human breast cancer cells. *Journal of Medicinal Food*, **13**: 298-305.

Yu, J.Q., Bao, W., and Lei, J.C. (2012). Emodin regulates apoptotic pathway in human liver cancer cells. *Phytotherapy Research*, (in press).

Zakhary, N.I., el-Aaser, A.A., Abdelwahab, A.A., Fathey, S.A., and Aboul-Ela, F. (1994). Effect of *Vicia faba* and bran feeding on nitrosamine carcinogenesis and formation. *Nutrition and Cancer*, **21**: 58-69.

Zhang, J., Yuan, K., Zhou, W.L., Zhou, J., and Yang, P. (2011). Studies on the active components and antioxidant activities of the extracts of *Mimosa pudica* Linn. from Southern China. *Pharmacognosy Magazine*, **7**: 35-39.

Zhao, Y., Liu, J.P., Lu, D., and Li, P.Y. (2007). A novel cycloartane triterpene glycoside from the seeds of *Cassia sophera* L. *Natural Product Research*, **21**: 494-499.

Zhou, Y., Liu, Y.E., Cao, J., Zeng, G., Shen, C., Li, Y., Zhou, M., Chen, Y., Pu, W., Potters, L., and Shi, Y.E. (2009). Vitexins, nature-derived lignin compounds, induce apoptosis and suppress tumor growth. *Clinical Cancer Research*, **15**: 5161-5169.

Zu, X.Y., Zhang, Z.Y., Zhang, X.W., Yoshioka, M., Yang, Y.N., and Li, J. (2010). Anthocyanins extracted from Chinese blueberry (*Vaccinium uliginosum* L.) and its anticancer effects on DLD-1 and COLO205 cells. *Chinese Medical Journal*, **123**: 2714-2719.

Traditional and Folk Herbal Medicine: Recent Researches Vol. 2 (2014) *Pages* **359–368**
Editor: **V.K. Gupta**
Published by: **DAYA PUBLISHING HOUSE, NEW DELHI**

11

Anti-inflammatory Activity of the Hydroethanol Extract of *Plantago erosa* in Various Models of Inflammation

C.C. Barua[1*], J.D. Roy[1], B. Buragohain[1], A.G. Barua[2] and Mangala Lahkar[3]

ABSTRACT

Plantago erosa ex Roxb (Plantaginaceae) a perennial herb, is of high ethnopharmacological relevance as the plant is traditionally used in folk remedy in inflammatory condition viz. insect bites, poison-ivy rashes, minor sores and boils, to cure snakebite and for coughs in rural areas of North east India. The present study focuses on the anti-inflammatory activity of hydroethanol extract of Plantago erosa (HEPE) in acute, sub-acute and chronic models of inflammation. The models used for this study are carageenan–induced paw edema in the rat, cotton pellet granuloma in rats and formaldehyde induced chronic polyarthritis in rats. It was observed that HEPE was effective in carageenan–induced edema in the rat as it showed maximum reduction at 2 and 5 h with a percent reduction of 68.42 per cent and 82.75 per cent respectively, at the dose 200 mg/kg p.o. in comparison to the standard drug-Indomethacin which showed a time dependent reduction of paw oedema volume from 6.25 per cent at 30 minutes to 94.74 per cent at 5 h. HEPE also showed a dose dependent reduction in weight of the pellet with percent reduction of 85.14 to 92.75 per cent at the dose 50 to 200 mg/

1 Department of Pharmacology and Toxicology, College of Veterinary Science, Khanapara, Guwahati – 781 022, Assam, India

2 Department of Veterinary Public Health, College of Veterinary Science, Khanapara, Guwahati – 781 022, Assam, India

3 Department of Pharmacology, Gauhati Medical College, Bhangagarh, Guwahati – 781 005, Assam, India

* *Corresponding author*: E-mail: chanacin@gmail.com; chanacin@satyam.net.in

kg p.o. in cotton pellet induced granuloma in rats. In formaldehyde mediated polyarthritis model in rats HEPE was the most effective extract at the dose 200mg/kg p.o. as it could significantly reduce the inflammation from 1st to 3rd day and again from 5th to 10th day. Hence it can be concluded that the HEPE exhibits potent anti-inflammatory activity in all the models of inflammation which could be due to the presence of steroids, triterpenes, tannins and diterpenes present therein.

Keywords: Anti-inflammatory, Hydroethanol, *Plantago erosa*, Carageenan, Cotton pellet granuloma, Formaldehyde, Polyarthritis.

Introduction

Inflammation is a pathophysiological response of mammalian tissues to a variety of hostile agents including infectious organisms, toxic chemical substances, physical injury or tumor growth leading to local accumulation of plasmic fluid and blood cells (Sobota *et al.*, 2000). Although a defence mechanism, the complex events and mediators involved in the inflammatory reaction can induce, maintain and aggravate many disorders. Most clinically important medicine for inflammation belongs to steroidal or non-steroidal anti-inflammatory chemical therapeutics. The use of non-steroidal anti-inflammatory drugs (NSAIDs) in the treatment of diseases associated with inflammatory reactions has adverse effects which pose a major problem in their clinical use. NSAIDs cause mild side effects that include stomach ulcers, bleeding ulcers, kidney dysfunction, constipation, dizziness and headaches. Hence, new anti-inflammatory drugs lacking such effects are being searched for as alternatives to NSAIDs (Afolayan *et al.*, 2010). Comparatively, the use of herbal and other plant based medicine has a long history with minimum or no side effects (Mishra and Allan, 2010). According to Kumara (2001), plant based drugs in traditional medicine are being paid much attention because of their minimal side effects, cheapness and also the fact that 80 per cent of the world population still relies on them.

Rheumatoid arthritis (RA) is a chronic, systemic inflammatory disease predominantly affecting the joints and peri-articular tissues. RA still remains a formidable disease, being capable of producing severe crippling deformities and functional disabilities. RA is classified as an inflammatory arthritis, the disease comprises of three basic inter-related processes like inflammation, synovial proliferation and joint tissue destruction (Lavanya *et al.*, 2010). In India, many Ayurvedic practitioners are using various indigenous plants for the treatment of different types of arthritic conditions. Although the application of these medicaments has a sound tradition and a rational background according to the Indian system of medicine, perhaps it is essential to investigate the rationality of their use in modern scientific terms. This is particularly important, firstly due to the gravity of the problem of rheumatism and arthritis and secondly due to the absence of the right type of drug of synthetic origin for its treatment. The presently available drugs provide only symptomatic relief and are not free from side effects (Shah *et al.*, 2006).

Plantago erosa ex Roxb (Plantaginaceae), a perennial herb is of high ethno-pharmacological relevance as the plant is traditionally used in folk remedy as astringent, antitoxic, antimicrobial, demulcent, expectorant and diuretic. A poultice of the leaves of *P. erosa* is used externally for insect bites, poison-ivy rashes, minor sores and boils which is a well known ethnic practice prevalent in native villages of North East India (Saklani and Jain, 1994). It is also known to cure snakebite. Different parts of the plant are used for coughs in rural areas of Assam. Recent research has shown *P. erosa* to be promising in lowering cholesterol and controlling diabetes. The aqueous extract of *P. major* is endowed with effective anti inflammatory and analgesic activities (Nunz *et al.*, 1997). *In vitro* cytotoxic, antiviral and immunomodulatory effects of *P. erosa* and *P. asiatica* are also reported (Chiang *et al.*, 2003). Its laxative effect after open haemorrhoidectomy showed that it shortens mean post operative hospital stay, expedite digestive function recovery and lessen the pain after stool (Atta *et al.*, 2005).

We have reported anti-inflammatory properties of the methanol extract of leaves of *Plantago eros*a (Barua *et al.*, 2010) and analgesic activity in hydroethanol extract of the plant (Barua *et al.*, 2011). These preliminary findings prompted us to study the anti inflammatory activity of the hydroethanol extract of *Plantago erosa* in acute, sub acute and chronic models of inflammation to explore the possibility of its use in arthritis.

Materials and Methods

Plant Materials

Leaves of *Plantago erosa* were collected during the month of July–Sept, 2010–2011 from the medicinal garden of the Department of Pharmacology and Toxicology, College of Veterinary Science, Khanapara, identified by taxonomist of NEIST, Jorhat, Assam and voucher specimens were deposited in the herbarium (*Plantago erosa* Specimen No. AAU/CVSC/PHT/04).

Preparation of Plant Extract

Leaves of *Plantago erosa* were shade dried and powdered. About 250 g of powdered leaves was soaked in 1000 ml of 50: 50 of aqueous ethanol for 72 h in a beaker and the mixture was stirred every 18 h using a sterile glass rod. Filtrate was obtained after passing through What man filter paper no 1 for three times and concentrated in Rotary evaporator (Equitron, Roteva) at 50°-60°C under reduced pressure. A dark brown hydroethanol extract of *Plantago erosa* (HEPE) was obtained and stored in air tight container at 4°C for further use. Recovery was 18.06 per cent (w/w) in terms of dry leaves.

Experimental Animals

Healthy adult albino mice weighing between 25-30 g and Wistar rats weighing between 150-200 g of either sex approximately of same age, were used for acute toxicity studies and anti-inflammatory activity respectively. The animals were group housed in polypropylene cages containing sterile paddy husk bedding under controlled conditions at 25° ± 3°C, RH 50 ± 5 per cent and 10/14 h of light/dark cycles. Food and water were provided *ad lib*. The study was conducted after obtaining approval from Institutional Animal Ethics Committee (No: 770/03/ac/CPCSEA/FVSc, AAU/IAEC/06/22).

Phytochemical Screening

HEPE was subjected to phytochemical screening as per the standard method for presence of different phytoconstituents (Harborne, 1991).

Acute Toxicity Studies

Acute oral toxicity of the plant extract was estimated by following up- and–down stair case method in mice as per OECD (TG 425) guidelines. Based on the toxicity study, the doses of plant extract for different parameters to be studied were selected.

Carageenan–Induced Edema in the Rat

This study was done by the method described by Winter *et al.*, 1962. Edema in one of the hind paws of the rats was induced by injecting 0.1 ml of 1 per cent solution of carageenan in normal saline solution into the plantar aponeurosis. HEPE (50, 100 and 200 mg/kg), standard drug-Indomethacin (10 mg/kg p.o.) or control were administered 1 h before the injection of the irritant. Volume of the injected paw was measured plethysmographically (Plethysmometer model: IITC 520 MR) after 30 mins of the injection and every 1 h up to 5 h. The difference between the two readings gives the edema developed in each rat. The mean of the group was calculated and compared with the mean edema developed in the control group. The percent inhibition produced by the test compound was compared with that produced by the standard drug. Percentage inhibitory activity is calculated by following equation.

$$\text{Per cent inhibition} = \frac{(C_t - C_0)\text{ control} - (C_t - C_0)\text{ treated}}{(C_t - C_0)\text{ control}} \times 100$$

where,

C_t is the paw volume at time t after carrageenan injection,

C_0 is the normal paw volume before carrageenan injection,

$(C_t - C_0)$ treated is the difference in the paw size before and after injection in treated group,

$(Ct - C_0)$ control is the difference in the paw size before and after injection in control group.

Cotton Pellet Granuloma in Rats

Sterilized cotton pellets weighing 50 mg each were subcutaneously implanted in the previously shaved back of rats aseptically (Winter and Porter, 1957). The animals were administered with HEPE (50, 100 and 200 mg/kg), standard drug-Indomethacin (10 mg/kg) or control orally from day 1 to 7. On 8th day, the animals were sacrificed and blood was collected for estimation of biochemical parameters. Pellets surrounded by granuloma tissue were dissected out carefully which was dried at 60°C in the hot air oven overnight to a constant weight. Mean weight of the granuloma tissue in each group was recorded and percentage inhibition was calculated by comparing the mean weight of the test group with the control group.

Formaldehyde Induced Chronic Polyarthritis in Rats

Arthritis is induced in the rat by injecting, 0.1 ml of 2.0 per cent formaldehyde solution on the first and third day of the experiment into the planter aponeurosis of hind paw (Seyle, 1949). HEPE (200 mg/kg), standard drug-Indomethacin (10 mg/kg) or control was administered orally from one to ten day of the experiment. The volume of the paw is measured plethysmographically daily for the next 10 days. The mean oedema developed in each group is calculated and compared with control group.

Statistical Analysis

All values were expressed as mean ± SE and analysed by using one way analysis of variance (ANOVA). P value less than 0.01 ($P < 0.01$) was the criteria for statistical significance.

Results and Discussion

In our previous study, we have reported that the hydroethanol extract of the plant exhibited potent analgesic activity (Barua *et al.*, 2011). In one of the model for analgesic activity *i.e.* formalin induced paw licking activity, the extract was found to completely abolish early phase and inhibit the late phase of paw licking dose dependently which indicates that it might cause complete inactivation of C- fiber in the early phase and the possibility that extract might act through activation of NMDA and non-NMDA receptors and NO cascade (Derard *et al.*, 1980; Davidson and Carlton, 1998). Based on these findings, the anti inflammatory activity of the hydroethanol extract of *Plantago eros*a was undertaken.

The phytochemical screening of the plant showed presence of steroids by Salkowski's and Lieberman Burchardt's test, diterpenes and triterpenes by Salkowski's and Lieberman Burchardt's test, tannins by ferric chloride and gelatin test. Oral administration of HEPE up to 2 g/kg did not produce any toxic effects in the normal behaviour of the mice. No mortality was observed and the extract was found to be safe at the given dose.

Carrageenan-induced rat paw edema is a suitable experimental animal model for investigating the anti-inflammatory effect of natural products and is believed to be biphasic; the first phase (1 h) involves the release of serotonin and histamine and the second phase (over 1 h) is mediated by prostaglandin, the cyclo-oxygenase products, and the continuity between the two phases is provided by kinins (Zhou *et al.*, 2008). The hydroethanol extract of *P. erosa* leaves showed maximum reduction of inflammation at 2 and 5 h with a percent reduction of 68.42 per cent and 82.75 per cent respectively, at the dose 200 mg/kg p.o whereas the standard drug, Indomethacin showed a time dependent reduction of paw oedema, volume from 6.25 per cent at 30 mins to 94.74 per cent at 5 h. HEPE inhibited both the phases in carageenan induced paw oedema in rats (Table 11.1). As compared to the hydroethanol extract, methanol extract of the plant could inhibit only the second phase of inflammation in

Table 11.1: Effect of HEPE in carageenan-induced paw oedema in rat.

Treatments	*Dose (mg/kgp.o)*	*Inflammatory Volume after Carrageenan Injection (ml) (Percent inhibition of inflammation)*					
		30 min	*1 h*	*2 h*	*3 h*	*4 h*	*5h*
Control	–	0.16 ± 0.01	0.18 ± 0.01	0.19 ± 0.01	0.22 ± 0.01	0.24 ± 0.01	0.29 ± 0.01
	50	0.15 ± 0.03 (6.25)	0.13 ± 0.01 (27.77)	0.11 ± 0.02 (6.25)	0.16 ± 0.01 (27.27)	0.14 ± 0.02 (41.16)	0.10 ± 0.01 (65.51)
HEPE	100	0.12 ± 0.01 (25.00)	0.10 ± 0.02 (44.44)	0.08 ± 0.01 (57.89)	0.14 ± 0.01 (36.36)	0.10 ± 0.02 (58.33)	0.07 ± 0.01 (75.86)
	200	0.11 ± 0.02 (31.25)	0.08 ± 0.01 (55.55)	0.06 ± 0.01 (68.42)	0.12 ± 0.01 (45.45)	0.08 ± 0.01 (66.66)	0.05 ± 0.01 (82.75)
Indomethacin	10	0.15 ± 0.01 (6.25)	0.15 ± 0.01 (16.66)	0.13 ± 0.01 (31.57)	0.11 ± 0.01 (50.00)	0.08 ± 0.01 (66.66)	0.02 ± 0.01 (93.10)

Values are Mean ± SE. N=6; P< 0.01.

carrageenan-induced rat paw edema in our previous study (Barua *et al.*, 2011), which might be due to different phytochemical ingredients or difference in the solvents (methanol or hydroethanol) used. The most effective dose in the methanol extract was 600 mg/kg p.o., whereas the hydroethanol extract inhibited both the inflammatory phases at 200 mg/kg p.o.

The inflammatory granuloma is a typical feature of established chronic inflammatory reaction (Ismail *et al.*, 1997). The dry weight of the pellets correlates well with the amount of granulomatous tissue (Swingle and Shideman, 1972). HEPE showed a dose dependent reduction in weight of the pellet with per cent reduction of 85.14 to 92.75 per cent at 50 to 200 mg/kg oral doses. The standard drug Indomethacin showed reduction of 81.77 per cent. The effect of HEPE at 100 and 200 mg/kg oral doses was better than the standard drug. The results are presented in Table 11.2.

Table 11.2: Effect of HEPE in cotton pellet granuloma in rats.

Treatments	*Dose (mg/kg p.o.)*	*Mean Weight of Pellets (mg)*	*Per cent Reduction*
Control	–	208.67 ± 0.00	0.00
HEPE	50	37.16 ± 1.27	82.19
	100	20.22 ± 2.22	90.31
	200	12.07 ± 2.34	94.21
Indomethacin	10	35.00 ± 3.16	81.77

Values are Mean ± SE. N=6; P< 0.01 vs vehicle.

Formaldehyde induced arthritis has been used to study the efficacy of a drug against the proliferative phase of inflammation *i.e.* chronic inflammation. Arthritis induced by formalin is a model used for evaluation of an agent with probable anti-proliferative activity (Banerjee *et al.*, 2002). In this model, there was gradual increase in percent reduction of inflammation with maximum reduction on 10th day (80.63) post treatment at 200 mg/kg oral dose which was the optimum dose in other two models of inflammation. The standard drug also showed similar pattern of reduction of inflammation (83.55 per cent) The results are presented in Table 11.3.

The present study revealed that the hydroethanol leaf extract of *Plantago erosa* (HEPE) was effective against acute (carrageenan induced paw oedema), sub acute (cotton pellet granuloma) as well as chronic (formaldehyde induced arthritis) types of inflammation. It might be possible that the various phytoconstituents present in HEPE might be responsible for inhibition of inflammation. A study on the anti-inflammatory activity of methanolic extract of *Solanum nigrum* suggests that steroids, alkaloids and triterpenoids might be responsible for anti-oedematous effect of the extract (Arunachalam *et al.*, 2009). The constituents of the rhizome of *Kaempferia galanga* includes a large amount of steroid and terpenoid and its alcoholic maceration has also been applied as liniment for rheumatism (Keys, 1976; Lieu, 1990). Likewise, phytoconstituents like steroids, triterpenes present in HEPE might play role in controlling inflammation in the present study. Further fractionation of the crude

Table 11.3: Effect of HEPE in formaldehyde induced chronic polyarthritis in rats.

Treatments	*Inflammatory Volume after Formaldehyde Injection (ml) (Per cent inhibition of inflammation)*										
	Dose (mg/kgp.o).	*1st day*	*2nd day*	*3rd day*	*4th day*	*5th day*	*6th day*	*7th day*	*8th day*	*9th day*	*10th day*
Control	–	4.36±0.72	3.12±0.52	2.52±0.42	4.52±0.75	2.38±0.39	2.27±0.37	2.19±0.36	2.01±0.33	2.76±0.46	3.77±0.62
HEPE	200	2.36±0.39 (45.87)	2.05±0.34 (34.29)	1.97±0.32 (21.82)	2.17±0.36 (51.99)	1.92±0.32 (19.32)	1.90±0.31 (16.29)	1.54±0.25 (29.68)	1.34±0.22 (33.33)	0.91±0.15 (67.02)	0.73±0.12 (80.63)
Indomethacin	10	2.14±0.35 (50.91)	1.87±0.31 (40.06)	1.58±0.26 (30.70)	2.13±0.35 (52.87)	1.97±0.32 (17.22)	1.46±0.24 (35.68)	1.39±0.23 (36.52)	1.14±0.19 (43.28)	0.84±0.14 (69.56)	0.62±0.10 (83.55)

Values are Mean ± SE. N=6. P< 0.01 vs vehicle.

extract and bioassay guided studies could help in identification of a particular fraction and/or compounds responsible for subsiding inflammation. In conclusion, HEPE possesses anti-inflammatory activity in acute, sub acute and chronic type of inflammation and provides a basis for further study to develop a suitable agent for treatment of various crippling inflammatory conditions like arthritis, rheumatism, osteoarthritis, spondylosis etc.

Acknowledgement

The authors are grateful to National Medicinal Plant Board, Govt. of India, New Delhi for providing financial assistance to carry out this work. Physical facility provided by the Director of Research (Vety), FVSc is also gratefully acknowledged.

References

Anthony, J. A., Oluseyi, M. A., Adeolu A. A., and Margaret O. S. (2010). Anti-inflammatory and analgesic activity of the methanol extract of *Malva parviflora* Linn (Malvaceae) in rats. *African Journal of Biotechnology*, **9:** 1225-1229.

Arunachalam, G., Subramanian, N., Perumal, P. G., Karunanithi, M., and Ravichandran, V. (2009). Evaluation of anti-inflammatory activity of methanolic extract of *Solanum nigrum* (Solanaceae). *Iranian Journal of Pharmaceutical Sciences*, **5:** 151-156

Atta, A. H., and Mouneir, S. M. (2005). Evaluation of some medicinal plants for anti diarrhoeal activity. *Phytotherapy Research*, **19:** 481-448.

Banerjee, S., Sur, T. K., Mandal, S., Chandra, D. P., and Sikdar, S. (2000). Assessment of the anti-inflammatory effects of *Swertia chirata* in acute and chronic experimental models in male albino rats. *Indian Journal of Pharmacology*, **32:** 21-24

Barua C.C., Buragohain, B., Roy J. D., Talukdar A., Barua A. G., Borah, P., and Lahon, L. C. (2011). Evaluation of analgesic activity of hydroethanol extract of *Plantago erosa* Ex Roxb. *Pharmacologyonline*, **2:** 86-95.

Barua, C.C., Pal, S. K., Buragohain, B., Roy J. D., Barua A. G., and Borah P. (2010). Studies on the anti-inflammatory properties of *Plantago erosa* leaf extract in rodents. *Journal of Ethnopharmacology*, **134:** 62-66.

Nikunja, K. M., and Allan, J. J. (2010). Evaluation of anti-inflammatory activity and potency of herbal formulation consists of different proportions of *Curcuma longa* and *Boswellia serrata* by using cotton pellet granuloma and xylene induced mice ear edema model. *International Journal of Pharmaceutical Techniques and Research*. **2:** 1855-1860.

Chiang, L. C., Chiang, W., Chang, M. Y., and Lin C. C. (2003). *In vitro* cytotoxic, anti-viral and immunomodulatory effects of *Plantago erosa* and *Plantago asiatica*. *Chinese Medicine*, **31:** 225-234.

Davidson, E. M., and Carlton S. M. (1998). Intraplanter injection of dextrophan, Ketamine or memantine attenuates formalin induced behaviors. *Brain Research*, **785:** 142.

Derard, R., Jougney, S., Delevalcee, F., and Falhout, M. (1980). Release of prostaglandins E and F in analgogenic reaction and its inhibition. *European Journal of Pharmacology*, **51:** 24.

Erciyas, E., Erkaleli. H. I., and Cosar, G. (1994). Antimicrobial evaluation of some styryl ketone derivatives and related thiol adducts. *Journal of Pharmacy Science*, **83:** 545-548.

Harborne, J. B. (1991). Phytochemical Screening –Guide to modern techniques of plant analysis. 2nd edition, Chapman and Hall, pp. 653.

Ismail, T.S., Gopalakrisan, S., Begum, V.H., and Elango, V. (1997). Anti-inflammatory activity of *Salacia oblonga* Wall and *Azima tetracantha* Lam. *Journal of Ethnopharmacology*, **56:** 145-152.

Keys, J. D. (1976). Chinese Herbs (Their Botany, Chemistry, and Pharmacodynamics). Charles E. Tuttle Company, Inc., Tokyo.

Koleva, I. I., Van Beek, T.A., Linssen, J. P., de Groot, A., and Evstapieva, L. N., 2002 *Phytochemical Analysis*, **13:** 8-17.

Kumara, N. K. V. M. R. (2001). Identification of strategies to improve research on medicinal plants used in Sri Lanka. In: WHO Symposium, University of Ruhuna, Galle, Sri Lanka.

Lieu, V.D. (1990). Medicinal Plants in Vietnam. Ha Noi. pp.239.

Nunz Guillen, M. E., Silva, D., Emim, J. A., Souccar, C., and Lapa, A. J. (1997). Analgesic and anti- inflammatory activities of the aqueous extract of *Plantago erosa*. *International Journal of Pharmacology*, **35:** 107.

Saklani, A., and Jain, S.K. (1994). *Cross cultural Ethnobotany of North East India*, pp. 97.

Seyle, H. 1949. Further studies concerning the participation of the adrenal cortex in the pathogenesis of arthritis. *British Medical Journal*, **2:** 1129-1135.

Sobota, R., Szwed, M., Kasza, A., Bugno, M., and Kordula, T. (2000). Parthenolide inhibits activation of signal transducers and activators of transcription (STATs) induced by cytokines of the IL- 6 family. *Biochemistry and Biophysics Research Communication*, **267:** 329-333.

Swingle, K.F., and Shideman, F.E. (1972). Phases of inflammatory response to subcutaneous implantation of cotton pellet and other modifications by certain anti-inflammatory agents. *Journal of Pharmacology and Experimental Therapeutics*, **183:** 226-234.

Winter, C.A., and Porter, C. A. (1962). Effect of alteration in side chain upon anti-inflammatory and liver glycogen activities of hydrocortisone esters. *Journal of American Pharmacological Association Science and Education*, **46:** 515-519.

Winter, C.A., Risley, E.A., and Nuss, C.W. (1962). Carrageenan- induced oedema in the hind paw of the rat as an assay for anti-inflammatory drugs. *Proceedings of the Society for Experimental Biology and Medicine*, **3:** 544-547.

Zhou, M., Wang, H., Suolangjiba, K. J., and Yu, B. (2008). Antinociceptive and anti-inflammatory activities of *Aquilaria sinensis* (Lour.) Gilg. Leaves extract. *Journal of Ethnopharmacology*, **117:** 345-350.

Traditional and Folk Herbal Medicine: *Pages* **369–386**
Recent Researches Vol. 2 (2014)
Editor: **V.K. Gupta**
Published by: **DAYA PUBLISHING HOUSE, NEW DELHI**

12

The Study on Chinese Patents upon Miao Ethnic Drug *Cynanchum wilfordii* (Maxim.) Hemsl and Bai/Yi Ethnic Drug *Geranium strictipes* R. Knuth

Tang Jun[1], Li Ting[1], Gao Pengfei[2*], Wu Xiumei[2], Geng Ling[2], Li Fengxian[2,3], Zhao Yu[2*] and Li Hui[2]

ABSTRACT

This paper analytically records Chinese patents during last two decades concerning the medicinal applications of Cynanchum wilfordii (Maxim.) Hemsl as a Miao minority herbal drug and Geranium strictipes R. Knuth as ethnic drug of Bai/Yi minorities. The two herbs at different families possess the same popular name of Ge Shan Xiao. Their origins, varieties, nomenclature, folk conventional usage, major functions, as well as their compatibility principles are compared and summarized in order to furnish more plentiful information and to afford suggestions on relative on-going R&D projects of these two herbal drugs.

Keywords: Patent study, Miao ethnic drug, Bai/Yi ethnic drug, *Cynanchum wilfordii, Geranium strictipes.*

Introduction

In China, ethnomedicine is composed of the traditional Chinese medicine, ethnic minority medicines and folk herbal medicines. Previously, systematic study majorly

1 School of Nursing, Dali University, Dali 671000, China

2 The Key Laboratory of Medicinal Insects and Spiders Resources for Development and Utilization, Yunnan Province,

3 The 4th People's Hospital in Yunnan Province, Dali University, Dali 671000, China

* *Corresponding author*: E-mail: 13013443093@126.com (Gao) or dryuzhao@126.com (Zhao)

focus on traditional Chinese medicine, whereas minority medicines including a large number of medical innovation resources with potential value receives less concern, and is worth researching for pharmaceutical investigators (Zhao and Hu, 2002; Gao *et al.*, 2011; Zhang and Zhao, 2003). In recent years, new drugs derived from ethnic minority medicines and local herbal medicines are constantly approved by Chinese SFDA access to the drug market, which not only created apparent economic benefits, but also spreaded cultures of ethnic minorities, and significantly achieved the purpose of promotion of a harmonious society.

Yunnan province is a multi-national region which composed of 26 ethnic minorities, which possess great treasure of folk medicines. It is important to collect, to inherit and to develop this treasury as soon as possible to advance together with chemical drugs and biopharmaceuticals (Woo *et al.*, 2007; Yu and Zhao, 2005). This paper afforded a brief analysis to the research progress and the status of patents applying of two ethnic medicinal herbs which own the same Chinese name of Ge Shan Xiao, and had a similar effect, whereas belongs to different plant families, utilized by different nations in various regions in southwestern China.

The Taxonomical Provenance differences between Miao Medicine *Cynanchum wilfordii* (Local Name: Ge Shan Xiao) and Bai/Yi Ethnic Drug *Geranium strictipes* (also named Ge Shan Xiao)

Miao medicine Ge Shan Xiao, belongs to Asclepiadaceae (genus of Cynanchum), is recorded as *Cynanchum wilfordii* (Maxim.) Hemsl in Dictionary of Traditional Chinese Medicine and Compendium of Materia Medica. It is also named as Ge Shan Qiao, Bai Shou Wu etc. It is a perennial herbal liana with a spindle-shaped flesh root in a grayish-brown color. The medicinal use roots are of 10 cm in length, 2 cm in diameter, which is born in hillside, valley, bushes and grass on the edge of the wood with an elevation of 800-1300 meters. This species is distributed in Hunan, Hubei, Shaanxi, Liaoning, Jiangsu, Sichuan, Shanxi, Anhui, Shandong, Xinjiang, Henan and Guizhou provinces, the most common herbal drugs are from wild resources. It was classified to spleen, stomach and kidney in Chinese meridian doctrine; with a bitter taste and a warm characteristic. The recorded efficacy of the title plant contains tonifying liver and kidney, strengthening the physique, invigorating spleen and stomach, as well as detoxification. It is a folk drug mainly used by Miao ethnic minority, and also applied by Buyei, Tuchia ethnic minorities. The drug "Ge Shan Xiao Ji Granule", which is developed and listed to the market successfully by Guizhou pharmarcertical company, –adopted *Cynanchum wilfordii* as major drug and its major function is disintegrating abdominal mass and promoting circulation of Qi. Indications for food retention, abdomen distends, apathy to food and acid regurgitation because of stomach disorders (Jiangsu New Medical School, 1977; State Drug Administration, 2002).

Moreover, another Ge Shan Xiao used by Bai and Yi ethnic minorities as folk medicine is actually the root of *Geranium strictipes* R. Knuth, which belongs to genus of Geranium L., Geraniaceae, and exists as the endemic plant of Yunnan province.

This perennial herb is 20 to 30 centimeters high, whose taproot is cone-shaped with a purple color and longer fibers, whose epidermis shows dark brown color, endothelium shows purple color, while the cut surface shows brownish yellow color. The raw materials owning yellow core and purple bark are preferred. Most of the title plants grow on the sunny slope. It is only distributed in western Yunnan province. The herbal drug contains tastes bitter mixed with light acerbity, and has a slight cold characteristic by traditional Chinese medicine classification. The recorded functions of the this herbal medicine concentrated on anti-inflammatory, hamostasia, restraining intestine; while the indications for enteritis, diarrhea, abdominal pain, internal bleeding, epistaxis, hematochezia, menorrhagia, post-partum hemorrhage, and injuries. Bai and Yi ethnic minorities often use it in the treatment of anorexia, enteritis, red and white diarrhea, indigestion disease etc (Li, 1990; Wang, *et al.*, 2008). The research group at Dali University carried out pharmacological and clinical investigations for a long time and more than 10 research papers has been published (Li, *et al.*, 2010; Lan and Zhang, 2010). Other Chinese group further performed the anti-HIV assay of this plant (Yang and Chen, 2007), which again revealed the potentiality of folk medicine for further development.

Case Studies on Patents Applications of *Cynanchum wilfordii* and *Geranium strictipes*

Summary of Patents Applying on Miao Drug *Cynanchum wilfordii*

There are 46 patents related to Miao drug *Cynanchum wilfordii* in the past two decades, including 43 invention patents and 3 utility model for patents. They can be classified as the following major items based on their different distinction between the clinical treatment of disease.

Warming the Kidney and Regulating the Function of the Liver and Spleen (One case)

A pharmaceutical company in Guangdong province applied the compound granules or tea bag, which is composed of thirteen kinds of medicinal materials such as *Cynanchum wilfordii*, Rou Cong Rong (*Cistanche deserticola* Ma), Gu Ya (Rice-grain Sprout), etc. Its functions are warming the kidney and fasting Ben, tonifying the spleen and helping digestion, relieving Qi Stagnancy in stomach, clearing the walls of the colon, clearing the intestine and detoxification, etc.

Treatment for Stomach Disorder/Gastroenteritis/Peptic Ulcer (13 Cases)

These patents are mainly applied from southwestern China such as Yunnan, Guizhou and Sichuan provinces. Guizhou Renhetong Pharmaceutical Co., Ltd applied a medicine for treatment of stomach disease composed of nine kinds of medicinal materials such as *Cynanchum wilfordii*, Ku Jin Pen (*Hemsleya macrosperma*), Mu Xiang (*Rosa banksiae*) and so on. Its function lies on treatment of deficiency of spleen and stomach, abdominal cold and pain, food retention and abdominal distension, abdominal and inflated chest rib, losing of appetite, and peptic ulcer, chronic gastritis, pancreatitis, cholecystitis (Guizhou Renhetong Pharmaceutical Co., Ltd, 2003).

The medicine applied by Guizhou Kai Li Pharmaceutical Co., Ltd, whose function is eliminating indigestion, has been on the market now (Guizhou kaili Pharmaceutical Co., Ltd, 2002). Guizhou Sanrentang Pharmaceutical Co, Ltd applied a traditional Chinese medicine patent, whose function is for stomach diseases and stomach bleeding and duodenal ulcer, which is made up of 12 kinds of medicinal materials such as *Cynanchum wilfordii*, Zhi Ke (*Poncirus trifoliata*), and Xiang Fu (*Cyperus rotundus* L.), etc. (Guizhou Sanrentang Pharmaceutical Co, Ltd, 2008)

Tian H Y applied a capsule preparation made up of nine kinds of medicinal materials such as *Cynanchum wilfordii*, Yi Zhi Huang Hua (*Solidago canadensis* L.), and Gua Zi Jin (*Carissa carandas*), etc. to treat with gastritis, indigestion, gastric ulcer, duodenal ulcer, abdominal mass and so on (Tian *et al.*, 2004).

Lu C R applied the preparation patent made up of four kinds of medicinal materials such as *Cynanchum wilfordii*, Ci Li Ye (Foliage of Rosa *Toxburghii Tratt. F. normalis* Rehd. et Wils.), Yang Liu Zhi (*Salix babylonica* L.) and San Qi (*Panax notoginseng*) to treat with atrophic gastritis, gastric ulcer, ulcer, etc. (Lu, 2002)

Wang C H applied a big compound patent made up of 27 kinds of medicinal materials such as *Cynanchum wilfordii*, Pu Gong Ying (*Taraxacum officinale*), Cao Xue Jie (*Polygonum paleaceum* Wall.) and so on to treat with atrophic gastritis. In his another authorized patent to treat with chronic gastritis, the applicant used 26 medicinal materials such as *Cynanchum wilfordii*, Chuan Lian Zi (Fructus Toosendan), Su Xin Hua (*Jasminum grandiflorum*) etc (Wang, 2007). Zhu W X made a powder medicine composed of 11 kinds of medicinal materials such as *Cynanchum wilfordii*, Qing Mu Xiang (*Saussurea costus* (Falc.) Lipech.), Shan Ci Gu (*Iphigenia indica*), and so on to treat with stomach disease (Zhu, 2005). Yuan Y declared a authorized patent, which is a small compound composed of *Cynanchum wilfordii* and Jin Qiao Mai (*Fagopyrum dibotrys* (D. Don) Hara) to treat with gastritis and gastrointestinal tract disease (Yuan, 2008). Zhang L L provided a compound medicine in the treatment of gastrointestinal diseases composed of 13 kinds of medicinal materials including *Cynanchum wilfordii* (Zhang, 1998). Xiao C S. applied a big compound made up of 11 kinds of folk medicines from Tuchia ethnic minority such as *Cynanchum wilfordii*, *Rosa banksiae* and so on, which is used in the treatment of gastric and duodenal ulcer (Xiao, 2004). Lian C D applied a compound medicine composed of 10 kinds of medicinal materials such as *Cynanchum wilfordii*, Wu Yao (*Lindera aggregate* (Sims) Kosterm.*var.playfairii* (Hemsl.) H.P. Tsui.), Ji Shi Teng (*Herba paederiae*), and so on to treat with superficial gastritis, erosive gastritis, gastric ulcer, and duodenal ulcer (Lian, 2008). Li F T applied a capsule made up of nine kinds of medicinal materials such as *Cynanchum wilfordii*, XiaoHuai Hua (*Desmodzum caudatum* (Thunb.) DC.), Yin Chen (*Artemisia scoparia* Waldstet Kit) and so on to treat with acute and chronic gastroenteritis (Li, 2003).

Niu Q D declared a compound medicine aiming at the treatment of gastroenteritis composed of nine kinds of medicinal materials such as *Cynanchum wilfordii*, XiaoHuai Hua (*Desmodzum caudatum* (Thunb.) DC.) and so on, which performing the functions of heat-clearing and detoxicating, as well as anti-inflammation and kills germs and stimulating the appetitetal (Niu, 2008).

Gastric Power Medicine, Invigorating the Stomach and Promoting Digestion Medicine and Health Products (Five cases)

Xiong L applied a patent named "new green gastric power medicine", which is composed of nine kinds of medicinal materials such as *Cynanchum wilfordii*, Huang Lian (*Rhizoma coptidis*), Zhi Ke (*Poncirus trifoliata*) and so on. It is wished to continually act on gastrointestinal and the digestive system, and functional linkage through different action mechanism sequentially, in order to strengthen the gastrointestinal local cholinergic nerve excitement, body liquid adjustment, senior central nervous comprehensive adjustment, so as to coordinate and strengthen the operation of gastrointestinal function (Xiong, 2004).

Guizhou Kedun Pharmaceutical Co., Ltd applied a small compound patent composed of *Cynanchum wilfordii*, Ma Lan Cao (*Herbia Wedeliae*) and Ci Li Ye (Foliage of *Rosa toxburghii* Tratt. *F. normalis* Rehd. et Wils.) for the treatment of children accumulating feed and loss of appetite (Guizhou Kedun Pharmaceutical Co., Ltd, 2002). Taiyihepu (Beijing) Academy of Traditional Chinese Medicine Co., Ltd applied a patent composed of 27 material medicines including *Cynanchum wilfordii*, which aims to treat with maldigestion (Taiyihepu (Beijing) Academy of Traditional Chinese Medicine Co., Ltd, 2009).

Li Y N applied a patent of pearl milk tea about its pharmaceutical technology, which is composed of 21 kinds of medicinal and food materials such as *Cynanchum wilfordii*, Dang Shen (*Salvia miltiorrhiza* Bunge), Shan Yao (*Rhizoma dioscoreae*) and so on and is to fill the blank of product whose function is for sub-health such as weak spleen and stomach, food lag in stomach and intestines in domestic food industry. This is the first case of bring *Cynanchum wilfordii* into Chinese tea culture of diet therapy and healthy drink productions (Li, 2008).

Microbiology Research Institute of China Science Academy applied a natural dealcoholic dosage to protect liver from suffering from AFLD, which also covers *Cynanchum wilfordii*, This is a case that used as a auxiliary of health food additives agent (Microbiology Research Institute of China Science Academy, 2007).

Treatment for Liver Cirrhosis, Liver Ascites (Three cases)

Yin H Y applied a patent of a compound composed of eight kinds of medicinal materials such as *Cynanchum wilfordii*, Ku Gua (*Momordica charantia* L.), Mu Gua (*Chaenomeles sinensis*), and so on. Mixed with pig bile by stirring, dried under sun and made into powder, to be taken with juice made from eels and white sugar in the treatment of liver cirrhosis, ascites and possess significant effect (Yin, 2002).

Xiao C S applied a patent of Chinese herbal medicine formula composed of 13 kinds of folk medicines from Tuchia ethnic minority including *Cynanchum wilfordii* in the treatment of hepatitis (Xiao, 2005).

Wang L applied a patent of powder composed of 31 kinds of medicinal materials from both animal and herbal resources such as *Cynanchum wilfordii*, Da Huang (*Radix et Rhizoma Rhei*), Jie Jin Cao (Huperzia spp.), and so on to treat with liver cirrhosis and ascites (Wang, 1998).

Treatment for Viral Hepatitis (One case)

Zhuo S R provided a kind of pills in the treatment of hepatitis B. Eight kinds of medicinal materials such as *Cynanchum wilfordii*, Di Er Cao (*Hypericum japonicum* Thunb. ex Murray), Ban Bian Lian (*Lobelia chinensis* Lour.), etc. were boiled with water and condensed, then mixed with juice of white radishes and brown sugar (Zhuo, 1999).

Treatment for Fatty Liver and Used as Anti-Alcoholic Medicines (Two cases)

Kong X J applied a Chinese medicine formula patent in the treatment of fatty liver, which is composed of 13 kinds of medicinal materials such as *Cynanchum wilfordii*, Chai Hu (Bupleurum spp.), Bai Shao (*Radix Paeoniae Alba*), and so on (Kong, 2010). Xie H B declared a kind of accompaniment drink used for anti-alcoholic and decreasing fire, which is composed of nine common kinds of medicinal materials and juice such as *Cynanchum wilfordii*, green tea, Yu Xing Cao (*Houttuynia cordata* Thunb.), and so forth. It is used for heat-clearing, decreasing fire, anti-alcoholic, reduce weight, antihypertensive, refresh spirit and invigorating spleen (Xie, 1992).

Treating for Cancer Diseases (Four cases)

Zhu G H was authorized a patent of a recipe to treat with tumor from ethnic minority Tuchia drugs. It is a large compound composed of 28 kinds of medicinal materials such as *Cynanchum wilfordii*, Zhu Jie Shen (*Rhizoma Panacis Japonica*), Ling Zhi (*Ganoderma lucidum* (Leyss.ex Fr.)Karst), and so on, which is used for the treatment of cervical cancer, nasopharyngeal carcinoma, lung cancer, etc. (Zhu, 2006)

Hu D L went public a kind of authorized patent concerning a capsule for the treatment of cancer, which is composed of 12 kinds of medicinal materials such as *Cynanchum wilfordii*, Wu Xing Cao (*Portulaca oleracea* L.), Ba Jiao Lian (*Dysosma versipellis* (Hance) M.Cheng ex Ying) etc. This formula is used to suppress tumor angiogenesis and to make tumor ischemia, hypoxia, hunger and atrophy and die, 80 cases of patients with tumor has been cured via this treatment (Hu, 2006). Chen Y X invented a kind of Chinese herbal medicinal wine cooked by 20 kinds of medicinal materials including *Cynanchum wilfordii*, Zang Hong Hua (*Crocus sativus* L.), Da Ji (*Euphorbia pekinensis* Rupr.),and so forth, which is for the treatment of leukemia (Chen, 1998).

Bai S R provided a kind of health tea composed of 10 kinds of medicinal materials including *Cynanchum wilfordii* for the cancer patients' rehabilitation and health protection (Bai, 1995).

Treatment for Leukoderma (1 case)

Yan Y H disclosed a medicine patent made up of six kinds of medicinal materials including *Cynanchum wilfordii*, Bu Gu Zhi (*Psoralea corylifolia* L.), Hei Gu Teng (*Biondia insignis* Tsiang), etc. The creams, gels and liquid membrane made from this compound drug were used to treat with 273 patients with vitiligo through externally curing internal disease, and 85 cases were recovered (Yan, 2008).

Treatment for Diabetes (1 case)

Zhou M Z was authorized a pharmaceutical compositions' patent composed of six kinds of medicinal materials including *Cynanchum wilfordii*, Bei Huang Qi (*Astragalus membranaceus* (Fisch.) Bunge), Huai Di Huang (*Rehmannia glutinosa* (Gaertn.) Li-Bosch), and so forth. It is used to treat diabetes and kidney disease, liver disease, encephalopathy and blood diseases related to it through various forms of preparations (Zhou, 2004).

Treatment for Orthopaedic Disease (Two cases)

Du Y Y applied a pharmaceutical compositions' patent composed of 13 kinds of rare medicinal materials such as *Cynanchum wilfordii*, Shan Ku Gua (*Trichosanthes cucumeroides* (Ser.) Maxim.), Zhui Feng Jian (*Patrinia heterophylla* Bunge), Nan Xing (*Rhizoma Arisaematis*) and so on, and it is used for fast treatment of bone fractures, which is proved to be a simple-prepared drug and release patients' pain (Du, 2005).

Yan C P in opened a patent of patch used for the treatment of bone hyperplasia, which was made up of six kinds of medicinal materials including *Cynanchum wilfordii*, Cuan Niu Xi (*Cyathula officinalis* Kuan) and so on, and auxiliary materials including sesame oil, rosin, beeswax, lubrication powder, zinc oxide, silicon nitrogen (Yan, 2011).

Therapy Rheumatism (1 case)

Chen Y X disclosed another patent of herbal drug wine for the treatment of rheumatism and blood poison. The wine was made by soaking 21 kinds of medicinal materials including *Cynanchum wilfordii* into rice wine, Zang Hong Hua (*Crocus sativus* L.), Da Ji (*Euphorbia* spp.), Niu Da Li (*Moghania philippinensis* (Merr. et Rolfe) Li.), Ren Shen (*Panax ginseng* C. A. Mey.), etc. and exhibited a satisfactory clinical effect (Chen, 1997).

Disordered Endocrine Therapy (1 case)

Zhang K J declared a authorized Chinese medicinal compositions' patent composed of 13 kinds of medicinal materials such as *Cynanchum wilfordii*, Ren Shen (*Panax ginseng* C. A. Mey.), Dang Shen (*Codonopsis pilosula* (Franch.) Nannf.), Yuan Shen (*Scrophularia ningpoensis* Hemsl.), and so forth. The compound possess the function of activating collaterals, getting rid of toxin and raising color, nourishing the liver and the stomach and balancing the Yang and Yin, which is therefore suitable to deal with the skin diseases, sexual dysfunction and endocrine disorders (Zhang, 2007).

The Effective Parts Extracting from the *Cynanchum wilfordii* (Seven cases)

The Second Military Medical University of the Chinese People's Liberation Army applied three patents of invention from *Cynanchum wilfordii*, *Cynanchum auriculatum* Royle ex Wight and *Cynanchum bungei* Decne. These three inventions involved two effective parts *viz.* total acetophenone, total saponins, while the contents of the two types of effective components both e"50 per cent in the extracted effective part. Among them the total acetophenone can be used in the treatment of liver disease and gastrointestinal disease, and total saponins can be used in the treatment of gastrointestinal diseases. The two patents used in the treatment of the gastrointestinal tract disease had been authorized (The Second Military Medical University of the Chinese People's Liberation Army, 2005; The Second Military Medical University of the Chinese People's Liberation Army, 2007; The Second Military Medical University of the Chinese People's Liberation Army and Guangdong Yipinhong Pharmaceutical Co., Ltd, 2004). Guizhou Normal University applied and got authorization about total saponins of *Cynanchum wilfordii* for the use of the prevention and control of the depression (Guizhou Normal University, 2007).

Jiangsu Academy of Traditional Chinese Medicine recently applied a patent of invention of degradation products of total C21 steroid saponins of *Cynanchum wilfordii*, which is evidenced to be active against tumor diseases (Jiangsu Academy of Traditional Chinese Medicine, 2010). Shandong Agricultural University has obtained the patent license for preparation method of the total polysaccharide of *Cynanchum wilfordii*, which has the immune adjustment function (Shandong Agricultural University, 2004). Beijing University of Chinese Medicine has been authorized a kind of preparation method of effective part from *Cynanchum wilfordii* by macroporous resin for anti-cancer use (Beijing University of Chinese Medicine, 2008).

The Patent of Pure Compounds Isolated from *Cynanchum wilfordii* (1 case)

To our knowledge, there is only one patent concerns the pure compound isolated from *Cynanchum wilfordii*, The diketone A applied by Zhangweidong from the Second Military Medical University of the Chinese People's Liberation Army. The patent also described that this acetophenone exhibited a strong protective effect to many kinds of brain damage on mice. There were also a remarkable stability and improve role on dementia patients and it could be used for the preparation of the drugs for prevention and cure various kinds of neurodegenerative diseases (The Second Military Medical University of the Chinese People's Liberation Army, 2008). It should be noted that there were others three patents of invention of effective parts by the same inventor, in which nine acetophenones and seven C21 steroidal saponins were included as pure compounds.

The Patents of *Cynanchum wilfordii* which was Used as an Alternative Materials (Two cases)

Korean Tianran Erduo Technology Company declared a kind of phytoestrogens composition used to prevent and cure menopause symptoms, the application disclosed the major medicinal effective ingredients of the composition—cinnamic acid was extracted from *Cynanchum wilfordii*.

Lu Z T was authorized applied patent concerning a kind of compound antidote of non-opioids composition, which was composed of non-opioids, analgesics, benzodiazepine drugs, anticholinergic drugs, opioid receptor antagonist, $\alpha 2$ receptor agonists, β receptor blockers, anti-depressant drugs and digestant according to a certain proportion. From the disclosed compound medicines, the digest tonic can be chosen from dry yeast, pepsin, Jixieteng, and *Cynanchum wilfordii*, etc. (Lu, 2000).

Feed Additives (1 case)

Besides the drugs for human beings, Yuanshijun applied a patent concerning usage of *Cynanchum wilfordii* as a kind of feed. Yuan S J filed for a patent of feed additives, which had been authorized, demonstrated using 13 kinds of medicinal materials including the malt, valley buds, *Cynanchum wilfordii* and so on as the green herb antibiotics to prevent and cure diseases of breeding animals and to improve its meat quality and taste (Yuan, 2000).

Patent for Utility Model Relevant to *Cynanchum wilfordii* (Three cases)

Mi X H filed for a patent of a cigarette holder reducing cigarette poison and diminishing pollution, whose packing filters part is composed of five kinds of medicinal materials including *Cynanchum wilfordii*, Chen Pi (*Pericarpium Citri Reticulatae*), Shan Zha Ye (leaves of *Crataegus pinnatifida Bge*), and so on which is used for neutralizing the nicotine in tobacco and attenuating the poison of cigarette (Mi, 2002).

Tao L H has applied two patents for utility model, one was the health preserving units and the other one was the health preserving boards, among them, *Cynanchum wilfordii*, which was useful for spleen and stomach, was always an optional programs of packing material for health preservation (Tao, 2010; Tao, 2010).

Summary of Applied Patents of Bai/Yi Folk Medicine *Geranium strictipes*

In the recent 20 years, there are 17 patents related to Bai/Yi folk medicine *Geranium strictipes*, no utility models were involved. The detailed investigations are listed below.

Treatment for Burn Wound and Other Skin Diseases (Five cases)

Liu S B applied a compound pharmaceutical patent composed of seven kinds of medicinal materials including charcoal of *Geranium strictipes*, Hua Zhang (*Rhiaoma Polygoni Cuspidati*), Huang Lian (*Coptis chinensis* Franch.), and so on to treat burn wound (Liu, 2010).

Xia X L in Lufeng of Yunnan province made a kind of compound composed of six kinds of medicinal materials such as Shigao, Dahuang, Cidiyu and so on in the treatment of common skin diseases as pruritus vulvae, lousy feet, scabies, sexual infection, burn wound, etc. (Xia, 2002).

In another patent, Mao H X provided a kind of liniment for skin composed of *Geranium strictipes*, Zi Cao (*Lithospermum erythrorhizon* Sieb. Et Zucc), and vegetable oil. The inventor declared that the liniment was not only effective to skin diseases such as skin acnes, acne, wrinkles rough, red blood silk, skin redness, chilblain, etc., but also had a certain curative effect to the diseases including psoriasis, skin cancer, burn wound, etc. (Mao, 2004).

Meng L fielded for a kind of topical drug to treat acute or chronic eczema, which was made up of 10 kinds of medicinal materials such as *Geranium strictipes*, Gan Cao (*Glycyrrhiza uralensis* Fisch), Jin Yin Hua (*Lonicera japonica* Thunb.), etc. according to the recipe, and made through the process of water extraction and concentration (Meng, 2008).

Guang D Y applied a traditional Chinese medicine in the treatment of fungal skin disease, which was composed of oral administration group A drugs and topical group B drugs, the medicine was a large compound made up of 28 kinds of medicinal materials such as *Geranium strictipes*, Bai Xian Pi (*Dictamnus dasycarpus* Turcz.), Di Fu Zi (*Kochia scoparia* (linn.) Schrad.), etc. (Guang, 2008).

Cool and Hemostatic Drugs (1 case)

Yang M J presented a patent describing one kind of nanometer granules, which is composed five kinds of medicinal materials such as *Geranium strictipes*, Da Hong Pao (*Clerodendrum bungei* Sterd.), *Senecio scandens* Buch.-Ham. ex D. Don etc. in nanometer levels. The nano-pharmacy apparently improved the bioavailability of listed granules made of these five raw material drugs (Yang, 2001).

Treatment for Gastrointestinal Diseases and Check Method of Drug Quality (Four cases)

Xia X L fileded for a patent of a kind of medicine in the treatment of digestive diseases, which was composed of eight kinds of medicinal materials such as *Geranium strictipes*, Da Huang (*Rheum palmatum* L.), Yuan Hu (*Corydalis turtschaninovii* Bess. *f. yahusuo* Y. H. Chou et C. C. Hsu), etc. (Xia, 2002).

Huang X afforded a patent for a traditional Chinese medicine, which was made up of seven kinds of medicinal materials such as *Geranium strictipes*, Deng Zhan Xi Xin (*Erigeron breviscapus* (Vaniot) Hand.-Mazz), Ye Men Guan (*Herba Lespedezae Cuneatae*), and so on, to treat gastrointestinal diseases (Huang, 2000).

Renhe(group) Development Co., Ltd applied a patent of one kind of traditional Chinese medicine composition and its preparation process and its quality control method. The composition was composed of six kinds of medicinal materials such as *Geranium strictipes*, Zi Jin Long (*Dactylicapnos scandens* (D. Don) Hutch.), Hai Piao Qiao (Endoconchasepiae), and so on in definite proportion by weight. It had significantly good therapeutic effect to the gastric ulcer, ulcer, chronic gastritis, gastric abscess, regurgitation of sourness, belching, gastric upset, black stools, etc. (Renhe(group) Development Co., Ltd, 2011).

Tang H Q fileded for a patent of quality check method of "Wei Shu capsules" which consisted of five herbal drugs like *Geranium strictipes*, Cao Guo (*Fructus Tsaoko*), Zhi Zhu Xiang (*Valeriana jatamansi* Jones), etc. (Tang, 2011).

Treatment for Viral Hepatitis B (1 case)

Li S H applied a compound patent of traditional Chinese medicine composed of 16 kinds of medicinal materials including *Geranium strictipes*, San Bai Cao (*Saururus chinensis* (Lour.) Baill.), Xuan Cao Gen (*Hemerocallis fulva* (Linn.) Linn.), and so on in the treatment of patients with viral hepatitis B (Li, 2009).

Treatment for Female Infertility (1 case)

Huang M H asked for patent of a kind of quick and efficient fertility dosage based on Chinese herbal medicine to treat with female infertility, which was composed of four parts of medicines as regulating the menstrual function, promoting pregnancy, protecting growth of the fetus and promoting the blood circulation. Among it, the medicine of regulating the menstrual function was a kind of large compound including *Geranium strictipes* (Huang, 1994).

Single Herb Preparations of *Geranium strictipes* (Three cases)

All of the three patents were applied by Yang R H (Yang, 2009; Yang, 2009; Yang, 2009). Firstly, the applicator applied the patent of *Geranium strictipes* process via lyophilization, packaging after frozen rapidly under -40 °C, so as to make the medicine and raw medicinal herbs in health product level which were saving more original colour and taste of grass. Then he fileded for *Geranium strictipes* as both an effervescent agent and a film agent. In the embodiments of the patents, the two preparation forms of *Geranium strictipes* could be used alone; or through compatibility of medicines with other commonly used traditional Chinese medicine. And it was pointed out that multi-layers film agent of *Geranium strictipes* can effectively solve the problem of drug compatibility.

Sickbed Cleaning Drum (1 case)

Sun X L invented a kind of sickbed cleaning drum composed of an axis and drum contained with drug ingredients. The production methods included soaking the drum in the extracting solution for 10 to 15 minutes. The extracted solution was made by mixing of *Geranium strictipes* with plaster, Da Huang (*Rheum palmatum* L.), and Huang Bai (*Phellodendron amurense* Rupr.) according to a certain proportion with water and decoction (Sun, 2005).

Feed Additives (Two cases)

Cheng Z B in Yunnan Agriculture Uuniversity applied a kind of Chinese herbal compound feed patent made up of 10 kinds of traditional Chinese medicinal materials such as *Geranium strictipes*, Huang Qi (*Astragalus membranaceus* (Fisch.)), Dang Gui (*Angelica sinensis* (Oliv.) Diels), and so on to treat piglets diarrhoea, and was used to improve piglets disease-resistant ability, intestinal development function, the growth performance, etc. (Yunnan Agriculture Uuniversity, 2007).

Zhang J X licensed a kind of feed additives of natural plant, whose basic formula including 16 kinds of medicinal materials such as Huang Qi (*Astragalus membranaceus* (Fisch.)), Dang Gui (*Angelica sinensis* (Oliv.) Diels), Bai Shao (), etc. And for adult pig feed, plus eight kinds of medicinal material including *Geranium strictipes*, Xia Ku Cao (*Prunella vulgaris* Linn.), Ru Xiang (*Boswellia carterii* Birdw.), etc. It can promote livestock rapid growth and stronger, and can effectively resist and prevent diseases, it can also replace antibiotics and chemical synthetic drug feed additives commonly used presently (Zhang, 2001).

Analysis and discussion

To sum up, there were some difference of emphasis in patent application status between Miao Ethnic Ge Shan Xiao (*Cynanchum wilfordii*) and Bai and Yi folk medicine Ge Shan Xiao (*Geranium strictipes*), which was reflected as the following trends.

Different Emphasis of their Function and Indication

It could be found from the above examples that *Geranium strictipes* produced in Yunnan province were mainly used for treatment of burn wound, dysentery, diarrhea, regulate the menstrual function, internal bleeding, gastralgia; this reflected its properties of "stopping bleeding and dysentery", which was relevant to its "convergence" function, thus it could be deduced that the "astrigent" should be to the typical characteristics of *Geranium strictipes*. The above clinical applications conformed to the records of indications application described in "Southern Yunnan Materia Medica", "Yunnan Chinese Herbal Medicine" and so on: dysentery, diarrhea, internal bleeding, menorrhagia, gastralgia " (Lan, 1396-1476; Yunnan Xinbin Health Department revolutionary committee, 1971). Comparatively, Miao folk medicine *Cynanchum wilfordii* mainly effects on nurishing liver and kidney, tonify the spleen and stomach, strengthens bones and tendons and detoxification. This could be found clearly from its frequency and content of patents applied: in the patent applications concerns *Cynanchum wilfordii*, there were five items in the treatment of liver disease, 20 items with the function of tonifying the stomach and helping digestion to treat with gastric diseases (including two utility model), five items in the treatment of bone diseases, one item used to regulate the menstrual function, and one item used as cool and hemostatic medicine. These uses of patents took up over half of the total number of 65, and were conformed to the recorded documents (Hunan food and drug administration, 2010; Ni and Xie, 2010; Wu and Zhou, 2006).

Moreover, it should be noticed that four patents in the treatment of cancer are not recorded in the pharmacopoeia such as Compendium of Materia Medica, the Traditional Chinese Medicine Dictionary, etc. Furthermore, the medicinal characteristic described in the patents upon the treatments of diabetes, neurodegenerative diseases and depression patent were also disclosed for the first time. As for the patents for animal feed, there were three items which reflected the trend that medicinal materials used for human had gradually replaced chemical antibiotics to use in the veterinary industry. Such green antibiotics will help to reduce indirect antibiotics abuse in meat food industry.

Single-Composition and Compound's Usage were Different

From the above summary of patents application status, it can be found that Bai/ Yi Ge Shan Xiao (*Geranium strictipes*) might be applied in the form of single herb medicine, whereas Miao Ge Shan Xiao (*Cynanchum wilfordii*) could only be filed in the form of compound, the most simple formula was composed of two kinds of medicinal materials, perhaps it reflected the difference of their using habits in folklore.

The Difference of Dosage Forms

From aspect of the dosage form, the applying patents of the two different Ge Shan Xiao had different features, it was clearly that there were more dosage forms in the patents of Bai/Yi Ge Shan Xiao (*Geranium strictipes*), besides oral forms, it tend to external use; whilst Miao Ge Shan Xiao (*Cynanchum wilfordii*) was used in oral forms without exception. Nevertheless, the ideas of "pearl milk tea" added *Cynanchum wilfordii* were worthy of emulation. Whether *Geranium strictipes* can also be developped into such kind of a popular drink or health food is apparently one of our future development direction.

Study Deepness of Two Ge Shan Xiao was Uneven

From the status analysis of the applied patents, there were not described activity of effective parts and single constituents during the research of *Geranium strictipes*.(Bai/Yi Ge Shan Xiao). However, *Cynanchum wilfordii* (Miao Ge Shan Xiao) already had related single constituents, and got pharmacological activities according to structurally different types of compound, furthermore, it had got corresponding patent protection of medicinal usage according to its specific pharmacological activities.

In our preliminary studies of *Geranium strictipes* (Bai/Yi Ge Shan Xiao), we had already got many purified components, which should be applied relevant patents against corresponding pharmacological activities screened out. Moreover, it could be found through the comparison that there were lack of systematic research of *Geranium strictipes* because it is endemically produced in Yunnan province, therefore the other investigators is hard to get this plant material. This is its inferiority of ethnic drug of Bai minority, whereas it exhibited also a very competitive advantage.

Through the above comparison and discussion, it could be found that Ge Shan Xiao of Yunnan province (*Geranium strictipes*) produced in Dali had a more unique resource advantages, and a clear development direction, which make it promising to be further developed to a new ethnic drugs to be approved by Chinese SFDA.

The research team at Dali University have been performing systemic and deep research on *Geranium strictipes*, it is promising to develop a new type of Bai/Yi folk medicine with independent intellectual property rights in the near future.

In the process of developing preparation, there are several points should be noted: firstly is to search and follow the local ethnic medicine treatment for specific clinical diseases in fixed prescription and dosage form; secondly is to comply with the requirement for new drugs by SFDA and to pay certain attention to the quality control and process selection of this new ethnic drug (Zhao, *et al.*, 2010); thirdly is to consider how to maximize the preservation of the original medicinal materials characteristics of purified effective parts; fourthly is to pay attention to wild resources reserves of this medicinal materials and its sustainable supply after industrialization.

It could not ensure to develop a kind of safety, effective, quality controlled, resources guaranteed creative ethnic drug in the treatment of digestive system diseases until above questions could be solved successfully.

Acknowledgements

This work was supported by IRTSTYN (2010-ZY-011), High Rank Talents Invited Project (2009CI121), and Programme for Yunnan Innovative Research Team.

References

Bai, S. R. Baojin tea: China, 95106977.2[P]. 1995-07-20.

Beijing University of Chinese Medicine. The extraction from Baishouwu used to treat with tumor and the drug composition including it: China, 200810055959.X[P]. 2008-01-03.

Chen, Y. X. A Chinese herbal wine for the treatment of leukemia: China, 98113258.8[P]. 1998-06-19.

Chen, Y. X. A Chinese herbal wine for the treatment of wind-damp and blood virus: China, 97114223.8[P]. 1997-08-12.

Dong, G. Y. A Chinese medicine for the treatment of mycotic skin disease: China, 200810097215.4[P]. 2008-05-06.

Du, Y. Y. A Chinese herbal medicine composition for the treatment of cataclasis: China, 200510032259.5[P]. 2005-10-15.

Gao, P., Xu, L., Xu, M., *et al.* (2011). Recent Progress in the Treatment of Liver Fibrosis by Natural Drug Products and their Active Chemical Entities, in China and Other Countries. *In: "Medicinal Plants: Phytochemistry, Pharmacology and Therapeutics"*. M/S Daya Publishing House, New Delhi, Vol 1, pp.177-214.

Dong, G. Y. A Chinese medicine for the treatment of mycotic skin disease: China, 200810097215.4[P]. 2008-05-06.Guizhou kaili Pharmaceutical Co., Ltd. A Chinese medicine for disintegrating abdominal mass and helping digesgtion.

Guizhou Kedun Pharmaceutical Co., Ltd. A pediatric digestive and appetizing drug and the process of the preparation thereof: China, 02127910.1[P]. 2002-11-27.

Guizhou Normal University. The usage of extraction of total saponins from Baishouwu: China, 200710200097.0[P]. 2007-01-23.

Guizhou Renhetong Pharmaceutical Co., Ltd. A kind of medicine for the treatment of gastric diseases: China, 02134069.2[P]. 2003-04-30.

Guizhou Sanrentang Pharmaceutical Co, Ltd. The Traditional Chinese medicine for the treatment of gastric diseases and stomach or duodenal ulcers: China, 200810068937.7[P]. 2008-10-14.

Hu, D. L. A Chinese herbal capsule for the treatment of reducing tumor and killing virus: China, 200610136820.9 [P]. 2006-12-08.

Huang, M. H. The preparation of Chengji for female with high speed and efficiency: China, 94112711.7[P]. 1994-12-26.

Huang, X. The Chinese medicine for treating gastric ailments: China, 00125922.9[P]. 2000-08-28.

Hunan Food and Drug Administration. (2010). "Chinese herbal medicine standards of Hunan province (2009 edition). Hunan Science and Technology Publishing House, Changsha.

Jiangsu Academy of Traditional Chinese Medicine. A degradation product of C21 total steroidal saponins from Baishouwu and its anti-tumor effect: China, 201010505530.3[P]. 2010-10-12.

Jiangsu New Medical School. (1977). Chinese medicine dictionary. Shanghai Science and Technology Press, Shanghai.

Kong, X. J. A Chinese medicinal formulae for the treatment of fatty liver: China, 201010529374.4[P]. 2010-10-21.

Lan, H., and Zhang, X. (2010). Effects of different *Geranium strictipes* R. Kunth extracts on the demineralization of bovine incisor. *Journal of Anhui Agricultural Sciences*, **38(32):** 18124-18125.

Lan, M. (Ming dynasty (1396-1476)). Diannan compendium. *In:* Vol. 2, Yunnan people's publishing house, Kunming, pp.275.

Li, F. T. A traditional Chinese medicine for the treatment of gastroenteritis: China, 01113034.2[P]. 2003-01-08

Li, S. H. The Chinese medicine for treating hepatitis B: China, 200910229969.5[P]. 2009-11-12.

Li, Y. N. A diet pearl milk tea preparation with the functions of nourishing stomach and tonifying spleen: China, 200810009719.6[P]. 2008-01-28.

Lian, C. D. A traditional Chinese medicine composition for the treatment of gastric diseases: China, 200810185488.4[P]. 2008-12-05.

Li, G. (1990). Yunnan Chinese medicine record. Yunnan science and technology press, Kunming.

Li, L., Guo, L., and Lan, H. (2010). Study on the bacteriostasis of *Geranium strictipes* to the oral cariogenic bacteria. *Journal of Anhui Agricultural Sciences*, **38(36):** 20600-20613.

Liu, S. B. A magic drug for the treatment of burns: China, 201010529191.2[P]. 2010-10-20.

Lu, C. R. Xiaopihewei capsule: China, 02134182.6[P]. 2002-11-25.

Lu, Z. T. The rehabilitation compound medicine combined from non opioid drug: China, 00112778.0[P]. 2000-03-15.

Mao, H. X. A skin linimentum and the process of preparation thereof: China, 200410021834.7[P]. 2004-02-13.

Meng, L. The external pharmaceutical preparation for treating acute or chronic eczema: China, 200810229605.2[P]. 2008-12-11.

Mi, X. H. Cigarette holder with the function of reducing smoking pollution: China, 02109995.2[P]. 2002-01-18.

Microbiology Research Institute of China Science Academy. Natural Juncao preparation for hangover alleviating and liver protecting: China, 200710198596.0[P]. 2007-12-14.

Ni, Y., and Xie, Y. (2010). Distribution of C21 steroidal glycosides in plants of Asclepiadaceae and their pharmacological activities. *Chinese Traditional and Herbal Drugs*, **41(1)**: 162-166.

Niu, Q. D. A Traditional Chinese Medicine for the treatment of gastroenteritis: China, 200810141364.6[P]. 2008-09-05.

Renhe (group) Development Co., Ltd. A Chinese medicine composition and the process of the preparation and the method of quality measurement thereof: China, 201110068391.7[P]. 2011-03-22.

Shandong Agricultural University. The extraction of Baishouwu polysaccharides and buccal tanblets preparation thereof: China, 200410024269.X[P]. 2004-06-11.

Sun, X. L. Clearing rotatum drum for hospital bed and the process of the preparation thereof: China, 200510050282.7[P]. 2005-04-18.

State Drug Administration. (2002). The national standards Assembly of Proprietary Chinese Medicines— spleen and stomach fascicles, Beijing.

Taiyihepu (Beijing) Academy of Traditional Chinese Medicine Co., Ltd. A Chinese medicine composition used to treat with diseases of disaccord of taste and flatulence and the process of the preparation thereof: China, 200910308864.9[P]. 2009-10-27

Tang, Q. H. A method of quality measurement for Changweishu capsul: China, 201110068391.7[P]. 2011-03-22.

Tao, L. H. health board: China, 201020132161.3[P]. 2010-03-16.

The Second Military Medical University of the Chinese People's Liberation Army. The application of saponins and total saponins from Baishouwu for the treatment of gastrointestinal problems: China, 200410099270.9[P]. 2004-12-29.

The Second Military Medical University of the Chinese People's Liberation Army. The application of Baishouwu or Baishouwu diketone A in the preparation of medicine for the treatment of neurodegenerative disease : China, 200710044238.4 [P]. 2008-03-12.

The Second Military Medical University of the Chinese People's Liberation Army. The application of total acetophenone and acetophenone from Baishouwu for the treatment of gastrointestinal problems: China, 200510027075.X [P]. 2005-06-23.

The Second Military Medical University of the Chinese People's Liberation Army. The application of total acetophenone and acetophenone from Baishouwu for the treatment of liver diseases: China, 200710172612.9[P]. 2007-12-20.

Tian, H. Y. A kind of Traditional Chinese medicine prescription for the treatment of gastric diseases: China, 200410022703.0[P]. 2004-06-01.

Wang, C. H. A kind of medicine for the treatment of stomach ailments: China, 200710066322.6[P]. 2007-10-26.

Wang, C. H. A kind of medicine for the treatment of stomach ulcers: China, 200710066329.8[P]. 2007-10-29.

Wang, L. A method for the preparation of drug used to treat with hepatocirrhosis ascites: China, 98112554.9[P]. 1998-07-11.

Wang, J., Li, X., Wang, L, *et al.* (2008). Pharmacognostical studies on root of *Geranium strictipes*. *Journal of Chinese Medicinal Materials*, **31(4):** 496-497.

Woo, A. Y. H., Zhao, Y., Zhang, R., *et al.* (2007). Angiogenesis and Chinese Medicinal Foods, *In:* Anti-Angiogenic, Functional, and Medicinal Foods,Ed By Jack N. Losso, CRC Press,New York, pp581–592.

Wu, Y., and Zhou, G. (2006). Research progress on the chemical components of Cynanchum. *Central South Pharmacy*, **4(5)**: 371-375.

Xia, X. L. A drug for the treatment of skin disease and the process of the preparation thereof: China, 02133655.5[P]. 2002-08-21.

Xiao, C. S. A Chinese herbal formulas for the treatment of hepatitis: China, 200410023075.8[P]. 2005-10-05.

Xiao, C. S. A Chinese herbal formulas for the treatment of stomach and duodenal ulcers: China, 200410023074.3[P]. 2004-03-31.

Xia, X. L. A drug for the treatment of digestive diseases and the process of the preparation thereof: China, 02133654.7[P]. 2002-08-21.

Xie, H. B. A method for the preparation of table beverage used to alleviate the hangover and to clear the inner heat: China, 92111990.9[P]. 1992-11-17.

Xiong, L. A dynamic drug for stomach: China, 200410045218.5[P]. 2004-05-29.

Yuan, S. J. Concentration feed additives from medicine: China, 00109916.7[P]. 2000-07-11.

Yan, C. P. A patch for the treatment of hyperosteogeny and the process of the preparation thereof: China, 201110195927.1[P]. 2011-07-13.

Yan, Y. H. A drug for the treatment of leukodermia: China, 200810305197.4[P]. 2008-10-27.

Yang, G., and Chen, D. (2007). *Geranium strictipes* chemical composition and its anti-AIDS virus activities. *Chinese Traditional and Herbal Drugs*, **38 (3):** 352-354.

Yang. M. J. Namo Qianzihong pharmaceutical preparation and the process of preparation thereof: China, 01101820.8[P]. 2001-02-05.

Yang, R. H. Zidiyu effervescing agent: China, 200910094585.7[P]. 2009-06-16.

Yang, R. H. Freezing Zidiyu: China, 200910094593.1[P]. 2009-06-16.

Yang, R. H. Zidiyu films: China, 200910094598.4[P]. 2009-06.

Yin, H. Y. A magic Chinese medicine for the treatment of liver diseases and the process of the preparation thereof: China, 02113473.1[P]. 2002-03-20.

Yu, R., and Zhao, Y. (2005). Modern biological technology in the application of the modernization of traditional Chinese medicine. China medical science and technology press, Beijing.

Yuan, Y. A medicine for the treatment of gastritis: China, 200810044224.7[P]. 2008-04-06

Yuan, S. J. Concentration feed additives from medicine: China, 00109916.7[P]. 2000-07-11.

Yunnan Agricultural University. Compound Chinese herbal medicinal feed additive for treating diarrhea in piglets: China, 200710066350.8[P]. 2007-11-05.

Yunnan Xinbin Health Department revolutionary committee. (1971). Yunnan Chinese herbal medicine. Kunming: Yunnan people's publishing house, Kunming.

Zhang, R., and Zhao, Y. (2003). *Chinese food and health products theory and practice*. Yunnan Science and Technology Press, Kunmin.

Zhao, Y., and Hu, J. (2002). Natural drug research and development trend of future development. *Journal of Zhejiang University (Medical Sciences)*, **31:** 479-487.

Zhang, L. L. A medicine for the treatment of gastroenteropathy: China, 97107572.7[P]. 1998-08-12

Zhang, K. J. A Chinese medicine composition for the treatment of endocrine disorders: China, 200710151238.4[P]. 2007-09-17.

Zhang, J. X. Natural herbal feed additive: China, 01129078.1[P]. 2001-11-16

Zhang rongping, and Zhao, Y.U. (2003). Chinese food and health products theory and practice [M]. Yunnan science and technology press, pp.7.

Zhao, Y.U., Hu jijiang. (2002). Natural drug research and development trend of future development [J]. *Journal of Zhejiang University (Medical Sciences)*, **31:** 479-487.

Zhao, Y., Liu, Q., Xu, Y., *et al*. (2010). Treatment of nonalcoholic and alcoholic fatty liver diseases by single natural drug and active natural products: research advances. *Journal ofIinternational Pharmaceutical Research*, **37(6):** 408-425.

Zhu, W. X. A Wushan Weitongsan for the treatment of stomach ailments: China, 200510048675.4 [P]. 2005-12-07

Zhu, G. H. A Tujia medical recipe for the treatment of tumor: China, 200610051018.X[P]. 2006-12-13.

Traditional and Folk Herbal Medicine: Recent Researches Vol. 2 (2014) *Pages* **387–405**
Editor: **V.K. Gupta**
Published by: **DAYA PUBLISHING HOUSE, NEW DELHI**

13

Cyclooxygenase Inhibitory, Anti-inflammatory and Anti-oxidant Effects of *Wrightia tinctoria* R.BR Leaves

A. Vijayalakshmi[1]*, V. Ravichandiran[2] and K. Masilamani[2]

ABSTRACT

Polyphenol contents, antioxidant and anti-inflammatory activities of Wrightia tinctoria (Roxb.) R.Br. leaves were investigated in this study in order to provide a scientific basis for the traditional use of plants to treat chronic inflammatory skin diseases such as psoriasis. The total phenolic content and flavonoid content was accessed using Folin-ciocalteu and $AlCl_3$ method respectively. The antioxidant potential of the samples was evaluated using inhibition of hydroxyl radical, 2,2-diphenyl-1-picrylhydrazyl (DPPH), and nitric oxide scavenging methods. The anti-inflammatory activity was evaluated using carageenan induced paw edema, cotton pellet granuloma pouch method in rats, membrane stabilizing potential and cyclooxygenase inhibition assay. The total phenolic content and flavonoid content was found to be 10.80 per cent w/w and 1.7 per cent w/w respectively. The HPTLC densitometric estimation of rutin was found to be 0.254 mg/gm. The ethanolic extract of Wrightia tinctoria with a concentration of 400 mg/kg b.w. showed maximum (56.25 per cent) inhibition on carrageenan induced rat paw edema, significantly inhibited granuloma formation ($p<0.01$) in cotton pellet granuloma model. The ethanol extract (at concentration range of 0.10-0.40 mg/ml) significantly protect the rat erythrocyte membrane against lysis induced by hypotonic solution. The ethanol extract was found to be a potent COX-2 inhibitor with an IC_{50} value of 64 μg/ml and a weak inhibitory effects on COX-1. The tested extract showed good dose-dependent free radical scavenging property in all the models. The IC_{50} values were found to be

1 College of Pharmacy, SRM University, Kottankulathur, Tamil Nadu, India

2 Department of Pharmacognosy, School of Pharmaceutical Sciences, VISTAS, Vels University, Chennai, Tamil Nadu, India

* *Corresponding author*: E-mail: aviji_1975@rediffmail.com

150 µg/ml, 160 µg/ml and 80 µg/ml respectively. The anti-inflammatory activity of W. tinctoria leaves could be at least in part due to COX-2 enzyme inhibition and free radical-scavenging activities which may be attributed to the presence of flavonoids and other polyphenols in the extract. These experimental results have established a pharmacological evidence for the folklore claim of the drug to be used as an anti inflammatory agent.

Keywords: COX-1/COX-2, Flavonoid, Inflammation, Polyphenol, Rutin, *Wrightia tinctoria*.

Introduction

Despite the progress made in medical research during the past decades, the treatment of many serious diseases is still problematic. Chronic inflammatory diseases remain one of the world's major health problems. Inflammation is the response of living tissues to injury involving a complex array of enzyme activation, mediator release, extravasations of fluid, cell migration, tissue breakdown and repair. The conventional drugs used to ameliorate this phenomenon are either too expensive or toxic and not commonly available to the rural folks that constitute the major populace of the world (Dharmasiri *et al.*, 2003).

The enzyme, Phospholipase A_2 converts membrane lipids into arachidonic acid, which is highly reactive and is rapidly metabolized by cyclooxygenase (prostaglandin synthase) to prostaglandins, which are major components that induce pain and inflammation. Cyclooxygenase enzymes (COXs, prostaglandin-endoperoxide synthases) catalyze two reactions, the first being a cyclooxygenase function consisting of the addition of molecular oxygen to arachidonic acid (AA) to form prostaglandin G2 (PGG2). The second is the conversion of PGG2 to PGH2 by a peroxidase function. Thus, COX is a bifunctional enzyme exhibiting both cyclooxygenase (from AA to PGG2) and peroxidase (from PGG2 to PGH2) activities (Hinz and Brune, 2002). COX-1 is constitutively expressed as a housekeeping enzyme in nearly all the tissues and mediates physiological responses (*e.g.*, cytoprotection of the stomach, platelet aggregation). On the other hand, COX-2 is expressed by cells that are involved in inflammation and has emerged as the isoform primarily responsible for the synthesis of prostanoids involved in acute and chronic inflammatory states of pathological processes (Hinz and Brune, 2002). Classical nonsteroidal anti-inflammatory drugs (NSAIDs) such as acetylsalicylic acid (aspirin) and diflunisal, which are grouped into the salicylate derivatives of NSAIDs, were shown to inhibit both COX-1 and COX-2 activities. None of the COX-2 selective inhibitors belonging to salicylates, which show selectivity for COX-2 inhibition with low concentrations, are reported to date (Warner *et al.*, 19999). Inhibition of COX-2-dependent PG synthesis accounts for the anti-inflammatory and analgesic effects of NSAIDs, whereas suppression of COX-1 can lead to many unwanted side effects (*e.g.*, gastrointestinal ulceration and bleeding, platelet dysfunctions). Thus, it has been thought that specific inhibitors for the COX-2–mediated reaction might have ideal therapeutic actions similar to those of classical NSAIDs with significantly less gastrotoxicity. So, there is a need of drugs which inhibit COX-2 without affecting COX-1 (selective COX-2 inhibitors). Baumann and coworkers were the first to report in a study that some dietary polyphenols inhibit arachidonic acid peroxidation (Baumann *et al.*, 1980). Since then several researches have reported that many dietary polyphenols possess COX-2 inhibitory or stimulatory effects (Luceri *et al.*, 2002). COX inhibition by polyphenols may account for anti-inflammatory effects, which reduce prostaglandin synthesis. Therefore, it should be noted that the concurrent use of polyphenols and NSAIDs could be beneficial or deleterious and thus necessitates constant attention by healthcare providers.

It is commonly accepted that in a situation of oxidative stress, reactive oxygen species (ROS) such as superoxide (O2•",OOH•), hydroxyl (OH•) and peroxyl (ROO•) radicals are generated. The ROS play an important role in the pathogenesis of various serious diseases, such as neurodegenerative disorders, cancer, cardiovascular diseases, atherosclerosis, cataracts, and inflammation (Aruoma, 1998). The mechanism of inflammation injury is attributed, in part, to release of reactive oxygen species from activated neutrophils and macrophages. This over production leads to tissue injury by damaging macromolecules and lipid peroxidation of membranes (Gutteridge, 1995). n addition, ROS propagate inflammation by stimulating release of cytokines such as interleukin-1, tumor necrosis factor-α, and interferon-γ, which stimulate recruitment of additional neutrophils and macrophages. Thus free radicals are important mediators that provoke or sustain inflammatory processes and, consequently, their neutralization by antioxidants and radical scavengers can attenuate inflammation (Delaporte *et al.*, 2002).

The use of traditional medicine is widespread and plants still represent a large source of natural antioxidants that might serve as leads for the development of novel drugs. Several anti-inflammatory, digestive, antinecrotic, neuroprotective, and hepatoprotective drugs have recently been shown to have an antioxidant and/or radical scavenging mechanism as part of their activity (Repetto and Llesuy, 2002). Several studies suggest that natural antioxidant and anti-inflammatory agents could be beneficial in the prevention and treatment of these pathologies.

Wrightia tinctoria (Roxb.) R.Br. commonly called as "Jaundice curative tree" in South India is a small deciduous tree of the family Apocynaceae distributed in Central India, Burma and Timor. The juice of the tender leaves is used efficaciously in jaundice. Crushed fresh leaves when filled in the cavity of decayed tooth relieve toothache. In Siddha system of medicine, it is used for chronic inflammatory diseases such as psoriasis and other skin diseases (Nadkarni, 1976). An oil 777 prepared out of the fresh leaves of the plant has been assigned to analgesic, anti-inflammatory and antipyretic activities, and to be effective in the treatment of psoriasis. It has been reported that the leaves of the plant contained beta-amyrin, steroids, indigo, indole alkaloid and flavonoid glycosides (Ram P Rastogi and Mehrotra, 1994). This study aimed to investigate the anti-inflammatory potential of standardized hydroalcoholic extracts (70 per cent v/v ethanol) of *Wrightia tinctoria* using *in-vivo* anti-inflammatory activity, COX inhibition assay and *in-vitro* anti-radical properties. Therefore an effort has been made to corroborate and establish scientific evidence for its ethnobotanical uses.

Methodology

Plant Material

Wrightia tinctoria (Roxb.) R.Br. is a small deciduous tree of the family Apocynaceae distributed in Central India, Burma and Timor. This plant is extensively used in the Indian system of medicine. The plants specimen for the proposed study was collected in Chennai, Tamilnadu. It was identified and authenticated by Dr. P. Jayaraman, Director, Plant Anatomy Research Centre (PARC), Tambaram, Chennai. A vocher specimen No. PARC/2011/882 has been deposited for further references.

Extraction and Fractionation

About 500 gm of the leaves of *Wrightia tinctoria* powder was extracted using a soxhlet apparatus with Ethanol (70 per cent v/v) (18 h). The extracted solution was filtered and concentrated in a rotary evaporator under reduced pressure (rotary vacuum flash evaporator). The total extract (100 g) thus obtained was subjected to chemical test for identification of various plant constituents and TLC using the mobile phase Toluene: ethyl acetate: formic acid (5: 4: 1).

Phytochemical Screening

Concentrated Ethanol (70 per cent) extract were screened for the presence of various phytoconstituents using standard procedure (Kokate *et al.*, 1997).

Thin Layer Chromatography (TLC)

To support preliminary chemical analysis, Ethanol (70 per cent) extract was subjected to TLC studies. A number of developing solvent systems were tried, but the satisfactory resolution was obtained in the solvent system Toluene: Ethyl acetate: Formic acid (5: 4: 1). The results were tabulated in Table 13.1 and shown in Figures 13.1 and 13.2.

Table 13.1: Thin layer chromatography of ethanol extract of *Wrightia tinctoria* leaves.

Test Extract	*Solvent System*	*Number of Spots*	*R_f Values*	*Detecting Agent*
Ethanol (70 per cent) extract	Toluene: Ethyl acetate: Ethanol (70 per cent) (7: 1: 0.5)	11	0.07	Vanillin in sulphuric acid
			0.13	
			0.15	
			0.23	
			0.36	
			0.43	
			0.53	
			0.57	
			0.72	
			0.85	
			0.94	

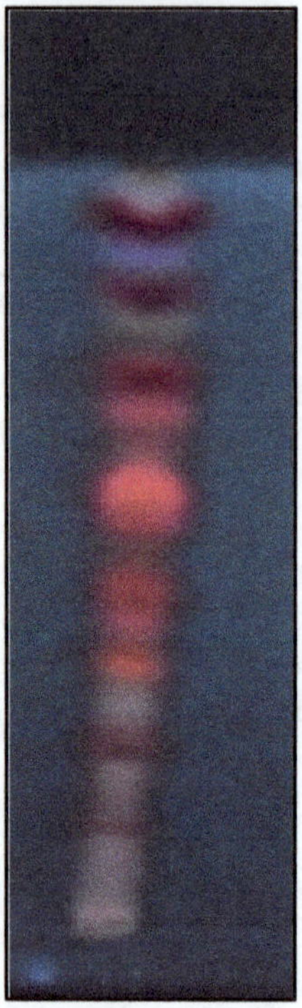

Figure 13.1: TLC fingerprint of ethanol extract.

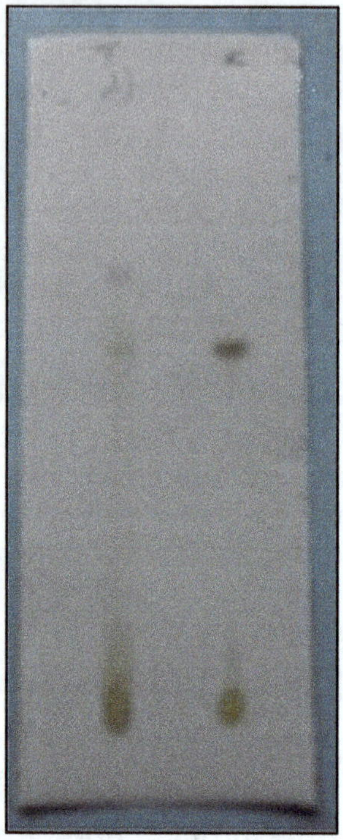

Figure 13.2: Rf value of test extract compared with that of standard rutin.

HPTLC Densitometric Estimation of Rutin

The Ethanol (70 per cent) extract of *Wrightia tinctoria* leaves was subjected to HPTLC for the conformation of the active constituents. HPTLC was performed on 10 cm × 10 cm aluminum backed plates coated with silica gel 60F254 (Merck, Mumbai, India). Standard solution of rutin and sample solution were applied to the plates as bands 8.0 mm wide, 30.0 mm apart, and 10.0 mm from the bottom edge of the same chromatographic plate by use of a Camag (Muttenz, Switzerland) Linomat V sample applicator equipped with a 100-μL Hamilton (USA) syringe. Ascending development to a distance of 80 mm was performed at room temperature (28 ± 2°C), with Toluene: Ethyl acetate: Formic acid (5 : 4: 1) (v/v/v), as mobile phase, in a Camag glass twin-trough chamber previously saturated with mobile phase vapor for 20 min. After development, the plates were dried with a hair dryer and then scanned at 254 nm with a Camag TLC Scanner with WINCAT software, using the deuterium lamp (Rich and Schibli). A stock solution of standard rutin (100μg/mL) was prepared in ethanol (70 per cent). Different volume of stock solution 1, 2 and 3μL, were spotted on to TLC plate to obtained concentration 100 ng, 200 ng and 300 ng/spot of rutin. Calibration curve range (100–300 ng/spot) for rutin was found to be linear. The data of peak areas, hcight and Rf values were shown in Table 13.2 and in Figure 13.3 and 13.4.

Table 13.2: HPTLC peak table.

Track	*Sample ID*	*Rf*	*Height*	*Area*	*Assigned Substance*
1	Reference	0.12	63.9	488.2	Rutin
2	Reference	0.12	61.7	478.0	Rutin
3	Reference	0.12	58.4	430.2	Rutin
4	Ethanol (70 per cent) extract	0.12	54.1	363.6	Rutin

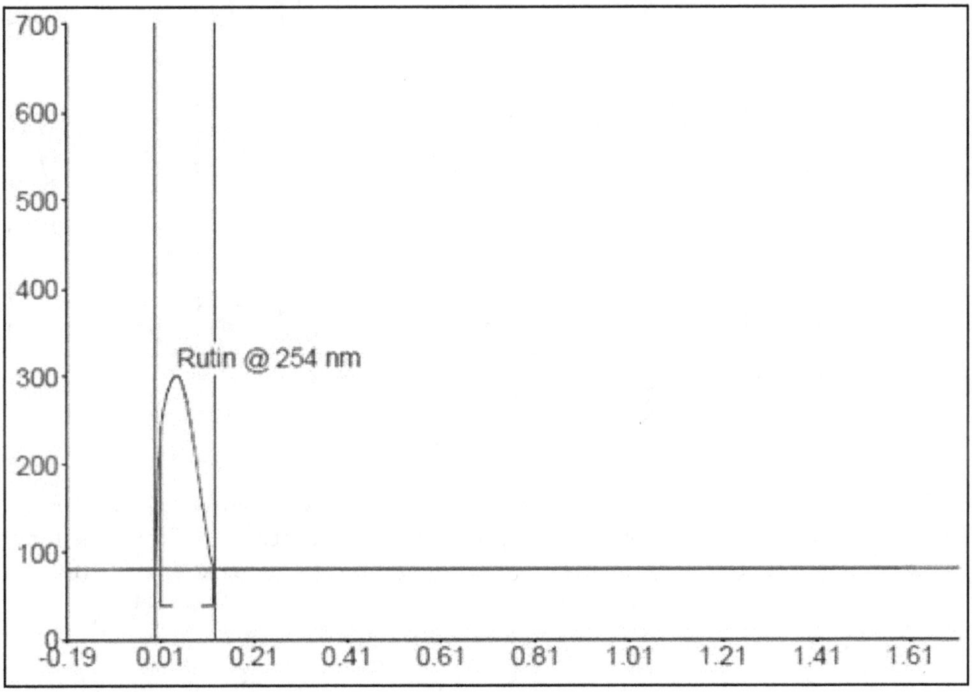

Figure 13.3: HPTLC densitometric scan at 254 nm.

Estimation of Phytoconstituents

Determination of Total Phenolic Content

The content of total phenolics in the powdered *Wrightia tinctoria* leaves was determined by using Folin- ciocalteu reagent. About 1 gram of the powder was extracted in an ultrasonic wave bath with 80 ml of aqueous ethanol solution (70 per cent v/v) for 2 hr. After cooling, the volume of the solution was adjusted to 100 ml. The final solution was centrifuged prior to the colorimetric determination. Tannic acid standards (10–110 mg/ml) were dissolved in 100 ml of aqueous ethanol solution (70 per cent v/v) respectively. About 10 ml of Folin- ciocalteu reagent was added to 1 ml of the extract solution and 1 ml of standard solution. After reacting for 3 min, 10 ml of 35 per cent sodium carbonate solution was added and the test solution was diluted

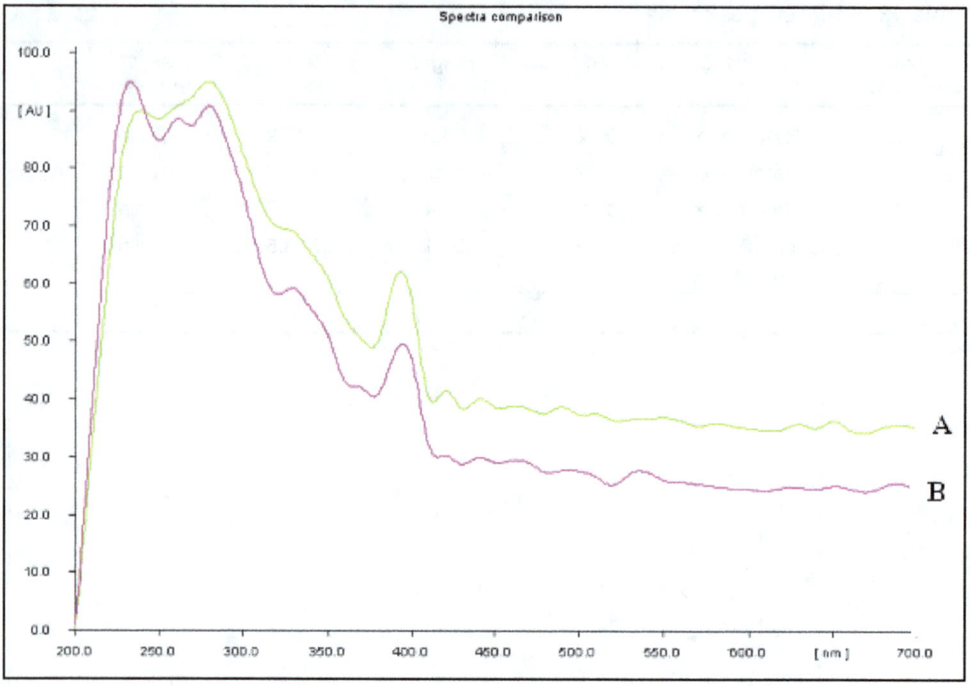

Figure 13.4: HPTLC densitometric scan at 327 nm

A: Standard (Rutin) B: Test sample (ethanol extract).

to 100 ml with water and mixed. After 45 min, an aliquot was centrifuged for 5 min. The absorption coefficient for the supernatant was measured at 745 nm. The total phenolic content of the extract was calculated using the mean regression coefficient from the standard (Yi-Chen Liu *et al.*, 2007).

Determination of Total Flavonoids

Total flavonoid content in dried aerial parts of the plant was estimated by spectrometric method (Perkin-Elmer UV-Vis spectrometer Lambda 16 (Germany). Dried powdered plant material (10 gm) was extracted by continuous mixing in 100 ml of 70 per cent ethanol, 24 hr at room temperature. After filtration, ethanol was evaporated until only water remained. Water phase was subsequently extracted with ethyl acetate. The extract was dried over anhydrous sodium sulphate, filtered and concentrated under vacuum up to a concentration of 1 gm/ml of extract. They were further diluted with ethyl acetate to obtain 0.01 gm/ml solutions. About 10 ml of the solution was transferred into a 25 ml volumetric flask, 1 ml of 2 per cent $AlCl_3$ was added and the solution was filled to volume with methanol-acetic acid and was kept aside for 30 min. The absorbance was measured at 390 nm against the same solution without $AlCl_3$ being blank. Rutin was used to construct the calibration curve in the concentration range 1.0-10.0 μg/ml (Kadifkova Panovska *et al.*, 2005).

Animals

Healthy Male Wistar rats (120–170 g) obtained from the animal house of Vels University, Tamilnadu, India was used for the study. Animals were housed in polypropylene cages and were left 7 days for acclimatization to animal room maintained under controlled condition (a 12 h light–dark cycle at 22 ± 2 %C) on standard pellet diet and water ad libitum. All animals were taken care of under ethical consideration as per the guidelines of CPCSEA with due approval from the Institutional Animal Ethics Committee (Registration No. XII/VELS/PCOG/04/2000/CPCSEA/IAEC/22.2.11). Animals were used for acute toxicity study and anti-inflammatory study.

Acute Toxicity

Acute toxicity study–up and down procedure was carried out as per the guidelines by Organization for Economic Co-operation and Development (OECD) 423. Rats (6/group) were divided into five groups. The first 2 groups received oral doses of 1000, and 2000 mg/kg of Ethanol (70 per cent) extract and 3rd group received saline (10 ml/kg) orally. Mortality was assessed 24 hours after administration. The animals were also observed for toxic symptoms and mortality was determined 24 hours after treatment (Lipnic *et al.*, 1995).

In-vitro Anti-inflammatory Activity

Membrane Stabilizing Activity

Preparation of Erythrocyte Suspension

Whole blood was obtained with heparinized syringes from rats through cardiac puncture. The blood was washed three times with isotonic buffered solution (154 mM NaCl) in 10 mM sodium phosphate buffer (pH 7.4). The blood was centrifuged each time for 10 minutes at 3000 g.

Hypotonic Solution-Induced Rat Erythrocyte Haemolysis

Membrane stabilizing activity of the extract was assessed using hypotonic solution-induced rat erythrocyte haemolysis. The test sample consisted of stock erythrocyte (RBC) suspension (0.50 ml) mixed with 5 ml of hypotonic solution (50 mM NaCl) in 10 mM sodium phosphate buffered saline (pH 7.4) containing the test sample (10, 25 and 50 mg/ml) or indomethacin (0.1 mg/ml). The control sample consisted of 0.5 ml of RBC mixed with hypotonic-buffered saline solution alone. The mixtures were incubated for 10 min at room temperature and centrifuged for 10 min at 3000 g and the absorbance of the supernatant was measured at 540 nm (Shinde *et al.*, 1999). The percentage inhibition of haemolysis or membrane stabilization was calculated by the formula:

$$100 \times (OD_1\text{-}OD_2/OD_1)$$

where,

OD_1: Optical density of hypotonic-buffered saline solution alone

OD_2: Optical density of test sample in hypotonic solution

COX-2 Inhibitory Activity

In vitro COX-2 inhibiting activities of the compounds have been evaluated using 'COX (ovine) inhibitor screening assay' kit with 96-well plates. Both ovine COX-1 and COX-2 enzymes were included. This screening assay directly measures PGF2 produced by SnCl2 reduction of COX-derived PGH2. COX-1, COX-2, initial activity tubes were prepared taking 950µl of reaction buffer, 10µl of heme and 10µl of COX-1 and COX-2 enzymes in respective tubes. Similarly, COX-1, COX-2 inhibitor tubes were prepared by adding 20µl of inhibitor (ethanol extract) in each tube in addition to the above ingredients. The inhibitory assays were performed in the presence of extracts at different concentrations (10 to 80 µg/mL). The background tubes correspond to inactivated COX-1 and COX-2 enzymes obtained after keeping the tubes containing enzymes in boiling water for 3 min. along with vehicle control. Reactions were initiated by adding 10µl of arachidonic acid in each tube and quenched with 50µl of 1M HCl. PGH2 thus formed was reduced to PGF2α by adding 100µl $SnCl_2$. The prostaglandin produced in each well was quantified using broadly specific prostaglandin antiserum that binds with major prostaglandins and reading the 96-well plate at 405 nm. The wells of the 96-well plate showing low absorption at 405 nm indicated the low level of prostaglandins in these wells and hence the less activity of the enzyme. Therefore, the COX inhibitory activities of the compounds could be quantified from the absorption values of different wells of the 96-well plate (Prabhjit Kaur *et al.*, 2010). The indomethacin (selective COX-1 inhibitor) and celecoxib (selective COX-2 inhibitor) were used as positive controls in the study. The test extract concentration causing 50 per cent inhibition of PGE2 release (IC_{50}) was calculated from the concentration-inhibition response curve by regression analysis.

In vivo Anti-inflammatory Activity

Carrageenan-Induced Paw Edema in Rats

The anti-inflammatory test was evaluated by the carrageenan-induced paw edema test in the rat, according to the method of Winter *et al.* (1962). Male Wistar rats (190-230 g) were briefly anesthetized with ethyl ether and injected subplantarly into the right hind paw with 0.1 mL of suspension of carrageenan (200 µg/mL) in isotonic saline. The left hind paw was injected with 0.1 mL of saline and used as a control. Paw volume was measured prior and 3 and 5 h after carrageenan administration, using a mercury plethysmograph (Ugo Basile, Italy). Ethanol extract (200 and 400 mg/kg) was orally administered 30 minutes prior to carrageenan injection. The control group received an equivalent volume of water. Indomethacin (5 mg/kg, *p.o.*) was used as the reference drugs. Results were expressed as percentage of inhibition of edema was calculated by the formula $(1 - Vt/Vc) \times 100$ where Vt and Vc are the mean paw volume in the treated and controlled groups, respectively.

Cotton Pellet Granuloma Pouch Method

Chronic inflammation was induced by cotton pellet granuloma. Rats (6 per group) were divided into four groups. The first 2 groups received oral doses of 200 and 400 mg/kg of the ethanol extract. The 3rd and 4th groups were treated orally with indomethacin (10 mg/kg) as a reference drug and saline (10 ml/kg) as control respectively. Autoclaved cotton pellet 50 ± 1 mg was implanted subcutaneously by making incision in the axilla and groin region of each rat under ether anesthesia. Drugs were administered orally for 7 consecutive days from the day of cotton pellet implantation. Animals were sacrificed on day 8th and the granuloma was dissected out, dried in an oven at 60° C for 24 hours and weighed. The increment in the dry weight of the pellet was taken as a measure of granuloma formation (Olajide *et al.*, 1999). The percentage of inhibition of granuloma was determined using the formula (1 – Wt/Wc) × 100 where Wt – Dry weight of the cotton in test animals and Wc–Dry weight of the cotton in control animals.

Antioxidant Activity

Hydroxyl Radical Scavenging Activity

The assay was performed by adding 0.1 ml EDTA, 0.01 ml of ferric chloride, 0.1 ml of hydrogen peroxide, 0.36 ml of deoxyribose, 1 ml of ethanol extract (25-400 µg/ml), 0.33 ml of phosphate buffer (50 mM. pH 7.4), 0.1 ml of ascorbic acid in sequence and incubated at 37°C for 1 hr. A 1.0 ml portion of the incubated mixture was mixed with 1.0 ml of 10 per cent tri chloro acetic acid and 1.0 ml of 0.5 per cent thio barbituric acid to develop the pink chromogen, which was measured at 532 nm (Annie Shirwaikar *et al.*, 2006).

$$\text{Percentage inhibition (per cent)} = \frac{(\text{Absorbance of control} - \text{Absorbance of test})}{\text{Absorbance of control}} \times 100$$

DPPH Radical Scavenging Activity

About 0.1 mM solution of DPPH in methanol was prepared and 1 ml of this solution was added to 3 ml of the different concentration (25-400 µg/ml) of ethanol extract and control (without the test compound, but with an equivalent amount of methanol) in different test tubes. The mixture was shaken and allowed to stand at room temperature for 30 min and the absorbance was measured at 517 nm using a spectrophotometer (Annie Shirwaikar *et al.*, 2006).

Nitric Oxide Scavenging Activity

Sodium nitroprusside (5 mM) in standard phosphate buffer saline solution (0.025 M, pH: 7.4) was incubated with different concentrations of ethanol extract (25-400 µg/ml), ascorbic acid as reference standard (25-400 µg/ml) dissolved in phosphate buffer saline (0.025 M, pH: 7.4) and the tubes were incubated at 25°C for 5 hr. Control experiments without the test compounds but equivalent amounts of buffer were conducted in an identical manner. After 5 hr, 0.5 ml of incubation solution was removed and diluted with 0.5 ml of Griess reagent. The absorbance of the chromophore was read at 546 nm (Annie Shirwaikar *et al.*, 2006).

Statistical Analysis

Level of significance of all the parameters was expressed as the arithmetic mean ± SEM and was analyzed by one-way analysis of variance (ANOVA), followed by Dunnett's "t" test. P value less than 0.05 ($P < 0.05$) was the critical criterion for statistical significance.

Results and Discussion

About 500 gms of the leaves of *Wrightia tinctoria* powder was extracted using a soxhlet apparatus with ethanol (70 per cent) and the yield of ethanol (70 per cent v/v) extract was about 6.4 per cent w/w.

Phytochemical Screening

Preliminary phytochemical screening of ethanol (70 per cent v/v) extract showed positive results for alkaloids, steroids, flavonoids, tannins, terpenoids and saponins.

Thin Layer Chromatography (TLC)

To support phytochemical screening, total ethanol (70 per cent v/v) extract was subjected to thin layer chromatography. The ethanol extract showed 11 well separated spots with R_f values 0.07, 0.13, 0.15, 0.23, 0.36, 0.43, 0.53, 0.57, 0.72, 0.85, 0.94 in Toluene: Ethyl acetate: Ethanol (70 per cent) (7: 1: 0.5) solvent system and vanillin in sulphuric acid as detecting agent (Table 13.1 and Figure 13.1). The Rf value of one spot matches with the R_f value of rutin (Figure 13.2).

High Performance Thin Layer Chromatography

In the current study, the flavonoid rutin was detected and quantified using HPTLC silica gel 60 F254 pre-coated plates with the mobile phase made of Toluene: Ethyl Acetate: Formic acid (5: 4: 1 v/v/v). It was conformed to the flavonoid rutin as one of the active constituent for the plant of *Wrightia tinctoria*. The detection of rutin was observed to be linear over a concentration range of 100 to 300 ngmL-1 with correlation coefficient of 0.96726. The concentration of rutin was found to be 0.254 mg/gm. The Rf values were correspondingly calculated and showed in the Table 13.2.

Estimation of Phytoconstituents

The chemical analysis of total phenolic and total flavonoid content were 10.80 per cent w/w and 1.7 per cent w/w respectively.

Acute Toxicity

The Ethanolic extract of *Wrightia tinctoria* leaves were found to be safe in the doses used and no abnormality in the gross behavioral studies also no mortality were noted in the doses of 1g/kg and 2 g/kg orally.

In-vitro Anti-inflammatory Activity

Effect on Erythrocyte Membrane Stability

The ethanol extract (at concentration range of 0.10-0.40 mg/ml) significantly protect the rat erythrocyte membrane against lysis induced by hypotonic solution. At a concentration of 0.40 mg/ml, the extract produced 36.84 per cent inhibition of RBC haemolysis as compared with 43.15 per cent produced by indomethacin. The compounds with membrane-stabilizing properties are well known for their ability to interfere with the early phase of inflammatory reactions, namely the prevention of the release of phospholipases that trigger the formation of inflammatory mediators and offer significant protection of cell membrane against injurious substance, thereby exhibit anti-inflammatory property (Perenz *et al.*, 1995). The ethanol extract demonstrate significant membrane stabilizing property (Table 13.3), contributing a significant role to its anti-inflammatory activity.

Table 13.3: Effect of ethanolic extract of *W. tinctoria* leaves on rat erythrocyte haemolysis induced by hypotonic solution.

Sample	*Concentration*	*Optical Density (OD)*	*Inhibition of Haemolysis (per cent)*
Hypotonic	50mM	0.95 ± 0.05	–
Ethanol extract	100 µg/ml	0.78 ± 0.02	17.89
Ethanol extract	200 µg/ml	0.69 ± 0.04**	27.36
Ethanol extract	400 µg/ml	0.60 ± 0.05**	36.84
Indomethacin	100 µg/ml	0.54 ± 0.07**	43.15

Values are mean ± SEM of 6 parallel measurements.

Statistical significant test for comparison was done by ANOVA, followed by Dunnet's 't' test (n=6).

**P< 0.01, when compared against control.

In vitro Inhibition of COX-2 Enzymatic Activity

Inhibition of enzymes of the arachidonic acid pathway may also contribute to the anti-inflammatory effects. The COX-2 is responsible for the biosynthesis of PGs under acute inflammatory conditions. So, the COX-2 plays an important role to be target enzyme on anti-inflammatory activity. The inhibitory effects of ethanolic extract of *Wrightia tinctoria* leaves on *in vitro* enzymatic activities were measured against COX-1/COX-2. Therefore, the ethanol extract were studied at a concentration of 10-80 µg/mL whether they inhibit cyclooxygenase-1 and -2. Interestingly, the ethanolic extract exhibited an inhibition of cyclooxygenase-1 and -2 of 6.70 per cent and of 62.9 per cent, respectively at highest concentration 80 µg/mL (Table 13.4).

Table 13.4: *In vitro* percentage inhibition for COX-1 and COX-2 enzymes by ethanol extract of *Wrightia tinctoria* leaves.

Sl.No.	*Drug*	*Concentration (µg/ml)*	*Per cent Inhibition*	
			COX-1	*COX-2*
1	Ethanol extract	10	10.8 ± 1.54**	5.50 ± 0.60*
		20	16.4 ± 0.79**	15.3 ± 1.06**
		40	9.3 ± 0.90**	43.37 ± 2.04**
		80	–	62.9 ± 0.80**
2	Indomethacin	30	96.8 ± 1.08**	56.66 ± 1.81**
3	Celecoxib	30	13.90 ± 0.38**	93.36 ± 0.90**

Values are mean ± SEM of 6 parallel measurements. Statistical significant test for comparison was done by ANOVA, followed by Dunnet's 't' test (n=6). **P< 0.01, when compared against control.

Flavonoids are the well studied class of polyphenols as COX-2 inhibitors. Kaempferol and quercetin exhibited anti-inflammatory activities by inhibiting iNOS and COX-2 protein levels in cultured human umbilical vein endothelial cells (Crespo *et al.*, 2008). Luteolin and galangin, well known flavonoid molecules were studied as first dietary polyphenols as inhibitors of arachidonic acid peroxidation (Baumann *et al.*, 1980). After this Chrysin and luteolin were considered as potent anti-inflammatory agents as they effectively suppressed COX-2 activity (Harris *et al.*, 2006). In 2008, Li and coworkers reported that a new molecule Malsudone along with known flavonoids luteolin, isoquercetin, 7-methoxyflavone and luteolin-7-O—glycoside possess potent inhibitory effect on COX-2 with moderate inhibition of COX-1. More recently, human clinical trials with COX-2 inhibitor drugs have shown similar anti-inflammatory and analgesic efficacy to traditional NSAIDs, yet with significantly less gastrotoxicity (Simon, 1999). The present investigation showed that *Wrightia tinctoria* leaves selectively inhibited COX-2 which may be due to the presence of flavonoid and rutin.

In-vivo Anti-inflammatory Activity

The carrageenan-induced paw edema model in rats is known to be sensitive to cyclooxygenase inhibitors and has been used to evaluate the effect of non-steroidal anti-inflammatory agents, which primarily inhibit the cyclooxygenase involved in prostaglandin synthesis. Carrageenan-induced rat paw oedema is used widely as a working model of inflammation in the search for new anti-inflammatory drug. The anti inflammatory activity of the ethanolic extract of *W. tinctoria* was evaluated by carrageenan-induced rat paw oedema method and the result is shown in Table 13.5. The extracts were tested at two different dose levels. The results showed that the ethanolic extract with a dose of 200 mg/kg b.w and ethanolic extract with a concentration of 400 mg/kg b.w. showed 46.55 and 53.44 per cent of inhibition on carrageenan induced rat paw edema at third hour. The results showed that the ethanolic extract with a dose of 200 mg/kg b.w and ethanolic extract with a concentration of 400 mg/kg b.w. showed 45.83 and 56.25 per cent of inhibition on carrageenan induced rat paw edema at fifth hour. The time course of edema development in carrageenan-induced paw edema model in rats is generally represented by a biphasic curve. The first phase, which occurs between 0 to 2.5 h after injection of the phlogistic agent, has been attributed to the release of histamine or serotonin. The edema volume reaches its maximum approximately 3 h post treatment and then begins to decline. The second phase of inflammatory reaction which is measured at 3 h is caused by the release of bradykinin, protease, prostaglandin and lysosome (Vinegar *et al.*, 1969). Therefore, it can be inferred that the inhibitory effect of ethanolic extract of *Wrightia tinctoria* on carrageenan-induced inflammation could be due to inhibition of the enzyme cyclooxygenase leading to inhibition of prostaglandin synthesis.

The effect of ethanolic extract of *Wrightia tinctoria* on cotton pellet induced granuloma in rats is shown in Table 13.6. In this the mean weights of the cotton pellets were determined. The weight of the granuloma for the control group of animals was found to be 57.2 ± 3.08 mg. Treatment with the ethanolic extract of *Wrightia*

Table 13.5: Effect of ethanolic extract of *W. tinctoria* leaves on carrageenan induced rat paw edema.

Groups	*Dose (mg/kg)*	*Initial Paw Volume (Mm/Hg)*	*Paw Volume*		*Per cent Inhibition*	
			3 H (Mm/Hg)	*5 H (Mm/Hg)*	*3 H*	*5 H*
Control	1.2	2.6	5.8 ± 2.078	4.8 ± 0.69**[a]	–	–
Ethanol extract	200	2.5	3.1 ± 0.621	2.6 ± 0.047**	46.55	45.83
	400	2.5	2.7 ± 0.238	2.1 ± 0.51**	53.44	56.25
Indomethacin (Standard)	10	2.5	1.9 ± 0.542	1.8 ± 0.080**	67.24	62.50

Values are mean ± SEM of 6 parallel measurements.

Statistical significant test for comparison was done by ANOVA, followed by Dunnet's 't' test (n=6). * $P< 0.05$, ** $P< 0.01$, when compared against control.

[a] $P< 0.01$ when compared against standard.

tinctoria (400 mg/kg, p.o.,) decreased the granuloma weight to 26.2 ± 0.89 mg. Treatment with indomethacin (10 mg/kg, p.o.,) produced a granuloma weight of 21.7 ± 0.65 mg. The ethanolic extract of *Wrightia tinctoria* and Indomethacin, both inhibited the granuloma tissue formation. The inhibition of the test extract (400 mg/kg) and standard drug was found to be 54.19 and 62.06 per cent respectively. This indicates that the test samples inhibit the increase in the number of fibroblasts, synthesis of collagen and mucopolysaccharides during granuloma tissue formation in the chronic inflammation stage.

Table 13.6: Effect of ethanolic extract of *W. tinctoria* leaves on cotton pellet induced granuloma in rats.

Treatment	*Dose (mg/kg, p.o.)*	*Granuloma wt. (mg)*	*Per cent inhibition*
Control (normal saline)	1 ml	57.2 ± 3.08[a]	–
Indomethacin	10	21.7 ± 0.65**	62.06**
Ethanolic extract	200	30.2 ± 2.39** [a]	47.20**
Ethanolic extract	400	26.2 ± 0.89**	54.19**

Values are mean ± SEM of 6 parallel measurements.

Statistical significant test for comparison was done by ANOVA, followed by Dunnet's 't' test (n=6).

***P*< 0.01, when compared against control. [a]*P*< 0.01 when compared against standard.

Anti-oxidant Activity

The oxygen free radicals liberated from phagocytes are important in the inflammation process as they implicate the activation of transcription factor, nuclear factor (NF)-κB which induces the formation of inflammatory cytokines and COX-2 (Winrow *et al.*, 1993). Hence, during the inflammatory response, the increased free radical generation is one of the tissue damaging factors. These undesirable effects of oxidative stress have been found to be controlled by the antioxidant and/or anti-inflammatory effects of compounds and plant extracts. Therefore, much attention has been given towards the antioxidant properties of plant extracts and identification of compounds responsible for these activities. This information encouraged us to evaluate the free radical scavenging effects of the 70 per cent ethanol extracts by employing three widely used hydroxyl, DPPH and nitric oxide free radicals. The results were reported in Figure 13.5.

Hydroxyl Radical Activity

The potentially reactive hydroxyl radicals can cause oxidative damage to DNA, lipids and proteins (Annie Shirwaikar *et al.*, 2006). The effect of ethanol extract of *Wrightia tinctoria* on the inhibition of free radical-mediated deoxyribose damage was assessed by means of iron (II)-dependent DNA damage assay, which showed significant results. The IC_{50} value being of 150 μg/ml.

Inhibition of DPPH Radical

DPPH assay is considered a valid and easy to evaluate scavenging activity of antioxidants, since the radical compound is stable and does not have to generated as

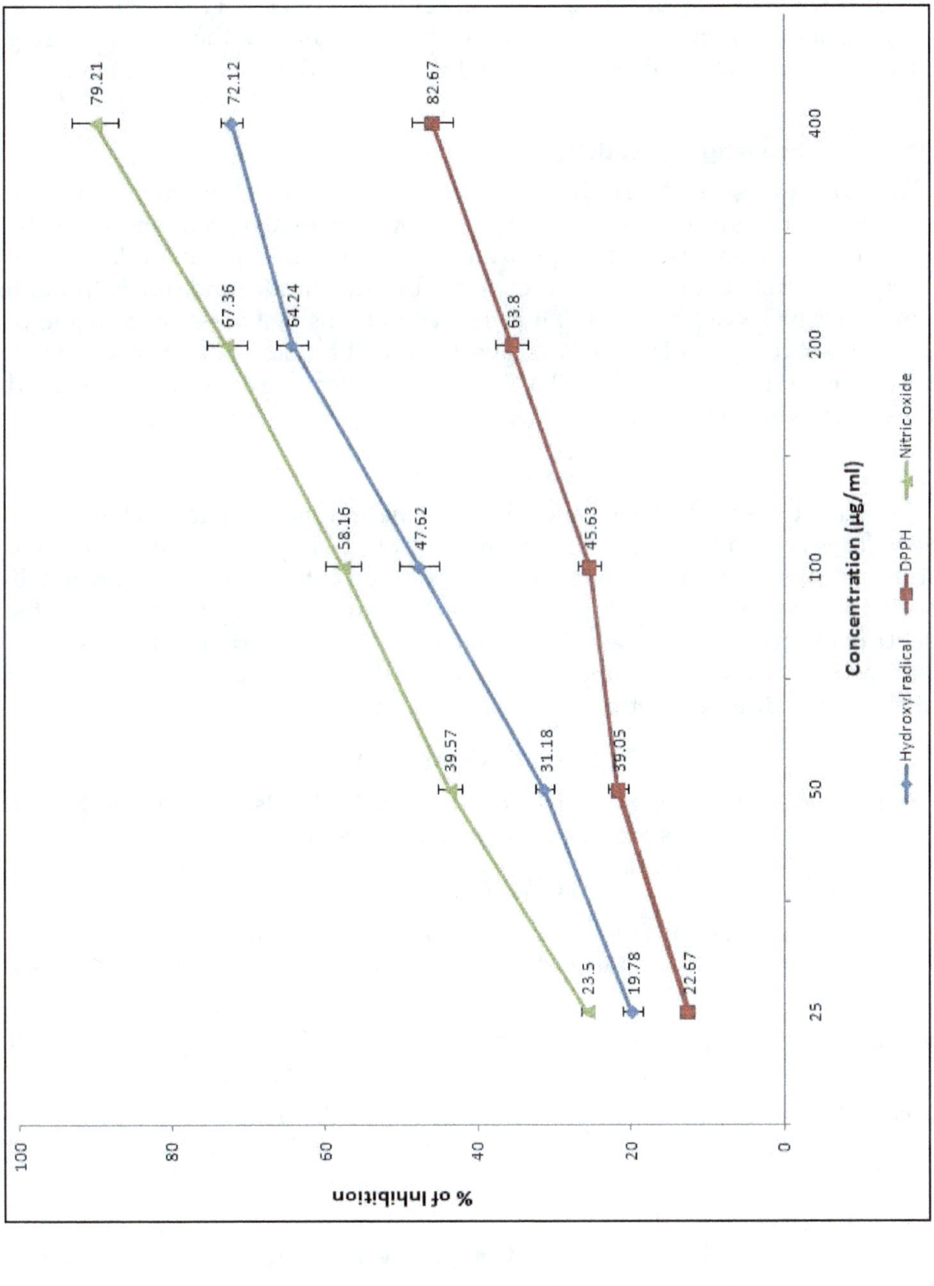

Figure 13.5: Free radical scavenging activity of ethanol extract of *Wrightia tinctoria* leaves.

in other radical assays. DPPH radicals react with suitable reducing agents and then electrons become paired off and the solution loses colour stochiometrically with the number of electrons taken up (Annie Shirwaikar *et al.*, 2006). Such reactivity has been widely used to test the ability of compound and plant extract to act as free radical scavengers. The potential decrease in the concentration of DPPH radical due to the scavenging ability of ethanol extract showed significant free radical scavenging activity of about 82.67 ± 1.35 per cent at higher doses with the IC_{50} value being 160 µg/ml.

Nitric Oxide Scavenging Activity

Nitric oxide (NO) is a free radical produced in mammalian cells, involved in the regulation of various physiological processes. However, excess production of NO is associated with several diseases. Nitric oxide is a very unstable species under aerobic conditions. It reacts with O^2 to produce stable product nitrate and nitrite through intermediates NO_2, N_2O_4 and N_3O_4. It is estimated by using Griess reagent and in presence of test compound which is a scavenger that the amount of nitrous acid will decrease (Annie Shirwaikar *et al.*, 2006). Significant scavenging activity was observed for the extract with an IC_{50} value of 80 µg/ml.

Conclusion

From the above study it was concluded that, the anti-inflammatory activity of *W. tinctoria* leaves could be at least in part due to COX-2 enzyme inhibition and free radical-scavenging activities which may be attributed to the presence of flavonoids and other polyphenols that are known to have direct antioxidant property due to the presence of hydroxyl groups, which can function as hydrogen donor. The results of this study provided a scientific support for the use of *W. tinctoria* leaves for the treatment of inflammatory diseases in ethnomedicine/Ayurveda.

Acknowledgements

Authors are extremely grateful to our Chancellor Dr. Ishari K. Ganesh, Vels University, for the facilities provided to complete this work successfully.

References

Annie Shirwaikar, S., Kirti Prabhu., and Punitha, I.S.R. (2006). *In vitro* antioxidant studies of *Sphaeranthus indicus* (Linn). *Indian Journal of Experiment Biology*, **44**: 993-996.

Aruoma, O.I. (1998). Free radicals, oxidative stress, and antioxidants in human health and disease. *Journal of the American Oil Chemists Society*, **75**: 199–212.

Baumann, J., Von Bruchhausen, F., and Wurm G. (1980). Flavonoids and related compounds as inhibitors of arachidonic acid peroxidation. *Prostaglandins*, **20**: 627-639.

Crespo, I., García-Mediavilla, M.V., Gutierrez, B., Sanchez-Campos, S., Tunon, M.J., and Gonzalez-Gallego, J.A. (2008). Comparison of the effects of kaempferol and quercetin on cytokine-induced pro-inflammatory status of cultured human endothelial cells. *The British Journal of Nutrition*, **100**: 968-976.

Delaporte, R.H., Sanchez, G.M., Cuellar, A.C., Giuliani, A., and Palazzo de Mello, J.C. (2002). Anti-inflammatory activity and lipid peroxidation inhibition of iridoid lamiide isolated from *Bouchea fluminensis* (Vell.) Mold. (Verbenaceae). *Journal of Ethnopharmacology*, **82**: 127–130.

Dharmasiri, J.R., Jayakody, A.C., Galhena, G., Liyanage, S.S.P., and Ratnasooriya, W,D. (2003). Antiinflammatory and analgesic activities of mature fresh leaves of *Vitex negundo*. *J. Ethnopharmacol.*, **87**: 199-206.

Gutteridge, J.M. (1995). Lipid peroxidation and antioxidants as biomarkers of tissue damage. *Clinical Chemistry*, **41**: 1819–1828.

Harris, G.K., Qian, Y., Leonard, S.S., Sbarra, D.C., and Shi, X. (2006). Luteolin and chrysin differentially inhibit cyclooxygenase-2 expression and scavenge reactive oxygen species but similarly inhibit prostaglandin-E2 formation in RAW-264.7 cells. *Journal of Nutrition*, **136**: 1517-1521.

Hinz, B., and Brune, K. (2002). Cyclooxygenase-2–10 years later. *J Pharmacol Exp Ther.*, **300**: 367–375.

Kadifkova panovska, T., Kulevanova, S., and Marina stefova. (2005). *In vitro* antioxidant activity of some *Teucrium* species (Lamiaceae). *Acta Pharm.*, **55**: 207-214.

Kokate, C.K., and Purohit, Gokhale, S.B. (1997). Text Book of Pharmacognosy. Vallabh Prakashan, New Delhi.

Li, X., Liu, Z., Zhang, X.F., Wang, L.J., Zheng, Y.N., Yuan, C.C., and Sun, G.Z. (2008). Isolation and characterization of phenolic compounds from the leaves of *Salix matsudana*. *Molecules*, **13**: 1530-1537.

Lipnic, R.L., Cotruvo, J.A., Hill, R.N., Bruce, R.D., Stitzel, K.A., Waker, A.P., *et al.* (1995). Comparison of the up and down conventional LD_{50} and fixed dose acute toxicity procedure. *Fundamental and Chemical Toxicology*, **3**: 223-231.

Luceri, C., Caderni, G., Sanna, A., and Dolara, P. (2002). Red wine and black tea polyphenols modulate the expression of cyclooxygenase-2, inducible nitric oxide synthase and glutathione-related enzymes in azoxymethane-induced f344 rat colon tumors. *Journal of Nutrition*, **132**: 1376-1379.

Nadkarni, A.K. (1976). Indian Materia Medica. Popular Prakashan, Bombay, pp. 245-249.

Olajide, O.A., Awe, S.O., and Makinde, J.M. (1999). Effect of the aqeous extract of *Bridelia ferruginea* stem bark on carrageenan-induced edema and granuloma tissue formation in rats and mice. *J Ethnopharmacol.*, **66**: 113–117.

Perenz, R.M., Perenz, S., Zavala, M.A., and Salazar, M. (1995). Anti-inflammatory activity of the bark of *Hipparatea excelsa*. *J. Ethnopharmacology*, **47**: 85-90.

Prabhjit Kaur, Satwinderjeet Kaur, Subodh Kuma, and Palwinder Singh. (2010). *Rubia cordifolia* L. and *Glycyrrhiza glabra* L. Medicinal Plants as Potential Source of COX-2 Inhibitors. *Am. J. Biomed. Sci.*, **2(2)**: 108-120.

Ram P Rastogi and Mehrotra, B.N. (1994). Compendium of Indian medicinal plants, Central Drug Research Institute Lucknow and National Institute of Science and Communication, New Delhi, pp. 1-5, 710, 684, 766, 898.

Repetto, M.G., and Llesuy, S.F. (2002). Antioxidant properties of natural compounds used in popular medicine for gastric ulcers. *Brazilian Journal of Medicine and Biological Research*, **35**: 523–534.

Rich E, Schibli A. HPTLC for the Analysis of Medicinal Plants. Thieme Medical Publisher, Inc, ISBN 13: 978-1-58890-409-6(US)

Shinde, V.A., Phadke, A.S., Naire, A.M., Mungantrivan, V.J., Dikshit and Saraf, V.O. (1999). Membrane stabilizing activity-a possible mechanism of action for anti-inflammation activity of *Cedrus deodara* wood oil. *Fitoterpia*, **70**: 251-257.

Simon, L.S. (1999). Anti-inflammatory and upper gastrointestinal effects of celecoxib in rheumatoid arthritis: a randomized controlled trial. *JAMA*, **282(20)**: 1921-8.

Vinegar, R., Schreiber, W. , and Hugo, R. (1969). Biphasic development of carrageenan oedema in rats. *J Pharmacol Exp Ther.*, **166**: 96-103.

Warner, T.D., Giuliano, F., Vojnovic, I., Bukasa, A., Mitchell, J.A., and Vane, J.R. (1999). Nonsteroid drug selectivities for cyclo-oxygenase-1 rather than cyclo-oxygenase-2 are associated with human gastrointestinal toxicity: a full *in vitro* analysis. *Proc Natl Acad Sci.*, **96**: 7563–7568.

Winrow, V.R., Winyard, P.G., Morris, C.J., and Blake, D.R. (1993) Free radicals in inflammation: second messangers and mediators of tissue destruction. *Br Med Bull.*, **49**: 506–22.

Winter, C.A., Risley, E.A., and Nuss, G.V. (1962). Carrageenan-induced edema in the hindpaw of the rat as an assay for anti-inflammatory drugs. *Proc Soc Exp Biol Med.*, **11**: 544-547.

Yi-Chen Liu, Jian-Guo Zeng, Bo Chen, and Shou-Zhou Yao. (2007). Investigation of Phenolic Constituents in *Echinacea purpurea* grown in China. *Planta Medica*, **73**: 73-89.

Traditional and Folk Herbal Medicine: Recent Researches Vol. 2 (2014) *Pages* **407–457**
Editor: **V.K. Gupta**
Published by: **DAYA PUBLISHING HOUSE, NEW DELHI**

14

Aphrodisiacs in Traditional Indian Phytotherapy: A Recent Perspective

Abhijit Dey[1]* and Anuradha Mukherjee[2]

ABSTRACT

Aphrodisiacs are the substances those enhance libido or sexual desire. Stress, diseases, medical therapy, diabetes, alcoholism, smoking, obesity and other lifestyle related health hazards are found to be responsible for decreased libido. Erectile dysfunction, premature ejaculation, sexual debility, oligospermia, sperm parameters such as sperm density and motility etc. are among the other related considerations. Human civilizations have been using several inorganic and organic compounds, medicinal plants, herbal formulations and drugs to stimulate sexual urge. Different ethnic groups residing in the rural parts of the world have been using herbal remedy as effective aphrodisiacs. The present review extensively compiles the traditional use of medicinal plants as aphrodisiacs used by various aboriginal groups of India. This state of the art review documents the recent literature published on utilization of aphrodisiac flora by the indigenous people of India mentioning the scientific names, plant families, reporting tribe, place of occurrence and other relevant information. Moreover, the present documentation may serve as an important start point which could eventually lead to the discovery of certain botanicals and/or isolated compounds having aphrodisiac property that might be evaluated scientifically in order to formulate such herbal preparations in drug discovery programs.

Keywords: Aphrodisiacs, Ethnobotany, Erectile dysfunction, India, Libido, Medicinal plants.

1 Assistant Professor, Department of Botany, Presidency University (Formerly Presidency College), 86/1, College Street, Kolkata – 700 073, West Bengal, India

2 Researcher and Assistant Teacher, MMHS, Dist: South 24 Paraganas, West Bengal, India

* *Corresponding author*: E-mail: abhijitbio25@yahoo.com

Introduction

Enhanced libido and sexual arousal have been considered as one of the eternal quest of mankind. It was found that 15 per cent of sexually active people are affected by infertility of which 50 per cent is being contributed by the male factor and few reports suggests a declining trend in sperm count among the male population (Pasqualotto *et al.*, 2004). Semen parameters in human are greatly affected by mental stress (Eskiocak *et al.*, 2005a, b, 2006). Stress along with anxiety and depression causes decrease in human libido (Arnett *et al.*, 1986). Interrelationship between erectile dysfunction and impotence are also related to diabetes mellitus (Rubin and Babbott, 1958; Nakanishi *et al.*, 2004). Moreover, it was noted that, hypogonadism and reduced level of testosterone are associated with obesity (Vermeulen, 1996). Furthermore, sexual debility is associated with using recreational agents such as cocaine, cigarette smoking and alcohol (Fahrner, 1987; Pasqualotto *et al.*, 2004). Readers are suggested to view the article written by Abel, 1980 for the influence of alcohol on libido. Infertility may be imposed by use of certain drugs and medications (Pasqualotto *et al.*, 2004). Roles of developmental defects such as hypospadias (Pasqualotto *et al.*, 2004; Mieusset and Soulié, 2005), cryptorchidism (Lipshultz, 1976; Hezmall and Lipshultz, 1982) and diseases like testicular cancer (Pasqualotto *et al.*, 2003) in infertility have also been investigated.

Natural products of plant and animal origin and synthetic chemical compounds have been considered as libido-lifters (Puri, 1971; Qureshi *et al.*, 1989; Waddell *et al.*, 1980; Dama *et al.*, 2010). Ginseng has been reported to possess remedy against erectile dysfunction in ancient China (Nair *et al.*, 2012). Some other compounds have been investigated for possessing aphrodisiac properties such as vitamin B12 (Sta-Mrad, 1955), hydrallazine (Davies, 1979), alkyl nitrite (Poulsen, 1983), yohimbine (Buffum, 1985), p-chlorophenylalanine (Gessa *et al.*, 1971), p-chlorophenylalanine methyl ester (Whalen and Luttge, 1970), (-) deprenyl (Yen *et al.*, 1982; Dalló *et al.*, 1990) some of which has yielded negative results. The eternal human quest for efficacious sexual stimulus (Borelli, 1960) to enhance libido (Kent, 1981) has enabled repeated investigations in search for new and authentic aphrodisiacs (Kent, 1975; Magrini, 1981). Herbal aphrodisiacs have been investigated pharmacologically *in vivo* for improvement of semen quality and motility and oligospermic conditions (Mahajan *et al.*, 2012). Herb based natural products might serve as a cure for erectile dysfunction and male sexual debility which are said to be related to the lack of libido (Pratap and Rajender, 2012). The age long human obsession to enhance sexual performance has rendered the human civilization to try various natural aphrodisiacs such as yohimbine and ginseng (Shamloul, 2010). Unfortunately such attempts are very limited when the use of aphrodisiacs for women is concerned (Shamloul, 2010). Male sexual debility is treated by popular herb based natural aphrodisiacs having the ability to enhance libido, increase sexual potency and/or pleasure (Malviya *et al.*, 2011a). Herbs have been reported for pro-sexual properties such as pro-ejaculatory (Carro-Juárez *et al.*, 2006) or enhancing ejaculatory capacity (Carro-Juárez *et al.*, 2009). These herbal remedies as aphrodisiacs are implicated to androgen increasing property (Gauthaman *et al.*, 2002), oxytocic activity (Carro-Juárez *et al.*, 2006), inhibition of ejaculation by urethral stimulation (Carro-Juárez *et al.*, 2009), vasorelaxant properties (Zamblé *et al.*,

2008), facilitating sexual arousal (Carro-Juárez *et al.*, 2004), pro-androgenic effect (Ang and Cheang, 2001), activity on the hypothalamus-pituitary-gonadal axis (Shukla *et al.*, 2009), rejuvenating male reproductive hormone balance (Gupta *et al.*, 2011), reducing stress and improving semen quality (Shukla *et al.*, 2010a), preventing premature ejaculation (Ratnasooriya and Dharmasiri, 2000) and decreased libido (Ratnasooriya and Fernando, 2008), facilitating increased and fast penile erections and elevating the level of serum testosterone (Ratnasooriya and Jayakody, 2006), acting by neurogenic mode (Ali and Rakkah, 2008) etc. Effects of herbal remedies on experimental animals have been investigated in terms of enhancement of sexuality in middle aged animals (Ang *et al.*, 2003), initiation of sexual performance in castrated male rats (Ang *et al.*, 2000), enhancing libido (Ang and Sim, 1997) etc.

Ethnobotany has been considered as a subject associated with man and plant relationship. Traditional phytotherapy has revealed the use of innumerable number of plant species (Dey and De, 2010, 2011, 2012a, b), either novel or has already been established in other systems of medicine. This branch of botany has provided the start-point of many pharmacological and pre-clinical tests of herbal preparations leading to the clinical trials and drug discovery programs. Traditional medicine of botanical origin has served as a rich reservoir of aphrodisiacs (Mendes, 2011; Kamatenesi-Mugisha and Oryem-Origa, 2005). Indigenous herbal remedies as pro-sexuals have been investigated in the laboratory and pharmacological investigations have yielded positive results by using experimental animals (Farnsworth *et al.*, 1967; Ang and Ngai, 2001; Adimoelja, 2000; Tempest *et al.*, 2008; Suresh *et al.*, 2009; Thakur *et al.*, 2009a). The readers may consult the recent review by Patel *et al.* (2011) for a compilation of botanical aphrodisiacs which have been pharmacologically investigated. However, the safety aspects of these herbal remedies have been questioned on several occasions (Venhuis *et al.*, 2008).

Enumeration

The authors have compiled a list of medicinal plants used as aphrodisiacs in the traditional Indian medicine. This review provides a compilation of recent literature published mostly the year 2000 onwards on the herbal aphrodisiacs used by different tribal and non-tribal communities residing in various parts of the country. The scientific names of the plants are provided alphabetically along with their respective families, part(s) used, used as/against particular types of sexual dysfunction, using people and locality of use/reporting area depending on the availability of information (Table 14.1). Regarding the use of specific plant parts, NM (not mentioned)/NS (not specified) have been used where either the information was absent or the plant part was not specifically mentioned. Some of the authors have mentioned other reference(s) to elucidate a plant's aphrodisiac activity, which have also been compiled. Polyherbal formulations have been mentioned considering a single herb's use as prosexual.

Discussion

The present study documents a total number of 219 ethnomedicinal plants belonging to 72 families claimed by the people residing in different parts of India to possess aphrodisiac activity. Author's citation and plant families were mostly

Table 14.1: A list of Indian ethnobotanicals used as aphrodisiac.

Plant	*Family*	*Plant Part Used*	*Used as/against*	*Using People/Locality of Use/ Reporting Area*	*Reference*
Abelmoschus crinitus Wall.	Malvaceae	Roots	Sexual disability	Bargarh district, Orissa	Sahu *et al.*, 2010
Abelmoschus esculentus (L.) Moench	Malvaceae	Root	To enhance sexual potentiality, cure impotency, spermaturia	Eastern Rajasthan	Upadhyay *et al.*, 2010
Abelmoschus manihot (L.) Medik.	Malvaceae	Root	Aphrodisiac	*Korku* tribe, Amravati district, Maharashtra	Jagtap *et al.*, 2008a
Abelmoschus moschatus Medik.	Malvaceae	Seeds	Aphrodisiac	Chandaka Denudated Forest Patches, Bhubaneswar, Orissa	Behera *et al.*, 2008
Abrus precatorius L.	Fabaceae	Seeds	Aphrodisiac	Gautala and Pitalkhora of Kannad, Aurangabad district, Maharashtra	Kshirsagar and Magar, 2011
		Roots	To enhance sexual vigor (men and women)	Southern Rajasthan	Jain *et al.*, 2004
		Seeds	Increase sperm count	*Irula* tribe, Hasanur hills, Erode district, Tamil Nadu	Revathi and Parimelaz-hagan, 2010
Abutilon indicum (L.) Sweet	Malvaceae	Whole plant	Aphrodisiac	Mahendergarh district, Haryana	Yadav *et al.*, 2010
		Seeds	Aphrodisiac	Sonaghati, Sonbhadra district, Uttar Pradesh	Singh *et al.*, 2002
Acacia arabica (Lam.) Willd.	Fabaceae	NS/NM	Male fertility disorders	*Saperas* community; Khetawas, Jhajjar district, Haryana	Panghal *et al.*, 2010
Acacia leucophloea Willd.	Fabaceae	NS/NM	To enhance sexual potency	*Meena* community, Rajasthan	Meena and Rao, 2010
		Gum	To increase sexual potency	Southern Rajasthan	Jain *et al.*, 2004
Acacia nilotica (L.) Willd. ex Delile	Fabaceae	Leaves	Sexual weakness	Rajasthan	Katewa *et al.*, 2004

Contd...

Table 14.1–*Contd...*

Plant	*Family*	*Plant Part Used*	*Used as/against*	*Using People/Locality of Use/ Reporting Area*	*Reference*
Acanthus ilicifolius L.	Acanthaceae	NS/NM	Aphrodisiac	Bhitarkanika wildlife sanctuary, Orissa	Pattanaik *et al.*, 2008b
Aconitum heterophyllum Wall. ex Royle	Ranunculaceae	Tuber	Aphrodisiac	*Vaidyas*, Ukhimath block, Uttarakhand	Semwal *et al.*, 2010
		Roots	Aphrodisiac	Nubra valley, Ladakh	Phani Kumar *et al.*, 2009
		Roots	Aphrodisiac	Alpine and Sub-Alpine zones, Western Himalaya	Bisht and Badoni, 2009
		Roots	Aphrodisiac	Indian Himalayas	Kala, 2005
Actiniopteris radiata (Sw.) Link	Pteridaceae	Whole plant	Aphrodisiac	Chitrakoot, Madhya Pradesh	Sikarwar *et al.*, 2008
Adiantum capillus-veneris L.	Pteridaceae	NS/NM	Aphrodisiac	Mizoram	Rai and Lalramnghinglova, 2010
		NS/NM	Aphrodisiac	Rajasthan	Parihar and Parihar, 2006
		NS/NM	Aphrodisiac	Arunachal Pradesh	Benniamin, 2011
Aegle marmelos (L.) Corrêa	Rutaceae	Fruits	Sexual debility	*Meena* community, Rajasthan	Meena and Rao, 2010
		Apical bud	Sexual debility (men and women)	Southern Rajasthan	Jain *et al.*, 2004
Albizia lebbeck (L.) Benth.	Fabaceae	Roots	Aphrodisiac	Kathua district, Jammu and Kashmir	Khan *et al.*, 2009
		NS/NM	Male fertility disorders	*Saperas* community; Khetawas, Jhajjar district, Haryana	Panghal *et al.*, 2010
Albizia stipulate DC. (Boivin)	Fabaceae	Tuber	To increase sperm count	North Maharashtra	Ahirrao *et al.*, 2011
Allium ascalonicum L.	Amaryllidaceae	Bulbs and Leaves	Aphrodisiac	*Meitei*; Manipur	Yumnam and Tripathi, 2012

Contd...

Table 14.1–*Contd...*

Plant	*Family*	*Plant Part Used*	*Used as/against*	*Using People/Locality of Use/ Reporting Area*	*Reference*
Allium hookeri Thwaites	Amaryllidaceae	Leaves	Aphrodisiac	*Meitei*; Manipur	Yumnam and Tripathi, 2012
Allium sativum L.	Amaryllidaceae	NS/NM	Aphrodisiac	*Pawara*, *Bhil* and *Pardhi* tribes, Satpuda region, Dhule and Jalgaon districts, Maharashtra	Jain *et al.*, 2010
		Bulb	Aphrodisiac	Gondia district, Maharashtra	Nimbekar *et al.*, 2011
Aloe vera (L.) Burm. f.	Xanthorrhoeaceae	Sucker	Sexual debility	*Malayali* tribes; Kolli hills, Tamil Nadu	Francis *et al.*, 2011
Alpinia galanga (L.) Willd.	Zingiberaceae	Rhizome	Aphrodisiac	*Malayalis*; Kollimalai hill tracts, Tamil Nadu	Anand *et al.*, 2006
		Bulb	Aphrodisiac	Sonebhadra district, Uttar Pradesh	Singh and Dubey, 2012
Alternanthera philoxeroides (Mart.) Griseb.	Amaranthaceae	Aerial parts	Aphrodisiac	*Meitei*; Manipur	Yumnam and Tripathi, 2012
Anacardium occidentale L.	Anacardiaceae	Seeds	Sexual debility	Puducherry	Padmavathy and Anbarashan, 2011
		Kernel	Aphrodisiac	*Chenchu* tribes; Andhra Pradesh	Reddy *et al.*, 1988
Argyreia speciosa (L. f.) Sweet	Convolvulaceae	NS/NM	Aphrodisiac	Chatara block, Sonebhadra district, Uttar Pradesh	Singh *et al.*, 2010
Aristolochia indica L.	Aristolochiaceae	Whole plant	Aphrodisiac	Madhya Pradesh	Malviya *et al.*, 2011b
Artemisia maritima L. ex Hook. f.	Asteraceae	NM/NS	Aphrodisiac	Indian trans-Himalaya	Kala, 2006

Contd...

Table 14.1–*Contd...*

Plant	*Family*	*Plant Part Used*	*Used as/against*	*Using People/Locality of Use/ Reporting Area*	*Reference*
Asparagus adscendens Roxb.	Asparagaceae	NS/NM	Aphrodisiac	Uttaranchal	Adhikari *et al.*, 2007
		Tuberous roots	Sexual debility	*Jaunsari* tribe; Garhwal Himalaya, Uttaranchal	Bhatt and Negi, 2006
		Roots	Sexual dysfunction	Kangra district, Himachal Pradesh	Uniyal *et al.*, 2011
		NS/NM	Aphrodisiac	Dehradun, Uttarakhand	Adhikari *et al.*, 2010
		Stem	Aphrodisiac	Kumaon region	Shah and Joshi, 1971
Asparagus curillus Buch.-Ham. ex Roxb.	Asparagaceae	NS/NM	Aphrodisiac	Uttaranchal	Adhikari *et al.*, 2007
Asparagus filicinus Buch.-Ham. ex D. Don	Asparagaceae	Tuberous roots	Aphrodisiac	*Jaunsaris*; Jaunsar-Bawar, Dehra Dun, Uttar Pradesh	Rana and Datt, 1997
		Roots	Aphrodisiac	Parvati valley; western Himalaya	Sharma *et al.*, 2004
Asparagus racemosus Willd.	Asparagaceae	Roots	Aphrodisiac	Thar desert, Rajasthan	Mohammed *et al.*, 2004
		Roots	Aphrodisiac	Mahendergarh district, Haryana	Yadav *et al.*, 2010
		Tuberous roots	Aphrodisiac	Sonebhadra district, Uttar, Pradesh	Singh and Dubey, 2012
		Roots	Aphrodisiac	*Mising* tribe; Assam	Sharma and Pegu, 2011
		Roots	Aphrodisiac	*Malayali*; Chitteri hills, Dharmapuri, Tamil Nadu	Kadhirvel *et al.*, 2010
		NS/NM	Aphrodisiac	Mornaula Reserve Forest, Kumaun, West Himalaya	Pant and Samant, 2010
		Roots	Aphrodisiac	Tuticorin district, Tamil Nadu	Muthukumar and Samuel, 2010
		NS/NM	Aphrodisiac	Madhya Pradesh	Dwivedi *et al.*, 2008
		Roots	Aphrodisiac	*Nagas*; Nagaland	Changkija, 1999

Contd...

Table 14.1–*Contd...*

Plant	*Family*	*Plant Part Used*	*Used as/against*	*Using People/Locality of Use/ Reporting Area*	*Reference*
		Roots	Aphrodisiac	Uttaranchal	Adhikari *et al.*, 2007
		Tuberous roots	Aphrodisiac	*Kandha* tribe; Orissa	Behera and Misra, 2005
		Tuber	Aphrodisiac	*Chenchus*; Nallamalais in Kurnool district, Andhra Pradesh	Sudhakar Reddy *et al.*, 2007
Asparagus racemosus Willd. var. *javanica* Baker	Asparagaceae	Roots	Aphrodisiac	Ratan Mahal Hills, Gujarat	Bedi, 1978
Avicennia alba Blume	Acanthaceae	NS/NM	Aphrodisiac	Bhitarkanika wildlife sanctuary, Orissa	Pattanaik *et al.*, 2008b
Avicennia officinalis L.	Acanthaceae	Roots	Aphrodisiac	Tuticorin district, Tamil Nadu	Muthukumar and Samuel, 2010
Balanophora dioica R. Br. ex Royle	Balanophoraceae	Inflorescence	Aphrodisiac	*Sangtam Naga* tribe; Kiphire district, Nagaland	Sangtam *et al.*, 2012
Barleria prionitis L.	Acanthaceae	Seeds	Increase sexual vigor	Medak district, Andhra Pradesh	Reddy *et al.*, 2010
Bauhinia vahlii Villar	Fabaceae	Roots	Aphrodisiac	Shivbari sacred grove of Himachal Pradesh	Jaryan *et al.*, 2010
Blepharis sindica Stocks ex T. Anderson	Acanthaceae	Seeds	Aphrodisiac	Thar desert, Rajasthan	Mohammed *et al.*, 2004
Boerhavia diffusa L.	Nyctaginaceae	Root tubers	Aphrodisiac	Rudrakod sacred grove, Nallamalais, Andhra Pradesh	Rao and Sunitha, 2011
Bombax ceiba L.	Malvaceae	Gum	Aphrodisiac	Western Mizoram	Lalfakzuala *et al.*, 2007
		Flower	Aphrodisiac	Hazaribag district, Jharkhand	Lal and Singh, 2012
		Roots	Aphrodisiac	Kathua district, Jammu and Kashmir	Khan *et al.*, 2009
		Roots	Sexual debility	Meerut district, Uttar Pradesh	Tomar, 2009

Contd...

Table 14.1–*Contd...*

Plant	*Family*	*Plant Part Used*	*Used as/against*	*Using People/Locality of Use/ Reporting Area*	*Reference*
		Stem bark	Aphrodisiac, impotency, to increase sperm content	*Kandha* tribe; Orissa	Behera and Misra, 2005
		Fruit	To cure female infertility	Rangia subdivision, Kamrup district, Assam	Choudhury *et al.*, 2011
Borassus flabellifer L.	Arecaceae	Fruits, Root	Aphrodisiac	Vattamalai hills, Namakkal district, Tamil Nadu	Kumar and Kumar, 2011
Bridelia scandens Gehrm.	Phyllantha-ceae	Bark	Aphrodisiac	*Kurichiar* tribes; of Tirunelli forest, Wayanad district, Kerala	Udayan *et al.*, 2008
Bryonia laciniosa L.	Cucurbitaceae	Seeds	Sterility in women	Satpura plateau region, Central India	Pandey and Shukla, 2008
Butea monosperma (Lam.) Taub.	Fabaceae	Fruits	Aphrodisiac	Nagzira wild life sanctuary, Gondia district, Maharashtra	Koche *et al.*, 2008
		Leaves	Aphrodisiac	Sonaghati, Sonbhadra district, Uttar Pradesh	Singh *et al.*, 2002
		Flower	Increase sperm count	Kinwat forest, Nanded district, Maharashtra	Biradar and Ghorband, 2010
Butea superb Roxb.	Fabaceae	Roots	Aphrodisiac	Garasia tribe; Sirohi district, Rajasthan	Meena and Yadav, 2010
Calanthe tricarinata Lindl.	Orchidaceae	Leaves, Bulb	Aphrodisiac	Manali wildlife sanctuary, Kullu district, Himachal Pradesh	Rana and Samant, 2011
Calotropis procera (Aiton) R. Br.	Apocynaceae	NS/NM	Male fertility disorders	*Saperas* community; Khetawas, Jhajjar district, Haryana	Panghal *et al.*, 2010
Campylandra aurantiaca Baker	Liliaceae	Rhizomes	Aphrodisiac	*Adi* tribe; Arunachal Pradesh	Srivastava and *Adi* community, 2009

Contd...

Table 14.1–*Contd...*

Plant	*Family*	*Plant Part Used*	*Used as/against*	*Using People/Locality of Use/ Reporting Area*	*Reference*
Cannabis sativa L.	Cannabaceae	Seeds	Sexual dysfunction	Kangra district, Himachal Pradesh	Uniyal *et al.*, 2011
		Leaves	Premature ejaculation	Western Uttar Pradesh	Khan and Khan, 2005
Capparis aphylla Hayne ex Roth.	Capparaceae	Bark	Aphrodisiac	Adilabad district, Andhra Pradesh	Omkar *et al.*, 2012
Capparis divaricata Lam.	Capparaceae	NM	Aphrodisiac	Sangli district, Maharashtra	Patil *et al.*, 2011
Carmona retusa (Vahl) Masam.	Boraginaceae	Leaves	Fertility inducer	Madurai District, Tamil Nadu	Ignacimuthu *et al.*, 2006
Carum carvi L.	Apiaceae	Seeds	Sexual dysfunction	Upper catchment of Dhauli Ganga, Nanda Devi Biosphere Reserve (NDBR), Central Himalaya	Kandari *et al.*, 2012
Cassia auriculata L.	Fabaceae	Seed, Root	Aphrodisiac	Vattamalai hills, Namakkal district, Tamil Nadu	Kumar and Kumar, 2011
		NS/NM	To increase sperm count	Satpura hills, Nandurbar district, Maharashtra	Patil, 2012
Cassia occidentalis L.	Fabaceae	NS/NM	Aphrodisiac	Chatara block, Sonebhadra district, Uttar Pradesh	Singh *et al.*, 2010
Carica papaya L.	Caricaceae	Fruits	Aphrodisiac	Vattamalai hills, Namakkal district, Tamil Nadu	Kumar and Kumar, 2011
		Fruits	Impotency	Western Uttar Pradesh	Khan and Khan, 2005
Caryota urens L.	Arecaceae	NS/NM	Aphrodisiac	Eastern Ghats, Andhra Pradesh	Naidu and Khasim, 2010
Celastrus paniculatus Willd.	Celastraceae	Seeds	Aphrodisiac	Salem district, Tamil Nadu	Mishra *et al.*, 2008
Celosia argentea L.	Amarantha-ceae	NS/NM	Aphrodisiac	Near Srinagar town of district Pauri Garhwal in Uttarakhand	Bahuguna *et al.*, 2011
Centella asiatica (L.) Urb.	Apiaceae	Whole plant	Spermaturia	Uttara Kannada district, Karnataka	Hegde *et al.*, 2007

Contd...

Table 14.1–*Contd...*

Plant	*Family*	*Plant Part Used*	*Used as/against*	*Using People/Locality of Use/ Reporting Area*	*Reference*
Ceropegia bulbosa Roxb.	Apocynaceae	Tuber	Aphrodisiac	Chhattisgarh	Kala, 2009
Chlorophytum arundinaceum Baker	Asparagaceae	Roots	Aphrodisiac	Satpura plateau region, Central India	Pandey and Shukla, 2008
Chlorophytum borivilianum Santapau and R.R. Fern.	Asparagaceae	Roots	Aphrodisiac	*Korku* tribe; Amravati district, Maharashtra	Jagtap *et al.*, 2008a
		Roots	Sexual weakness	Chandauli district, Uttar Pradesh	Singh and Singh, 2009
		Roots	Aphrodisiac	*Pawra* tribe; Nandurbar district, Maharashtra	Jagtap *et al.*, 2008b
Chlorophytum tuberosum (Roxb.) Baker	Asparagaceae	Roots	Aphrodisiac	Chandaka Denudated forest patches, Bhubaneswar, Orissa	Behera *et al.*, 2008
		Tubers	Aphrodisiac	Dhamtari district, Chhattisgarh	Acharya *et al.*, 2012
Cissus quadrangularis L.	Vitaceae	Plant	Aphrodisiac	Vattamalai hills, Namakkal district, Tamil Nadu	Kumar and Kumar, 2011
Citrullus vulgaris Schrad.	Cucurbitaceae	Seeds	Aphrodisiac	Kathua district, Jammu and Kashmir	Khan *et al.*, 2009
Clitoria ternatea L.	Fabaceae	Root, Seed	Aphrodisiac	Vattamalai hills, Namakkal district, Tamil Nadu	Kumar and Kumar, 2011
		Root	To cure female infertility	Rangia subdivision, Kamrup district, Assam	Choudhury *et al.*, 2011
Coccinia indica Wight and Arn.	Cucurbitaceae	Leaves	To increase semen production	*Kani* tribe; Tirunelveli hills, Tirunelveli and Kanyakumari districts, Tamil Nadu	Ayyanar and Ignacimuthu, 2011
Cocculus pendulus (J.B. Först. and G. Forst.) Diels	Menispermaceae	Whole plant	Sexual debility	Eastern Rajasthan	Upadhyay *et al.*, 2010

Contd...

Table 14.1–*Contd...*

Plant	*Family*	*Plant Part Used*	*Used as/against*	*Using People/Locality of Use/ Reporting Area*	*Reference*
Cocos nucifera L.	Arecaceae	Spadix	Aphrodisiac	Coastal and sand dune, Orissa	Pattanaik *et al.*, 2008a
		Fruit	Aphrodisiac	Kurukshetra district, Haryana	Lal and Yadav, 1983
Coelogyne cristata Lindl.	Orchidaceae	Bulbs	Aphrodisiac	Mountain region, Uttarakhand	Phondani, 2011
Colchicum luteum Baker	Colchicaceae	Corms	Aphrodisiac	Kathua district, Jammu and Kashmir	Khan *et al.*, 2009
		NS/NM	Aphrodisiac	Jammu and Kashmir	Pant and pant, 2011
Corchorus depressus (L.) C. Chr.	Malvaceae	Whole plant	Sexual vitality	Southern Rajasthan	Jain *et al.*, 2004
Coriandrum sativum L.	Apiaceae	NS/NM	Aphrodisiac	Hazaribag district, Jharkhand	Lal and Singh, 2012
Costus speciosus (J. König) Sm.	Costaceae	Rizomes	Aphrodisiac	Gondia district, Maharashtra	Kamble *et al.*, 2011
Crocus sativus L.	Iridaceae	NS/NM	Aphrodisiac	Kishtwar, Jammu and Kashmir	Kumar *et al.*, 2009
Croton bonplandianum Baill	Euphorbiaceae	NS/NM	Aphrodisiac	Nandurbar district, Maharashtra	Tayade and Patil, 2005
Cucurbita maxima Duchesne	Cucurbitaceae	Seeds	Enhance sexual vigor	Tinsukia district, Assam	Buragohain, 2011
Curculigo orchioides Gaertn.	Hypoxidaceae	Rhizome	Aphrodisiac	*Malayalis*; Kollimalai hill tracts, Tamil Nadu	Anand *et al.*, 2006
		Leaves	Aphrodisiac	Niyamgiri hill ranges, Orissa	Dash *et al.*, 2008
		Roots	Aphrodisiac	Satpura plateau region, Central India	Pandey and Shukla, 2008
		Tubers	Aphrodisiac	Rudrakod sacred grove, Nallamalais, Andhra Pradesh	Rao and Sunitha, 2011
		Roots	Aphrodisiac	Chhatarpur district, Madhya Pradesh	Arjariya and Chaurasia, 2009
		Roots	Aphrodisiac	Adilabad district, Andhra Pradesh	Omkar *et al.*, 2012

Contd...

Table 14.1–*Contd...*

Plant	*Family*	*Plant Part Used*	*Used as/against*	*Using People/Locality of Use/ Reporting Area*	*Reference*
		Roots	Impotency, aphrodisiac	Almora district, Uttarakhand	Kumari *et al.*, 2011
		NS/NM	Sexual stimulant	West and South district of Tripura	Sen *et al.*, 2011
		NS/NM	Control semen ejaculation	South and West district of Tripura	Majumdar and Datta, 2004
		Tuber	Promoter of sexual desire	Chhattisgarh	Shukla *et al.*, 2008
		Leaves	Increase sperm count	*Irula* tribe, Hasanur hills, Erode district, Tamil Nadu	Revathi and Parimelazhagan, 2010
		Roots	Aphrodisiac, impotency, to increase sperm content	*Kandha* tribe; Orissa	Behera and Misra, 2005
		Rhizome	To improve semen production	*Valaiyans*; Karandamalai, Dindigul district, Tamil Nadu	Kottaimuthu, 2008
Curcuma amada Roxb.	Zingiberaceae	Rhizome	Aphrodisiac	Madhya Pradesh	Malviya *et al.*, 2011b
Curcuma aromatic Salisb.	Zingiberaceae	Rhizome	Aphrodisiac	Madhya Pradesh	Malviya *et al.*, 2011b
Cynoglossum lanceolatum Forssk.	Boraginaceae	Whole plant	Aphrodisiac	Indian trans-Himalaya	Kala, 2006
		NS/NM	Aphrodisiac	Dehradun, Uttarakhand	Adhikari *et al.*, 2010
Dactylorhiza hatagirea (D. Don) Soó	Orchidaceae	Tuber	Aphrodisiac	Vaidyas, Ukhimath block, Uttarakhand	Semwal *et al.*, 2010
		Tuber	Aphrodisiac	Indian trans-Himalaya	Kala, 2006
		Tuber	Sexual dysfunction	Upper catchment of Dhauli Ganga, Nanda Devi Biosphere Reserve (NDBR), Central Himalaya	Kandari *et al.*, 2012
		Roots	Aphrodisiac	Parvati valley; Western Himalaya	Sharma *et al.*, 2004

Contd...

Table 14.1–*Contd...*

Plant	*Family*	*Plant Part Used*	*Used as/against*	*Using People/Locality of Use/ Reporting Area*	*Reference*
		NS/NM	Aphrodisiac	Uttarkashi district, Garhwal Himalaya	Rawat and Jalal, 2011
		Tubers	Aphrodisiac	Uttarakhand	Jalal *et al.*, 2008
		Tubers	Aphrodisiac	Uttarakhand	Jalal and Rawat, 2009
		Tubers	Aphrodisiac	Indian Himalayas	Kala, 2005
		Tuber	Aphrodisiac	Indian Himalayas	Kala, 2005
Dalbergia sissoo Roxb. ex DC.	Fabaceae	Wood	Aphrodisiac	Chatara block, Sonebhadra district, Uttar Pradesh	Singh *et al.*, 2010
Datura metel L.	Solanaceae	NS/NM	Male fertility disorders	*Saperas* community; Khetawas, Jhajjar district, Haryana	Panghal *et al.*, 2010
Decaschistia cuddapahensis T.K. Paul and M.P. Nayar	Malvaceae	Roots	Aphrodisiac	Seshachalam hills, Andhra Pradesh	Sudhakar Reddy *et al.*, 2009
		Roots	Aphrodisiac	Eastern Ghats	Pattanaik *et al.*, 2009
Dendropthoe falcate (L.f) Ettingsh	Loranthaceae	NS/NM	Aphrodisiac	Kerala	Shanavaskhan *et al.*, 2012
Dioscorea bulbifera L.	Dioscoreaceae	NS/NM	Aphrodisiac	Chatara block, Sonebhadra district, Uttar Pradesh	Singh *et al.*, 2010
Dioscorea hispida Dennst.	Dioscoreaceae	Tubers	Aphrodisiac	Satpura plateau region, Central India	Pandey and Shukla, 2008
Dioscorea oppositifolia L.	Dioscoriaceae	Tuber	To increase sperm number	Rapur forest division, Nellore district, Andhra Pradesh	Neelima *et al.*, 2011
Diospyros melanoxylon Roxb.	Ebenaceae	Flowers	Aphrodisiac	Chatara block, Sonebhadra district, Uttar Pradesh	Singh *et al.*, 2010
Diospyros peregrine Gürke	Ebenaceae	Fruits	Aphrodisiac	Western Ghats region, Maharashtra	Deshmukh and Waghmode, 2011

Contd...

Table 14.1–*Contd...*

Plant	*Family*	*Plant Part Used*	*Used as/against*	*Using People/Locality of Use/ Reporting Area*	*Reference*
Diplazium polypodioides Blume	Woodsiaceae	Fronds	Aphrodisiac	Deoprayag; Garhwal Himalaya	Gaur and Bhatt, 1994
Dracaena terniflora Roxb.	Asparagaceae	Roots	Spermaturia	Uttara Kannada district, Karnataka	Hegde *et al.*, 2007
Drypetes roxburghii (Wall.) Hurus.	Putranjivaceae	Fruit	Aphrodisiac	Sonebhadra district, Uttar Pradesh	Singh and Dubey, 2012
Eclipta alba (L.) Hassk.	Asteraceae	Plant	Aphrodisiac	Manipur	Jain *et al.*, 2011
		Leaves	Sexual debility	Eastern Rajasthan	Upadhyay *et al.*, 2010
Elephantopus scaber L.	Asteraceae	Flower	Aphrodisiac	Vattamalai hills, Namakkal district, Tamil Nadu	Kumar and Kumar, 2011
Epipactis veratrifolia Boiss. and Hohen.	Orchidaceae	Rizomes	Aphrodisiac	Uttarakhand	Dangwal *et al.*, 2010
Eruca sativa Mill.	Brassicaceae	Whole plant	Aphrodisiac	Indian trans-Himalaya	Kala, 2006
Erythrina variegatai L.	Fabaceae	Flower	Aphrodisiac	Eastern Ghats, Andhra Pradesh	Naidu and Khasim, 2010
Eulophia campestris Wall. ex Lindl.	Orchidaceae	Rhizome	Aphrodisiac	North Eastern states	Medhi and Chakrabarti, 2009
Eulophia spectabilis (Dennst.) Suresh	Orchidaceae	Tuber	Aphrodisiac	Niyamgiri hill ranges, Orissa	Dash *et al.*, 2008
Eulophia dabia (D. Don) M.S. Balakr.	Orchidaceae	Tubers	Aphrodisiac	Uttarakhand	Jalal *et al.*, 2008
Eulophia ochreata Lindl.	Orchidaceae	NS/NM	Aphrodisiac	Satpura hills, Maharashtra	Jagtap *et al.*, 2010
Euphorbia hirta L.	Euphorbiaceae	NS/NM	Semen debility, aphrodisiac	*Pawara*, *Bhil* and *Pardhi* tribes; Satpuda region, Dhule and Jalgaon districts, Maharashtra	Jain *et al.*, 2010
		Fruit	Spermatorrhea, aphrodisiac	Gondia district, Maharashtra	Nimbekar *et al.*, 2011

Contd...

Table 14.1–*Contd...*

Plant	*Family*	*Plant Part Used*	*Used as/against*	*Using People/Locality of Use/ Reporting Area*	*Reference*
Evolvulus alsinoides (L.) L.	Convolvulaceae	Plant	Aphrodisiac	*Malayalis*; Kollimalai hill tracts, Tamil Nadu	Anand *et al.*, 2006
		Plant	Aphrodisiac	Sonaghati, Sonbhadra district, Uttar Pradesh	Singh *et al.*, 2002
		Plant	Aphrodisiac	Sonebhadra district, Uttar Pradesh	Singh and Dubey, 2012
Ficus benghalensis L.	Moraceae	NS/NM	Sexual potency	*Meena* community, Rajasthan	Meena and Rao, 2010
		Leaf latex	Aphrodisiac	*Kandha* tribe; Orissa	Behera and Misra, 2005
		Latex	Aphrodisiac	*Korku* tribe; Amravati district, Maharashtra	Jagtap *et al.*, 2006
		Latex	Aphrodisiac	Uttaranchal	Kala, 2004
		Latex	Aphrodisiac	Kurukshetra district, Haryana	Lal and Yadav, 1983
		NS/NM	Male fertility disorders	*Saperas* community; Khetawas, Jhajjar district, Haryana	Panghal *et al.*, 2010
Ficus hispida L. f.	Moraceae	Fruits	Aphrodisiac	*Kondareddis*; Khammam district, Andhra Pradesh	Reddy *et al.*, 2008
Ficus racemosa L.	Moraceae	Stem	Aphrodisiac, to increase sperm content	*Kandha* tribe; Orissa	Behera and Misra, 2005
		Latex	Aphrodisiac	Sonaghati, Sonbhadra district, Uttar Pradesh	Singh *et al.*, 2002
Ficus religiosa L.	Moraceae	Bark	Aphrodisiac	Kumaragiri hills, Salem district, Tamil Nadu	Alagesaboopathi, 2009
		Bark, Fruit	Aphrodisiac	Vattamalai hills, Namakkal district, Tamil Nadu	Kumar and Kumar, 2011
Flemingia nana Roxb. ex Aiton	Fabaceae	Roots	Aphrodisiac	Dhamtari district, Chhattisgarh	Acharya *et al.*, 2012

Contd...

Table 14.1–*Contd...*

Plant	*Family*	*Plant Part Used*	*Used as/against*	*Using People/Locality of Use/ Reporting Area*	*Reference*
Flemingia strobilifera (L.) R. Br.	Fabaceae	Roots	Aphrodisiac	Balaghat district, Madhya Pradesh	Jain *et al.*, 2011
Flickingeria nodosa (Dalzell) Seidenf.	Orchidaceae	NS/NM	Aphrodisiac	Kerala	Shanavaskhan *et al.*, 2012
Gloriosa superb L.	Colchicaceae	Tuber	Aphrodisiac	Kanjamalai hills, Salem district, Tamil Nadu	Alagesaboopathi, 2011
		Tuber	Aphrodisiac	Kumaragiri hills, Salem district, Tamil Nadu	Alagesaboopathi, 2009
Glycyrrhiza glabra L.	Fabaceae	Stem bark	Aphrodisiac	South Kerala	Ijinu *et al.*, 2011
Gmelina arborea Roxb. ex Sm.	Lamiaceae	Fruit	To cure male infertility	Rangia subdivision, Kamrup district, Assam	Choudhury *et al.*, 2011
Hemidesmus indicus (L.) R. Br	Apocynaceae	Roots	To increase sperm count	North Maharashtra	Ahirrao *et al.*, 2011
		Roots	Spermaturia	Uttara Kannada district, Karnataka	Hegde *et al.*, 2007
Heracleum wallichii DC.	Apiaceae	Roots	Aphrodisiac	Manali wildlife sanctuary, Kullu district, Himachal Pradesh	Rana and Samant, 2011
Hibiscus rosa-sinensis L.	Malvaceae	NS/NM	Aphrodisiac	*Pawara*, *Bhil* and *Pardhi* tribes, Satpuda region, Dhule and Jalgaon districts, Maharashtra	Jain *et al.*, 2010
		Roots	Spermaturia	Uttara Kannada district, Karnataka	Hegde *et al.*, 2007
		Bark	Aphrodisiac	*Pawara*, *Bhil* and *Pardhi* tribes, Satpuda region, Dhule and Jalgaon districts, Maharashtra	Jain *et al.*, 2010
		Bark	Aphrodisiac	Nagzira wild life sanctuary, Gondia district, Maharashtra	Koche *et al.*, 2008

Contd...

Table 14.1–*Contd...*

Plant	*Family*	*Plant Part Used*	*Used as/against*	*Using People/Locality of Use/ Reporting Area*	*Reference*
		NS/NM	Aphrodisiac	Hazaribag district, Jharkhand	Lal and Singh, 2012
		Ovary	Aphrodisiac	Gondia district, Maharashtra	Nimbekar *et al.*, 2011
Hippophae rhamnoides L.	Elaeagnaceae	NS/NM	Aphrodisiac	Jammu and Kashmir	Pant and pant, 2011
		NS/NM	Sexual dysfunction	Upper catchment of Dhauli Ganga, Nanda Devi Biosphere Reserve (NDBR), Central Himalaya	Kandari *et al.*, 2012
		Fruit	Aphrodisiac	Indian Himalayas	Kala, 2005
		Fruits	Aphrodisiac	Uttaranchal	Adhikari *et al.*, 2007
Hygrophila auriculata Heine	Acanthaceae	NS/NM	Aphrodisiac	Dehradun, Uttarakhand	Adhikari *et al.*, 2010
		Seeds	Aphrodisiac, impotency, to increase sperm content	*Kandha* tribe; Orissa	Behera and Misra, 2005
Hygrophila schulli (Buch.-Ham.) M.R.Almeida and S.M. Almeida	Acanthaceae	Seeds	Sexual weakness	Buldhana district, Maharashtra	Ahirrao and Patil, 2010
Hypoxis aurea Lour.	Hypoxidaceae	Roots	Aphrodisiac	Manali wildlife sanctuary, Kullu district, Himachal Pradesh	Rana and Samant, 2011
Ichnocarpus frutescens (L.) W.T. Aiton	Apocynaceae	Roots	Seminal weakness, sexual disability	Ayyakarkoil Forest Province, South-western Ghats, Tamil Nadu	Rajendran *et al.*, 2004
Ipomoea digitata L.	Convolvula-ceae	Tuberous root	To increase sperm count	Rapur forest division, Nellore district, Andhra Pradesh	Neelima *et al.*, 2011
Ipomoea mauritiana Jacq.	Convolvula-ceae	Seeds	Aphrodisiac, to increase sperm content	*Kandha* tribe; Orissa	Behera and Misra, 2005
Ipomoea pes-tigridis L.	Convolvulaceae	Seeds	Premature ejaculation	Southwest Coast of India	Nazar *et al.*, 2008
Jatropha curcasl L.	Euphorbiaceae	Seed	Aphrodisiac	Vattamalai hills, Namakkal district, Tamil Nadu	Kumar and Kumar, 2011

Contd...

Table 14.1–*Contd...*

Plant	*Family*	*Plant Part Used*	*Used as/against*	*Using People/Locality of Use/ Reporting Area*	*Reference*
Juglans regia L.	Juglandaceae	Fruits	Aphrodisiac	*Sangtam Naga* tribe; Kiphire district, Nagaland	Sangtam *et al.*, 2012
Leea macrophylla Roxb. ex Hornem.	Vitaceae	Roots	Male sexual debility	Sitamata wildlife sanctuary, Rajasthan	Jain *et al.*, 2005
		Tuber	Male sexual debility	Aravalli hills, Rajasthan	Swarnkar and Katewa, 2008
Leptadenia reticulata (Retz.) Wight and Arn.	Apocynaceae	Whole plant	Seminal weakness	Saraswati River region, Patan district, North Gujarat	Seliya and Patel, 2009
Leucas aspera (Willd.) Link	Lamiaceae	Leaves	Aphrodisiac, impotency, to increase sperm content	*Kandha* tribe; Orissa	Behera and Misra, 2005
Lilium polyphyllum D. Don	Liliaceae	Bulb	Aphrodisiac	Manali wildlife sanctuary, Kullu district, Himachal Pradesh	Rana and Samant, 2011
Mallotus philippensis (Lam.) Müll. Arg.	Euphorbiaceae	Fruits	Aphrodisiac	Shivbari sacred grove of Himachal Pradesh	Jaryan *et al.*, 2010
Malvastrum coromande-lianum (L.) Garcke	Malvaceae	Seeds	Premature ejaculation	Rewa district, Madhya Pradesh	Shukla *et al.*, 2010b
Melia azedarach L.	Meliaceae	Leaves	To increase sperm production	*Kani* tribe; Tirunelveli hills, Tirunelveli and Kanyakumari districts, Tamil Nadu	Ayyanar and Ignaci-muthu, 2011
Merremia emarginata (Burm. f.) Hallier f.	Convolvulaceae	Seeds	Sexual debility	Rajasthan	Jain *et al.*, 2008
Mimosa pudica L.	Fabaceae	Roots	Aphrodisiac	*Gond* tribe; Bhandara district, Maharashtra	Gupta *et al.*, 2010
		Roots	Aphrodisiac	Villupuram district, Tamil Nadu	Sankaranarayanan *et al.*, 2010

Contd...

Table 14.1–*Contd...*

Plant	*Family*	*Plant Part Used*	*Used as/against*	*Using People/Locality of Use/ Reporting Area*	*Reference*
		Roots	Aphrodisiac	Tuticorin district, Tamil Nadu	Muthukumar and Samuel, 2010
		NS/NM	Aphrodisiac	Madhya Pradesh	Dwivedi *et al.*, 2008
		Roots	Sexual weakness	Kanyakumari district, Tamil Nadu	Jeeva *et al.*, 2006
		Leaves	Aphrodisiac	Kurukshetra district, Haryana	Lal and Yadav, 1983
		Whole plant	Aphrodisiac	Hazaribag district, Jharkhand	Lal and Singh, 2012
		Roots	Enhance sexual vigor	Tinsukia district, Assam	Buragohain, 2011
Mimosa hamata Willd.	Fabaceae	NS/NM	Male sexual debility	Sitamata wildlifesanctuary, Rajasthan	Jain *et al.*, 2005
Mirabilis jalapa L.	Nyctaginaceae	Roots	Aphrodisiac	Mizoram	Rai and Lalramng-hinglova, 2011
Marsilea minuta L.	Marsileaceae	NS/NM	Aphrodisiac	Western Ghats	Benjamin and Manickam, 2007
Momordica charantia L.	Cucurbitaceae	Fruits	Sexual debility	Bargarh district, Orissa	Sahu *et al.*, 2010
		Fruits	Sexual debility	Similipal Bioreserve, Orissa	Behera, 2006
Momordica dioica Roxb. ex Willd.	Cucurbitaceae	NS/NM	Male fertility disorders	*Saperas* community; Khetawas, Jhajjar district, Haryana	Panghal *et al.*, 2010
Moringa oleifera Lam.	Moringaceae		Aphrodisiac	*Siddha* healers, Virudhunagar district, Tamil Nadu	Mutheeswaran *et al.*, 2011
		Flower	To increase sperm production	*Kani* tribe; Tirunelveli hills, Tirunelveli and Kanyakumari districts, Tamil Nadu	Ayyanar and Ignaci-muthu, 2011
		Flower	To increase sperm production	Kancheepuram District of Tamil Nadu	Muthu *et al.*, 2006

Contd...

Table 14.1–*Contd...*

Plant	*Family*	*Plant Part Used*	*Used as/against*	*Using People/Locality of Use/ Reporting Area*	*Reference*
Mucuna pruriens (L.) DC.	Fabaceae	Seeds, pods	Aphrodisiac	*Malayalis*; Kollimalai hill tracts, Tamil Nadu	Anand *et al.*, 2006
		Seeds	Aphrodisiac	*Pawara, Bhil* and *Pardhi* tribes; Satpuda region, Dhule and Jalgaon districts, Maharashtra	Jain *et al.*, 2010
		Seeds	Aphrodisiac	South Kerala	Ijinu *et al.*, 2011
		Seeds	Aphrodisiac	*Siddis*; Uttara Kannada district, Karnataka	Bhandary *et al.*, 1995
		Seeds	Aphrodisiac, male sterility	Satpura plateau region, Central India	Pandey and Shukla, 2008
		NS/NM	Aphrodisiac	Chhatarpur district, Madhya Pradesh	Arjariya and Chaurasia, 2009
		Seeds	Aphrodisiac	Gautala and Pitalkhora of Kannad, Aurangabad district, Maharashtra	Kshirsagar and Magar, 2011
		NS/NM	Aphrodisiac	Nilgiri Hills, Western Ghats, South India	Rajan *et al.*, 2002
		Seeds	Fertility improver	*Malasar* tribe; Coimbatore district, Tamil Nadu	Venkataswamy *et al.*, 2010
Murraya paniculata (L.) Jack	Rutaceae	NS/NM	Sexual stimulant	West and South district of Tripura	Sen *et al.*, 2011
		Roots	Sexual stimulant	Tripura	Majumdar *et al.*, 2006
Myristica fragrans Houtt.	Myristicaceae	Seeds	Aphrodisiac	*Malayalis*, Kollimalai hill tracts, Tamil Nadu	Anand *et al.*, 2006
Nerium indicum Mill.	Apocynaceae	Roots	Aphrodisiac	Chatara block, Sonebhadra district, Uttar Pradesh	Singh *et al.*, 2010

Contd...

Table 14.1–*Contd...*

Plant	*Family*	*Plant Part Used*	*Used as/against*	*Using People/Locality of Use/ Reporting Area*	*Reference*
Nymphaea alba L.	Nymphaea-ceae	Stalks	Aphrodisiac	*Meitei*; Manipur	Yumnam and Tripathi, 2012
Nymphoides indica (L.) Kuntze	Menyantha-ceae	Stalks and Flowers	Aphrodisiac	*Meitei*; Manipur	Yumnam and Tripathi, 2012
Ocimum americanum L.	Lamiaceae	Seeds	Sexual potency	*Meena* community, Rajasthan	Meena and Rao, 2010
Ocimum gratissimum L.	Lamiaceae	Leaves	Aphrodisiac	Mizoram	Rai and Lalramnghin-glova, 2011
Ocimum sanctum L.	Lamiaceae	Seeds	Aphrodisiac	Satpura plateau region, Central India	Pandey and Shukla, 2008
		NM/NS	Male fertility disorders	*Saperas* community; Khetawas, Jhajjar district, Haryana	Panghal *et al.*, 2010
		Seeds	Aphrodisiac	Chhatarpur district, Madhya Pradesh	Arjariya and Chaurasia, 2009
Oenanthe javanica (Blume) DC.	Apiaceae	Aerial parts	Aphrodisiac	*Meitei*; Manipur	Yumnam and Tripathi, 2012
Orchis latifolia L.	Orchidaceae	Tubers, Roots	Aphrodisiac	North Eastern states	Medhi and Chakrabarti, 2009
		Tubers, Roots	Aphrodisiac	Eastern Himalaya, Darjeeling district, west Bengal	Hussain and Hore, 2007
Paederia foetida L.	Rubiaceae	Whole plant	Aphrodisiac	Gorakhpur district, Uttar Pradesh	Singh *et al.*, 1997
Panax ginseng C.A. Mey.	Araliaceae	Roots	Aphrodisiac	*Nagas*; Nagaland	Changkija, 1999
Panax psuodogiseng Wall.	Araliaceae	Rhizome	Aphrodisiac	Senapati district, Manipur	Pfoze *et al.*, 2012
Pandanusf ascicularis Lam.	Pandanaceae	NS/NM	Aphrodisiac	Jogimatti forest, Chitradurga district, Karnataka	Hiremath *et al.*, 2010

Contd...

Table 14.1–*Contd...*

Plant	*Family*	*Plant Part Used*	*Used as/against*	*Using People/Locality of Use/ Reporting Area*	*Reference*
Pedalium murex L.	Pedaliaceae	NS/NM	Aphrodisiac	Sariska and Siliserh regions, Alwar district, Rajasthan	Jain *et al.*, 2009
		Plant	Aphrodisiac	Vattamalai hills, Namakkal district, Tamil Nadu	Kumar and Kumar, 2011
		NS/NM	Male fertility disorders	*Saperas* community; Khetawas, Jhajjar district, Haryana	Panghal *et al.*, 2010
Peganum harmala L.	Nitrariaceae	Whole plant	Aphrodisiac	Thar desert, Rajasthan	Mohammed *et al.*, 2004
		Seeds	Aphrodisiac	Indian trans-Himalaya	Kala, 2006
		Stem	Aphrodisiac	Nubra valley, Ladakh	Phani Kumar *et al.*, 2009
Peucedanum nagpurense (Cl.) Prain	Apiaceae	Whole plant	Promoter of sexual desire	Chhattisgarh	Shukla *et al.*, 2008
Pimpinella tirupatiensis Ball	Apiaceae	Tuber	Sexual debility	Tirumala Hills of Eastern Ghats	Sudhakar *et al.*, 2011
Piper betle L.	Piperaceae	Leaves	Aphrodisiac	Villupuram district, Tamil Nadu	Sankaranarayanan *et al.*, 2010
Piper nigrum L.	Piperaceae	NS/NM	Aphrodisiac	Rewa district, Madhya Pradesh	Shukla *et al.*, 2010b
Phoenix sylvestris (L.) Roxb.	Arecaceae	Roots	Aphrodisiac	Kathua district, Jammu and Kashmir	Khan *et al.*, 2009
		Fruits	Male fertility	*Kani* tribe; Tirunelveli hills, Tirunelveli and Kanyakumari districts, Tamil Nadu	Ayyanar and Ignacimuthu, 2011
Phyla nodiflora (L.) Greene	Verbenaceae	Roots	Enhance female sexual desire	Bargarh district, Orissa	Sahu *et al.*, 2010
Phyllanthus amarus Schumach. and Thonn.	Phyllanthaceae	Whole plant	Aphrodisiac	Northern India	Kala *et al.*, 2006

Contd...

Table 14.1–*Contd...*

Plant	*Family*	*Plant Part Used*	*Used as/against*	*Using People/Locality of Use/ Reporting Area*	*Reference*
Phyllanthus fraternus G.L. Webster	Phyllantha-ceae	Whole plant	Spermaturia	Uttara Kannada district, Karnataka	Hegde *et al.*, 2007
Polygonatum multiflorum (L.) All.	Asparaga-ceae	Roots	Aphrodisiac	Manali wildlife sanctuary, Kullu district, Himachal Pradesh	Rana and Samant, 2011
		Roots	Aphrodisiac	Indian Himalayas	Kala, 2005
Polygonatum verticillatum (L.) All.	Asparaga-ceae	Tuber	Aphrodisiac	Vaidyas, Ukhimath block, Uttarakhand	Semwal *et al.*, 2010
		Tuber	Aphrodisiac	Manali wildlife sanctuary, Kullu district, Himachal Pradesh	Rana and Samant, 2011
		NS/NM	Aphrodisiac	Mornaula Reserve Forest, Kumaun, West Himalaya	Pant and Samant, 2010
		Roots	Aphrodisiac	Indian Himalayas	Kala, 2005
Psoralea corylifolia L.	Fabaceae	NS/NM	Premature ejaculation	Amarkantak region, Madhya Pradesh	Srivastava *et al.*, 2012
		NS/NM	Premature ejaculation	Koraput District, Odisha	Dhal *et al.*, 2011
Pterocarpus marsupium Roxb.	Fabaceae	NS/NM	Aphrodisiac	*Madugga* tribe; Siruvani forest, Coimbatore district, Tamil Nadu	Soudahmini *et al.*, 2005
Pueraria tuberosa (Roxb. ex Willd.) DC.	Fabaceae	Shoot	Aphrodisiac	Amarakantak region, Madhya Pradesh	Kumar *et al.*, 2004
Raphanus sativus L.	Brassicaceae	Seeds	Sexual debility	Bargarh district, Orissa	Sahu *et al.*, 2010
Rhododendron anthopogon D. Don	Ericaceae	NS/NM	Aphrodisiac	Indian Himalayas	Kala, 2005
		NS/NM	Aphrodisiac	Indian trans-Himalaya	Kala, 2006
Rhododendron lepidotum Wall. ex G. Don	Ericaceae	NS/NM	Aphrodisiac	Indian Himalayas	Kala, 2005

Contd...

Table 14.1–*Contd...*

Plant	*Family*	*Plant Part Used*	*Used as/against*	*Using People/Locality of Use/ Reporting Area*	*Reference*
Ricinus communis L.	Euphorbiaceae	NS/NM	Aphrodisiac	*Pawara*, *Bhil* and *Pardhi* tribes, Satpuda region, Dhule and Jalgaon districts, Maharashtra	Jain *et al.*, 2010
Rivea hypocrateriformis Choisy	Convolvula-ceae	Whole plant	Sexual weakness	*Bhilla* tribe; Maharashtra	Kamble *et al.*, 2010
		NS/NM	Sexual weakness	Dehradun, Uttarakhand	Adhikari *et al.*, 2010
Scindapsus officinalis (Roxb.) Schott	Araceae	NS/NM	Aphrodisiac	Dehradun, Uttarakhand	Adhikari *et al.*, 2010
Securinega virosa (Roxb. ex Willd.) Pax and K. Hoffm.	Phyllanthaceae	Roots	Aphrodisiac	Uttarakhand	Adhikari *et al.*, 2007
		Shoots	Aphrodisiac	Kappat hills, Gadag district, Karnataka	Harihar and Kotresha, 2010
Sesamum orientale L.	Pedaliaceae	NS/NM	Aphrodisiac	*Vaidyas*; Alaknanda Catchment Uttarakhand	Phondani *et al.*, 2009
		Seeds	Aphrodisiac	Tehri Garhwal district, Uttarakhand	Kala, 2010
Sida cordata (Burm. f.) Borss. Waalk.	Malvaceae	Roots	Aphrodisiac	Rewa district, Madhya Pradesh	Shukla *et al.*, 2010b
Sida cordifolia L.	Malvaceae	Seeds	Aphrodisiac	Thar desert, Rajasthan	Mohammed *et al.*, 2004
		NS/NM	Aphrodisiac	Sariska and Siliserh regions, Alwar district, Rajasthan	Jain *et al.*, 2009
		Plant	Aphrodisiac	Vattamalai hills, Namakkal district, Tamil Nadu	Kumar and Kumar, 2011
		NS/NM	Sexual weakness	Dehradun, Uttarakhand	Adhikari *et al.*, 2010
Sida rhombifolia L.	Malvaceae	Leaves, Roots	Aphrodisiac	Satpura plateau region, Central India	Pandey and Shukla, 2008

Contd...

Table 14.1–*Contd...*

Plant	*Family*	*Plant Part Used*	*Used as/against*	*Using People/Locality of Use/ Reporting Area*	*Reference*
Smilax macrophylla Willd.	Smilacaceae	Roots	Seminal emission	Satpura plateau region, Central India	Pandey and Shukla, 2008
Smilax zeylanica L.	Smilacaceae	Roots	Aphrodisiac	Adilabad district, Andhra Pradesh	Omkar *et al.*, 2012
		Roots	Aphrodisiac, impotency, to increase sperm content	*Kandha* tribe; Orissa	Behera and Misra, 2005
		Roots	Sexual stimulant	Tripura	Majumdar *et al.*, 2006
Solanum trilobatum L.	Solanaceae	Flower	Male fertility	*Kani* tribe; Tirunelveli hills, Tirunelveli and Kanyakumari districts, Tamil Nadu	Ayyanar and Ignaci-muthu, 2011
Spermacoce hispida L.	Rubiaceae	Seeds	Aphrodisiac	Rudrakod sacred grove, Nallamalais, Andhra Pradesh	Rao and Sunitha, 2011
Sphaeranthus indicus L.	Asteraceae	Plant	Aphrodisiac	Satpura plateau region, Central India	Pandey and Shukla, 2008
		NM	Aphrodisiac	Durgapur, West Bengal	Banerjee *et al.*, 2012
Sterculia villosal Roxb.	Malvaceae	Root	Aphrodisiac, erectile problem	Amarakantak region, Madhya Pradesh	Kumar *et al.*, 2004
Talinum portulacifolium (Forssk.) Asch. ex Schweinf.	Talinaceae	Leaves	To increase sperm count	Rapur forest division, Nellore district, Andhra Pradesh	Neelima *et al.*, 2011
Tamarindus indica L.	Fabaceae	NS/NM	Aphrodisiac	*Pawara*, *Bhil* and *Pardhi* tribes, Satpuda region, Dhule and Jalgaon districts, Maharashtra	Jain *et al.*, 2010
		NS/NM	Sexual weakness	Mizoram	Hazarika *et al.*, 2012
Tamarix dioica Roxb. ex Roth	Tamaricaceae	Bark	Aphrodisiac	Shekhawati region, Rajasthan	Katewa and Galav, 2006
Taxus baccata L.	Taxaceae	Leaves	Aphrodisiac	*Jaintia* tribe; Meghalaya	Jaiswal, 2010
		Leaves	Aphrodisiac	Meghalaya	Dolui *et al.*, 2004

Contd...

Table 14.1–*Contd...*

Plant	*Family*	*Plant Part Used*	*Used as/against*	*Using People/Locality of Use/ Reporting Area*	*Reference*
		Leaves	Aphrodisiac	Kathua district, Jammu and Kashmir	Khan *et al.*, 2009
Terminalia arjuna (Roxb. ex DC.) Wight and Arn.	Combretaceae	Roots	Wet dream	Rangia subdivision, Kamrup district, Assam	Choudhury *et al.*, 2011
Terminalia bellirica (Gaertn.) Roxb.	Combretaceae	Kernel	Aphrodisiac	Mizoram	Rai and Lalramnghinglova, 2010
Terminalia catappa L.	Combretaceae	Seeds	Sperm production	Southwest Coast of India	Nazar *et al.*, 2008
Terminalia chebula Retz.	Combretaceae	Fruits	Stimulate sexual power in men	*Kani* tribe; Tirunelveli hills, Tirunelveli and Kanyakumari districts, Tamil Nadu	Ayyanar and Ignacimuthu, 2011
Thunbergia coccinea Wall.	Acanthaceae	Tubers	Aphrodisiac	*Nagas*; Nagaland	Changkija, 1999
Tinospora cordifolia (Willd.) Miers	Menispermaceae	Stem	Aphrodisiac	Eastern Himalaya, Darjeeling district, west Bengal	Hussain and Hore, 2007
		Stem	Aphrodisiac	*Nyishi* (*Daffla*) tribe; Arunachal Pradesh	Srivastava and *Nyishi* Community, 2010
		NS/NM	Aphrodisiac	Sonebhadra district, Uttar Pradesh	Singh and Dubey, 2012
		Root	Aphrodisiac	Eastern Ghats, Andhra Pradesh	Naidu and Khasim, 2010
		Whole plant	Aphrodisiac	Mornaula Reserve Forest, Kumaun, West Himalaya	Pant and Samant, 2010
		Stem	Aphrodisiac	Mizoram	Rai and Lalramnghinglova, 2010
Tinospora sinensis (Lour.) Merr.	Menispermaceae	Stem	Aphrodisiac	*Nagas*; Nagaland	Changkija, 1999
Trachyspermum ammi Sprague	Apiaceae	Seeds	Aphrodisiac	South Kerala	Ijinu *et al.*, 2011

Contd...

Table 14.1–*Contd...*

Plant	*Family*	*Plant Part Used*	*Used as/against*	*Using People/Locality of Use/ Reporting Area*	*Reference*
Tribulus lanuginosus L.	Zygophyllaceae		To increase fertility and sperm production	*Malayali*; Chitteri hills, Dharmapuri, Tamil Nadu	Kadhirvel *et al.*, 2010
Tribulus terrestris L.	Zygophyllaceae	Fruits	Aphrodisiac	Thar desert, Rajasthan	Mohammed *et al.*, 2004
		Leaf	Aphrodisiac	*Bej*; Sonapur, Kamrup district, Assam	Sharma and Sharma, 2010
		Fruits	Aphrodisiac	Satpura plateau region, Central India	Pandey and Shukla, 2008
		Roots, Fruit	Aphrodisiac	Vattamalai hills, Namakkal district, Tamil Nadu	Kumar and Kumar, 2011
		Fruits, Whole plant	Premature ejaculation	Western Uttar Pradesh	Khan and Khan, 2005
		Fruits, leaves	Impotency, to increase sperm number	Rajasthan	Katewa *et al.*, 2004
		Fruits	Sexual inadequacy	Puducherry	Padmavathy and Anbarashan, 2011
Trigonella foenum-graecum L.	Fabaceae	NS/NM	Aphrodisiac	*Pawara*, *Bhil* and *Pardhi* tribes, Satpuda region, Dhule and Jalgaon districts, Maharashtra	Jain *et al.*, 2010
		Whole plant	Aphrodisiac	Gondia district, Maharashtra	Nimbekar *et al.*, 2011
Urena lobata L.	Malvaceae	Roots	Aphrodisiac	Mizoram	Rai and Lalramnghinglova, 2010
Valeriana jatamansi Jones	Caprifoliaceae	Roots	Aphrodisiac	*Jaunsari* tribe; Garhwal Himalaya, Uttaranchal	Bhatt and Negi, 2006

Contd...

Table 14.1–*Contd...*

Plant	*Family*	*Plant Part Used*	*Used as/against*	*Using People/Locality of Use/ Reporting Area*	*Reference*
Verbascum thapsusl L.	Scrophulariaceae	NM/NS	Aphrodisiac	Shivbari sacred grove of Himachal Pradesh	Jaryan *et al.*, 2010
Vigna mungo (L.) Hepper	Fabaceae	Seed	Aphrodisiac	Vattamalai hills, Namakkal district, Tamil Nadu	Kumar and Kumar, 2011
Vigna unguiculata (L.) Walp.	Fabaceae	Pods	To enhance sexual vigor	Saraswati river region, Patan district, North Gujarat	Seliya and Patel, 2009
Viola betonicifolia Sm.	Violaceae	Whole plant	Aphrodisiac	Nanda Devi Biosphere Reserve, Chamoli district, Uttarakhand	Dangwal *et al.*, 2011
Viola canescens Wall.	Violaceae	Whole plant	Aphrodisiac	Nanda Devi National Park (NDNP) Uttarakhand	Rana *et al.*, 2010
Viscum articulatum Burm. f.	Santalaceae	NS/NM	Aphrodisiac	Kerala	Shanavaskhan *et al.*, 2012
Vitis vitiginea (L.) W.L. Theob.	Vitaceae	Tuber	To increase potency (in men)	Southern Rajasthan	Jain *et al.*, 2004
Withania somnifera (L.) Dunal	Solanaceae	Root, leaves	Aphrodisiac	*Malayalis*, Kollimalai hill tracts, Tamil Nadu	Anand *et al.*, 2006
		Root	Aphrodisiac	Gulbarga district, Karnataka	Ghatapanadi *et al.*, 2011
		Root, leaves	Aphrodisiac	Madugga tribe; Siruvani forest, Coimbatore district, Tamil Nadu	Soudahmini *et al.*, 2005
		Root, leaves	Aphrodisiac	Aurangabad district, Maharashtra, India	Mali and Bhadane, 2011
		Leaves	Aphrodisiac	Niyamgiri hill ranges, Orissa	Dash *et al.*, 2008
		Plant	Aphrodisiac	Gautala and Pitalkhora of Kannad, Aurangabad district, Maharashtra	Kshirsagar and Magar, 2011

Contd...

Table 14.1–*Contd...*

Plant	*Family*	*Plant Part Used*	*Used as/against*	*Using People/Locality of Use/ Reporting Area*	*Reference*
		Leaves, Roots	Aphrodisiac	Sonebhadra district, Uttar Pradesh	Singh and Dubey, 2012
		Roots	Aphrodisiac	Parvati valley; western Himalaya	Sharma *et al.*, 2004
		NS/NM	Aphrodisiac	Chandauli District, Uttar Pradesh	Singh and Singh, 2009
		NS/NM	Sexual weakness	Churu district, Rajasthan	Khan and Singh, 2012
Wrightia tinctoria R. Br.	Apocynaceae	Bark, Seeds	Aphrodisiac	Kalrayan and Shervarayan Hills, Eastern Ghats, Tamil Nadu	Kadavu and Dixit, 2009
Xanthium strumarium L.	Asteraceae	Roots	Aphrodisiac	Gulbarga district, Karnataka	Ghatapanadi *et al.*, 2011
Zanthoxylum limonella Alston	Rutaceae	Fruits	Sexual stimulant	*Tripuri* and *Reang* tribes, Tripura	Das *et al.*, 2009
Zingiber officinale Roscoe	Zingiberaceae	Rhizomes	Aphrodisiac	Coimbatore city, Tamil Nadu	Nisha and Rajesh-kumar, 2010

represented on the basis of the information provided in www.tropicos.org. Among the aphrodisiac representatives, most popular plants were found to be *Asparagus racemosus* (13), *Curculigo orchioides* (13), *Withania somnifera* (10), *Dactylorhiza hatagirea* (9), *Mucuna pruriens* (9), *Mimosa pudica* (8), *Tribulus terrestris* (7), *Bombax ceiba* (6), *Ficus benghalensis* (6), *Hibiscus rosa-sinensis* (6) and *Tinospora cordifolia* (6) where the number in the parenthesis indicates the number of ethnobotanical reports of the plants as aphrodisiacs (The plants with more than 5 reports have been considered only) (Figure 14.1). Among the plant families, Fabaceae (27) was found to be the predominant one followed by Malvaceae (16), Asparagaceae (11), Orchidaceae (10), Acanthaceae and Apiaceae (both 8), Apocynaceae and Convolvulaceae (both 7), Cucurbitaceae (6) and Asteraceae, Euphorbiaceae and Lamiaceae (all 5) where the number in the parenthesis indicates the number of plants reported from each families (Figure 14.2). Although most of the reports describes the herbs as aphrodisiacs, some specific activities such as seminal weakness, enhancement of sexual vigor, promotion of sexual desire, use against sexual debility or weakness, impotency, erectile dysfunction, premature ejaculation, spermatorrhea, spermaturia etc. have also been mentioned. Interestingly, most of the reports involve male sexual enhancement or the treatment of male sexual debility, whereas the reports on the effects of herbal pro-sexuals on women were found to be of little importance. This reflects the male dominating society of rural India where the sexuality of women represents a big taboo in the gender-biased society.

Among the above-mentioned botanicals claimed as pro-sexuals, aphrodisiac and or androgenic activities *in vitro* and *in vivo* experimental models have been described in *Mucuna pruriens* (Ahmad *et al.*, 2008; Suresh *et al.*, 2009; Shukla *et al.*, 2009, 2010a; Gupta *et al.*, 2011; Mahajan *et al.*, 2012); *Chlorophytum borivillianum* (Kenjale *et al.*, 2008; Thakur *et al.*, 2009a, b; Mahajan *et al.*, 2012); *Allium tuberosum* (Guohua *et al.*, 2009); *Crocus sativus* (Hosseinzadeh *et al.*, 2008); *Bombax ceiba* (Bhargava *et al.*, 2011); *Myristica fragrans* (Tajuddin *et al.*, 2003; 2005); *Zingiber officinale* (Kamtchouing *et al.*, 2002a); *Panax ginseng* (Nocerino *et al.*, 2000); *Tribulus terrestris* (Gauthaman *et al.*, 2002, 2003; Gauthaman and Adaikan, 2005; Tyagi *et al.*, 2008; Singh *et al.*, 2012); *Curculigo orchioides* (Chauhan *et al.*, 2007; Thakur *et al.*, 2009a); *Asparagus racemosus* (Thakur *et al.*, 2009a; Wani *et al.*, 2011a); *Mimosa pudica* (Pande and Pathak, 2009); *Tinospora cordifolia* (Wani *et al.*, 2011b); *Abelmoschus manihot* (Rewatkar *et al.*, 2010); *Dactylorhiza hatagirea* (Thakur and Dixit, 2007); *Pedalium murex* (Balamurugan *et al.*, 2010) etc. Spermatogenic effect of *Withania somnifera* has also been reported (Abdel-Magied *et al.*, 2001). Pharmacological efficacy against sexual debility of certain herbs not mentioned in the list such as *Eurycoma longifolia* (Ang *et al.*, 2000, 2001, 2003; Ang and Ngai, 2001); *Diodia scandens* (Onuaguluchi and Nwafor, 1999); *Eulophia campestris* (Mahajan *et al.*, 2012); *Camellia sinensis* (Ratnasooriya and Fernando, 2008); *Mondia whitei* (Lampiao *et al.*, 2008); *Butea frondosa* (Ramachandran *et al.*, 2004); *Piper guineense* (Kamtchouing *et al.*, 2002b); *Orchis malculata* (Ageel *et al.*, 2006); *Argyreia nervosa* (Subramoniam *et al.*, 2007) etc. have also been tested in the laboratory.

In the end, it can be said that the present review is a comprehensive account of Indian plant based aphrodisiacs. Secondly, since many consider the ethnobotanical explorations as the starting point of herbal drug discovery, the present documentation

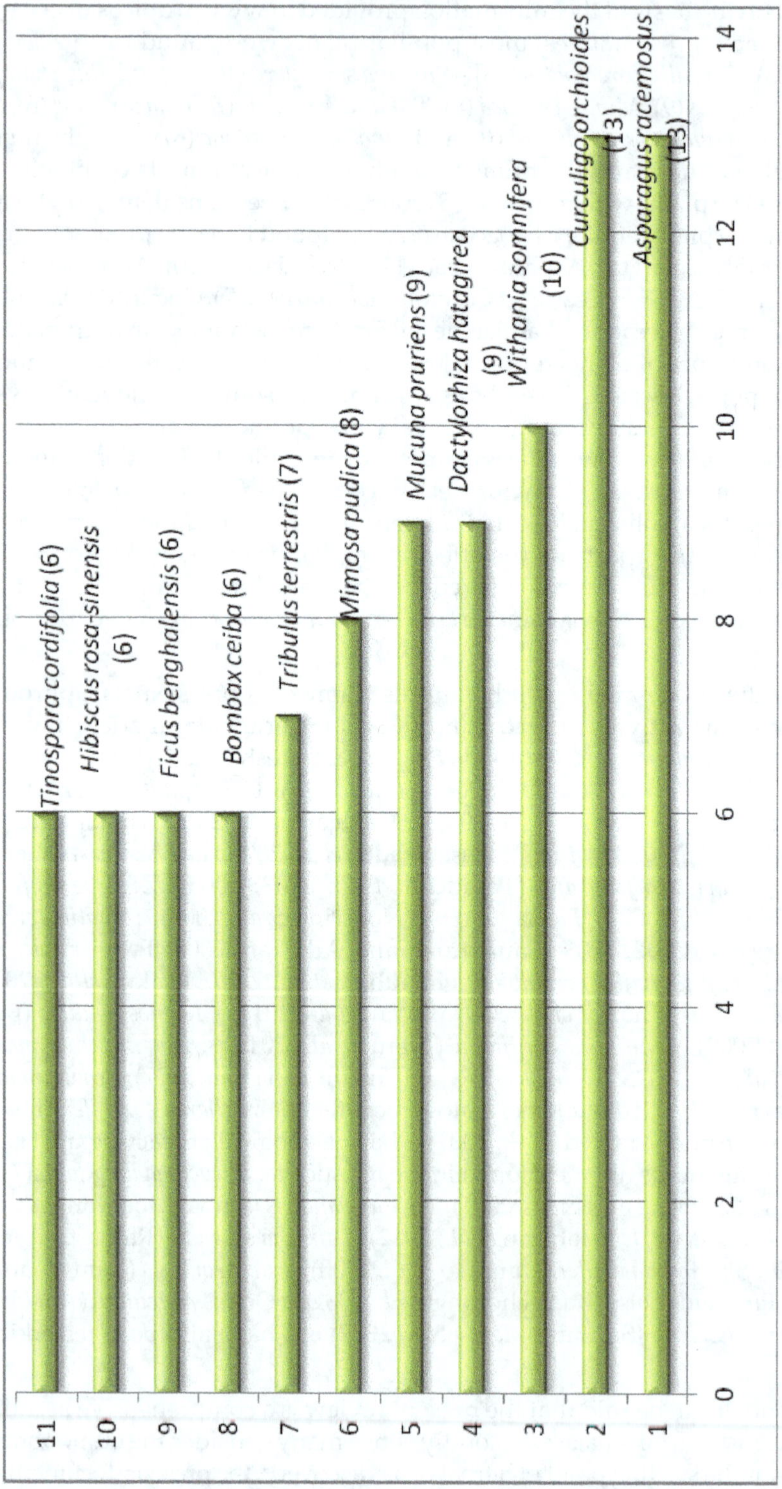

Figure 14.1: Distribution of predominat aphrodisiac plants (digits indicate the number of times reported).

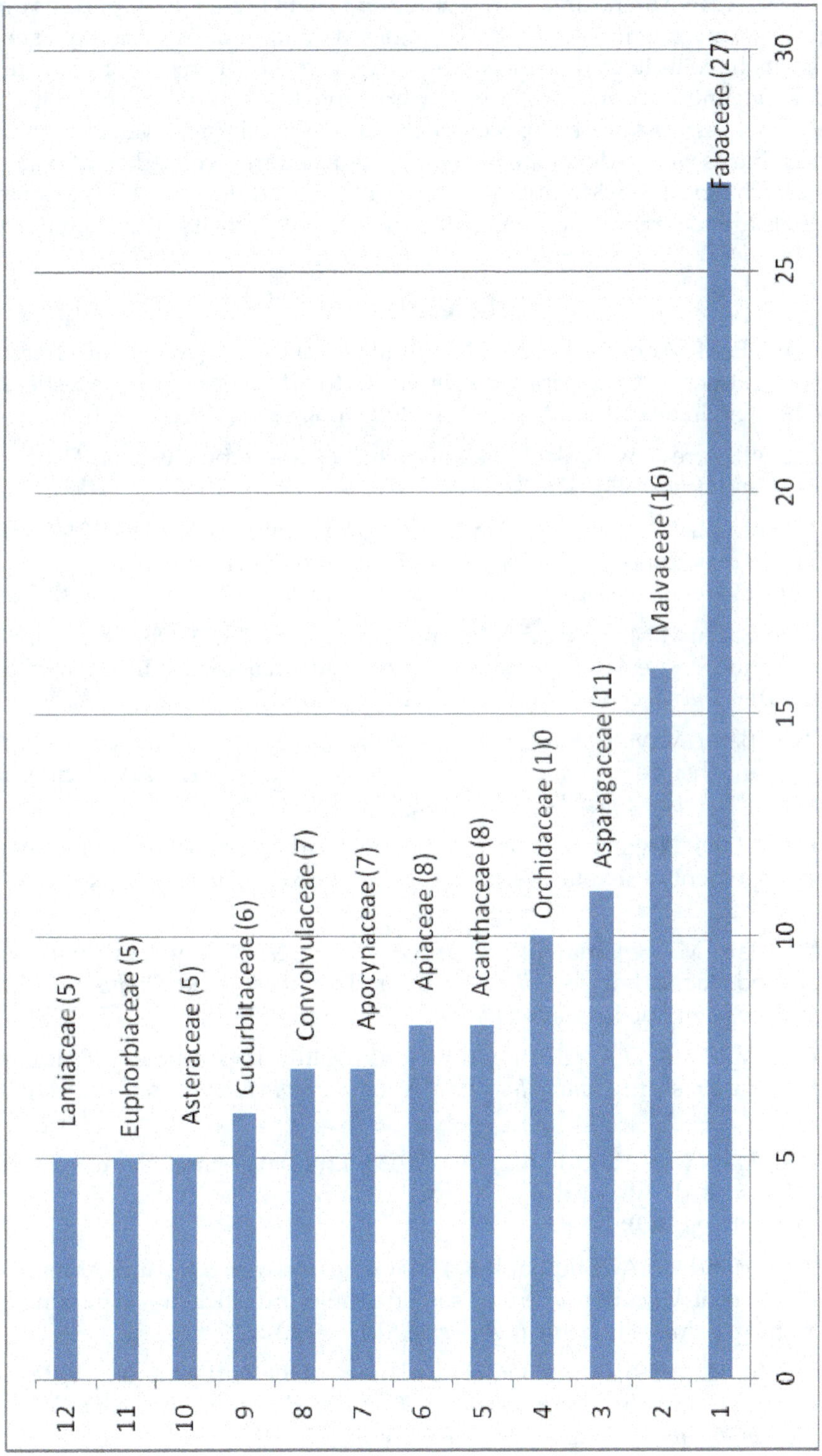

Figure 14.2: Distribution of predominat families containing aphrodisiac plants (number indicates the plant species present within each family).

may serve as a valuable literature. Thirdly, the plants with higher use reports as pro-sexuals were also investigated for pharmacological activity as aphrodisiac and some positive results have indicated encouraging correlation between ethnobotany and lab based experimentation. Fourthly, considering the safety aspects, lack of development of drug resistance and price aspect, herbal remedies are given importance in the developing as well as developed countries as an alternative medicinal system. Lastly, the toxicological considerations of the herbal libido-lifters must be evaluated and the tribals are to be informed regarding the possible toxicity of certain herbal preparations.

References

Abdel-Magied, E. M., Abdel-Rahman, H. A., and Harraz, F. M. (2001). The effect of aqueous extracts of *Cynomorium coccineum* and *Withania somnifera* on testicular development in immature Wistar rats. *J Ethnopharmacol.*, **75(1)**: 1-4.

Abel, E.A. (1980). A review of alcohol's effects on sex and reproduction. *Drug and Alcohol Dependence*, **5(5):** 321–332.

Acharya, V., Sharma, V., Patra P.K., Naik, M.L., and Kanungo, V.K. (2012). Plants used by kamar, gond and halba tribe of Dhamtari district of Chhattisgarh for relief of sickle cell disease. *Recent Research in Science and Technology*, **4(3):** 1-3.

Adhikari, B.S., Babu, M.M., Saklani, P.L., and Rawat, G.S. (2007). Distribution, use pattern and prospects for conservation of medicinal shrubs in Uttaranchal State, India. *Journal of Mountain Science*, **4(2):** 155-180.

Adhikari, B.S., Babu, M.M., Saklani, P.L., and Rawat, G.S. (2010). Medicinal plants diversity and their conservation status in Wildlife Institute of India (WII) campus, Dehradun. *Ethnobotanical Leaflets*, **14**: 46-83.

Adimoelja, A. (2000). Phytochemicals and the breakthrough of traditional herbs in the management of sexual dysfunctions. *International Journal of Andrology*, **23 (Suppl. 2)**: 82–84.

Ageel, A. M., Islam, M. W., Ginawi, O. T., and Al-Yahya, M. A. (1994). Evaluation of the aphrodisiac activity of *Litsea chinensis* (Lauraceae) and *Orchis malculata* (Orchidaceae) extracts in rats. *Phytother. Res.*, **8(2)**: 103–105.

Ahirrao, Y.A., and Patil, D.A. (2010). Indigenous healthcare practices in Buldhana district (Maharashtra). *Indian Journal of Natural Products and Resources*, **1(1):** 85-88.

Ahirrao, Y.A., Patil, M.V., and Patil, D.A. (2011). Ethnomedicinal claims of some botanicals sold by vendors in North Maharashtra (India). *Journal of Ecobiotechnology*, **3(9):** 1-3.

Ahmad, M.K., Mahdi, A.A., Shukla, K.K., Islam, N., Jaiswar, S.P., and Ahmad, S. (2008). Effect of *Mucuna pruriens* on semen profile and biochemical parameters in seminal plasma of infertile men. *Fertil Steril.*, **90(3):** 627-635.

Alagesaboopathi, C. (2009). Ethnomedicinal plants and their utilization by villagers in Kumaragiri hills of Salem district of Tamil Nadu, India. *Afr. J. Traditional, Complementary and Alternative Medicines*, **6(3):** 222–227.

Alagesaboopathi, C. (2011). Ethnobotanical studies on useful plants of Kanjamalai hills of Salem district of Tamil Nadu, Southern India. *Archives of Applied Science Research*, **3(5):** 532-539.

Ali, S. T., and Rakkah, N. I. (2008). Probable neuro sexual mode of action of *Casimiroa edulis* seed extract versus [correction of verses] sildenafil citrate (Viagra(tm)) on mating behavior in normal male rats. *Pak J Pharm Sci.*, **21(1):** 1-6.

Anand, R.M., Nandakumar, N., Karunakaran, L., Ragunathan, M., and Murugan, V. (2006). A survey of medicinal plants in Kollimalai hill tracts, Tamil Nadu. *Natural Product Radiance*, **5(2):** 139-143.

Ang, H.H., and Cheang, H.S. (2001). Effects of *Eurycoma longifolia* jack on laevator ani muscle in both uncastrated and testosterone-stimulated castrated intact male rats. *Arch Pharm Res.*, **24(5):** 437-440.

Ang, H.H., and Ngai, T.H. (2001). Aphrodisiac evaluation in non-copulator male rats after chronic administration of *Eurycoma longifolia* Jack. *Fundam Clin Pharmacol.*, **15(4):** 265-268.

Ang, H.H., and Sim, M.K. (1997). *Eurycoma longifolia* Jack enhances libido in sexually experienced male rats. *Exp Anim.*, **46(4):** 287-290.

Ang, H.H., Cheang, H.S., and Yusof, A.P. (2000). Effects of *Eurycoma longifolia* Jack (Tongkat Ali) on the initiation of sexual performance of inexperienced castrated male rats. *Exp Anim.*, **49(1):** 35-38.

Ang, H.H., Ikeda, S., and Gan, E.K. (2001). Evaluation of the potency activity of aphrodisiac in *Eurycoma longifolia* Jack. *Phytother Res.*, **15(5):** 435-436.

Ang, H.H., Ngai, T.H., and Tan, T.H. (2003). Effects of *Eurycoma longifolia* Jack on sexual qualities in middle aged male rats. *Phytomedicine*, **10(6-7):** 590-593.

Arjariya, A., and Chaurasia, K. (2009). Some medicinal plants among the tribes of Chhatarpur district (M.P.) India. *Ecoprint.*, **16**: 43-50.

Arnett, J.L., Toews, H., and John, A. (1986). Loss of libido due to stress. *Medical Aspects of Human Sexuality*, **20(1):** 140-148.

Ayyanar, M., and Ignacimuthu, S. (2011). Ethnobotanical survey of medicinal plants commonly used by Kani tribals in Tirunelveli hills of Western Ghats, India. *J Ethnopharmacol.*, **134(3):** 851-864.

Bahuguna, Y.M., Sharma, J., and Gairola, S. (2011). Phytodiversity in the submergence area of the Srinagar hydroelectric power project in Garhwal Himalaya, Uttarakhand, India. *International Journal of Environmental Sciences*, **1(7):** 1448-1458.

Balamurugan, G., Muralidharan, P., and Polapala, S. (2010). Aphrodisiac activity and curative effects of *Pedalium murex* (L.) against ethanol-induced infertility in male rats. *Turk J Biol.*, **34**: 153-163.

Banerjee, S., Kar, D., Banerjee, A., and Palit, D. (2012). Utilization of some aquatic macrophytes in Borobandh-a lentic water body in Durgapur, West Bengal,

India: Implications for socio-economic upliftment of local stakeholder. *Indian J. Applied and Pure Bio.*, **27(1):** 83-92.

Bastiaan, J. V., Leonore, B. T., and de Kaste, D. (2008). Designer drugs in herbal aphrodisiacs. *Forensic Science International.*, **177(2–3):** 25–27.

Bedi, S.J. (1978). Ethnobotany of the Ratan Mahal Hills, Gujarat, India. *Economic botany*, **32(3):** 278-284.

Behera, K.K. (2006). Ethnomedicinal plants used by the tribals of Similipal bioreserve, Orissa, India: A pilot study. *Ethnobotanical Leaflets*, **10**: 149-173.

Behera, K.K., Sahoo, S., and Patra, S. (2008). Floristic and medicinal uses of some plants of Chandaka denudated forest patches of Bhubaneswar, Orissa, India. *Ethnobotanical Leaflets*, **12**: 1043-1053.

Behera, S.K., and Misra, M.K. (2005). Indigenous phytotherapy for genito-urinary diseases used by the Kandhatribe of Orissa, India. *Journal of Ethnopharmacology*. **102(3):** 319–325.

Benjamin, A., and Manickam, V.S. (2007). Medicinal pteridophytes from the Western Ghats. *Indian Journal of Traditional Knowledge*, **6(4):** 611-618.

Benniamin, A. (2011). Medicinal ferns of North Eastern India with special reference to Arunachal Pradesh. *Indian Journal of Traditional Knowledge*, **10(3):** 516-522.

Bhandary, M.J., Chandrashekar, K.R., and Kaveriappa, K.M. (1995). Medical ethnobotany of the Siddis of Uttara Kannada district, Karnataka, India. *J Ethnopharmacol.*, **47(3):** 149-158.

Bhargava, C., Thakur, M., and Yadav, S. K. (2011). Effect of *Bombax ceiba* L. on spermatogenesis, sexual behaviour and erectile function in male rats. *Andrologia*, **44(Suppl. 1):** 474-478.

Bhatt, V.P., and Negi, G.C.S. (2006). Ethnomedicinal plant resources of Jaunsari tribe of Garhwal Himalaya, Uttaranchal. *Indian Journal of Traditional Knowledge*, **5(3):** 331-335.

Biradar, S.D., and Ghorband, D.P. (2010). Ethnomedicinal wisdom of tribals of Kinwat forest of Nanded district (Maharashtra). *Indian Journal of Natural Products and Resources*, **1(2):** 254-257.

Bisht, C., and Badoni, A. (2009). Medicinal strength of some alpine and sub-alpine zones of Western Himalaya, India. *New York Science Journal*, **2(5):** 41-46.

Borelli, S. (1960). The effectiveness of sexual stimulants. Article in German. *Arch Klin Exp Dermatol.*, **211**: 147-149.

Buffum, J. (1985). Pharmacosexology update: yohimbine and sexual function. *J Psychoactive Drugs*, **17(2):** 131-132.

Buragohain, J. (2011). Ethnomedicinal plants used by the ethnic communities of Tinsukia district of Assam, India. *Recent Research in Science and Technology*, **3(9):** 31-42.

Carro-Juárez, M., Alcazar, C., Ballesteros-Polvo, E., and Villalobos-Peñalosa, P. (2009). Increase of ejaculatory capacity by systemic administration of the oquichpatli (*Senecio cardiophyllus*) aqueous crude extract in male rats. *J Ethnopharmacol*, **126(3):** 506-511.

Carro-Juárez, M., Cervantes, E., Cervantes-Méndez, M., and Rodríguez-Manzo, G. (2004). Aphrodisiac properties of *Montanoa tomentosa* aqueous crude extract in male rats. *Pharmacol Biochem Behav*., **78(1):** 129-134.

Carro-Juárez, M., Lobatón, I., Benítez, O., and Espíritu, A. (2006). Pro-ejaculatory effect of the aqueous crude extract of cihuapatli (*Montanoa tomentosa*) in spinal male rats. *J Ethnopharmacol*., **106(1):** 111-116.

Changkija, S. (1999). Folk medicinal plants of the Nagas in India. *Asian Folklore Studies*, **58**: 205-230.

Chauhan, N.S., Rao, C. V., and Dixit, V. K. (2007). Effect of *Curculigo orchioides* rhizomes on sexual behaviour of male rats. *Fitoterapia*, **78(7-8):** 530-534.

Choudhury, N., Mahanta, B., and Kalita, J.C. (2010). An ethnobotanical survey on medicinal plants used in reproductive health related disorders in Rangia Subdivision, Kamrup district, Assam. *International Journal of Science and Advanced Technology*, **1(7):** 154-159.

Dalló, J., Yen, T.T., and Knoll, J. (1990). The aphrodisiac effect of (-)deprenyl in male rats. *Acta Physiol Hung*., **75**: 75-76.

Dama, M. S., Akhand, P. S., and Rajender, S. (2010), Nature versus nurture–plant resources in management of male infertility. *Front Biosci (Elite Ed).*, **2**: 1001-1014.

Dangwal, L.R., Rana, C.S., and Sharma, A. (2011). Ethno-medicinal plants from transitional zone of Nanda Devi Biosphere Reserve, district Chamoli, Uttarakhand (India). *Indian Journal of Natural Products and Resources*, **2(1)**: 116-120.

Dangwal, L.R., Sharma, A., Kumar, N., Rana, C.S., and Sharma, U. (2010). Ethno-medico botany of some aquatic angiospermae from North-West Himalaya. *Researcher*, **2(4):** 49-54.

Das, H.B., Majumdar, K., Datta, B.K., and Ray, D. (2009). Ethnobotanical uses of some plants by Tripuri and Reang tribes of Tripura. *Natural Product Radiance*,**8(2):** 172-180.

Dash, P.K., Sahoo, S., and Bal, S. (2008). Ethnobotanical studies on orchids of Niyamgiri hill ranges, Orissa, India. *Ethnobotanical Leaflets*, **12**: 70-78.

Davies, I.B. (1979). Hydrallazine as an aphrodisiac? *Br Med J.*, **2(6201)**: 1367.

Deshmukh, B.S., and Waghmode, A. (2011). Role of wild edible fruits as a food resource: Traditional knowledge. *International Journal of Pharmacy and Life Sciences*, **2(7):** 919-924.

Dey, A., and De, J.N. (2010). Ethnoveterinary uses of medicinal plants by the aboriginals of Purulia district, West Bengal, India. *International Journal of Botany*, **6(4):** 433-440.

Dey, A., and De, J.N. (2011). Traditional use of medicinal plants in pediatric and maternal care practiced by the ethnic groups of Purulia district, West Bengal, India. *International Journal of Medicinal and Aromatic Plants*, **1(3)**: 189-194.

Dey, A., and De, J.N. (2012a). Anti snake venom botanicals used by the ethnic groups of Purulia district, West Bengal, India. *Journal of Herbs, Spices and Medicinal Plants*, **18(2)**: 152-165.

Dey, A., and De, J.N. (2012b). Ethnomedicinal plants used by the tribals of Purulia district, West Bengal, India against gastrointestinal disorders. *Journal of Ethnopharmacology*, **143(1):** 68-80.

Dhal, Y., Sahu, R.K., and Deo, B. (2011). Ethnomedicinal survey of Koraput district, Odisha : An update. *Journal of Pharmacy Research*, **4(11):** 4142-4145.

Dolui, A.K., Sharma, H.K., Marein, T.B., and Lalhriatpuii, T.C. (2004). Folk herbal remedies from Meghalaya. *Indian Journal of Traditional Knowledge*, **3(4)**: 358-364.

Dwivedi, S., Dwivedi, A., and Dwivedi, S.N. (2008). Folk lore uses of some plants by the tribes of Madhya Pradesh with special reference to their conservation. *Ethnobotanical Leaflets*, **12**: 763-771.

Eskiocak, S., Gozen, A. S., Taskiran, A., Kilic, A. S., Eskiocak, M., and Gulen, S. (2006). Effect of psychological stress on the L-arginine-nitric oxide pathway and semen quality. *Braz J Med Biol Res.*, **39(5):** 581-588.

Eskiocak, S., Gozen, A.S., Kilic, A.S., and Molla, S. (2005a). Association between mental stress and some antioxidant enzymes of seminal plasma. *Indian J Med Res.*, **122(6):** 491-496.

Eskiocak, S., Gozen, A.S., Yapar, S.B., Tavas, F., Kilic, A.S., and Eskiocak, M. (2005b). Glutathione and free sulphydryl content of seminal plasma in healthy medical students during and after exam stress. *Hum Reprod.*, **20(9):** 2595-2600.

Fahrner, E.M. (1987). Sexual dysfunction in male alcohol addicts: Prevalence and treatment. *Archives of sexual behavior*, **16(3):** 247-257.

Farnsworth, N. R., Hess, M., Fisher, A. E., Martello, R. E., and Cammarato, L. V. (1967). Medicinal folklore evaluation I. Alleged androgenic and aphrodisiac action of pega palo (*Rhynchosia pyramidalis*). *J. Pharm. Sci.*, **56(8):** 967–970.

Francis, X.T., Freeda, R.A., and Dhivyaa, M. (2011). Ethnomedcinal survey of Malayali tribes in Kolli hills of eastern ghats of Tamil Nadu, India. *Indian Journal of Traditional Knowledge*, **10(3)**: 559-562.

Gaur, R.D., and Bhatt, B.P. (1994). Folk utilization of some Pteridophytes of Deoprayag area in Garhwal Himalaya: India. *Economic Botany*, **48(2)**: 146-151.

Gauthaman, K., and Adaikan, P. G. (2005). Effect of *Tribulus terrestris* on nicotinamide adenine dinucleotide phosphate-diaphorase activity and androgen receptors in rat brain. *J Ethnopharmacol.*, **96(1-2):** 127-132.

Gauthaman, K., Adaikan, P.G., and Prasad, R.N. (2002). Aphrodisiac properties of *Tribulus Terrestris* extract (Protodioscin) in normal and castrated rats. *Life Sci.*, **71(12):** 1385-1396.

Gauthaman, K., Ganesan, A. P., and Prasad, R. N. (2003). Sexual effects of puncturevine (*Tribulus terrestris*) extract (protodioscin): an evaluation using a rat model. *J Altern Complement Med.*, **9(2):** 257-265.

Gessa, G.L., Tagliamonte, A., and Tagliamonte, P. (1971). Aphrodisiac effect of p-chlorophenylalanine. *Science,* **171(3972)**: 706.

Ghatapanadi, S.R., Johnson, N., and Rajasab, A.H. (2011). Documentation of folk knowledge on medicinal plants of Gulbarga district, Karnataka. *Indian Journal of Traditional Knowledge*, **10(2):** 349-353.

Guohua, H., Yanhua, L., Rengang, M., Dongzhi, W., Zhengzhi, M., and Hua, Z. (2009). Aphrodisiac properties of *Allium tuberosum* seeds extract. *J Ethnopharmacol.*, **122(3):** 579-582.

Gupta, A., Mahdi, A.A., Ahmad, M.K., Shukla, K.K., Bansal, N., Jaiswer, S.P., and Shankhwar, S.N. (2011). A proton NMR study of the effect of *Mucuna pruriens* on seminal plasma metabolites of infertile males. *J Pharm Biomed Anal.*, **55(5):** 1060-1066.

Gupta, P. K., Gupta, S., Murugan, P. M., and Singh, S. B. (2009). Ethnobotanical studies of Nubra valley–A cold arid zone of Himalaya. *Ethnobotanical Leaflets,* **13:** 752-765.

Gupta, R., Vairale, M.G., Deshmukh, R.R., Chaudhary, P.R., and Wate, S.R. (2010). Ethnomedicinal uses of some plants used by Gond tribe of Bhandara district, Maharashtra. *Indian Journal of Traditional Knowledge*, **9(4):** 713-717.

Harihar, N.S., and Kotresha, K. (2010). Wild medicinal plants of Kappat hills, Gadag district, Karnataka. *Research and Reviews in Biomedicine and Biotechnology*, **1(2):** 111-118.

Hazarika, T.K., Lalramchuana and Nautiyal, B.P. (2012). Studies on wild edible fruits of Mizoram, India used as ethno-medicine. *Genetic Resources and Crop Evolution,* **59(8):** 1767-1776.

Hegde, H.V., Hegde, G.R., and Kholkute, S.D. (2007). Herbal care for reproductive health: Ethno medicobotany from Uttara Kannada district in Karnataka, India. *Complementary Therapies in Clinical Practice*, **13(1):** 38–45.

Hezmall, H.P., and Lipshultz, L.I. (1982). Cryptorchidism and infertility. *Urol Clin North Am.*, **9(3):** 361-369.

Hiremath, V.T., Vijaykumar, M.M.T., and Taranath, T.C. (2010). Survey on ethno medicinal plants of Jogimatti forest Chitradurga district, Karnataka, India. *Environment and We An International Journal of Science and Technology,* **5**: 223-233.

Hosseinzadeh, H., Ziaee, T., and Sadeghi, A. (2008), The effect of saffron, *Crocus sativus* stigma, extract and its constituents, safranal and crocin on sexual behaviors in normal male rats. *Phytomedicine*, **15(6-7):** 491-495.

Hussain, S., and Hore, D.K. (2007). Collection and conservation of major medicinal plants of Darjeeling and Sikkim Himalayas. *Indian Journal of Traditional Knowledge*, **6(2):** 352-357.

Ignacimuthu, S., Ayyanar, M., and Sivaraman, S.K. (2006). Ethnobotanical investigations among tribes in Madurai district of Tamil Nadu (India). *Journal of Ethnobiology and Ethnomedicine*, **2**: 25.

Ijinu, T.P., Anish, N., Shiju, H., George, V., and Pushpangadan, P. (2011). Home gardens for nutritional and primary health security of rural poor of South Kerala. *Indian Journal of Traditional Knowledge,* **10(3):** 413-428.

Jagtap, S., Deokule, S., Mukherjee, S., Kuvalekar, A., Devkar, S, Harsulkar, A., and Pawar, P. (2010). Assessment of nutritional value of some wild edible plants from Satpura hills of Maharashtra, India. *Journal of Herbal Medicine and Toxicology,* **4(1):** 77-82.

Jagtap, S.D., Deokule, S.S., and Bhosle, S.V. (2006). Ethnopharmacological communication. Some unique ethnomedicinal uses of plants used by the Korku tribe of Amravati district of Maharashtra, India. *Journal of Ethnopharmacology,* **107(3):** 463–469.

Jagtap, S.D., Deokule, S.S., and Bhosle, S.V. (2008a). Ethnomedicobotanical uses of endemic and RET plants utilised by the Korku tribe of Amravati district, Maharashtra. *Indian Journal of Traditional Knowledge*, **7(2):** 284-287.

Jagtap, S.D., Deokule, S.S., and Bhosle, S.V. (2008b). Ethnobotanical uses of endemic and RET plants by Pawra tribe of Nandurbar district, Maharashtra. *Indian Journal of Traditional Knowledge*, **7(2):** 311-315.

Jain, A., Katewa, S.S., Galav, P.S., and Sharma, P. (2005). Medicinal plant diversity of Sitamata wildlife sanctuary, Rajasthan, India. *Journal of Ethnopharmacology*, **102(2):** 143–157.

Jain, A., Katewa, S.S., Chaudhary, B.L., and Galav, P. (2004). Folk herbal medicines used in birth control and sexual diseases by tribals of southern Rajasthan, India. *J Ethnopharmacol*., **90(1):** 171-177.

Jain, A., Katewa, S.S., Galav, P., and Nag, A. (2008). Some therapeutic uses of biodiversity among the tribals of Rajasthan. *Indian Journal of Traditional Knowledge*, **7(2):** 256-262.

Jain, D.L., Baheti, A.M., Jain, S.R., and Khandelwal, K.R. (2010). Use of medicinal plants among tribes in Satpuda region of Dhule and Jalgaon districts of Maharashtra–An ethnobotanical survey. *Indian Journal of Traditional Knowledge,* **9(1):** 152-157.

Jain, S.C., Jain, R., and Singh, R. (2009). Ethnobotanical Survey of Sariska and Siliserh regions from Alwar district of Rajasthan, India. *Ethnobotanical Leaflets*, **13**: 171-188.

Jain, S.P., Srivastava, S., Singh, J., and Singh, S.C. (2011). Traditional phytotherapy of Balaghat district, Madhya Pradesh, India. *Indian Journal of Traditional Knowledge,* **10(2):** 334-338.

Jaiswal, V. (2010). Culture and ethnobotany of Jaintia tribal community of Meghalaya, Northeast India–A mini review. *Indian Journal of Traditional Knowledge*, **9(1):** 38-44.

Jalal, J.S., and Rawat, G.S. (2009). Habitat studies for conservation of medicinal orchids of Uttarakhand, Western Himalaya. *African Journal of Plant Science*, **3(9):** 200-204.

Jalal, J.S., Kumar, P. and Pangtey, Y.P.S. (2008). Ethnomedicinal orchids of Uttarakhand, Western Himalaya. *Ethnobotanical Leaflets*, **12:** 1227-1230.

Jaryan, V., Uniyal, S.K., Singh, G.R.D., Lal, B., Kumar, A., and Sharma, V. (2010). Role of traditional conservation practice: highlighting the importance of Shivbari sacred grove in biodiversity conservation. *The Environmentalist*, **30(2):** 101-110.

Jeeva, S., Kiruba, S., Mishra, B.P., Venugopal, N., Dhas, S.S.M., Regini, G.S., Kingston, C., Kavitha, A., Sukumaran, S., Raj, A.D.S., and Laloo, R.C. (2006). Weeds of Kanyakumari district and their value in rural life. *Indian Journal of Traditional Knowledge*, **5(4):** 501-509.

Kadavu, K., and Dixit, A.K. (2009). Ethnomedicinal studies of the woody species of Kalrayan and Shervarayan hills, Eastern Ghats, Tamil Nadu. *Indian Journal of Traditional Knowledge*, **8(4):** 592-597.

Kadhirvel, K., Ramya, S., Sudha, T.P.S., Ravi, A.V., Rajasekaran, C., Selvi, R.V., and Jayakumararaj, R. (2010). Ethnomedicinal survey on plants used by tribals in Chitteri hills. *Environment and We An International Journal of Science and Technology*, **5**: 35-46.

Kala, C.P. (2005). Indigenous uses, population density, and conservation of threatened medicinal plants in protected areas of the Indian Himalayas. *Conservation Biology*, **19(2):** 368–378.

Kala, C.P. (2006). Medicinal plants of the high altitude cold desert in India: Diversity, distribution and traditional uses. *International Journal of Biodiversity Science and Management*, **2(1):** 43-56.

Kala, C.P. (2009). Aboriginal uses and management of ethnobotanical species in deciduous forests of Chhattisgarh state in India. *Journal of Ethnobiology and Ethnomedicine*, **5**: 20.

Kala, C.P. (2010). Status of an indigenous agro-forestry system in changing climate: A case study of the middle Himalayan region of Tehri Garhwal, India. *Journal of Forest Science*, **56(8):** 373–380.

Kala, C.P., Dhyani, P.P., and Sajwan, B.K. (2006). Developing the medicinal plants sector in northern India: challenges and opportunities. *Journal of Ethnobiology and Ethnomedicine*, **2**: 32.

Kamble, R. B., Mungole, A., Maske, M., Chaturvedi, A., and Chaturvedi, A. (2011). Indigenous traditional knowledge (ITK) from forest dwellers of Gondia district, Maharashtra. *Journal of Ecobiotechnology*, **3(10):** 14-18.

Kamble, S.Y., Patil, S.R., Sawant, P.S., Sawant, S., Pawar, S.G., and Singh, E.A. (2010). Studies on plants used in traditional medicine by Bhilla tribe of Maharashtra. *Indian Journal of Traditional Knowledge*, **9(3):** 591-598.

Kamtchouing, P., Fandio, A. S. M., Dimo, T., and Jatsa, H. B. (2002), Evaluation of androgenic activity of *Zingiber officinale* and *Pentadiplandra brazzeana* in male rats. *Asian J Androl.*, **4(4)**: 299-301.

Kamtchouing, P., Mbongue, G. Y., Dimo, T., Watcho, P., Jatsa, H. B., and Sokeng, S. D. (2002). Effects of *Aframomum melegueta* and *Piper guineense* on sexual behaviour of male rats. *Behav Pharmacol.*, **13(3):** 243-247.

Kandari, L.S., Phondani, P.C., Payal, K.C., Rao, K.S., and Maikhuri, R.S. (2012). Ethnobotanical study towards conservation of medicinal and aromatic plants in upper catchments of Dhauli Ganga in the central Himalaya. *Journal of Mountain Science*, **9(2):** 286-296.

Katewa, S.S., and Galav, P.K. (2006). Additions to the traditional folk herbal medicines from Shekhawati region of Rajasthan. *Indian Journal of Traditional Knowledge*, **5(4):** 494-500.

Katewa, S.S., Chaudhary, B.J., and Jain, A. (2004). Folk herbal medicines from tribal area of Rajasthan, India. *Journal of Ethnopharmacology*, **92(1):** 41–46.

Kenjale, R., Shah, R., and Sathaye, S. (2008). Effects of *Chlorophytum borivilianum* on sexual behaviour and sperm count in male rats. *Phytother Res.*, **22(6):** 796-801.

Kent, S. (1975). On the trail of an authentic aphrodisiac. *Geriatrics*, **30(12):** 96-99.

Kent, S. (1981). Drugs to boost sexual potency. *Geriatrics*, **36(7):** 158-166.

Khan, J.B., and Singh, G.P. (2012). Ethno-medicinal diversity of Churu district traditionally used by rural people. *Indian Journal of Fundamental and Applied Life Sciences*, **2(1):** 109-114.

Khan, M., Kumar, S., and Hamal, I.A. (2009). Medicinal plants of Sewa River Catchment area in the Northwest Himalaya and its implication for conservation. *Ethnobotanical Leaflets*, **13**: 1113-1139.

Khan, V.A., and Khan, A.A. (2005). Herbal folklores for male sexual disorders and debilities in western Uttar Pradesh. *Indian Journal of Traditional Knowledge*, **4(3):** 317-324.

Koche, D.K., Shirsat, R.P., Imran, S., Mohd. N., Zingare, A.K., and Donode, K.A. (2008). Ethnobotanical and ethnomedicinal survey of Nagzira Wild Life Sanctuary, District Gondia (M.S.) India- Part I. *Ethnobotanical Leaflets*, **12:** 56-69, 2008.

Kottaimuthu, R. (2008). Ethnobotany of the Valaiyans of Karandamalai, Dindigul district, Tamil Nadu, India. *Ethnobotanical Leaflets*, **14**: 920-941.

Kshirsagar, A. A., and Magar, B. K. (2011). Morphotaxonomic authentification of ethno-medicinal plants from Gautala and Pitalkhora of Kannad, District Aurangabad Maharashtra. *Asian Journal of Plant Science and Research*, **1(4)**: 17-24.

Kumar, M., Paul, Y., and Anand, V.K. (2009). An ethnobotanical study of medicinal plants used by the locals in Kishtwar, Jammu and Kashmir, India. *Ethnobotanical Leaflets*, **13**: 1240-1256.

Kumar, R., Suman, N.R., and Dash, S.S. (2004). Traditional uses of plants by tribals of Amarakantak region, Madhya Pradesh. *Indian Journal of Traditional Knowledge*, **3(4):** 383-390.

Kumari, P., Joshi, G.C., and Tewari, L.M. (2011). Diversity and status of ethno-medicinal plants of Almora district in Uttarakhand, India. *International Journal of Biodiversity and Conservation*, **3(7):** 298-326.

Lal, H.S., and Singh, S. (2012). Study of plant biodiversity of Hazaribag district Jharkhand, India and its medicinal uses. *Bioscience Discovery*, **3(1):** 91-96.

Lal, S.D., and Yadav, B.K. (1983). Folk medicines of Kurukshetra district (Haryana), India. *Economic botany*, **37(3)**: 299-305.

Lalfakzuala, R., Lalramnghinglova, H., and Kayang, H. (2007). Ethnobotanical usages of plants in western Mizoram. *Indian Journal of Traditional Knowledge*, **6(3)**: 486-493.

Lampiao, F., Krom, D., and Plessis, S. S. D. (2008). The *in vitro* effects of *Mondia whitei* on human sperm motility parameters. *Phytother. Res.*, **22(9)**: 1272–1273.

Lipshultz, L.I. (1976). Cryptorchidism in the subfertile male. *Fertil Steril.*, **27(6):** 609-620.

Magrini, G. (1981). [Anything new on aphrodisiacs? (author's transl). *Ther Umsch.*, **38(11):** 1082-1088.

Mahajan, G.K., Mahajan, A.Y., and Mahajan, R.T. (2012). Efficacy of aphrodisiac plants towards improvement in semen quality and motility in infertile males. *J Complement Integr Med.*, **9(1):** 1553-3840.

Majumdar, K., and Datta, B.K. (2007). A study on ethnomedicinal usage of plants among the folklore herbalists and Tripuri medical practitioners: Part-II. *Natural Product Radiance*, **6(1):** 66-73.

Majumdar, K., Saha, R., Datta, B.K., and Bhakta, T. (2006). Medicinal plants prescribed by different tribal and non-tribal medicine men of Tripura state. *Indian Journal of Traditional Knowledge*, **5(4):** 559-562.

Mali, P.Y., and Bhadane V.V. (2011). Ethno-medicinal wisdom of tribals of Aurangabad district (M.S.), India. *Indian Journal of Natural Products and Resources*, **2(1)**: 102-109.

Malviya, N., Jain, S., Gupta, V.B., and Vyas, S. (2011a). Recent studies on aphrodisiac herbs for the management of male sexual dysfunction—a review. *Acta Pol Pharm.*, **68(1):** 3-8.

Malviya, N., Jain, S., Gupta, V.B., and Vyas, S. (2011b). Indigenous herbal remedies used by tribals of Madhya Pradesh for improving their sexual performance and problem associated with sexuality. *International Journal of Research in Ayurveda and Pharmacy*, **2(2):** 399-402.

Medhi, R.P., and Chakrabarti, S. (2009). Traditional knowledge of NE people on conservation of wild orchids. *Indian Journal of Traditional Knowledge*, **8(1):** 11-16.

Meena, A.K., and Rao, M.M. (2010). Folk herbal medicines used by the Meena community in Rajasthan. *Asian Journal of Traditional Medicines,* **5(1):** 19-31.

Meena, K.L., and Yadav, B.L. (2010). Studies on ethnomedicinal plants conserved by Garasia tribes of Sirohi district, Rajasthan, India. *Indian Journal of Natural Products and Resources,* **1(4):** 500-506.

Mendes, F. R. (2011), Tonic, fortifier and aphrodisiac: adaptogens in the Brazilian folk medicine. *Rev. bras. farmacogn.,* **21(4):** 754-763.

Mieusset, R., and Soulié, M. (2005). Hypospadias: Psychosocial, sexual, and reproductive consequences in adult life. *Journal of Andrology,* **26(2):** 163-168.

Mishra, S.B., Dwivedi, S., Shashi, A., and Prajapati, K. (2008). Ethnomedicinal uses of some plant species by ethnic and rural peoples of the Salem district of Tamil Nadu with special reference to the conservation of vanishing species. *Ethnobotanical Leaflets,* **12:** 873-887.

Mohammed, S., Kasera, P.K., and Shukla, J.K. (2004). Unexploited plants of potential medicinal value from the Indian Thar desert. *Natural Product Radiance,* **3(2):** 69-74.

Mugisha, M. K., and Origa, O. M. (2005). Traditional herbal remedies used in the management of sexual impotence and erectile dysfunction in western Uganda. *African Health Sciences,* **5(1):** 40-49.

Mutheeswaran, S., Pandikumar, P., Chellappandian, M., and Ignacimuthu, S. (2011). Documentation and quantitative analysis of the local knowledge on medicinal plants among traditional Siddha healers in Virudhunagar district of Tamil Nadu, India. *J Ethnopharmacol.,* **137(1):** 523-533.

Muthu, C., Ayyanar, M., Raja, N., and Ignacimuthu, S. (2006). Medicinal plants used by traditional healers in Kancheepuram District of Tamil Nadu, India. *Journal of Ethnobiology and Ethnomedicine,* **2:** 43.

Muthukumar, K., and Samuel, A.S. (2010). Traditional herbal medicines of the coastal diversity in Tuticorin District, Tamil Nadu, India. *Journal of Phytology,* **2(8):** 38–46.

Naidu, K.A., and Khasim, S.M. (2010). Contribution to the floristic diversity and ethnobotany of Eastern Ghats in Andhra Pradesh, India. *Ethnobotanical Leaflets,* **14:** 920- 941.

Nair, R., Sellaturay, S., and Sriprasad, S. (2012). The history of ginseng in the management of erectile dysfunction in ancient China (3500-2600 BCE). *Indian J Urol.,* **28(1):** 15-20.

Nakanishi, S., Yamane, K., Kamei, N., Okubo, M., and Kohno, N. (2004). Erectile dysfunction is strongly linked with decreased libido in diabetic men. **7(2):** 113-119.

Nazar, S., Ravikumar, S., and Williams, G.P. (2008). Ethnopharmacological Survey of Medicinal Plants along the Southwest Coast of India. *Journal of Herbs, Spices and Medicinal Plants.* **14(3-4):** 219-239.

Neelima, M., Prasad, G.P., Sudarsanam, G., Penchala Pratap, G., and Jyothi, B. (2011). Ethnobotanical studies in Rapur forest division of Nellore district in Andhra Pradesh. *Life Sciences Leaflets*, **11**: 333-345.

Nimbekar, T. P., Katolkar, P. P., Wanjari, B. E., Nema, M. V., and Patil, A. T. (2011). Traditional knowledge of medicinal plants of gondia district (in maharashtra state): an ethno-botanical survey. *Journal of Herbal Medicine and Toxicology*, **5(2):** 9-17.

Nisha, M.C., and Rajeshkumar, S. (2010). Survey of crude drugs from Coimbatore city. *Indian Journal of Natural Products and Resources*, **1(3):** 376-383.

Nocerino, E., Amato, M., and Angelo, A. I. (2000). The aphrodisiac and adaptogenic properties of ginseng. *Fitoterapia*, **71(1):** 1–5.

Omkar, K., Suthari, S., Alluri, S., Ragan, A., and Vatsavaya, S.R. (2012). Diversity of NTFPs and their utilization in Adilabad district of Andhra Pradesh, India. *Journal of Plant Studies*, **1(1):** 33-46.

Onuaguluchi, G., and Nwafor, P. (1999). Pharmacological basis for the use of the antivenene water soluble extract of *Diodia scandens* as a laxative, oxytocic agent and a possible aphrodisiac in traditional medicine practice in eastern Nigeria. *Phytother Res.*, **13(6):** 459-463.

Padmavathy, A., and Anbarashan, M. (2011). Phytomedicinal study of coastal sand dune floras in Puducherry. *Journal of Medicinal Plants Research*, **5(12)**: 2566-2571.

Pande, M., and Pathak, A. (2009). Aphrodisiac activity of roots of *Mimosa pudica* Linn. ethanolic extract in mice. *International Journal of Sciences and Nanotechnology Pharmaceutical*, **2(1)**: 477-486.

Pandey, A. K., and Shukla, P. K. (2008). The role of medicinal plants in the health care and rural economy in the tribals of Satpura plateau region of central India. *Indian Forester*, **134(11):** 1438-1446.

Panghal, M., Arya, V., Yadav, S., Kumar, S., and Yadav, J.P. (2010). Indigenous knowledge of medicinal plants used by Saperas community of Khetawas, Jhajjar district, Haryana, India. *Journal of Ethnobiology and Ethnomedicine*, **6**: 4.

Pant, S., and Pant, V.S. (2011). Status and conservation management strategies for threatened plants of Jammu and Kashmir. *Journal of Phytology*, **3(7):** 50-56.

Pant, S., and Samant, S.S. (2010). Ethnobotanical observations in the Mornaula Reserve Forest of Komoun, West Himalaya, India. *Ethnobotanical Leaflets*, **14**: 193-217.

Parihar, P., and Parihar, L. (2006). Some pteridophytes of medicinal importance from Rajasthan. *Natural Product Radiance*, **5(4):** 297-301.

Pasqualotto, F.F., Lucon, A.M., Sobreiro, B.P., Pasqualotto, E.B., and Arap, S. (2004). Effects of medical therapy, alcohol, smoking, and endocrine disruptors on male infertility. *Rev. Hosp. Clin.*, **59(6):** 375-382.

Pasqualotto, F.F., Pasqualotto, E.B., Agarwal, A., and Thomas, A.J. Jr. (2003). Detection of testicular cancer in men presenting with infertility. *Rev. Hosp. Clin.*, **58(2):** 75-80.

Patel, D.K., Kumar, R., Prasad, S.K., and Hemalatha, S. (2011). Pharmacologically screened aphrodisiac plant-A review of current scientific literature. *Asian Pacific Journal of Tropical Biomedicine*, **1(1):** 131-138.

Patil, H.M. (2012). Ethnobotanical notes on Satpura hills of Nandurbar district, Maharashtra, India. *Research Journal of Recent Sciences*, **1**: 326-328.

Patil, S.B., Nilofar, S., Naikwade, Chandrakant, S.M., and Awale, V.B. (2011). Some medicinal plants used by people of Sangli district, Maharashtra. *Asian J. Pharm. Res.*, **1(2):** 42-43.

Pattanaik, C., Reddy, C.S., and Dhal, N.K. (2008a). Phytomedicinal study of coastal sand dune species of Orissa. *Indian Journal of Traditional Knowledge*, **7(2):** 263-268.

Pattanaik, C., Reddy, C.S., Dhal, N.K., and Das, R. (2008b). Utilisation of mangrove forests in Bhitarkanika wildlife sanctuary, Orissa. *Indian Journal of Traditional Knowledge*, **7(4):** 598-603.

Pattanaik, C., Reddy, S., and Reddy, K.N. (2009). Ethno-medicinal survey of threatened plants in Eastern Ghats, India. *Our Nature*, **7**: 122-128.

Pfoze, N.L., Kumar, Y., and Myrboh, B. (2012). Survey and assessment of ethnomedicinal plants used in Senapati district of Manipur State, Northeast India. *Phytopharmacology*, **2(2):** 285-311.

Phondani, P.C. (2011). Worth of traditional herbal system of medicine for curing ailments prevalent across the mountain region of Uttarakhand, India. *Journal of Applied Pharmaceutical Science*, **1(9):** 81-86.

Phondani, P.C., Maikhuri, R.K., and Bisht, N.S. (2009). Medicinal plants used in the health care system practiced by traditional vaidyas in Alaknanda catchment of Uttarakhand, India. *Ethnobotanical Leaflets*, **13:** 1453-1467.

Poulsen, P.A. (1983). Alkyl nitrite as an aphrodisiac. Article in Danish. *Ugeskr Laeger*, **145(29):** 2213-2215.

Pratap, S.A., and Rajender, S. (2012). Potent natural aphrodisiacs for the management of erectile dysfunction and male sexual debilities. *Front Biosci (Schol Ed).*, **4**: 167-180.

Puri, H. S. (1971). Vegetable Aphrodisiacs of India. **11(2):** 1742-1753.

Qureshi, S., Shah, A. H., Tariq, M., and Ageel, A. M. (1989). Studies on herbal aphrodisiacs used in Arab system of medicine. *Am J Chin Med.*, **17(1-2):** 57-63.

Rai, P.K., and Lalramnghinglova, H. (2010). Ethnomedicinal plant resources of Mizoram, India: Implication of traditional knowledge in health care system. *Ethnobotanical Leaflets*, **14**: 274-305.

Rai, P.K., and Lalramnghinglova, H. (2011). Ethnomedicinal plants of India with special reference to an Indo-Burma hotspot region: An overview. *Ethnobotany Research and Applications*, **9**: 379-420.

Rajan, S., Sethuraman, M., and Mukherjee, P. K. (2002). Ethnobiology of the Nilgiri hills, India. *Phytother. Res.*, **16(2)**: 98–116.

Rajendran, S.M., Agarwal, S.C., and Sundaresan, V. (2004). Lesser known ethnomedicinal plants of the Ayyakarkoil forest province of Southwestern Ghats, Tamil Nadu, India–Part I. *Journal of Herbs, Spices and Medicinal Plants,* **10(4):** 103-112.

Ramachandran, S., Sridhar, Y., Sam, S. K., Saravanan, M., Leonard, J. T., Anbalagan, N., and Sridhar, S. K. (2004). Aphrodisiac activity of *Butea frondosa* Koen. ex Roxb. extract in male rats. *Phytomedicine,* **11(2-3):** 165-168.

Rana, M.S., and Samant, S.S. (2011). Diversity, indigenous uses and conservation status of medicinal plants in Manali wildlife sanctuary, North western Himalaya. *Indian Journal of Traditional Knowledge,* **10(3):** 439-459.

Rana, T.S., and Datt, B. (1997). Ethnobotanical observation among Jaunsaris of Jaunsar-Bawar, Dehradun(U.P.), India. **35(5):** 371-374.

Ranaa, C. S., Sharma, A., Kumar, N., Dangwal, L. R., and Tiwari, J. K. (2010), Ethnopharmacology of some important medicinal plants of Nanda Devi National Park (NDNP) Uttarakhand, India. *Nature and Science,* **8(11):** 82-85.

Rao, B.R.P., and Sunitha, S. (2011). Medicinal plant resources of Rudrakod sacred grove in Nallamalais, Andhra Pradesh, India. *J Biodiversity,* **2(2):** 75-89.

Ratnasooriya, W.D., and Dharmasiri, M.G. (2000). Effects of *Terminalia catappa* seeds on sexual behaviour and fertility of male rats. *Asian J Androl.,* **2(3):** 213-219.

Ratnasooriya, W.D., and Fernando, T.S. (2008). Effect of black tea brew of *Camellia sinensis* on sexual competence of male rats. *J Ethnopharmacol.,* **118(3):** 373-377.

Ratnasooriya, W.D., and Jayakody, J.R. (2006). Effects of aqueous extract of *Alpinia calcarata* rhizomes on reproductive competence of male rats. *Acta Biol Hung.,* **57(1):** 23-35.

Rawat, V.S., and Jalal, J.S. (2011). Sustainable utilization of medicinal plants by local community of Uttarkashi district of Garhwal, Himalaya, India. *European Journal of Medicinal Plants,* **1(2):** 18-25.

Reddy, K.N., Reddy, C.S., and Raju, V.S. (2008). Ethnomedicinal observations among the Kondareddis of Khammam district, Andhra Pradesh, India. *Ethnobotanical Leaflets,* **12:** 916-926.

Reddy, K.N., Trimurthulu, G., and Reddy, C.S. (2010). Medicinal plants used by ethnic people of Medak district, Andhra Pradesh. *Indian Journal of Traditional Knowledge,* **9(1):** 184-190.

Reddy, M.B., Reddy, K.R., and Reddy, M.N. (1988). *A Survey of Medicinal Plants of Chenchu Tribes of Andhra Pradesh, India.* **26(4):** 189-196.

Revathi, P., and Parimelazhagan, T. (2008). Traditional knowledge on medicinal plants used by the Irula tribe of Hasanur hills, Erode district, Tamil Nadu, India. *Ethnobotanical Leaflets,* **12:** 195-203.

Rewatkar, K. K., Shahzad, N., Ahmed, A., Khan, Md. I., and Ganesh, N. (2010). A landmark approach to aphrodisiac property of *Abelmoschus manihot* (L.). *International Journal of Phytomedicine,* **2:** 312-319.

Rubin, A., and Babbott, D. (1958). Impotence and diabetes mellitus free. *Jama.*, **168(5):** 498-500.

Sahu, A.R., Behera, N., and Mishra, S.P. (2010). Use of ethnomedicinal plants by natives of Bargarh district of Orissa, India. *Ethnobotanical Leaflets*, **14**: 136-160.

Sangtam, T.L., Jamir, N.S., Deb, C.R., and Jamir, S. (2012). A Study on the medicinal plants used by the Sangtam Naga tribe in Kiphire district, Nagaland, India. *International Journal of Ayurvedic and Herbal Medicine*, **2(2):** 267-275.

Sankaranarayanan, S., Bama, P., Ramachandran, J., Kalaichelvan, P.T., Deccaraman, M., Vijayalakshimi, M., Dhamotharan, R., Dananjeyan, B., and Bama, S.S. (2010). Ethnobotanical study of medicinal plants used by traditional users in Villupuram district of Tamil Nadu, India. *Journal of Medicinal Plants Research*, **4(12):** 1089-1101.

Seliya, A.R., and Patel, N.K. (2009). Ethnomedicinal uses of climbers from Saraswati river region of Patan district, North Gujarat. *Ethnobotanical Leaflets*, **3**: 865-872.

Semwal, D.P., Saradhi, P.P., Kala, C.P., and Sajwan, B.S. (2010). Medicinal plants used by local Vaidyas in Ukhimath block, Uttarakhand. *Indian Journal of Traditional Knowledge*, **9(3):** 480-485.

Sen, S., Chakraborty, R., De, B., and Devanna, N. (2011). An ethnobotanical survey of medicinal plants used by ethnic people in West and South district of Tripura, India. *Journal of Forestry Research*, **22(3):** 417-426.

Shah, N.C., and Joshi, M.C. (1971). An ethnobotanical study of the Kumaon region of India. *Economic botany*, **25(4):** 414-422.

Shamloul, R. (2010). Natural aphrodisiacs. *J Sex Med.*, **7(1):** 39-49.

Shanavaskhan, A.E., Sivadasan, M., Alfarhan, A.H., and Thomas, J. (2012). Ethnomedicinal aspects of angiospermic epiphytes and parasites of Kerala, India. *Indian Journal of Traditional Knowledge*, **11(2):** 250-258

Sharma, P.K., Chauhan, N.S., and Lal, B. (2004). Observations on the traditional phytotherapy among the inhabitants of Parvati valley in western Himalaya, India. *Journal of Ethnopharmacology*, **92(2–3):** 167–176.

Sharma, R., and Sharma, H.K. (2010). Ethnomedicines of Sonapur, Kamrup district, Assam. *Indian Journal of Traditional Knowledge*, **9(1):** 163-165.

Sharma, U.K., and Pegu, S. (2011). Ethnobotany of religious and supernatural beliefs of the Mising tribes of Assam with special reference to the 'Dobur Uie'. *Journal of Ethnobiology and Ethnomedicine*, **7**: 16

Shukla, A.N., Srivastava, S., and Rawat, A.K.S. (2010b). An ethnobotanical study of medicinal plants of Rewa district, Madhya Pradesh. *Indian Journal of Traditional Knowledge*, **9(1)**: 191-202.

Shukla, K.K., Mahdi, A.A., Ahmad, M.K., Jaiswar, S.P., Shankwar, S.N., and Tiwari, S.C. (2010a). *Mucuna pruriens* reduces stress and improves the quality of semen in infertile men. *Evid Based Complement Alternat Med.*, **7(1):** 137-144.

Shukla, K.K., Mahdi, A.A., Ahmad, M.K., Shankhwar, S.N., Rajender, S., and Jaiswar, S.P. (2009). *Mucuna pruriens* improves male fertility by its action on the hypothalamus-pituitary-gonadal axis. *Fertil Steril.*, **92(6):** 1934-1940.

Shukla, R., Chakravarty, M., and Gautam, M.P. (2008). Indigenous medicine used for treatment of gynecological disorders by tribal of Chhattisgarh, India. *Journal of Medicinal Plants Research*, **2(12):** 356-360.

Sikarwar, R.L.S., Pathak, B., and Jaiswal, A. (2008). Some unique ethnomedicinal perceptions of tribal communities of Chitrakoot, Madhya Pradesh. *Indian Journal of Traditional Knowledge*, **7(4)**: 613-617.

Singh, A., and Dubey, N.K. (2012). An ethnobotanical study of medicinal plants in Sonebhadra district of Uttar Pradesh, India with reference to their infection by foliar fungi. *Journal of Medicinal Plants Research*, **6(14):** 2727-2746.

Singh, A., and Singh, P.K. (2009). An ethnobotanical study of medicinal plants in Chandauli district of Uttar Pradesh, India. *Journal of Ethnopharmacology*, **121(2):** 324–329.

Singh, A.K., Raghubanshi, A.S., and Singh, J.S. (2002). Medical ethnobotany of the tribals of Sonaghati of Sonbhadra district, Uttar Pradesh, India. *Journal of Ethnopharmacology*, **81(1)**: 31–41.

Singh, P.K., Kumar, V., Tiwari, R.K., Sharma, A., Rao, C.V., and Singh, R.H. (2010). Medico-ethnobotany of 'Chatara' block of district Sonebhadra, Uttar Pradesh, India. *Advances in Biological Research*, **4 (1):** 65-80.

Singh, S., Nair, V., and Gupta, Y.K. (2012). Evaluation of the aphrodisiac activity of *Tribulus terrestris* Linn. in sexually sluggish male albino rats. *J Pharmacol Pharmacother.*, **3(1):** 43-47.

Singh, V.K., Ali, Z.A., and Siddiqui, M.K. (1997). Medicinal plants used by the forest ethnics of Gorakhpur district (Uttar Pradesh), India. **35(3):** 194-206.

Soudahmini, E., Senthil, G.M., Panayappan, L., and Divakar, M.C. (2005). Herbal remedies of Madugga tribes of Siruvani forest, South India. *Natural Product Radiance*, **4(6):** 492-501.

Srivastava, A., Patel, S.P., Mishra, R.K., Vashistha, R.K., Singh, A., and Puskar, A.K. (2006). Ethnomedicinal importance of the plants of Amarkantak region, Madhya Pradesh, India. *International Journal of Medicinal and Aromatic Plants*, **2(1):** 53-59.

Srivastava, R.C. and Adi community. (2009). Traditional knowledge of Adi tribe of Arunachal Pradesh on plants. *Indian Journal of Traditional Knowledge*, **8(2):** 146-153.

Srivastava, R.C., and Nyishi Community. (2009). Traditional knowledge of Nyishi (Daffla) tribe of Arunachal Pradesh. *Indian Journal of Traditional Knowledge*, **9(1):** 26-37.

Sta-mrad, A. (1955). Aphrodisiac effects of high doses of vitamine B12. Article in French. *Tunis Med.*, **43(2):** 147-148.

Subramoniam, A., Madhavachandran, V., Ravi, K., and Anuja, V. S. (2007). Aphrodisiac property of the elephant creeper *Argyreia nervosa*. *J Endocrinol Reprod.*, **11(2):** 82–85.

Sudhakar Reddy, C., Reddy, K.N., Murthy, E.N., and Raju, V.S. (2009). Traditional medicinal plants in Seshachalam hills, Andhra Pradesh, India. *Journal of Medicinal Plants Research*, **3(5)**: 408-412.

Sudhakar Reddy, S., Reddy, K.N., Thulsi Rao, K., and Pattanaik, C. (2007). Ethnobotanical studies on medicinal plants used by the Chenchus of Nallamalais in Kurnool district, Andhra Pradesh, India. *Research Journal of Medicinal Plant*, **1(4)**: 128-133.

Sudhakar, A., Ramesh, C., Nagaraju, N., and Murthy, N. S. R. (2011). Pharmacognostical studies on the root tuber of *Pimpinella tirupatiensis* Bal. and Subr.–An endemic to Tirumala Hills of Eastern Ghats, India. *International Journal of Pharmacognosy and Phytochemical Research*, **3(3):** 56-60.

Suresh, S., Prithiviraj, E., and Prakash, S. (2009). Dose- and time-dependent effects of ethanolic extract of *Mucuna pruriens* Linn. seed on sexual behaviour of normal male rats. *J Ethnopharmacol.*, **122(3):** 497-501.

Swarnkar, S., and Katewa, S.S. (2008). Ethnobotanical observation on tuberous plants from tribal area of Rajasthan (India). *Ethnobotanical Leaflets*, **12**: 647-666.

Tajuddin, A. S., Latif, A., and Qasmi, I. A. (2003). Aphrodisiac activity of 50 per cent ethanolic extracts of *Myristica fragrans* Houtt. (nutmeg) and *Syzygium aromaticum* (L) Merr. and Perry. (clove) in male mice: a comparative study. *BMC Complement Altern Med.*, **3**: 6.

Tajuddin, A. S., Latif, A., Qasmi, I.A., and Amin, K. M. (2005). An experimental study of sexual function improving effect of *Myristica fragrans* Houtt. (nutmeg). *BMC Complement Altern Med.*, **5**: 16.

Tayade, S.K., and Patil, D.A. (2005). Hitherto untapped plantlore from Nandurbar district (Maharashtra). *Natural Product Radiance*, **4(1):** 46-50.

Tempest, H. G., Homa, S. T., Routledge, E. J., Garner, A., Zhai, X. P., and Griffin, D. K. (2008). Plants used in Chinese medicine for the treatment of male infertility possess antioxidant and anti-oestrogenic activity. *Syst Biol Reprod Med.*, **54(4-5)**: 185-195.

Thakur, M., and Dixit, V. K. (2007). Aphrodisiac activity of *Dactylorhiza hatagirea* (D.Don) Soo in male albino rats. *Evid Based Complement Alternat Med.*, **4(1)**: 29-31.

Thakur, M., Bhargava, S., Praznik, W., Loeppert, R., and Dixit, V. K. (2009). Effect of *Chlorophytum borivilianum* Santapau and Fernandes on sexual dysfunction in hyperglycemic male rats. *Chin J Integr Med.*, **15(6)**: 448-453.

Thakur, M., Chauhan, N. S., Bhargava, S., and Dixit, V. K. (2009). A comparative study on aphrodisiac activity of some ayurvedic herbs in male albino rats. *Arch Sex Behav.*, **38(6):** 1009-1015.

Tomar, A. (2009). Folk medicinal uses of plant roots from Meerut district, Uttar Pradesh. *Indian Journal of Traditional Knowledge*, **8(2)**: 298-301.

Tyagi, R. M., Aswar, U. M., Mohan, V., Bodhankar, S. L., Zambare, G. N., and Thakurdesai, P. A. (2008). Study of furostenol glycoside fraction of *Tribulus terresteris* on male sexual function in rats. *Pharmaceutical Biology*, **46(3)**: 191-198.

Udayan, P.S., Harinarayanan, M.K., Tushar, K.V., and Balachandran, I. (2008). Some common plants used by Kurichiar tribes of Tirunelli forest, Wayanad district, Kerala in medicine and other traditional uses. *Indian Journal of Traditional Knowledge*, **7(2)**: 250-255.

Uniyal, S.K., Sharma, V., and Jamwal, P. (2011). Folk medicinal practices in Kangra district of Himachal Pradesh, Western Himalaya. *Human Ecology*, **39(4)**: 479-488.

Upadhyay, B., Parveen, Dhaker, A.K., and Kumar, A. (2010). Ethnomedicinal and ethnopharmaco-statistical studies of Eastern Rajasthan, India. *Journal of Ethnopharmacology*, **129(1)**: 64–86.

Venkataswamy, R., Mubarack, H.M., Doss, A., Ravi, T.K., and Sukumar, M. (2010). Ethnobotanical study of medicinal plants used by Malasar tribals in Coimbatore district of Tamil Nadu (South India). *Asian J. Exp. Biol. Sci.*, **1(2)**: 387–392.

Vermeulen, A. (1996). Decreased androgen levels and obesity in men. *Annals of Medicine*, **28(1)**: 13-15.

Waddell, T. G., Jones, H., and Keith, A. L. (1980). Legendary chemical aphrodisiacs. *J. Chem. Educ.*, **57(5)**: 341.

Wani, J. A., Achur, R. N., and Nema, R. K. (2011). Phytochemical screening and aphrodisiac activity of *Asparagus racemosus. International Journal of Pharmaceutical Sciences and Drug Research*, **3(2)**: 112-115.

Wani, J. A., Achur, R. N., and Nema, R. K. (2011). Phytochemical screening and aphrodisiac property of *Tinospora cordifolia. International Journal of Pharmaceutical and Clinical Research*, **3(2):** 21-26.

Whalen, R.E., and Luttge, W.G. (1970). P-chlorophenylalanine methyl ester: an aphrodisiac? *Science*, **169(3949):** 1000-1001.

Yadav, S., Yadav, J.P., Arya, V., and Panghal, M. (2010). Sacred groves in conservation of plant biodiversity in Mahendergarh district of Haryana. *Indian Journal of Traditional Knowledge*, **9(4):** 693-700.

Yen, T.T., Dalló, J., and Knoll, J. (1982). The aphrodisiac effect of low doses of (-) deprenyl in male rats. *Pol J Pharmacol Pharm.*, **34(5-6):** 303-308.

Yumnam, J.Y., and Tripathi, O.P. (2012). Traditional knowledge of eating raw plants by the Meitei of Manipur as medicine/nutrient supplement in their diet. *Indian Journal of Traditional Knowledge*, **11(1):** 45-50.

Zamblé, A., Martin-Nizard, F., Sahpaz, S., Hennebelle, T., Staels, B., Bordet, R., Duriez, P., Brunet, C., and Baillcul, F. (2008). Vasoactivity, antioxidant and aphrodisiac properties of *Caesalpinia benthamiana* roots. *J Ethnopharmacol.*, **116(1):** 112-119.

Traditional and Folk Herbal Medicine: Recent Researches Vol. 2 (2014) *Pages* **459–476**
Editor: **V.K. Gupta**
Published by: **DAYA PUBLISHING HOUSE, NEW DELHI**

15

Andrographis paniculata and its Major Diterpenoid–Andrographolide: Traditional Uses, Phytochemistry and Pharmacological Properties–A Review

Kavitha Sankar P.C.[1] and Beena Joy[1*]

ABSTRACT

Andrographis paniculata (Nees.) has been used for centuries as a medicinal herb in Asia and Europe for the treatment of upper gastrointestinal tract and upper respiratory infections, fever, herpes and other chronic diseases. It is widely used in Ayurveda as a popular remedy for the treatment of various liver disorders. This chapter summarizes a review of phytochemical and pharmaceutical potentials of the plant, Andrographis paniculata. Phytochemical investigation of this well studied herb showed that it is a rich source of bioactive flavonoids and diterpenoids, major constituent is andrographolide. The structures of these compounds have been established on the basis of various spectral studies. The present chapter compiles consequential compendium of pharmacological benefits of the plant as well as the major diterpenoids and the structure- activity relationship. Detailed research regarding the phytochemical and pharmacological aspects of the A. paniculata leads to the fact that it is a valuable source of unique natural products for various medicinal applications.

Keywords: *Andrographis paniculata,* Traditional uses, Phytochemistry, Pharmaceutical potentials.

1 Agroprocessing and Natural Products Division, National Institute for Interdisciplinary Science and Technology (NIIST-CSIR), Thiruvananthapuram, Kerala, India

* *Corresponding author*: E-mail: bjoy1571@gmail.com

Introduction

Medicinal plants are the major sources for the therapeutic remedies of various ailments. Their active phytocostituents are mainly responsible for these potential medicinal effects. The genus *Andrographis* (family Acanthaceae) includes 28 species of small shrubs occurring in tropical Asian countries. Of these *Andrographis paniculata* NEES is the most important medicinal plant and having been widely used in Chinese traditional medicines (Li *et al.*, 2007). It is an annual and branched plant with lanceolate green leaves and attains heights of 60-70 cm (Mishra *et al.*, 2007). It grows abundantly in Asian countries like India, Sri Lanka, Pakistan, China, Java, Malaysia and Indonesia. In India it is commonly known as 'Kalmegh' and in China as 'Chuanxinlian' or 'Yijianxi', mainly found in the plains of the country and is one of the commonly used medicinal plants in Ayuevedic and Unani systems of medicines. The plant is also known as the 'king of bitters' (Kabeeruddin *et al.*, 1937; Shahid, 2011) because of extreme bitter taste in every part of plant body. On the basis of literature survey it has been observed that the aerial parts (leaves and stems) of the plants are most commonly used to extract the active phytochemicals, however the whole plant or roots are mentioned to a limited extend (Shahid, 2011). Diterpenoids and flavanoids are the major class of secondary metabolites isolated from this plant. Therefore these active constituents are believed to be responsible for the pharmacological properties of *A. paniculata* (Tang and Eisenbrand, 1992; Saxena *et al.*, 1998). This review is aimed to summarize the knowledge concerning the phytochemical constituents isolated up to date from the plant *A. paniculata*. The traditional uses, pharmacological aspects of the plant and its significance in treating various ailments are well documented.

Traditional Uses

A. paniculata is one of the herbs mentioned in all ancient scriptures of Ayurveda. Extensive research has revealed that *A. paniculata* has a broad spectrum of pharmacological effects and some of them are extremely beneficial such as antimicrobial, antifungal, antioxidant, antiinflammatory, antihelmintic, antiplasmodial, antipyretic, antimalarial, anticancer and antidiarrhoeal effects (Trivedi, 2004; Jain and Tarafder, 1963; Dua *et al.*, 1999; Puri *et al.*, 1993; Matsuda *et al.*, 1994; Kumar *et al.*, 2004; Saxena *et al.*, 1998). According to Unani system of medicine it is useful in the treatment of chronic hepatitis. It is one of the most studied herbs due to its unique ability to treat various ailments like hypoglycemia, leprosy, gonorrhea, scabies, skin eruptions, hypertension, neoplasia, diuresis, dyspepsia, influenza etc. (Handa and Sharma, 1990). Leaves and roots are the mostly used plant parts in traditional medicines and has been used in many European countries as herbal supplement for health promotion; in Asian countries such as China, Malaysia, Indonesia and Thailand for the treatment of gastric disorders, infectious diseases, common colds for many years (Gamble, 1956). The herb *A. paniculata* which has been used in Indian medicinal practice mainly for the treatment of diabetes, dysentery, herpes, enteritis, peptic ulcer, skin infections; as an antidote against snake bites etc (Chang, 1987). The plant is official in Indian Pharmacopoeia as a prominent constituent of atleast 26 Ayurvedic formulations used to treat liver disorders and can be widely used to treat neoplasm as mentioned in ancient Ayurvedic literature (Balachandran and Govindarajan, 2005; Khare, 2007; Handa *et al.*, 1986; Kapil *et al.*, 1993). Decoction of fresh leaves of the plant is given to infants to relieve griping and loss of appetite (Chopra *et al.*, 1982; Khory and Katrak, 1984). Due to its blood purifying activity it is believed to be a good remedy for the treatment of leprosy, gonorrhea, scabies, boils, skin eruptions and chronic and seasonal fevers (Chopra, 1980). In addition to that pharmacological and clinical studies have demonstrated that *A. paniculata* possesses cardiovascular, immunostimulatory, adaptogenic, astringent, carminative and cytotoxic activities (Jarukamjorn and Nemoto, 2008).

Phytochemical Properties

The active phytoconstituents prodominently reported from the plant include diterpene lactones, flavonoids and polyphenols (Pandey *et al.*, 2011). More than 20 diterpenoids and about 10 flavonoids have been reported from this species over the past three dacades (Koteswara Rao *et al.*, 2004; Li *et al.*, 2007). The areal parts of the plant are mainly used to extract the active phytochemicals. Detailed phytochemical investigations on the chemical composition of *A. paniculata* showed that it is a rich source of 2´-oxygenated flavonoids (Govindachari *et al.*, 1969; Jalal *et al.*, 1979; Gupta *et al.*, 1983; Gupta *et al.*, 1996; Kuroyanagi *et al.*, 1987; Koteswara Rao *et al.*, 2004; Wenkui *et al.*, 2007), labdane diterpenoids (Kleipool, 1952; Chan *et al.*, 1971; Balmain and Connolly, 1973; Fujita *et al.*, 1984; Puri *et al.*, 1993; Matsuda *et al.*, 1994; Jantan *et al.*, 1994; Munta *et al.*, 2003; Li *et al.*, 2007) polyphenols and steroids (Siripong *et al.*, 1992; Ali *et al.*, 2002). Recent studies conducted to investigate the pharmacological activities of *A. paniculata* and/or its chemical constituents, revealed andrographolide, the prime constituent mainly attributed for its therapeutic potentials (Rajani *et al.*, 2000; Lomlim *et al.*, 2003). Bicyclic diterpenoid lactone andrographolide [$C_{20}H_{30}O_5$; (3-[2-{decahydro-6-hydroxy-5-(hydroxymethyl)-5,8α-dimethyl-2-methylene-1-naphthalenyl} ethylidene]dihydro-4-hydroxy-2(3H)-furanone) is mainly concentrated in leaves and can be easily isolated from the crude plant extracts as crystalline solids (Figure 15.1). It has a very bitter in taste and was first reported from the plant in 1950's. It has been reported to possess significant antitumor activity (Charkravarthy *et al.*, 1951; Cava *et al.*, 1962; Sheeja and Kuttan, 2007; Munta *et al.*, 2003; Kumar *et al.*, 2004; Rajagopal and Kumar, 2003; Cheung *et al.*, 2005); anti HIV (Calabrese *et al.*, 2000), cardioprotective (Yoopan *et al.*, 2007), hepatoprotective properties (Handa *et al.*, 1990; Sarawat *et al.*, 1995; Trivedi *et al.*, 2007). GC-MS analysis of ethanolic leaf extract of *A. paniculata* observed the presence of fatty acids, fatty acid esters, di-triterpenes, steroids etc (Kalaivani *et al.*, 2012).

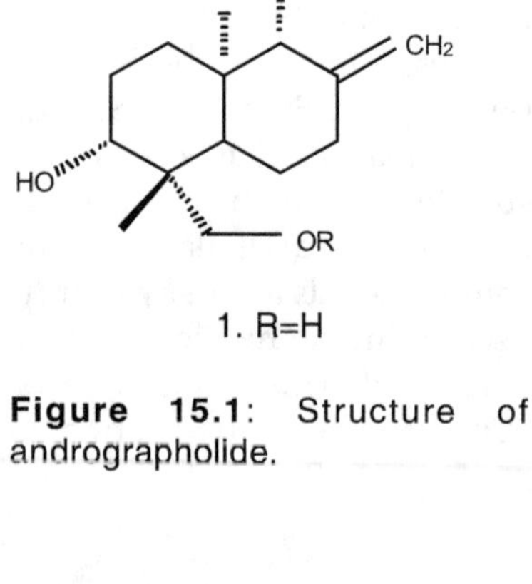

1. R=H

Figure 15.1: Structure of andrographolide.

Figure 15.2: Structures of flavonones and diterpenoids isolated from *A. paniculata*.

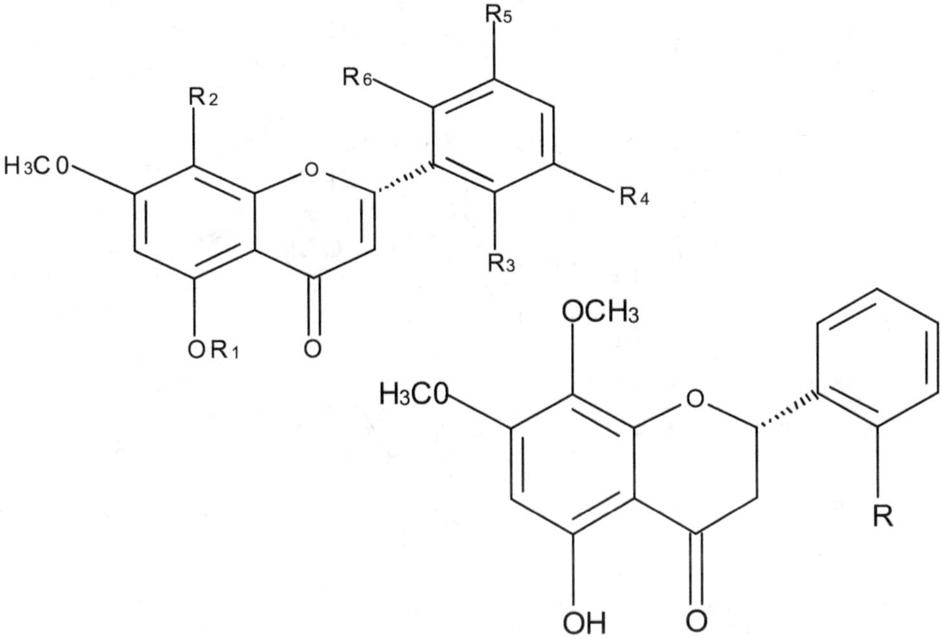

	R_1	R_2	R_3	R_4	R_5	R_6		R
1	H	OCH_3	H	H	H	H	8	H
2	H	OCH_3	OCH_3	H	OCH_3	H	9	OH
3	H	OCH_3	OCH_3	OCH_3	H	H		
4	H	H	OCH_3	H	H	OCH_3		
5	H	OCH_3	OCH_3	H	H	H		
6	glc	OCH_3	H	H	H	H		
7	H	OCH_3	Oglc	H	H	H		

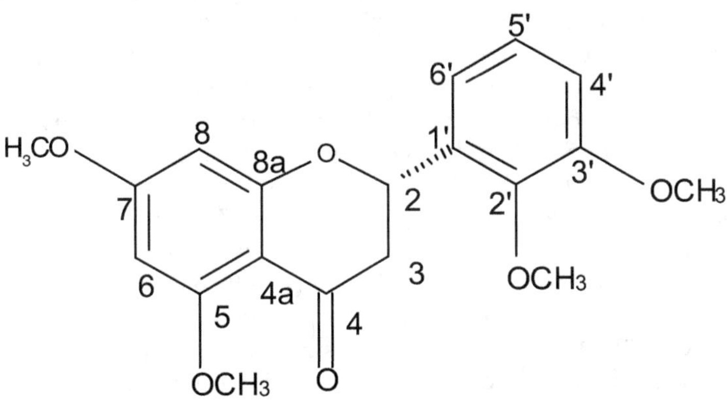

Contd...

Figure 15.2–*Contd...*

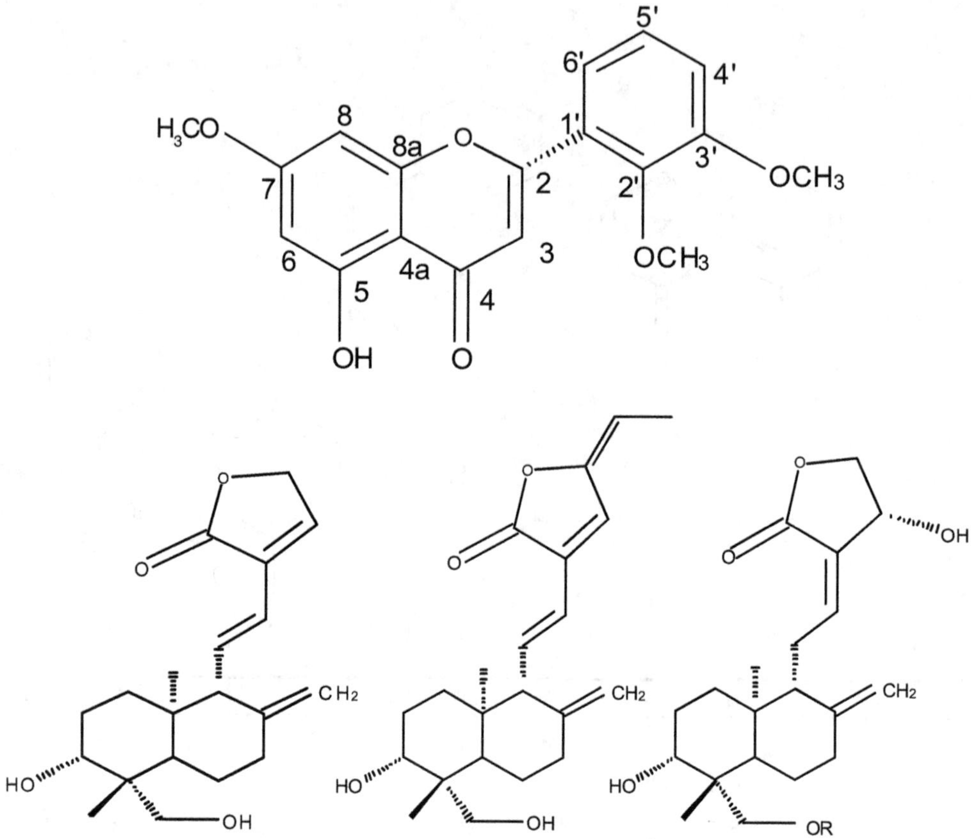

Besides andrographolide, other major compounds reported from the plant include andrographolide derivatives such as 14-deoxyandrographolide, 14-deoxy-11-oxo-andrographolide, 14-deoxy-11,12-didehydroandrographolide, andrographosterol, neoandrographolide, homoandrographolide, andrographone, andrographane, andrographin, andrographosterin, andrograpanin, angrographiside, isoandrographolide and bisandrographolide flavonids: 5-hydroxy-7,8,2',3'-tetramethoxy flavone, 5,5'-dihydroxy-7,8,2'-trimetroxyflavone, 5-hydroxy-7,8,2',6'-tetramethoxyflavone, 5,3'-dihydroxy-7, 8,4'-trimethoxyflavone, 2'-hydroxy-5,7,8-trimethoxyflavone, 5-hydroxy-7,8,2',3',4'-pentamethoxyflavone, 5,2',6'-trihydroxy-7-methoxyflavone 2'-O-beta-D-glucopyranoside, 5,7,8,2'-tetramethoxyflavone, 5-hydroxy-7,8-dimethoxyflavanone, 5-hydroxy-7,8-dimethoxyflavone, 5,2'-dihydroxy-7,8-dimethoxyflavone, 5-hydroxy-7,8,2',5'-tetramethoxyflavone, 5-hydroxy-7,8,2'-trimethoxyflavone, 5,4'-dihydroxy-7,8,2',3'-tetramethoxyflavone, 5,2'-dihydroxy-7,8-dimethoxyflavone-2'-O-beta-D-glucopyranoside and wightin, two phenylpropanoids: trans-cinnamic acid and 4-hydroxy-2-methoxycinnamaldehyde, oleanolic acid, β-sitosterol, stigmasterol, apigenin-7,4'-di-O-methyl ether etc (Chen *et al.*, 2006; Chong *et al.*, 2011; Niranjan *et al.*, 2010) (Figure 15.2). Ent-labdane type diterpenoids were also reported from the aerial parts of the plant *A. paniculata viz.* 3-O-β-D-glucopyranosyl-14, 3-O-β-D- glucopyranosyl-andrographolide, 19-O-[β-D-apiofuranosyl(1-2)-β-D-glucopyranoyl]3,14-dideoxy-andrographolide, 19-dideoxyandrographolide, 12S-hydroxyangrographolide, 7R-hydroxy-14-deoxy-andrographolide, 7S-hydroxy-14-deoxy-andrographolide, 12S,13S-hydroxy-andrographolide, 12R,13R-hydroxy-andrographolide, andrographatoside. These compounds showed significant antimicrobial activity against various bacterial and fungal strains (Shen *et al.*, 2006; Zhang *et al.*, 2006; Chen *et al.*, 2008; Dua *et al.*, 2004). Dua *et al.*(2004), reported some xanthone type compounds from the roots of *A. paniculata* i.e 1,8-dihydroxy-3,7-dimethoxy xanthone, 4,8-dihydroxy-2,7-dimethoxy xanthone, 1,2-dihydroxy-6,8-dimethoxy xanthone, 1-hydroxy-3,7,8-trimethoxy xanthone, which exhibited antimalarial activity against Plasmodium berghei infections. Neoandrographolide has been reported to exhibit anti-HIV activity (Khan and Khanum Atiya, 2003). Besides andrographolide, another constituent of *A. paniculata*, isoandrographolide was found to exhibit cytotoxicity in *in vitro* KB cell lines (Wenkui *et al.*, 2007).

Pharmacological Properties

Hepatoprotective Activity

A. paniculata and the diterpenoids are the main constituents in traditional indigeneous medicine to treat various liver disorders such as hepatitis (Trivedi and Rawal, 2000; Trivedi and Rawal, 2001) or damage induced by galactosamine (Visen *et al.*, 2007), paracetamol (Visen *et al.*, 1993) and carbon tetrachloride (Handa and Sharma, 1990; Choudhary and Poddar, 1984). The choleretic potential of the major diterpenoids as a stimulant for proper functioning of gall bladder in anesthetized guinea pigs and conscious rats was reported by Shukla *et al.*, in 1992. The efficacy of andrographolide and neoandrographolide against ethanol and acetaminophen induced liver damage was comparable to silymarin, the standard hepatoprotective drug (Singha *et al.*, 2007). However previous studies were carried out in this area suggested (Kapil *et al.*, 1993), besides andrographolide several other constituents in the extract may be responsible for the hepatoprotective activity of the plant. The hepatoprotective activity is due to influence of certain metabolic enzymes. The plant extract and its main active constituent diterpenoids were investigated for its influence on liver metabolizing enzyme such as cytochrome P450, glutathione S-transferase etc (Singh *et al.*, 2001; Jarukamjorn *et al.*, 2006). Hepatoprotective studies on administration of aqueous plant extract to lymphoma bearing mice revealed that the catalase, superoxide dismutase, glutathione-S- transferase enzyme activity are enhanced while lactase dehydrogenase activity get reduced (Verma, 2008). The alcohol and aqueous extract are found to increase CYP1A1 and CYP2B without altering the P450 and inhibit CYP1A2 and CYP2C in rat and human liver chromosomes (Pekthong *et al.*, 2008; Pekthong *et al.*, 2009). Further clinical research is recommended to draw conclusion on the effect of the *A. paniculata* and its constituents on hepatic metabolizing enzymes.

Antioxidant Activity

Aqueous extract of *A.paniculata* were investigated for their radical scavenging and antioxidant activity in brain and liver organs of animal models and the observed activities were summarized due to the presence of flavonoids in the extract (Das *et al.*, 2009; Lin *et al.*, 2009; Verma *et al.*, 2008). The aqueous extract was found to be more effective radical scavenger than alcoholic extract in nicotine induced inhibition of mitochondrial electron chain process. Usually nicotine inhibition process lead to generation of nitric oxide free radicals which are scavenged effectively by *A. paniculata* extract and/or andrographolide. The methanol extract of the plant was also found to be effective in scavenging reactive oxygen species (ROS·) and Lipopolysaccharide (LPS)- stimulated nitric oxide (NO·) radicals (Sheeja *et al.*, 2006; Batkhuu *et al.*, 2002; Liu *et al.*, 2007). Structure activity relationship of diterpenoids as free radical scavengers are reported (Kamdem *et al.*, 2002). Intake of *A. paniculata* extract help to maintain the balance of nitric oxide/endothelin in the tissues (Wang *et al.*, 1997). Studies throw light on the exploring the potential of the plant and the diterpenoid in treating neurodegenerative disease such as Parkinson's disease (Abu-Ghefreh *et al.*, 2009; Liu *et al.*, 2007).

Anti inflammatory Activity

The methanol extract of *A.paniculata* and/or andrographolide inhibited carrageenan induce inflammation (Sheeja *et al.*, 2006) owing to its radical scavenging activity. Studies by Deng (1978); Yin and Guo (1993) reported that diterpenoids reduce inflammation induced by histamine/adrenaline in the order dehydroandrographolide, neoandrographolide and andrographolide. These diterpenoids stimulate the brain to the release of andrenocorticotrophic hormones which signals the release of anti inflammatory hormones namely cortisol from adrenal glands. Different mechanisms have been proposed on anti-inflammatory activity of andrographolide involving modulating macrophage and neutrophil activity. The mechanism of suppression of nitric oxide synthase, mRNA and protein expression in RAW 264.7 cells induced with lipopolysaccharides and interferon-gamma by andrographolide in a dose dependent manner was reported by Chiou *et al.*, in 2000. Similar dose dependent inhibition of nitric oxide and prostaglandin by neoandrographolide was reported by Liu *et al.* (2007).

Cardiovacsular Protective Property

Intake of plant extract increase blood clotting duration; decreasing arterial narrowing due to injury or high fat diet; heart muscle damage after myocardial infraction in animals (Zhao *et al.*, 1990; Zhao *et al.*, 1991); activate fibrinolysis and antihypertensive effects. Hence plant extracts can be administered in pre- and post-treatments in angioplasty. Thisoda *et al.* (2006) suggested that the major diterpenoid andrographolide and 14-deoxy-11, 12-didehydroandrographolide in the aqueous of the plant are mainly contributing to antiplatelet aggregation (Thisoda *et al.*, 2006).

Anticytotoxic and Anti-HIV Activity

The potential of *A. paniculata* in treating cancer, HIV infections were largely explored (See *et al.*, 2002; Sheeja *et al.*, 2007; Zhao *et al.*, 2008). Andrographolide also reported to possess anticancer and immune stimulating potential (Kumar *et al.*, 2004; Rajagopal *et al.*, 2003; Li *et al.*, 2007). The plant extract as well as the andrographolide were reported to induce myeloid leukemia cell differentiation in mice (Cheung *et al.*, 2005). Nakanishi *et al.* (1965) reported antitumor activity of methanolic extract of leaves in rats. Studies done by Siripong *et al.* (1992) reported that the methanolic extract of *A. paniculata* leaves exhibited significant cytotoxicity against KB cells (human epidermoid carcinoma of the nasopharynx) and P388 (mouse lymphocytic leukemia) cells. Further bioassay guided isolation of the methanolic extract of leaves yielded diterpenoids such as andrographolide, 14-deoxy-11,12-didehydroandrographolide, neoandrographolide and *in vitro* cytotoxicity against KB and P388 cell lines revealed andrographolide to possess significant cytotoxic activity (Siripong *et al.*, 1992). Jiang *et al.* (2007) studies reported the anticancer potential of andrographolide by suppressing the adhesion of gastric cancer cells. *In vitro* and *in vivo* studies in detrimental T cells, reported the potential of andrographolide as T cells obstructor (Iruretagoyena *et al.*, 2005). Reddy *et al.* (2005) studies reported that besides the diterpenoids and flavanones, Bis-andrographolide ether isolated from areal parts exhibit anti-HIV and cytotoxic activity. Talukdar *et al.* (1968) reported that the andrographolide modify the cellular signal transduction in HIV infected and prevent the transmission of HIV virus. Many clinical studies have reported the anti-HIV activity of the constituents of the plant.

Antidiabetic Activity

Hypoglycemic studies on *A. paniculata* administered to non-diabetic rabbits reported that the aqueous extract was effective in reducing hyperglycemia in oral glucose fed rabbits (Borhanuddin *et al.*, 1994) and the plasma glucose levels in streptozotocin induced diabetic rats in dose dependent manner but not in adrenaline injected animals. Ethanolic extract administered to streptomycin induced diabetic rats exhibited lowered levels of fasting serum glucose, thiobarbituric acid stimulative substances in liver with gain in body weight and found to enhance the activity of superoxide dismutase and catalase enzymes (Zhang and Tan, 2000). Reyes *et al.* (2006) studies in alloxan induced diabetic rats reported the restoration of estrous cycle in rats by intake of *A. paniculata*. Andrographolide administered to streptomycin–induced diabetic rats and normal rats was found to lower plasma glucose levels in dose dependent manner. However more pronounced effect was noted in normal rats (Yu *et al.*, 2003). Yu *et al.* (2003) study is contradictory to studies in rats by Dandu and Inamdar (2009), in which aqueous extract of the plant was not effective in lowering glucose levels. Yu *et al.* (2003) studies with intravenous administration of andrographolide to diabetic rats suggested that the glucose lowering effect may be due to more glucose consumption by skeletal muscles, noted by increase in mRNA and GLUT4 enzyme levels in skeletal tissues. However studies by Subramanian *et al.* (2008) proposed that the hypoglycemic activity of alcoholic extract of *A.paniculata*/ andrographolide was due to α-amylase and α-glucosidase inhibitory effects in the intestine. Another *in vitro* study suggested that insulin released from β-cells through ATP modulated potassium channels contribute to hypoglycemic effect (Wibudi *et al.*, 2008). Studies on water and alcohol extract of *A. paniculata* and andrographolide proposed several mechanistic way for explaining the hypoglycemic activity but still further work needed for identifying the active constituents (Wibudi *et al.*, 2008; Subramanian *et al.*, 2008).

Antimicrobial and Antiviral Activity

The crude aqueous extract was found to possess antimicrobial activity against some strains such as *S. aureus*, *P. aeruginosa* but no significant activity against *E. coli*, *Salmonella*, *Staphylococcus aureus* (Singha *et al.*, 2003; Zaidan *et al.*, 2005). The antibacterial activity may be due to andrographolides and arabinogalactan proteins in the aqueous extract. Voravuthikunchai *et al.* (2006) also reported that the ethanol extract was inactive against different strains of *E. coli.* The major diterpenoids reported to exhibit viricidal activity against herpes simplex virus (HSV-1) (Wiart *et al.*, 2005).

Antimalarial and Antipyretic Activity

Antimalarial potential of *A. paniculata* against *Plasmodium berghei*, parasite that transmit malaria and *Plasmodium falciparum* were attributed to reactivation of superoxide dismutase, an antioxidant enzyme in the former and presence of xanthones in the extracts in the latter species (Rahman *et al.*, 1999; Mishra *et al.*, 2009; Dua *et al.*, 2004). Xanthones from the plant are also reported to possess protozoa growth inhibitory potential (Dua *et al.*, 2009). *A. paniculata* and its phytochemicals are active constituents in formulations to treat filariasis (Dutta and Sukul, 1982; Kapil *et al.*, 1993). Efficacy of *A. paniculata* extract as central nervous system depressant drugs has been proved in 2001 (Mandal *et al.*, 2001). However extensive research to optimize the concentration for clinical feasibility is now demanded. *A. paniculata* was found to be used in traditional medicines for lowering fever. The chief phytochemical constituent, andrographolide reported to possess antipyretic, antiulcerogenic and analgesics activity which is comparable to that of aspirin (Madav *et al.*, 1995; Madav *et al.*, 1996). Oral inake of *A. paniculata* plant as well as the standardized dried extract of *A. paniculata* (SHA-10) efficiency was comparable to acetaminophen in relieving symptoms from pharyngotonsillitis (Cáceres *et al.*, 1999).

Anti-diarrheal and Upper Respiratory Tract Infections

Formulations containing andrographolide and neoandrographolide were reported to be effective against bacillary dysentery and diarrhea compared to standard drugs chloramphinicol/furazolidine and loperamide (Imodium) (Chang and But, 1987). Several reports on *A. paniculata* extract and in combination with Eleutherococcus senticosus in treating uncomplicated upper respiratory tract infections (UTRI) in adults and children in the age group 4-11 yrs are available (Cáceres *et al.*, 1999; Gabrielian *et al.*, 2002; Spasov *et al.*, 2004; Poolsup *et al.*, 2004; Coon *et al.*, 2004). However more clinical trials should be performed on *A. paniculata* to test its efficacy in UTRI as the results on combinations with *E. senticosus* was reported to be effective.

Antivenom Activity

A. paniculata and its phytochemicals are active constituents in formulations to treat snake venom (Chang *et al.*, 1987). Swamy *et al.* (2008) studied the anti viper potency of the plant extract as well as purified fractions in mice. The *in vitro* and *in vivo* studies reported that isolated fractions (2.4mg/kg body weight) more effective to combat snake venom compared to oral intake of three-fold this quantity is needed for crude plant extracts.

Antifertility

Traditional medicines such as Ayurveda advocates avoiding the consumption of the plant during pregnancy. Studies on antifertility effects of *A. paniculata* and its main constituent, andrographolide in animal models revealed the antispermatogenic and ovulation hindering effects (Akbarsha and Murugaian, 2000; Janarthanan, 1990). Another study with mice of both sex observed that female mice fed with *A.paniculata* when mated with male mice did not conceive pregnancy while in case of mating between plant administered male mice- normal female mice, pregnancy was reported (Zoha *et al.*, 1989). Chang *et al.* (1987) reported abortion in rabbits fed with *A. paniculata* and also observed the suppression of human placental chorionic trophoblastic cell growth. However *A. paniculata* fed to pregnant rats did not affect progesterone levels during initial gestation period (Panossian *et al.*, 1999). Though many explanation for the mechanistic pathways of antifertility activity such as uterine relaxation by obstructing calcium channels (Burgos *et al.*, 2001); inhibition of spermatogenisis; degenerative changes in epididymis, seminiferous tubules, seminal vesicles have been suggested, most of the trials were carried out for a short duration. So the choice of *A. paniculata* as an alternative resource in medical therapy for contraception requires further active research involving the isolation of its phytoconstituents and longer clinical trials.

Conclusions

The major bioactive compound 2' oxygenated flavonoids which occur rarely in nature, in addition to andrographolide diterpenoids from *A. paniculata* confined the chemotaxonomic importance of *Andrographis* species in the Acanthaceae family. In traditional Chinese medicine the pretreatment with various extracts and constituents of *A. paniculata* against hepatotoxicity are very consistent. The plant is found to be very effective in a number of polyherbal formulations for the treatment of liver ailments. Andrographolide, the primary medical component of *A. paniculata* in terms of bioactive properties and abundance has been reported for its anticancer and anti HIV effects leads to explore the relevance of plant in modern medicine as a potent chemotherapeutic agent. Neoandrographolide another active component of *A. paniculata* has also been reported to show anti HIV activity. The ent-labdane diterpenoids reported from the species exhibited significant antimicrobial effect. Among the andrographolide analogues, 14-deoxy-11,12-didehydro andrographolide is immunostimulatory, antiinfectve and antiatherosclerotic and 14-deoxy-andrographolide is immunomodulatory and antiatherosclerotic. Among the less abundant constituents of the plant andrograpanin and isoandrographolide are anti-inflammatory and tumor suppressive.

References

Abu-Ghefreh, A.A., Canatan, H., and Ezeamuzie, C.I. (2009). *In vitro* and *in vivo* anti-inflammatory effects of andrographolide. *Int Immunopharmacol.*, **9:** 313-318.

Akbarsha, M.A., and Murugaian, P. (2000). Aspects of the male reproductive toxicity/ male antifertility property of andrographolide in albino rats: Effect on the testis and cauda epididymidal spermatozoa. *Phytother Res.*, **14:** 432-435.

Ali, M.S., Saleem, M., Ahmad, W., Parvez, M., and Yamdagni. (2002). A chlorinated monoterpene ketone, acylated p-sitosterol glycosides and a flavanone glycoside from *Mentha longifolia* (Lamiaceae). *Phytochem.*, **59:** 889–895.

Balachandran, P., and Govindarajan, R. (2005). Cancer–an ayurvedic perspective. *Pharmacol* Balmain, A., and Connolly, J.D. (1973). Minor diterpenoid constituents of *Andrographis paniculata* Nees. *J Chem Soc., Perkin Trans 1.*, **1:** 1247–1251.

Batkhuu, J., Hattori, K., Takano, F., *et al.* (2002). Suppression of NO production inactivated macrophages *in vitro* and *ex vivo* by neoandrographolide isolatedfrom *Andrographis paniculata*. *Biol Pharm Bull.*, **25:** 1169-1174.

Borhanuddin, M., Shamsuzzoha, M., and Hussain, A.H. (1994). Hypoglycemia effects of *Andrographis paniculata* Nees on non-diabetic rabbit. *Bangladesh Med Res Counc Bull.*, **20:** 24-26.

Burgos, R.A., Aguila, M.J., Santiesteban, E.T., *et al.* (2001). *Andrographis paniculata* (Nees) induces relaxation of uterus by blocking voltage operated calcium channels and inhibits Ca^{+2} influx. *Phytother Res.*, **15:** 235-239.

Cáceres, D.D., Hancke, J.L., Burgos, R.A., *et al.* (1999). Use of visual analogue scale measurements (VAS) to assess the effectiveness of standardized *Andrographis paniculata* extract SHA-10 in reducing the symptoms of common cold. *Phytomed.*, **6:** 217-223.

Calabrese, C., Berman, S.H., Babish, J.G., Ma, X., Shinto, L., Dorr, M., Wells, K., Wenner, G.A., and Standish, L.J. (2000). A phase I trial of andrographolide in HIV positive patients and normal volunteers. *Phytother Res.*, **14:** 333-338.

Cava, M.R., Chan, W.R., Haynes, J., Johnson, L.F., and Weinstein, B. (1962). The structure of andrographolide. *Tetrahedron*, **18:** 397-403.

Chan, W.R., Taylor, D.R., Willis, C.R., Bodden, R.L., and Fehlhaber, H,W. (1971). The structure and stereochemistry of neoandrographolide, a diterpene glycoside from *Andrographis paniculata* Nees. *Tetrahedron*, **27:** 5081–5091.

Chang, H.M., But, P.P.H. (1987). *Pharmacology and Applications of Chinese Materia Medica*. English translation by Shem Chang-Shing Yeung, Sih Cheng-Yao and Lai-Ling Wang (Chinese Medicinal Material Research Centre, "e Chinese University of Hong Kong), Singapore: World Scientific Publishing Co. Pte. Ltd; 2: 918-928.

Charkravarthy, R.N., and Charkravarthy, D. (1951). Andrographolide the active constituent of *Andrographis paniculata* NEES, *Ind Med Gazette*, **86:** 96-97.

Chen, L-X., Qu, G-X., and Qiu, F. (2006). Studies on diterpenoids from *Andrographis paniculata*. *China journal of Chinese materia medica*, **31:** 1594-1597.

Chen, L., Zhu, H., Wang, R., Zhou, K., Jing, Y., and Qiu, F. (2008). Ent-Labdane diterpenoid lactone stereoisomers from *Andrographis paniculata*. *J Nat Prod.*, **71:** 852-855.

Cheung, H.Y., Cheung, S.H., Li, J., Cheung, C.S., Lai, W.P., Fong, W.F., and Leung, F.M. (2005). Andrographolide isolated from *Andrographis paniculata* induces cell cycle arrest and mitochondrial mediated apotosis in human leukemia HL-60 cells. *Planta Med.*, **71:** 1106-1111.

Chiou, W.F., Chen, C.F., and Lin, J.J. (2000). Mechanism of suppression of inducible nitric oxide synthase (iNOS) expression in RAW 264.7 cells by andrographolide. *Br J Pharmacol.*, **129:** 1553-1560.

Chong, X., and Wang, Z-T. (2011). Chemical constituents from roots of *Andrographis paniculata*. *Acta Pharm Sinic.*, **46:** 317-321.

Chopra, R.N., Nayer, S.L., and Chopra, I.C. (1980). Glossary of Indian Medicinal Plants, CSIR, New Delhi, pp. 18.

Choudhary, B.R., and Poddar, M.K. (1984). Andrographolide and kalmegh (*Andrographis paniculata*) extract: *in vivo* and *in vitro* effect on hepatic lipid peroxidation. *Method Find Exp Clin.*, **6:** 481-485.

Coon, J.T., and Ernst, E. (2004). *Andrographis paniculata* in the treatment of upper respiratory tract infections: a systematic review of safety and efficacy. *Planta Med.*, **70:** 293-298.

Dandu, A.M., and Inamdar, N.M. (2009). Evaluation of beneficial effects of antioxidant properties of aqueous leaf extract of *Andrographis paniculata* in STZ-induced diabetes, *Pak J Pharm Sci.*, **22:** 49-52.

Das, S., Gautam, N., Dey, S.K., *et al.* (2009). Oxidative stress in the brain of nicotine-induced toxicity: Protective role of *Andrographis paniculata* Nees and vitamin E. *Appl Physiol Nutr Metab.*, **34:** 124-135.

Deng, W.L. (1978). Outline of current clinical and pharmacological research on *Andrographis paniculata* in China. *Newsletters of Chinese Herbal Med.*, **10:** 27-31.

Dua, V.K., Ojha, V.P., Roy, R., Joshi, B.C., Valecha, N., Devi, C.U., Bhatnagar, M.C., Sharma, V.P., and Subbarao, S.K. (2004). Antimalarial activity of some xanthones isolated from the roots of *Andrographis paniculata*. *J Ethnopharmacol.*, **95:** 247-251.

Dua, V.K., Verma, G., and Dash, A.P. (2009). *In vitro* antiprotozoal activity of some xanthonesisolated from the roots of *Andrographis paniculata*. *Phytother Res.*, **23:** 126-128.

Dutta, A., and Sukul, N.C. (1982). Filaricidal properties of a wild herb, *Andrographis paniculata*. *J Helminthol.*, **56:** 81-84.

Fujita, T., Fujitani, R., Takeda, Y., Yakaishi, Y., Yamada, T., Kido, M., and Miura, I. (1984). On the diterpenoids of *Andrographis paniculata*: X-ray crystallographic analysis of andrographolide and structure determination of new minor diterpenoids. *Chem Pharm Bull.*, **32:** 2117–2125.

Gabrielian, E.S., Shukarian, A.K., Goukasova, G.I., *et al.* (2002). A double blind, placebo-controlled study of *Andrographis paniculata* fixed combination Kan Jang in the treatment of acute upper respiratory tract infections including sinusitis. *Phytomed.*, **9:** 589-597.

Gamble, J.S. (1956). *In*: Flora of the Presidency of Madras, Vol. 2, Botanical Survey of India, Calcutta, pp. 1048-1052.

Govindachari, T.R., Pai, B.R., Srinivasan, M., and Kalyanaraman, P.S. (1969). Chemical investigation of *Andrographis paniculata*. *Ind J Chem.*, **7:** 306.

Gupta, K.K., Taneja, S.C., Dhar, K.L., and Atal, C.K. (1983). Flavonoids of *Andrographis paniculata*. *Phytochem.*, **22:** 314–315.

Gupta, K.K., Taneja, S.C., Dhar, K.L. (1996). Flavonoid glycoside of *Andrographis paniculata*. *Ind J Chem.*, **35B:** 512–513.

Handa, S.S., and Sharma, A. (1990). Hepatoprotective activity of andrographolide from *Andrographis paniculata* against carbon tetrachloride. *Ind J Med Res.*, **92:** 276-283.

Iruretagoyena, M.I., Tobar, J.A., Gonzalez, P.A., Sepulveda, S.E., Figueroa, C.A., Burgos, R.A., Hancke, J.L., and Kalergis, A.M. (2005). Andrographolide interferes with T cell activation and reduces experimental autoimmune encephalomyelitis in the mouse. *J Pharmacol Exp Ther.*, **312:** 366-372.

Jalal, M.A.F., Overton, K.H., and Rycroft, D.S. (1979). Formation of three new flavones by differentiating callus cultures of *Andrographis paniculata*. *Phytochem.*, **18:** 149–151.

Janarthanan, S. (1990). Antifertility effects of andrographolide in rats. *J Ecobiol.*, **2:** 235-329.

Jantan, I., and Waterman, P.G. (1994). Ent-14-b-hydroxy-8(17), 12-labdadien-16,15-olide-3b,19-oxide: a diterpene from the aerial parts of *Andrographis paniculata*. *Phytochem.*, **37:** 1477–1479.

Jarukamjorn, K., Don-in, K., Makejaruskul, C., Laha, T., Daodee, S., Pearaksa, P., and Sripanidkulchai, B. (2006). Impact of *Andrographis paniculata* crude extract on mouse hepatic cytochrome P450enzymes. *J Ethnopharmacol.*, **105:** 464-467.

Jarukamjorn, K., and Nemoto, N. (2008). Pharmacological aspects of *Andrographis paniculata* on health and its major diterpenoid constituent andrographolide. *Journal of Health Science*, **54:** 370-381.

Kabeeruddin, M. (1937). Kitabul Advia, Vol 2, Aligarh Barqi Press, Delhi, India, pp.148.

Kalaivani, C.S., Sathish, S.S., Janakiraman, N., and Johnson, M. (2012). GC-MS studies on *Angrographis paniculata* (Burm.f.) Wall. Ex Nees – A medicinally important plant. *Int J Med Arom plants,* **2:** 69-74.

Kamdem, R.E., Sang, S., and Ho, C.T. (2002). Mechanism of the superoxide scavenging activity of neoandrographolide-a natural product from *Andrographis paniculata* Nees. *J Agric Food Chem.*, **50:** 4662-4665.

Kapil, A., Koul, I.B., Banerjee, S.K., and Gupta, B.D. (1993). Antihepatotoxic effects of major diterpenoid constituents of *Andrographis paniculata*. *Biochem Pharmacol.*, **46:** 182-185.

Khare, C.P. (2007). Indian Medicinal Plants: An Illustrated Dictionary, Springer, Berlin, Germany.

Kleipool, R.J.C. (1952). Constituents of *Andrographis paniculata* Nees. *Nature,* **169:** 33–34.

Koteswara Rao, Y., Vimalamma, G., Rao, C.V., and Tzeng, Y-M. (2004). Flavonoids and andrographolides from *Andrographis paniculata*. *Phytochem.,* **65:** 2317–2321.

Kumar, R.A., Sridevi, K., Kumar, N.V., Nanduri, S., and Rajagopal, S. (2004). Anticancer and immunostimulatory compounds from *Andrographis paniculata*. *J Ethanopharmacol.,* **92:** 291-295.

Kuroyanagi, M., Sato, M., Ueno, A., and Nishi, K. (1987). Flavonoids from *Andrographis paniculata*. *Chem Pharm Bull.,* **35:** 4429–4435.

Li, W., Xu, X., and Zhang, H. (2007). Secondary metabolites from *Andrographis paniculata*. *Chem Pharm Bull.,* **55:** 455–458.

Lin, F.L., Wu, S.J., Lee, S.C., and Ng, L.T. (2009). Antioxidant, antioedema and analgesic activities of *Andrographis paniculata* extracts and their active constituent andrographolide. *Phytother Res.,* **23:** 958-964.

Liu, J., Wang, Z.T., Ji, L.L., and Ge, B.X. (2007). Inhibitory effects of neoandrographolide on nitric oxide and prostaglandin E2 production in LPS-stimulated murine macrophage. *Mol Cell Biochem.,* **298:** 49-57.

Liu, J., Wang, Z.T., and Ji, L.L. (2007). *In vivo* and *in vitro* anti-inflammatory activities of neoandrographolide. *Am J Chin Med.,* **35:** 317-328.

Madav, H.C., Tripathi, T., and Mishra, S.K. (1995). Analgesics, antipyretic and antiulcerogenic effects of andrographolide. *Ind J Pharm Sci.,* **57:** 121-125.

Madav, S., Tandan, S.K., Lal, J., and Tripathi, H.C. (1996). Anti-inflammatory activity of andrographolide. *Fitoterapia,* **67:** 452-458.

Mandal, S.C., Dhara, A.K., and Maiti, B.C. (2001). Studies on psychopharmacological activity of *Andrographis paniculata* extract. *Phytother Res.,* **15:** 253-256.

Matsuda, T., Kuroyanagi, M., Sugiyama, S., Umehara, K., Ueno, A., and Nishi, K. (1994). Cell differentiation inducing diterpenes from *Andrographis paniculata* Nees. *Chem Pharm Bull.,* **42**: 1216–1225.

Mishra, S.K., Sangwan, N.S., and Sangwan, R.S. (2007). *Andrographis paniculata* (Kalmegh): A review. *Pharmacognosy Rev.,* **1:** 283-298.

Mishra, K., Dash A.P., Swain, B.K., and Dey, N. (2009). Anti-malarial activities of *Andrographis paniculata* and *Hedyotis corymbosa* extracts and their combination with curcumin. *Malar J.,* **8:** 26.

Munta, K.R., Reddy, M.V.B., Gunasekar, D., Murthy, M.M., Caux, C., and Bodo, B. (2003). A flavone and an unusual 23-carbon terpenoid from *Andrographis paniculata*. *Phytochem.,* **62:** 1271–1275.

Nakanishi, K., Sasaki, S., Kiang, A.K., Goh, J., Kakaisawa, H., Ohashi, M., Watanabe, J., Yokntuni, H., Masumara, C., and Toyashi, M. (1965). Phytochemistry survey of Malaysian plants preliminary chemical and pharmacological screening. *Chem. Pharm. Bull.,* **13:** 882-896.

Niranjan, A., Tewari, S.K., and Lehri, A. (2010). Biological activities of Kalmegh (*Andrographis paniculata* Nees) and its active principles – A review. *Ind J Nat Prod Res.,* **1:** 125-135.

Pandey, M.K., Sing, G.N., Sharma, R.K., and Latha, S. (2011). Physiochemical standardization of *Andrographis paniculata* (NEES): An ayurvedic drug. *Int J Pharm Res Dev.,* **3:** 81-89.

Panossian, A., Kochikian, A., Gabrielian, E., *et al.* (1999). Effect of *Andrographis paniculata* extract on progesterone in blood plasma of pregnant rats. *Phytomed.,* **6:** 157-161.

Pekthong, D., Martin, H., Abadie, C., *et al.* (2008). Differential inhibition of rat and human cytochrome P450 by *Andrographis paniculata* extract and andrographolide. *J Ethnopharmacol.,* **115:** 432-440.

Pekthong, D., Blanchard, N., Abadie, C., *et al.* (2009). Effects of *Andrographis paniculata*extract and Andrographolide on hepatic cytochrome P450 mRNA expression and monooxygenase activities after *in vivo* administration to rats and *in vitro* in rat and human hepatocyte cultures. *Chem Biol Interact.,* **79:** 247-255.

Poolsup, N., Suthisisang, C., Prathanturarug, S., *et al.* (2004). *Andrographis paniculata* in the symptomatic treatment of uncomplicated upper respiratory tract infection: Systemic review of randomized controlled trials. *J Clin Pharm Ther.,* **29:** 37-45.

Puri, A., Saxena, R., Saxena, R.P., Saxena, K.C., Srivastava, V., and Tandon J.S. (1993). Immunostimulant agents from *Andrographis paniculata*. *J Natur Prod.,* **56:** 995–999.

Rahman, N.N.N.A., Futura, T., Kojima, S., Takane, K., and Mohd, M.A. (1999). Antimalarial activity of extracts of Malaysian medicinal plants. *J Ethanopharmacol.,* **64:** 249-254.

Rajagopal, S., Kumar, R.A., Deevi, D.S., Sathyanarayana, C., and Rajagopalan, R. (2003). Andrographolide, a potential cancer therapeutic agent isolated from *Andrographis paniculata*. *J Exp Ther Oncol.,* **3:** 147-158.

Rajani, M., Shrivastava, N., and Ravishankara, M.N. (2000). A rapid method for isolation of andrographolide from *Andrographis paniculata* Nees (Kalmegh). *Pharm Biol.,* **38:** 204–209.

Reddy, V.L., Reddy, S.M., Ravikanth, V., Krishnaiah, P., Goud, T.V., Rao, T.P., Ram, T.S., Gonnade, R.G., Bhadbhade, M., and Venkateswarlu, Y. (2005). A new bis-andrographolide ether from *Andrographis paniculata* Nees and evaluation of anti-HIV activity. *Nat Prod Res.,* **19:** 223-230.

Reyes, B.A., Bautista, N.D., Tanquilit, N.C., Anunciado, R.V., Leung, A.B., Sanchez, G.C., Magtoto, R.L., Castronuevo, P., Tsukamura, H., and Maeda, K.I. (2006). Anti diabetic potentials of *Momordica charantia* and *Andrographis paniculata* and their effects on estrous cyclicity of alloxan-induced diabetic rats. *J Ethanopharmacol.,* **105:** 196-200.

Sarawat, B., Visen, P.K.S., Patnaik, G.K., and Dhawan, B.N. (1995). Effect of andrographolide against galactosamine induced hepatotoxicity. *Fitoterapia,* **66:** 415-420.

Saxena, S., Jain, D.C., Bhakuni, R.S., and Sharma, R.P. (1998). Chemistry and pharmacology of *Andrographis* species. *Indian Drugs,* **35:** 458-467.

See, D., Mason, S., and Roshan, R. (2002). Increased tumor necrosis factor alpha (TNF-alpha) and natural killer cell (NK) function using an integrative approach in late stage cancers. *Immunol Invest.,* **31:** 137-153.

Shahid, A. (2011). *Andrographis paniculata*: A review of pharmacological activities and clinical effects. *Alternat Med Rev.,* **16:** 66-77.

Sheeja, K., Shihab, P.K., and Kuttan, G. (2006). Antioxidant and anti-inflammatoryactivities of the plant *Andrographis paniculata* Nees. *Immunopharmacol Immunotoxicol.,* **28:** 129-140.

Sheeja, K., and Kuttan, G. (2007). Activation of cytotoxic T lymphocyte responses and attenuation of tumor growth *in vivo* by *Andrographis paniculata* extract and andrographolide. *Immunopharm Immunot.,* **29:** 81-93.

Shen, Y-H., Li, R-T., Xiao, W-L., Xu, G., Lin, Z-W., Zhao, Q-S., and Sun, H-D. (2006). Ent-labdane diterpenoids from *Andrographis paniculata*. *J Nat Prod.,* **69:** 319-322.

Shukla, B., Visen, P.K., Patnaik, G.K., and Dhawan, B.N. (1992). Choleretic effect of andrographolide in rats and guinea pigs. *Planta Med.,* **58:** 146-149.

Singh, R.P., Banerjee, S., and Rao, A.R. (2001). Modulatory influence of *Andrographis paniculata* on mouse hepatic and extrahepatic carcinogen metabolizing enzymes and antioxidant status. *Phytother Res.,* **15:** 382-390.

Singha, P.K., Roy, S., and Dey, S. (2003). Antimicrobial activity of *Andrographis paniculata*. *Fitoterapia,* **74:** 692-694.

Singha, P.K., Roy, S., and Dey, S. (2007). Protectiveactivity of andrographolide andarabinogalactan protcins from *Andrographis paniculala* Nees againstethanol-induced toxicity in mice. *J Ethnopharmacol.,* **111:** 13-21.

Siripong, P., Kongkathip, B., Preechanukool, K., Picha, P., Tunsuwan, K., and Taylor, W.C. (1992). Cytotoxic diterpenoid constituents from *Andrographis paniculata* Nees. Leaves. *J. Sci. Soc. Thailand,* **18:** 187-194.

Spasov, A.A., Ostrovskij, O.V., Chernikov, M.V., and Wikman, G. (2004). Comparative controlled study of *Andrographis paniculata* fixed combination, Kan Jang and an Echinacea preparation as adjuvant, in the treatment of uncomplicated respiratory disease in children. *Phytother Res.,* **18:** 47-53.

Subramanian, R., Asmawi, M.Z., and Sadikun, A. (2008). *In vitro* alpha-glucosidase and alpha-amylase enzyme inhibitory effects of *Andrographis paniculata* extract and andrographolide. *Acta Biochim Pol.,* **55:** 391-398.

Samy, R.P., Thwin, M.M., Gopalakrishnakone, P., and Ignacimuthu, S. (2008). Ethanobotanical survey of folk plants for the treatment of snakebites in southern part of Tamil Nadu, India. *J Ethanopharmacol.*, **115:** 302-312.

Talukdar, P.B., and Banerjee, S. (1968). Studies on the stability of andrographolide. *Indian J Chem.*, **6:** 252-254.

Tang, W., and Eisenbrand, G. (1992). Chinese Drugs of Plant Origin Chemistry, Pharmacology and Use in Traditional and Modern Medicine. Springer Verlag, Berlin, Germany, pp. 97-103.

Thisoda, P., Rangkadilok, N., Pholphana, N., *et al.* (2006). Inhibitory effect of *Andrographis paniculata* extract and its active diterpenoids on platelet aggregation. *Eur J Pharmacol.*, **553:** 39-45.

Trivedi, N., and Rawal, U.M. (2000). Hepatoprotective and toxicological evaluation of *Andrographis paniculata* on severe liver damage. *Ind J Pharmacol.*, **32:** 288-293.

Trivedi, N., and Rawal U.M. (2001). Hepatoprotective and antioxidant property of *Andrographis paniculata* (Nees) in BHC induced liver damage in mice. *Ind J Exp Biol.*, **39:** 41-46.

Trivedi, N.P., Rawal, U.M., and Patel, B.P. (2007). Hepatoprotectve effects of andrographolide against hexachlorocyclohexane-induced oxidative injury. *Intgr Cancer Ther.*, **6:** 271-280.

Verma, N., and Vinayak, M. (2008). Antioxidant action of *Andrographis paniculata* on lymphoma. *Mol Biol Rep.*, **35:** 535-540.

Visen, P.K.S., Shukla, B., Patnaik, G.K., and Dhawan, B.N. (1993). Andrographolide protects rat hepatocytes against paracetamol induced liver damage. *J Ethanopharmacol.*, **40:** 131-136.

Visen, P.K.S., Sarawat, B., Vuksan, V., and Dhawan, B.N. (2007). Effect of andrographolide on monkey hepatocytes against galactosamine nduced cell toxicity: An *in-vitro* study. *J Comp Integr Med.*, **4:** 10.

Voravuthikunchai, S.P., and Limsuwan, S. (2006). Medicinal plant extracts as anti-*Escherichia coli* O157: H7 agents and their effects on bacterial cell aggregation. *J Food Prot.*, **69:** 2336-2341.

Wang, H.W., Zhao, H.Y., and Xiang, S.Q. (1997). Effects of *Andrographis paniculata* component on nitric oxide, endothelin and lipid peroxidation in experimental atherosclerotic rabbits. *Zhoung-guo Zhong Xi Yi Jie He Za Zhi.*, **17:** 547-549.

Wiart, C., Kumar, K., Yosuf, M.Y. *et al.* (2005). Antiviral properties of ent-labdene diterpenes of *Andrographis paniculata* (Nees), inhibitors of herpes simplex virus type 1. *Phytother Res.*, **19:** 1069-1070.

Wibudi, A., Kiranadi, B., Manalu, W., *et al.* (2008). The traditional plant, *Andrographis paniculata* (Sambiloto), exhibits insulin-releasing actions *in vitro. Acta Med Indones.*, **40:** 63-68.

Yoopan, N., Thisoda, P., Rangkadilok, N., Sarasitiwat, S., Pholphana, N., Ruchirawat, S., and Satayavivad, J. (2007). Cardiovascular effects of 14 deoxy 11, 12 didehydro andrographolide and *Andrographis paniculata* extracts. *Planta Med.*, **73:** 503-511.

Yu, B.C., Hung, C.R., Chen, W.C., and Cheng, J.T. (2003). Antihyperglycemic effect of andrographolide in streptozotocin-induced diabetic rats. *Planta Med.*, **69:** 1075-1079.

Zaidan, M.R., Noor Rain, A., Badrul, A.R., *et al.* (2005). *In vitro* screening of five local medicinal plants for antibacterial activity using disc diffusion method. *Trop Biomed.*, **22:** 165-170.

Zhang, X.F., and Tan, B.K. (2000). Antihyperglycemic and antioxidant properties of *Andrographis paniculata* in normal and diabetic rats. *Clin Exp Pharmacol Physiol.*, **27:** 358-363.

Zhang, X-Q., Wang, G-C., Ye, W-C., Li, Q., Zhou, G-X., and Yao, X-S. (2006). New diterpenoids from *Andrographis paniculata* (Burm. f.) Nees. *J Integr Plant Biol.*, **48:** 1122"1125.

Zhao, H.Y., and Fang, W.Y. (1990). Protective effects of *Andrographis paniculata* Nees. On post-infarction myocardium in experimental dogs. *J Tongji Med Univ.*, **10:** 212-217.

Zhao, H.Y., and Fang, W.Y. (1991). Antithrombotic effects of *Andrographis paniculata* Nees in preventing myocardial infarction. *Chinese Med J (Engl).*, **104:** 770-775.

Zhao, F., He, E.Q., Wang, L., and Liu, K. (2008). Anti-tumor activities of andrographolide, a diterpene from *Andrographis paniculata*, by inducing apoptosis and inhibiting VEGF level. *J Asian Nat Prod Res.*, **10**: 467-473.

Zoha, M.S., Hussain, A.H., and Choudhary, S.A. (1989). Antifertility effect of *Andrographis paniculata* in mice. *Bangladesh Med Res Counc Bull.*, **15:** 34-37.

Traditional and Folk Herbal Medicine: *Pages* **477–528**
Recent Researches Vol. 2 (2014)
Editor: **V.K. Gupta**
Published by: **DAYA PUBLISHING HOUSE, NEW DELHI**

16

Phytopharmacological Overview on Controversial Drug: *Murva*

Varsha Tiwari[1]*, Anita Singh[1] and Abhishek Tiwari[1]

ABSTRACT

Ayurveda system of medicine provides a vast range of drugs for a number of diseases without side effects. Murva is one of such important ayurvedic drug used in the treatment of various disorders like intermittent fever, abdominal colic, laxative, urinary diseases, pruritus and diabetes mellitus. Murva is a controversial drug also known as "Sandigdha dravays" is a term used for medicinal plants having various sources. Eleven different plants are considered as Murva in different books and in throughout India. This comprehensive review provides the latest information on eleven plants (considered as murva) with regards to their pharmacological and phytochemical findings.

***Keywords*:** Controversial drugs, *Murva*, Pharmacological, Phytochemical, Diabetes mellitus.

Introduction

"Sandigdha dravays" is a term used for medicinal plants having controversial botanicals as various sources. The Ayurvedic and Sanskrit literature describes a herb with names, which do not precisely indicate the botanical source but many a times attribute to therapeutic utility of the plant. India is a country having a variety of languages and population dependent on different tribal and folklore medicine. The variation in the language sometimes is responsible for confusion in the nomenclature of different plants having similar name. Moreover, the descriptions of a plant in the ancient literature are found in verses having ample use of synonyms. These synonyms

1 Department of Pharmacognosy, Devsthali Vidyapeeth College of Pharmacy, Lalpur, Rudrapur (Uttarakhand).

* *Corresponding author*: E-mail: varshat1983@gmail.com

have caused controversy in the identification of plants and hence the correct source sometime is misleading with a fictious plant. It has become an important task to generate parameters of identification as well as differentiation among different plant sources having a similar name. Since herbal products are prepared using the extracts of plants known for a particular activity, the controversial source sometimes leads to inefficacious preparations (Sethiya *et al.*, 2009, 2010).

Plant-based medicines form a very important component of total medicines available for treating various human and veterinary diseases. The increased demand and reduced availability has reportedly led to even adulterations of many plant-based formulations. Hence, a need has been felt to study the market and other associated processes, such as market demand, supply, harvesting, processing, and use. This requires coordinated and focused research on these aspects of medicinal value of all the plants and their actual uses so that the medicinal value of all the plants can be scientifically verified along with the documentation of traditional knowledge (Hashmi *et al.*, 1994).

Groups of Drugs Used in Ayurveda

In *Ayurveda*, medicinal plants have been classified according to pharmacological action. *Charaka* has described drugs in groups for alleviating diseases. *Arshoghana* contains 10 drugs having beneficial effect in hemorrhoids; the name is derived from Sanskrit words ARSH (Blood loss) and GHNA (killer or remover). Another method adopted by *Charaka* is based on collection of three or more plants having identical properties in one group signifies as *Gana* (means group) (Singh, 2008). Based on the physical forms of the different botanical sources, drugs are grouped in different texts of Ayurveda and are classified as follows:

(a) *Brahatpanchmula*

It is a combination of roots of five medicinal plants, including Bilva (*Aegle marmelos*), Agnimantha (*Premna integrifolia*), Shoynaka (*Oroxylum indicum*), Gambhari (*Gmelina arborea*), and Patla (*Sterospermum suaveolens*).

(b) *Dashmula*

It is a combination of roots of ten medicinal plants, including Bilva (*Aegle marmelos*), Agnimantha (*Premna integrifolia*), Shoynaka (*Oroxylum indicum*), Gambhari (*Gmelina arborea*), Patla (*Sterospermum suaveolens*), Shalaparni (*Desmodium gangeticum*), Prishnaparni (*Uraria picta*), Brahati (*Solanum indicum*), Kantkari (*Solanum xanthocarpum*), and Gokshura (*Tribulus terrestris*).

(c) *Triphala*

It is a combination of fruits of three medicinal plants, including Haritaki (*Terminalia chebula*), Vibithaka (*Terminalia belerica*), and Amalaki (*Emblica officinalis*).

(d) *Murva*

It is a combination of roots of eleven medicinal plants, including *Marsdenia tenacissima, Marsdenia roylei, Bauhinia tomentosa, Helicteres isora, Maerua oblongifolia, Clematis triloba, Chonemorpha fragrans, Salvadora persica, Wattakaka volubilis, Senseviera roxburghiana, Argyreia nervosa.*

Since the plant drugs are collected generally from wild sources sometimes by skilled but many a times by unskilled collectors and due to improper identification of correct species of plant/organ and collection methods lead to a state of confusion, and thus a controversy is cropped up (Table 16.1) (Vaidya, 1982; Kumar, 2007).

Table 16.1: Shows the controversial plants with their important ayurvedic formulations.

Sl.No.	*Name of the Formulation*	*No. of Plants*	*Confusing Plants*
1.	Daruharidra	02	*Berberis aristata* (Berberidaceae), *Coscinium fenestratum* (Menispermaceae)
2.	Brahmi	03	*Bacopa moniera* (Scrophulariaceae), *Hydrocotyle asiatica* (Umbellifereae), *Moniera cuneate*
3.	Amaravala	02	*Cascuta reflexa* (convolvulaceae), *Cassyatha filiformis* (Lauraceae)
4.	Punarnava	02	*Trianthema partulacastrum* (Ficoidaceae), *Boerhavia diffusa* (Nyctaginaceae)
5.	Pasanabheda	11	*Saxifraga ligulata* (saxifragaceae), *Aerva lanata, Aerva javanica* (Amaranthaceae), *Ammania baccifera* (Lythraceae), *Bergenia ligulata* (saxifragaceae), *Bryophyllum calcynum* (Crassulaceae), *Coleus aromaticum* (Labiatae), *Rotuala aquatica* (Boraginaceae), *Bridelia Montana* (Euphorbiaceae), *Homania riparia* (Euphorbiaceae), *Ocimum basilicum* (Labiatae)
6.	Jivanti	03	*Leptadenia reticulata* (Asclepiadaceae), *Desmotrichun fimbritum* (Orchiaceae), *Cimicifuga foetida* (Ranunculaceae)
7.	Shankhpushpi	04	*Convolvulus pluricaulis* (Convolvulaceae), *Evolvulus alsinoides* (Convolvulaceae), *Canscora decussate* (Gentianaceae), *Clitorea ternatea* (Papilonaceae)
8.	Rasna	10	*Vanda tessellate* (Orchidaceae), *Alpinia galangal* (Scitaminaceae), *Pluchea lanceolata* (Compositae), *Viscum album* (Loranthaceae), *Withania coagulans* (solanaceae), *Aristolochia indica* (Aristolochiaceae), *Inula racemosa* (Compositae), *Rauwolfia serpentiana* (Apocynaceae), *Lochnera rosea* (Apocynaceae), *Enicostema littorale* (Gentianaceae)
9.	Nagakesaram	03	*Mesua ferrea* (Guttifereae), *Ochrocarpus longifolius* (Guttifereae), *Dillenia pentagyna* (Dilleniaceae)
10.	Brahma-dandi	03	*Argemone maxicana* (Papaveraceae), *Tricholepis glaberrima* (Compositae), *Echinops echinatus* (Echinaceae)
11.	Ashoka	03	*Polyalthia longifolia* (Anonaceae), *Saraca indica* (Caesalpiniaceae), *Shorea robusta* (Dipterocarpaceae)
12.	Twak	03	*Cinnamomum tamala* (Lauraceae), *Cinnamomum zeylanicum* (Lauraceae), *Cinnamomum cassia* (Lauraceae).
13.	Murva	11	*Marsdenia tenacissim* (Asclepiadaceae), *Marsdenia roylei* (Asclepiadaceae), *Bauhinia tomentosa* (Leguminoseae), *Helicteres isora* (Sterculiaceae), *Maerua oblongifolia* (Capparaceae), *Clematis triloba* (Ranunculaceae), *Chenomorpha fragrans* (Apocynaceae), *Salvadora persic* (Salvadoraceaa), *Wattakaka volubilis* (Asclepiadaceae), *Senseviera roxburghiana* (Agavaceae), *Argyreia nervosa* (Convolvulaceae).

There is a huge surge in Ayurvedic plants and world over it has become a subject of intensive research for various aspects. There have also been substantial efforts to standardize the Ayurvedic crude drugs as well as finished Ayurvedic medicines. However, these initiatives would imperatively need establishing correct identity of the plant drug. The long history of safe usage of Ayurvedic medicines can be extrapolated only when the botanical identity of the plant going into those medicines is established and standardized. Hence proper nomenclature of all crude drugs and establishing their exact botanical origin is must.

Current practice of ayurvedic physician and Industry is to use one of the several alternative herbs that are considered equivalent to ayurvedic plant drugs. Some amount of legal sanction exists for use of plants as one can use only plant drugs mentioned in one of the 56 authorized books by Drugs and Cosmetic Act, 1940 of India. Hence even though, one would use the alternative herbs, one would restrict to only those which are mentioned in the ayurvedic authorized text books. However, for global acceptance as well as providing safe and effective ayurvedic products, it would be required to identify which of the particular botanical entities is to be used to have regulatory compliance. Ayurvedic scholars, plant taxonomists and regulators need to work together on this (Gupta, 2003; Anonymus, 1978).

Murva is a controversial drug and 11 plants are considered as *murva* in different parts of India. One of the synonyms of this plant is *'Dha-nurgunopayogya'* meaning 'the plant whose bark is being used for the bow-strings'. The plant which has tough fibres is the *Murva*. There are many such fibre yielding plants are found in the vegetable kingdom. The plants considered as *Murva* are:

- *Marsdenia tenacissima* (Roxb.) Moon
- *Marsdenia roylei* Wight
- *Bauhinia tomentosa* L.
- *Helicteres isora* L.
- *Maerua oblongifolia* (Forsk.) A. Rich
- *Clematis triloba* Heyne ex Roth
- *Chonemorpha fragrans* Moon (*C. macrophylla* Roxb. G.Don (Apocynaceae)
- *Salvadora persica* L.
- *Wattakaka volubilis* (Linn. f.) Stapf
- *Senseviera roxburghiana* Schult and Schult.f.(*S. zeylanica* Roxb.)
- *Argyreia nervosa* (Burm.f.) convolvulaceae (other names are convolulus nervosus (Burm.f.) *Argyreia speciosa L.f.(Sweet), Lettsomia nervosa* (Burm.f.) Roxb.

Murva is used in several Ayurvedic formulations as an ingredient (Tables 16.2 and 16.3) for the treatment of intermittent fever, abdominal colic, laxative, urinary diseases, pruritus, diabetes mellitus, epilepsy, piles, typhoid, sterility, rigidity in lower limbs and skin diseases. *Marsdenia tenacissima* (Asclepiadaceae) is an accepted botanical source of *Murva. H. isora* is also considered as *Murva* in Punjab, *Maerua arneria* is used as *Murva* in Bihar, *Chonomorpha fragrans* used as *Murva* in Kerala,

Clematis triloba and *Bauhinia tomentosa* used sa *Murva* in West Bengal, *Senseviera roxburghiana* Schult and Schult.f.(*S. zeylanica* Roxb.) in west Bengal, *Salvadora persica* L. and *Wattakaka volubilis* (Linn. f.) Stapf in south India; whereas *Marsdenia tenacissimais* is an accepted source of *Murva* (Singh, 2002; Kulkarni and Shahida, 2004).

Table 16.2: Antiobesity formulations of *Murva.*

Sl.No.	*Name of the Formulation*	*Manufactured by*	*Indications*
1.	OBENYL Tabs	Charak	Obesity, obesity in patients with diabetes, hypertension, heart disease.
2.	R-EXEX Capsule	Ajmera	Exogenous obesity, abnormalities in carbohydrate metabolism.

Table 16.3: Antipyretic formulations of *Murva.*

Sl.No.	*Name of the Formulation*	*Manufactured by*	*Indications*
1.	FEVEREND syrup	Chirayu	Fever
2.	MAHASUDHARSHAN GHANBATI Tabs	Baidyanath	All types of Fevers.
3.	PYRICIN Tabs	Manil	Common cold, fever, bodyache, headache, cough.
4.	M.P.6 Capsule	Vasu	Malaria, refractory malaria, hyperpyrexia, pain and bodyache.

Description of Plants

Marsdenia tenacissima (Roxb.) Moon (An accepted source of Murva)

Synonyms: Rajmahal- creeper, Jiti, maruabel

Family: Asclepiadaceae.

Description

It is a perennial climber with grey bark. Leaves are 7-15 cm long and 7-10 cm wide, broadly ovate, acuminate, deeply cordate at the base with rounded lobes; both surfaces are densely velvety tomentose when young, but become almost glabrous above when old (Figure 16.1). Flowers are greenish yellow, arranged in much branched, corymbose cymes. Fruit is a follicle, about 10-15 cm long (Kumar, 2007).

Murva is used in several ayurvedic formulations as an ingredient for the treatment of intermittent fever, abdominal colic, laxative, urinary diseases, pruritus, diabetes mellitus, epilepsy, piles, typhoid, sterility, rigidity in lower limbs and skin diseases.

Figure 16.1: Shows *Marsdenia tenacissima* plant.

Figure 16.2: Habitat of *Marsdenia tenacissima* plant.

Distribution

This species is globally distributed in India, Burma, Indo-China and west China (Figure 16.2). Within India, it is found in the Himalays from Kumaon to Assam up to an altitude of 1500 m. (Yoganarasimhan, 2000; Anonymus, 2004).

Phytochemical Studies

11α, 12β-Acetonyltenacissoside F acetone solvate: a new polyoxypregnane glycoside was isolated from the stems of *Marsdenia tenacissima* (Li *et al.*, 2006).

Further polyoxypregnanes were isolated from *Marsdenia tenacissima* (Qiu *et al.*, 1996).

A new C_{21} steroid glycoside was isolated from *Marsdenia tenacissima* (Jing *et al.*, 2008).

Two new Polyoxypregnanes from *Marsdenia tenacissima* were isolated and their structures were also elucidated (Luo *et al.*, 1993).

Polyoxypregnane glycosides were isolated from the stems of *Marsdenia tenacissima* by high-performance liquid chromatography/tandem mass spectrometry (Juanjuan *et al.*, 2008).

Five glycosides from the Chinese drug 'tong-guang-san' (stems of *Marsdenia tenacissima*) were isolated (Miyakawa *et al.*, 1986).

Polyoxypregnanes from the stems of *Marsdenia tenacissima* have been isolated (Li *et al.*, 2007).

Eight polyoxypregnane glycosides, marsdenosides A-H, were isolated and their structures were also established on the basis of spectroscopic analysis and chemical evidence from the $CHCl_3$ soluble fraction of the ethanol extract of the *Marsdenia tenacissima* (stem) (Deng *et al.*, 2005).

Four Pregnane glycosides named marstenacissides A (1), B (2), C (3) and D (4), have been isolated and their structures were established on the basis of chemical and spectral methods from the *Marsdenia tenacissima* (stem) (Xia *et al.*, 2004).

Fingerprinting of *Marsdenia tenacissima* by Capillary Electrophoresis Compared to HPLC and this method has been compared with that of conventional HPLC method (Luhua *et al.*, 2010).

Three New Polyoxypregnane Glycosides from *Marsdenia tenacissima* were isolated and their from the stem structures elucidated by means of chemical evidence (Deng *et al.*, 2005).

Cissogenin, a pregnane genin was isolated rom *Marsdenia tennacissima* (Singhal *et al.*, 1980).

Tenasogenin, a pregnane ester was isolated from *Marsdenia tennacissima* (Singhal *et al.*, 1980).

Pharmacological Studies

Antitumor potential of *Marsdenia tenacissima* has been evaluated. The herb possessed extensive antitumor effects in experimental and clinical applications (Qian *et al.*, 2009).

Tenacigenin B Derivatives from the *Marsdenia tenacissima* have been shown to increased the sensitivity of HepG2/Dox cells towards the antitumour drugs doxorubicin, vinblastine, puromycin, and paclitexel (Figure 16.3).

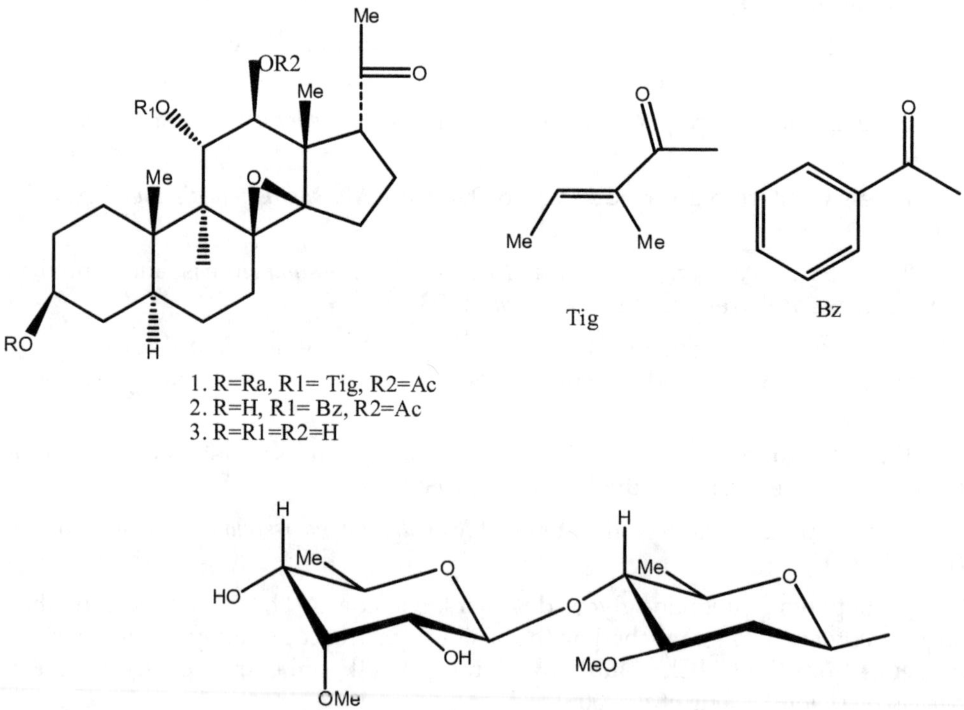

1. R=Ra, R1= Tig, R2=Ac
2. R=H, R1= Bz, R2=Ac
3. R=R1=R2=H

Figure 16.3: Shows the structure of Marsdenosides.

Important Ayurvedic Formulations

Ayaskrti, Dadhika ghrata, Mahapanchagarya ghrata, Sudarshana churna, Candana Balakasad taila, Prameha Mihira taila, Gandharva churna and *Sidha Praneswar-rasa.*

Marketed Formulations

Murva is used as one of the ingredient in *Mahamanjistadi kvatha, Sudarshana churna, Aravindaasava Mahayogaraj guggulu.* Few Commercial formulations of *Murva* are given in table 1 and 2 (Yoganarasimhan, 2000; Nayak, 2007).

Marsdenia roylei Wight

Figure 16.4: Shows *Marsdenia roylei* plant.

Synonyms: Muurvaa (var.) Maruaa-bel, Khaarchu (Garhwal), Dudilaro

Family : Asclepiadaceae.

Description

The plant can grow in semi-shade (light woodland) or no shade (Figures 16.4–16.6). The flowers are hermaphrodite. The juice of the stem is used in the treatment of gastric troubles and peptic ulcers. A silky-white strong fibre is obtained from the stems. It is used for making ropes (Anonymous, 2005).

Figure 16.5: Shows *Marsdenia roylei* twig.

Figure 16.6: Shows *Marsdenia roylei* herbarium.

Distribution

Plant is a shrub can be found as climbing at altitudes of 900–2400 m from Pakistan to Arunachal Pradesh.

Phytochemical Studies

A new pregnane glycoside has been isolated from *Marsdenia roylei* as potential antioxidant and antidyslipidemic agents (Sethi *et al.*, 2008).

Two pregnane oligoglycosides have been isolated from *Marsdenia roylei* (Gupta *et al.*, 2009).

Two oligosaccharides have been isolated from *Marsdenia roylei* (Kumar *et al.*, 1999).

Trisaccharides have been isolated from *Marsdenia roylei* (Kumar *et al.*, 1999).

Pregnanes and pregnane glycosides from *Marsdenia roylei* (Gupta *et al.*, 2003).

Marsin from *Marsdenia roylei* (Gupta *et al.*, 2003) (Figure 16.7).

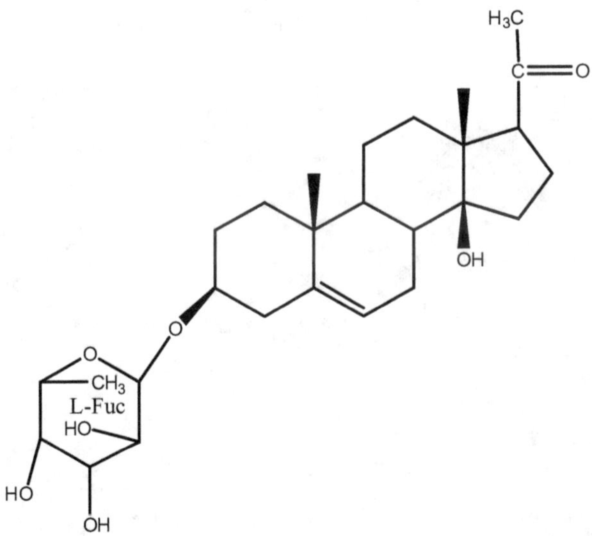

Figure 16.7: Marsin structure.

A new compound Marsin was isolated from *Marsdenia roylei* (Khare *et al.*, 1997).

Structural studies of pregnane and their glycosides from *Marsdenia roylei* have been done (Sethi *et al.*, 2008).

Pharmacological Studies

A new pregnane glycoside roylenine (**1**) was isolated from the $CHCl_3$-EtOH (4: 1) extract of *Marsdenia roylei* and its structure was established on the basis of spectroscopic studies. (Gupta *et al.*, 2004).

Bauhinia tomentosa Linn.

Synonyms: Yellow Bauhinia, Yellow Kachnar.

Family: Leguminoseae (Fabaceae).

Description

This is a medium to large shrub to a small tree, up to 4m in height. Leaves are divided into two lobes, light green in colour, with a leathery texture, carried on branches that are often drooping. It produces large bell-shaped, bright yellow flowers with a black to deep maroon coloured centre from December to March. The species name "*tomentosa*" means hairy and it refers to the velvety/hairy pods (Bhattacharjee, 1998; Singh and Panda, 2005; John, 2001; Sikarwar *et al.*, 2003).

According to Kirtikar and Basu (1993), the bruised bark is applied externally to tumors and wounds. Decoction of the root bark is administered for inflammation of the liver and as vermifuge (Figures 16.8–16.10).

Distribution

This plant can be found through out India and Sri Lanka.

Figure 16.8: Shows *Bauhinia tomentosa* plant.

Figure 16.9: Shows *Bauhinia tomentosa* flower.

Figure 16.10: Shows *Bauhinia tomentosa* fruit.

Pharmacological Activities

Pharmacognostical and preliminary phytochemical studies of *Bauhinia tomentosa* flower has been done (Manivannan, *et al.*, 2010).

Colouring matter of the flower petals of *Bauhinia tomentosa* (Linn) has been isolated (Row and Viswanadham, 1954).

Bauhinia tomentosa stimulates the immune system and scavenges free radical generation *in vitro* (Kannan and Renitta, 2010).

Antioxidant potential of ethanolic extract of *Bauhinia tomentosa* (Linn) flower has also been evaluated (Vijayakumar *et al.*, 2010).

Antimicrobial activity of *Bauhinia tomentosa* and *Bauhinia vahlii* roots has been done on the hexane, ethylacetate and methanol extracts of roots (Dugasani *et al.*, 2010).

Helicteres isora L.

Figure 16.11: Shows *Helicteres isora* plant.

Synonyms: *Murva*, Avartani, Indian screw tree

Family : Sterculiaceae

Figure 16.12: Shows *Helicteres isora* flower.

Figure 16.13: Shows *Helicteres isora* fruit.

Description

Helicteres isora (Sterculiaceae) is a woody shrub or small tree 2-3 m tall, young shoots stellate hairy, bark thin and strong (Figures 16.11–16.13). The fruits of *Helicteres isora* Linn (Sterculiaceae) have been used in the indigenous system of medicine in India for the treatment of griping bowels and diarrheal diaseases (Sharma, 2005). The roots and the bark are expectorant, demulcent, hypoglycemic and useful in colic, scabies, gastropathy, diabetes, diarrhoea and dysentery (Anonymous, 2005; Gamble, 2005; Saldanha and Nicolson, 1976).

The Plant wood is twice as hard as teak, is extensively used for making boats, cards, carriages, firewoods, bows of violin, planking, tool -handles, beams and fence posts. Hence the plant is also known as *Murva* in Punjab and *Avartani* in Sanskrit is proved to be a very important medicinal plant (Vaidyaratnam, 1995).

Distribution

It is distributed from Jhelum Eastwards to Nepal, Bihar, West Bengal, Central, Western and Southern India.

Pharmacological Activities

- ☆ Antioxidant and anti-diabetic activity of the plant has been done (Suthar *et al.*, 2009).
- ☆ Anti-hyperglycemic activity of the plant has been done (Venkatesh *et al.*, 2003).
- ☆ Glucose uptake activity of the plant has been done (Gupta *et al.*, 2009).
- ☆ Effect of *H. isora* bark extract on Protein metabolism and marker enzymes in streptozotocin induced diabetic rats has been done (Kumar *et al.*, 2007).
- ☆ Antispasmodic activity of the plant has been done (Pohocha and Grampurohit, 2001).
- ☆ Influence of *H. isora* bark extract on Glycaemic control and Renoprotective activity in Streptozotocin -induced diabetic rats has been studied (Kumar *et al.*, 2008).
- ☆ Antifungal activity of the plant has been done (*Badgujar, 2009)*.
- ☆ Effect of saponins from *Helicteres isora* on lipid and glucose metabolism regulating genes expression has been studied (Bhavsar *et al.*, 2009).
- ☆ *In vitro* α-Amylase inhibition and Anti-hyperglycemic activity of *H. isora* in Streptozotocin-induced rats has been studied (Alok *et al.*, 2009).
- ☆ Phytochemical and antibacterial potential of *Helicteres isora L.* fruits against enteric bacterial pathogen has been studied (Tambekar *et al.*, 2008).
- ☆ Preliminary biochemical screening of *H. isora* stem bark in CCl4 induced toxicity in rats has been done (Dehvi *et al.*, 2008).
- ☆ *In vitro* antioxidant activity of hot aqueous extract of *H. isora* fruits has been done (Basniwal *et al.*, 2009).

- Attenuation of *H. isora* L. bark extracts on streptozotocin-induced alterations in glycogen and carbohydrate metabolism in albino rats has been done (Kumar *et al.*, 2009).
- Effect of *Helicteres isora* bark extract on blood glucose and hepatic enzymes in experimental diabetes has been studied (Kumar *et al.*, 2006).
- Hypoglycaemic effect of *Helicteres isora* bark extract in rats has been studied (Kumar *et al.*, 2006).
- Influence of *Helicteres isora* administration for diabetes mellitus: Its effect on erythrocyte membrane and antioxidant status has been studied (Kumar *et al.*, 2009).
- Antimicrobial activity of *Helicteres isora* root has been studied (Venkatesh S *et al.*, 2007).
- Antinociceptive activity of the plant has been studied (Venkatesh *et al.*, 2007).
- Hypolipidaemic activity of *Helicteres isora* L. bark extracts in streptozotocin induced diabetic rats has been studied (Kumar and Murugesan, 2008).
- Effect of *Helicteres isora* bark extracts on heart antioxidant status and lipid peroxidation in streptozotocin diabetic rats has been studied (Kumar *et al.*, 2008).
- Antipyretic activity of alcohol and aqueous extract of *Helicteres isora* L. root has been studied (Tiwari *et al.*, 2010).

Phytochemical activity

Quantitative determination of β- sitosterol from stem bark of *Helicteres isora* L has been studied (Jain *et al.*, 2009).

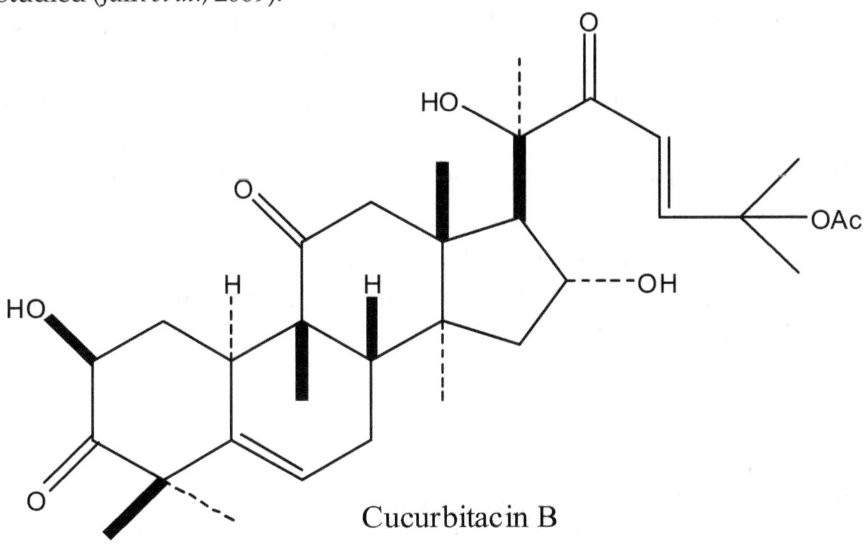

Figure 16.14: Cucurbitacin B structure.

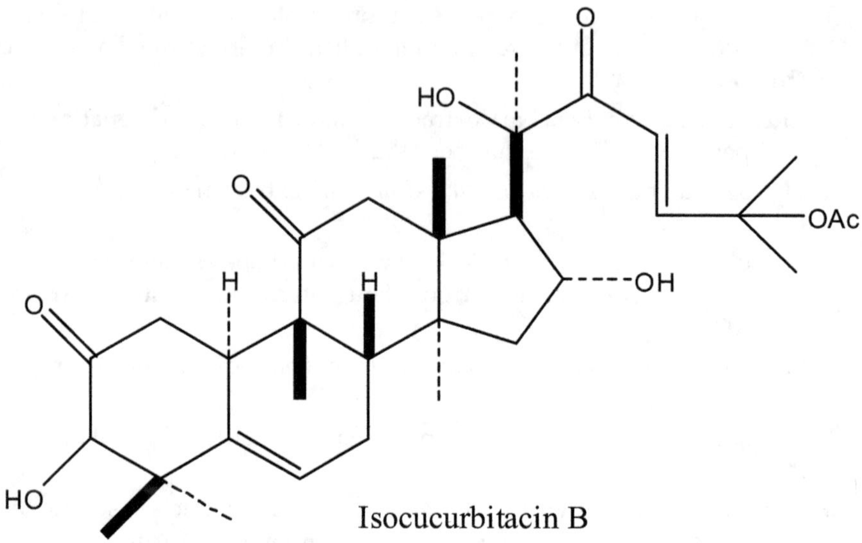

Figure 16.15: Isocucurbitacin B structure.

Isolation and characterization of bioactive terpene from the fruits has been studied (Sandhya and Grampurohit, 2008).

Cucurbitacin B and Isocucurbitacin B: Cytotoxic components of *Helicteres isora* has been isolated (Bean *et al.*, 1985) (Figures 16.14 and 16.15)

Helisterculins A and B, Two New (7.5', 8.2')-Neolignans, and Helisorin, the First (6.4', 7.5', 8.2') Neolignan, from *Helicteres isora* has been isolated (Tezuka *et al.*, 1999).

Flavonoid glucuronides from *Helicteres isora* has been isolated (Kamiya *et al.*, 2001).

Studies on the constituents of fruits of *Helicteres isora* L have been done (Deshmukh and Pandit 1968).

Preliminary Phytochemical analysis, HPTLC studies of the plant has been done (Tiwari *et al.*, 2010).

Maerua oblongifolia (FORSK.) A. Rich:

Synonyms: Murva, Morata, *Maerua angustifolia* A.Rich., *Streblocarpus oblongifolius* (Forssk.), *Maerua racemosa* Lanza.

Family : Capparaceae

Description

This plant is a rare woody bushy under-shrub, sometimes scandent, with thick rootstock and thick leaves, flowers strongly scented. The plant survives annual burning by throwing up shoots from its thick rootstock. It is recorded as used by a Wolof medicineman in Senegal along with eight other drug-plants to treat syphilis (Figures 16.16–16.18).

Figure 16.16: Shows *Maerua oblongifolia* plant.

Figure 16.17: Shows *Maerua oblongifolia* twig.

Figure 16.18: Shows *Maerua oblongifolia* flower.

Distribution

The plant is distributed in savanna woodland from Senegal to Nigeria, and in Sudan to the Red Sea and Arabia.

Phytochemical Studies

Three lupane triterpenoid were isolated and characterized from *Maerua oblongifolia* by one and two dimensional NMR, MS and IR spectra, as lup-20(29)-en-3β,30-diol (Figure 16.19) (Mogib, 1999).

The *in vitro* antioxidant, antimicrobial and cytotoxic activities of the plant have been done (Fatimi *et al.*, 2007).

Antipyretic activity studies of two botanical sources of the drug *Murva* have been done (Madhavan *et al.*, 2010).

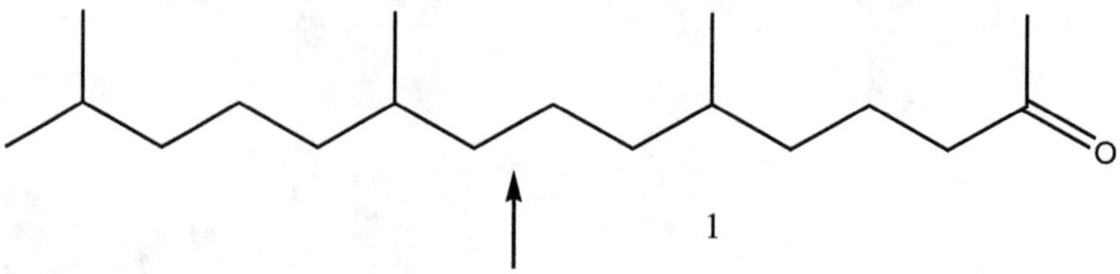

1

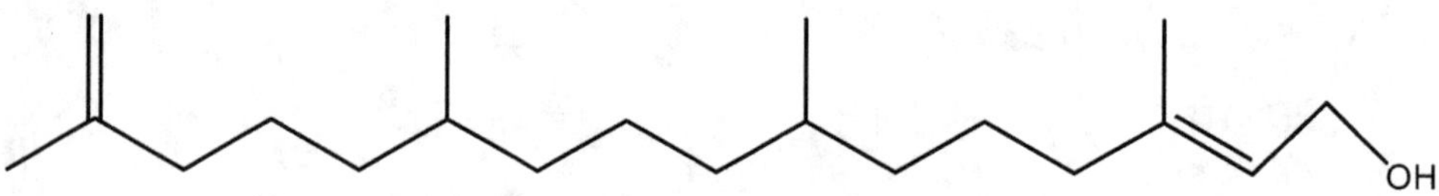

2

1. R' $=CH_2OH$, R" $= CH_3$

2. R' $=CHO$, R" $= CH_3$

3. R' $=CH_3$, R"$= CH_2OH$

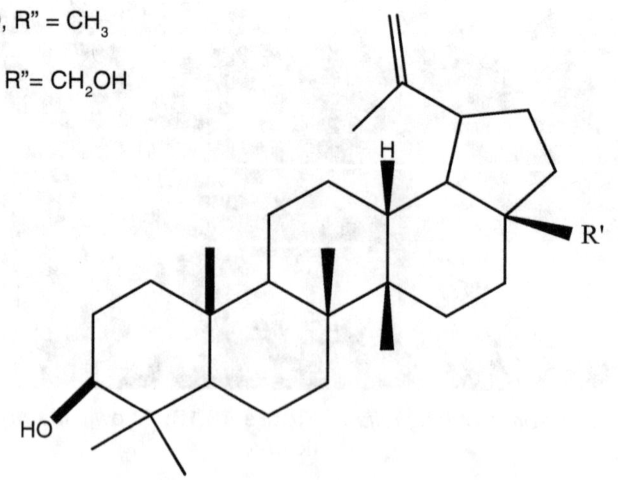

Figure 16.19: Lupane triterpenoid structure.

Figure 16.20: Shows *Clematis triloba* plant.

Clematis triloba Heyne ex Roth:

Synonyms: *Clematis kassneri* Engler, Thunbergii, Deccan Clematis, Murhar, Morvel, Laghuparnika.

Family: Ranunculaceae

Description

Figure 16.21: *Clematis triloba* plant with flower.

It is a less common and extensive climber. Sepals are white 4-6, membranous. Many species of Clematis are grown as ornamentals for their colourful flowers. Other members of the family are grown as hardy annuals, biennials, and perennials, as aquatics, and as alpines, the following genera being most commonly represented in cultivation (Chopra, 1995; Kirtikar and Basu, 1996; Anonymous, 1988). Presence of Oleanolic acid and ursolic acid in *Clematis triloba* are also present (Fai and Tao, 2009) (Figures 16.20–16.22).

Figure 16.22: *Clematis triloba* flower.

Distribution

The plant is distributed worldwide, but richly represented in the temperate regions, particularly in East Asia. The plant is distributed throughout India growing in moist forests (Wencai Wang *et al.*,) (Figures 16.23 and 16.24).

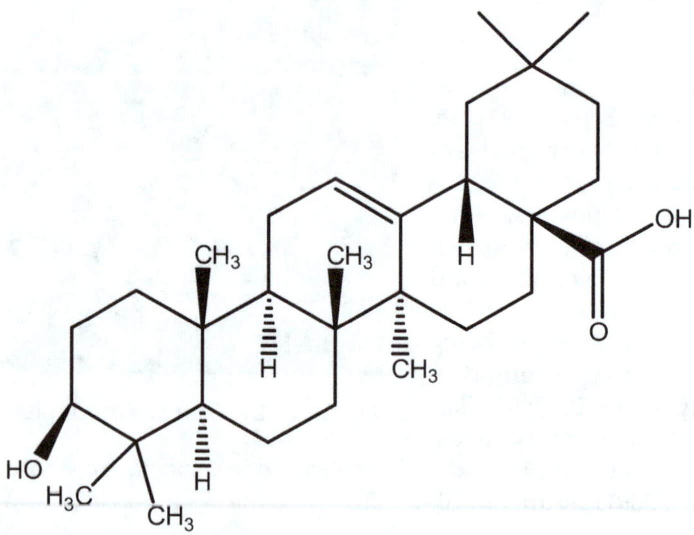

Figure 16.23: Structure of Oleanolic acid.

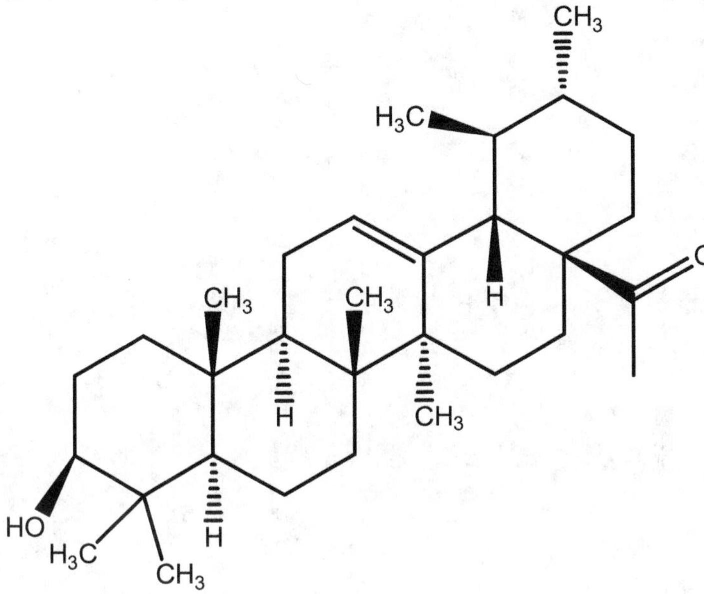

Figure 16.24: Structure of Ursolic acid.

Pharmacological Activity

Anthelmintic activity of a polyherbal preparation containing *Thespesia populnea* (bark), *Terminalia alata* (bark), *Clematis triloba* (roots) and *Ceratophyllum demersum* (leaves) has been done (Dwivedi *et al.*, 2009).

Antipyretic activity of root of *Climatis triloba* heyne ex. roth in rats has been done (Hatapakki *et al.*, 2010).

Preclinical evaluation of antidiabetic activity of root extracts of *Clematis triloba* Heyne in wistar rats has been done. The antidiabetic activity of metahnol root extract showed a significant decrease in blood glucose in STZ and STZ-NIC induced diabetic rats (Shah and Gohil, 2012).

Chonemorpha fragrans Moon (*C. macrophylla* Roxb. G.Don):

Synonyms: Frangipani Vine, Wood vine, Moorva, Garbhedaro, Murva, Morata.

Family: Apocynaceae

Description

This beautiful vine is a stunning climbing plant, flowering profusely from May–July. The flowers deceptively resemble the Frangipani. It is one of the powerful climbers of the Indian and Malayan forests, climbing to the tops of the tallest trees. Flowers are pure white with a yellow center, and have a delicious rich fragrance. Even without the leaves, the vine is eye-catching with large shiny leaves with prominent veins.

Traditional literature claims it as antidiabetic (Kirtikar and Basu, 1996) (Figures 16.25–16.27).

Figure 16.25: Shows *Chonemorpha fragrans* plants.

Figure 16.26: *Chonemorpha fragrans* flower bunch.

Figure 16.27: *Chonemorpha fragrans* flower.

Distribution

It is distributed on southern part of India (Anonymous, 1988).

Phytochemicals Studies

Chonemorphine, stigmasterol, and ecdysterone: Steroids has been isolated through bioassay-directed plant screening programs (Shah *et al.*, 1989) (Figures 16.28–16.30).

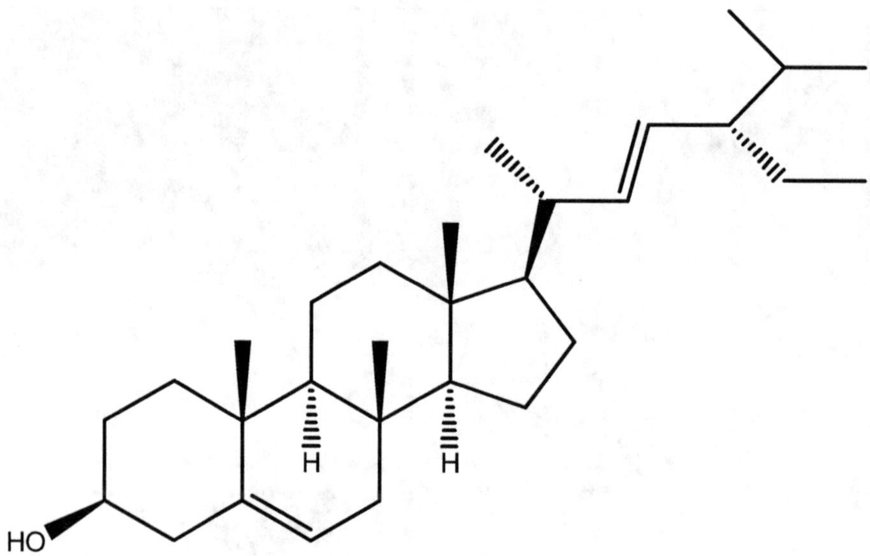

Figure 16.28: Structure of stigmasterol.

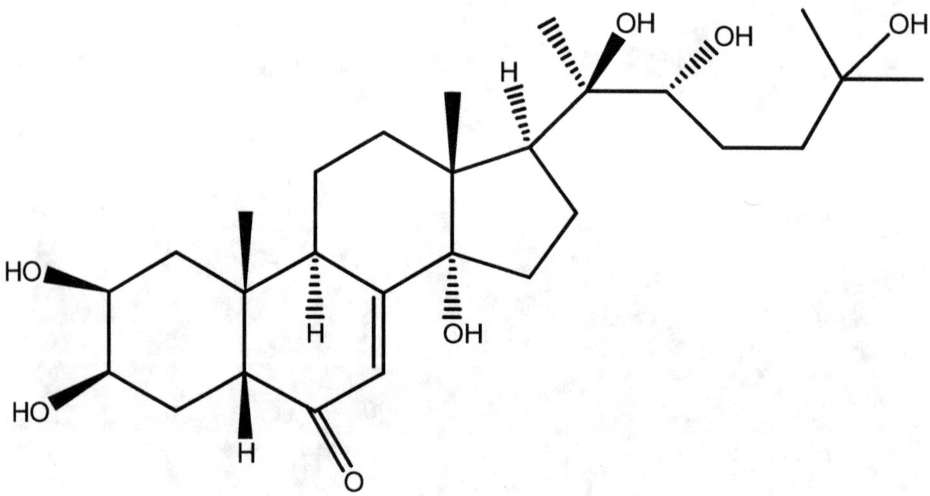

Figure 16.29: Structure of ecdysterone.

Chonemorphine and rapanone–Antipara-sitic agents from plant source has been reported (Shah *et al.*, 1987).

Occurrence of Funtumafrine C in *Chonemorpha macrophylla* has been reported (Chatterjee and Banerjee, 1972).

Pharmacological Studies

Alcohol extract of *Chonemorpha fragrans* root at the dose of 200 mg/kg per oral (p.o.) was reported to produced significantly reduction in blood glucose level in fasted normal rat and against glucose over loaded rats (Shende *et al.*, 2009).

Figure 16.30: Three dimensional structure of ecdysterone.

Skeletal muscle relaxant effect of *Chonemorpha macrophylla* in experimental animals (Roy *et al.*, 2005).

Antiamoebic activity of chonemorphine, a steroidal alkaloid, in experimental models (Chatterjee, 1987).

Salvadora persica L.

Synonyms: Miswak (tooth brush), Kharijal

Family: Salvadoraceae

Description

It is an upright evergreen small tree or shrub, seldom more than 1 ft in diameter reaching a maximum height of 3 m. The fresh leaves are eaten as salad and are used in traditional medicine for treatment of cough, asthma, scurvy, rheumatism, piles, and other diseases. The use of miswak is a pre-Islamic custom, which was adhered to by the ancient Arabs to get their teeth white and shiny (Bhandari, 1990; Craene and Wanntorp, 2009; Sadhan and Almas, 1999). The beneficial effects of miswak in respect of oral hygiene and dental health are partially due to its mechanical action and partially due to pharmacological action.

Fruits are sweet and edible. A fermented drink is reported to be made from the fruits. Fruits possess deobstruent, carminative, diuretic, lithontriptic, and stomachic properties and are used in biliousness and rheumatism. Seeds are used as purgative, diuretic and tonic seed oil is applied on the skin in rheumatism (Atassi, 2002; Bukar *et al.*, 2004).

Figure 16.31: Shows *Salvadora persica* plants.

Figure 16.32: *Salvadora persica* plant with fruits.

Figure 16.33: *Salvadora persica* flower.

Figure 16.34: *Salvadora persica* fruits.

Destribution

It is widely distributed in the arid regions of India and often on saline soils (Danfillo *et al.*, 2004).

Phytochemical Studies

- Chemical constituents (octacosanol, 1-triacantanol, β-sitosterol,β-sitosterol-3-O-β-d-glucopyranoside) from the stem of *Salvadora persica* has been isolated (Jain and saxena,1984)
- History of the toothbrush has been studied shows the presence of essential oil contained [α- and β-thujones, camphor, cineole, β-cymene, limonene, β-myrcene, borneol, linalool, and bornyl acetate] and nonvolatile fraction contained humulene, caryophyllene, β-santatol, andfarnesol (Almas, 2001).
- Benzylamide has been isolated and identified as butanediamide, N1, N4-bis (phenylmethyl)-2(S)-hydroxy-butanediamine (1), *N*-benzyl-benzamide (2), *N*-benzyl-2-phenylacetamide (3), and benzyl urea (4). Compound 3 was found to posses a significant inhibitory effect on human collagen-induced platelet aggregation, and a moderate antibacterial activity against *Escherichia coli* (Khalil, 2006).
- Heavy metals contents are present in the leaf galls of the plant, which posses some growth promoting principles. Metal contents in the control as well as in the galls are prepared and presented in (Table 16.4). The data showed an enhanced content of Cu, Ni, Mn, V, Ti, and Mo in the galls when compared with the control (Raj *et al.*, 1979).
- The effects of extracts of chewing sticks (*Salvadora persica*) on healthy and periodontally involved human dentine has been studied (Almas, 2001; Nawal *et al.*, 2007).

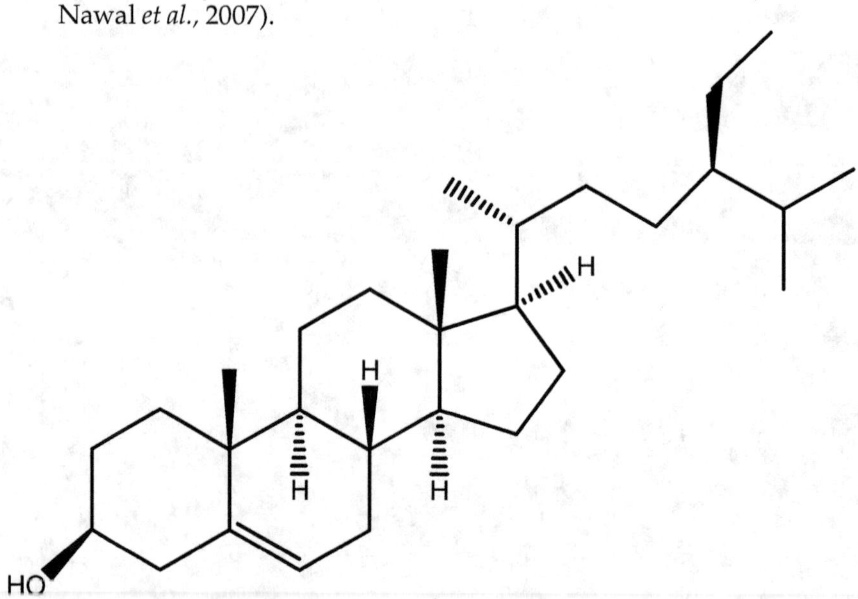

Figure 16.35: Structure of β-sitosterol.

Table 16.4: Heavy metal analysis in the leaf galls of *Salvodra persica*.

Metal	*Control*	*Galls*
Cu	6.0	8.0
Zn	5.0	5.0
Ni	8.0	10.0
Mn	20.0	25.0
Co	15.0	15.0
Pb	0.20	0.20
Cd	0.20	0.20
Ce	0.005	0.004
V	0.005	0.20
Vi	–	0.005
Mo	0.30	0.55
Hg	–	–
Fe	5.0	55.0

Pharmacological Studies

The antimicrobial effects of extracts of *Azadirachta indica* and *Salvadora persica* chewing sticks have been studied. The best antimicrobial effect for *Salvadora persica* extract was noticed at 15 per cent (Almask *et al.*, 2005).

- ☆ The miswak (chewing stick) and oral health studies on oral hygiene practices of urban Saudi Arabians have been done. The effects of prolonged administration of a lyophilized stem decoction of *Salvadora persica* were evaluated in diet induced rat hypercholesterolemic. The results showed that the plant decoction significantly lowered cholesterol and LDL plasma levels in the rats (Galati *et al.*, 1999).
- ☆ Antiulcer activity of this plant has been evaluated, plant possessed significant protective action against ethanol and stress-induced ulcers (Sanogo *et al.*, 1999).
- ☆ Anticonvulsant activity of the plant has been evaluated shows the potentiation of sodium pentobarbital activity and on generalized tonic-clonic seizure pentylentertazol (PTZ) on the rat is (Monforte *et al.*, 2002).
- ☆ Antifertility activity of the plant has been done. The results indicate that miswak has adverse effects on male and female reproduction systems and fertility (Darmani *et al.*, 2003).
- ☆ Antibacterial activity of the plant has been done. The results found that the strong antibacterial effects against all bacteria tested is due to the presence of a volatile active antibacterial compounds (Edi and Selim, 1994).
- ☆ The effects of the extracts of *Salvadora persica* and derum were examined on the proliferation of Balb/C 3T3 of fibroblast and viability of carcinogenic bacteria. Results shows that miswak and derum have adverse effects on

the growth of cariogenic microorganisms, with derum as more active than miswak (Darmine and Al-Hiyasat, 2006).

- ✰ Antimycotic activity of the plant has been done. Aqueous extracts of miswak could be used to reduce the growth of *Candida albicans*. Such inhibition lasts for up to 36 h at concentrations of 15 per cent and above (Sarvesh Paliwal *et al.*, 2007).
- ✰ Release of calcium and chloride into saliva of the plant has been done. They reported that miswak produced significant increases in calcium (22-fold) and chloride (6-fold), and significant decreases in phosphate and pH, saturation of saliva with calcium inhibits demineralization and promotes demineralization of tooth enamel, whereas high concentration of chloride inhibits calculus formation (Kaur *et al.*, 2004).
- ✰ Analgesic effect of the plant has been done. They found that miswak was more effective against thermal stimuli than against chemical stimuli and also acts as an analgesic (Mansour *et al.*, 1996).
- ✰ The cytotoxic potential of plant on gingival and other periodontal structures, using the agar overlay method was studied. Results showed no cytotoxic effect by a freshly cut and freshly used miswak. The cytotoxicity in this study became evident only after 24 h because the agar overlay method depends on the diffusion of the medicament to the agar material (Baeshen *et al.*, 2008; Kubota *et al.*, 1988; Norton and Addy, 1989; Abo and Al-Bagieh, 1996).
- ✰ The effects of prolonged administration of a lyophilized stem decoction of *plant* were evaluated in diet induced rat hypercholesterolemic. The results showed that plant decoction was inactive at 18 h after treatment, whereas at 27 h it was able to reduce cholesterol and LDL plasma levels; in all the experiments HDL and triglycerides were unchanged (Galati *et al.*, 1999).

Wattakaka volubilis (Linn. f.) Stapf

Synonyms: Green milkweed climber, *Dregea volubilis* (Linn. f.) Benth ex. Hook f., Jukti, Cotton milk plant, Green wax flower, Haemajeevanthi, Haemalatha, Naakochhikni.

Family: Asclepiadaceae

Description

It is a tall woody climber or herb/shrub usually twinning the stem sometimes succulent. Leaves opposite and whorled and rarely alternate. Flowers are hermaphrodite, regular, solitary or many toghter usally in auxillary. Roots of this plant are considered as murva and are used in the treatment of several disorders like intermittent fever, laxatives, urinary diseases, pruritus, diabetes mellitus, epilepsy, expectorant, demulsant, piles typhoiod, sterility, shin diseases astringent, abdominal coliac and a cure for scabies. The juice of juvenile leaves are used in the treatement of pyoderma and fever in children.The roots are used as an emetic, also used in the

Figure 16.36: Shows *Wattakaka volubilis* plant.

Figure 16.37: *Wattakaka volubilis* flower.

Figure 16.38: Habitat of *Wattakaka volubilis.*

treatment of kidney diseases. Plant is reported to posses mild CNS-depressant, antispasmodic, anthelmintic, antimutagenic, and anticancer activities.

Distribution

It occurs throughout the hotter parts of India and bar Nicobar Islands ascending to an altitude of 1500m (Hooker BJD, 1883).

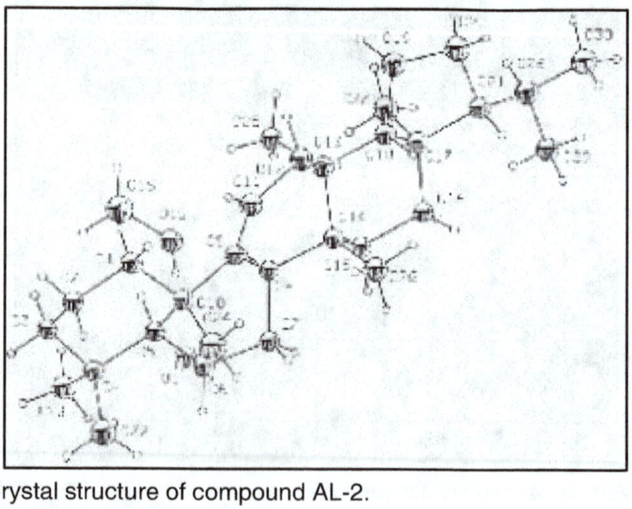

Figure 16.39: Crystal structure of compound AL-2.

Phytochemical Studies

Unusual novel triterpenoid ether, multiflor-7-en-12,13 ether and a new multifloe-7-en-12a-ol from plant has been reported (Reddy *et al.*, 2002).

Compound named AL-2 and two new triterpenoids derivatives WV1 and WV2 and an acetylated derivative WV1a were derived from this plant. (Reddy and Venkateshwarlu, 2003).

Polyoxy pregnane glycosides from the flowers of *Dregea volubilis* (Sahu *et al.*, 2002).

Reichstein and co- workers 2 studied the components of the seeds of the plant and deduced the structure of drevogenins A, B, D and P (Rastogi and Mahrotra, 1969).

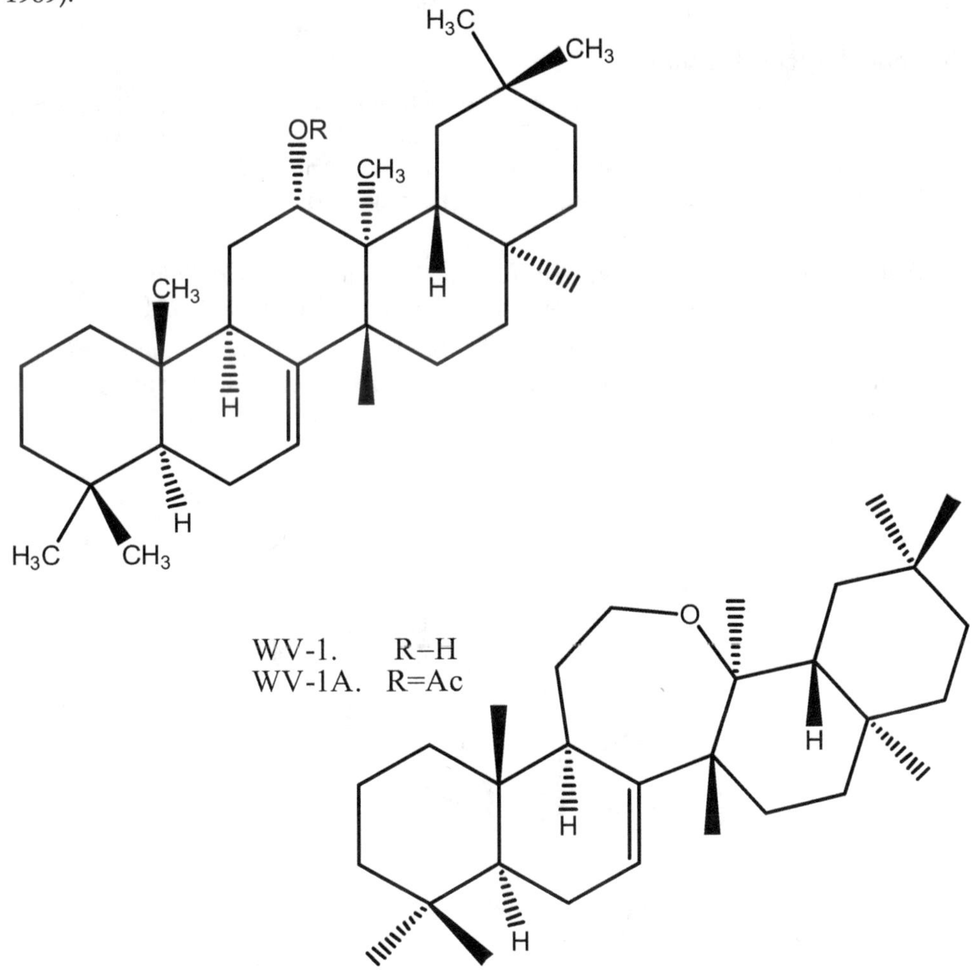

WV-1. R–H
WV-1A. R=Ac

WV-2

Figure 16.40: Structure of Isolated compounds (WV1 and WV2).

Polyoxy pregnane glycosides from the flowers of Dregea Volubilis has been isolated and characterization of twelve polyhydroxy C/D cis-pregnane glycosides from the same plant collected from Thailand has been reported (Panda *et al.*, 2003).

A quantitative estimation of aeridin in *Wattakaka volubilis* by HPTLC has been done and its structure was elucidated by 1D and 2D NMR analyses (Mridula *et al.*, 2009).

A new polyoxypregnane glycoside from the roots of *Dregea volubilis* (L.f) Benth.ex Hook. f and its chondroprotective effect has been evaluated (Sanyacharernkul *et al.*, 2009).

Isolation of β sitosterol, kaempherol-3-galactoside, a 2-deoxy sugar, drevogenin A, drevogenin P, D–cymarose and L-olendrose from the plant has also been reported (Biswas and Samit, 2010).

Pharmacological Activity

In vitro anti-leishmanial and anti-tumour activities of a pentacyclic triterpenoid compound isolated from the fruits of *Dregea volubilis* Benth *Asclepiadaceae* has been reported. Two pregnane glycosides, dregeosides Ap1 and A01, isolated from this plant showed antitumor activities against melanoma B-16 in mice (Biswas and Samit, 2010).

Anti-hyperlipidemic and hypoglycemic activity of *Wattakaka volubilis* methanol extract in alloxan-induced diabetic rats has been done (Kumar *et al.*, 2010).

Antidiabetic, antihyperlipidaemic and antioxidant activity of *Wattakaka volubilis* (l. f) stapf leaves in alloxan induced diabetic rats has been reported (Maruthupandian *et al.*, 2010).

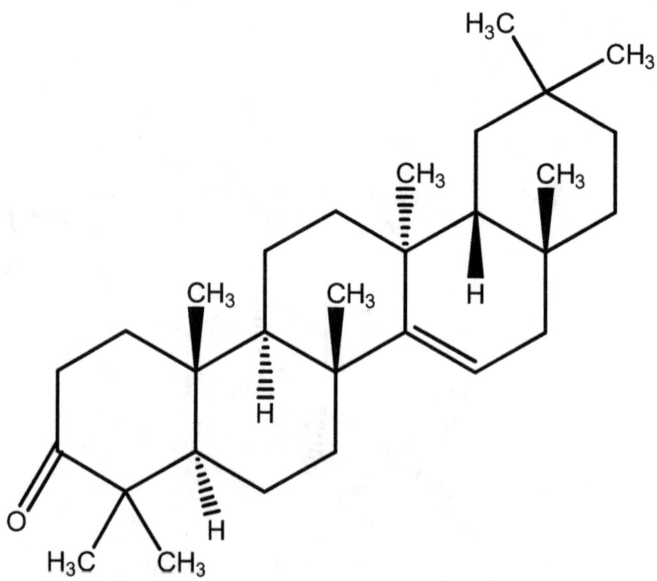

Figure 16.41: Structure of taraxerone (D-friedoolean-14-en, 3 one).

Antiinflammatory, analgesic and anti- lipid peroxidative properties of *Wattakaka volubilis* (Linn. f.) Stapf has been done (Divya *et al.*, 2009).

Preliminary phytochemical evaluation of certain anticancer crude drugs including *Wattakaka volubilis* used by Adivasis of Rayalaseema region, Andhra Pradesh, India has been reported (Krishnaiah *et al.*, 2008).

Hepatotoxicity of *Nigella sativa* seeds and *Dregea volubilis* leaves has been reported (Kamani *et al.*, 1991).

Acute and sub-chronic toxicity study of *Dregea volubilis* fruit in mice has been done (Biswas and Biswakanth, 2010).

Antitumor effect of *Dregea volubilis* fruit in Ehrlich ascites carcinoma bearing mice has been reported (Biswas and Samit, 2010).

Senseviera roxburghiana schult and Schult.F. (*S. zeylanica* ROXB.)

Synonyms: *Murva, Sansevieria zeylanica* Roxb., *Cordyline hyacinthoides* E. F. Wight, Buntot-tigre, Aspi-aspi (Pamp.), Baniat (Is.), Bowstring hemp (Engl.)

Family: Liliaceae

Figure 16.42: *Senseviera roxburghiana* plant.

Description

It is a herbaceous perennial with short stem and a rootstock that is very stout, branching and stoliniferous. Flowers are pale-colored, numerous, in facsicles of 3 to 6, sweet-scented, 2.5 to 3 cm long, with the perianth segments nearly twice as long as the tube. Fruit is sparingly produced, globose, about 8 mm indiameter. Plant contains an active constituent sansevierine (alkaloid). The roasted leaves used as an emollient, rootstocks used for cough, juice of tender shoots used in children to clear their throats of phlegm. It is also used as febrifuge, tonic and purgative, tender roots and rhizome used as expectorant.

Distribution

It is a native of tropical Asia and is cultivated in many parts of the Philippines.

Pharmacological Sudies

Antitumor activity of *Sansevieria roxburghiana* rhizome against Ehrlich ascites carcinoma in mice. Study of hydroalcohol extract of plant for antitumor activity against EA carcinoma exhibited remarkable antitumor activity attributed to its augmentation of endogenous antioxidant mechanisms (Haldar *et al.*, 2010).

Antidiabetic activity and modulation of antioxidant status of plant rhizome inStreptozotocin-induced diabetic rats has been reported. Hydroalcohol extract of plant rhizome in streptozotocin-induced diabetic rats demonstrated remarkable antidiabetic activity possibly through its modulation of endogenous antioxidant status (Haldar *et al.*, 2010).

Invitro cytotoxicity and anticancer activity of the methanol extracts was evaluated. The minimum effective concentration of methanol extracts of leaves that were nontoxic to 3T3 cells but toxic to HepG2 cells (IC_{50}) was recorded at a concentration lesser than 100µg/ml of the extract (Philip *et al.*, 2011).

Argyreia nervosa (Burm.f.) Bojer var. *speciosa*

Synonyms: Hawaiian Baby Woodrose, Adhoguda, Vidhara, Elephant Creeper, Woolly Morning Glory, Samudra-sok, Vridha daraka, Murva.

Family: Convolvulaceae

Figure 16.43: *Argyreia nervosa* plant.

Description

Argyreia nervosa invites attention of the researchers worldwide for its pharmacological activities ranging from aphrodisiac to nematicidal activities. Traditionally, leaves were used by Rajasthani tribe to prevent conception. Chemical analysis revealed the presence of triterpenoids, flavanoids, steroids and lipids. Many polyherbal formulations containing this plant parts are available in the market.This plant is a rare example of a plant whose hallucinogenic properties have only recently been discovered by non-Hawaiians (Praveen *et al.*, 1990; Babber *et al.*, 1978; Shukla *et al.*, 1999; Mishra *et al.*, 1978; George *et al.*, 1949; Khan and Khanum, 2005; Gamble,1956).

Figure 16.44: *Argyreia nervosa* plants with fruits.

Distribution

It is a climbing shrub with woody tomentose stem, found mainly in Deccan, Karnataka and East slopes of the West Ghats at an altitude of 900 meters.

Phytochemical Studies

Two new flavones glycosides were reported from the leaves of the plant (Ahmad *et al.*, 1993).

Seeds of the plant were found to posses glycerides of alminate, stearic, linoleic, linolenic, oleic acids (Biswas *et al.*, 1947; Batra and Mehta, 1985).

Presence of ergoline alkaloids (clavine type) have also been reported from the seed. Two novel aryl esters, Tetradecanyl palminate 5, 8-oxidotetraco-san-10-one were isolated and characterized from the roots (Nair *et al.*, 1987; Shukla *et al.*, 1999).

Fruits were reported to posses n-tricontanol, β-sitosterol, p-hydroxy cinnamoyloctadecanolate and caffeic acid (Gamble, 1956).

Pharmacological Studies

The root and flower were exploited for the aphrodisiac activity. Alcohol extract of the root was found to posses aphrodisiac activity (Subramonium *et al.*, 2007).

Figure 16.45: *Argyreia nervosa* flower.

Figure 16.46: *Argyreia nervosa* fruits.

Immunomodulatory activity on the root extract of the plant has been evaluated. Reults showed enhanced production of antibodies in mice, but failed to show any effect on macrophage phagocytosis (Gokhale *et al.*, 2003).

Hepatoprotective activity of ethanol and ethyl acetate extract of root has been evaluated.These extracts were found to be capable of enhancing and maintaining the activity of hepatic enzymes (Habbu *et al.*, 2008).

n-hexane, chloroform, ethyl acetate and water extracts of root were found to posses a significant CNS depressant activity (Galani and Patel, 2009).

Methanol extract of stem was found to posses dose dependent percentage blood glucose reduction in normal as well as in diabetic rats (Hemet *et al.*, 2008).

Aqueous extract of plant foud to posses a significant nootropic activity in both young and aged mice (Hanumanthachar *et al.*, 2007).

Alcohol extract of the root was found to posses a significant anti-inflammatory activity against granuloma formation techniques in albino rats (Srivastava and Shukla, 1998).

Hydroalcoholic extract of the plant at the dose of 200 and 400 mg/kg body weight significantly delayed the latency to the onset of first clones as well as onset of death in unprotected mice and exhibited protection in 16.66 and 33.33 per cent of pentylenetetrazolen treated mice respectively (Vyavhare and Bodhankar, 2009).

Conclusion

Ayurveda is the science of living, hence demand for ayurvedic medicines is ever increasing tremendously over the allopathic medicines due to fewer side effects. The major problem with ayurvedic drugs are identification and standardization. In Ayurveda some drug are called as controversial drugs like *Pashanuveda*, *Murva* etc. *Murva* is one that has tough fibres hence eleven plants are used as *Murva* in different parts of India. Acceptable source of *Murva* is *Marsdenia tennacissima*. *Murva (Marsdenia tennacissima)* is used in the treatment of various ailments as described above.

In this review we have presented information on all eleven plants are considered as *Murva* in different parts of the India with regards to their phytochemicals and pharmacological work done till now.This review will help in the identification of correct botanical source of *Murva* as well as provides a lot of latest information regarding other plants.

A little work has been done on these plants as given above. There is an intense need of attention and research on these plants and ensure whether they are equivalent, inferior or superior with that of *Murva* (*Marsdenia tennacissima*) regarding phytochemical and therapeutic point of view.

References

Almas, K. (1993). Miswak (chewing stick) and its oral health. *Postgrad Dentist*, **3**: 214.

Amin, T.T., and Al-Abad, B.M. (2008). Oral hygiene practices, dental knowledge, dietary habits and their relation to caries ong male primary school children in Al Hassa, Saudi Arabia. *Int J Dent Hyg.*, **6**: 361-70.

Ayyanar, M., Sankara, S.K., and Ignacimuthu, S. (2008). Traditional herbal medicines used for the treatment of diabetes among two major tribal groups in South Tamil Nadu, India. *Ethno Botanical Leaflet*, **47**: 389-394.

Almas, K. (2001). The effects of extracts of chewing sticks (*Salvadora persica*) on healthy and periodontally involved human dentine: a SEM study. *Indian J Dent Res.*, **12**: 127-32.

Almas, K.(1999). The antimicrobial effects of extracts of *Azadirachta indica* (Neem) and *Salvadora persica* (Arak) chewing sticks. *Indian J Dent Res.*, **10**: 23-6.

Al-Otaibi, M. (2004). The miswak (chewing stick) and oral health. Studies on oral hygiene practices of urban Saudi Arabians. *Swed Dent J Suppl.*, **167:** 2-75.

Almas, K. (2001). The antimicrobial effects of seven different types of Asian chewing sticks. *Odontostomatol Trop.*, **24**: 17-20.

Almask, K., Skang, N., and Ahmad, I. (2005). An *in vitro* antimicrobial comparison of Miswak extract with commercially available non-alcohol mouth rinses. *Int J Dent Hyg.*, **3(1):** 189-194.

Alok, S., Vijayakumar,M., Raj, V.A., Talib, M.D., Unnikrishnan, M.K., and Rao, C.V. (2009). *In vitro* α-amylase inhibition and antihyperglycemic activity of *Helicteres isora* in streptozotocin-induced rats. *I J Pharma. clinical res.*, **1(1):** 15-18.

Almas, K., and Al-Zeid, Z. (2004). The immediate antimicrobial effect of a toothbrush and miswak on cariogenic bacteria: a clinical study. *J Contemp Dent Pract.*, **5**: 105-14.

Almas, K., Al –Bagieh, N., and Akpata, E.S. (1997). *In vitro* antibacterial effect of freshly cut 1-month-old Miswak extracts. *Biomed Lett.*, **56**: 145-9.

Al-Bagieh, N., and Almas, K. (1997). *In-vitro* antibacterial effects of aqueous and alcohol extracts of Miswak (chewing sticks). *Cairo Dent J.*, **13:** 221-4.

Anonymous. (2005). The Wealth of India, Raw Materials.vol.5. New Delhi: CSIR, pp. 279.

Abo, Al-Samh D., and Al-Bagieh, N. (1996). A Study of antibacterial activity of the miswak extract *in vitro*. *Biomed lett.*, **53:** 225–38.

Anonymus. (2004). Ayurvedic Pharmacoepia of India. Ist ed. New Delhi: Govt. of India, Ministry of health and welfare, Department of Ayush., Part I, Vol.III.p178-80.

Ahmad, M., Jain, N., Khan, M.S., and Shafiullah, Iliad, M. (1993). Two new flavones glycosides were reported from the leaves of the *Argyreia nervosa*. *Journal of Chemical research,* **(S)7**: 248.

Alali, F., and Lafi, T. (2003). GC-MS analysis and bioactivity testing of the volatile oil from the leaves of the tooth brush tree *Salvadora persica L. Nat Prod Res.,* **17(3)**: 189-94.

Badgujar,V.B., Jain, P.S., and Atassi, F. (2002). Oral home care and the reasons for seeking dental care by individuals on renal dialysis. *J Contemp Dent Prac.,* **3**: 31-41.

Badgujar, S.V. (2009). Antifungal activity of stem bark of *Helicteres isora* Linn. *Drug Invention Today,* **1(2):** 135-136.

Bukar, A., Danfillo, I.S., Adeleke, O.A., and Ogunbodede, E.O. (2004). Traditional oral health practices among Kanuri women of Borno State, Nigeria. *Odontostomatol Trop.,* **27**: 25-31.

Baeshen, H.A., Kjellberg, H., Lingstr, A.P., and Birkhed, D. (2008). Uptake and release of fluoride from fluoride-impregnated chewing sticks *in vitro* and *in vivo*. *Caries Res.,* **42**: 363-8.

Bean, M.F., Antoun, M., Abramson, D., Chang, C.J., Jerry, L., and McLaughlin. (1985). Cucurbitacin B and Isocucurbitacin B: cytotoxic components of *Helicteres isora*. *J Nat Prod.,* **48(3):** 500.

Bhavsar, S.K., Singh, S., Giri, S., Jain, M.R., and Santani, D.D. (2009). Effect of saponins from *Helicteres isora* on lipid and glucose metabolism regulating genes expression. *J Ethnopharmacol.,* **124(3):** 426-33.

Bhattacharjee, S.K. (1998). Handbook of Medicinal Plants. Jaipur (India): Pointer publishers, pp 56-7,1998.

Bhandari, M.M. (1990). Flora of the Indian desert. 1st ed. Dhrti printers, NewDelhi; 1990.

Babber,O.P., Joshi, M.N., and Madan, A.R. (1978). Evaluation of plants for antiviral activity. *Indian J Med Res.*, **76**: 54-65.

Baeshen, H.A., Kjellberg, H., Lingstr, A.P., and Birkhed, D. (2008). Uptake and release of fluoride from fluoride-impregnated chewing sticks *in vitro* and *in vivo. Caries Res.*, **42:** 363–8.

Biswas, B., Tiwari, L.D., and Dutt, S. (1947). Chemical composition of the fixed oil from the seeds of *Argyreia nervosa. Indian soap journal*, **13**: 51-54.

Basniwal, P.K., Suthar, M., Rathore, G.S., Gupta, R., Kumar, V., and Pareek, A. (2009). *In vitro* antioxidant activity of hot aqueous extract of *H. isora* fruits. *NPR.*, **8(5):** 483-487.

Batra, A., and Mehta, B.K. (1985). Chromatographic analysis and antibacterial activity of the seed oil of *Argyreia nervosa.Fitoterapia,* **56**: 357-59.

Bachhav, R.S., Gulecha, V.S., and Upasni, C.D. (2009). Analgesic and anti-inflammatory activity of *Argyreia nervosa* root. *Indian J Pharmacol.*, **41(4):** 158-161.

Biswas, M., and Biswakanth, K. (2010). Acute and sub-chronic toxicity study of *Dregea volubilis* fruit in mice. *Journal of Phytology*, **2(8):** 06–10.

Biswas, M., and Samit, B. (2010). Antitumor effect of *Dregea volubilis* fruit in ehrlich ascites carcinoma bearing mice. *Global Journal of Pharmacology*, **4(3):** 102-1-6.

Chopra, R.N. (1955). A review of work on Indian Medicinal Plants. Indian council of medical research, New Delhi.

Chatterjee, A., and Banerjee, J. (1972). Occurrence of Funtumafrine C in *Chonemorpha macrophylla. Indian J Chem.*, **10**: 1197.

Chatterjee, D.K., Iyer, N., and Ganguli, B.N. (1987). Antiamoebic activity of chonemorphine, a steroidal alkaloid, in experimental models. *Parasitol Res.*, **74(1):** 30-3

Cooke, T. (2006). The Flora of Presidency of Bombay.2nd ed.vol.2.Dehradun: BSI., : 164-167.

Craene, R.D., and Wanntorp, L. (2009). Floral development and anatomy of Salvadoraceae. *Ann Bot.*, **104**: 913-23.

Darout, I.A., Albandar, J.M., and Skaug, N. (2000). Periodontal status of adult Sudanese habitual users of miswak chewing sticks or toothbrushes. *Acta Odontol Scand.*, **58**: 5-30.

Darmine, H., Nusayr, T., and Al-Hiyasat, A.S. (2006). The effects of extracts of *Salvadora persica* also examined on proliferation Balb/C 3T3 of fibroblast and viability of carcinogenic bacteria. *Pharmacog Review*, **4**: 62-6.

Darmani, H., Al-Hiyasat, A.S., Betieha, A.M., and Alkofahi, A. (2003). The effect of an extract of *Salvadora persica* (Meswak chewing stick) on fertility of male and female mice. *PhytoMedicine*, **10**: 63-5.

Dwivedi, A., Dwivedi, S., Sitoke, A. K., Patel, R., and Jhade, D. (2009). Anthelmintic activity of a polyherbal preparation. *Ethnobotanical Leaflets,* **13**: 259-62.

Darout, I. A., Christy, A.A., Skaug, N., and Egeberg, K.P. (2000). Identification and quantification of some potentially antimicrobial anionic component in Miswak extract. *Indian Journal of Pharmacology*, **32**: 11-14.

Deng, J., Liao, Z., Chen, D., and Marsdenosides, A.H. (2005). Polyoxypregnane glycosides from *Marsdenia tennacissima. Phytochemistry*, **66(90):** 1040-1051.

Deng, J., Liao, Z., and Chen, D. (2005). Three new polyoxypregnane glycosides from *Marsdenia tenacissima. Helvetica Chimica Acta.*, **88(10):** 2675-2682.

Dehvi, R., Gayathri, K., Shabi, M.M., Subashini, U., Dubey, G.P., and Rajamanickam, G.V. (2008). A preliminary biochemical screening of *Helicteres isora* L. stem bark in carbon tetrachloride induced toxicity in rats. *Bulgarian J of Veterinary Medicine,* **11(4)**: 235-42.

Divya, T.S., Latha, P.G., Usha, K., Anuja, G.I., Suja, S.R., and Shymal, S. *et al.* (2009). Antiinflammatory, analgesic and anti- lipid peroxidative properties of *Wattakaka volubilis* (Linn. f.) Stapf. *NPR.*, **8(2):** 137-141.

Dugasani, S., Balijepalli, M.K., Tandra, S., and Pichika, M.R. (2010). Antimicrobial activity of *Bauhinia tomentosa* and *Bauhinia vahlii* roots. *Pharmacogn Mag.*, **6(23):** 204–207.

Deshmukh, V.K., and Pandit, U. (1968). A study of fruits of *Helicteres isora* L. *Indian J Pharm.*, **30**: 283.

Edi, M.A., and Selim, H.A. (1994). Retrospecitve study on the relationship between Miswak chewing stick and periodontal health. *Egyptian Dent J.*, **40:** 589-92.

Fatimi, M.A., Wurster, M., Schröder, G., and Lindequist, U. (2007). Antioxidant, antimicrobial and cytotoxic activities of selected medicinal plants from Yemen. *Journal of Ethnopharmacol.*, **111(3):** 657-666.

Fai, Y.M., and Tao, C.C. (2009). A review of presence of oleanolic acid in natural products. *Sample Review for Natura Proda Medica,* **4**: 4-6.

Galati, E.M., Monforte, M.T., Forestieri, A.M., Miceli, N., Bader, A., and Trovato, A. (1999). *Salvadora persica* hypolipidemic activity on experimental hypercholesterolemia in rat. *Phytomedicine,* **6**: 181-5.

Gupta, A.K., Sharma, M., and Tandon, N. (2004). Indian Medicinal Plants. vol.4. New Delhi: ICMR,India, pp 102-25.

George, M., and Pandalai, K.M. (1949). Investigations on plant antibiotics Part IV Further search for antibiotic substances in Indian medicinal plants. *Indian J Med Res.*, **37**: 169-181.

Galani, V.J., and Patel, B.G. (2009). Central nervous system activity of *Argyreia nervosa* in mice. *Research journal of Pharmaceutical Technology*, **2(2):** 331-34.

Galati, E.M., Monforte, M.T., Forestieri, A.M., Miceli, N., Bader, A., and Trovato, A. (1999). *Salvadora persica* hypolipidemic activity on experimental hypercholesterolemia in rat. *Phytomedicine,* **6**: 181–5.

Gupta, A.K. (2003). Quality standards of Indian medicinal plants. vol. 1. ICMR, India.

Gupta, R.N., Pareek, A., Suthar, M., Rathore, G.S., Basniwal, P.K., and Jain, D. (2009). Study of glucose uptake activity of *Helicteres isora* Linn. fruits in L-6 cell lines. *Int J Diabetes Dev Ctries.*, **29(4):** 170–173.

Gample, G.S. (2005). The flora of presidency of Madras.vol.2.Calcutta: BSI (repr.); pp: 821-22.

Gupta, V.S., Kumar, A., Deepak, D., Khare, A., and Khare, N.K. (2003). Pregnanes and pregnane glycosides from *Marsdenia roylei. Phytochemistry,* **64(8):** 1327-1333.

Gokhale, A.B., Damre, A.S., and Saraf, M.N. (2003). Investigation into evaluation of anti-inflammatory and antiarthriic activity of *S. lappa. Phytomedicine,* **9(5):** 433-437.

Gupta, V., Kumar, A., Khare, A., and Khare, N. K. (2009). Two pregnane oligoglycosides from *Marsdenia roylei. NPR.,* **23(14):** 1342-1354.

Habbu, P.V., Shastry, R.A., Mahadevan, K.M., Hanumanthachar, Joshi., and Das, S.K. (2008). Hepatoprotective effects of *Argyreia nervosa* in rats. *Aft. J. Trad. CAM.,* **5(2):** 158-164.

Hatapakki, B.C., and Hukkeri, V.I. (2010). Antipyretic activity of root of *Climatis triloba* heyne ex. roth in rats. *Pharmacie Globale (IJCP),* **2(06):** 65-69.

Haldar, P.K., Kar, B., Bala, A., Bhattacharya, S., and Mazumder, U.K.(2010). Antitumor activity of *Sansevieria roxburghiana* rhizome against Ehrlich ascites carcinomainmice. *Pharm Biol.,* **48(12):** 1337-43.

Hemet, L.E., Sataynarayan,T., Ramesh, A., Prasad, D., Routhu,Y., and Srinivas, K.V. (2008). Hypoglycemic and antihyperglycemic effect of *Argyreia nervosa* Sweet in normal and in alloxan induced diabetic rats.*Journal of Natural remedies,* **8(2):** 203-208.

Hanumanthachar, J., Navneet, K., and Jyotibala, C. (2007). Evaluation of nootropic effect of *Argyreia nervosa* in mice. *J Health Sci.,* **53(4):** 382-388.

Habbu, P.V., Shastry, R.A., Mahadevan, K.M., Hanumantachar Joshi, and Das, S.K. (2008). Hepatoprotactive effects of *Argyreia nervosa* in rats. *African journal of Traditional CAM.,* **5(2):** 158-164.

Haldar, P. K., Kar, B., Bhattacharya, S., Bala, A., and Kumar, R. B. S. (2010). Antidiabetic activity and modulation of antioxidant status by *Sansevieria roxburghiana* rhizome in streptozotocin-induced diabetic rats. *Diabetologia Croatica,* 39-4.

Hashmi, S., and Ghouse, A.K. (1994). Analysis of Jadwar, a controversial Herbal Drug. *Hamdard Med.,* **37**: 79-84.

Hooker, B.J.D. (1883). Brit FI, Flora of china, 16: 250.

Hu, Y.J., Shen, X.L., Lu, H.L., Zhang,Y.H., Huang, X.A., and Fu, L.C. (2008). Hypoglycemic and anti-hyperglycemic effect of *Argyreia nervosa* Sweet in normal and in alloxan induced diabetic rats. *J. Nat. Rem.,* **8(2):** 203-208.

Galani, V.J., and Patel, B.G. (2009). Central nervous system activity of *Argyreia nervosa* roots in mice. *Research. J. Pharm and Tech.*, **2(2):** 331-334.

Gokhle, A.B., Damre, A.S., and Saraf, M.N. (2003). Investigation in to the Immunomodulatory activity of *Argyreia nervosa. J Ethanopharmacol.*, **84(1):** 109-114.

Jain, M., and Saxena, V.K. (1984). Chemical constituents of the stem of *Salvadora persica. Acta Ciencia Indica*, **10**: 127.

Joy, P.P., Thomas, J., Mathew, S., and Skari, B.P. (1998). Medicinal plants. Kerala Agricultural University, Kerala: Aromatic and medicinal plants research station, **(85):** 182-184.

Jain, P.S., Badgujar, V.B., Patil, R.R., Haswani, N.G., and Chaudhari, S.G. (2009). Quantitative determination of β-sitosterol from stem bark of *Helicteres isora* Linn. *Journal of Pharmacy Research*, **2(9):** 1397-1398.

Jing, L., Zhi, B.Y., Yun, H.Y., and Ya, W.Z. (2008). A new C_{21} steroidal glycoside from *Marsdenia tenacissima. Chinese Chemical Letters*, **19(4):** 444-446.

John, A. (2001). Healing plants of Peuinsular India. Pasrotta: CABI, pp 310-24.

Juanjuan, C., Xiaoyu, L.i., Cuirong, S., Yuanjiang, P., and Peter, S. (2008). Identification of Polyoxypregnane glycosides from the stems of *Marsdenia tenacissima* by high-performance liquid chromatography/tandem mass spectrometry. *Talanta*, **77(1):** 152-159.

Joshi, S.G. (2008). Medicinal plants. New Delhi: Oxford and IBH Publishing Company Pvt. Ltd., pp 69.

Katkar, K., Mridula, G., Shreedhara, C., Patil, P., Naik, V., and Bhamar, S., *et al.* (2009). Aeridin, a phenanthropyran derivative from *Wattakaka volubilis* l. *f. RJPP.*, **1(3):** 131-32.

Khalil, A.T. (2006). Benzylamides from *Salvadora persica. Arch Pharm Res.*, **29**: 952-6.

KeshavaMurthy, K.R., and Yoganarasimhan, S.N. (1990). Flora of Coorg (Kodagu) Karnataka, India. Vismat Publishers, India.

Kubota, K., Tanaka, T., Murata, Y., and Hirasawa, M. (1988). Effect of tannic acid on adherence of Candida to denture base. *J Dent Res.*, **67**: 183.

Kumar, G., Banu, G.S., Murugesan, A.G., and Pandian, M.R. (2007). Effect of *Helicteres isora* bark extract on protein metabolism and marker enzymes in streptozotocin-induced diabetic rats. *Iranian Journal of Pharmaceutical Research*, **6(2):** 123-129.

Kumar, G., Banu, G.S., and Murugesan, A.G. (2008). Influence of *Helicteres isora* L. bark extracts on glycaemic control and renoprotective activity in streptozotocin-induced diabetic rats. *International Journal of Pharmaceutical Sciences and Nanotechnology*, **1(3):** 122-127.

Kannan, N., Renitta, R.E., and Guruvayoorappan, C. (2010). *Bauhinia tomentosa* stimulates the immune system and scavenges free radical generation *in vitro. J Basic Clin Physiol Pharmacol.*, **21(2):** 157-168.

Kumar, A., Khare, A., and Khare, N.K. (1999). Two oligosaccharides from *Marsdenia roylei*. *Phytochemistry*, **50(8):** 1353-1357.

Kumar, G., Banu, G.S., Murugesan, A.G. (2009). Attenuation of *Helicteres isora* L. bark extracts on streptozotocin-induced alterations in glycogen and carbohydrate metabolism in albino rats. *Hum Exp Toxicol.*, **28(11):** 689-696.

Kumar, G., Murugesan,A.G., and Pandian, M. R. (2006). Effect of *Helicteres isora* bark extract on blood glucose and hepatic enzymes in experimental diabetes. *Pharmazie*, **61(4):** 353-5.

Kamiya, K., Saiki, Y., Hama, T., Fujimoto, Y., Endang, H., and Umar, M. (2001). Flavonoid glucuronides from *Helicteres isora*. *Phytochem.*, **57(2):** 297-301.

Kumar, G., Sharmila, B.G., Murugesana, A.G., and Pandian, M.R. (2006). Hypoglycaemic effect of *Helicteres isora* bark extract in rats. *J Ethnopharmacol.*, **107(2):** 304-307.

Kumar, G., Banu, S., and Murugesan, A.G. (2009). Influence of *Helicteres isora* administration for diabetes mellitus: Its effect on erythrocyte membrane and antioxidant status. *Food Chem Toxicol.*, **47(8):** 1803-1809.

Kumar, A., Khare, A., and Khare, N.K. (1999). Trisaccharide from *Marsdenia roylei*. *Phytochemistry*, **52(4):** 675-9.

Khare, A., Gupta, V.S., Khare, N.K. (1999). Structural studies of Pregnane and their glycosides from *Marsdenia roylei*. *Phytochemistry*, **52(4):** 675-679.

Kolammal, M. (1978). Pharmacognosy of Ayurvedic drugs, series-1 Trivendrum, Ayurveda research institute. pp: 1.

Kubota, K., Tanaka, T., Murata,Y., and Hirasawa, M. (1988). Effect of tannic acid on adherence of Candida to denture base. *J Dent Res.*, **67:** 183.

Kumar, C.D. (2007). Pharmacognosy can help minimize accidental misuse of herbal medicine. *Curr Sci.*, **93**: 1356-8.

Kulkarni, P.H., and Shahida, A. (2004). The Ayurvedic Plants. 1st ed. New Delhi: Sri Satguru Publications, Indological and Oriental publishers, pp. 177.

Kumar, G., and Murugesan,A.G. (2008). Hypolipidaemic activity of *Helicteres isora* L. bark extracts in streptozotocin induced diabetic rats. *J Ethnopharmacol.*, **116(1):** 161-6.

Kumar, G., Gani, S.B., and Arunachalam, G.M. (2008). The effect of *Helicteres isora* bark extracts on heart antioxidant status and lipid per-oxidation in streptozotocin diabetic rats. *J Appl Biomed.*, **6**: 89-95.

Kumar, A., Ahmed, A.B., Vekateshvaran, A., and Panagal, Mani. (2010). Anti-hyperlipidemic and hypoglycemic activity of *Wattakaka volubilis* methanol extract in alloxan-induced diabetic rats. *J P R.*, **3(8):** 1913-1915.

Kirtikar, K.R., and Basu, B.D. (1996). *Indian medicinal plant 500 species*. Oriental enterprises Deharadun, Vol. II, pp 263.

Luhua, Z., Bingren, X., and Xiying, T. (2010). Fingerprinting of *Marsdenia tenacissima* by capillary electrophoresis compared to HPLC. *J of Chromatographic Science*, **48(5):** 417-424.

Li, Q.F., Wang, X.L., Yu, K.B., Peng, S.L., and Ding, L.S. (2006). 11 α, 12β-Acetonyltenacissoside F acetone solvate: a new polyoxypregnane glycoside from the stems of *Marsdenia tenacissima. Acta Crystallographica Section E*, **62(11):** 216-225.

Luo, S.Q., Lin, L.Z., Cordell, G.A., Xue, L., and Johnson, M.E. (1993). Polyoxypregnanes from *Marsdenia tennacissima*. *Phytochemistry*, **34(6):** 1615-162.

Li, Q.F., Wang, X.L., Ding, L.S., and Chang, Zhang. (2007). Polyoxypregnanes from the stems of *Marsdenia tenacissima. Chinese Chemical Letters*, **18(7):** 831-34.

Muthu, C., Ayyanar, M., Raja, N., and Ignacimuthu, S. (2006). Medicinal plants used by traditional healers in Kancheepuram district of Tamil Nadu, India. *Journal of Ethnobiology and ethnomedicine*, **2**: 43-52.

Krishnaiah,P.V., Ratnam,K.V., and Raju, R.R. (2008). Preliminary phytochemical evaluation of certain anticancer crude drugs used by adivasis of Rayalaseema region, Andhra Pradesh, India. *Ethnobotanical Leaflets*, **12**: 693-97.

Kamani, H., Tennekoon, and Jeevathayaparan, S. (1991). Possible hepatotoxicity of *Nigella sativa* seeds and *Dregea volubilis* leaves. *Journal of Ethnopharmacol.*, **31(3):** 283-9.

Nawal, A.K., Sabawi, A.I., Khalik, A.K., Sheikh, A., and Mahmoud, Y.T. (2007). The antimicrobial activity of *Salvadora persica* solution (Miswak-Siwak) as root canal irrigant (a comparative study). *Journal of Pure and Applied Science*, **4(3):** 69-91.

Maruthupandian, A., Mohan,V.R., and Sampathraj, R. (2010). Antidiabetic, antihyperlipidaemic and antioxidant activity of *Wattakaka volubilis* (l. f) stapf leaves in alloxan induced diabetic rats. *IJPSR.*, **1(11):** 83-90.

Moulish, B., Bikash, M.N., Partha, P., Kumar, G.A., Sukdeb, B., and Kanti, H.P. (2009). *In vitro* anti-leishmanial and anti-tumour activities of a pentacyclic triterpenoid compound isolated from the fruits of *Dregea volubilis* Benth Asclepiadaceae. *Tropical Journal of Pharmaceutical Research*, **8(2):** 127-131.

Mridula, G., Shreedhara, C.S., Suthar, K.K., Chauhan, A., and Singh, V. A. (2009). Quantitative estimation of aeridin in *Wattakaka volubilis* by HPTLC. *International Journal of Pharmaceutical and Clinical Research*, **1(3):** 115–118.

Silija, V.P., Samitha Varma, K., and Mohanan, K.V. (2008). Ethnomedicinal plant knowledge of the Mullukuruma tribe of Wayanad district, Kerala. *Indian Journal of Traditional Knowledge*, **7:** 604-612.

Monforte, M.T., Trovato, A., Rossitto, A., Forestieri, A.M., Daquino, A., and Miceli, N. (2002). Anticonvulsant and sedative effects of *Salvadora persica* stem extracts. *Phytother Res.*, **16:** 395-7.

Mogib, M.A. (1999). A lupane triterpenoid from *Macrua oblongifolia*. *Phytochemistry*, **51(3):** 445-448.

Mishra, S.H., and Chaturvedi, S.C. (1978). Antibacterial and antifungal activity of the oil and unsaponifiable matter of *Argyreia speciosa* Sweet. Indian Drugs. *Pharmaceut Ind.,* **13(5):** 29-31.

Madhavan, V., Shukla, A.S., Murali, A., Usha, M., and Yoganarasimhan, S.N. (2010). Antipyretic activity studies of two botanical sources of the drug *Murva. Asian journal of traditional medicines,* **5(5):** 171-180.

Manivannan, R., Vijaykumar, M., Balasubramaniam, A., Baby, E., and Kumar, N.S. (2010). Pharmacognostical and preliminary phytochemical studies of *Bauhinia tomentosa* flower. *Journal of Pharmacy Research,* **3(3):** 502-505.

Miyakawa, S., Yamaura, K., Hayashi, K., Kaneko, K., and Mitsuhashi, H. (1986). Five glycosides from the Chinese drug 'Tong-guang-san' from the stems of *Marsdenia tennacissima. Phytochemistry,* **25(12):** 2861-2865.

Mansour, M.I., Khateeb, T.L., and Mazraoo, A.A. (1996). The analgesic effects of Miswak. The *Saudi Dental Journal,* **08(2):** 87-91.

Nair, G.G., Daniel, M., and Sabnis, S.D. (1987). Ergolines in the seeds of Convolvulaceae.*IJPS.,* **49:** 100-102.

Nayak, B. (2007). Ayurvedline. 9th ed. Bangalore: Ayurvedline, part 2. p. 498-667.

Norton, M.R., and Addy, M. (1989). Chewing sticks versus tooth brushes in West Africa. *Clin Prev Dent.,* **11**: 11–37.

Natarajan, B., and Paulsan, B.S. (2000). An Ethno pharmacological study from Thane district, Maharashtra, India: Traditional knowledge compared with modern biological sciences. *Pharmaceu Biol,* **38(2):** 139 – 151.

Panda, N., Banerjee, S., Mandal, N.B., and Sahu, N.P. (2006). Pregnane glycosides. *Nat Prod Commun.,* **4**: 665-669.

Panda, N., Mandal, N.B., Banerjee, S., Sahu, N.P., Koike, K., Nikaidio, T., Weber, M., and Luger, P. (2003). Polyhydroxy pregnanes from *Dregea volubilis. Phytochemistry,* **61:** 8400-8405.

Pandikumar, P., Ayyanar, M., and Ignacimuthu, S. (2007). Medicinal plants used by *Malasar* tribes of Coimbatore district, Tamil Nadu. *Indian Journal of Traditional Knowledge,* **6**: 579-582.

Pullaiah,T., Ramakrishnaiah, V., Rani, S., and Rao, P.N. (2000). Flora of Guntur District, Andhra Pradesh, India. L Regency publication. India, pp 145-46.

Paliwal, S., Chauhan, R., Siddiqui, A., Paliwal, S., and Sharma, J. (2004). Evaluation of antifungal activity of *Salvodra Persica* leaves. *NPR.,* **6(5):** 372-74.

Philip, D., Kaleena, P.K., and Valivittan, K. (2011). *In vitro* cytotoxicity and anticancer activity of *Sansevieria roxburghiana. International Journal of Current Pharmaceutical Research,* **3(3):** 71-73.

Pohocha, N., and Grampurohit, N.D. (2001). Anti-spasmodic activity of the fruits of *Helicteres isora* L. *Phytotherapy Res.,* **15(1):** 49-52.

Philipa, D., Kaleena, P. K.B., and Valivittan, K. (2011). *In vitro* cytotoxicity and anticancer activity of *Sansevieria roxburghiana. International Journal of Current Pharmaceutical Research,* **3(3):** 71-71.

Praveen, N., Khan,N.U., and Singhal, K.C. (1990). Anti-filarial activity of *Argyreia nervosa* against *Setaria cervi in vitro. Phytother Res.,* **4:** 162-164.

Qiu, S.X., Luo, S.Q., Lin, L.Z., and Cordell, G.A. (1996). Further polyoxypregnanes from *Marsdenia tennacissima. Phytochemistry,* **41(5):** 1385-1388.

Qian, J., Hua, H., and Qin, S. (2009). Progress on antitumor effects of *Marsdenia tenacissima. Zhongguo Zhong Yao Za Zhi.,* **34(1):** 11-3.

Roy, R.K., Ray, N.M., and Das, A.K. (2005). Skeletal muscle relaxant effect of *Chonemorpha macrophylla* in experimental animals. *Indian J pharmacol.,* **37(2):** 116-19.

Row, L.R., and Viswanadham, N. (1954). Colouring matter of the flower petals of *Bauhinia tomentosa* Linn. *Proceedings Mathematical Sciences,* **39(5):** 240-242.

Rastogi, R.P., and Mahrotra, B.N. (1960-1969): Compendium of Indian medicinal plants,vol-1.Lucknow CDRI and New Delhi NISC: 433-34.

Rajadurai, M., Vidhya, V.G., Ramya, M., and Bhaskar, A. (2009). Ethno-Medicinal plants used by the traditional healers of Pacchamalai hills, Tamil Nadu, India. *Ethnomedicine,* **3**: 39-41.

Raj, K.P., and Agarwal, Y.K. (1979). Heavy metal contents of the leaf-gall. *Sci Cult.,* **45**: 35.

Saldanha, C.J., and Nicolson, D.H. (1976). Flora of Hassan District, Karnataka. vol. 1. New Delhi: Amerind Publishing Co., pp 138-44.

Sanyacharernkul, S., Itghiarbha, A., Kongtawelert, P., and Meepowpan, P. A. (2009). New Polyoxypregnane glycoside from the roots of *Dregea volubilis* (L.f) Benth.ex Hook. f and its chondroprotective effect. *American Journal of Biochemistry and Biotechnology,* **5(4):** 202-209.

Sahu, N.P., Panda, N., Mandal, N.B., Banerjee, S., Koike,K., and Nikaido, T. (2002). Polyoxy pregnane glycosides from the flowers of *Dregea Volubilis. Phytochemistry,* **61**: 383-88.

Sanogo, R., Monforte, M.T., Daquino, A., Rossitto, A., Maur, D.D., and Galati, E.M. (1999). Antiulcer activity of *Salvadora persica* structural modifications. *Phytomedicine,* **6:** 363-6.

Sethiya, N.K., Nahata, A., Mishra, S.H., and Dixit, V.K. (2009). An update on Shankhpushpi, a cognition boosting Ayurvedic medicine. *Zhong Xi Yi Jie He Xue Bao.,* **7**: 1001-22.

Sethiya, N.K., Trivedi, A., Patel, M.B., and Mishra, S.H. (2010). Comparative pharmacognostical investigation on four ethanobotanicals traditionally used as Shankhpushpi in India. *J Adv Pharm Tech Res.,* **1**: 388-95.

Samh, A.L., and Bagleh, A.N. (1996). A Study of antibacterial activity of the miswak extract *in vitro. Biomed let.,* **53:** 225-38.

Singh, A. (2008). Contribution of dravyaguna to herbal materia medica. *Ethnobotanical Leaflets*, **12**: 599-602.

Singh, U.K., Govil, J.N., and Singh, G. (2002). Recent progress in Medicinal Plants. vol.1.USA: SCI Tech Publishing LLC', pp 3.

Singhal, S., Khare, M.P., and Khare, A. (1980). Cissogenin, a pregnane genin from *Marsdenia tennacissima*. *Phytochemistry*, **19(11):** 2427-2430.

Singhal, S., Khare, M.P., and Khare, A. (1980). Tenasogenin, a pregnane ester from *Marsdenia tennacissima*. *Phytochemistry*, **19(11):** 2431-2433.

Sethi, A., Paswan, S., Srivastava, S., Khare, N.K., Bhatia, A., and Kumar, A. (2008). A new pregnane glycoside from *Marsdenia roylei* as potential antioxidant and antidyslipidemic agents. *J Asian Nat Prod Res.*, **10(11-12):** 1023-8.

Singh, M.P., and Panda, H. (2005). Medicinal Herbs with their formulations vol.1. Delhi: Daya Publishing House,pp 159-160.

Shah, V.C., Adolf, S.D., and Souza,N.J. (1989). Chonemorphine, stigmasterol, and ecdysterone: Steroids isolated through bioassay-directed plant screening programs. *Steroids*, **53(3-5)**: 559-565.

Shah, V., Sunder, R., and Desouza, N.J. (1987). Chonemorphine and rapanone–Antipara-sitic agents from plant source. *American Chemical society*, **50:** 730-731.

Shende, V.S., Sawant, V.A., Turuskar, A.O., Chatap, V.K., and Vijaya, C. (2009). Evaluation of hypoglycemic and antihyperglycemic effects of alcoholic extract of *Chonemorpha fragrans* root in normal and alloxan induced diabetic rats. *Pharmacog Mag.*, **5(19):** 36-41.

Sadhan, A.L., and Almas. (1999). Miswak (chewing stick)–A cultural and scientific heritage. *Saudi Dent J.*, **11:** 81.

Sikarwar, R., Kumar, V., Rawat, A., and Mehrotra, S. (2003). Folk use of plants in veterinary medicine in central india. 2nd World Congress on " Biotechnology developements of Herbal Medicine" NBRI, Lucknow,UP,India, p. 105, february 20-22.

Sharma, P.V., and Dravyaguna, Vijnana. (2005). vol.2. Varanasi: Chaukhambha Bharti Academy, pp 480-81.

Suthar, M., Rathore, G.D., and Pareek, A. (2009). Antioxidant and antidiabetic activity of *Helicteres isora* (L.) fruits. *Indian J Pharm Sci.*, **71(6):** 695–699.

Subramonium, A., Madhavachandran, V., Ravi, K., and Anuja, V.S. (2007). Aphrodisiac property of the elephant creeper *Argyreia nervosa. Journal of Endocrinology*, **11(2):** 82-85.

Srivastava, A., and Shukla, Y.N. (1998). Aryl esters and a coumarin from *Argyreia speciosa.Indian journal of Chemistry*, **37(B):** 192-194.

Sandhya, P., and Grampurohit, N.D. (2008). Isolation and characterization of bioactive terpene from the fruits of *Helicteres isora* L. *Phcog Mag.*, **4(14):** 107-11.

Saldanha, C.J., and Nicolson, D.H. (1976). Flora of Hassan District of Karnataka. New Delhi: Amerind Publishing Co, pp 455.

Shukla, Y.N., Srivstava, A., Kumar, S., and Kumar, S. (1999). Phytotoxic and antimicrobial constituents of *Argyreia nervosa* and *Oenothera biennis*. *J. Ethnopharmacol.*, **67**: 241-245.

Subramonium, A., Madhavachandran, V., Ravi, K., and Anuja, V.S. (2007). Aphrodisiac property of the *Argyreia nervosa* Elephant Creeper. *J Endocrinal Report.*, **11(2):** 82-85.

Tambekar, D.H., Khante, B.S., Panzade, B.K., Dahikar.S.B., and Banginwar,Y.S. (2008). Evaluation of phytochemical and antibacterial potential of *Helicteres isora* L. fruits against enteric bacterial pathogens. *African J Trad, Comp and Alt Med.*, **5(3):** 290-293.

Tiwari, V., Tiwari, A., and Madhavan,V. (2010). Preliminiary phytochemical analysis, HPTLC studies and antipyretic activity of alcohol and aqueous extract of *Helicteres isora* L. root. *IJPPS.*, **2(2):** 74-79.

Tezuka, Y., Terazono, M., Kusumoto, T.I., Kawashima, Y., Hatanaka, Y., and Kadota, S. (1999). Helisterculins A and B, two new (7.5′, 8.2′)-neolignans, and helisorin, the first (6.4′,7.5′,8.2′)-neolignans, from the Indonesian medicinal plant *Helicteres isora. Helv Chimica Acta,* **82(3):** 408-417.

Vijayakumar, M., Balasubramaniam, A., Manivannan, R., and Kumar, N.S. (2010). Antioxidant potential of ethanolic extract of *Bauhinia tomentosa* (Linn) flower. *RJPBCS,* **1(2):** 143.

Vaidya, B.L. (1982). Some controversial drugs in Indian medicine. Chaukhambha Orientalia. Varanasi.

Venkatesh, S., Reddy, G.D., and Reddy, B.M. (2003). Anti-hyperglycemic activity of *Helicteres isora* root in alloxan induced diabetic rats. *Pharmaceu Biol.,* **41(5):** 347-50.

Venkatesh, S., Sailaxmi, K., Reddy, B.M., and Ramesh, M. (2007). Antimicrobial activity of *Helicteres isora* root. *Indian J Pharm Sci.,* **69**: 687-9.

Vaidyaratnam, P.S. (1995). Indian Medicinal Plants. vol. 3. Orient Longman Ltd. pp 132-35.

Venkatesh, S., Laxmi, K.S., Reddy, B.M., and Ramesh, M. (2007). Antinociceptive activity of *Helicteres isora. Fitoterapia,* **78(2):** 146-48.

Vyavhare, N.S., and Bodhankar, S.L. (2009). Anticonvulsant activity of *Argyreia speciosa* in mice. *Indian jounal of Pharmaceutical science,* **71(2):** 131-134.

Wiart, C. (2006). Medicinal plants used in Asia and the Pacific. London: Taylor and Francis.

Xia, Z.H., Xing, W.X., Mao, S.L., Lao, A.N., Uzawa, J., Yoshida, S., and Fujimoto, Y. (2004). Pregnane glycosides from the stems of *Marsdenia tenacissima. J Asian Nat Prod Res.,* **6(2):** 79-85.

Yoganarasimhan, S.N. (2000). Medicinal Plants of India vol.1: Tamil Nadu. Bangalore Cybermedia, pp 263.

Yoganarasimhan, S.N. (2000). Medicinal Plants of India vol.2: Tamil Nadu. Bangalore Cybermedia, pp 263.

Traditional and Folk Herbal Medicine: *Pages* **529–542**
Recent Researches Vol. 2 (2014)
Editor: **V.K. Gupta**
Published by: **DAYA PUBLISHING HOUSE, NEW DELHI**

17

Evaluation of Clinical Efficacy of Selected Hair Care Herbs in Ayurvedic Hair Oil for Treatment of Dandruff, Hair Fall and Stress

Farhat S. Daud[1*], Sarang J. Deshpande[1], Shailesh B. Malekar[1] and Shubhangi D. Kamble[1]

ABSTRACT

Dandruff and Hair fall are common disorders affecting the scalp characterizing itching and white scales. Apart from several infectious pathologies, stress is also a very important etiological factor for causing dandruff. Ayurvedic hair oil was so formulated that it counteracts dandruff, hair fall and stress by using some hair care herbs like Amla, Brahmi, Jatamansi and Neem. The anti microbiological activity of this oil against causative organism of dandruff P. ovale cultured from the white scale of scalp was first confirmed. The Clinical Efficacy of the hair oil in the management of dandruff, hair fall and stress was then evaluated through an open, non-comparative, non-randomized, phase III clinical trial, conducted as per the ethical guidelines and approved by Institutional Ethics Committee. 36 patients of both sexes, from the age group of 18-35 years, suffering from Dandruff were enrolled in the study. There was a significant reduction ($p<0.05$) in the mean score for itching, white scales, hair fall and stress at the end of 2 weeks. There were no clinically significant adverse reactions, either reported or observed. Therefore, it was concluded that the formulated Ayurvedic hair oil is effective and safe in the management of dandruff, hair fall and stress.

Keywords: Ayurvedic hair oil, Clinical efficacy, Dandruff, Hair fall, Herbs, Stress.

---

1 Unijules Life Sciences Limited, 1505/1, "Universal Square", Shantinagar, Nagpur – 440 002, M.S., India

* *Corresponding author*: E-mail: fairydaud@yahoo.com

Introduction

Dandruff ("Pityriasis simplex") is a most common embarrassing hair disorder, characterized by itching of scalp, hair loss, scaling etc. If not treated properly it may lead to serious disorders like dermatitis, psoriasis, inflammation of scalp etc. Dandruff affects 5 per cent of the global population and mostly occurs after puberty (between ages of 20 and 30 years). (Schrader and Domsch, 2005)

Pityrosporum ovale is the causative microorganism of dandruff. It feeds on the dermal lipids and proteins and facilitates lipase activity, which releases proinflammatory free fatty acids causing dermal inflammation and tissue damage. The lipase activity indicates that in addition to hypersensitivity, *P. ovale* releases toxic chemicals, which contribute to the development of a fungal infection. Also reoccurrence is the commonest problem with the currently available synthetic medicaments for dandruff which may be due to poor efficacy, non compliance etc. (Berth and Holden, 1998).

According to the National Alopecia Areata Foundation, alopecia areata is a common autoimmune skin disease resulting in the loss of hair on the scalp and elsewhere. It affects approximately two percent of the population overall. While it affects both men and women equally, it is diagnosed more often in women, since they are more likely to seek treatment (Petukhova *et al.*, 2010). Apart from scalp disease there are several lifestyle factors which contribute to male and female hair loss including stress, smoking, heavy drinking, and sun exposure. Thus the herbs were selected by considering all these aspects and were incorporated in a Til oil base using Traditional Ayurvedic preparation method.

Materials and Methods

The method used to evaluate efficacy of the selected hair care herbs in the formulated Ayurvedic hair oil was by both subjective (by patient) and objective (by Doctor) evaluation while nature and acceptance of Ayurvedic hair oil was by subjective evaluation.

This trial was a prospective, open, non comparative, phase III trial conducted at Clinics of Ayurvedic Physicians with proper ethical consideration for the period of two weeks. At least 30 patients suffering from dandruff and hair fall were enrolled to get sufficient data for statistical consideration.

Selection and Pharmacology of Herbs

Four herbs with proven efficacy in hair problems were selected namely –Neem Amla, Brahmi and Jatamansi.

Neem

It consists of all aerial parts of plant known as *Azardirachta indica*, family *Meliaceae*. (Kokate *et al.*, 2005).

Parts Usually Used

Seeds, Bark, Leaves, Fruits

Description of Plant

Neem is known as free tree of India as it is found almost everywhere in India. It is considered as a magic tree, which has properties that not only relieves but also cures from illness (Bhowmik, *et al.*, 2010). Neem is evergreen, tall, and fast growing tree. It is native to India and well established in at least 30 countries world-wide, in South East Asia, Africa, West Indies, Central and South America (Singh *et al.*, 2008).

Chemical Constituents

Neem plants contain several thousands of chemical constituents, of special interest are the terpenoids known from different parts of neem plant. Of its biological constituents the most active and well studied compound is Azardirachtion. Besides the terpenoids neem also contains more than 20 sulphurous compounds responsible for characteristics smell of seeds and neem oil (Singh *et al.*, 2008).

Neem has been highly successfully against harmful fungi, parasites, and viruses. It has been most helpful in treating a variety of skin and scalp problems and diseases including psoriasis eczema and other persistent conditions (Bhowmik,. *et al.*, 2010).

Figure 17.1: *Azardirachta indica* (Neem).

Nimbidin Margolone, margolonone and isomargolonone are the main constituents of Neem leaves responsible for anti bacterial activity (Biswas *et al.*, 2002) and responsible for controlling dandruff and hair fall (Figure 17.1).

Amla

Amla consists of dried as well as fresh fruits of the plant *Emblica offincinalis* Gaerth (*Phyllanthus emblica* Linn.) belonging to family *Euphorbiaceae*.

Parts Used

Amla Fruits

Description of Plant

A small medium sized tree, found in mixed deciduous forests, ascending to 1300m on hills and native plan of India, mostly collected in winter season after ripening (Ayurvedic Pharmacopoeia of India, 2001).

Chemical Constituents

Amla fruit is rich natural source of vitamin C and contains 600 to 750 per hundred g of the fresh pulp. Fruits also contain about 0.5 per cent fat, phyllemblin and 5 per

Figure 17.2: *Emblica offincinalis* (Amla).

cent tannin. Amla fruit also rich in minerals like phosphorus, iron and calcium. It contains appreciable amount of pectin.

The fruits are dehydrated and stored. Vitamin content of dried fruits is not lost considerably due to presence of tannins which retards oxidation of vitamin C. Thus possess the property of promoting hair growth and prevents premature graying of hair (Kokate *et al.*, 2005) (Figure 17.2)

Brahmi

It consists of the fresh leaves and the stems of the plant known as *Bacopa moniera* Linn. (*Herpestis moniera*) belonging to family *Scrophulariaceae* (Kokate *et al.*, 2005).

Description of Plant

Brahmi is a small herb with light purple flowers. It grows in wet and sandy areas and near streams in Indian tropical regions. It is a creeping herb with numerous branches and small fleshy, oblong leaves. Flowers and fruits appear in summer. The whole plant is medicinally useful (Bammidi *et al.*, 2011).

Chemical Constituents

The pharmacological effects of *Bacopa monniera* are attributed to the presence of a number of biologically active compounds, including alkaloids, saponins and sterols. The compounds responsible for the memory enhancing effects of Bacopa monniera are triterpenoid saponins called "Bacosides" (Bammidi *et al.*, 2011).

Figure 17.3: *Bacopa moniera* (Brahmi).

Brahmi is found to contain alkaloids as main constituents which are responsible for hair growth (Banerjee *et al.*, 2009) (Figure 17.3).

Jatamansi

Jatamansi consists of dried rhizomes of *Nardostachys jatamansi*, D.C. from family Valerianaceae (Kokate *et al.*, 2005).

Parts Used

Roots and Rhizomes

Description of Plant

Nardostachys jatamansi DC. is a small, perennial, dwarf, hairy, rhizomatous, herbaceous, endangered and most primitive species within family *Valerianaceae*. The plant is distributed in the Himalayas from Pakistan, India (Jammu and Kashmir, Himachal Pradesh, Uttarakhand, Sikkim) to Nepal, Tibet and China between 3300 to 5000 m asl.

Chemical Constituents

Chemically, the dried roots contain sesquiterpenes like jatamansone (valeranone), spirajatamol, patchouli alcohol, norseychelanone, jatamol A and B, lignans and

Figure 17.4: *Nardostachys Jatamansi* (Jatamansi).

neolignans like virolin, pinoresinol, jatamansic acid, terpenic coumarins like oroselol and jatamansin. The essential oil is composed of sesquiterpeniods and coumarins. Jatamansone is the principle sesquiterpenoid and renders majority of the biological activity (Sakina *et al.*, 2012).

It contains compounds like Nardal, Jatamansic acid, Nardin responsible for hair growth (Rao *et al.*, 2011). The 70 per cent hydroalcoholic extract of *Nardostachys jatamansi* rhizome, contains good amounts of polyphenols and flavonoids thereby exhibiting high antioxidant and free radical scavenging activities. The extract was found to chelate iron, scavenge superoxide radical and has a high reducing power. It inhibited protein oxidation, DNA damage and decreased malondialdehyde level. Hence, these invitro assays support the idea that *Nardostachys jatamansi* can be a good source of natural antioxidants thereby alleviating oxidative stress mediated anxiety (Razack, S. *et al.*, 2012) (Figure 17.4).

Method for Preparation of Ayurvedic Hair oil

Ayurvedic hair oil was formulated by the Ayurvedic 'Siddha Taila' (Shastri, 1984) procedure using the herbs Neem, Amla, Bramhi, and Jatamansi in Sesame oil. This oil help in better absorption through the scalp, nourishes the scalp tissues, helps in strengthening hair and prevents hair fall.

Procedure for Preparation of Siddha Taila (Shastri, 1984)

1. Raw herb of Jatamansi, Bramhi, Neem and Amla (in crude form mesh size 40) were weighed and added to extraction vessel.
2. Water (16 times of herb) was added and allowed to boil till the quantity reaches to 1/4 th amount.
3. Extract was filtrated with muslin cloth.
4. Til oil was taken in another heating vessel and above herbal extract was added into it.
5. Entire mixture was heated at slow rate until whole water is evaporated from the oil.
6. Finally oil was filtered with muslin cloth. This is called Siddha taila.

Procedure of Preparation of Ayurvedic Hair Oil

1. In above Siddha taila, Light liquid paraffin and Vitamin E was added in appropriate quantity and allowed to mix properly.
2. The oil was finally filtered with muslin cloth and stored at cool, dry place away from direct light (Figure 17.5).

Confirmation of Activity of Ayurvedic Hair Oil with a known Standard Sample (Leeming and Notman, 1987)

Neem Extract known for its antimicrobial activity was selected as **Standard**. It was procured from Ghaziabad Aromatics, Ghaziabad. Microbial Analysis of the Standard (Neem extract) and Ayurvedic hair oil (Product sample) was carried out with fungus isolated from human scalp *P.ovale* on Sabouraud Dextrose Agar medium.

Figure 17.5: Ayurvedic Hair oil.

It was found that the growth of fungi was absent in both Standard and Sample (Figure 17.6).

Clinical Study of Formulated Ayurvedic Hair Oil

Inclusion criteria: A total of 30 patients, ages between and including 18 to 35 who were diagnosed as suffering from mild to moderate form of dandruff and hair fall, and who were willing to give informed written consent were included in the study.

Exclusion criteria: Patients with severe scalp infection, Psoriatic patch, history of hypersensitivity to oils, cosmetics, children below eighteen years of age, patients with severe systemic disease, patients with genetic and endocrinal disorders, and those patients who refused to give informed written consent were excluded from the study. Pregnant and lactating women were also excluded from the study.

All enrolled patients underwent a thorough clinical examination of scalp skin for dandruff. All patients were advised to apply 10 ml of Ayurvedic Hair Oil twice

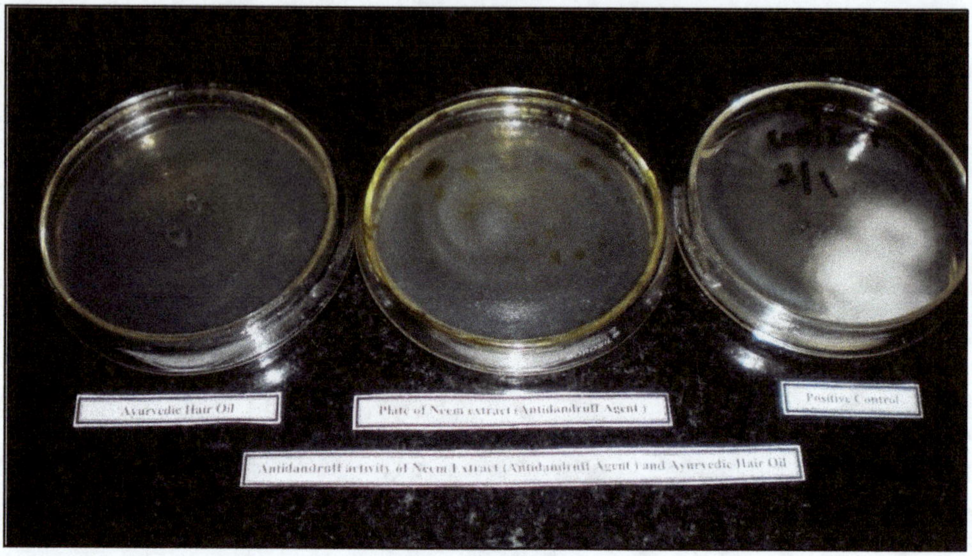

Figure 17.6: Anti-dandruff activity of neem extract (standard) and ayurvedic hair oil (sample).

daily for a period of 2 weeks with gentle massage to the entire scalp and were advised to leave the Ayurvedic hair oil on the scalp for a contact period of minimum 2-3 hours after application. Patients were told to restrict themselves to Ayurvedic hair oil as the only treatment for their dandruff. No other active treatment intervention during the study period was allowed. Safety evaluation was done by assessment of any undesired effect observed during and after the completion of study period.

Follow-Up and Assessment

All patients were followed for a period of 2 weeks and at each follow-up visit, they were asked about the frequency of the application of Hair Oil and also for nature and acceptance in order to check the compliance to the oil. Clinical assessment of dandruff was done objectively (by doctor). Thorough scalp examination was done after completion of 1 week and at the end of the study. The severity of the dandruff symptoms (itching, white scales and hair fall) was recorded on a score scale from '0 to 3' (0=Nil, 1=Mild, 2=Moderate and 3=Severe). Subjective assessment for nature and acceptance of hair oil and also for efficacy and stress was set as an analogue scale grading to 'Poor, below average, Average, Good, Very good and Excellent' which was done by Subject.

The predefined primary efficacy endpoints were rapid clinical improvement and symptomatic control of the dandruff. The predefined secondary safety endpoints for short- and long-term were assessed by incidence of adverse events and patient compliance to the therapy.

Statistical Analysis

Statistical analysis was done according to intention-to-treat principles. The changes in various parameters from baseline values and the values after 2 weeks were evaluated by "Paired't' Test" by using 'Instat statistical software'. The minimum level of significance was fixed at 95 per cent confidence limit and a 2-sided p value of <0.05 was considered as significant.

Results

An Ayurvedic hair oil containing four known hair care herbs was formulated to counter Dandruff, Hair fall and Stress and was evaluated for its clinical efficacy through an open, non-comparative, non-randomized, phase III clinical trial, conducted as per the ethical guidelines and approved by Institutional Ethics Committee. Total 44 subjects were enrolled for the treatment out of which 36 subjects completed the 2 weeks treatment period.

Subjective evaluation for efficacy of hair oil showed that 88.88 per cent response was observed for antidandruff property, whereas 100 per cent response was for anti hair-fall activity and 94.44 per cent response was for hair-strengthening property (Table 17.1).

Table 17.1: Subjective evaluation of efficacy of Ayurvedic hair oil.

Grading	*Subjects*	*Effect*							
		Antidandruff Property		*Anti Hairfall Property*		*Hair Strengthing Property*		*Stress Relieving Property*	
Excellent	Responders	2	32	0	36	0	34	20	32
V. Good		6		16		18		8	
Good		24		20		16		4	
Average	Non responders	4	4	0	0	2	2	4	4
Below Avg.		0		0		0		0	
Poor	0		0		0		0		

* $p<0.05$

Objective evaluation done for anti-dandruff activity by a physician showed that it significantly reduced the itching of scalp as $p<0.05$ after 2 weeks of application. White scales on scalp and hair fall was also significantly reduced ($p<0.05$) after 2 weeks of treatment (Figures 17.6–17.8).

Subjective evaluation for nature and acceptability of product by subject self assessment, it was observed that 88.88 per cent subjects are responders for appearance, 100 per cent response is for fragrance, 88.88 per cent responded for ease of application where as 38.88 per cent subjects responded for ease of washing (Table 17.2).

Discussion

Dandruff generally gets aggravated by humidity, season, trauma, emotional disturbances etc. It is characterized by desquamation of scalp which is manifestation of seborrheic dermatitis caused by *P. ovale*. The age at which sebaceous activity is at

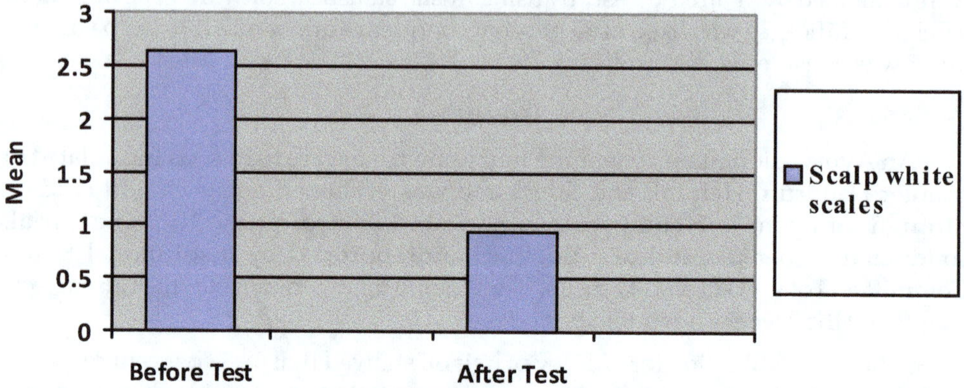

Figure 17.6: Significant reduction in Scalp White scales $p<0.05$.

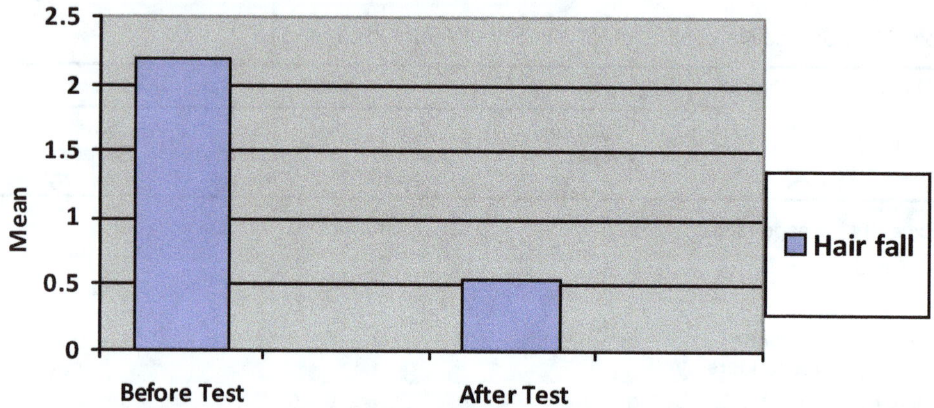

Figure 17.7: Significant reduction in Hair fall $p<0.05$.

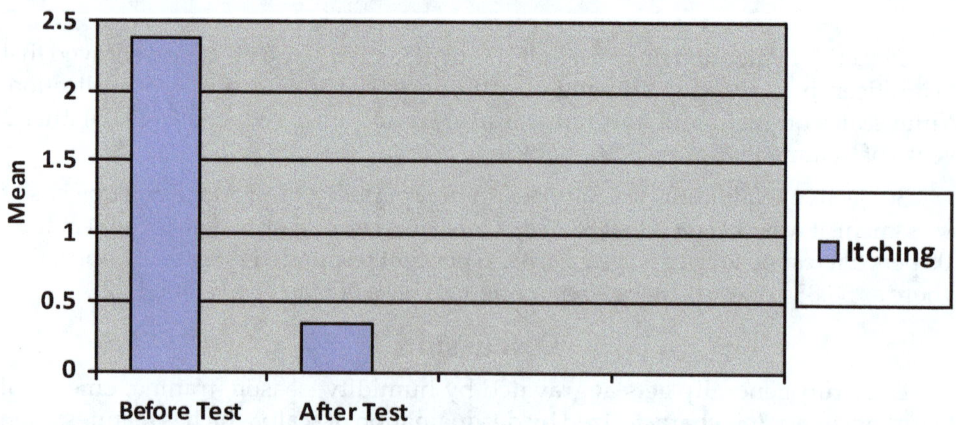

Figure 17.8: Significant reduction in Itching $p<0.05$.

peak level is the onset of dandruff disorders. It is improved in summer as *P. ovale* counteracts by ultraviolet rays of sunlight and gets worse during winter as other organisms are also involved and sebum level is also increased (Mc Grath and Murphy, 1991).

Table 17.2: Nature and acceptance of Ayurvedic hair oil.

Grading	*Subjects*	*Effect*							
		Appearance		*Fragrance*		*Ease of Application*		*Ease of Washing*	
Excellent	Responders	0	32	22	36	2	32	0	14
V. Good		8		14		8		6	
Good		24		0		22		8	
Average	Non responders	4	4	0	0	4	4	20	22
Below Avg.		0		0		0		2	
Poor		0		0		0		0	

* p<0.0.

Dandruff is often visible (grayish white flakes) on hair, shoulders and is associated with itching, hair fall which may lead to severe eczema, can affect the skin around nose, eyebrows, ear, face etc.

These studies aims towards reducing the causative organisms and provide symptomatic relief from itching, scaling of white scalp and hair fall and prevent further complications. The principal ingredients such as Neem, Bhramhi, Jatamansi, and Amla are of proven efficacy in hair care. They exert anti-inflammatory, antifungal activity, antioxidant and also have stress relieving properties.

With this Polyherbal Ayurvedic hair oil a significant reduction was observed in the mean scores of Itching, white Scalps and Hair fall which itself proves its anti-inflammatory, antifungal and immunostimulatory action of its ingredients. Microbiological study has also proved its activity against *P. ovale*. Subjective evaluation also states its nature, acceptance, efficacy and stress relieving property significantly.

Conclusion

This study observed a significant reduction in the mean scores of itching, white scales and hair fall observed by the physician.

Subjective evaluation for nature and acceptability revealed remarkable response. Response for efficacy of hair oil is also remarkable. Thus it can be concluded that this Ayurvedic hair oil has excellent antidandruff action due to the synergistic antifungal, anti-inflammatory and local immunostimulatory actions of its ingredients. There were no clinically significant adverse reactions, either reported or observed, during the entire study period and overall compliance to the treatment was excellent. Therefore, it may be concluded that, Ayurvedic hair oil is effective and safe in the management of Dandruff, Hair fall and Stress.

References

Bammidi, S.R., Volluri, S.S., Chippada, S.C., Avanigadda, S., and Vangalapati, M. (2011). A review on pharmacological studies of *Bacopa monniera*. *Journal of Chemical, Biological and Physical Sciences,* **1(2, Sec. B):** 250-259.

Banerjee, P., Sharma, M., and Nema, R. (2009). Preparation, evaluation and hair growth stimulating activity of herbal hair oil. *Journal of Chemical and Pharmaceutical Research,* **1(1):** 261-267. Available online from http: / /jocpr.com/ first-issue / article-26.html

Berth, J.L., and Holden, C.A. (1998). Eczema Lichenification, Prurigo and Erythroderma, in Rook's Text book of dermatology, 6th edn, Vol. I, Ed. by Burns, T., Breathnach, S., Cox, N., and Griffiths, C. Wiley Blackwell, New Jersey, pp. 638-642.

Bhowmik, D., Chiranjib, Yadav, J., Tripathi, K.K., and Kumar, S. (2010). Herbal remedies of *Azardirachta Indica* and its medicinal application. *J.Chem.Pharm. Res*., **2(1):** 62-72.

Biswas, K., Chattopadhyay, I., Banerjee, R. K., and Bandopadhyay. U. (2002). Biological activities and medicinal properties of neem (*Azadirachta indica*). *Current Science Association,* **82(11)**: 1336-1345.

Chauhan, R.S., Nautiyal, M. C., and Arun Kumar. (2011). Analysis of variabilities in populations of *Nardostachys jatamansi* DC. in Garhwal Himalaya, India. *Journal of Plant Breeding and Crop Science,* **3(9):** 190-194.

Government of India (2001). Ayurvedic pharmacopoeia of India, The controller of Publications, p. p 4.

Kokate, C. K., Purohit, A. P., and Gokhale, S. B. (2005). Pharmacognosy, 31st edn. Nirali prakashan, Pune, 219.

Kokate, C. K., Purohit, A. P., and Gokhale, S. B. (2005).Pharmacognosy, 31st edn, Nirali prakashan, Pune, pp. 344.

Kokate, C. K., Purohit, A. P., and Gokhale, S., B. (2005). Pharmacognosy, 31st edn, Nirali prakashan, Pune, pp. 581.

Kokate, C. K., Purohit, A. P., and Gokhale, S., B. (2005). Pharmacognosy, 31st edn, Nirali prakashan, Pune, pp.262, 263.

Leeming, J., and Notman, F. (1987). Improved methods for isolation and enumeration of *Melassezia furfur* from human skin. *Journal of Clinical Microbiology, American Society for Microbiology,* **25 (10):** 2017-2019.

Mc, Grath J., and Murphy, G. M. (1991). The control of seborrheic dermatitis and dandruff by antipityrosporal drug. *Drugs,* **41(2)**: 178-184.

Petukhova, L., Duvic, M., Hordinsky, M., Norris, D., Price, V., Shimomura, Y., Kim, H., Singh, P., Lee, A., Chen, W.V., Meyer, K., Paus, R., Jahoda, C., Amos, C., Gregersen, P., and Christiano, A. (2010). Genome-wide association study in alopecia areata implicates both innate and adaptive immunity. *Nature,* **466 (7302):** 113 DOI: 10.1038/nature09114

Rao, V. G., Annamalai, T., and Mukhopadhyay, T. (2011). Phytochemical investigation and hair growth studies on the rhizomes of *Nardostachys jatamansi* DC. *Pharmacognosy Magazine*, **7(26)**: 146–150. Available from http: // www.ncbi.nlm.nih.gov / pmc / articles / PMC3113354/

Razack, S., Khanum, F., and Ilaiyaraja, N.(2012). *Nardostachys jatamansi*, an anti anxiety herb: *In vitro* antioxidant potency and protective efficacy against biomolecules oxidation. Available online on: www.apjtb.com / press / 2012 / B177.docx

Razack, S., Khanum, F., and Ilaiyaraja, N. (*2012*.) *Nardostachys jatamansi*, an anti anxiety herb: *In vitro* antioxidant potency and protective efficacy against biomolecules oxidation. *Biochemistry and Nutrition discipline*, Defence Food Research Laboratory, Mysore.pp.1-15.

Schrader, K., and Domsch, A. (2005). Cosmetology- Theory and practice, Vol II, Verlag fur chemische Industrie, Augsburg, pp. 2-4.

Shastri, D. (1984). Sharangdhar samhita. 3rd edition, Madhyam Khanda, Baidyanath Prakashan, 321.

Tomar, A., and Singh, K.K. (2008). Neem: An Introduction. *In:* Neem A Treatise, Ed. By Singh, K.K, Phogat, S., Tomar, A., Dhillon, R.S., I K International, New Delhi, pp.4,7.

Traditional and Folk Herbal Medicine: Recent Researches Vol. 2 (2014) *Pages* **543–564**
Editor: **V.K. Gupta**
Published by: **DAYA PUBLISHING HOUSE, NEW DELHI**

18

Biological and Functional Properties of Bee Products for Medicinal Purposes

Danielle N.A. Silva[1], João J.S. Sousa[2] and Maria G. Campos[1*]

ABSTRACT

Due to the possible therapeutic applications of bee products in veterinary or traditional medicine, it is important to be aware of their potential registration as "Herbal Medicinal Products" and/or "Traditional Herbal Medicinal Products" which would increase their value in the near future.

For this registration to be possible, quality control of crude materials collected by beekeepers need an accurate follow-up and the purification of extracts will need to adhere to the European directives established by the European Medicines Agency.

The use of Bee Products for therapeutic and food purposes is ancestral and, over the last decade several studies have been developed in this direction. The investigated bioactivities in these raw materials have shown a strong indication for viability in different applications. Variations in floral sources could determine the type of the therapeutic range of bioactivities. The antioxidant activity, prevention of prostate cancer, increase bone mass, anti-inflammatory bioactivity in rheumatoid arthritis and antimicrobial action are some of the examples.

From the research developed, markers will be established for quantitative and qualitative analysis of each of the products and/or extracts, once they vary depending on the floral source, for instance, bee pollen. Their inclusion in pharmaceutical formulations, even in a simplified registration, needs to be supported with a monograph that safeguards the Public Health and maintains its bioactivities in-depth knowledge and is consistent. Despite this enormous

1 Drug Discovery Group, Center for Pharmaceutical Studies, Faculty of Pharmacy, University of Coimbra. Heath Science Campus,3000-548. Coimbra, Portugal.

2 Drug Development, Center for Pharmaceutical Studies, Faculty of Pharmacy, University of Coimbra. Heath Science Campus, 3000-548 Coimbra, Portugal

* *Corresponding author*: E-mail: mgcampos@ff.uc.pt

potential and the fact that work tends to be systematized, the standardization of quality control standards for the more bioactive bee products, such as bee venom, propolis, royal jelly and bee pollen require toxicity studies, which are still poorly developed. In this chapter some of these questions will be discussed as a starting point for the development of these crude materials.

Further reliable studies with this goal can provide a short-term development of a document that assures to the consumer a quality product and a risk assessment that is consistent with the legislation.

The determination of toxic contaminants such as heavy metals and pesticides, among others, should be included in the final documents. Another unneglectable aspect is the contamination of genetically modified plants whose impact is still unpredictable, and this still needs research because there are few results available concerning this issue.

This opens an important contribution to the profitability of bee products and also allows the consumer the confidence that is crucial to guarantee good products on the market.

Keywords: Apitherapy, Bee pollen, Bee products, Royal jelly, Honey, Apitoxin, Propolis.

Introduction

Research points out the utility of natural sources as potential patterns that can be screened for bioactivity leading to better and cheaper options in the treatment of many diseases (Newman and Cragg, 2012). Areas such as cancer and multidrug-resistant bacteria and/or fungi, focus the attention of chemical discovery for new targets of medicines.

Another important issue relates to the need for more therapeutic tactics in the diseases related to stress and/or increase of life span. Currently, diseases such as Parkinson's, psoriasis and atopic dermatitis and others that cause different changes in the human immune system (for instance, derived from cancer or stress) require immunomodulatory approaches that presume new medicines.

As an example of the importance of the natural sources in drug discovery, Newman and Cragg (2012) show figures related to the total 118 approved antibacterial drugs where only 26 are totally synthetic. Aside from the 14 vaccines all the others are from natural products (10), derived from natural product generally means a semisynthetic modification (67), made completely by synthesis but the pharmacophore is/was from natural products. A similar approach can be done for anticancer medicines with a total of 128 approved, from which only 20 are completely synthetic; the remained are natural products and/or derived from them.

In this chapter, we demonstrate that bee products have the potential to be explored for therapeutic application for diseases that lack bioactive molecules.

This is due to the fact that, on the one hand, they have an ancestral use that dates back more than 6000 years in the traditional medicine in ancient Egypt, and, on the other hand, in latter publications they have been screened for important bioactivities with very positive results.

However, the use of bee products in conventional medicine is limited to certain indications where they have shown effects which are "equal to" or "better than" those of standard treatments, for example, in treating wounds and burns (Molan, 2006) and fighting arthritis (Lee *et al.*, 2005).

Apitherapeutic societies claim that bee products are efficacious in several circumstances (Munstedt and Franke, 2005; Munstedt and Hackthal, 2006) however these claims of effectiveness are not supported by published reports. Access to health care and to medications in particular, is, consequently, a global concern for all, with the possible exception of the very wealthy countries.

In the paper "Natural Products and Traditional Medicine: Turning on a Paradigm" the authors alert to the fact that what is urgently needed is a Global Alliance for Natural Product Development and the necessary role in health care, which involves traditional medicines. This was a mechanism to bring representatives together to discuss the global issues and implications in new strategic terms, for the global development of natural products based medicinal agents for health care (Cordell and Colvard, 2012).

Strategies are needed to encourage and rethink the responsible use of natural resources prudently and to consider how drugs will be derived synthetically, semisynthetically, and naturally in the future. Hence, the first global paradigm shift is the assumption that resources will be available in the future to provide access to needed medicines, either synthetic or natural. This involves bringing the concept of sustainability to drug discovery and development (Cordell, 2011; Cordell *et al.*, 2012).

Apart from the others already cited above, one of the situations reported relates to urinary tract infections (UTIs). This is a widespread problem (Foxman, 2003; Bjerklund *et al.*, 2007) whose recurrence is common in women, resulting in considerable morbidity, multiple antibiotic treatments and increased expenditures. The increasing resistance of *E. coli*, bacteria most frequently isolated during UTIs to diverse antibiotics is an international public health problem (Wagenlehner and Naber, 2006; Coque *et al.*, 2008).

Therefore, it is important to create new strategies for public health prevention to treat UTIs for instance. Propolis' antimicrobial activities are well documented against different bacteria and also in this situation. *In vitro* propolis may act directly on microorganisms, and *in vivo* it may stimulate the immune system, activating bacterial death. Propolis may show additional effects with antimicrobial drugs– fluoroquinolones and b-lactams (Sforcin and Bankova, 2011).

In this chapter we will show the potential of bee products moving from Traditional Medicine to important research carried out in medical applications and also discuss the importance of the standardization of crude extracts in a future perspective if they will be submitted under the EU legislation on pharmaceutical products for human use that applies in general to traditional herbal medicines using the simplified registration procedure introduced in 2004. "This simplified registration procedure is intended for herbal medicinal products with a long tradition, which do not fulfill the requirements for a marketing authorisation, in particular those requirements whereby an applicant can demonstrate by detailed references to published scientific literature that the constituent or the constituents of the medicinal products has or have a well-established medicinal use with recognised efficacy and level of safety (so-called "well established use") (http: / /ec.europa.eu / health / human-use / herbal-medicines /).

Materials and Methods

To collect data on the biological and functional properties of bee products for medicinal purposes we conducted a structured review of literature in electronic databases, using model parameters from the systematic review of Cook *et al*. (1995). This model applies scientific strategies to limit bias in the critical evaluation and synthesis of all relevant studies on a specific topic in order to minimize the error in the conclusions. The search was conducted using keywords such as: apitherapy, bee products, bioactivities, flavonoids, phenolic compounds, honey, propolis, bee venom, and bee pollen.

The literature researched is composed of articles published in indexed sources about bee products from the period from 1968 to 2012. This period was purposively extended showing that many years ago there was an interest in scientific proof of bee products, however today we are still finding an "empty" meaning in relation to validation of these therapeutic products. In this chapter, the characterization studies applied to the scientific validation of bee products were established according to the criteria of inclusion and exclusion, and are important in the process of identification and evaluation of relevant results that substantiate the therapeutic importance of bee products. Therefore we have:

Inclusion Criteria

- ✰ Evidence-based clinical practices.
- ✰ Preclinical studies: that guide to the type of pharmacological action.
- ✰ Clinical studies that demonstrate what are the beneficial or harmful implications for the human body.
- ✰ Experience of beekeepers in the use of bee products.

Exclusion Criteria

- ✰ Studies that do not address therapeutic effects.

The execution of a scientific review is based on systematic methods in a sequence of steps recommended by the Centre for Reviews and Dissemination (CRD) Report (Khan *et al.*, 2001), published by the Centre for NHS Review and Dissemination, University of York (England) and another publication Cochrane–Cochrane Handbook (Higgins and Green, 2009).

The Cochrane Collaboration is an international organization that aims to prepare, maintain and provide access to systematic reviews on the effects of interventions in health. This was established in the UK in 1993.

The objective of starting with a systematic review is that it can present conclusive results whether favorable or unfavorable, or be inconclusive in the case that there are no studies of sufficient quality to endorse or refute any information.

Results

The results obtained for the selected biological and functional properties of some bee products are developed in the text that follows:

Honey

Definition and Main Constituents

Honey is defined in the European Directive 2001/110/EC, 20 December 2001, as "The natural sweet substance produced by *Apis mellifera* bees from the nectar of plants or from secretions of living parts of plants or excretions of plant-sucking insects on the living parts of plants, which the bees collect, transform by combining with specific substances of their own, deposit, dehydrate, store and leave in honeycombs to ripen and mature."

When placed on the market as honey or used in a product intended for human consumption, honey must meet the composition criteria set out in Annex II of this Directive, and must, as much as possible, be free from organic or inorganic matters foreign to its composition.

There is an exception of point 3 of Annex I – in the same Directive, "must not have any foreign tastes or odours, have begun to ferment, have an artificially changed acidity or have been heated in such a way that the natural enzymes have been either destroyed or significantly inactivated."

Honey is naturally a very concentrated solution of several sugars, mainly carbohydrates such as glucose and fructose in which it contains some of both; Blossom honey not less than 60g/100g and honeydew honey, blends of honeydew honey with blossom honey not less than 45g/100g. Blossom honey contains no less than 60g/100g (glucose and fructose) this ratio is the same in honeydew honey. Blends of honeydew honey and blossom honey contain no less than 45g/100g.

European unifloral honeys can be distinguished by their typical flavonoid profile (Tomás-Barberán *et al.*, 2001).

The chemical composition of honey is variable and depends on regional and climatic conditions and on the type of flowers visited by the bees, thus, it is classified as unifloral or polyfloral (Fredes, 2004; Muñoz *et al.*, 2007).It has also been suggested that the presence of variable amounts of heavy metals can be related to the geographical and botanical origin of this product (Fredes, 2004; Fredes and Montenegro, 2006).It is important to highlight, specific compounds such as chrysin, galangin, isorhamnetin, kaempferol, myricetin, naringenin, pinobanksin, pinocembrin, quercetin, and rutin when regarding flavonoids indentified in Chilean honeys (Muñoz *et al.*, 2007; Montenegro *et al.*, 2009).Neverthelss, these compounds are in very low amounts and the probability of their contribution to the bioactivity of the product is scarce. They can be used for the origin identification and standardization of these crude materials.

Selected Biological and Functional Properties For Medicinal Purposes

Immunomodulatory Potential

The immune system is involved in many diseases and its immunomodulation can alleviate them strengthening and suppressing the host immune response. Many researchers have focused on identifying plant derived natural substances that can serve this purpose (Saravanan *et al.*, 2012) and honey was one of them. Mesaik *et al.* (2008) reported an immunomodulatory potential of this crude material in the course of phagocytosis, mainly by modifying the oxidative burst process through the inhibition of phagocytic myeloperoxidase activity. It also had an inhibitory effect exerted by the major sugar constituents of honey on exocytsis-associated ROS formation catalyzed by myeloperoxidase, providing a perpective for a future exploration of this bioactivity. The flora was not discussed, probably because the experiment revealed that the sugars could be the major source of the involved bioactivity in this case, all the honeys have a similar amount of them.

Treatments of Wounds

There is a scarcity of medications available for the treatment of surgical and burn wounds, mainly due to bacterial resistance such as metycilin-resistan *Staphylococcus aureus* and vancomycin-resistant *Enterococcus faecium* (Brudzinsky and Lannigan, 2012).

This lack of antibiotic treatments for these resistant bacteria involves an urgency to develop new molecules that can help alleviate infections developed by them.

Peter Molan, Professor of Biological Sciences and Director of the Honey Research Unit University of Waikato, Hamilton, New Zealand was one of the first authors to publish results that prove the involvement of other substances besides sugars and hydrogen peroxide in the healing of wounds. He developed a new regard for the therapeutic effects of honey:

"There has been a renaissance in recent times for the use of this bee product, in wound dressing, for the treatment of wounds, burns, and skin ulcers. In the past decade there have been many reports of case studies, experiments using animal models, and randomized controlled clinical trials that provide a large body of very convincing evidence for its effectiveness, and biomedical research that explains how honey produces such good results. As a dressing on wounds, honey provides a moist healing environment, rapidly clears infection, deodorizes,and reduces inflammation, edema, and exudation. Also, it increases the rate of healing by stimulation of angiogenesis,granulation, and epithelialization, making skin grafting unnecessary and giving excellent cosmetic results. On burns its antibacterial and anti-inflammatory properties allow a moist healing environment to be maintained that protects the wounds from deterioration and fibrosis" (Molan, 2001a).

In Tonks *et al.* (2003), the authors explain that using monocytic cell line cultures, honey has been proved to stimulate the release of inflammatory cytokines, such as tumor necrosis factor-alpha, interleukin-1β and interleukin-6 which are also involved in triggering a number of immune responses to infection. Thus, they are helpingthe understanding of this complex issue (Molan, 2001b).

Hemorrhoids and Anal Fissures

Health problems as hemorrhoids and anal fissures cause pain and discomfort to many patients and doctors look for better options in clinical practice. A mixture of honey (50 per cent), olive oil (29 per cent) and beewax (21 per cent) has shown to be safe and effective in a prospective pilot study conducted to evaluate the therapeutic effect of topical application in the above cited situations. Patients were treated with a 12h application of a spoon size of the mixture around the skin twice daily. The therapeutic effect of topical application of the mixture on patients with anal fissures or hemorrhoids was carried out in 13 males and 2 females, mean age of 45 years. The honey mixture significantly reduced bleeding and relieved itching in these patients with hemorrhoids. Patients with anal fissures had a significant reduction in pain in 3 weeks, bleeding (4 weeks), and itching (2 weeks) after the treatment. In these cases no side effects were reported with use of the studied mixture (Al-Waili *et al.*, 2006). However, clinical evaluation of these complaints must be carefully carried out since tumors can exist in this region of the body with similar symptoms.

Labial and Genital Herpes Simplex Lesions

Another important experiment that was conducted with topical application of honey was carried out with the treatment of labial and genital *Herpes simplex* (zoster) lesions.

The authors performed a clinical trial of an efficient treatment with honey for labial herpes and genital herpes. The study was compared to systemic acyclovir in immunocompromized patients with zoster to prevent secondary bacterial skin infections and to accelerate healing of the herpetic lesions. No side effects were reported indicating that the treatment was safe and it effectively managed the signs and symptoms of recurrent lesions (Al-Waili, 2004).

The majority of the research involving bee products in this clinical situation was made with propolis, mainly from Brazilian botanical sources, and this approach with honey is another possibility for this complicated treatment.

Propolis

Definition and Main Constituents

Propolis is a resinous material collected by bees from exudates and buds of plants, then mixed with wax and bee enzymes and its colour, consistency and chemical composition are intimately related to the flora visited by the bees and to the season during which it is collected (Bankova, 2005).

At present, more than 300 compounds, mainly polyphenols, have been identified as constituents of propolis (Salatino *et al.*, 2005). Most polyphenols are flavonoids, followed by phenolic acids, esters, aldehydes, ketones, and so forth (Orsolic *et al.*, 2005). However, propolis contains some compounds which are toxic and induce hypersensitive reactions, especially in topic applications. Skin is the main target organ for treatments with this product, with contact dermatitis being a common manifestation (Munstediet *et al.*, 2007).

Selected Biological and Functional Properties for Medicinal Purposes

Treatment of Burns

Following Senedese *et al.* (2011), a topically applied product based on Brazilian green propolis was developed for the treatment of burns. For such substance to be used more safely in future clinical applications, this study evaluated the mutagenic potential of topical formulations supplemented with the cited extract (1.2, 2.4 and 3.6

per cent) based on the analysis of chromosomal aberrations and of micronuclei. Under the present conditions, topical formulations containing different concentrations of green propolis used for the treatment of burns showed no mutagenic effect in either test system, but 3.6 per cent propolis gel was found to be cytotoxic in the *in vitro* test.Thus, these results conclude the reinforcement of hypersensitivity of reactions cited by Munstediet *et al.* (2007), and the development of these questions need to be completed with further research, in a order to allow the possibility of a safety formulation of theses extracts. The application for other floral sources of propolis is not linear.The bioactive compounds involved will be different depending on botanical sources involved. Therefore, each type of propolis requires specified analysis to prove its efficacy and security.

Prostate Cancer

Prostate cancer is one of the most commonly diagnosed cancers in men, and is the second leading cause of cancer related death in men after lung cancer. It has shown that nutritional supplements rich in polyphenolic compounds such as propolis play a significant role in prostate cancer chemoprevention. However, there is a long road ahead to realize how we can help to reduce this disease.

Prostate cancer therapy usually includes radiotherapy, hormone therapy, and chemotherapy (Mazharand Waxman, 2008). Localized prostate cancer is sensitive to conventional radiotherapy using megavoltage photons (X-rays), but residual disease often causes clinical relapse in a large proportion of patients (Zietman *et al.*, 1993). Hormone therapy is also commonly used alone or in combination with other therapies in advanced or high-risk prostate cancer (Lam *et al.*, 2006), but hormone therapy will eventually fail and the prostate cancer will continue to grow and spread (Vijjanand Dubey, 2007).

Recently, new strategies are applied in order to provide a better understanding of this cancer. For example, to facilitate proteomic analysis, such as, the surface-enhanced laser desorption ionization time-of-flight mass spectrometry (SELDI-TOF-MS) has been introduced (Zhang *et al.*, 2006; Malik *et al.*, 2005). A study in Turkey evaluated the "Effect of Turkish propolis extracts on proteome of prostate cancer cell line" (Barlak *et al.*, 2011). In this paper the authors used the proteome approach because it is more accurate. The term "proteome" describes "all proteins expressed by the genome of a cell, a tissue or an organism" and was introduced by Marc Wilkins (Clarke *et al.*, 2003), for a more effective evaluation of the data. The aim of this study was to evaluate if the protein expression profile in PC-3 prostate cancer cell lines could be differentiated when incubated with dimethyl sulfoxide and water extracts of Turkish propolis.

The antioxidant potentials of dimethyl sulfoxide and water extracts of propolis were found in correlation with the amount of total phenolic compounds. Dimethyl sulfoxide and water extracts of propolis (20 μg/mL) reduced the cell viability to 24.5 per cent and 17.7 per cent, respectively. It was concluded that dimethyl sulfoxide and water extracts of Turkish propolis may have antiproliferative activity through differentiating protein expression profiles in PC-3 prostate cancer cell lines along with their antioxidant capacity (Barlak *et al.*, 2011).

Diet and dietary factors play an important role in preventing and treating chronic diseases including cancer (Výuda-Martos *et al.*, 2008; Orsolic and Basic, 2005).Thus, propolis is thought to improve health and prevent diseases such as inflammation and cancer because of its antioxidant potential (Banskota *et al.*, 2001) and the anti-tumoral activity might be attributed to a single substance, or to synergistic effects of several compounds, or to potential metabolites (Carballo *et al.*, 2008) providing new possibilities for further research. These are some of the questions that still are on the table for new research but an important result can be achieved with this natural resource.

Urinary Tract Infections

As explained in the introduction recurrent urinary tract infections are a significant cause of discomfort mainly for women.

The study conducted by Lavigne *et al.* (2011), in the article "Propolis can potentialise the anti-adhesion activity of proanthocyanidins on uropathogenic *Escherichia coli* in the prevention of recurrent urinary tract infections", reveals that free forms of galanginand pinocembrin were present in very low amounts in urine indicating that these compounds are excreted mainly in a conjugated form. In fact these are two main flavonoids (potential active compounds) found in the extract of propolis from various sources (Volpi and Bergonzini, 2006).

The concentration of the conjugates of pinocembrin in urine was relatively constant during the first 12 h, and reached its highest level 24 h after administration. In contrast, the conjugates of galangin in urine were high at 4-6 h after administration, and appeared to be maintained during the next collecting times. In addition, pinocembrin seems to be excreted more efficiently.This study shows that the dose-dependent and time-dependent effects are also observed when propolis was added to proanthocyanidins (PACs).The additional effect of PACs was clearly noted. The authors could speculate that in addition to the anti-adhesion activity of PACs, propolis killed the bacteria, decreasing the number of adherent bacteria and facilitating the activity of proanthocyanidins (Zafriri *et al.*, 1989; Foo *et al.*, 2000).

Some of the activities described above (anti-tumoral, mutagenic effect, anti-inflammatory), can be related with the antioxidant activity of this versatile product, for example, the correlation observed between the antioxidant activity and chemical composition of its different fractions, with special emphasis on the presence of flavonoids and *p*-coumaric acid derivatives. The authors concluded that the components of propolis act through different mechanisms sequestering reactive oxygen species. Artepillin C (3,4-diprenyl-*p*-coumaric acid), a major constituent of green propolis, is also an excellent scavenger of free radicals similar to catechins (Nakanishi *et al.*, 2003) and in this case probably further research can be focused only on the main component or the semisynthetic derivatives as hypothesis for new drugs.

Bee Venom (Apitoxin)

Definition and Main Constituents

The Bee venom is a colourless liquid, the active portion of the venom is a complex mixture of proteins, which causes local inflammation and acts as an anticoagulant.

This venom is produced in the abdomen of worker bees from a mixture of acidic and basic secretions; however this product is acidic (pH 4.5 to 5.5).

Unfortunately, some people are potentially allergic to bee stings and instead of developing immunity, they become severely allergic after a sting. This causes a serious generalized reaction, including skin irritation, a change in heart rate with a falling blood pressure, difficulty in breathing, fainting; they require urgent medical attention because they can die of anaphylactic shock.

Bee venom (BV) contains a variety of different peptides, including melittin (a major component), apamin, adolapin and mast cell degranulating peptide (Price *et al.*, 1983; Jentsch and Mucke, 1977).

One of the treatments made with this product is "BV acupuncture" that involves the injection of dilute bee venom into acupuncture points, used in disorders such as pain, arthritis, rheumatoid diseases, cancer, and skin diseases (Nam *et al.*, 2003; Son *et al.*, 2007).

Selected Biological and Functional Properties For Medicinal Purposes

Parkinson's Disease

Parkinson's disease (PD) is the second most common neurodegenerative disorder and is characterized by the selective loss of dopaminergic neurons of the substantia nigra pars compacta, resulting in reduced striatal dopamine. Levodopa is the drug of choice in PD due to its ability to initially improve core symptoms by increasing basal ganglia dopamine activity.

However, after years of therapy, patients have motor complications and the benefit from each dose becomes weaker ("wearingoff" fluctuations), more unpredictable ("oneoff" fluctuations), and associated with involuntary movements (dyskinesias). In addition, patients continue to suffer from fluctuations in motor function that are inherent to the disease itself (Verhagen, 2002).

Research carried out in Korea, reported an effectiveness of both acupuncture and bee venom acupuncture for Parkinson's disease (Cho, 2011). In this pilot study, both acupuncture and bee venom acupuncture showed promising results as adjuvant therapies for Parkinson's disease. Recently, the anti-neuroinflammatory effect of bee venom has been investigated, and the possibility of its use in the treatment of neurodegenerative disorders has been suggested (Yang *et al.*, 2010), corroborating, to some extent, the above cited study. Studies involving the toxic doses and the possible side effects still need further and more and effective research before the ponderation of this natural source of bioactive compounds can be candidate to a future application in the treatment of this disease.

Prostate Cancer

As stated previously, the increase of lifespan brings a new overview for this clinical situation involving many men.

Park *et al.* (2004), in Korea, studied bee venom for "Activation of Caspase Pathway via Inactivation of NF-κB" in prostate cancer. This product and its mainly constituent, melittin, decreased anti-apoptotic proteins but induced pro-apoptotic proteins. In this study, it was demonstrated that bee venom and its major compound, melittin, inhibited cancer cell growth in prostate carcinoma LNCaP, DU145, and PC-3 cells through induction of apoptotic cell death *in vitro* as well as *in vivo* xenograft model via activation of caspase pathway through suppression of constituted activation of NF-kB activity. This is an important effect that can be added to other results and used in some of the possibilities to treat this cancer with the exception of the toxicological assessment for which there is still little information.

Neuroinflammatory Effects

The study conducted by Yang *et al.* (2010), suggests that Bee venom (BV) could be a potential therapeutic agent for antineuroinflammatory effects in an animal model of Amyotrophic lateral sclerosis (ALS), a disease affecting the central nervous system and causing the death of motor neurons.The effects of BV treatment on anti-neuroinflammation of hSOD1G93A mice were assessed via immunoreactions using Iba 1 as a microglia marker and TNF-a antibody. The BV treatment increases survival rate and motor activity in symptomatic hSOD1G93A transgenic mice.To determine the effects of BV on the survival and motor activity of hSOD1G93A mice, BV (0.1 $\mu g/g$) was bilaterally administered (subcutaneously) at ST36.The administration of BV at a symptomatic stage of disease progression resulted in increased motor activity and a prolonged life.

The present study illustrated that BV treatment prevented mitochondrial disruption and served a neuroprotective role *in vivo* via the activation of cell survival signal transduction pathways, such as the PI3K and ERK pathways, which subsequently protected against the death of motor neurons in symptomatic hSOD1G93A mice.

Royal Jelly

Definition and Main Constituents

Royal jelly (RJ) that is produced by the hypopharingeal and mandibular glands of worker honeybees is well known to be a necessary food for the growth of the queen honeybee. This is because it contains a variety of free amino acids, sugars, minerals, vitamins (Park *et al.*, 2012) and organic acids (Isidorov *et al.*, 2009). Besides its usage as a dietary supplement, a number of biological and immunopotentiating actions have been reported for RJ. These include antitumor activities (Tamura *et al.*, 1987) and protective activity against hemopoietic dysfunction in X-irradiated mice (Emori *et al.*, 1998), however more recently proofs are scarce and the research with this natural source still needs more validation. Important research is carried out in order to assure the quality control of the product even including the possibility for a geographical identification using the ratio between the Carbone and Azote isotopes (Daniele *et al.*, 2011) which will be an important tool to be used if included in pharmaceutical formulations.

Selected Biological and Functional Properties for Medicinal Purposes

Atopic Dermatitis

Atopic dermatitis (AD) that occurs primarily in childhood is a chronic inflammatory skin disease characterized by pruritic and eczematous skin lesions.

The incidence of AD is steadily increasing, especially in industrialized countries, and population studies suggest that 10–15 per cent of the population is affected by AD during childhood (Coleman *et al.*, 1997; Leung, 1997).

Oral administration of Royal Jelly suspensions to the PiCltreated NC/Nga mice inhibited the development of *AD-like* skin lesions, as judged from the significant decrease in both the clinical symptom scores of the dorsal skin and the total scores obtained from the histological findings (Matsumoto *et al.*, 1999) and this can be explored for a better treatment of these situations.

Other Skin Effects

For cosmetic reasons, the demand for effective and safe skin-whitening agents is high and research in this field is scarce. Han *et al.*(2011) study the effect of royal jelly in topic application since the key enzyme in the melanin synthetic pathway is tyrosinase, many depigmenting agents in the treatment of hyperpigmentation act as tyrosinase inhibitors. The authors investigated the hypo-pigmentary mechanism of royal jelly in a mouse melanocyte cell line, B16F1. They found that mRNA levels of tyrosinase were reduced by royal jelly suggesting that royal jelly reduces melanin synthesis by down-regulation of tyrosinase mRNA transcription and serves as a new candidate in the design of new skin-whitening or therapeutic agents.

Aging remains to be an important issue in cosmetics and the reduced amount in estrogen increase constantly with a high contribution to this situation. Park *et al.* (2012) carried out animal models of skin aging in the context of estrogen deficiency and assessed the antiaging effects of royal jelly on skin. In this *in vivo* model of skin aging, bilateral ovariectomies were performed in 12-week-old virgin female Sprague-Dawley rats. Results show that the level of procollagen type I protein was increased in the dorsal skin of ovariectomized rats fed with a dietary supplement containing 1 per cent royal jelly extract; the amount of collagen recovered was close to the normal level. With these preliminary results the authors discuss the possibility that Royal Jelly may protect against skin aging by enhancing collagen production in rats with ovariectomy-induced estrogen deficiency.

Bee Pollen

Definition and Main Constituents

Bee pollen is the result of the agglutination of flower pollens, made by worker honeybees, with nectar (and/or honey) and salivary substances, and collected at the hive entrance.The male gametophyte of flowers, mainly collected by the honeybee is *Apis mellifera* for the purpose of feeding its larvae in the early stages of development (Campos *et al.*, 2008).Bee pollen can also be defined as a fine powder-like material produced by flowering plants pollen, mixed with nectar and bee secretions (Leblanc *et al.*, 2009).Commercially traded pollen is mainly pollen collected by the honeybee *Apis mellifera* for the purpose of feeding its larvae in the early stages of development. Collected flower pollen is accumulated as corbicular pellets in pollen baskets on the rear legs of the honeybee and it is a mixture of these pellets that comprises bee pollen (Campos *et al.*, 2008).

It is known that bee pollen contains lipids, sugars, proteins, amino acids, vitamins, carotenoids, polyphenolics such as flavonoids and carbohydrates (Humanand and Nicolson, 2006).

A high concentration of reducing sugars, essential amino acids, unsaturated and saturated fatty acids, and the presence of Zn, Cu, Fe, and high K/Na ratio makes bee pollen very important for human diets (Serra-Bonvehi and Escola-Jorda, 1997; Villanueva *et al.*, 2002; Bastos *et al.*, 2004; Almeida-Muradian *et al.*, 2005).

The phenolic composition of pollen predominantly consists of flavones and flavonol glycosides and of hydroxycinnamic acids derivatives (Campos,1997; Campos *et al.*, 1997). This composition tends to be *species-specific* and very important for floral origin identification (Campos *et al.*, 1997; Almaraz-Abarca *et al.*, 2004). A wide range of flavonoid types has been determined previously from pollen samples of different geographic origin among which mainly flavones (Wiermann, 1968), flavonols (Meurer *et al.*, 1988), dihydroflavonols and chalcones (Wollenweber *et al.*, 1979). Benzoic acid derivatives and hydroxycinnamic acids have been also identified from pollen (Serra-Bonvehí *et al.*, 2001). Flavonoids are widely recognized as reliable chemotaxonomic markers not only of plants but also of bee pollen and honey (Markham and Campos, 1996; Tomas-Lorente *et al.*, 1992).

The bee pollen has been related to the therapeutic properties antibiotic, antineoplastic, antidiarrheal and antioxidant (Leblanc *et al.*, 2009; Campos *et al.*, 2003; Carpes *et al.*, 2009). There is also evidence supporting the use of pollen in prostate diseases. These studies were conducted with rye pollen and showed positive results (Freckmann, 1979).

Selected Biological and Functional Properties for Medicinal Purposes

In a recent publication, Campos and Frigerio (2010) reviewed the more important therapeutic possibilities of bee pollen from various floral origins, in this chapter we report the more recent ones.

For example, the content in iron was correlated to the potential of this crude material to be used for anaemia and the hemopoietic system.The antioxidant activity of many types of pollen was evaluated and a full overview was published by Lopes *et al.* (2011). This was an important contribution for the possibilities that can be explored from this bioactivity (Almaraz-Abarca *et al.*, 2004).

Other bioactivities as anti inflammatory activity (*Pinus densiflora* correlation to flavonoids content), antitumoral activity (*Brassica napus* correlated to a *polysaccharide* LBPP), antiartherosclerotic activity (fat-soluble pollen extract), activity on osteoporosis (*Cistus ladaniferus* involving its constituents withPM < 1000), activity in disorders of the prostate and the hidroxamic acid as the possible reason for the positive results are some of the situations developed in Campos and Frigerio (2010).

Antioestrogenic and Antigenotoxic Activity

Between the poluants and the same natural compounds as isoflavones the linkage to the estrogen receptors is of concern (Campos *et al.*, 2006). There is concern with the linkage of estrogen receptors with pollutants and the same natural compounds such as isoflavones.

In some situations this affinity can be involved in hormonal-like tumours and all the caution is required to avoid these products.

This is new research published by Pinto *et al.* (2010) with bee pollen samples that was not screened in the above cited review which was published more or less at the same time. The authors carried out assays with *Cystus incanus* and *Salix alba* in a way to evaluate by yeast estrogen screen and micronucleous in human lymphocites. Both samples showed a marked inhibitory effect on the activity of 17- beta- estradiol and failed to cause estrogenic activity and chromosome damage. One prepation of the derivative extracts made of each *specie* showed a marked antigenotoxic effect against the action of anticancer drugs as mytomicin C, bleomycin and vincristine. Thus, this study supported their potential use as safe future drugs that need more toxicological essays but they are a good possibility in a near future as crude materials at least. Thus, this study supports their potential as safe future drugs, however, further toxicological assays are required, but there is a good possibility in the near future for the use as crude materials.

Although more toxicological tests are needed in further research, conlusions from Pinto *et al.* (2010) are a good starting point to support the use of pollen with potential for future safe drugs.

Toxicity

The toxicological approach to these beehive products is almost nule. However, for bee pollen pirroliziolidinic alkaloids were found in certain species as *Echium vulgare, Senecio ovatus* and *S. jacobaea* (Boppré *et al.*, 2008). It is known that these compounds are hepatotoxic mainly by chronic consumption. The authors discuss in this paper how eating during the drying process of bee pollen to reduce the moisture could affect the amount and the structures of these compounds, providing a different point of view to this toxicity.

The issue is not yet clarified and further research will be carried out in a near future assuring the safety or risk of bee pollen from these floral origins.

Herbal-Drug Interactions

The Herbal-Drug Interactions are a hot topic (www.oipm.uc.pt) and there has been a relevant interaction published associated with bee pollen and warfin (Hurren and Lewis, 2010). This drug is consumed for blood thickening and the risk of hemorragic episodes is frequent. Consumption of Warfrin is associated with thinning blood and has a frequent risk of hemorrhagic episodes. To assure the correct dose, patients taking this compound are asked to do a narrow screen of their Internacional Normalized Ratio (INR). INR has accurate values for each situation but with a normal therapeutic, it ranges between 2.0 to 3.0.This is one of the drugs where these herbal-drug interactions are frequently discussed. This is exactly because this exam is done to avoid the risk of the intake and any changes lead to the determination of the potential cause.

In the situation reported by Hurren and Lewis (2010) a 71 year old Caucasian man arrived for routine warfarin monitoring and the INR was 7.1 and in the previous 9 months, it was 1.9-3.3. The discrepancy was related to the intake of bee pollen and after he stopped his consumption of bee pollen the levels of INR decreased to 3.7, an accepted clinical value. The medical evaluation of this herbal-drug interaction was done with various details that are unnecessary for this discussion however, the situation itself is pertinent.

The intake of medicines and herbs and/or natural products as supplements needs to be thoughtful and all the occurred interactions signalized for further investigations and alerts avoiding future similar situations.

The intake of medicines and herbs and/or natural products such as, supplements needs to be thoughtfully considered and those interactions that have occurred need to be signaled for further investigation. Alerts to patients and health care professionals should be made to avoid similar situations in the future.

Discussion

The development of drugs of natural origin presents an overview of some aspects of the possible future roles of chemistry and traditional medicine focusing on the development of new strategies for the health worldwide. These are not separate issues, since we are one large, deeply interwoven, totally vulnerable organism (Cordell and Colvard, 2012).Over time, countries, or consortia of countries, will develop the infrastructures to produce their own sustainable medicinal agents from natural sources based on the quality of their natural products. International aid programs will be needed to assist countries to potentiate their infrastructure and resources, including their facilities and their scientists.

The sustainability of the heath systems and the green chemistry are oriented nowadays for new medical targets. Bee products as honey, propolis, bee venom (apitoxin), royal jelly and bee pollen have proven to have good potential as crude materials for new medicines in diseases as resistant bacteria, virus, hemorrhoids and anal fissures (honey), prostate cancer (propolis and bee venom), psoriasis and urinary tract infections (propolis), atopic dermatic (royal jelly), parkinson´s disease and neuroinflammatoty situations (bee venom).

The potential benefit of bee products is supported by several studies and, in the majority of them phenolic compounds are present in the form of flavonoids (Viuda-Martos *et al.*, 2008) with multiple benefits to human health–anti-inflammatory, antioxidant, antiallergic, antithrombotic, normolipidemic, hepato-protective, antiviral and anticarcinogenic activities (Havsteen, 2002; Middleton *et al.*, 2000; Pérez-Trueba, 2003). The positive findings from the meta-analysis on honey and wound healing show that some tests may well prove to have interesting results (Molan, 2006).

Insect pollination, mainly by bees, plays a crucial role in the human food supply chain. Yet, the global bee population has declined alarmingly in recent years, putting a huge strain on the world's food resources. One promising option would be sustainable management of bee colonies. Pollination increases planet's biodiversity and, consequently, causes an increase of natural resources to be exploited by scientists in order to validate these products for conventional medicine and/or given interesting new molecules that can lead to future new drugs.

Acknowledgements

The authors thank José Alexandre Rodrigues, Northeastern University, Boston–USA, for reviewing the English version.

References

Almaraz-Abarca, N., Campos M.G., Avila-Reyes J.A., Naranjo-Jimenez N., Herrera-Corral J., and Gonzalez-Valdez L.S. (2004).Variability of antioxidant activity among honeybee-collected pollen of different botanical origin.*Interciencia*, **29:** 574-578.

Almeida-Muradian, L. B., Pamplona, L. C., Coimbra, S., and Barth, O. M. (2005). Chemical composition and botanical evaluation of dried bee pollen pellets. *Journal of Food Composition and Analysis***18**(1): 105 -111.

Al-Waili, Noori, S., Saloom, K. S., and Al-Waili, T. N.(2006). The safety and efficacy of a mixture of honey,olive oil, and beeswax for the management of hemorrhoids and anal fissure: A pilot study. *The Scientific World Journal* 6, 1998–2005, TSW *Holistic Health and Medicine* ISSN 1537-744X; DOI 10.1100/tsw.333.

Al-Waili.(2004). Topical honey application *vs.* acyclovir for the treatment of recurrent *herpes simplex* lesions.*Med Sci Monit.*, **10** (8): 94-98.

Bankova, V. (2005). Recent trends and important developments in propolis research. *Evidence-Based Complementary andAlternative Medicine*, **2** (1): 29–32.

Banskota, A., Tezuka Y., and Kadota S. (2001). Recent progress in pharmacological research of propolis. *Phytother Res.*, **15:** 561-571.

Barlak, Y., Deðer O., Çolak M., Karataylý, S. C., Bozdayý, A. M., and Yücesan, F. (2011). Effect of Turkish propolis extracts on proteome of prostate cancer cell line. *Proteome Science*, **9:** 74-84.

Bastos, D. H. M., Barth, M. O., Rocha, C. I., Cunha, I. B. S., Carvalho, P. O., Torres, E. A. S., and Michelan, M. F. (2004). Acid composition and palynological analysis of bee (*Apis*) pollen loads in the states of São Paulo and Minas Gerais, Brazil. *Journal of Apicultural Research*, **43**(2): 35-39.

Bjerklund, J.T.E., Cek M, Naber K., Strachounski L, Svendsen M.V., Tenke P., Pep and Peap (2007). Study investigators, European Society of Infections in Urology: Prevalence of hospital-acquired urinary tract infections in urology departments. *EurUrol.*, **51:** 1100-1111.

Boppré M., Colegate S.M., Edgar J.A., and Fischer O. W. (2008). Hepatotoxic pirrolizidinic alkaloids in pollen and drying-related implications for commercial processing of bee-pollen. *J. of Agricultural and Food Chemistry*, **56** (14): 5662–5672.

Brudzynski, K., and Lannigan, R. (2012). Mechanism of honey bacteriostatic actionagainst MRSA and VRE envolveshydroxylradicals generated from honey's hydrogenperoxide. *Frontiers in microbiology*, **3:** 1-8.

Campos, M.G. (1997). Caracterização do polenapicola pelo seu perfil em compostos fenolicos e pesquisa de algumas actividades biologicas. PhDthesis. FacultyofPharmacy, Universityof Coimbra, Portugal.

Campos, M. G., Bogdanov, S., Almeida-Muradian L. B., Szczesna T., Mancebo Y., Frigerio, C., and Ferreira F. (2008). *Journal of Apicultural Research* inc. *Bee World*, **47** (2): 154-161.

Campos, M.G., and Frigerio, C. (2010). Bee-Pollen Therapeutical Value. *In:* Book series Comprehensive Bioactive Natural Products, Vol 1 – Potential and Challenges. Ed. V. K. Gupta. published by M/S. Studium Press LLC, USA., Chapter 11, pp257-277. ISBN: 1-933699-51-5

Campos, M.G., Markham K.R., Mitchell K.A., and Proenca da Cunha A. (1997). An approach to the characterization of bee pollens via their flavonoid/phenolic profiles. *Phytochem Anal.*, **8:** 181-185.

Campos, M. G., Paranhos António, H., Matos Miguel, P., Câmara Maria, T, Cunha Margarida, M., Pinto, P., Armando, J. Silvestre, Amado F., and Neto, P. C. (2006). Comparative analysis of over the counter tablet preparations of isoflavones extracted from soy available in Portugal. *Natural Products Communications*, **1** (11): 973-980.

Campos, M.G., Webby R.F., Markham K.R., Mitchell K.A., and Cunha A.P. (2003). Age-induced diminution of free radical scavenging capacity in bee pollens and the contribution of constituent flavonoids. *J Agric Food Chem.*, **51:** 742-745.

Carballo, D.D., Malak S., Bardenheuer W., Freistuehler M., and Reusch H.P. (2008). The contribution of plukenetione A to the anti-tumoral activity of Cuban propolis. *Bioorg Med Chem.*, **16** (22): 9635-9643.

Carpes, S.T., Mourao, G.B., Alencar, S.M., and Masson, M.L. (2009). Chemical composition and free radical scavenging activity of *Apis mellifera* bee pollen from Southern Brazil. *Brazilian J Food Technol.*, **12:** 220-229.

Cho, I.H.(2011). Bee venom reduces neuroinflammation in the MPTP–induced model of Parkinson's disease. *Int J Neurosci.*, **121**(4): 209e17.

Clarke, W., Zhang, Z., and Chan, D.W. (2003).The application of clinical proteomics to cancer and other diseases.*ClinChem Lab Med.*, **41** (12): 1562-1570.

Coleman, R., Trembath, R.C., and Harper, J.I. (1997). Genetic studies of atopy and atopic dermatitis. *Br. J. Dermatol.*, **136:** 1 – 5.

Cook, D. J., Sackett, D. L., and Spitzer, W. O. (1995). Methodologic guidelines for systematic reviews of the randomized control trials in health care from the potsdam consultation on meta-analysis. *J Clin Epidemiol.*, **48** (1): 167-171.

Coque, T.M., Novais, A., Carattoli, A., Poirel, L., Pitout, J., Peixe, L., Baquero, F., Canton, R., and Nordmann P. (2008). Dissemination of clonally related *Escherichia coli* strains expressing extended-spectrum beta-lactamase CTX-M-15. *Emerg Infect Dis.*, **14:** 195-200.

Cordell, G. A. (2011). Sustainable medicines and global health care. *Planta Med.* **77:** 1129"1138.

Cordell,G.A., and Colvard, M.D. (2012). Natural products and traditional medicine: Turning on a paradigm. *Journal of Natural Products*,**75:** 514-535.

Cordell, G. A., Martinez, J. L., Marinoff, M., and Rastrelli, L. (2012). Eds. Science Publishers: Enfield, NH; pp 1"45.

Council Directive 2001/110/EC, 20 December (2001). *Legislation relating to honey*, European Union.

Daniele, G., Wytrychowski, M., Batteau, M., Guibert, S., and Casabianca, H. (2011). Stable isotopes ratio measurements of royal Jelly samples for controlling production procedures: impact of sugar feeding. *Rapid Commun.Mass Spectrom.*, **25:** 1929-1932

Emori, Y., Oka H., Ohya, O., Tamaki, H., and Hayashi, H. (1998). The protective effect of royal jelly against the hemopoiesis dysfunction in X-irradiated mice. *Biot herapy*, (Jpn), **12:** 313– 9.

Foo, L.Y., Lu, Y., Howell, A.B., and Vorsa, N. (2000). A-type proanthocyanidintrimers from cranberry that inhibit adherence of uropathogenic P-fimbriated *Escherichia coli*. *J Nat Prod.*, **63:** 1225-1228.

Foxman, B. (2003). Epidemiology of urinary tract infections: incidence, morbidity, and economic costs. *Dis Mon.*, **49:** 53-70.

Freckmann, K. (1979). ImkereiimRheinland und in der Pfalz. Cologne: Rheinland-Verlag. 81–3.

Fredes, C., and Montenegro, G. (2006). Contenido de metales pesados y otros elementos traza en mieles de abeja en Chile. *Ciencia e Investigación Agraria*, **33:** 57-66.

Fredes, C.P. (2004). Relaciones entre el origen geográfico y botánico de mieles chilenas y el contenido de metales pesados. Tesis para optar al grado de Magíster en Ciencias Vegetales. Facultad de Agronomía e Ingeniería Forestal, Pontificia Universidad Católica de Chile. Santiago, Chile. pp. 33.

Han S. M., Yeo, J. H., Cho, Y.H., and Pak, S. C. (2011). Royal Jelly reduces melanin synthesis through down-regulation of tyrosinase expression. *Am J. Chin. Med.*, **39**(6): 1253-1260.

Havsteen, B.H. (2002). The biochemistry and medical significance of the flavonoids. *Pharmacology and Therapeutics*, **96:** 67-202.

Higgins, J.P.T., and Green, S. (2009). Cochrane handbook for systematic reviews of interventions. *The Cochrane Collaboration*, Version 5.0.

Humanand, H., and Nicolson, S W. (2006). Nutritional content of fresh, bee-collected and stored pollen of *Aloe greatheadii* var. *davyana* (Asphodelaceae). *Phytochemistry*, **67:** 1486–1492.

Hurren K.M., and Lewis, C.L. (2010). Probable interaction between warfarin and bee pollen. *Am. J. Health-Syst Pharm.*, **67:** 2034-2037.

Jentsch, J., and Mucke, H.W. (1977). Bee venom peptides. XVIII. Peptide-m and mcd-peptide: isolation and characterization. *Int J Pept Protein Res.*, **9:** 78–9.

Khan, K.S., TerRiet, G., Glanville, J., Sowden, A.J., and Kleijnen, J. (editors). (2001). Undertaking Systematic Reviews of Research on Effectiveness: CRD's Guidance for those Carrying Out or Commissioning Reviews (CRD Report Number 4) (2nd edition). York (UK): NHS Centre for Reviews and Dissemination, University of York.

Isidorov, V. A., Czyzewska, U., Isidorova, A. G., and Bakier, S. (2009). Gas chromatography and mass spectrosmetric characterization of organic acids extracted from some preparations containing lyophilized royal jelly. *Journal of Chromatography B,* **877:** 3776-3780.

Lam, J.S., Leppert, J.T., Vemulapalli, S.N., Shvarts, O., and Belldegrun, A.S. (2006). Secondary hormonal therapy for advanced prostate cancer. *J Urol.*, **175**(1): 27–34).

Lavigne, J., Vitrac, X., Bernard, L, Bruycre,F., and Sotto, A. (2011). Propolis can potentialise the anti adhesion activity of proanthocyanidins on uropathogenic *Escherichia coli* in the prevention of recurrent urinary tract infections. *BMC Research Notes*, **4:** 522.

Leblanc, B.W., Davis, O.K., Boue, S., Delucca, A., and Deeby, T. (2009). Antioxidant activity of Sonoran desert bee pollen. *Food Chem.*, **115:** 1299-1305.

Lee, J.D., Park, H.J., Chae, Y., and Lim, S. (2005). An overview of bee venom acupuncture in the treatment of arthritis. *Evid Based Complement Alternat Med.*, **2:** 79–84.

Leung, D.Y.M. (1997). Atopic dermatitis: immunobiology and treatment with immune modulators. *ClinExp Immuno.*, **107**: 25– 30.

Lopes, J., Stanciu, O.G., Campos, M.G., Almaraz-Abarca, N., Muradian, L.B., and Marghitas, L.A. (2011). Bee pollen antioxidant activity–a review: achievements and further challenges. *Journal of Pharmacognosy*, **2** (2): 25-38.

Malik, A., Afaq, F., Sarfaraz, S., Adhami, V.M., Syed, D.N., and Mukhtar, H. (2005). Pomegranate fruit juice for chemoprevention and chemotherapy of prostate cancer. *Proc Natl Acad Sci USA*. **102**(41): 14813– 14818).

Markham, K.R., and Campos, M.G. (1996).7 and 8-O-methylherbacetin-3-O-sophorosides from bee pollen and some structure/activity observations. *Phytochemistry*, **43:** 763-767.

Matsumoto, M., Ra, C., Kawamoto, K., Sato, H., and Itakura, A. (1999).Ig Ehyperproduction through enhanced tyrosine phosphorylation of Janus kinase 3 in NC/Nga mice, a model for human atopic dermatitis. *J Immunol.*, **162:** 1056 –63.

Mazhar, D., and Waxman, J. (2008). Early chemotherapy in prostate cancer. *Nat Clin Pract Urol.*, **5** (9): 486–493.

Mesaik, M.A., Azim, M.K., and Mohiuddin, S. (2008). Honey modulates oxidative burst of Honey. *Advances in Food Research*, **24:** 287-375.

Meurer, B., Wiermann, R., and Strack, D. (1988). Phenylpropanoid patterns in fagales pollen and their phylogenetic relevance. *Phytochemistry*, **27:** 823-828.

Middleton, E., Kandaswami, C., and Theoharides.T.C. (2000). The effects of plants flavonoids on mammalian cells: implications for inflammation, heart disease and cancer. *Pharmacological Reviews*, **52:** 673-751.

Molan, P.C. (2001a). Why honey is effective as a medicine. 2. The scientific explanation of its effects. *Bee World*, **82:** 22-40.

Molan, P. C. (2001b). Potential of Honey in the Treatment of Wounds and Burns, *Am J ClinDermatol.*, **2**(1): 13-9.

Molan, P.C.(2006).The evidence supporting the use of honey as a wound dressing. *Int J Low Extrem Wounds*.**5:** 40 – 54 Erratum in: *Int J Low Extrem Wounds*, 5: 122.

Montenegro, G., Salas, F., Peña, R.C., and Pizarro, R. (2009). Actividad antibacteriana y antifúngica de mieles monoflorales de *Quillaja saponaria*, especie endémica de Chile. *Oyton*, **78:** 141-146.

Munstedt, K., and Franke, F.E. (2005). Pollen therapy and its scientific evidence. *Am Bee J.*, **145**: 511–3.

Munstedt, K., and Hackethal,A. (2006). Propolis–NeueDatenzumEinsatz in der Medizin. *Biene.*, **142**: 20–2.

Munstedt, K., Hellner, M., Hackethal, A., Winter, D., and Von Georgi, R. (2007). Contact allergy to propolis in beekeepers. *Allergologiaet Immunopathologia*, **35** (3): 95–100.

Munõz, O., Copaja, S., Speisky, H., Peña, R.C., and Montenegro, G. (2007).Contenido de flavonoides y compuestos fenólicos de mieles chilenas e índice antioxidante. *Quimica Nova*, **30**: 848-851.

Nakanishi, I., Uto, Y., and Ohkubo, K. (2003). Efficient radical scavenging ability of artepillin C, a major component of Brazilian propolis, and the mechanism. *Organic and Biomolecular Chemistry*, **1(9):** 1452–1454.

Nam, K.W., Je, K.H., Lee, J.H., Han, H.J., Lee, H.J., and Kang, S.K. (2003). Inhibition of COX-2 activity and proinflammatory cytokines (TNF-alpha and IL-1beta) production by water-soluble sub-fractionated parts from bee (*Apismellifera*) venom. *Arch. Pharm. Res.*, **26**(5): 383-388.

Newman and Cragg. (2012). Natural products as sources of new drugs over 30 years from 1981 to 2010. *Journal of Natural Products*,**75:** 311-335.

Orsolic, N., and Basic I. (2005). Water-soluble derivative of propolis and its polyphenolic compounds enhance tumoricidal activity of macrophages. *J Ethnopharmacol.*, **102:** 37-45.

Orsolic, N., Terzic, S., Mihaljevik, Z., Sever, L., and Basic I. (2005). Effects of local administration of propolis and its polyphenolic compounds on tumor formation and growth. *Biological and Pharmaceutical Bulletin*, **28:** 1928–1933.

Park, H. M., Cho, M. H., Cho, Y., and Kim, S. Y. (2012). Royal Jelly increases collagen production in rat skin after ovariectomy. *J. Med. Food*, **15** (6): 568-575.

Park, H.J., Lee, S.H., Son, D.J., Oh, K.W., Kim, K.H., Song, H.S., Kim, G.J., Oh, G.T., Yoon, D.Y., and Hong, J.T. (2004). Antiarthritic effect of bee venom: Inhibition of inflammation mediator generation by suppression of NF-kappa B through interaction with the p50 subunit. *Arthritis Rheum.*, **50**(11): 3504–3515).

Pérez-Prueba, G. (2003). Los flavonoides: antioxidantes o prooxidantes. *Revista Cubana de Investigación Biomédica,* **22:** 48-57.

Pinto, B. Caciagli, F., Riccio, E., Reali, D., Saric, A., Balog, T., Likic, S., and Scarpato, R. (2010). Antiestrogenic and antigenotoxic activity of bee pollen from *Cystus incanus* and *Salix alba* as evaluated by yeast estrogen screen and the micronucleousassy in human lymphocites. *European J. of Medical Chemistry*, **45:** 4122-4128.

Price, J.H., Hillman, K.S., Toral, M.E., and Newell, S. (1983). The public's perceptions and misperceptions of arthritis. *Arthritis Rheum.*, **26:** 1023–8.

Salatino, E., Teixeira, W., Negri, G., and Message, D. (2005). Origin and chemical variation of Brazilian propolis. *Evidence-BasedComplementary and Alternative Medicine*, **2** (1): 33– 38.

Saravanan,S., Prakash, B.N., Pandikumar, P., Karunai, R.M., Gabriel, P.M., and Ignacimuthu, S.(2012). Imunomodulatory potential of *Enicostema axillare* (Lam.) A. Raynal, a traditional medicinal plant. *J. of Ethopharmacology,* **140:** 239-246.

Senedese, J., Rodrigues, A., Furtado, M., Faustino, V., Berretta, A., Marchetti,J., and Tavares, D. (2011). Assessmentof themutagenicActivity of Extracts of Brazilian Propolisin Topical Pharmaceutical Formulations on Mammalian Cells *In Vitro* and *InVivo*. *Evidence-Based Complementary and Alternative Medicine*, Article ID 315701, doi: 10.1093/ecam/nen049.

Serra-Bonvehi, J., and Escola Jorda, R. (1997). Nutrient composition and microbiological quality of honey bee collected pollen in Spain. *Journal of Agriculture and Food Chemistry,* **45:** 725-732.

Serra-Bonvehí, J., SolivaTorrentó, M., and Centelles Lorente, E. (2001). Evaluation of polyphenolic and flavonoid compounds in honeybee-collected pollen produced in Spain. *J Agric Food Chem.,* **49:** 1848-1853.

Sforcin, J.M., and Bankova, V. (2011). Propolis: Is there a potential for the development of new drugs? *J Ethnopharmacol.,* **133:** 253-260.

Son, D.J., Lee, J.W., Lee, Y.H., Song, H.S, Lee, C.K., and Hong, J.T. (2007).Therapeutic application of anti-arthritis, pain-releasing, and anti-cancer effects of bee venom and its constituent compounds. *Pharmacol Ther.,* **115**(2): 246e70.

Tamura, T., Fujii A., and Kuboyama N. (1987). Antitumor effects of royal jelly (RJ). *Folia Pharmacol* (Jpn)., **89:** 73– 80.

Tomás-Barberán, F A., Martos, I., Ferreres, F., Radovic, B S., and Anklam, E. (2001). HPLC flavonoid profiles as markers for the botanical origin of European unifloral honeys. *Journal of theScience of Food and Agriculture,* **81** (5): 485-496.

Tomas-Lorente, F., Garcia-Grau, M.M., Nieto, J.L., and Tomas-Barberan, F.A. (1992). Flavonoids from *Cistus ladanifer* bee pollen. *Phytochemistry*, **31:** 2027-2029.

Tonks, A.J., Cooper, R.A., Jones, K.P., Blair, S., Parton, J., and Tonks. A. (2003). Honey stimulates inflammatory cytokine production from monocytes. *Cytokine*, **21:** 242-247.

Verhagen, M. L. (2002). Recognition and treatment of response fluctuations in Parkinson's disease: review article. *Amino Acids.*, **23** (1e3): 141-145.

Vijjan, V., and Dubey, D. (2007). New therapeutic targets in the treatment of prostate cancer. *Indian. J Urol.*, **23**(1): 61–66).

Villanueva, M. T. O., Marquina, A. D., Serrano, R B., and Abellán, G. B.(2002).The importance of bee-collected pollen in the diet: a study of its composition. *International Journal of Food Sciences and Nutrition*, **53**(3): 217-224.

Viuda-Martos, M., Ruiz-Navajas, Y., Fernández-López, J., and Pérez-Alvarez, J.A. (2008). Functional properties of honey, propolis, and royal jelly. *Journal of Food Science*, **73:** R117-R124.

Viuda-Martos, M., Ruýz-Navajas, Y., Fern'andez-L'opez, J., and P'erez-'Alvarez, JÁ.(2008). Functional properties of honey, propolis and royal jelly. *J Food Sci.*, **73**(9): 116-124.

Volpi, N, and Bergonzini, G. (2006). Analysis of flavonoids from propolis by on-line HPLC-electrospray mass spectrometry. *J. Pharm. Biomed. Anal.*, **42:** 354-361.

Wagenlehner, F.M., and Naber, K.G. (2006). Treatment of bacterial urinary tract infections: presence and future. *Eur Urol.*, **49:** 235-244.

Wiermann, R. (1968). Phenylpropanoid metabolism by pollen. I. Survey of flavonoid components isolated from gymnosperms and angiosperms. *Ber Dent Bot Ges.*, **81:** 3-16.

Wollenweber, E., and Wiermann, R. (1979). On the pigmentation of the pollen of *Nothofagus antarctica* (Forst.) Oerst. (Fagaceae). *Z Naturforsch.*, **34c:** 1289-1291.

Yang, E.J., Jiang, J.H., Lee, S.M., Yang, S.C., Hwang, H.S., Nothofagusantarctica Lee, M.S. (2010). Bee venom attenuates neuroinflammatory events and extends survival in amyotrophiclateral sclerosis models. *J Neuroinflammation*, **7:** 69.

Zafriri, D., Ofek, I., Adar, R., Pocino, M., Sharon, N. (1989). Inhibitory activity of cranberry juice on adherence of type I and type P fimbriated *Escherichia coli* to eucaryotic cells. *Antimicrob Agents Chemother.*, **33:** 92-98.

Zhang, H., Kong, B., Qu, X., Jia, L., Deng, B., and Yang, Q. (2006). Biomarker discovery for ovarian cancer using SELDI-TOF-MS. *GynecolOncol.*, **102:** 61-66.

Zietman, A. L., Shipley, W. U., and Willett, C.G.(1993). Residual disease after radical surgery or radiation therapy for prostate cancer.Clinical significance and therapeutic implications.*Cancer*, **71** (3 Suppl): 959–969.

Traditional and Folk Herbal Medicine: Recent Researches Vol. 2 (2014) *Pages* **565–572**
Editor: **V.K. Gupta**
Published by: **DAYA PUBLISHING HOUSE, NEW DELHI**

19

Aloe vera: A Miracle Plant

Archana Peshin Raina[1*]

ABSTRACT

Aloe vera *is one of the oldest known medicinal plant which is regarded as a valuable plant source in both traditional medicine and modern drug development. The characteristic bitter taste of* Aloe vera *leaves possesses many medicinal properties. It bears multinutritional qualities, showing multiple actions that helps to prevent diseases, maintains and restores health and promotes healing. Aloin is the principal active constituent of aloes, which is a mixture of glycosides used as purgative principle. Barbaloin is the main glycoside present in aloin that varies from 3.8 per cent to 22.1 per cent in Curacao aloe. The mucilaginous parts of the leaves contain sugars; glucose, galactose, mannose and galactouronic acid and is used in various types of skin disorders. In India,* Aloe vera *is grown commercially for its high demand in cosmetic industries as well as Indian Systems of Medicine.*

Keywords: *Aloe vera*, Aloe gel, Aloin, Barbaloin, *Curacao aloe*.

Introduction

Aloe vera Linn. (Family-Liliaceae) is a genus found in tropics and introduced in India for ornamental and medicinal purposes. The word Aloe is said to be derived from Arabic word "Alloeh" which means "bitter and shining substance". In India, Aloe is known by several names such as Ghrit Kumari, Kunvar Pathu, Musabbar, Ghee Kanwar and Indian Aloe. There are more than 300 species of aloe (Reynold, 1985) through out the world, but only few are of pharmaceutical importance. Three commercially important species which are the main sources of the drug are: *A. barbadensis* (*A. vera*); which yields Curacao aloe or Indian aloe which is produced in

1 Germplasm Evaluation Division, National Bureau of Plant Genetic Resources, New Delhi – 110 012, India

* *Corresponding author*: E-mail: aprraina@yahoo.co.in

the West Indies, *A. ferox* and its hybrids which yield Cape aloe, produced in South Africa and *A. perryi* which yields the Socotrine aloe. Other aloe species (*A. africana* and *A. spicata*) yield aloes of lesser importance. Four species of this genus are introduced in India for ornamental and medicinal purposes, out of them *Aloe barbadensis* (Synonym: *Aloe vera* Linn) has become naturalized in all parts of the country.

Plant Origin and Distribution

Aloe is native to Africa, Canary Islands, Spain and Mediterranean countries and has spread to East and West Indies, Southern USA, Central America, India, China and other countries. From semi arid areas to hilly (up to 1500 m, height) and coastal areas, it can be seen growing in gardens, wastelands and forests. It is also grown as ornamental plants in the kitchen gardens, farmhouses and offices. The plant is a boon to arid zone or the drought prone areas. It is a salt tolerant crop and can be grown in salt affected soils also. Successful cultivation of aloe in coastal areas of Kerala and Tamilnadu in India is an excellent example of its salt tolerance nature. It can resist water stress as well. In India, the species is distributed in Maharashtra, Tamil Nadu, Kerala, Andhra Pradesh, Karnataka, Rajasthan, Gujarat and Madhya Pradesh.

Plant Description

Aloe vera is a short stem perennial plant with leaves succulent, fleshy, sessile with spiny-margins, erect, lanceolate, tapering, 10 cm wide at base and can grow up to 60 cm in length with pointed a apex (Figure 19.1). The leaf surface is quite smooth and shining, dark glaucous green or pale green. Leaves contain a clear jelly-like mucilaginous pulp inside. It also contains an erect stout stalk arising out from the

Figure 19.1: Field view of *Aloe vera*.

center of the plant, which may be up to 6m long. The flowers are pendulous, imbricate, tubular, yellow to orange red in colour, drooping, borne in cylindrical terminal racemes on central flower stalk. The fruits are oval capsule containing numerous seeds.

Chemical Composition and their Uses for Mankind

Aloe has been used as a folk medicine for 3000 years, it is also now used in the modern medicine in many parts of the world and is called "a pharmacy in a plant" (Pharmacopoeia of India, 1966). From the leaves three types of commercial products can be obtained: the dried exudates, excreted from the aloin cells present in the zone of the vascular bundle; the gel, a mucilaginous juice present in the center of the leaf (hydrenchyma); and the oil, extracted by organic solvents. The purgative activity of aloe is due to presence of anthraquinone compounds, aloin (Fairbairn, 1949).

The plant produces at least six antiseptic agents such as lupeol, salicylic acid, urea nitrogen, phenols, cinnamonic acid and sulphur. They kill or control fungus, mold, bacteria and viruses. Lupeol and salicylic acid present in the juice are very effective pain-killer. It contains at least three anti-inflammatory fatty acids, cholesterol, campesterol and beta-sitosterol. They are highly effective in treatment for burns, cuts, scrapes, abrasions, allergic reactions, rheumatic fevers, acid indigestion, and ulcers. Its drink contains more than 70 essential ingredients of nutritional importance. It shows excellent immuno, antiseptic, antifungal, anti-inflammatory and cooling properties. Aloe because of its nutritional qualities, anti-oxidant and medicinal properties, is often called the "miracle plant" natural healer.

In India, it is grown commercially for its high demand in cosmetic industry as well as Indian Systems of Medicine. The succulent leaves are the economic part of this plant. The gel produced from its mature leaves is used in cosmetic industry for preparation of shampoo, soap, face creams and moisturizing agents. The harmful effect of synthetic cosmetics on the tender skin has been established through several research findings all over the world. Therefore, several herbal based beauty care products have come in the market, aloe gel is one of them. Several research reports indicate that aloe gel contains anti wrinkle and anti -ageing properties. *Aloe vera* is nature's miraculous gift for complete everyday skincare (as well as for hair) and is known as the "healing plant". The number of skin care products based on *Aloe vera* and aloe gel is growing daily. It presents great commercial opportunities among various medicinal plants.

The leaf gel or polysaccharides have been known to possess important biological properties such as anti-inflammatory, antibacterial, antitumor, antiallergic, infected wound healing by immuno enhancement and general tonic effect and because of these pharmacological properties of *A. barbadensis* extract and gel/polysaccharides are gaining a lot of attention from both clinical and commercial angle. Based on these biological properties a number of formulations are widely available in the market, especially in the USA, for the treatment of skin disorders and as general skin care cosmetic and health food. Because of these commercial products, growing of Aloe plants and the processing of its gel has become a big industry in some countries.

Chemistry of Aloes

Leaf exudates of aloe contain aloins, a small amount of volatile oil, resin gum, emodin anthraquinone derivatives, chrysophanic acid and traces of coumarins. The curacao aloes obtained from *A. barbadensis* is known to contain four major constituents *viz.* aloin A also known as barbaloin (Figure 19.2), its 10–epimer aloin B (Manitto *et al.*, 1990), both having anthrone skeleton and C- glucoside substituents; aloesin, formerly known as aloeresin B and aloeresin A. The Curacao aloes also contain a small quantity of free anthraquinone, aloe-emodin–anthrone. In Curacao aloe, barbaloin is accompanied by isobarbaloin. Socotrine and Zanzibar aloes do not contain isobarbaloin whereas Cape aloe contains traces only. Recently a number of new C-glycosyl –chromones such as neoaloesin-A, isoaloresin D and aloresin E have been

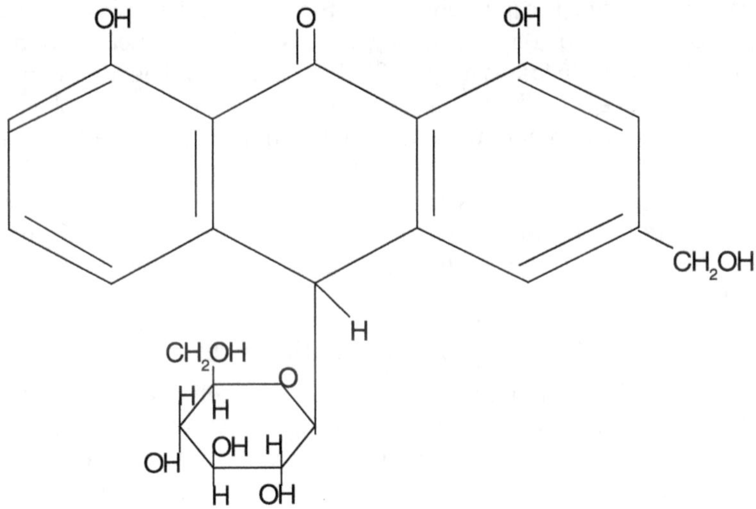

Figure 19.2: Structure of Barbaloin.

isolated (Joshi, 1998). The percentage of barbaloin, the principal glycoside of the aloes responsible for its cathartic activity, varies from source to source and season to season. The Indian aloes is reported to contain barbaloin in the range of 2.52-13.39 per cent (Bhavsar *et al.*, 1990). There is more barbaloin in exudates of young mature leaves just below the apex and the level decreases in older leaves towards the bas of the plant (Groom and Reynold, 1987). Aloin content was found positively correlated with short, broad and thick leaf types of Indian aloe (Srivastava *et al.*, 1990). It is also reported that anthraquinones present in the plant were highest before flowering and then fell during growth process.

Aloes/aloin are in use, in many countries, as cathartic agent, in traditional medicine for over two centuries and products containing aloes/aloin have been in market in major western countries. In India, it has been used in laxative formulations along with other components.

Varieties of *Aloe barbedensis*

National Bureau of Plant Genetic Resources, Pusa Campus, New Delhi has identified some accessions having high aloin and gel content. The high aloin content accessions are IC111267, IC111269, IC111271, IC111273, IC111279, IC111280 and the high gel content accessions are IC111266, IC111267, IC111272, IC111277, IC111280.

Chemistry of Aloe Gel

The mucilaginous jelly obtained from *A. barbadensis* leaves is reported to be a mixture of partially acetylated glucomannans that differ in their glucose to mannose ratio and also in the degree of acetylation. The composition of polysaccharides seems to vary significantly from source to source and also from season to season, although the principal polysaccharide present in all the analyzed sample is essentially acetylated glucomannan with change in degree of acetylation, linearity or branching of polysaccharide and molecular weight of polysaccharides. The mucilaginous jelly isolated from fleshy leaves of *A. barbedensis* is reported to be a mixture of polysaccharides containing mainly a glucogalactomannan with linear monosaccharides units linked by 1,4 bonds. The polysaccharides constituents 0.2 –0.3 per cent of fresh gel (0.8 to 1.2 per cent of the dry matter content). Along with the polysaccharides, the leaf extract is reported to contain arachidonic acid, a potential precursor for the prostanoids synthesis and essential amino acids arginine, histidine, phenylalanine, valine, leucine and isoleucine.

Processing of *Aloe vera*

The term "Aloe" used in medicine stands for the dried juice, which flows from the transversely cut bases of leaves. For processing of "Aloe" the juice is allowed to drain from the cut leaves into vessels and then concentrated by evaporation, either spontaneously or frequently by boiling. The juice is colorless or yellow when obtained fresh from leaves but becomes dark brown due to evaporation and boiling. When the juice is concentrated rapidly over a fire, the product obtained on cooling is amorphous and semi transparent and is called "glassy" or "vitreous" aloe.

Gel Extraction

It is well known that the aloe gel contains anti-wrinkle and anti-ageing properties. Most commercial brands of *Aloe vera* contain inorganic substances either in the form of fillers or as preservative or other conditioners, which may have adverse effect on the skin. However, use of fresh, instant aloe gel is natural choice for keeping the skin tender for a long time. It requires skill and experience to extract the gel out of the leaves. The leaves left over after draining of aloe juice are cut open and mucilage is scraped out with a blunt edges knife for isolation of gel. The extracted mucilage so obtained is stirred vigorously in a blender to make it homogenous mixture. The mixture is strained with the help of a muslin cloth and then filtered. The gel is precipitated from the extract by slowly adding acetone while stirring. The whole content is kept overnight and gel is isolated by centrifugation. This gel is redissolved in slightly warm water and transferred to a tube of known weight. It is dried at a high temperature (below 100ºC). For maximum yield the crop is cut after 9-10 months of planting. During this period mature leaves are more juicy, weighty and fresh. Too much old leaves do not give good quality gel, its quantity is also decreased. Very new and immature small leaves also contain least quantity of gel. For production of juice mature leaves of appropriate size are cut from the base and are hung for 5-6 hr keeping cut portion downside. The Aloe gel of superior quality should be transparent without any mould and for this the extraction is done under dark

An excision is made transversely on the leaves, which are heaped on elevated pillars allowing the gel to flow down. The gel is collected in a trough or plastic container. It takes one or two days to get the gel completely drained from the leaves. The gel obtained is opaque and waxy extract is called as "hepatic or livery aloe." Rapid concentration of the extract over a strong fire gives amorphous, semi transparent mass called 'Vitreous Aloe' (Chopra *et al.*, 1958). This is used externally to cure jaundice, constipation, acidity etc. If the whole leaf is slightly roasted over a gently fire, it becomes easier to extract the gel from the leaves.

Aloe Based Commercial Products

- ☆ Herbal dietary supplement
- ☆ Nutraceutical beverages
- ☆ Cosmetic
- ☆ Personal care

Pharmacological Uses

It is stomachic, tonic in small doses, purgative in large doses and anthelmintic. Aloe is used mainly in traditional medicine as a cathartic agent and dried juice is given in constipation, while fresh juice has cooling effect on body. Aloe is useful in functional sterility and disturbed menstrual function, root useful in colic. Undiluted fresh leaf juice has, anti-asthmatic, anti-bacterial, anti-burn, anti-hair loss, emollient and analgesic properties in case of insect stings, while dried leaf juice has anti-fungal, anti- X-ray burn, cosmetic and hair conditioner properties.

Wound Healing

The *Aloe vera* leaf has polysaccharide in the parenchymal cells which is used by the plant for energy. Mucilage is a different storage form. All polysaccharides are not the same. When mucilage is placed over a wound, the wound remains moist and does not drop as does dry wounds. Epidermal and fibroblast growth factors come from the mucilage and stimulate the fibroblast directly for growth and repair. The cells as a result migrate within the wound in a proper manner to increase wound healing.

As Dietary Supplement

The internal use of *Aloe vera* is well documented. It helps in maintaining normal healthy stomach lining, digestion, cell growth, and as a general overall tonic. Studies in animals suggest possible beneficial effects on maintenance of normal healthy function of the kidneys, heart muscle, clot formation, general cell growth and support of normal healthy cell proliferation, blood sugar levels, the immune system and even longevity. On the consumer level, *Aloe vera* has become of wide-spread interest to athletes, sports participants and other highly active people interested in supporting and maintaining their overall activity levels.

As Laxative

Since ancient times, drug aloe (prepared from the bitter yellow juice of the leaf) and its derivatives are used extensively today as active ingredients in commercial laxative preparations, most often in combination with other botanical laxatives such as cascara sagrada bark and senna leaves or pods. Aloe leaves are cut at the base to release the juice, which is then heated to evaporate the water. The remaining dark brown mass is drug aloe. Commercial aloin is a refined form of drug aloe that contains high concentrations of barbaloin, aloe's main laxative constituent.

For Everyday Use

Aloe gel can be applied on the all over body including face, neck and hands. First it turns into a thin flaky white layer and after some time it will slowly become a moisturous, transparent and invisible film that supplies nutrients to skin. This film protects skin from exposing to harmful dust, sunrays, and smoke etc. This facial preparation cleanses and softens the skin and improves its texture. *Aloe vera* can be used as a hair shampoo and conditioner. *Aloe vera* is an excellent natural hair conditioner and can completely eliminate the need to use other chemical conditioners.

Economics

The farmers harvest the leaves three times in a year. In each harvest, 6-8 ton of fresh leaves are obtained in an acre. Out of the total yield of 18 to 24 tones of fresh leaves, the farmers are able to generate a gross income of Rs.40, 000 to Rs. 60,000 fresh leaves with an average yield of 20 t/a in one acre. Such farmers generally make marketing tie up with buyers to whom the leaves are sold either as fresh or dried. This way, farmers make Rs. 25,000 to Rs. 40,000 net income in one acre. It should be noted that Aloes are grown in poor or marginal soils and mostly under high alkaline soils where no other annual crops could generate this amount of income.

Scope

There has been a growing need for cultivation of medicinal and herbal plants in the developed/developing countries. Products of plants parts, extracts and essential oils either used in the traditional system of medicine or in the modern system of medicine, are finding larger usage in human health. The species is in considerable demand for its medicinal properties with specific utility in the cosmetic industry. Dried Aloe leaves or powders are often exported to the European countries. The price of dried Aloes in our country ranges from Rs.600 to Rs.1000 per kg depending upon the 'aloin' content and color of the dried leaves. Aloe gels are expensive herbal cosmetics.

References

Anonymous (1985). The Wealth of India, Raw Materials. CSIR, New Delhi, Volume A, pp 191-193.

Bhavsar G.C., Chauhan, M.G., and Upadyay, U.M. (1990). Evaluation of Indian Aloes. *Indian J. Nat. Prod.,* **6** : 11-13.

Chopra, R.N., Chopra, I.C., Handa, K.L., and Kapoor, L.D.(1958). Indigenous drugs in India. U.N. Dhar and Sons, Calcutta, 2 nd Edition

Groom, Q.J., and Reynold, T. (1987). Barbaloin in *Aloe* species. *Planta Med.,* **53**(4): 345-348.

Fairbairn, J.W. (1949). The active constituents of the vegetable prugatives containing anthracene derivartives Part 1. Glycosides and aglycones. *J. Pharm Pharmacol.,* **1** : 683-694.

Joshi, S. P. (1998). Chemical constituents and biological activity of *Aloe barbadensis*–a review. *Journal of Medicinal and Aromatic Plant Sciences*, **20**(3): 768-773.

Manitto, P., Monti, D., and Speranza, G. (1990). Studies on aloe. Part 6; Confirmation and absolute configuration of aloin A and B and related 10-C glucosyl-9-anthrones. *J. Chem. Soc. Perk.,* I : 1297-1300.

Pharmacopoeia of India, (Ministry of Health, N. Delhi) 1966, 36.

Reynolds. T. (1985a). The compounds in *Aloe* leaf exudates: A review. *Bot. J. Linn. Soc.*, **90**: 157-177.

Reynold, T. (1985b). Observation on the phytochemistry of the *Aloe* leaf exudate compounds. *Botanical J. Linn. Soc.*, **90:** 179-199.

Reynolds, T., and Dweck, C. (1999). *Aloe vera* leaf gel. *Journal of Ethnopharmacology*, **68**: 1-3, 37

Satyabrata, Maiti and Ram Chandra.(2002).Cultivation of *Aloe vera* (Extension Bulletin) NRC for Medicinal and Aromatic Plants, Boriavi, Anand, 99 1-6.

Srivastava, V.K., Singh, B.M., Gupta, V., Chaudhari, N.K., and Gupta, R. (1990).Variability in Indian Aloe. *Indian J. Pl. Genet. Resources*, **3**(2): 93-96.

Singh, B.M., Srivastava, V.K., Kidwai, M.A., Gupta, V., and Gupta, R. (1995). Aloe, Psoralea and Mucuna. *Advances in Horticulture* Vol. 11-Medicinal and Aromatic plants. Eds. K. L. Chadha and Rajendra Gupta. Malhotra Publishing House, New Delhi-110064; pp 513-525.

Waller, G.R., Mangiafico, S., Ritchey, C.R., and Cumberland, C.D. (1978). Natural products from *Aloe barbadensis* Miller. *Lloydia*, **41**(6): 648.

Previous Volumes Contents

— Volume 1 —

2012, xvii+407p., col. plts., figs., tabls., ind., 25 cm Rs. 2400
ISBN 978-81-7035-766-7

Index

B

C

D

E

M

N

O

P

Q

R

T

U

V